W9-AXU-326

Texts and Monographs in Physics

W. Greiner B. Müller J. Rafelski

Quantum Electrodynamics of Strong Fields

With an Introduction into Modern Relativistic Quantum Mechanics

With 258 Figures

Springer-Verlag
Berlin Heidelberg New York Tokyo

Professor Dr. Walter Greiner
Professor Dr. Berndt Müller
Institut für Theoretische Physik der Johann-Wolfgang-Goethe Universität
D-6000 Frankfurt/Main, Fed. Rep. of Germany

Professor Dr. Johann Rafelski
Institute of Theoretical Physics and Astrophysics, University of Cape Town
Rondebosch 7700, South Africa

Editors

ISBN 3-540-13404-2 Springer-Verlag Berlin Heidelberg New York Tokyo
ISBN 0-387-13404-2 Springer-Verlag New York Heidelberg Berlin Tokyo

Library of Congress Cataloging in Publication Data. Greiner, Walter, 1935–. Quantum electrodynamics of strong fields. (Texts and monographs in physics) Includes bibliographies and index. 1. Quantum electrodynamics. 2. Heavy ion collisions. 3. Field theory (Physics) I. Müller, Berndt. II. Rafelski, Johann. III. Title. IV. Title: Strong fields. V. Series. QC680.G73 1985 537.6 84-26824

Typesetting: K + V Fotosatz, Beerfelden
Offset printing and bookbinding: Konrad Triltsch, Graphischer Betrieb, Würzburg
2153/3130-543210

Preface

The fundamental goal of physics is an understanding of the forces of nature in their simplest and most general terms. Yet there is much more involved than just a basic set of equations which eventually has to be solved when applied to specific problems. We have learned in recent years that the structure of the ground state of field theories (with which we are generally concerned) plays an equally fundamental role as the equations of motion themselves. Heisenberg was probably the first to recognize that the ground state, the vacuum, could acquire certain properties (quantum numbers) when he devised a theory of ferromagnetism. Since then, many more such examples are known in solid state physics, e. g. superconductivity, superfluidity, in fact all problems concerned with phase transitions of many-body systems, which are often summarized under the name synergetics.

Inspired by the experimental observation that also fundamental symmetries, such as parity or chiral symmetry, may be violated in nature, it has become widely accepted that the same field theory may be based on different vacua. Practically all these different field phases have the status of more or less hypothetical models, not (yet) directly accessible to experiments. There is one magnificent exception and this is the change of the ground state (vacuum) of the electron-positron field in superstrong electric fields. A whole new area of physics has developed when it became clear in the late 1960s and early 1970s that the vacuum of QED in supercritical fields becomes charged, i. e. it undergoes a phase change from a neutral to a charged vacuum. Most important is the fact that experiments have been systematically proposed and performed to test these fundamental ideas in the laboratory. The proposed existence of superheavy quasimolecules and finally of giant nuclear systems were milestones in this development, culminating in the discovery of sharp positron lines which probably signal the spontaneous decay of the vacuum and – at the same time – the existence of rather long-lived giant nuclei. It was a breathtaking development both in theory and experiments over the course of the last two decades.

We shall trace the essential steps of this scientific endeavour in this book. Starting with an overview of the physical, historical and philosophical implications (Chap. 1), a review of necessary elementary theoretical prerequisites (Chaps. 2, 3, 4), the Klein paradox in its original historical version (Chap. 5), we then develop systematically quantum electrodynamics of strong fields, always focussing on the non-perturbative aspect of the issues (Chaps. 6, 8 – 10). To attract the attention of graduate students also, Chap. 7 includes a synopsis of "ordinary" quantum electrodynamics, i. e. QED of weak fields.

Formulation of the theory of quasimolecules and heavy-ion collision dynamics can be found in Chaps. 11 und 12. No theory stands by itself in physics, but must be proven right or wrong by experiments. The courageous and ingenious experiments performed over the course of many years are summarized and discussed in Chap. 13.

Theory is treated further in Chaps. 14–16. The difficult problem of vacuum polarization is discussed in detail in Chaps. 14 und 15, and many-body effects in strong fields, including the self-energy in high Z atoms, in Chap. 16.

If bosons are bound in strong and supercritical potentials the phenomena change quantitatively and to some extent also conceptually. Chaps. 17–19 are devoted to these topics. Finally, in Chaps. 20 and 21, supercritical phenomena in other areas of physics, like superstrong gluon fields and supercritical gravitational fields, are presented to give an impression of the richness and variety of the applications of the idea of a supercritical vacuum.

We also tried to exhibit a number of details of the theoretical developments in order to ease the learning process for students. Such sections are indicated by a vertical grey line in the margin.

The last 15 years during which this research has been going on were most rewarding. We started with vague theoretical ideas which became more and more concrete and quantitative until they finally became real physics through experiments which were then compared in detail to theory. It is a pleasure to thank friends and colleagues, above all J. Reinhardt and G. Soff, and also P. Gärtner, U. Heinz, J. Kirsch, U. Müller, W. Pieper, P. G. Reinhard, T. de Reus, P. Schlüter, R. K. Smith, D. Vasak, K. W. Wietschorke and several others who made important contributions. Without their energy, enthusiasm and reliable work, the complex theory of these involved processes would not be as developed as it is.

We should equally like to thank a number of experimentalist friends and colleagues, in particular J. Greenberg, P. Kienle, and W. Meyerhof, and also H. Backe, H. Bokemeyer, K. Bethge, F. Bosch, T. Cowen, A. Gruppe, E. Kankeleit, C. Kozhuharov, D. Liesen, D. Schwalm, P. Vincent and their associates for their prodigous work and dedication. They followed a courageous path against many stumbling blocks, set backs and outside discouragements. The close contact with them was of mutual benefit and most gratifying.

Finally we should like to thank our technical assistant B. Utschig and our secretaries R. Lasarzig, L. Schubert and J. Parsons for their patience and steady help in preparing the manuscript. One of us (B. M.) would also like to thank the secretarial staff of the Department of Physics at Vanderbilt University and of the Kellogg Radiation Laboratory at the California Institute of Technology, where part of the manuscript was written, for their help.

Frankfurt am Main and Cape Town, May 1985

Walter Greiner
Berndt Müller · Johann Rafelski

Contents

1. Introduction

The structure of the vacuum is one of the most important topics in modern theoretical physics. In the best understood field theory, Quantum Electrodynamics (QED), a transition from the neutral to a charged vacuum in the presence of strong external electromagnetic fields is predicted. This transition is signalled by the occurrence of spontaneous e^+e^- pair creation. The theoretical implications of this process as well as recent successful attempts to verify it experimentally using heavy ion collisions are discussed. A short account of the history of the vacuum concept is given. The role of the vacuum in various areas of physics, like gravitation theory and strong interaction physics is reviewed.

1.1 The Charged Vacuum

Our ability to calculate and predict the behaviour of charged particles in weak electromagnetic fields is primarily due to the relative smallness of the fine-structure constant $\alpha \approx 1/137$. However, physical situations exist in which the coupling constant becomes large, e.g. an atomic nucleus with Z protons can exercise a much stronger electromagnetic force on the surrounding electrons than could be described in perturbation theory, and hence it is foreseeable that the new expansion parameter $(Z\alpha)$ can quite easily be of the order of unity. In such cases non-perturbative methods have to be used to describe the resultant new phenomena, of which the most outstanding is the massive change of the ground-state structure, i.e. of the vacuum of quantum electrodynamics.

In fact, the fundamental understanding of the vacuum, its properties and phases is of more general interest. During the last decade in field theories the structure of the vacuum has been considerably elucidated. But the conjecture that the vacuum is not just an empty and trivial space domain and its philosophical implications have a very long history dating back to the Greek philosophers of the Eleatic school and to Aristotle in particular. We shall come back to this interesting development and describe the historical roots below. Only with the advent of a quantum field theory did it become clear that a vacuum is an object capable of being subjected to physical experiments.

The most fascinating aspect of the vacuum of quantum field theory, which is discussed extensively in this book, is the possibility that it allows for the creation of real particles in strong (static) external fields: the normal vacuum state is unstable and decays into a new vacuum that contains real particles. This in itself is a deep philosophical and physical insight. But it is more than an academic problem, for two reasons. Firstly, very strong electric fields are available for laboratory experiments and secondly, here is an example of a more general phenomenon encountered in strongly interacting quantum field theory, namely, analogous physical phenomena are encountered again when discussing the interaction of other fundamental particles and fields.

Quantum electrodynamics is commonly, but falsely, identified with its perturbative expansion in powers of the fine-structure constant $\alpha = e_0^2/\hbar c$. Hence, the perturbative radiative corrections are normally referred to as the testing ground of QED. However, this traditional view ignores the importance of understanding the interaction of electrons with a strong source such as a heavy nucleus. Here the coupling constant becomes $Z\alpha$, where Ze is the electric charge of the nucleus. Since $Z\alpha$ is not generally small, and also because understanding bound states requires non-perturbative treatment, we must also consider other phenomena than radiative phenomena. That there are effects of zeroth order, i.e. 'yes or no' effects, which are most fundamental aspects of QED, does not seem to be well known. The most profound of these effects is the change of the vacuum (more specifically, the electron-positron vacuum) in over-critical fields.

Let us first illustrate here what is meant by these statements. One of the basic features of (relativistic) quantum field theory is the intrinsic many-body aspect of the theory. It becomes important as soon as a quantum field is considered in a strong external (classical, i.e. unquantized) field: particles may be produced spontaneously when there is a change of the vacuum caused by sufficiently strong and either strongly space- or time-dependent external fields. In the static case, even below the critical potential strengths, virtual particle pairs are generated as fluctuations of the vacuum, but because of the uncertainty relation the lifetime of the virtual pair is only of the order $\Delta t \sim \hbar/m_0 c^2$. However, if such a virtual pair is separated during this time by more than a Compton wavelength and if it has gained more kinetic energy than twice its rest mass, the pair may become real, as for overcritical potentials. Hence the strong background field has to extend over a sufficiently large region to lead to observable critical behaviour. Indeed, a potential difference of more than 1 MV in a Van de Graff machine, that is, over macroscopic distances, has a negligible probability to lead to particle production. However, if both the potential and its gradient are sufficiently strong, the particle creation process continues until either the potential differences in the external field are reduced substantially or the Pauli principle prevents further particle creation. Very often the new ground state will be charged, thus it has dynamically broken charge symmetry. The new ground state is called 'supercritical vacuum' or 'charged vacuum' [Mü 72].

A precise definition of the vacuum in quantum electrodynamics is obtained considering the single-particle spectrum (Fig. 1.1) of electrons in an external field.

The spectrum contains positive and negative energy states separated by an energy gap of $2m_0c^2$, containing the localized bound states. The negative energy states are ψ_{n}, the positive energy states, ψ_{p}. The quantum field operator at time $t = 0$ is defined as

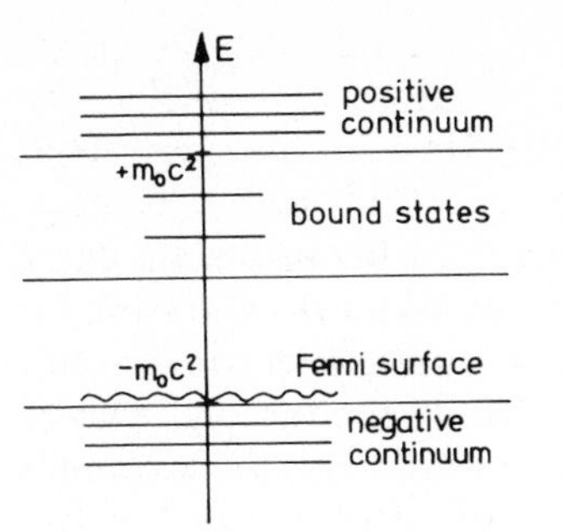

Fig. 1.1. Single-particle spectrum of the Dirac equation in an attractive external field

$$\hat{\Psi}(\boldsymbol{x}, t=0) = \sum_p \hat{b}_p \psi_p(\boldsymbol{x}) + \sum_n \hat{a}_n^+ \psi_n(\boldsymbol{x}) , \tag{1.1}$$

where the operators $\hat{b}_p$ are the annihilation operators for electrons and $\hat{a}_n^+$ the creation operators for positrons. This definition guarantees that the energy $\hat{H}_D$ of the electron-positron field is positive definite, i.e. (Chap. 9)

$$\hat{H}_0 = \sum_p E_p \hat{b}_p^+ \hat{b}_p + \sum_n |E_n| \hat{a}_n^+ \hat{a}_n . \tag{1.2}$$

The vacuum $|0\rangle$, characterized by the Fermi energy E_F, i.e. by the states $|n\rangle$ with $E_n < E_F$ filled with electrons, is best illustrated using hole theory (Fig. 1.2). A hole in this "Dirac sea" is then interpreted as a positron. The position of the Fermi surface dividing the single particle states into 'electronic' and 'positronic' states can, for a given physical system, be determined experimentally by observing, e.g., the threshold for e^+e^- pair production. Normally, such a vacuum $|0\rangle$ is neutral. Calculating the vacuum expectation value of the charge operator (Chap. 9)

$$\hat{\varrho}(\boldsymbol{x}) = \frac{e}{2} [\hat{\Psi}^+(\boldsymbol{x}, 0), \hat{\Psi}(\boldsymbol{x}, 0)]_- , \tag{1.3}$$

which is the zero component of the current four-vector

$$\hat{j}_\mu = \frac{e}{2} [\hat{\bar{\Psi}}, \gamma_\mu \hat{\Psi}]_- , \tag{1.4}$$

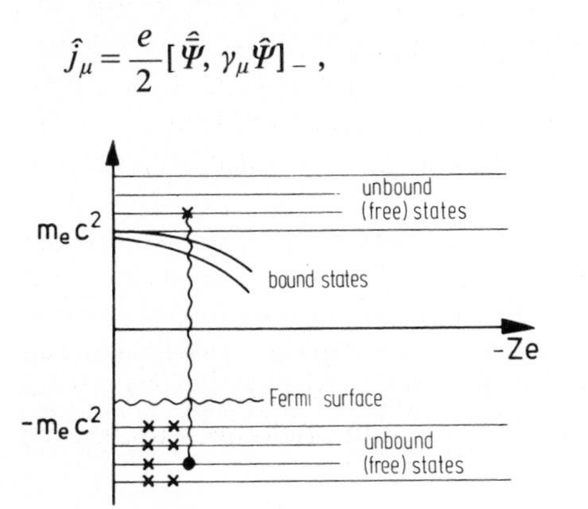

Fig. 1.2. The neutral QED vacuum can be viewed as negative energy states filled with electrons. The infinite charge is renormalized to zero

gives

$$\langle 0 | \hat{\varrho} | 0 \rangle \equiv \varrho_{\text{pol}}^{\text{vac}}(\boldsymbol{x}) = \frac{1}{2} e \left[\sum_{n} \psi_{\text{n}}^{+} \psi_{\text{n}}(\boldsymbol{x}) - \sum_{p} \psi_{\text{p}}^{+} \psi_{\text{p}}(\boldsymbol{x}) \right] . \tag{1.5}$$

This expression vanishes by symmetry for the field-free case because of the equal number and structure of n and p states. It leads to the well-known expression for the vacuum-polarization charge $\varrho_{\text{pol}}^{\text{vac}}(\boldsymbol{x})$ when these interactions are present. Vacuum polarization is now a very well understood phenomenon as, e.g., part of the Lamb shift, energy shifts in muonic atoms; even in weak fields the vacuum is a polarizable medium, characterized by occurrence of a displacement charge density $\varrho_{\text{pol}}^{\text{vac}}(\boldsymbol{x})$, which for weak fields does not contain a net charge:

$$\int \varrho_{\text{pol}}^{\text{vac}}(\boldsymbol{x})\, d^3x = 0 . \tag{1.6}$$

In Fig. 1.3 a qualitative illustration of this phenomenon is represented by the remarkable fact that upon renormalization the displacement charge density is positive nearer to the positive charge source.

If the charge of the central nucleus is continuously increased, the spectrum of electrons near a point-like and an extended source looks as presented in Fig. 1.4. For point nuclei the so-called fine-structure formula results, which has no s states beyond $Z = 137$. This puzzling behaviour is postponed to the so-called *critical charge* $Z_{\text{cr}} \approx 173$ for extended nuclei, where the $1s$ state "dives", i.e. joins the negative continuum. For even higher central charge at $Z'_{\text{cr}} \approx 183$, the $2p$ state dives, etc. The significance of this is the following.

In the supercritical case the dived state is degenerate with the (occupied) negative electron states. Hence *spontaneous* e^+e^- *pair creation* becomes possible, where an electron from the Dirac sea occupies the additional state, leaving a hole in the sea which escapes as a positron while the electron's charge remains near the

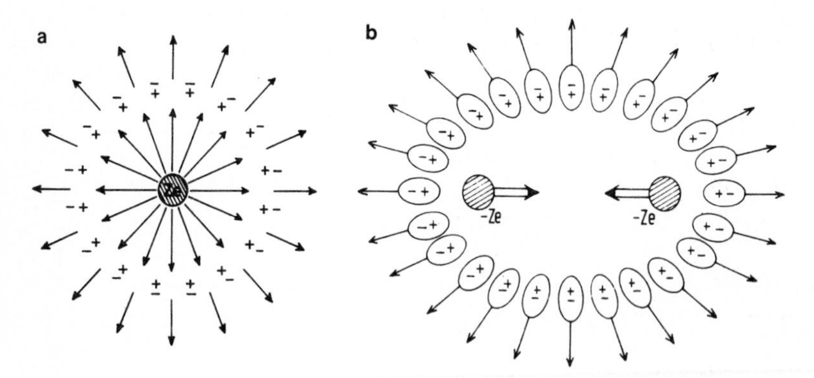

Fig. 1.3a, b. The vacuum polarization charge around a central nucleus (**a**) and around two colliding heavy nuclei (**b**). In (**a**) the static s electrons are shifted in energy somewhat due to modification of the Coulomb potential by $\varrho_{\text{pol}}^{\text{vac}}$ (this is part of the Lamb shift); in (**b**) the vacuum polarization charge is partially stripped off because of ionic motion. The vacuum polarization charge cannot readjust fast enough to the permanently changing position of the moving nuclei [So 77a]. Here "e" denotes the electron charge

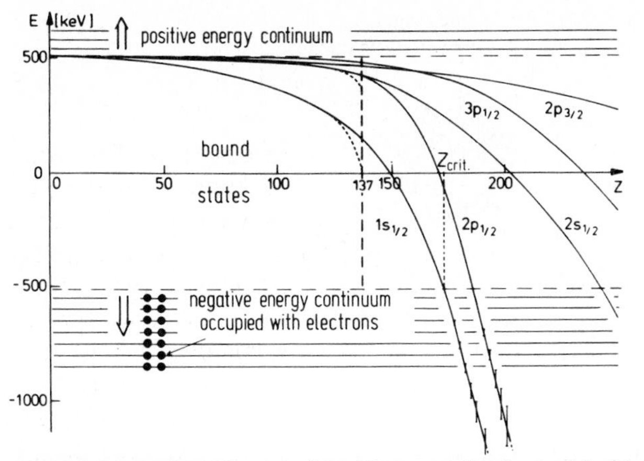

Fig. 1.4. Lowest bound states of the Dirac equation for nuclei with charge Z. While the Sommerfeld fine-structure energies (– – –) for $\varkappa = -1$ (s states) end at $Z = 137$, the solutions for extended Coulomb potentials (———) can be traced down to the negative-energy continuum reached at the critical charge Z_{cr} for the $1s$ state. The bound states entering the continuum obtain a spreading width as indicated [Mü 72a]

source [Pi 69]. This is a *fundamentally new process*, whereby the *neutral vacuum* of QED *becomes unstable in supercritical electric fields. It decays within about* 10^{-19} s *into a charged vacuum* [Mü 72a – c].

The charged vacuum is now stable due to the Pauli principle, that is the number of emitted particles remains finite. The vacuum is first charged twice because two electrons with opposite spins can occupy the $1s$ shell. After the $2p_{1/2}$ shell has dived beyond $Z'_{cr} = 185$, the vacuum is charged four times, etc. This change of the vacuum structure is *not a perturbative effect*, as are the radiative QED effects (vacuum polarization, self-energy, etc.), but has all the features of a *phase transition*. In the new charged vacuum the polarization charge contains a real component, not only a displacement charge density of net zero charge:

$$Q^{vac} = \int \langle \text{charged vacuum} | \hat{\varrho} | \text{charged vacuum} \rangle \, d^3x = 2e, 4e, \ldots \; . \tag{1.7}$$

When displaying the vacuum charge as a function of the proton number Z of the source, the usual discontinuity associated with a phase transition is evident. The dashed line in Fig. 1.5 is smoothed, ignoring the atomic shell structures. We rec-

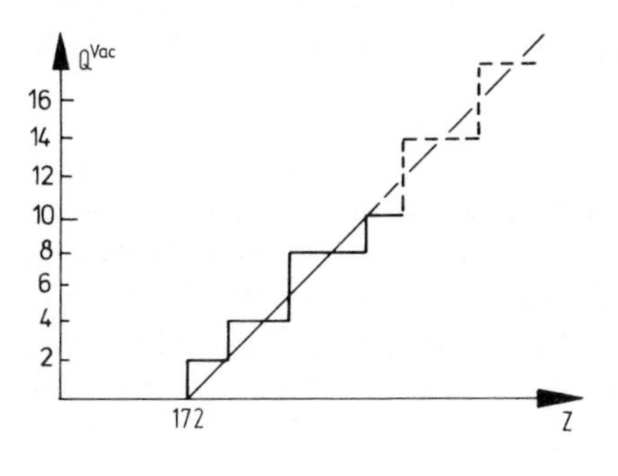

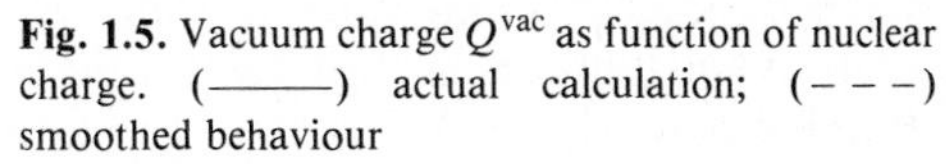

Fig. 1.5. Vacuum charge Q^{vac} as function of nuclear charge. (———) actual calculation; (– – –) smoothed behaviour

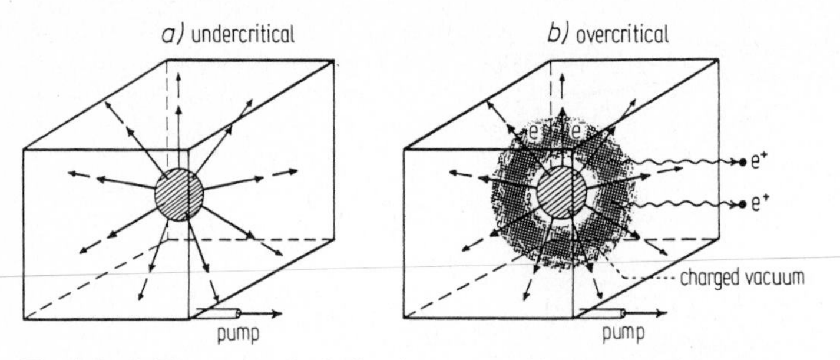

Fig. 1.6 a, b. The vacuum in the box is penetrated by the electric field of a central nucleus with charge $-Ze$; e being the charge of an electron. The nucleus is solely a spectator, furnishing the source of the electric field

ognize a 2nd-order transition. The full lines are the actual results with strong shell effects. One can view Q^{vac} as the order parameter of the transition.

The physical significance of the vacuum charge is further vividly illustrated in Fig. 1.6. The space in the box is thought to be pumped empty by an elementary particle pump. (Theoreticians can "invent" such a tool.) Only the central nucleus can act as a source for the electric field left as a spectator. The empty space in the box represents the neutral vacuum. In the undercritical case (a) it is stable. In the supercritical case (b), however, two positrons are emitted into free states, travelling around in the box and simultaneously two electron charges are localized around the nucleus. The positrons, being in free states, are easily pumped away. The electron cloud left over is the charged vacuum, representing the stable ground state in the supercritical field case. With great effort (i.e., substantial supply of energy) the two vacuum charges can also be pumped out, but within about 10^{-19} s again two positrons are created and the vacuum charge cloud reestablishes itself. In the supercritical situation only the charged vacuum is stable.

With these non-perturbative concepts well established, one still has to show that what has become of the radiative corrections is a negligible and well controllable effect. In this context the following questions can be raised.

1) *Can vacuum polarization hinder the vacuum decay,* i.e. prevent diving of the bound levels? The answer is *no.* The energy shift due to vacuum polarization is of the order of a -10 keV (about one percent of the binding energy) at the diving point [Gyu 75, Ri 75].

2) *Can self-energy prevent diving?* The answer is *also no.* Self-energy corrections shift the energy approximately $+10$ keV at the critical charge as *Soff* et al. have established [So 82b].

3) *Can non-linear effects in the electromagnetic field or in the Dirac field eventually become so large in supercritical fields that vacuum decay is prevented* or, at least, substantially shifted toward higher central charges? The answer is again *no,* if the excellent agreement between QED of weak fields and high precision experiments (Lamb shift, μ atoms, etc.) is not to be destroyed [So 73, Wa 74].

4) What happens in the *point limit,* i.e. when the radius of an overcritical charge distribution $R \to 0$. This very interesting problem was solved several years

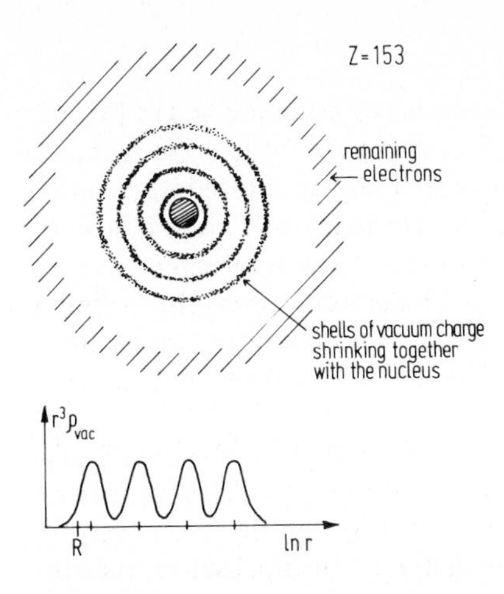

Fig. 1.7. As the radius R of the central charge shrinks to zero, the vacuum has an onion-shell type structure and becomes higher and higher charged, shielding the central charge. Thus point charges higher than 137 are prevented in QED

ago [Gä 81], indicating that as $R \to 0$, more and more bound levels with $\varkappa = \pm 1$ dive. The vacuum becomes higher and higher charged; the charge distribution of the vacuum in which the $(ns, (n+1)p^2)$ shells are arranged like onion skins shrinks to a point, so that in the limit $R \to 0$ the effective charge Z_{eff} of the system, consisting of the central charge Z minus the vacuum charge Z_{vac} approaches 137:

$$Z_{\text{eff}} = Z - Z_{\text{vac}} \xrightarrow[R \to 0]{} 137 .$$

This is a most interesting result indicating that QED *does not allow point* charges with charge larger than 137 (Fig. 1.7). In other words, the coupling constant of QED (for point particles) can never become larger than one. If a charged fundamental (point-like) spin-0 boson existed, the limiting charge of a point-like nucleus would be reduced to 137/2, because of the Klein-Gordon equation [Ba 81c].

1.2 From Theory to Experimental Verification

The crucial question is: Is the decay of the neutral vacuum an observable event? The question can be answered positively, but first the following concepts need to be established.

i) In heavy-ion collisions near the Coulomb barrier, the electrons behave as if the source were the combined Coulomb field of both nuclei. Since the electrons' motion is similar to the binding electron in a molecule this is a quasi-molecular state. However, the electronic binding is much too weak to keep the nuclei together.

ii) Under certain conditions the nuclear interactions keep the colliding ions from separating again for 10^{-19} s or more. Then the vacuum has time to decay if

iii) there has been a substantial probability for establishing the neutral vacuum in the entrance channel, i.e. for a hole in the strongly bound K shell.

We now discuss these three points in more detail.

1.2.1 Superheavy Quasimolecules

Quasi-stable islands of superheavy nuclei were predicted around $Z = 114$ [Me 67, Ni 69a, Mo 68] and around $Z = 164$ [Mo 69], but there was little hope from the beginning that they could be produced in sufficient quantity. More importantly, the predicted proton number (and hence the electric field) was not sufficiently large to lead to supercriticality. The new idea was to use the slowness of nuclear motion as compared to the motion of inner-shell electrons when the colliding nuclei are close to each other during a collision. Thus a superheavy molecule or atom is simulated (Fig. 1.8) [Ra 71, Mü 72c]. Hence these intermediate systems were called superheavy quasimolecules.

The adiabaticity conditions particularly are also fulfilled for the highly relativistic inner-shell electrons in such systems [Ra 72]. *Scheid* and *Greiner* [Sche 70, Gr 70] began to study *nuclear quasimolecules*, in 1967 at the University of Virginia (where Pieper and Greiner worked on electrons in superheavy nuclei at the same time) to explain the nuclear molecular resonances observed by *Bromley* and collaborators [Br 60]. It was then a small step from the nuclear quasimolecule to the electronic quasimolecule.

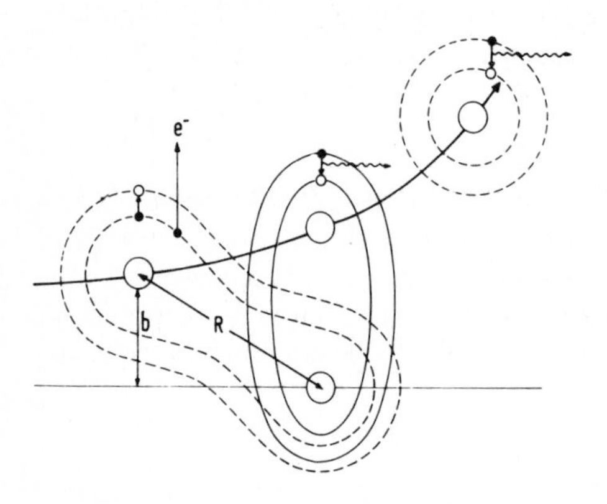

Fig. 1.8. Basic idea for forming superheavy quasimolecules in heavy-ion collisions. Since the relative motion of the two nuclei is slow ($V_{\rm ion} \approx c/20$) compared to the motion of the inner electrons ($V_{\rm e} \approx c$), the latter are expected to orbit around both centres

This proposal of the Frankfurt school had its precursors. An alternative to the theoretical method for calculating ion-atom collisions in Born approximation is the perturbed stationary state method [Mo 33], known for a longtime. *Mott* described in his excellent paper already fifty years ago the elements of the time-dependent molecular perturbation theory, using the eigenfunctions of the quasimolecule, similar to the so-called Born-Oppenheimer approach. About the same time, *Coates* [Co 34] observed what are now called quasi-molecular x-rays: he interpreted the x-ray bumps observed as radiation stemming from an intermediate quasi-molecular collision system. As with many untimely discoveries, these results remained obscure, as the measurements did not have independent importance within the precision obtainable at that time.

On investigating the vacuum stability of quantum electrodynamics in 1969/1970, the Frankfurt school first proposed that supercritical quasimolecules would be created in heavy-ion collisions and would be essential in understanding

QED of strong fields [Gr 69, Ra 71, Mü 72a–c]. Similar ideas were also expressed by *Gershtein* and *Zel'dovich* [Ge 70] but were not pursued further much beyond a general proposal. The need to understand relativistic particles in two centre potentials led the Frankfurt group to further work [Mü 73a, 76b] which has provided a basis for a quantitative description of the physical phenomena.

Some milestones establishing the formation of superheavy quasimolecules are the following.

a) *Saris* et al. [Sa 72] observed continuous x-ray spectra in various ion–atom collisions, which were interpreted as quasi-molecular L x-rays. Nearly simultaneously, the GSI-Cologne group of *Armbruster* and colleagues [Ar 72] observed quasi-molecular M x-rays from the superheavy I + Au system.
b) *Meyerhof* et al. measured broad continuous x-ray spectra in Br – Br collisions, which were interpreted as K-molecular x-rays [Me 73b]. These results were confirmed shortly thereafter by *Davies* and *Greenberg* [Da 74b].
c) *Greenberg* et al. [Gr 74] discovered the asymmetry of the quasi-molecular spectrum in Ni + Ni collisions, theoretically predicted by *Müller* and *Greiner* [Mü 74a, b, Gr 76]. Asymmetry of M x-rays was simultaneously found by *Kraft* et al. [Kr 74]. These measurements constituted the basis of the final identification of quasi-molecular x-rays by *Meyerhof* [see e)], and for the systematic spectroscopy of quasiatoms by *Stoller* et al. [St 77].
d) *Kaun* and his group observed K-quasi-molecular x-rays from superheavy systems [Ka 74].
e) *Meyerhof* et al. rigorously confirmed the existence of quasi-molecules [Me 75] by measuring the Doppler shift of quasi-molecular x-rays (1975) using the asymmetry of the x-ray spectrum.
f) *O'Brien* et al. isolated selected MO transitions to vacant $1s\sigma$ states using the cascade coherence between MO and K x-rays [O'Br 80].

1.2.2 Nuclear Sticking

While studying adiabaticity and the effects of the electron binding in nuclear collisions, the idea emerged that nuclear forces could also be crucially important in delaying the motion of colliding heavy ions. This phenomenon is well known in light nuclear systems, where nuclear molecules were first discovered by *Bromley* et al. in 1960 [Al 60, Br 74] and theoretically founded by the nuclear two-centre shell model [Ho 69, Ma 72, Pr 70, Pa 81]. This development led us to suggest in 1977 that indeed the nuclear time delay would greatly influence the spontaneous positron spectrum from vacuum decay [Ra 78c]. If the two nuclei stick for one reason or another, the spontaneous positron decay line due to the decay of the neutral to the charged vacuum should be clearly visible. The longer the sticking time T, the more pronounced the line structure should become, hence unambiguously proving the formation of the charged vaccuum.

The evolution of the $1s$ and $2p$ states is shown in Fig. 1.9 qualitatively as a function of time, assuming that between times t_1 and t_2 the nuclei stick, leading to pronounced lines of positrons (insert) in the spectrum. Occasionally, when

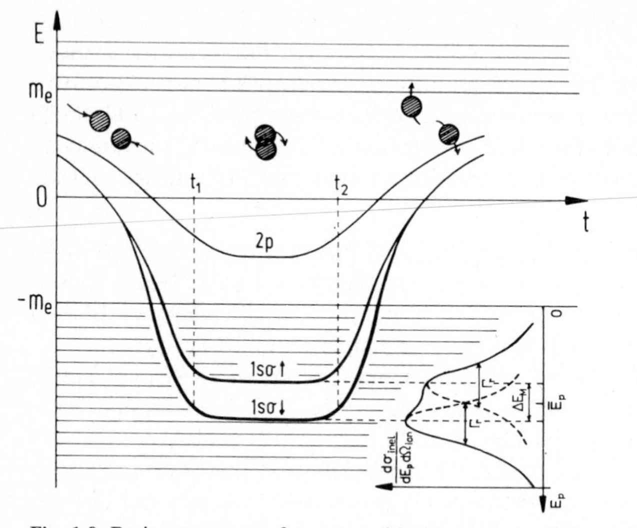

Fig. 1.9. Positron spectrum from overcritical heavy-ion collisions with sticking (formation of superheavy nuclear systems). The colliding, sticking and separating nuclei are indicated by black circles (●). The $1s$ and $2p$ quasi-molecular levels are drawn as a function of time

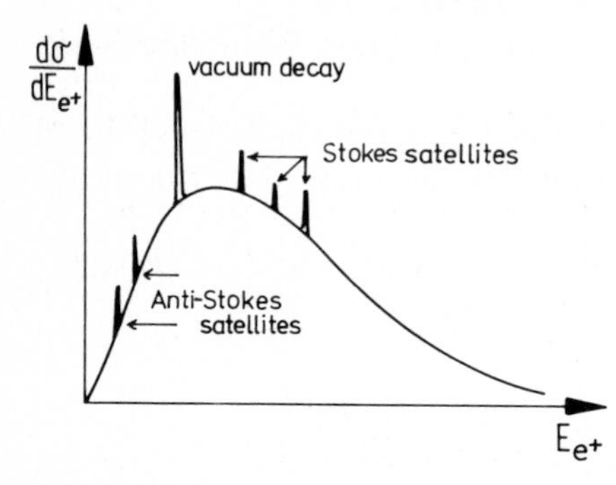

Fig. 1.10. Schematic positron spectrum from a long-living giant nuclear system. The expected Stokes and anti-Stokes lines are due to excited (mostly collective) states of the giant nuclear system. They are shown on top of the smooth dynamic spectrum

nuclear rotation is substantial, one may even notice the magnetic Zeeman splitting as shown.

Reinhardt extended these investigations and incorporated them in a general formulation of positron production processes in sub- and supercritical heavy-ion collisions [Re 81b]. In addition, it has been suggested that the giant nuclear system eventually formed in such sticking processes might itself be excited and decay by supercritical internal conversion [Re 81a, Mü 83c]. The internal conversion rate in such giant systems increases by many orders of magnitude compared to that in ordinary heavy nuclei. Thus additional positron lines might appear at higher energies of the spectrum, yielding information on the nuclear structure of the giant nuclear system formed in the collision. A schematic form of the expected spectrum is shown in Fig. 1.10. The dynamically induced part of the spectrum is due to induced and direct positron production (shake-off of the vacuum polarization cloud).

1.2.3 *K*-Shell Ionization

The experimental signature for the charged vacuum, viz. spontaneous positron creation, requires the existence of a vacancy in the supercritical bound state. In principle, one can produce an uranium nucleus with only one or even no electron(s) in the *K* shell by sending the beam through a thin stripping foil. However, the beam energy must exceed several hundred MeV per nucleon, far too much for nuclear collision experiments at energies of 5 – 6 MeV per nucleon further down the beam line. (With the SIS 18 accelerator such experiments may become possible in a few years; the fully stripped heavy ions would be cooled in a storage ring and the naked heavy ion beams split and later crossed, as was proposed by W. Greiner and Ch. Schmelzer [Be 75].) The important step was to recognize that *K*-shell vacancy could be produced with sufficient probability by ionization in the same collision prior to positron creation [Mü 72c]. This was disputed for several years, with estimates ranging from a probability of 10^{-5} or less [Me 74, Fo 76] to 10^{-1} [Bu 74], until a full-scale calculation predicted a *K* vacancy probability of several percent [Be 76b]. This prediction has been confirmed by experiment [Gr 77]. Today, the dependence of the *K*-shell ionization probability on impact parameter has been measured for various superheavy quasimolecules, where it agrees excellently with theoretical calculations [Li 82]. In particular, observation of the tremendously enhanced *K*-vacancy creation in superheavy quasimolecules [Gr 77, Ma 78a, An 78b] provided experimental proof that collisions of very heavy ions are an ideal testing ground for QED of strong fields.

Use of the *K*-vacancy measurements to determine binding energies (suggested by *Müller* et al. [Mü 78, So 78b]) was experimentally realized by *Behncke* et al. [Be 78, Li 80]. The first positron experiments with heavy ions establishing their non-nuclear origin and their energy dependence were carried out by *Backe* and collaborators in 1978 [Ba 78].

Coincidence experiments of positrons and heavy ions [Ba 78, Ko 79b] showed that positron creation is a non-perturbative phenomenon depending on nuclear charge like Z^{20} ("shake-off" of vacuum polarization). Observation of peaks in the coincidence positron spectrum in 1981 by *Kienle* et al. [Ki 83] and by *Bokemeyer, Greenberg, Schwalm* et al. [Bo 83a, Gr 83b] was a breakthrough in the search for spontaneous positron creation. The second group was the first to perform experiments with most kinematic parameters fully determined, which seems essential for reproducing the positron spectra. At about the same time, *Backe* and collaborators [Ba 83b] studied positron emission in heavy-ion collisions in which a deep inelastic nuclear reaction occurs. Triggering with fission products, they were able to observe time-delay effects in the positron spectrum due to nuclear "sticking" of the order of 10^{-21} s, thus also contributing to the observation of the spontanous decay of the vacuum.

The last year (1983) has witnessed the repeated observation and full confirmation of the narrow line structure in the positron spectrum of various supercritical collision systems (U + U, U + Th, U + Cm) by *Kienle, Clemente, Bosch, Kozhuharov* et al. [Cl 84], and by *Schwalm, Backe, Greenberg* and associates

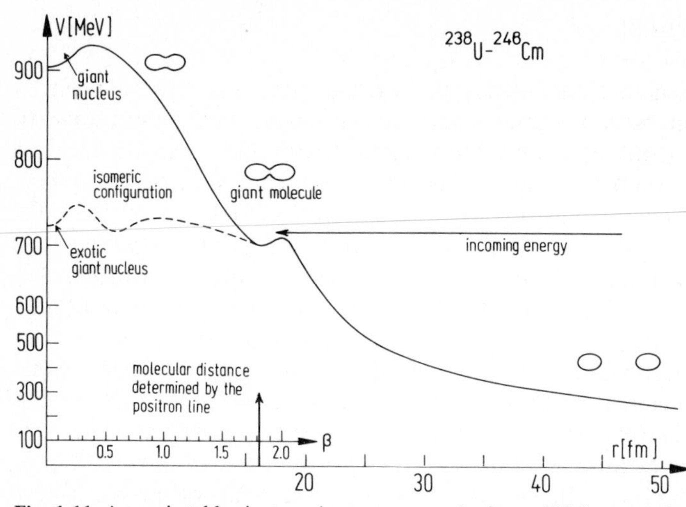

Fig. 1.11. A quasi-stable giant nuclear system can be formed if the attractive nuclear force leads to a pocket in the heavy-ion potential. Here r is the separation of the two nuclei

[Schw 83]. The absence of these structures in undercritical systems (Pb + U) is an important check that the positron lines indeed originate from the giant system and not from standard conversion processes in the uranium atoms, as Kienle's group determined. Further such checks are necessary.

The decisive test that the narrow line structure in the positron spectra does not originate from some ordinary processes (like pair conversion, monoenergetic pair conversion) lies in measuring the δ-electron and x-ray spectra. Both Greenberg's and Kienle's groups measured these spectra for U + Cm and U + U systems, respectively, during summer 1983, and found that these spectra are smooth. *Cowan, Greenberg* et al. even measured the Doppler shift of the positrons from the resonance structure and showed that those positrons are emitted from a system moving with the center-of-mass velocity of the compound system. This unseems to prove that the positron lines do not come from conversion of excited nuclear states.

Once this possibility is excluded, analysis of the experimental data shows that the positron lines must come from a long-lived giant nuclear system with a lifetime of $T > 5 \times 10^{-20}$ s. The line intensity requires only a rather small cross-section of a few millibarn to form the giant nuclear system in collision. The long lifetime of the combined nuclear system may be tentatively understood if the attractive nuclear force leads to a binding pocket in the internuclear potential (Fig. 1.11) [Se 84, Gr 83c]. The giant system can be formed only in the immediate neighbourhood of the barrier [He 83b].

This is precisely the behaviour found in experiments where there is a narrow beam energy window of only 0.2 – 0.3 MeV per nucleon for observing the positron line structures.

The positive proof that the origin of the observed positron lines is, indeed, the decay of the vacuum in the strong electric field of a giant nuclear system can be further confirmed by a Doppler shift analysis of the line structures. If the emitting system moves with the velocity of the centre-of-mass of the two ions, the origin of radiation from the combined atomic system is proven. First experiments have been carried out, with the result that the line broadening is compatible with Doppler broadening from the centre-of-mass motion, but not from the motion of the separated nuclei [Bo 84]. Again, further such investigations deem necessary.

1.3 Theoretical Developments

Although Sect. 1.4 treats the historical evolution of the idea that particle-antiparticle pairs can be created in strong fields, we note already here that this idea emerged repeatedly between 1930 and 1965, but was never considered worthy of further pursuit. This attitude changed from 1968–1971, when the theoretical schools in Frankfurt [Gr 69, Pi 69, Ra 71, Mü 72c] and Moscow [Ge 70, Ze 72] became aware that sufficiently strong electric fields could be realized under laboratory conditions in collisions of two very heavy atoms. Both groups were, in part, motivated in their studies by plans at Berkeley, Darmstadt and Dubna to build machines that could accelerate even the heaviest atoms, bringing them as close together as their nuclear radii. Vice versa, these ideas have particularly motivated the scientists in Germany to build the Unilac accelerator at GSI so that uranium can be lifted over the Coulomb barrier of the heaviest elements.

An essential step in understanding the decay of the neutral vacuum into the charged vacuum was made by the description of the decay as an autoionization process of positrons [Mü 72a–c, Ra 74]. Charged vacuum properties, such as the real vacuum polarization, energy spectrum of the emitted positrons, could be investigated. Fano's autoionization theory [Fa 61] also enabled simple treatment of the time-dependence of the decay of the neutral vacuum during a heavy-ion collision [Mü 72c, Ra 74, Re 81b]. This is particulary important because supercritical electric fields exist only for a finite time in heavy-ion collisions.

When solutions of the two-centre Dirac equation became available [Mü 73a, b], positron creation in heavy-ion collisions could be investigated in realistic situations. The probability of making a K vacancy in the course of the collision could be reliably calculated; it turned out to be of the order of several percent. Predictions about the background of dynamically induced positrons were later quantitatively confirmed by experiment.

A statistical description of the charged, supercritical vacuum by *Müller* and *Rafelski* [Mü 75] enabled the decay of the neutral vacuum to be viewed as the prototype of a phase transition in quantum field theory. The expectation value of the total charge in the vacuum state is the order parameter for the phase change. Further developments and considerable improvements of this path later allowed the limiting case of a point-like supercritical charge to be treated, which is shielded by the charge of the vacuum [Gä 81].

The fully dynamical calculations showed that heavy-ion collisions below the Coulomb barrier are so short that spontaneous positron emission is hidden in the background of dynamic processes of pair-production. Our suggestion that nuclear reactions could provide the time needed for the charged vacuum state to establish itself [Ra 78c], which was initially ridiculed, has become the central idea for theoretical work. New methods to treat supercritical heavy-ion collisions developed by *Reinhardt* et al. [Re 81b] enabled quantitative predictions for positron spectra with nuclear delay to be made [Re 81a, Mü 83c], which are crucial for comparisons with experiments. As the importance of nuclear reactions is increasingly recognized, present theoretical work tries to put this influence on a solid basis in the framework of a quantum mechanical reaction theory [He 83a, 84a, b, Re 83, To 83, 84].

1.4 Historical Annotations on the Vacuum

1.4.1 The Concept of Vacuum

Since the early Greek philosophers our view of the physical world has been dominated by certain paradigms, i.e. specific pictures, of selected physical entities. Such entities are space, time and matter as the basis of natural philosophy or, more specifically, of physics. Therefore, it is no surprise that our concept of "vacuum", intimately connected with the picture of space, time and matter, is among the most fundamental issues in the scientific interpretation of the world.

The picture of a vacuum has undergone perpetual modifications during the last twenty-five centuries as the available technologies have changed; often old, abandoned ideas have been resurrected when new information became accessible. Many aspects of today's concept of the vacuum date back to ancient Greek philosophy, but have only recently been established by modern experiments.

Two concepts alternatively formed the ancient Greek point of view concerning the vacuum: (i) "vacuum" is the void separating material objects and allowing their relative motion; and (ii) "empty space" is not possible, i.e. the vacuum was conceived as a medium. In the Pythagorean point of view, "air" was identified with the void as a seat of pure numbers.

In the atomism of Democritos (400 BC), the vacuum manifests itself as "intervals that separate atom from atom and body from body, assuring their discreteness and possibility of motion". Lucretius, (60 BC), writes in *De Rerum Natura:* "All nature then, as it exists, by itself is founded on two things: there are bodies and there is a void in which these bodies are placed and through which they move about". For Plato (380 BC), a physical body was merely a part of space limited by geometrical surfaces containing nothing but empty space. For Plato physics was merely a geometry, a point of view, which finally led to Einstein's geometric vision of the world and the modern theory of spaces of fibre bundles.

Opposed to that attitude was the opinion of Parmenides (480 BC) and Melissos (350 BC), "according to whom the universe was a compact plenum, one continuous unchanging whole" [Me 71]. Aristotle (350 BC) also rejected the

Platonic, Democritian and Pythagorean concepts of space, and, avoiding the notion of empty space, he spoke of space as the total sum of all places occupied by bodies, and of place (*topos*) as that part of space whose limits coincide with the limits of the occupying body. Rejecting the vacuum, Aristotle insisted that the containing body has to be everywhere in contact with that which is "contained". Empedokles performed an experiment with a "vacuum" around 450 BC, using the so-called clepsydra or water thief, a bronze sphere with an open neck and small holes in the bottom, used as a kitchen ladle at that time. Consequently, the idea of a natural "horror vacui" was developed, i.e. the belief that nature tries to avoid the formation of vacuum. However, with the invention of the air pump (around 1640) by Otto v. Guericke, and of the barometer by Toricelli (1643), it became clear that air can be removed from the interior of a vessel. The question remained as to what kind of vacuum is formed if all air molecules are pumped out of the vessel.

Since then many different concepts of vacuum were developed by scientists, different vacua as carriers for various kinds of physical phenomena. Newton's laws of mechanics require an absolute space for the principle of inertia to make sense. He writes in the *Principia*: "Absolute space, owing to its nature, remains the same and fixed regardless of any relationship to any substance". Mach considered Newton's absolute space an empty philosophical notion which should be replaced by an operational device to construct inertial frames. That idea developed into the requirement that local inertial coordinate systems should be determined by the energy-momentum distribution in the universe (Mach's principle). In Einstein's theory of gravity it means that the bundle of local inertial frames of reference is determined by Einstein's field equation when combined with suitable initial values of energy momentum and geometry on a space-like hypersurface. As an experimental consequence, e.g., it is predicted that near the surface of the earth a free gyroscope, with its axis oriented perpendicular to the rotational axis of the earth, should precess at a rate of 0.05 seconds of arc per year (Lense-Thirring effect). Attempts are underway to measure this precession by a gyroscope launched into orbit around the earth.

When the wave nature of light had been firmly established, the hypothesis of the vacuum as an elastic medium, the "ether", was developed parallel to the theory of elasticity worked out in the early nineteenth century by Navier, Cauchy, Poisson and others. From this point of view the ether had to be a very firm, elastic body to account for the high velocity of light, but it had to impose practically no friction on planetary motion. Michelson's experiment (1881) showed that the velocity of light is not influenced by the motion of the earth with respect to the ether by quantities of order $(v/c)^2$, and thus rejected the notion of ether as an absolute system of reference. This provided the basis for Einstein's principle of special relativity, whereby all inertial systems are physically equivalent. Newton's "absolute space" has lost most of its attributes, yet is not completely abandoned in Einstein's geometrodynamics. Unaccelerated frames are still "preferred" coordinate systems, although locally, within an infinitesimally small space-time volume, a freely falling system in a gravitational field may not be distinguished from a local inertial frame. When larger regions of space-time are considered, a true gravitational field is revealed by tidal forces and the radiation field of an "accelerated" charge can always be detected in the far zone.

Relativistic quantum mechanics and quantum field theory with the possibility of pair creation laid the grounds for our present concept of the nature of a vacuum. In today's language, a vacuum consists of a polarizable gas of *virtual* particles, fluctuating randomly. It is found that in the presence of strong external fields, the vacuum may even contain "real" particles. The paradigm of "virtual particles" not only expresses a philosophical notion, but directly implies observable effects:

1) The occurrence of spontaneous radiative emission from atoms and nuclei can be attributed to the action of the fluctuations of the virtual gas of photons.

2) The virtual particles cause effects of zero-point motion as in the Casimir effect. (Two conducting, uncharged plates attract each other in a vacuum environment with a force varying like the inverse fourth power of their separation.) Hawking's pair formation effect by a collapsing body may also be understood as an overcritical "decay of the vacuum" in strong gravitational fields or as a gravitational Casimir effect.

3) The electrostatic polarizability of virtual fluctuations can be measured in the Lamb shift and Delbrück scattering. The electron in the hydrogen atom is subject not only to the Coulomb potential ϕ of the nucleus but also to the fluctuation field of the virtual particles. If L is the typical dimension of the system, fluctuations in the electrostatic field strength are of the order

$$\Delta \varepsilon \sim \frac{(\hbar c)^{1/2}}{L^2} .$$

This field slightly displaces the electron by Δx from the usual classical orbit, leading to a shift of atomic energy levels of the order ΔE:

$$\Delta E \cong \frac{(\Delta x)^2}{2} \langle \nabla^2 \phi \rangle_{\text{average}} .$$

Precise calculations give a difference of 1057.9 MHz between the energies of the $2s_{1/2}$ and $2p_{1/2}$ levels in the hydrogen atom, in good agreement with experiment. Photon scattering by an external electric field (Delbrück scattering) caused by interaction of the photons with virtual electrons and positrons was first measured by *Wilson* [Wi 53].

4) The magnetic polarizability of the vacuum can be measured by observing the rotation of the polarization of laser light travelling through strong magnetic fields. Experiments are in progress to determine this effect [Ia 79].

1.4.2 The Vacuum in Strong Fields

The first indication that something strange would happen in strong electric fields came when *O. Klein* studied the reflection and penetration of electron waves on a potential barrier V_0, immediately after the first formulation of the Dirac

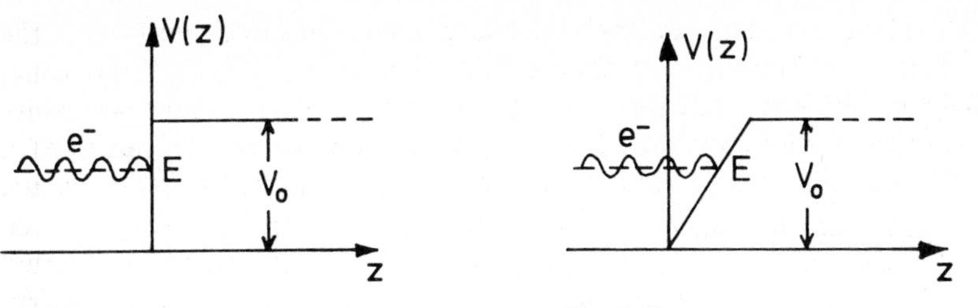

Fig. 1.12 **Fig. 1.13**

Fig. 1.12. The Klein paradox. If the barrier V_0 becomes very high ($V_0 \to \infty$), 83% of the electron wave will penetrate into the potential

Fig. 1.13. Sauter studied the Klein paradox for smoothed-out potential barriers

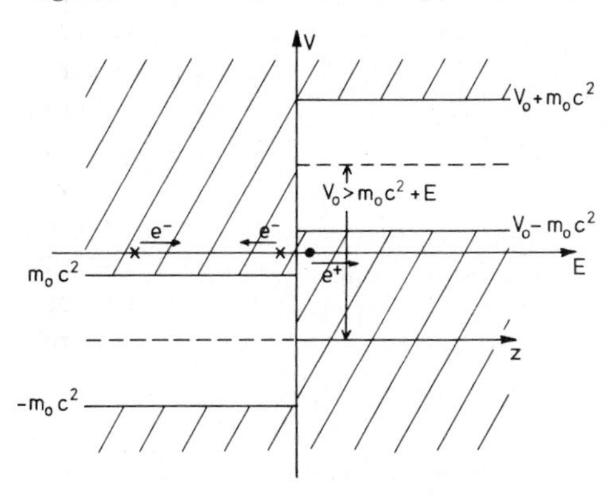

Fig. 1.14. By introducing the quantized-field aspected using hole theory, Hund explained the Klein paradox. The incoming electrons could knock out additional electrons at the barrier and positrons would move inside the potential

equation in 1928 [Kl 29]. He found that for very high potential barriers an unusually large number of electrons penetrate into the wall (about 83% for $V_0 \to \infty$). Moreover, Klein observed that these electrons had negative energy. This was later called *Klein's paradox*, Fig. 1.12. *Sauter* [Sa 31] found essentially the same results in the more general case where the barrier is not sudden, but smoothed out, Fig. 1.13. If the potential step of order m_0c^2 appears over a distance of the Compton wavelength $d \geq \lambda_e$, penetration of electrons into the classically forbidden region ceases.

In 1940 *Hund* [Hu 41, 54] introduced the quantized field aspect by discussing Klein's paradox in the framework of hole theory, Fig. 1.14. The particles of negative energy in the high potential domain are then electrons from the vacuum. In this picture there can be more reflected electrons than incoming ones, because the incoming electrons can knock out an electron-positron pair at the surface of the wall, the electron joining the reflected electrons and the positron (hole) travelling inside the high potential. Particle creation by strong, infinitely extended constant electrostatic fields (which is related to the Klein paradox) was quantitatively predicted by *Heisenberg* and *Euler* in 1936 [He 36].

Clearly, pair creation for Klein's paradox is stimulated by the incoming electron beam. This introduces some time-dependence, which is responsible for knocking out pairs from the barrier. It is similar to induced pair creation,

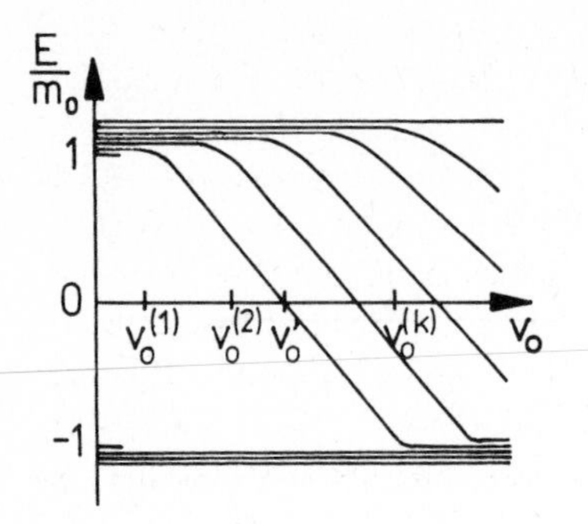

Fig. 1.15. Energy spectrum of a Dirac particle in a deep square well potential

appearing in the time-dependent heavy-ion collision (shake-off of the vacuum polarization discussed above). It may be possible to create pairs by the time-dependence of an external gauge field since a quantized field can be viewed as consisting of infinitely many harmonic oscillators ("modes"). If the quantized field is then excited from its ground state (i.e. from the vacuum state), real particles are produced. In some cases these two possibilities of pair creation by external gauge fields are connected by a gauge transformation: a globally static external gauge field may represent a dynamic field if only the gauge is suitably chosen. One simple example is the constant electrostatic field described by the static potential $A_\mu = (\boldsymbol{E} \cdot \boldsymbol{x}, \boldsymbol{O})$ and also by $A_\mu = (O, \boldsymbol{E} t)$.

Shortly before Hund's contribution in 1939, *Schiff, Snyder*, and *Weinberg*, in Oppenheimer's institute, discussed particles in a deep well [Schi 40]. They found an energy spectrum similar to the behaviour as we understand it today, shown in Fig. 1.15 for the Dirac equation.

When the depth of the square well potential exceeds $2 m_e c^2$, the energy of the lowest bound state becomes equal to $-m_e c^2$, and the bound level joins the states of the Dirac sea. They recognized that beyond this "critical" potential depth, the Dirac sea, i.e., the vacuum, contains more states than before. If these are all occupied by electrons, the ground state has non-zero charge. That these bound states "dive" into the negative energy continuum, keeping their identity to a large extent (up to a spreading) width, was not recognized.

In 1945, *Pomeranchuk* and *Smorodinsky* [Po 45], in Landau's institute, showed that a similar phenomenon occurs in the deep Coulomb potential of a superheavy nucleus with charge $Z = 175$ (at a nuclear radius of 8 fm). They accepted the interpretation given by *Schiff* et al. [Schi 40] that the ground state becomes charged, but they failed to recognize that this could occur only with the accompanying positron production. They write in their article: "If an electron happens to be in such a well (or in the field of such a charge), it will jump on this unoccupied level. The vacuum acquires a charge $-$e, and the electron, since this is the level of an unobservable vacuum, will pass into an unobservable state". This is close to ideas of a charged vacuum, but that was not clearly recognized then. The ideas were still rather confused.

The same is true for the investigation of *Werner* and *Wheeler* [We 58], who studied superheavy nuclei, more or less hypothetically from the present point of view. They also solved the Dirac equation for extended high-Z nuclei up to $Z = 170$ and found that the K-binding energy can reach $2m_ec^2$. The fact that beyond $Z = 170$ something fundamental happens was not noticed by them. The spontaneous positron emission in such a supercritical situation was first recognized in 1961 by *Voronkov* and *Koleznikov* [Vo 61]. This paper was rediscovered around 1977 by Armbruster who had been approached by *Koleznikov* at a conference in Dubna. Even though this short paper contained for the first time the correct interpretation of what happens at the critical point, it did not influence the later physical developments, because the idea of how to realize sufficiently strong fields in an atomic collision was missing and the theoretical concepts had not been further worked out. A relative step back was the paper dealing with the Klein paradox as regards atoms by *Beck* et al. [Be 63], in which they stated that if the binding energy for electrons in a potential becomes larger than $2m_ec^2$, those states "will be occupied". That they did not understand the physical implications of this phenomenon is hidden in a footnote in which they remark "that this occupation is – of course – not in the sense of a physical process, but according to some definition" they give.

1.5 The Vacuum in Modern Quantum Physics

In modern quantum field theories the vacuum necessarily has to have certain structures, most of which are not yet completely understood. Let us review here the best known cases [Le 81].

a) In some ϕ^4-*theories* the vacuum is degenerate, because the $V(\phi)$ potential looks as depicted in Fig. 1.16, and degenerate ground states are obviously possible with $\langle\phi\rangle = \pm\phi_0$.

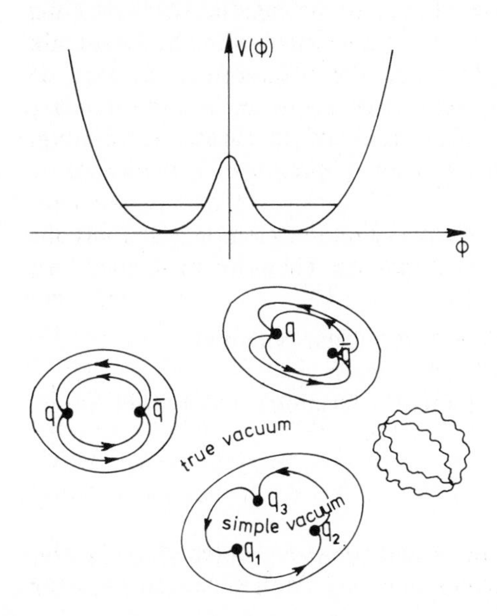

Fig. 1.16. The ground state (vacuum) in a ϕ^4 theory is degenerate

Fig. 1.17. The true vacuum of QCD is liquid-like, with virtual gluon and quark balls of a certain size forming and decaying. Real quarks can exist only within the simple vacuum; they are caught inside the bubbles

b) In gauge theories of weak and electromagnetic interactions, it must be assumed that the vacuum contains *Higgs fields* to produce masses for the particles in a gauge-invariant way. Such Higgs fields have non-vanishing expectation values of fields in the ground state, such as (a). It is difficult to judge the physical reality of these constructions, but we do not know any better way to construct gauge-invariant, renormalizable theories allowing the gauge particles to obtain mass. However, it has been suggested that the Higgs fields express dynamic symmetry breaking.

c) In quantum chromodynamics, the theory of strong interactions, two vacua have been discussed:

i) *The true vacuum* expels chromodynamic field lines similarly to a superconductor repelling magnetic field lines. Hence, an assembly of real quarks, which are the source of chromoelectric field lines, forms a bubble (bag) within the true vacuum.
ii) For a *perturbative vacuum* the true vacuum cannot exist in this bubble because of the presence of chromoelectric fields. It is likely that the bag contains particles behaving as if they were in structureless simple vacuum, subject to the laws of perturbative quantum field theory, Fig. 1.17.

d) Grand unified theories of all interactions require that a very large symmetry group is broken sequentially, leading to the richness of the physical world around us. It is generally believed that the symmetry breaking may, at least in part, be dynamic and the vacuum structure is correspondingly complex and unexplored.

The problem with the vacuum ideas described under (a) and (b) is that the structure of the vacuum is a priori predetermined by the choice of the Lagrangian. In cases (c) and (d) the contrary applies. There is probably a more fundamental Lagrangian, but we have no tools to explore the vacuum structures in a relatively easy and controllable way. But "experimenting" with the vacuum is necessary to substantiate the underlying ideas, or, put differently, to convert the models and vague ideas into physical pictures of consequence in understanding the world. This has been achieved by the description of the vacuum structure in the field theory best known today, i.e. *the vacuum of quantum electrodynamics*, which is the central issue of this book.

The intense study of the problem of electrons bound to strong potentials has stimulated interest in a number of related problems. They can be divided into three groups:

a) bosons bound to strong electromagnetic and nuclear potentials;
b) fermions and bosons in strong gravitational fields;
c) the vacuum structure of strongly interacting fermions and bosons (overcriticality in QCD).

1.5.1 Pion Condensation

Let us first discuss the theory of bosons bound to strong potentials. Clearly, some conceptual features must be different since the Pauli exclusion principle

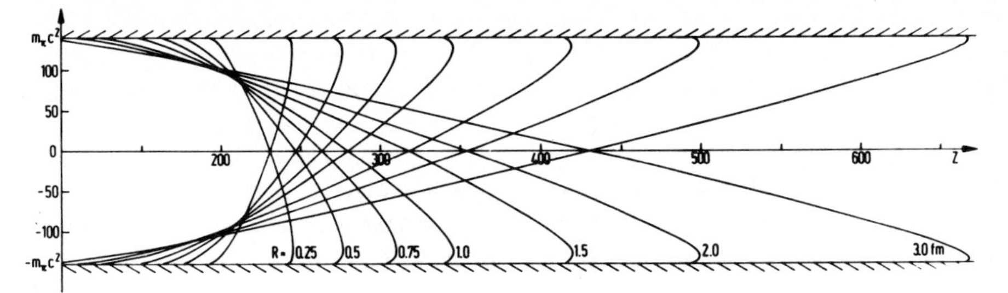

Fig. 1.18. Qualitative features of eigenvalues from the Klein-Gordon equation for a square well potential

stabilizes the charged vacuum for fermions in overcritical fields; without it continued positron production would ensue. This is the central problem when considering *bosons* in strong external potentials. The first step here is to understand the single-particle spectrum of the Klein-Gordon equation.

As already mentioned, the earliest investigation of the solutions of the Klein-Gordon equation with a strong external potential was carried out by *Schiff* et al. [Schi 40]. They solved the problem of the square well potential and found that the spectrum behaves qualitatively different from that of the Dirac equation as the potential strength is varied. They discovered that for a given state, there are two critical points; at a value V, an antiparticle state with the same quantum numbers emerges from the negative energy continuum, while at V_{cr} the particle and antiparticle states meet each other. No particle or antiparticle state with these particular quantum numbers is found above V_{cr}, Fig. 1.18.

Although this behaviour of the spectrum is suggestive for any potential, a different result was found later by *Popov* (1971) and discussed by *Bawin* and *Lavine* (1975) for long-range potentials. In particular, they found that the Coulomb potential of a finite size charge has an eigenvalue spectrum similar to that of the Dirac equation [Po 71, Ba 75].

Although the superbound pion system is theoretically a very interesting problem, there does not seem to be a prospect of experimental tests in atomic collisions. This can be seen by noting that an external potential comparable in strength to the mass of the pion $R m_\pi \alpha^{-1} \sim 2000$ is needed. Here R describes the nuclear size. This formula is valid only if $R m_\pi > 1$ and therefore does not apply to electrons bound in the nuclear Coulomb field. *Synder* and *Weinberg* successfully introduced a second quantization of the theory for potentials smaller than V_{cr} but made no attempt to treat the overcritical case [Sn 40].

The modern development of the subject which led to present understanding of the supercritical state for pions was stimulated by the work on the overcritical Dirac equation. *Migdal* [Mi 72] argued that to stabilize the vacuum in the overcritical Bose case, some further interaction effect must be included in the Hamiltonian, such as a $\lambda \phi^4$ term (see above). This is based on the consideration that as V approaches V_{cr}, the energy necessary to make a meson pair vanishes, allowing, in principle, an infinite number of pairs to be produced. To stabilize the vacuum, a further positive definite part in the Hamiltonian is needed to stop the production of the condensate when a certain meson density is reached. In this context introduction of an arbitrary interaction is, however, quite unsatisfactory as the condensate and its inherent stability must follow from a picture of dynamic symmetry breaking.

This much more difficult problem was first approached by *Klein* and *Rafelski* [Kl 75a, b, 77, Ra 78b], who showed, in particular, that it is possible to consider the charge of the condensate itself as the stabilizing agent. They demonstrated that the Coulomb repulsion, which is always present in any charge distribution of the meson condensate, suffices to stabilize the condensate. Their treatment was formally complete for both the under- and overcritical cases. A full relativistic quasi-particle formalism was developed for $V > V_{cr}$ and equations determining the Bose condensates were given. At that time it had not yet been realized that the $1s$ state of the Klein-Gordon equation joins the lower continuum for long-range potentials, much in the same way as for Dirac states [Ba 75]. Recognition of this feature of the Klein-Gordon spectrum led [Kl 75b] to the conjecture that a *charged condensate* develops in the long-range supercritical Coulomb field, which was further substantiated and the properties of the charged condensate described in [Kl 77, Ra 78b]. In accordance with charge conservation, a large number of free antimesons would be produced in this process. Furthermore, in view of the possibility of weak decay of the charged vacuum, other mechanisms leading to pion condensates in external fields have also been described.

Another form of "pion condensation" arises because nuclear matter itself may have excited states with the quantum numbers of a pion, i.e. $O^- - T = 1$ states. Now, as a function of nuclear density the effective nucleon–nucleon interaction generally changes. It turns out that such excited states (in normal nuclear matter) may become the ground state in dense nuclear matter. However, intensive searches for this effect were negative [Os 82], so that one must conclude that the critical density for this second type of pion condensation is beyond experimental reach, or as has been argued very convincingly in recent years by *Faessler*, that such an effect does not exist at all [Fä 83]. Therefore these developments are not further discussed in this volume.

1.5.2 Strong Gravitational Fields

The success of QED of strong fields raises the question whether further instabilities of the Dirac vacuum, similar to the electrostatic instability, can occur in nature. Although many instabilities of the Dirac vacuum can be visualized via energy level diagrams based on solutions of classical field equations, they should ultimately be described in the framework of quantum field theory. For this purpose one needs a manifestly covariant formulation of the quantized Dirac field in curved space-time enabling description of particle creation. Such a formalism has, for example, been suggested by *Rumpf* [Ru 76], whereby pair creation by external fields can be described by introducing four (instead of two) classes of field modes, viz. ingoing and outgoing particle and antiparticle modes. This formalism, however, is not applicable to situations where a new stable ground state of the supercritical Dirac system develops. Then one has to rely on energy level diagrams. Stabilization of the supercritical vacuum can then be described by statistical methods first suggested for the highly charged supercritical vacuum in quantum electrodynamics [Mü 75]. For example, in the framework of generalized gravitation theory with spatial torsion, the Dirac vacuum develops a non-zero expectation value of spin in strong graviational torsion fields [So 79c, 82c].

Let us sketch the general procedure for studying Dirac particles in gauge fields. One starts with the free Dirac action of spin-$\frac{1}{2}$ particles that have (internal) degrees of freedom with values on a semisimple Lie group. Background fields are introduced as gauge fields describing the parallel transport of the spinor from one space point to the next. Typical gauge groups are $U(1)$, forming the basis of QED, the Poincaré group P, underlying the Einstein-Cartan theory of gravitation, and the $SU(n)$ groups, playing such an important role in modern theories of elementary interactions (Yang-Mills theories). It turns out that the nature of this group profoundly influences the stability or instability of the Dirac vacuum in a strong, globally static gauge field.

In general, non-Abelian gauge theories such as the Einstein-Cartan theory or Yang-Mills theories introduce a complicated and rich vacuum structure into the Dirac system [So 82c]. The complications arise from the gauge freedom: gauge non-equivalent potentials may lead to identical field strengths but to different behaviour of the Dirac field. On the other hand, locally gauge-equivalent potentials may lead to different vacuum structures distinguishable by global, topological properties. For example, a uniformly accelerated observer could measure particles in the Minkowski vacuum. Here, the transformation from an observer at rest to a uniformly accelerated observer is a gauge transformation in the sense of Poincaré symmetry of the theory, but leads to a new vacuum structure, the vacuum of the accelerated observer.

The new vacuum structure of the accelerated observer means that the Fock space of those states actually measured by a physical detector depends on the coordinate system of the observer and on the topology of the submanifold of the entire space-time, which is causally connected with the observer. Thus there is the apparent paradox that the gravitating part of matter can be described by a covariant energy-momentum tensor $T_{\mu\nu}$, but that, on the other hand, the trajectory of the detector enters into the detection of field quanta, which is therefore a coordinate- (or gauge-)dependent phenomenon [Un 76]. This behaviour is particularly evident in the theory of fermions in Rindler space, as shown by *Soffel, Müller* and *Greiner* [So 80b].

It turns out that the observer moving with a constant acceleration $\mathcal{g}$ in Minkowski space experiences a thermal flux of all kinds of fermions with effective temperature $T = \mathcal{g}\hbar/2ck$ (k is Boltzmann's constant). This result, known for some time for particles with integer spin [Fu 73, Un 76], depends crucially upon the fact that the uniformly accelerated observer in Minkowski space has an event horizon: there are parts of Minkowski space-time with which the observer cannot communicate [Fu 73, Un 76]. Thus the geometric and topological structure of the submanifold naturally connected with the observer's state of motion leads to the appearance of the thermal particle spectrum.

1.5.3 Vacuum Structure of Strongly Interacting Fermions and Bosons

Consider that in a *Gedankenexperiment* the fine-structure constant is increased arbitrarily. At a certain point, if the bound positronium state reaches zero mass, an inherent instability of QED could occur [Ra 78a]. For $\alpha \to \alpha_{cr} \sim 1$ muons, the heavy partners of electrons can no longer be produced as charged bare quanta, since the spontaneous production of an e^+e^- pair would then become possible, leading to a neutralized muon state with the muon charge carried away by the electron field. At even greater values of α (probably about 1.5) one could argue that the electron field also develops an inherent instability and a complex vacuum structure develops. This has been explicitly shown in the context of the so-called Schwinger model [Schw 62]. In one-dimensional space this problem can be solved, but the instability appears already for *any* arbitrarily small coupling constant. However, as this example shows, the method developed for studying supercritical fields can be applied successfully to gauge theories, particularly to studying the structure of the QCD vacuum. Here is simultaneously a strongly coupled theory with the bose field having both attractive *and* repulsive self-interaction. The inherent instability of the perturbative vacuum of this theory has been widely discussed [Ma 77a, b, 78b, Mi 78, Ma 79a], but only recently has the new ground state based on the perturbative vacuum from gauge field condensates around a Fermi source been constructed [Mü 82, Ca 83]. This new ground state, which must be considered in a satisfactory theory, requires substantial screening of the colour charge of the source.

From this study one learns that the essential aspect of the interaction is the supercritical attractive force between the vector gluons and the (average) colour magnetic field of the condensate when both are antiparallel. This property can be further exploited to construct a new, non-perturbative global vacuum state, based on an approximate evaluation of the zero point energy of the interacting gauge fields and in particular on the associated 'effective Lagrangian'. In this context it was shown that the supercritical binding of vector gluons in a colour-magnetic field lowers the energy density of the interacting ground state below that of the perturbative state *globally* [Sa 77, Ni 78]. Using this qualitative model quantitatively, one obtains for the first time the true vacuum state which is indeed quite different from the perturbative (naive) vacuum in which (perturbative) quarks are found. An ultimate description of the true QCD vacuum state in terms of the perturbative fields, perhaps similar in idea to the BCS method, but certainly quite different in detail is still to be given. This problem is intrinsically related with the colour confinement problem of QCD. The general problems associated with the quark-gluon ground state are discussed in Chap. 21, but of course remain incomplete.

Our intention is primarily to develop a modern view of the properties of the vacuum of particles in various strong fields. In particular, we want to convey that related phenomena are connected with a change in the ground state for many different types of interaction. All these phenomena have in common that some critical strength of the interaction exists at which it becomes energetically more favourable to produce particles out of the old vacuum during the transition to a new ground state. Depending on the geometry of the fields, either one particle species is emitted to infinity while the quantum numbers of the other species are retained in the interaction region, or both 'particles' and 'antiparticles' (i.e., their charges) remain localized, providing a real vacuum polarization. In either case, the phenomenon screens the interaction strength. These considerations are particularly important for theories supercritical for any value of the coupling strength, such as QCD, for which it is therefore difficult to guess the correct ground-state wave function.

The necessary condition under which the vacuum state (phase transition) can change is that the field distinguishes the particles according to some quantum number, such as electric charge, spin, colour. This condition may be expressed more formally by saying that the interaction must dynamically break some symmetry of the non-interacting theory. Consequently, the changed ground state is characterized by the spontaneous global (over the typical domain of the interaction) breaking of the symmetry, i.e. it is charged or carries spin, colour, etc. Therefore, the underlying subject of this volume is really dynamic symmetry breaking in strong (gauge) fields. But we emphasize that this phenomenon is typical for most interactions and not characteristic for a special one. Of course, *decay* into a charged vacuum state in strong electromagnetic fields plays a distinct role as the one example amenable to laboratory experiment.

Bibliographical Notes

The fundamental ideas concerning the change of the vacuum state in strong electric fields are discussed in the following review articles: [Mü 76a, Re 77, 84, Ra 78b, Br 78].

These ideas have been extended to other gauge fields, particularly to gravitational fields, in the review [So 82c].

A report on the status of experimental investigations up to 1977 is also contained in [Re 77]. A more recent overview of the field has been given by *Greiner* in [Gr 83c].

Those interested in more details about these topics should consult *Quantum Electrodynamics of Strong Fields* (Proceedings of the Lahnstein conference 1981) [Gr 83a]. It contains a detailed account of the historical development and a survey of the status of experiments in the field up to 1981.

For more general reading [Re 76a, Fu 79, Gr 80, 82] can be recommended.

2. The Wave Equation for Spin-1/2 Particles

Following Dirac's original derivation the four-component relativistic wave equation describing spin-1/2 particles is introduced. The energy spectrum, which exhibits two continua for both negative and positive energies, as well as the corresponding free spinor solutions are discussed. Finally we study in detail the transformation properties of Dirac-spinors under the Lorentz transformations and the construction of bilinear covariants.

2.1 The Dirac Equation

In 1928, *Dirac* searched for a covariant relativistic wave equation of Schrödinger form [Di 28]

$$\mathrm{i}\hbar\frac{\partial\Psi}{\partial t}=\hat{H}\Psi\,. \tag{2.1}$$

Because of the difficulties encountered then in interpreting the Klein-Gordon equation

$$\hat{E}^2\Phi=(\hat{p}^2c^2+m_0^2c^4)\,\Phi\qquad\text{or}\qquad -\hbar^2\frac{\partial^2}{\partial t^2}\Phi=(-\hbar^2c^2\nabla^2+m_0^2c^4)\,\Phi\,, \tag{2.2}$$

which did not allow for a positive definite probability density, Dirac required that (2.1) take a positive definite probability density $\varrho\sim\Psi^*\Psi$. The charge-density interpretation for the Klein-Gordon case was not yet known and, besides, would have made little physical sense, because the π^+ and π^- mesons as charged spin-0 particles had not yet been discovered [Gr 81b, Bj 65].

Since an equation of the form (2.1) is linear in the time derivative, linearity in the space derivatives is also required (homogeneity in space-time) and hence (2.1) must more specifically be of the form

$$\mathrm{i}\hbar\frac{\partial\Psi}{\partial t}=\left[\frac{\hbar c}{\mathrm{i}}\left(\hat{\alpha}_1\frac{\partial\Psi}{\partial x^1}+\hat{\alpha}_2\frac{\partial\Psi}{\partial x^2}+\hat{\alpha}_3\frac{\partial\Psi}{\partial x^3}\right)+\hat{\beta}m_0c^2\Psi\right]\equiv\hat{H}_f\Psi\,. \tag{2.3}$$

The as yet unknown coefficients $\hat{\alpha}_i$ and $\hat{\beta}$ cannot be c numbers, because otherwise (2.2) would not be form invariant, not even against simple space rotations. Probably $\hat{\alpha}_i$ and $\hat{\beta}$ are matrices (operators), indicated by the circumflex (^) above the symbols. Therefore Ψ cannot be a scalar function, but must be a column vector

$$\Psi(\boldsymbol{x}, t) = \begin{pmatrix} \psi_1(\boldsymbol{x}, t) \\ \psi_2(\boldsymbol{x}, t) \\ \vdots \\ \psi_N(\boldsymbol{x}, t) \end{pmatrix} . \tag{2.4}$$

Let us already call it a spinor, even though the precise definition of a spinor is given below (2.90 – 96).

The positive definite probability density is then of the form

$$\varrho(\boldsymbol{x}, t) = \Psi^\dagger \Psi(\boldsymbol{x}, t) = (\psi_1^*, \psi_2^* \ldots \psi_N^*) \begin{pmatrix} \psi_1 \\ \psi_2 \\ \vdots \\ \psi_N \end{pmatrix} = \sum_{i=1}^{N} \psi_i^* \psi_i(\boldsymbol{x}, t) . \tag{2.5}$$

To prove that $\varrho(\boldsymbol{x}, t)$ of (2.5) can be interpreted as a probability density, $\varrho(\boldsymbol{x}, t)$ must be the time component of a current four vector, which obeys the continuity equation. The dimension N of the spinor is determined below. From (2.3, 4) we conclude that $\hat{\alpha}_i$ and $\hat{\beta}$ have to be $N \times N$ square matrices, so that on both sides of (2.3) a column vector with N components appears. This simply means that the Schrödinger-type equation (2.3) is, in fact, a system of N coupled linear differential equations for the spinor components $\psi_\sigma(\boldsymbol{x}, t)$, $\sigma = 1, 2 \ldots N$. Using this, (2.3) can be written as

$$\begin{aligned} \mathrm{i}\hbar \frac{\partial \psi_\sigma}{\partial t} &= \frac{\hbar c}{\mathrm{i}} \sum_{\tau=1}^{N} \left(\hat{\alpha}_1 \frac{\partial}{\partial x^1} + \hat{\alpha}_2 \frac{\partial}{\partial x^2} + \hat{\alpha}_3 \frac{\partial}{\partial x^3} \right)_{\sigma\tau} \psi_\tau + m_0 c^2 \sum_{\tau=1}^{N} (\hat{\beta})_{\sigma\tau} \psi_\tau \\ &\equiv \sum_{\tau=1}^{N} (\hat{H}_f)_{\sigma\tau} \psi_\tau , \qquad \sigma = 1, 2 \ldots N . \end{aligned} \tag{2.6}$$

Equation (2.3) is simply a short-hand notation of (2.6). To proceed, one naturally requires

a) the correct energy-momentum relation for a relativistic particle, i.e.

$$E^2 = c^2 p^2 + m_0^2 c^4 , \tag{2.7}$$

b) the Hamiltonian $\hat{H}_f$ of (2.3) should be hermitian,
c) the continuity equation for the density (2.5), and
d) Lorentz covariance (i.e. Lorentz form invariance) for (2.3), or respectively, (2.6).

To fulfil requirement (a), every component ψ_σ of the spinor Ψ has to obey the Klein-Gordon equation (2.2), i.e.

$$-\hbar^2 \frac{\partial^2 \psi_\sigma}{\partial t^2} = (-\hbar^2 c^2 \nabla^2 + m_0^2 c^4) \psi_\sigma . \tag{2.8}$$

Now, iteration of (2.3) yields

$$-\hbar^2\frac{\partial^2\Psi}{\partial t^2}=-\hbar^2c^2\sum_{i,j=1}^{3}\frac{\hat{\alpha}_i\hat{\alpha}_j+\hat{\alpha}_j\hat{\alpha}_i}{2}\frac{\partial^2\Psi}{\partial x^i\partial x^j}$$

$$+\frac{\hbar m_0c^3}{\mathrm{i}}\sum_{i=1}^{3}(\hat{\alpha}_i\hat{\beta}+\hat{\beta}\hat{\alpha}_i)\frac{\partial\Psi}{\partial x^i}+\hat{\beta}^2m_0^2c^4\Psi \tag{2.9}$$

and comparing (2.8, 9) leads directly to the following conditions for the matrices $\hat{\alpha}_i$ and $\hat{\beta}$:

$$\begin{aligned}\hat{\alpha}_i\hat{\alpha}_j+\hat{\alpha}_j\hat{\alpha}_i&=2\delta_{ij}\mathbb{1},\\ \hat{\alpha}_i\hat{\beta}+\hat{\beta}\hat{\alpha}_i&=0\\ \hat{\alpha}_i^2&=\hat{\beta}^2=\mathbb{1}\,.\end{aligned}\tag{2.10}$$

These anticommutation relations define an algebra for the four matrices. In order to fulfil condition (b) these matrices have to Hermitian:

$$\begin{aligned}\hat{\alpha}_i^\dagger&=\hat{\alpha}_i\,,\\ \hat{\beta}^\dagger&=\hat{\beta}\,.\end{aligned}\tag{2.11}$$

Hence the eigenvalues of the matrices $\hat{\alpha}_i$ and $\hat{\beta}$ are real, and since $\hat{\alpha}_i^2=\mathbb{1}$ and $\hat{\beta}^2=\mathbb{1}$ we conclude that the eigenvalues can only be ± 1. This is most easily seen in the eigenrepresentation of the matrices, where the matrices are diagonal. For example, $\hat{\alpha}_i$ has then the form

$$\hat{\alpha}_i=\begin{pmatrix}A_1^i & 0 & \dots & 0\\ 0 & A_2^i & & 0\\ \vdots & & & 0\\ 0 & 0 & \dots & 0\,A_N^i\end{pmatrix},$$

where A_ν^i $(\nu=1,2\dots N)$ are the eigenvalues of $\hat{\alpha}_i$. Because of

$$\hat{\alpha}_i^2=\mathbb{1}=\begin{pmatrix}1 & 0 & 0\dots & 0\\ 0 & 1 & 0 & \vdots\\ \vdots & 0 & 1 & 0\\ 0 & \dots & \dots 0 & 1\end{pmatrix}=\begin{pmatrix}(A_1^i)^2 & 0 & \dots & 0\\ 0 & (A_2^i)^2 & & \vdots\\ \vdots & & & 0\\ 0 & \dots & \dots & 0\,(A_N^i)^2\end{pmatrix},$$

then

$$(A_\nu^i)^2=1\;\Leftrightarrow\;A_1^i=\pm 1\,. \tag{2.12}$$

Furthermore, the anticommutation relations of (2.10)

$$\hat{\alpha}_i=-\hat{\beta}\hat{\alpha}_i\hat{\beta}$$

indicate that the trace of the matrices $\hat{\alpha}_i$ and $\hat{\beta}$ must vanish. Since $\mathrm{tr}\,\hat{A}\hat{B}=\mathrm{tr}\,\hat{B}\hat{A}$

$$\mathrm{tr}\,\hat{\alpha}_i=+\mathrm{tr}\,\hat{\beta}^2\hat{\alpha}_i=+\mathrm{tr}\,\hat{\beta}\hat{\alpha}_i\hat{\beta}=-\mathrm{tr}\,\hat{\alpha}_i$$

and hence $\mathrm{tr}\,\hat{\alpha}_i = 0$. Similarly, $\mathrm{tr}\,\hat{\beta} = 0$. Now, since the trace of a matrix equals the sum of its eigenvalues and because of (2.12), each of the matrices $\hat{\alpha}_i$ and $\hat{\beta}$ has to have an equal number of positive $(+1)$ and negative (-1) eigenvalues and the dimension of these matrices has to be even. The smallest even dimension $N=2$ must be excluded, because in two dimensions only 3 anticommuting matrices exist, namely the 3 Pauli matrices $\hat{\sigma}_i$. We therefore try $N=4$ and easily find an explicit representation for the $\hat{\alpha}_i$ and $\hat{\beta}$ matrices, obeying all conditions (2.10):

$$\hat{\alpha}_i = \begin{pmatrix} 0 & \hat{\sigma}_i \\ \hat{\sigma}_i & 0 \end{pmatrix}, \quad \hat{\beta} = \begin{pmatrix} \mathbb{1} & 0 \\ 0 & -\mathbb{1} \end{pmatrix}, \tag{2.13}$$

where the three well-known Pauli matrices $\hat{\sigma}_i\,(i=1,2,3)$ and the 2×2 unit matrix $\mathbb{1}$ are

$$\hat{\sigma}_1 = \begin{pmatrix} 0 & 1 \\ 1 & 0 \end{pmatrix}, \quad \hat{\sigma}_2 = \begin{pmatrix} 0 & -i \\ i & 0 \end{pmatrix}, \quad \hat{\sigma}_3 = \begin{pmatrix} 1 & 0 \\ 0 & -1 \end{pmatrix}, \quad \mathbb{1} = \begin{pmatrix} 1 & 0 \\ 0 & 1 \end{pmatrix} \tag{2.14}$$

and obey the anticommutation relations [Bj 64, Gr 81b]:

$$\hat{\sigma}_i\hat{\sigma}_j + \hat{\sigma}_j\hat{\sigma}_i = 2\delta_{ij}\mathbb{1}\,. \tag{2.15}$$

Next we construct the current density and verify its continuity equation. Therefore (2.3) is multiplied from the left by the adjoint spinor $\Psi^\dagger = (\psi_1^*, \psi_2^*, \psi_3^*, \psi_4^*)$, which gives

$$\mathrm{i}\hbar\,\Psi^\dagger\frac{\partial}{\partial t}\Psi = \frac{\hbar c}{\mathrm{i}}\sum_{k=1}^{3}\Psi^\dagger\hat{\alpha}_k\frac{\partial}{\partial x^k}\Psi + m_0c^2\,\Psi^\dagger\hat{\beta}\Psi\,. \tag{2.16}$$

The Hermitian conjugate of (2.3) is

$$-\mathrm{i}\hbar\frac{\partial}{\partial t}\Psi^\dagger = -\frac{\hbar c}{\mathrm{i}}\sum_{k=1}^{3}\frac{\partial\Psi^\dagger}{\partial x^k}\hat{\alpha}_k^\dagger + m_0c^2\,\Psi^\dagger\hat{\beta}^\dagger\,. \tag{2.17}$$

Multiplied from the right by Ψ, it gives

$$-\mathrm{i}\hbar\frac{\partial\Psi^\dagger}{\partial t}\Psi = -\frac{\hbar c}{\mathrm{i}}\sum_{k=1}^{3}\frac{\partial\Psi^\dagger}{\partial x^k}\hat{\alpha}_k\Psi + m_0c^2\,\Psi^\dagger\hat{\beta}\Psi\,, \tag{2.18}$$

where the Hermiticity of the Dirac matrices ($\hat{\alpha}_k^\dagger = \hat{\alpha}_k$, $\hat{\beta}^\dagger = \hat{\beta}$) has been used. Subtracting (2.18) from (2.16) then yields

$$+\mathrm{i}\hbar\frac{\partial}{\partial t}(\Psi^\dagger\Psi) = \frac{\hbar c}{\mathrm{i}}\sum_{k=1}^{3}\frac{\partial}{\partial x^k}(\Psi^\dagger\hat{\alpha}_k\Psi) \tag{2.19}$$

or

$$\frac{\partial\varrho}{\partial t} + \mathrm{div}\,\boldsymbol{j} = 0 \tag{2.20}$$

if we identify

$$\varrho = \Psi^\dagger \Psi \quad \text{and} \quad j^k = c\Psi^\dagger \hat{\alpha}_k \Psi \tag{2.21}$$

as the *probability density* and *current density* respectively. The former is positive definite, as required. From the continuity equation (2.20) follows immediately

$$\frac{\partial}{\partial t}\int_V d^3x\, \Psi^\dagger \Psi = -\int_V \operatorname{div} \boldsymbol{j}\, d^3x = -\int_F \boldsymbol{j} \cdot d\boldsymbol{F} \Rightarrow 0\,, \tag{2.22}$$

where F is the surface of the volume V. The last step in this equation is valid for infinite volumes with the surface at infinity where the wave functions $\Psi(\boldsymbol{r}\to\infty)$ vanish. Hence (2.22) expresses the conservation of total probability with time.

We have tacitly assumed that $\boldsymbol{j} = \{j^1, j^2, j^3\}$ is a three-vector, i.e. transforms under three dimensional rotations like a vector in 3-space. This, however, must still be proven. One also expects that

$$j^\mu = \{c\varrho, \boldsymbol{j}\} = \{c\varrho, j^1, j^2, j^3\} \tag{2.23}$$

forms a four-vector, whose components transform via a Lorentz transformation from one inertial frame to another. This, as well as the form invariance of the Dirac equation (which is called *covariance*) under Lorentz transformations must still be demonstrated before we can accept the Dirac equation as a proper relativistic wave equation, Sect. 2.5.

Note that (2.13) gives a special representation of the Dirac matrices and hence of the Dirac equation. The choice of the matrices $\hat{\alpha}_k$, $\hat{\beta}$ of (2.13) is not unique, however. This can be seen by realizing that any unitary transformation $\hat{S}$ leads to new matrices

$$\hat{\alpha}_i' = \hat{S}\hat{\alpha}_i\hat{S}^{-1}\,, \quad \hat{\beta}' = \hat{S}\hat{\beta}\hat{S}^{-1}\,, \tag{2.24}$$

which also fulfil (2.10). As an example we calculate the first of the anticommutators (2.10) and find

$$\begin{aligned}
\hat{\alpha}_i'\hat{\alpha}_j' + \hat{\alpha}_j'\hat{\alpha}_i' &= \hat{S}\hat{\alpha}_i\hat{S}^{-1}\hat{S}\hat{\alpha}_j\hat{S}^{-1} + \hat{S}\hat{\alpha}_j\hat{S}^{-1}\hat{S}\hat{\alpha}_i S^{-1} \\
&= \hat{S}(\hat{\alpha}_i\hat{\alpha}_j + \hat{\alpha}_j\hat{\alpha}_i)\hat{S}^{-1} = \hat{S}(2\delta_{ij}\mathbb{1})\hat{S}^{-1} \\
&= 2\delta_{ij}\mathbb{1}\,.
\end{aligned}$$

2.2 The Free Dirac Particle

To get a preliminary understanding of the physical content of the Dirac equation, let us investigate the solutions of the free equation (2.3), i.e. of the equation without potentials. We rewrite it in the form

$$\mathrm{i}\hbar \frac{\partial \Psi}{\partial t} = (c\hat{\alpha} \cdot \hat{p} + m_0 c^2 \hat{\beta})\, \Psi \equiv \hat{H}_f \Psi \,, \tag{2.25}$$

where $\hat{p} = \{\hat{p}^1, \hat{p}^2, \hat{p}^3\} = -\mathrm{i}\hbar\{(\partial/\partial x^1), (\partial/\partial x^2), (\partial/\partial x^3)\}$ is the momentum operator. Stationary solutions of (2.25) are found with the ansatz

$$\Psi(\boldsymbol{x}, t) = \Psi(\boldsymbol{x}) \exp\left(-\frac{\mathrm{i}}{\hbar}\varepsilon t\right) \tag{2.26}$$

with which (2.25) becomes

$$\varepsilon\, \Psi(\boldsymbol{x}) = \hat{H}_\mathrm{f} \Psi_\mathrm{f} \,. \tag{2.27}$$

The quantity ε characterizes the time development of the stationary wave function $\Psi(\boldsymbol{x}, t)$. For the Dirac equation it is identical with the energy [Gr 81b].

For many applications it is useful to split up the four-spinor (2.4) into two two-spinors φ and χ according to

$$\Psi = \begin{pmatrix} \psi_1 \\ \psi_2 \\ \psi_3 \\ \psi_4 \end{pmatrix} = \begin{pmatrix} \varphi \\ \chi \end{pmatrix} . \tag{2.28}$$

Using now the explicit form (2.13) for the Dirac matrices, (2.27) becomes

$$\varepsilon \begin{pmatrix} \varphi \\ \chi \end{pmatrix} = c \begin{pmatrix} 0 & \hat{\sigma} \\ \hat{\sigma} & 0 \end{pmatrix} \cdot \hat{p} \begin{pmatrix} \varphi \\ \chi \end{pmatrix} + m_0 c^2 \begin{pmatrix} \mathbb{1} & 0 \\ 0 & -\mathbb{1} \end{pmatrix} \begin{pmatrix} \varphi \\ \chi \end{pmatrix} \tag{2.29}$$

or

$$\begin{aligned} \varepsilon \varphi &= c\hat{\sigma} \cdot \hat{p}\, \chi + m_0 c^2 \mathbb{1}\, \varphi \,, \\ \varepsilon \chi &= c\hat{\sigma} \cdot \hat{p}\, \varphi - m_0 c^2 \mathbb{1}\, \chi \,. \end{aligned} \tag{2.30}$$

Since $[\hat{p}, H_\mathrm{f}]_- = 0$, states with well-defined linear momentum may arise

$$\begin{pmatrix} \varphi \\ \chi \end{pmatrix} = \begin{pmatrix} \varphi_0 \\ \chi_0 \end{pmatrix} \exp\left(\frac{\mathrm{i}}{\hbar} \boldsymbol{p} \cdot \boldsymbol{x}\right) , \tag{2.31}$$

which, after insertion into (2.30), yield the same equations as (2.30) for (φ_0/χ_0) after replacing the operator $\hat{p}$ by the momentum $\boldsymbol{p}$ (which is a c number). Ordering the resulting equations yields

$$\begin{aligned} (\varepsilon - m_0 c^2)\, \mathbb{1}\, \varphi_0 - c\hat{\sigma} \cdot \boldsymbol{p}\, \chi_0 &= 0 \\ -c\hat{\sigma} \cdot \boldsymbol{p}\, \varphi_0 + (\varepsilon + m_0 c^2)\, \mathbb{1}\, \chi_0 &= 0 \,, \end{aligned} \tag{2.32}$$

and leads to non-trivial solutions only if the determinant

$$\begin{vmatrix} (\varepsilon - m_0c^2)\mathbb{1} & -c\hat{\sigma}\cdot p \\ -c\hat{\sigma}\cdot p & (\varepsilon + m_0c^2)\mathbb{1} \end{vmatrix} = 0 \tag{2.33}$$

vanishes. Using now the well-known relation for Pauli matrices [Gr 81b]

$$(\hat{\sigma}\cdot A)(\hat{\sigma}\cdot B) = A\cdot B\,\mathbb{1} + \mathrm{i}\hat{\sigma}\cdot(A\times B) \tag{2.34}$$

gives

$$\begin{aligned} &\varepsilon^2 = m_0^2c^4 + c^2p^2\,, \quad \text{or} \\ &\varepsilon = \pm E_\mathrm{p} = \lambda E_\mathrm{p} \quad \text{with} \\ &E_\mathrm{p} = +\sqrt{c^2p^2 + m_0^2c^4}\,, \quad \lambda = \pm 1\,. \end{aligned} \tag{2.35}$$

The two signs $\lambda = \pm 1$ correspond to *positive and negative energy solutions* for the Dirac equation. For given $\varepsilon = \lambda E_\mathrm{p}$ one obtains from (2.32)

$$\chi_0 = \frac{c(\hat{\sigma}\cdot p)}{m_0c^2 + \varepsilon}\varphi_0\,. \tag{2.36}$$

Thus χ_0 is determined, if φ_0 is known. We may choose most generally

$$\varphi_0 = \begin{pmatrix} u_1 \\ u_2 \end{pmatrix} \equiv u \tag{2.37}$$

with the normalization

$$\varphi_0^\dagger\varphi_0 = u^\dagger u = u_1^* u_1 + u_2^* u_2 = 1\,.$$

Here u_1, u_2 are c numbers. The complete solutions of the free Dirac equation, which are also eigenstates of the momentum operator $\hat{p}$, now read

$$\Psi_{p\lambda}(x,t) = N_\lambda \begin{pmatrix} u \\ \dfrac{c\hat{\sigma}\cdot p}{m_0c^2 + \lambda E_\mathrm{p}}u \end{pmatrix} \frac{\exp\left[\dfrac{\mathrm{i}}{\hbar}(p^2\cdot x - \lambda E_\mathrm{p}t)\right]}{\sqrt{2\pi\hbar}^{\,3}}\,. \tag{2.38}$$

Here $\lambda = \pm 1$ characterizes the positive and negative energies, respectively ($\varepsilon = \lambda E_\mathrm{p}$), and N_λ is a normalization factor, which is readily determined from

$$\int \Psi_{p\lambda}^\dagger(x,t)\,\Psi_{p'\lambda'}(x,t)\,d^3x = \delta_{\lambda\lambda'}\,\delta(p - p') \tag{2.39}$$

to be

$$N_\lambda = \sqrt{\frac{m_0c^2 + \varepsilon}{2\varepsilon}} = \sqrt{\frac{m_0c^2 + \lambda E_\mathrm{p}}{2\lambda E_\mathrm{p}}}\,. \tag{2.40}$$

Note that the radicand is always positive. The energy spectrum corresponding to the solutions $\Psi_{p\lambda}(x,t)$ is given by

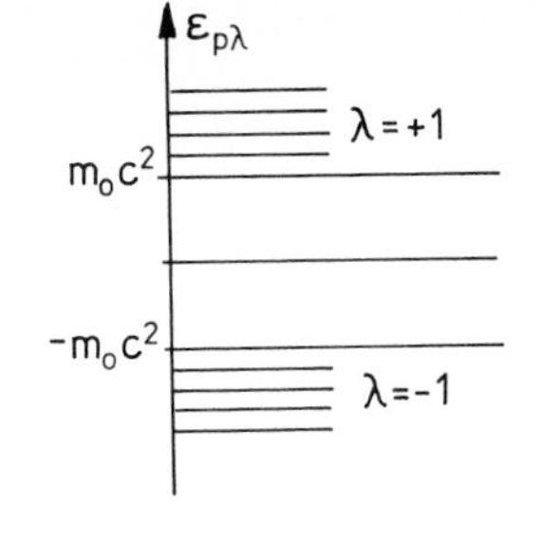

Fig. 2.1. Energy spectrum of the free Dirac equation

$$\varepsilon_{p\lambda} = \lambda E_p = \lambda\sqrt{m_0^2c^4 + c^2\boldsymbol{p}^2} \tag{2.41}$$

and is depicted in Fig. 2.1.

It shows a gap of $2m_0c^2$ separating the positive energy continuum ($\varepsilon_{p1} = +\sqrt{m_0^2c^4+c^2\boldsymbol{p}^2}$) from the negative energy continuum ($\varepsilon_{p,-1} = -\sqrt{m_0^2c^4+c^2\boldsymbol{p}^2}$). Interpretation of the states with $\lambda = -1$ is considered in Sect. 2.3. Here we note that all states (2.38) are obviously also eigenfunctions of the linear momentum operator $\hat{\boldsymbol{p}}$:

$$\hat{\boldsymbol{p}}\,\Psi_{p\lambda} = \boldsymbol{p}\,\Psi_{p\lambda}\,. \tag{2.42}$$

For each momentum $\boldsymbol{p}$ two kinds of states exist, namely those with positive energy ($\lambda = +1$) and those with negative energy ($\lambda = -1$), both having the same absolute energy $|\varepsilon_{p\lambda=1}| = |\varepsilon_{p\lambda=-1}| = E_p$.

We shall now show that still another quantum number, the *helicity*, exists, which helps classify the free solutions completely. First note that the operator

$$\hat{\boldsymbol{\Sigma}}\cdot\hat{\boldsymbol{p}} = \begin{pmatrix} \hat{\boldsymbol{\sigma}} & 0 \\ 0 & \hat{\boldsymbol{\sigma}} \end{pmatrix}\cdot\hat{\boldsymbol{p}} \tag{2.43}$$

commutes with the free Dirac Hamiltonian $\hat{H}_f$ [cf. (2.25)] and with the momentum operator

$$\begin{aligned} [\hat{H}_f, \hat{\boldsymbol{\Sigma}}\cdot\hat{\boldsymbol{p}}]_- &= 0\,, \\ [\hat{\boldsymbol{p}}, \hat{\boldsymbol{\Sigma}}\cdot\hat{\boldsymbol{p}}]_- &= 0\,. \end{aligned} \tag{2.44}$$

Now

$$\hat{\boldsymbol{S}} = \frac{\hbar}{2}\hat{\boldsymbol{\Sigma}} = \frac{\hbar}{2}\begin{pmatrix} \hat{\boldsymbol{\sigma}} & 0 \\ 0 & \hat{\boldsymbol{\sigma}} \end{pmatrix} \tag{2.45}$$

can be considered as the four-dimensional generalization of the spin operator $\hat{\boldsymbol{S}} = \hbar\hat{\boldsymbol{\sigma}}/2$ of the Pauli theory. The helicity operator $\hat{\Lambda}_s$ is now defined as the projection of $\hat{\boldsymbol{S}}$ along the momentum $\hat{\boldsymbol{p}}$ (Fig. 2.2), i.e.

$$\hat{\Lambda}_s = \frac{\hbar}{2}\hat{\boldsymbol{\Sigma}}\cdot\frac{\hat{\boldsymbol{p}}}{|\hat{\boldsymbol{p}}|} = \hat{\boldsymbol{S}}\cdot\frac{\hat{\boldsymbol{p}}}{|\boldsymbol{p}|}\,. \tag{2.46}$$

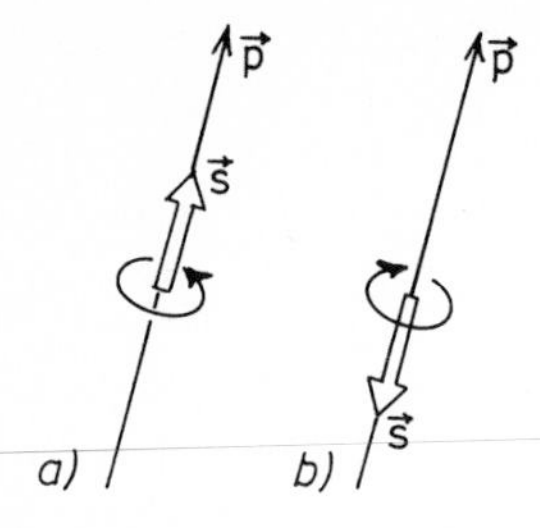

Fig. 2.2a, b. Electrons with **(a)** positive and **(b)** negative helicity. The double arrow symbolizes spin, and ↷ rotation of the spinning electron

With (2.44) it is easily verified that $\hat{\Lambda}_s$ also commutes with $\hat{H}_{\mathrm{f}}$ and $\hat{\boldsymbol{p}}$:

$$[\hat{\Lambda}_s, \hat{H}_{\mathrm{f}}]_- = 0\,, \qquad [\hat{\Lambda}_s, \hat{\boldsymbol{p}}]_- = 0\,. \tag{2.47}$$

Without loss of generality we can assume that the spinor wave moves in the z direction. Then

$$\boldsymbol{p} = \{0, 0, p\} \quad \text{and}$$

$$\hat{\Lambda}_s = \hat{S}_z = \frac{\hbar}{2}\hat{\Sigma}_z = \frac{\hbar}{2}\begin{pmatrix} \hat{\sigma}_z & 0 \\ 0 & \hat{\sigma}_z \end{pmatrix} = \frac{\hbar}{2}\begin{pmatrix} 1 & 0 & 0 & 0 \\ 0 & -1 & 0 & 0 \\ 0 & 0 & 1 & 0 \\ 0 & 0 & 0 & -1 \end{pmatrix} \tag{2.48}$$

with the eigenvalues $\pm\hbar/2$. The eigenvectors of $\hat{\Lambda}_s$ are immediately recognized to be

$$\begin{pmatrix} u_1 \\ 0 \end{pmatrix}, \begin{pmatrix} u_{-1} \\ 0 \end{pmatrix}, \begin{pmatrix} 0 \\ u_1 \end{pmatrix}, \begin{pmatrix} 0 \\ u_{-1} \end{pmatrix}, \tag{2.49}$$

where $u_1 = \begin{pmatrix} 1 \\ 0 \end{pmatrix}$, $u_{-1} = \begin{pmatrix} 0 \\ 1 \end{pmatrix}$ are the standard eigenvectors of $\hat{\sigma}_z$. The classification of the free states (2.38) of the Dirac equation can be completed by noting explicitly that

$$\Psi_{p_z, \lambda, \sigma} = N_\lambda \begin{pmatrix} u_\sigma \\ \dfrac{c\hat{\sigma}_z p}{m_0c^2 + \lambda E_{\mathrm{p}}} u_\sigma \end{pmatrix} \exp\left[\frac{\mathrm{i}}{\hbar}(p_z z - \lambda E_{\mathrm{p}} t)\right], \tag{2.50}$$

$$\lambda = \pm 1\,, \qquad \sigma = \pm 1\,,$$

and that states with $\sigma = \pm 1$ have positive and negative helicity, respectively. Obviously

$$\langle \Psi_{p_z \lambda \sigma} | \Psi_{p'_z \lambda' \sigma'} \rangle = \delta_{\lambda\lambda'} \delta_{\sigma\sigma'} \delta(p_z - p'_z) . \tag{2.51}$$

2.3 Single-Particle Interpretation of Plane (Free) Dirac Waves

Here we shall briefly discuss the single-particle interpretation of the Dirac equation and its solutions. It is already evident from the occurrence of negative energy solutions that a single-particle interpretation in its full significance is not possible.

Let us first note that the time-development factor $\varepsilon = \lambda E_p$ of (2.35) must be interpreted as energy. This can be proven by establishing the Lagrangian density

$$\mathscr{L} = \mathrm{i}\hbar\, \Psi^\dagger \frac{\partial}{\partial t} \Psi + \mathrm{i}\hbar c\, \Psi^\dagger \nabla \cdot \hat{\alpha} \Psi - m_0 c^2 \Psi^\dagger \hat{\beta} \Psi , \tag{2.52}$$

from which the Dirac equation (2.25) can be derived in the standard way by applying the Euler-Lagrange field equations. From (2.52) also the energy-momentum tensor $T_{\mu\nu}$ and in particular its T_{00} component and hence the energy of the system can be calculated. One finds [Gr 81b], see also Sect. 8.1,

$$E = \int T_{00}(\Psi_{p,\lambda,\sigma})\, d^3x = \lambda E_p = \varepsilon . \tag{2.53}$$

This clearly indicates that the negative time-development factor $\varepsilon = -E_p$ cannot be reinterpreted as positive energy (as for the Klein-Gordon equation) but must be negative energy. A single-particle theory cannot explain this observation at all. Dirac's hole theory solves this dilemma. Assume that real electrons are described only by the positive energy states of (2.38), i.e. by the states $\Psi_{p,\lambda=+1,\sigma}$.

The states with negative energy shall all be occupied by electrons, Fig. 2.3. Because of the Pauli principle this also prevents an electron in a positive energy state from losing energy by some mechanism (e.g. radiation) and so dropping deeper and deeper into negative energy states.

Naturally the question arises about the physical meaning of this *Dirac sea.* It is shown below (Chap. 4) that a hole in this sea represents a positron (antiparticle of an electron) and the sea itself characterizes the vacuum. It is, however,

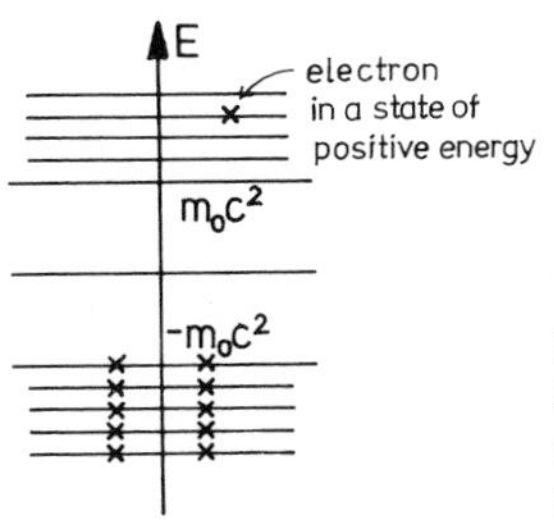

Fig. 2.3. The states with negative energy ($E < -m_0c^2$) are occupied by electrons and form the so-called Dirac sea, representing the vacuum, because observable electrons occur generally only with positive energy

already clear that the interpretation of the negative energy solutions of the Dirac equation leads us away from the single-particle to the many-body interpretation. Figure 2.3 demonstrates this aspect vividly. Prediction of antiparticles has been one of the outstanding triumphs of relativistic quantum theory.

2.4 The Dirac Particle Coupled to Electromagnetic Fields – Non-Relativistic Limits and Spin of the Dirac Equation

The electromagnetic fields are described by the four-potential

$$A^{\mu}(x) = \{A_0(x), \boldsymbol{A}(x)\}\,, \tag{2.54}$$

which is coupled minimally to the particle through the substitution

$$\hat{p}_{\mu} \rightarrow \hat{p}_{\mu} - \frac{e}{c} A_{\mu} \equiv \hat{\pi}_{\mu} \tag{2.55}$$

in order to ensure gauge invariance. Hence the Dirac equation with electromagnetic potentials becomes

$$c\left(\mathrm{i}\hbar \frac{\partial}{\partial ct} - \frac{e}{c} A_0(x)\right) \Psi(x) = \left[c\hat{\boldsymbol{\alpha}} \cdot \left(\hat{\boldsymbol{p}} - \frac{e}{c}\boldsymbol{A}(x)\right) + \hat{\beta} m_0 c^2\right] \Psi(x) \tag{2.56}$$

or

$$\begin{aligned} \mathrm{i}\hbar \frac{\partial \Psi}{\partial t} &= \left[c\hat{\boldsymbol{\alpha}} \cdot \left(\hat{\boldsymbol{p}} - \frac{e}{c}\boldsymbol{A}(x)\right) + eA_0(x) + \hat{\beta} m_0 c^2\right] \Psi(x) \\ &\equiv (\hat{H}_{\mathrm{f}} + \hat{H}_{\mathrm{int}})\, \Psi(x)\,. \end{aligned}$$

The particle and field interaction is contained in the interaction Hamiltonian

$$\hat{H}_{\mathrm{int}} = -\frac{e}{c} c\hat{\boldsymbol{\alpha}} \cdot \boldsymbol{A}(x) + eA_0(x) = -\frac{e}{c} \boldsymbol{v} \cdot \boldsymbol{A}(x) + eA_0(x)\,, \tag{2.57}$$

where

$$\begin{aligned} \boldsymbol{v} &= \frac{d\boldsymbol{x}}{dt} = \frac{1}{\mathrm{i}\hbar}[\boldsymbol{x}, \hat{H}]_- = \frac{1}{\mathrm{i}\hbar}[\boldsymbol{x}, \hat{H}_{\mathrm{f}}]_- \\ &= \frac{1}{\mathrm{i}\hbar}[\boldsymbol{x}, c\hat{\boldsymbol{\alpha}} \cdot \hat{\boldsymbol{p}} + \hat{\beta} m_0 c^2]_- = \frac{c}{\mathrm{i}\hbar}[\boldsymbol{x}, \hat{\boldsymbol{\alpha}} \cdot \hat{\boldsymbol{p}}]_- \\ &= c\hat{\boldsymbol{\alpha}} \end{aligned} \tag{2.58}$$

is the relativistic velocity operator. Expression (2.57) is identical with the classical interaction of a point particle with electromagnetic fields.

Let us now study the non-relativistic limit of (2.56) by again decomposing the four-spinor Ψ into the two spinors $\tilde{\varphi}$ and $\tilde{\chi}$ according to

$$\Psi = \begin{pmatrix} \tilde{\varphi} \\ \tilde{\chi} \end{pmatrix}, \tag{2.59}$$

and rewriting (2.56) as

$$\mathrm{i}\hbar \frac{\partial}{\partial t} \begin{pmatrix} \tilde{\varphi} \\ \tilde{\chi} \end{pmatrix} = c\hat{\boldsymbol{\sigma}} \cdot \hat{\boldsymbol{\pi}} \begin{pmatrix} \tilde{\chi} \\ \tilde{\varphi} \end{pmatrix} + eA_0(x) \begin{pmatrix} \tilde{\varphi} \\ \tilde{\chi} \end{pmatrix} + m_0 c^2 \begin{pmatrix} \tilde{\varphi} \\ -\tilde{\chi} \end{pmatrix}. \tag{2.60}$$

With the ansatz

$$\begin{pmatrix} \tilde{\varphi} \\ \tilde{\chi} \end{pmatrix} = \begin{pmatrix} \varphi \\ \chi \end{pmatrix} \exp\left(-\mathrm{i} \frac{m_0 c^2}{\hbar} t \right)$$

the rest energy is split off and (2.60) becomes

$$\mathrm{i}\hbar \frac{\partial}{\partial t} \begin{pmatrix} \varphi \\ \chi \end{pmatrix} = c\hat{\boldsymbol{\sigma}} \cdot \hat{\boldsymbol{\pi}} \begin{pmatrix} \chi \\ \varphi \end{pmatrix} + eA_0 \begin{pmatrix} \varphi \\ \chi \end{pmatrix} - 2m_0 c^2 \begin{pmatrix} 0 \\ \chi \end{pmatrix}. \tag{2.61}$$

Now consider the lower component of this equation first and assume that the kinetic energy is much smaller than the rest energy, i.e. $|\mathrm{i}\hbar\, \partial\chi/\partial t| \ll |m_0 c^2 \chi|$, and also that the Coulomb energy is much smaller than the rest energy: $|eA_0(x)\,\chi| \ll |m_0 c^2 \chi|$. Then

$$\chi \approx \frac{\hat{\boldsymbol{\sigma}} \cdot \hat{\boldsymbol{\pi}}}{2m_0 c} \varphi, \tag{2.62}$$

indicating that in the non-relativistic limit the χ component of the four-spinor is much smaller than the φ component:

$$|\chi| \sim \frac{1}{2} \frac{v}{c} |\varphi|.$$

Inserting (2.62) into the equation for the upper component of (2.61) gives

$$\mathrm{i}\hbar \frac{\partial \varphi(x)}{\partial t} = \frac{\hat{\boldsymbol{\sigma}} \cdot \hat{\boldsymbol{\pi}}\ \hat{\boldsymbol{\sigma}} \cdot \hat{\boldsymbol{\pi}}}{2m_0} \varphi(x) + eA_0(x)\,\varphi(x). \tag{2.63}$$

With (2.34) the first term on the right-hand side is readily simplified:

$$\begin{aligned}
\hat{\boldsymbol{\sigma}} \cdot \hat{\boldsymbol{\pi}}\ \hat{\boldsymbol{\sigma}} \cdot \hat{\boldsymbol{\pi}} &= \hat{\boldsymbol{\pi}}^2 + \mathrm{i}\hat{\boldsymbol{\sigma}} \cdot (\hat{\boldsymbol{\pi}} \times \hat{\boldsymbol{\pi}}) \\
&= \left(\hat{\boldsymbol{p}} - \frac{e}{c} \boldsymbol{A}(x) \right)^2 + \mathrm{i}\hat{\boldsymbol{\sigma}} \cdot \left[\left(-\mathrm{i}\hbar\nabla - \frac{e}{c}\boldsymbol{A}(x) \right) \times \left(-\mathrm{i}\hbar\nabla - \frac{e}{c}\boldsymbol{A}(x) \right) \right] \\
&= \left(\hat{\boldsymbol{p}} - \frac{e}{c} \boldsymbol{A}(x) \right)^2 - \frac{e}{c} \hbar \hat{\boldsymbol{\sigma}} \cdot (\nabla \times \boldsymbol{A}) \\
&= \left(\hat{\boldsymbol{p}} - \frac{e}{c} \boldsymbol{A}(x) \right)^2 - \frac{e\hbar}{c} \hat{\boldsymbol{\sigma}} \cdot \boldsymbol{B}.
\end{aligned}$$

Thus (2.63) becomes

$$\mathrm{i}\hbar\frac{\partial\varphi(x)}{\partial t}=\left[\frac{\left(\hat{p}-\frac{e}{c}A\right)^2}{2m_0}-\frac{e\hbar}{2m_0c}\hat{\sigma}\cdot B+eA_0(x)\right]\varphi(x)\,. \tag{2.64}$$

This is the well-known *Pauli equation* [Gr 81b]. The two components of φ describe the two spin degrees of freedom. From non-relativistic quantum mechanics it is known that the Pauli equation describes particles with spin 1/2. Hence we can conclude that also *the Dirac equation – as the relativistic generalization of the Pauli equation – has to describe spin-1/2 particles.* This is especially so, since the spin, as an intrinsic property, should be present independently whether the particle moves with relativistic or non-relativistic velocities.

2.5 Lorentz Covariance of the Dirac Equation

A proper relativistic theory must be Lorentz covariant, i.e. invariant in form when changing from one inertial frame to another. Lorentz transformations are briefly recapitulated, and for a deeper group theoretical discussion [Bj 64, Gr 81b] are recommended.

Two observers A and B sitting in different inertial frames describe the same physical event in their respective space-time coordinates. The coordinates observer A uses to describe the event are x^μ, those of observer B for the same event are $x^{\mu'}$. Both coordinates are connected by a Lorentz transformation

$$(x^\nu)'=\sum_{\mu=0}^{3}a^\nu_\mu x^\mu\equiv a^\nu_\mu x^\mu\,,\qquad x^\mu=\{x^0,x^1,x^2,x^3\}=\{ct,x,y,z\}\,, \tag{2.65}$$
$$x_\mu=\{x_0,x_1,x_2,x_3\}=\{ct,\,-x,\,-y,\,-z\},$$

which is a linear, homogeneous transformation and the coefficients a^ν_μ depend only on the relative velocity and on the spatial orientation of the two coordinate systems. The Lorentz transformations (2.65) preserve the distance between two space-time points, which can be expressed differentially by invariance of the length element

$$ds^2=dx^\mu dx_\mu=g_{\mu\nu}dx^\mu dx^\nu\,,\qquad g_{\mu\nu}=\mathrm{diag}\{1,-1,-1,-1\}=g^{\mu\nu}\,, \tag{2.66}$$

as

$$dx^\mu dx_\mu=dx^{\mu\prime}dx'_\mu\,. \tag{2.67}$$

This yields

$$dx^{\mu\prime}dx'_\mu=a^\mu_\nu a^\sigma_\mu dx^\nu dx_\sigma\overset{!}{=}dx^\nu dx_\nu=\delta^\sigma_\nu dx^\nu dx_\sigma\,,$$

and hence

$$a^\mu_\nu a^\sigma_\mu=\delta^\sigma_\nu\,. \tag{2.68}$$

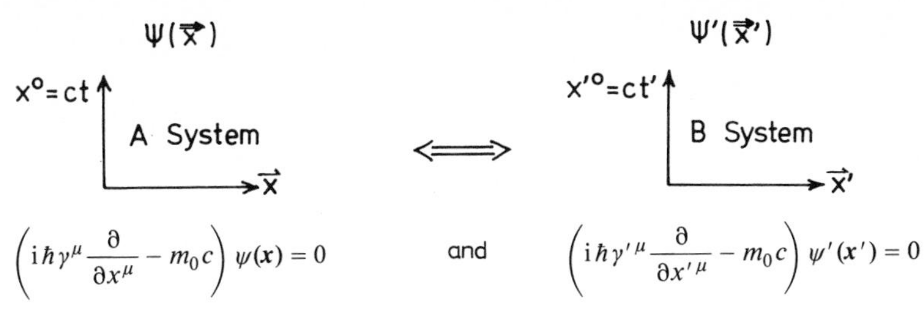

$$\left(\mathrm{i}\hbar\gamma^\mu\frac{\partial}{\partial x^\mu}-m_0c\right)\psi(x)=0 \qquad \text{and} \qquad \left(\mathrm{i}\hbar\gamma'^\mu\frac{\partial}{\partial x'^\mu}-m_0c\right)\psi'(x')=0$$

Fig. 2.4. Both A and B observe the same form of the Dirac equation for their respective spinors $\psi(x)$ and $\psi'(x')$

These are the orthogonality relations for the Lorentz transformations (L.T.). *Proper* and *improper* Lorentz transformations are distinguished; because of

$$(\det a^\mu_\nu)^2=1\ ,$$

which follows from (2.68), one can conclude that either

$$\begin{aligned}\det(a^\mu_\nu)&=+1\quad\text{(proper L.T.)}\quad\text{or}\\ \det(a^\mu_\nu)&=-1\quad\text{(improper L.T.)}\ .\end{aligned}\tag{2.69}$$

The former are the *proper L.T.'s* and the latter *improper.* The proper Lorentz transformations consist of the group of those coordinate transformations from one inertial frame into another one which can be constructed from the identity by continuous infinitesimal operations (rotations, translations, etc.). The improper L.T.'s contain either a (discrete) space inversion or a (discrete) time inversion. Such discrete transformations cannot be constructed from the identity by successive infinitesimal transformations. The determinant of their transformation coefficients is therefore negative.

Our task is now to connect the measurements of observer A with those of observer B. Each observer performs the measurements in her own inertial frame. More precisely, a relation (transformation law) between $\psi'(x')$ and $\psi(x)$ is required (Fig. 2.4).

For a given $\psi(x)$ for A, the as yet unknown transformation law must enable $\psi'(x')$ of observer B to be calculated. The Lorentz covariance requirement now means that $\psi(x)$ in the A system and $\psi'(x')$ in the B system have to obey the respective Dirac equations in these systems. The Dirac equations in both systems must have the same form to ensure that both systems are equal in the sense of the principle of relativity.

In the following it is more convenient to denote the Dirac equation in four-dimensional form. In this way the symmetry between the time coordinate $x^0=ct$ and the space coordinates is better exhibited. Starting with (2.25), we multiply it from the left by $\hat{\beta}/c$, which gives for $\psi(x)$

$$\left(\hat{\beta}\mathrm{i}\hbar\frac{\partial}{\partial ct}+\sum_{k=1}^{3}\hat{\beta}\hat{\alpha}_k\mathrm{i}\hbar\frac{\partial}{\partial x^k}-m_0c\right)\psi(x)=0\ ,$$

and with the definitions

$$\gamma^0 = \hat{\beta}\,, \quad \gamma^i = \hat{\beta}\hat{\alpha}_i\,, \quad (i = 1,2,3) \tag{2.70}$$

finally

$$\mathrm{i}\hbar\left(\gamma^0 \frac{\partial}{\partial x^0} + \gamma^1 \frac{\partial}{\partial x^1} + \gamma^2 \frac{\partial}{\partial x^2} + \gamma^3 \frac{\partial}{\partial x^3}\right)\psi - m_0 c\,\psi = 0\,. \tag{2.71}$$

Even though the γ^ν are matrices, they are denoted without the operator sign (ˆ), because confusion is not possible. Also, the four-vector x^ν is simply denoted by x, as before,

$$x^\nu = \{x^0, x^1, x^2, x^3\} \equiv x\,, \tag{2.72}$$

so that the notation is simplified. The γ^μ matrices allow a more elegant formulation of the anticommutation relations (2.10), namely

$$\gamma^\mu\gamma^\nu + \gamma^\nu\gamma^\mu = 2g^{\mu\nu}\mathbb{1}\,, \tag{2.73}$$

where $\mathbb{1}$ is the 4×4 unit matrix. The γ^i $(i = 1,2,3)$ are unitary $((\gamma^i)^\dagger = (\gamma^i)^{-1})$ and anti-Hermitian $(\gamma^{i\dagger} = -\gamma^i)$. These statements follow from

$$(\gamma^i)^2 = -\mathbb{1} = -\gamma^i\gamma^{i\dagger}$$

and hence

$$(\gamma^i)^\dagger = (\gamma^i)^{-1} = -\gamma^i\,. \tag{2.74}$$

On the other hand, the γ^0 matrix is Hermitian $(\gamma^{0\dagger} = \gamma^0)$, as can be seen from

$$(\gamma^0)^2 = \mathbb{1} = \gamma^0\gamma^{0\dagger}\,. \tag{2.75}$$

The γ^μ can also be denoted explicitly using (2.13, 70):

$$\gamma^i = \begin{pmatrix} 0 & \hat{\sigma}^i \\ -\hat{\sigma}^i & 0 \end{pmatrix}, \quad \gamma^0 = \begin{pmatrix} \mathbb{1} & 0 \\ 0 & -\mathbb{1} \end{pmatrix}. \tag{2.76}$$

A further shorthand notation used from time to time is the Feynman dagger, defined through

$$\not{A} \equiv \gamma^\mu A_\mu = g_{\mu\nu}\gamma^\nu A^\mu = \gamma^0 A^0 - \sum_{i=1}^{3} \gamma^i A^i = \gamma^0 A^0 - \boldsymbol{\gamma}\cdot\boldsymbol{A}\,,$$

and the Nabla dagger

$$\not{\nabla} \equiv \gamma^\mu \frac{\partial}{\partial x^\mu} = \gamma^0 \frac{\partial}{\partial ct} + \sum_{i=1}^{3} \gamma^i \frac{\partial}{\partial x^i} = \frac{\gamma^0}{c}\frac{\partial}{\partial t} + \boldsymbol{\gamma}\cdot\nabla\,. \tag{2.77}$$

With that the free Dirac equation (2.71) can be simply written as

$$(\mathrm{i}\hbar \nabla\!\!\!/ - m_0 c)\,\psi(x) = 0\,, \tag{2.78}$$

or with the four-momentum operator $\hat{p}_\mu = \mathrm{i}\hbar\,\partial/\partial x^\mu$:

$$(\hat{p}\!\!\!/ - m_0 c)\,\psi(x) = 0\,. \tag{2.79}$$

The Dirac equation (2.56) with minimal coupling to the electromagnetic four-potential A^μ reads in this notation

$$\left(\hat{p}\!\!\!/ - \frac{e}{c}A\!\!\!/ - m_0 c\right)\psi(x) = 0\,. \tag{2.80}$$

Both $\hat{p}^\mu$ and A^μ are four-vectors, and consequently this is also true for the difference $p^\mu - eA^\mu/c$. Therefore, in the following discussion of covariance, we restrict ourselves to the free Dirac equation (2.79) without loss of generality.

2.5.1 Formulation of Covariance (Form Invariance)

Covariance of the Dirac equation (2.79) has two requirements.

1) A prescription allowing observer B to construct her own spinor $\psi'(x')$ from spinor $\psi(x)$ reported by observer A must exist. The $\psi'(x')$ of B has to describe the same physical state as the $\psi(x)$ of A.
2) According to the principle of relativity the physical principles, i.e. the basic physical equations, are the same in every inertial system. Hence $\psi'(x')$ must be the solution of a Dirac equation, which in x' coordinates has the same form as (2.79), namely

$$\left(\mathrm{i}\hbar\,\gamma'^\mu \frac{\partial}{\partial x^{\mu\prime}} - m_0 c\right)\psi'(x') = 0\,. \tag{2.81}$$

According to the relativity principle the γ'^μ must obey the same anticommutation relations (2.73), otherwise the inertial frames of A and B would not be equivalent. Thus

$$\gamma'^\mu\gamma'^\nu + \gamma'^\nu\gamma'^\mu = 2g^{\mu\nu}\mathbb{1}\,, \tag{2.82}$$

and also

$$\begin{aligned}(\gamma'^0)^\dagger &= \gamma'^0\,,\\ (\gamma'^i)^\dagger &= -\gamma'^i\,,\end{aligned} \tag{2.83}$$

which means that for the same reasons the same Hermiticity and anti-Hermiticity relations for the Dirac matrices must be valid for both observers as in (2.74). This, by the way, also assures Hermiticity of the free Dirac Hamiltonian (2.25) in the x' system [Gr 81b, Go 55]

$$\hat{H}_{\mathrm{f}}' = -c\gamma'^0\gamma'^k\left(\mathrm{i}\hbar\frac{\partial}{\partial x'^k}\right) + \gamma'^0 m_0 c^2\,, \tag{2.84}$$

i.e.

$$(\hat{H}_{\mathrm{f}}')^\dagger = \hat{H}_{\mathrm{f}}'\,. \tag{2.85}$$

In a longer algebraic proof [Go 55, Gr 81b] one can show that all 4×4 matrices γ'^μ, which obey (2.82, 83), are equivalent up to a unitary transformation $\hat{U}$, i.e.

$$\gamma'^\mu = \hat{U}^\dagger\gamma^\mu\hat{U}\,, \qquad \hat{U}^\dagger = \hat{U}^{-1}\,. \tag{2.86}$$

Now, since unitary transformations do not change the physics, without loss of generality we can use in the reference frame of observer B *the same γ matrices* as in the reference frame of A. Hence we can set $\gamma'^\mu = \gamma^\mu$, and (2.81) becomes

$$(\not{p}' - m_0 c)\,\psi'(x') = 0\,, \tag{2.87}$$

where now

$$\not{p}' = \mathrm{i}\hbar\gamma^\nu\frac{\partial}{\partial x'^\nu}\,. \tag{2.88}$$

Let us now turn to determining the transformation between $\psi'(x')$ and $\psi(x)$. It must be linear, because Dirac equations (2.79, 87) are linear in the wave functions. Hence

$$\psi'(x') = \psi'(\hat{a}x) = \hat{S}(\hat{a})\,\psi(x) = \hat{S}(\hat{a})\,\psi(\hat{a}^{-1}x')\,, \tag{2.89}$$

with $\hat{a}$ denoting the Lorentz transformation matrix $\hat{a} = (a_\mu^\nu)$. Here $\hat{S}(\hat{a})$ is a 4×4 matrix depending on a_μ^ν and acts on the four components of the spinor $\psi(x)$. Via a_ν^μ it depends on the relative velocities and relative orientations of the coordinate systems of A and B. The inverse operator $\hat{S}^{-1}(\hat{a})$ has to exist, because A must also be able to construct her spinor $\psi(x)$ from the spinor $\psi'(x')$ of B (relativity principle: equal opportunity for both physicists). Therefore

$$\psi(x) = \hat{S}^{-1}(\hat{a})\,\psi'(x') = \hat{S}^{-1}(\hat{a})\,\psi'(\hat{a}x)\,. \tag{2.90}$$

Because of (2.89) it follows that

$$\psi(x) = \hat{S}(\hat{a}^{-1})\,\psi'(x') = \hat{S}(\hat{a}^{-1})\,\psi'(\hat{a}x)\,, \tag{2.91}$$

and therefore

$$\hat{S}^{-1}(\hat{a}) = \hat{S}(\hat{a}^{-1})\,. \tag{2.92}$$

This is a condition for $\hat{S}$, which can now be constructed. To that purpose we start with the Dirac equation (2.78) of A:

$$\left[\mathrm{i}\hbar\gamma^\mu\frac{\partial}{\partial x^\mu} - m_0 c\right]\psi(x) = 0$$

and insert $\psi(x)$ from (2.90), which gives

$$\left[\mathrm{i}\hbar\,\gamma^{\mu}\hat{S}^{-1}(\hat{a})\frac{\partial}{\partial x^{\mu}}-m_0 c\hat{S}^{-1}(\hat{a})\right]\psi'(x')=0\,.$$

Multiplication with $\hat{S}(\hat{a})$ from the left gives

$$\left[\mathrm{i}\hbar\,\hat{S}(\hat{a})\,\gamma^{\mu}\hat{S}^{-1}(\hat{a})\frac{\partial}{\partial x^{\mu}}-m_0 c\right]\psi'(x')=0\,, \tag{2.93}$$

using $\hat{S}(\hat{a})\,\hat{S}^{-1}(\hat{a})=\mathbb{1}$. Now we express $\partial/\partial x^{\mu}$ in terms of x' coordinates (2.65):

$$\frac{\partial}{\partial x^{\mu}}=\frac{\partial x'^{\nu}}{\partial x^{\mu}}\frac{\partial}{\partial x'^{\nu}}=a^{\nu}_{\mu}\frac{\partial}{\partial x'^{\nu}} \tag{2.94}$$

and insert this into (2.93):

$$\left[\mathrm{i}\hbar(\hat{S}(\hat{a})\,\gamma^{\mu}\hat{S}^{-1}(\hat{a})\,a^{\nu}_{\mu})\frac{\partial}{\partial x'^{\nu}}-m_0 c\right]\psi'(x')=0\,. \tag{2.95}$$

Because of the form invariance of the Dirac equation, (2.95) has to be identical with (2.87). Comparing both leads us to conclude

$$\hat{S}(\hat{a})\,\gamma^{\mu}\hat{S}^{-1}(\hat{a})\,a^{\nu}_{\mu}=\gamma^{\nu} \tag{2.96}$$

or, equivalently,

$$a^{\nu}_{\mu}\,\gamma^{\mu}=\hat{S}(\hat{a})\,\gamma^{\nu}\hat{S}^{-1}(\hat{a})\,. \tag{2.97}$$

This is the most important equation for determining the $\hat{S}(\hat{a})$ operator. In fact, determining $\hat{S}(\hat{a})$ gives the solution of (2.97). Condition (2.97) holds for all Lorentz transformations, i.e. for proper and improper L.T.'s because in this deduction det $(a^{\mu}_{\nu})=\pm 1$ has not been used. As soon as we have shown that a solution of (2.97) exists, and as soon as we have constructed $\hat{S}(\hat{a})$, we have proven the Lorentz covariance of the Dirac equation. It will be shown that the group property $\hat{S}(\hat{a}\hat{a}')=\hat{S}(\hat{a})\,\hat{S}(\hat{a}')$ also holds.

At this stage a spinor can be precisely defined, up to now introduced only as a column vector with four components. *In general a wave function*

$$\psi(x)=\begin{pmatrix}\psi_1(x)\\ \vdots\\ \psi_4(x)\end{pmatrix}$$

is called a Lorentz spinor with four components, if $\psi(x)$ *transforms according to* (2.90) *with the condition* (2.97). Such a 4 spinor is also called a bispinor, because it is built up of two 2-spinors:

$$\psi(x) = \begin{bmatrix} \begin{pmatrix} \psi_1(x) \\ \psi_2(x) \end{pmatrix} \\ \\ \begin{pmatrix} \psi_3(x) \\ \psi_4(x) \end{pmatrix} \end{bmatrix} .$$

For $\hat{S}(\hat{a})$ one has to expect novel structures, i.e. structures uncommon to, e.g., tensor analysis, because bilinear forms of the spinor field $\psi(x)$ such as the four-current (2.21)

$$\{c\varrho, j^k\} = \{c\psi^\dagger\psi, c\psi^\dagger\hat{\alpha}^k\psi\} \quad (k = 1,2,3)$$

have to be four-vectors. The following shows how this is achieved.

In general it is easier to construct the various group operators of a symmetry group for infinitesimal transformations. The operators for finite transformations (rotations, translations, etc.) are then built up from consecutive infinitesimal operations. This procedure is adapted to construct $\hat{S}(\hat{a})$ and, therefore, to investigate first infinitesimal Lorentz transformations, given by the matrix

$$a^\nu_\mu = \delta^\nu_\mu + \Delta\omega^\nu_\mu , \quad \text{where} \tag{2.98}$$

$$\Delta\omega^{\nu\mu} = -\Delta\omega^{\mu\nu} , \tag{2.99}$$

as can be easily concluded from the orthogonality relations (2.68) by neglecting terms of order $(\Delta\omega^{\mu\nu})^2$:

$$\begin{aligned} a^\mu_\nu a^\sigma_\mu = \delta^\sigma_\nu &= (\delta^\mu_\nu + \Delta\omega^\mu_\nu)(\delta^\sigma_\mu + \Delta\omega^\sigma_\mu) \\ &\approx \delta^\mu_\nu\delta^\sigma_\mu + \delta^\mu_\nu\Delta\omega^\sigma_\mu + \delta^\sigma_\mu\Delta\omega^\mu_\nu \\ &= \delta^\sigma_\nu + \Delta\omega^\sigma_\nu + \Delta\omega^\sigma_\nu . \end{aligned}$$

Because of (2.99) 6 independent elements $\Delta\omega^{\mu\nu}$ exist. Each of these group parameters (rotation angles) generates an infinitesimal Lorentz transformation. We investigate two examples here, after which it is straight-forward to proceed similarly for other cases:

a) $\Delta\omega^{10} = -\Delta\omega^{01} \equiv -\Delta\beta \neq 0 ; \quad$ all other $\quad \Delta\omega^{\mu\nu} = 0 .$

We then have

$$\begin{aligned} \Delta\omega^0_1 = g_{1\sigma}\Delta\omega^{\sigma 0} = g_{11}\Delta\omega^{10} &= -\Delta\omega^{10} = +\Delta\beta \\ = \Delta\omega^{01} = \Delta\omega^1_0 &= -\Delta\omega^0_1 \quad \text{and} \end{aligned}$$

$$\Delta\omega^1_i = g_{i\sigma}\Delta\omega^{\sigma 1} = 0 \quad \text{for all} \quad i = 1,2,3, \quad \text{etc.}$$

Because of (2.65)

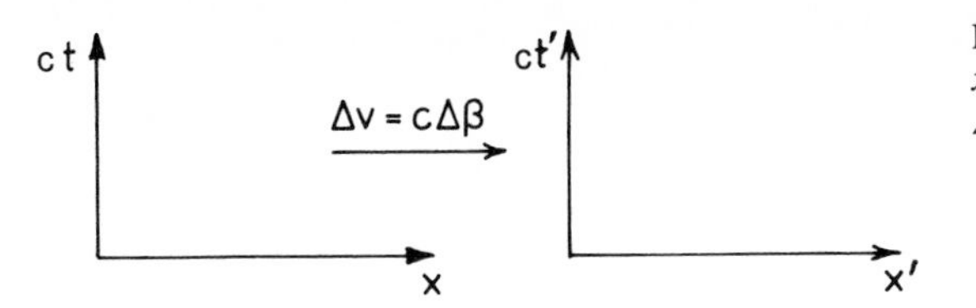

Fig. 2.5. Illustration of (2.100). The x' system moves with velocity $\Delta v = c\Delta\beta$ relative to the x system

$$
\begin{aligned}
(x^\nu)' &= (\delta^\nu_\mu + \Delta\,\omega^1_0\delta^\nu_1\delta^0_\mu + \Delta\,\omega^0_1\delta^\nu_0\delta^1_\mu)x^\mu \\
&= (\delta^\nu_\mu - \Delta\,\beta\delta^\nu_1\delta^0_\mu - \Delta\,\beta\delta^\nu_0\delta^1_\mu)x^\mu ,
\end{aligned}
$$

which is explicitly (Fig. 2.5)

$$
\begin{aligned}
(x^0)' &= x^0 - \Delta\,\beta x^1 = x^0 - \frac{\Delta\,v}{c}x^1 , \\
(x^1)' &= x^1 - \Delta\,\beta x^0 = x^1 - \frac{\Delta\,v}{c}x^0 , \\
(x^2)' &= x^2 , \\
(x^3)' &= x^3 .
\end{aligned}
\tag{2.100}
$$

The inertial frame of observer B (x' system) therefore moves against the inertial frame of A (x system) with velocity $\Delta\,v = c\Delta\,\beta$ along the positive x^1 axis, which follows from

$$
(x^1)' = 0 \Rightarrow \frac{x^1}{x^0} = \frac{x^1}{ct} = \frac{\Delta\,v}{c} = \Delta\,\beta .
$$

b) Secondly,

$$
\Delta\,\omega^1_2 = -\Delta\,\omega^{12} = \Delta\,\omega^{21} \equiv \Delta\,\varphi ; \qquad \text{all other} \quad \Delta\,\omega^{\mu\nu} = 0 .
$$

Here the Lorentz transformation (2.65) reads

$$
(x^\nu)' = (\delta^\nu_\mu + \delta^\nu_1\delta^2_\mu\Delta\,\varphi + \delta^\nu_2\delta^1_\mu(-\Delta\,\varphi))x^\mu
$$

or explicitly

$$
\begin{aligned}
(x^0)' &= x^0 , \\
(x^1)' &= x^1 + \Delta\,\varphi x^2 , \\
(x^2)' &= x^2 - \Delta\,\varphi x^1 , \\
(x^3)' &= x^3 .
\end{aligned}
\tag{2.101}
$$

This transformation is immediately recognizable as a rotation through the angle $\Delta\,\varphi$ around the z axis, which would read for finite angles

$$
\begin{aligned}
(x^1)' &= \cos\varphi x^1 + \sin\varphi x^2 , \\
(x^2)' &= -\sin\varphi x^1 + \cos\varphi x^2 .
\end{aligned}
\tag{2.102}
$$

2.5.2 Determining the $\hat{S}(\hat{a})$ Operator for Infinitesimal Lorentz Transformations

Let us now return to determining the operator

$$\hat{S}(\hat{a}) = \hat{S}(\Delta\omega^{\mu\nu}) .$$

We expand $\hat{S}$ in a power series of $\Delta\omega^{\mu\nu}$ and stop after linear terms, so that

$$\begin{aligned} \hat{S}(\Delta\omega^{\mu\nu}) &= \mathbb{1} - \frac{\mathrm{i}}{4}\hat{\sigma}_{\mu\nu}\Delta\omega^{\mu\nu} , \\ \hat{S}^{-1}(\Delta\omega^{\mu\nu}) &= \mathbb{1} + \frac{\mathrm{i}}{4}\hat{\sigma}_{\mu\nu}\Delta\omega^{\mu\nu} , \\ \hat{\sigma}_{\mu\nu} &= -\hat{\sigma}_{\nu\mu} , \\ \gamma_0\hat{\sigma}^{\dagger}_{\mu\nu}\gamma_0 &= \hat{\sigma}_{\mu\nu} . \end{aligned} \tag{2.103}$$

The factor i/4 is chosen to allow for a simple structure of the yet unknown coefficients $\hat{\sigma}_{\mu\nu}$. One should clearly note that *each of the six quantities $\hat{\sigma}_{\mu\nu}$ is itself a 4×4 matrix!* The same is true, of course, for the operator $\hat{S}$. Inserting now (2.103, 98) into condition (2.97) gives a system of equations for $\hat{\sigma}_{\mu\nu}$:

$$(\delta^{\nu}_{\mu} + \Delta\omega^{\nu}_{\mu})\gamma^{\mu} = \left(\mathbb{1} - \frac{\mathrm{i}}{4}\hat{\sigma}_{\alpha\beta}\Delta\omega^{\alpha\beta}\right)\gamma^{\nu}\left(\mathbb{1} + \frac{\mathrm{i}}{4}\hat{\sigma}_{\alpha\beta}\Delta\omega^{\alpha\beta}\right)$$

or

$$\Delta\omega^{\nu}_{\mu}\gamma^{\mu} = -\frac{\mathrm{i}}{4}\Delta\omega^{\alpha\beta}(\hat{\sigma}_{\alpha\beta}\gamma^{\nu} - \gamma^{\nu}\hat{\sigma}_{\alpha\beta}) . \tag{2.104}$$

In the last step terms quadratic in $\Delta\omega^{\mu\nu}$ have been neglected. Using the antisymmetry equation (2.99) $\Delta\omega^{\mu\nu} = -\Delta\omega^{\nu\mu}$ it is straightforward to derive from (2.104) the relation

$$[\hat{\sigma}_{\alpha\beta}, \gamma^{\nu}]_{-} = -2\mathrm{i}(g^{\nu}_{\alpha}\gamma_{\beta} - g^{\nu}_{\beta}\gamma_{\alpha}) , \tag{2.105}$$

which determines $\hat{\sigma}_{\alpha\beta}$. Since $\hat{\sigma}_{\alpha\beta}$ has to be antisymmetric in α and β, see (2.103), it is plausible to try the ansatz

$$\hat{\sigma}_{\alpha\beta} = \frac{\mathrm{i}}{2}[\gamma_{\alpha}, \gamma_{\beta}]_{-} , \tag{2.106}$$

which indeed solves (2.105), as a small calculation demonstrates [Gr 81b]. With that the $\hat{S}(\hat{a})$ operator for infinitesimal Lorentz transformations is now determined and according to (2.103) reads

$$\hat{S}(\Delta\omega^{\mu\nu}) = \mathbb{1} - \frac{\mathrm{i}}{4}\hat{\sigma}_{\mu\nu}\Delta\omega^{\mu\nu} = \mathbb{1} + \frac{1}{8}[\gamma_{\mu}, \gamma_{\nu}]_{-}\Delta\omega^{\mu\nu} . \tag{2.107}$$

2.5.3 The $\hat{S}(\hat{a})$ Operator for Finite Lorentz Transformations

We now construct $\hat{S}(\hat{a})$ for finite, proper Lorentz transformations. To obtain the finite Lorentz transformation (2.65) from the infinitesimal one (2.98), first $\Delta\omega_\mu^\nu$ is denoted in the form

$$\Delta\omega_\mu^\nu = \Delta\omega(\hat{I}_n)_\mu^\nu , \tag{2.108}$$

where $\Delta\omega$ is the infinitesimal parameter of the Lorentz group ("infinitesimal rotation angle") around an axis in $\boldsymbol{n}$ direction. The 4×4 matrix (space and time components) for a unit Lorentz rotation around the $\boldsymbol{n}$ axis is $(\hat{I}_n)_\mu^\nu$. For example, the Lorentz rotation (2.100) is a Lorentz transformation with a velocity along the x axis $\Delta v = c\Delta\beta$ which can easily be brought to the form (2.108):

$$a_\mu^\nu = \delta_\mu^\nu + \Delta\omega_\mu^\nu = \delta_\mu^\nu + \Delta\omega(\hat{I}_x)_\mu^\nu = \delta_\mu^\nu - \Delta\beta(\delta_1^\nu\delta_\mu^0 + \delta_0^\nu\delta_\mu^1) ,$$

and thus with $\Delta\omega = -\Delta\beta$

$$(\hat{I}_x)_\mu^\nu = -(\delta_1^\nu\delta_\mu^0 + \delta_0^\nu\delta_\mu^1) = \begin{pmatrix} 0 & -1 & 0 & 0 \\ -1 & 0 & 0 & 0 \\ 0 & 0 & 0 & 0 \\ 0 & 0 & 0 & 0 \end{pmatrix} . \tag{2.109}$$

Obviously only the matrix elements

$$(\hat{I}_x)_1^0 = (\hat{I}_x)_0^1 = -(\hat{I}_x)^{01} = +(\hat{I}_x)^{10} = -1$$

are non-vanishing. The following relations also hold:

$$(\hat{I}_x)^2 = \begin{pmatrix} 1 & 0 & 0 & 0 \\ 0 & 1 & 0 & 0 \\ 0 & 0 & 0 & 0 \\ 0 & 0 & 0 & 0 \end{pmatrix} \quad \text{and} \quad (\hat{I}_x)^3 = (\hat{I}_x) . \tag{2.110}$$

2.5.4 Finite, Proper Lorentz Transformations

The algebraic properties (2.110) are useful for determining according to (2.65, 98) finite, proper L.T.'s from infinitesimal ones (2.108). From (2.100)

$$\begin{aligned} (x^\nu)' &= \lim_{N\to\infty}\left(\mathbb{1} + \frac{\omega}{N}\hat{I}_x\right)^\nu_{\alpha_1}\left(\mathbb{1} + \frac{\omega}{N}\hat{I}_x\right)^{\alpha_1}_{\alpha_2} \dots x^{\alpha_N} \\ &= \lim_{N\to\infty}\left[\left(\mathbb{1} + \frac{\omega}{N}\hat{I}_x\right)^N\right]^\nu_\mu x^\mu = (\mathrm{e}^{\omega\hat{I}_x})^\nu_\mu x^\mu = [\cosh(\omega\hat{I}_x) + \sinh(\omega\hat{I}_x)]^\nu_\mu x^\mu \\ &= \left[\left(\mathbb{1} + \frac{(\omega\hat{I}_x)^2}{2!} + \frac{(\omega\hat{I}_x)^4}{4!} + \dots\right) + \left(\frac{(\omega\hat{I}_x)}{1!} + \frac{(\omega\hat{I}_x)^3}{3!} + \dots\right)\right]^\nu_\mu x^\mu \end{aligned}$$

$$= \left[\left(\mathbb{1} + \frac{\omega^2}{2!}(\hat{I}_x)^2 + \frac{\omega^4}{4!}(\hat{I}_x)^4 + \dots \right) + \left(\frac{\omega}{1!} + \frac{\omega^3}{3!} + \dots \right)(\hat{I}_x) \right]^\nu_{\ \mu} x^\mu$$

$$= [\mathbb{1} - (\hat{I}_x)^2 + (\cosh\omega)(\hat{I}_x)^2 + (\sinh\omega)(\hat{I}_x)]^\nu_{\ \mu} x^\mu .$$

This reads explicitly

$$\begin{pmatrix} x^{0\prime} \\ x^{1\prime} \\ x^{2\prime} \\ x^{3\prime} \end{pmatrix} = \begin{pmatrix} \cosh\omega & -\sinh\omega & 0 & 0 \\ -\sinh\omega & \cosh\omega & 0 & 0 \\ 0 & 0 & 1 & 0 \\ 0 & 0 & 0 & 1 \end{pmatrix} \begin{pmatrix} x^0 \\ x^1 \\ x^2 \\ x^3 \end{pmatrix} ,$$

or

$$\begin{aligned} x^{0\prime} &= (\cosh\omega)x^0 - (\sinh\omega)x^1 = \cosh\omega(x^0 - (\mathrm{tgh}\,\omega)x^1) , \\ x^{1\prime} &= -(\sinh\omega)x^0 + (\cosh\omega)x^1 = \cosh\omega(x^1 - (\mathrm{tgh}\,\omega)x^0) , \\ x^{2\prime} &= x^2 , \\ x^{3\prime} &= x^3 . \end{aligned} \tag{2.111}$$

The finite Lorentz rotation angle ω can now easily be expressed in terms of the relative velocity $v_x = c\Delta\beta_x$ of the two inertial systems. Namely, the origin of the x' systems is described in the x system by

$$x^{1\prime} = 0 = \cosh\omega(x^1 - (\mathrm{tgh}\,\omega)x^0) ,$$

from which

$$\frac{x^1}{x^0} = \frac{x^1}{ct} = \frac{v_x}{c} = \mathrm{tgh}\,\omega = \beta \tag{2.112}$$

follows. Hence

$$\cosh\omega = \frac{\cosh\omega}{\sqrt{\cosh^2\omega - \sinh^2\omega}} = \frac{1}{\sqrt{1 - \mathrm{tgh}^2\omega}} = \frac{1}{\sqrt{1-\beta^2}} , \tag{2.113}$$

and therefore (2.111) becomes

$$\begin{aligned} x^{0\prime} &= \frac{x^0 - \beta x^1}{\sqrt{1-\beta^2}} , \\ x^{1\prime} &= \frac{x^1 - \beta x^0}{\sqrt{1-\beta^2}} , \\ x^{2\prime} &= x^2 , \\ x^{3\prime} &= x^3 , \end{aligned} \tag{2.114}$$

which is the Lorentz transformation for finite velocities $v = c\beta$. This procedure for constructing finite Lorentz transformations from infinitesimal ones can be generalized to motions and rotations along an arbitrary axis. Six matrices (generators) $(\hat{I}_n)^\mu_{\ \nu}$ exist corresponding to the six independent Lorentz transformations. They are the four-dimensional generalizations of the well-known generators of three-dimensional space rotations (angular momentum operators) known from non-relativistic quantum mechanics [Bj 64, Gr 81b].

2.5.5 The $\hat{S}$ Operator for Finite Lorentz Transformations

Now, finally, we establish the spinor transformation operator for finite "rotation angles" starting from (2.103, 108). We shall perform infinitely many such infinitesimal transformations, one after the other, so that

$$
\begin{aligned}
\psi'(x') = \hat{S}(\hat{a})\,\psi(x) &= \lim_{N\to\infty}\left(\mathbb{1} - \frac{\mathrm{i}}{4}\,\frac{\omega}{N}\,\hat{\sigma}_{\mu\nu}(\hat{I}_n)^{\mu\nu}\right)^N \psi(x) \\
&= \exp\left[-\frac{\mathrm{i}}{4}\,\omega\,\hat{\sigma}_{\mu\nu}(\hat{I}_n)^{\mu\nu}\right]\psi(x)\,.
\end{aligned}
\tag{2.115}
$$

For the special "rotation" (2.109), which corresponds to a Lorentz transformation along the x axis

$$
\begin{aligned}
\psi'(x') &= \exp\left\{-\frac{\mathrm{i}}{4}\,\omega[\hat{\sigma}_{01}(\hat{I}_x)^{01} + \hat{\sigma}_{10}(\hat{I}_x)^{10}]\right\}\psi(x) \\
&= \exp\left\{-\frac{\mathrm{i}}{4}\,\omega[\hat{\sigma}_{01}(+1) + \hat{\sigma}_{10}(-1)]\right\}\psi(x) \\
&= \exp\left\{-\frac{\mathrm{i}}{2}\,\omega\,\hat{\sigma}_{01}\right\}\psi(x) \\
&\equiv \hat{S}_{\mathrm{L}}\,\psi(x)\,,
\end{aligned}
\tag{2.116}
$$

because, according to (2.103) $\hat{\sigma}_{10} = -\hat{\sigma}_{01}$ and according to (2.109) $(\hat{I}_x)^{01} = +1$ and $(\hat{I}_x)^{10} = -1$.

Similarly, for the rotation around the z axis using of (2.101)

$$
\Delta\omega^{\nu}_{\ \mu} = \Delta\varphi(\hat{I}_z)^{\nu}_{\ \mu}\,,
$$

$$
(\hat{I}_z)^{\nu}_{\ \mu} = \begin{pmatrix} 0 & 0 & 0 & 0 \\ 0 & 0 & 1 & 0 \\ 0 & -1 & 0 & 0 \\ 0 & 0 & 0 & 0 \end{pmatrix},
$$

$$
\begin{aligned}
\psi'(x') &= \exp\left[-\frac{\mathrm{i}}{4}\,\varphi\,\hat{\sigma}_{\mu\nu}(\hat{I}_z)^{\mu\nu}\right]\psi(x) \\
&= \exp\left\{-\frac{\mathrm{i}}{4}\,\varphi[\hat{\sigma}_{12}(\hat{I}_z)^{12} + \hat{\sigma}_{21}(\hat{I}_z)^{21}]\right\}\psi(x)
\end{aligned}
$$

$$
\begin{aligned}
&= \exp\left\{-\frac{\mathrm{i}}{4}\varphi[\hat{\sigma}_{12}(-1)+\hat{\sigma}_{21}(+1)]\right\}\psi(x) \\
&= \exp\left[\frac{\mathrm{i}}{2}\varphi\hat{\sigma}_{12}\right]\psi(x) = \exp\left(\frac{\mathrm{i}}{2}\varphi\hat{\sigma}^{12}\right)\psi(x) \\
&\equiv \hat{S}_{\mathrm{R}}\,\psi(x)\,.
\end{aligned}
\tag{2.117}
$$

With the explicit $\hat{\sigma}^{\alpha\beta}$ (2.106) we easily find

$$
\begin{aligned}
\hat{\sigma}^{12} &= \frac{\mathrm{i}}{2}[\gamma^1,\gamma^2]_- = \frac{\mathrm{i}}{2}\left[\begin{pmatrix}0 & \hat{\sigma}_1\\ -\hat{\sigma}_1 & 0\end{pmatrix}\begin{pmatrix}0 & \hat{\sigma}_2\\ -\hat{\sigma}_2 & 0\end{pmatrix} - \begin{pmatrix}0 & \hat{\sigma}_2\\ -\hat{\sigma}_2 & 0\end{pmatrix}\begin{pmatrix}0 & \hat{\sigma}_1\\ -\hat{\sigma}_1 & 0\end{pmatrix}\right] \\
&= \frac{\mathrm{i}}{2}\left[\begin{pmatrix}-\hat{\sigma}_1\hat{\sigma}_2 & 0\\ 0 & -\hat{\sigma}_1\hat{\sigma}_2\end{pmatrix} - \begin{pmatrix}-\hat{\sigma}_2\hat{\sigma}_1 & 0\\ 0 & -\hat{\sigma}_2\hat{\sigma}_1\end{pmatrix}\right] \\
&= -\frac{\mathrm{i}}{2}\begin{bmatrix}\hat{\sigma}_1\hat{\sigma}_2-\hat{\sigma}_2\hat{\sigma}_1 & 0\\ 0 & \hat{\sigma}_1\hat{\sigma}_2-\hat{\sigma}_2\hat{\sigma}_1\end{bmatrix} \\
&= -\frac{\mathrm{i}}{2}\begin{pmatrix}2\mathrm{i}\hat{\sigma}_3 & 0\\ 0 & 2\mathrm{i}\hat{\sigma}_3\end{pmatrix} = \begin{pmatrix}\hat{\sigma}_3 & 0\\ 0 & \hat{\sigma}_3\end{pmatrix},
\end{aligned}
\tag{2.118}
$$

where the commutation relations (2.15) were used for the 2×2 Pauli matrices.

The similarity between the spinor transformation (2.117) and the transformation of two-component Pauli spinors under space rotations [Gr 81b] should be noted:

$$
\varphi'(x') = \exp\left(\frac{\mathrm{i}}{2}\boldsymbol{\omega}\cdot\hat{\boldsymbol{\sigma}}\right)\varphi(x)\,. \tag{2.119}
$$

Accordingly, the $\omega^{\mu\nu}$ can be interpreted as *covariant angle-variables*, because they enter the Lorentz transformations (2.98) in a similar manner as the components of the rotation vector $\boldsymbol{\omega}$ enter the three-dimensional space rotations. The occurrence of half the angles in the transformation law (2.117) causes the doubled value of the spinor transformations. A spinor transforms into itself after a rotation through 4π (not 2π!). Because of this, physical observables in a spinor theory (like the Dirac theory) have to be bilinear in ψ (or of even power in ψ). Only in this way can an observable turn into itself under a rotation through 2π, as experience shows.

According to (2.115) the transformation operator $\hat{S}_{\mathrm{R}}$ for space rotations of spinors is

$$
\hat{S}_{\mathrm{R}}(\omega_{ij}) = \exp\left(-\frac{\mathrm{i}}{4}\hat{\sigma}_{ij}\,\omega^{ij}\right), \qquad i,j=1,2,3\,. \tag{2.120}
$$

It is a unitary operator, because the $\hat{\sigma}_{ij}$ $(i,j=1,2,3)$ are Hermitian and

$$
\hat{S}_{\mathrm{R}}^{\dagger} = \exp\left(\frac{\mathrm{i}}{4}\hat{\sigma}_{ij}^{+}\,\omega^{ij}\right) = \exp\left(\frac{\mathrm{i}}{4}\hat{\sigma}_{ij}\,\omega^{ij}\right) = \hat{S}_{\mathrm{R}}^{-1}\,. \tag{2.121}
$$

For true Lorentz transformations ("Lorentz boosts", i.e. transformations to a moving coordinate system) this does not hold. This can be seen, for example, for the Lorentz transformation operator $\hat{S}_{\rm L}$ as it appears in (2.116):

$$\hat{S}_{\rm L}^{\dagger} = \exp\left(-\frac{\rm i}{2}\omega\hat{\sigma}_{01}\right) = \exp\left(+\frac{\omega}{2}\hat{\alpha}_1\right) = \hat{S}_{\rm L}^{\dagger} \mp \hat{S}_{\rm L}^{-1} . \tag{2.122}$$

Here we used

$$\hat{\sigma}_{01} = \frac{\rm i}{2}(\gamma_0\gamma_1 - \gamma_1\gamma_0) = \frac{\rm i}{2}(\hat{\beta}\hat{\beta}\hat{\alpha}_1 - \hat{\beta}\hat{\alpha}_1\hat{\beta}) = \frac{\rm i}{2}(\hat{\alpha}_1 + \hat{\alpha}_1) = {\rm i}\,\hat{\alpha}_1 .$$

It is not difficult to show that in this case

$$\hat{S}_{\rm L}^{-1} = \gamma_0\hat{S}_{\rm L}^{\dagger}\gamma_0 , \tag{2.123}$$

and that this relation holds both for true Lorentz transformations and space rotations, i.e. that in general [Bj 64, Gr 81b]

$$\hat{S}^{-1} = \gamma_0\hat{S}^{\dagger}\gamma_0 . \tag{2.124}$$

This formula will be used repeatedly.

2.5.6 The Four-Current Density

In (2.20 – 23) the four-current density was introduced

$$\{j^{\mu}\} = \{j^0, \boldsymbol{j}\} = \{c\psi^{\dagger}\psi, c\psi^{\dagger}\hat{\boldsymbol{\alpha}}\psi\} = \{c\psi^{\dagger}\gamma^0\gamma^{\mu}\psi\} ,$$

or simply

$$j^{\mu}(x) = c\psi^{\dagger}(x)\gamma^0\gamma^{\mu}\psi(x) = c\bar{\psi}(x)\gamma^{\mu}\psi(x) , \qquad \bar{\psi} = \psi^{\dagger}\gamma^0 , \tag{2.125}$$

but it was not shown that it is really a four-vector. We can do this now by demonstrating that (2.125) transforms under Lorentz transformations as

$$\begin{aligned} j'^{\mu}(x') &= c\psi'^{\dagger}(x')\gamma^0\gamma^{\mu}\psi'(x') \\ &= c\psi^{\dagger}(x)\hat{S}^{\dagger}\gamma^0\gamma^{\mu}\hat{S}\psi(x) \\ &= c\psi^{\dagger}(x)\gamma^0\hat{S}^{-1}\gamma^{\mu}\hat{S}\psi(x) \qquad \text{[because of (2.124)]} \\ &= ca^{\mu}_{\ \nu}\psi^{\dagger}(x)\gamma^0\gamma^{\nu}\psi(x) \qquad \text{[because of (2.97)]} \\ &= a^{\mu}_{\ \nu}j^{\nu}(x) . \end{aligned} \tag{2.126}$$

This is indeed the transformation law of a four-vector as in (2.65). Also the continuity equation (2.20) can now be denoted in Lorentz covariant form as

$$\frac{\partial j^{\mu}(x)}{\partial x^{\mu}} = 0 . \tag{2.127}$$

This proves in particular that the probability density

$$j^0(x) = c\varrho(x) = c\,\psi^\dagger\psi(x)$$

is indeed the time component of a four vector.

In the following it turns out to be useful to abbreviate the often occurring combination

$$\bar{\psi} \equiv \psi^\dagger\gamma^0 . \tag{2.128}$$

Here $\bar{\psi}$ is called the *adjoint spinor.* Its transformation law under Lorentz transformations follows from (2.89, 123):

$$\begin{aligned}\bar{\psi}'(x') &= \psi'^\dagger(x')\gamma^0 = (\hat{S}\psi(x))^\dagger\gamma^0 = \psi^\dagger(x)\hat{S}^\dagger\gamma^0 \\ &= \psi^\dagger(x)\gamma^0\hat{S}^{-1} = \bar{\psi}(x)\hat{S}^{-1} .\end{aligned} \tag{2.129}$$

2.6 Spinor Under Space Inversion (Parity Transformation)

Next we consider the improper Lorentz transformation of space inversion defined by

$$\begin{aligned}\boldsymbol{x}' &= -\boldsymbol{x} , \\ t' &= t .\end{aligned} \tag{2.130}$$

The corresponding transformation matrix is

$$a_\mu^\nu = \begin{pmatrix} 1 & 0 & 0 & 0 \\ 0 & -1 & 0 & 0 \\ 0 & 0 & -1 & 0 \\ 0 & 0 & 0 & -1 \end{pmatrix} = g^{\mu\nu} . \tag{2.131}$$

Also in this case the Dirac equation must be covariant, because (2.130) is nothing else than a special case of the general Lorentz transformation (2.65). Therefore, all the general considerations from the last section can be taken over, except those based on infinitesimal transformations. The space inversion (2.130) cannot be constructed by a sequence of infinitesimal operations out of the identity.

The corresponding operator for spinor transformation is called $\hat{P}$ (from "parity"), i.e. in this special case $\hat{S} \equiv \hat{P}$. The general condition (2.97) also holds now and reads

$$a_\mu^\nu\gamma^\mu = \hat{P}\gamma^\nu\hat{P}^{-1}$$

or

$$a_\nu^\sigma a_\mu^\nu\gamma^\mu = \hat{P}a_\nu^\sigma\gamma^\nu\hat{P}^{-1} \Leftrightarrow \delta_\mu^\sigma\gamma^\mu = \hat{P}\sum_{\nu=0}^{4} g^{\sigma\nu}\gamma^\nu\hat{P}^{-1} \Leftrightarrow \hat{P}^{-1}\gamma^\sigma\hat{P} = g^{\sigma\sigma}\gamma^\sigma .$$

Setting $\sigma = \nu$, this finally becomes

$$\hat{P}^{-1}\gamma^\nu\hat{P} = g^{\nu\nu}\gamma^\nu . \tag{2.132}$$

One should note that no summing over ν occurs on the rhs. This equation has the simple solution

$$\hat{P} = \mathrm{e}^{\mathrm{i}\varphi}\gamma^0 , \quad \hat{P}^{-1} = \mathrm{e}^{-\mathrm{i}\varphi}\gamma^0 , \tag{2.133}$$

where φ is still an arbitrary phase. In analogy to the proper Lorentz transformations (2.118), where only a rotation through 4π yields the identity transformation for the spinor, we request that *four inversions reproduce the spinor*, i.e.

$$\hat{P}^4 \psi = \psi = \mathrm{e}^{\mathrm{i}4\varphi}(\gamma^0)^4 \psi = \mathrm{e}^{\mathrm{i}4\varphi}\psi , \tag{2.134}$$

then

$$(\mathrm{e}^{\mathrm{i}\varphi})^4 = 1 .$$

Hence

$$\mathrm{e}^{\mathrm{i}\varphi} = \pm 1, \ \pm \mathrm{i} . \tag{2.135}$$

The operator $\hat{P}$ of (2.133) is unitary:

$$\hat{P}^{-1} = \mathrm{e}^{-\mathrm{i}\varphi}\gamma^0 = \hat{P}^{\dagger} , \tag{2.136}$$

and also obeys (2.124)

$$\hat{P}^{-1} = \gamma^0 \hat{P}^{\dagger} \gamma^0 . \tag{2.137}$$

The spinor transformation under space inversion reads explicitly

$$\psi'(x') = \psi'(-\boldsymbol{x}, t) = \hat{P}\psi(x) = \mathrm{e}^{\mathrm{i}\varphi}\gamma^0 \psi(\boldsymbol{x}, t) . \tag{2.138}$$

In the non-relativistic limit ψ reduces to $\psi \rightarrow \begin{pmatrix} \varphi \\ 0 \end{pmatrix}$ and therefore becomes an eigenstate of $\hat{P}$. In general, with

$$\begin{aligned} \psi &= \begin{pmatrix} \psi_1(x) \\ \psi_2(x) \\ \psi_3(x) \\ \psi_4(x) \end{pmatrix} , \\ \hat{P} \begin{pmatrix} \psi_1(x) \\ \psi_2(x) \\ \psi_3(x) \\ \psi_4(x) \end{pmatrix} &= \mathrm{e}^{\mathrm{i}\varphi} \begin{pmatrix} \psi_1(x) \\ \psi_2(x) \\ -\psi_3(x) \\ -\psi_4(x) \end{pmatrix} . \end{aligned} \tag{2.139}$$

Thus, the eigenstates with positive energy in the non-relativistic limit have opposite $\hat{P}$ eigenvalues to those of negative energy. One speaks, therefore, of *opposite "inner" parity* of these states.

Other improper Lorentz transformations still exist, the most important of which is *time reversal*, Sect. 4.3.

2.7 Bilinear Covariants of Dirac Spinors

Sixteen linearly independent 4×4 matrices must exist, here denoted by

$$(\hat{\Gamma}^n)_{\alpha\beta}\,, \qquad n = 1, 2 \ldots 16\,. \tag{2.140}$$

It turns out that it is possible to construct 16 such linearly independent matrices from the Dirac matrices and their products. They are

$$\begin{aligned} &\hat{\Gamma}^{\mathrm{S}} = \mathbb{1}\,, \quad \hat{\Gamma}^{\mathrm{V}}_{\mu} = \gamma_\mu\,, \quad \hat{\Gamma}^{\mathrm{T}}_{\mu\nu} = \hat{\sigma}_{\mu\nu} = -\hat{\sigma}_{\nu\mu}\,, \\ &\quad [1] \qquad\qquad [4] \qquad\qquad\qquad [6] \\ &\hat{\Gamma}^{\mathrm{P}} = \mathrm{i}\gamma^0\gamma^1\gamma^2\gamma^3 \equiv \gamma_5 \equiv \gamma^5\,, \quad \hat{\Gamma}^{\mathrm{A}}_{\mu} = \gamma_5\gamma_\mu\,. \\ &\quad [1] \qquad\qquad\qquad\qquad\qquad\qquad [4] \end{aligned} \tag{2.141}$$

The numbers in brackets [$\cdots$] indicate the number of a particular type of matrices. The superscripts S, V, T, P, A stand for scalar, vector, tensor, pseudoscalar and axial vector, respectively. This nomenclature will become transparent below. These matrices (2.141) have some important properties.

a) For each $\hat{\Gamma}^n$

$$(\hat{\Gamma}^n)^2 = \pm\mathbb{1}\,. \tag{2.142}$$

This is obvious for $\hat{\Gamma}^{\mathrm{S}}$, also for $(\hat{\Gamma}^{\mathrm{V}}_{\mu})^2 = \gamma_\mu^2 = g^{\mu\mu}$ (because of (2.73)), and can be checked straightforwardly for all other matrices (2.141).

b) For each $\hat{\Gamma}^n$ (except $\hat{\Gamma}^{\mathrm{S}}$) at least one $\hat{\Gamma}^m$ exists such that

$$\hat{\Gamma}^n\hat{\Gamma}^m = -\hat{\Gamma}^m\hat{\Gamma}^n\,. \tag{2.143}$$

This is illustrated just for $\hat{\Gamma}^{\mathrm{V}}_{\mu}$, for which $\hat{\Gamma}^{\mathrm{V}}_{\mu}\hat{\Gamma}^{\mathrm{V}}_{\nu} = -\hat{\Gamma}^{\mathrm{V}}_{\nu}\hat{\Gamma}^{\mathrm{V}}_{\mu}$ for $\nu \neq \mu$ because of the anticommutation relations (2.73). Also $\hat{\Gamma}^{\mathrm{T}}_{\mu\nu}\hat{\Gamma}^{\mathrm{P}} = -\hat{\Gamma}^{\mathrm{P}}\hat{\Gamma}^{\mathrm{T}}_{\mu\nu}$ and similarly for all other matrices. Relation (2.143) is important, because from it and (2.142) follows

$$\pm\hat{\Gamma}^n = -\hat{\Gamma}^m\hat{\Gamma}^n\hat{\Gamma}^m = +\hat{\Gamma}^n(\hat{\Gamma}^m)^2\,,$$

and therefore, since $\mathrm{tr}\,\hat{A}\hat{B} = \mathrm{tr}\,\hat{B}\hat{A}$,

$$\pm\,\mathrm{tr}\,\hat{\Gamma}^n = -\,\mathrm{tr}\,\hat{\Gamma}^n(\hat{\Gamma}^m)^2 = \mathrm{tr}\,\hat{\Gamma}^n(\hat{\Gamma}^m)^2 = 0\,. \tag{2.144}$$

In other words, *all the matrices* (2.141) *except* $\hat{\Gamma}^{\mathrm{S}}$ *have vanishing trace.*

c) For given $\hat{\Gamma}^a$ and $\hat{\Gamma}^b$ with $a \neq b$ one can always find a $\hat{\Gamma}^n \neq \hat{\Gamma}^{\mathrm{S}}$ such that

$$\hat{\Gamma}^a\hat{\Gamma}^b = f^{ab}_n\hat{\Gamma}^n\,, \qquad \text{(no summation over } n!) \tag{2.145}$$

where f^{ab}_n is a c number. This is proved by a few examples:

$$\hat{\Gamma}^{\mathrm{V}}_{\mu}\hat{\Gamma}^{\mathrm{V}}_{\nu} = \gamma_\mu\gamma_\nu = \frac{\gamma_\mu\gamma_\nu + \gamma_\nu\gamma_\mu}{2} = -\mathrm{i}\,\hat{\sigma}_{\mu\nu} = -\mathrm{i}\hat{\Gamma}^{\mathrm{T}}_{\mu\nu}\,.$$

The f factor is in this case $f = -i$. Similarly, one finds

$$\hat{\Gamma}^{\mathrm{V}}_{\mu}\hat{\Gamma}^{\mathrm{T}}_{\sigma\tau} = \frac{\mathrm{i}}{2}\gamma_\mu(\gamma_\sigma\gamma_\tau - \gamma_\tau\gamma_\sigma) = \begin{cases} -\mathrm{i}\gamma_\sigma g_{\mu\mu} = -\mathrm{i}g_{\mu\mu}\hat{\Gamma}^{\mathrm{V}}_{\sigma} & \text{for} \quad \mu = \tau \\ +\mathrm{i}\gamma_\tau g_{\mu\mu} = \mathrm{i}g_{\mu\mu}\hat{\Gamma}^{\mathrm{V}}_{\tau} & \text{for} \quad \mu = \sigma \\ \pm\mathrm{i}\hat{\Gamma}^{\mathrm{A}}_{\varkappa}, \quad \varkappa \neq \mu, \sigma, \tau & \text{for} \quad \mu \neq \sigma \neq \tau , \end{cases}$$

etc. Properties (a – c) allow the linear independence of the covariants (2.141) to be proved quite elegantly. Suppose a relation of the form

$$\sum_n a_n \hat{\Gamma}^n = 0 \tag{2.146}$$

exists. Then, after multiplication with $\hat{\Gamma}^m \neq \hat{\Gamma}^{\mathrm{S}}$ and taking the trace,

$$\begin{aligned} 0 = \sum_n a_n \operatorname{tr} \hat{\Gamma}^n \hat{\Gamma}^m &= a_m \operatorname{tr}(\hat{\Gamma}^m)^2 + \sum_{n \neq m} a_n \operatorname{tr} f_r^{nm} \hat{\Gamma}^r \\ &= a_m(\pm 1) \quad + 0 , \end{aligned}$$

because of (2.142, 144). Hence $a_m = 0$ for all $m \neq s$. For $\hat{\Gamma}^m = \hat{\Gamma}^s$ similarly

$$\begin{aligned} 0 = \operatorname{tr} \sum_n a_n \hat{\Gamma}^s \hat{\Gamma}^n &= a_s \operatorname{tr}(\hat{\Gamma}^s)^2 + \sum_{n \neq s} a_n \operatorname{tr} \hat{\Gamma}^n \\ &= 4a_s \qquad + 0 , \end{aligned}$$

and thus also $a_s = 0$. Hence all coefficients a_n vanish necessarily if (2.146) holds, which means that $\hat{\Gamma}^n$ are linearly independent. In other words, each 4×4 matrix (2.141) cannot be expressed through the others. Next we investigate the behaviour of bilinear forms of the type

$$\bar{\psi}(x)\hat{\Gamma}^n \psi(x) \tag{2.147}$$

under Lorentz transformations. We shall need the obvious relation

$$\gamma^\mu \gamma_5 + \gamma_5 \gamma^\mu = 0 \tag{2.148}$$

from which follows

$$[\gamma_5, \hat{\sigma}_{\mu\nu}]_- = -\frac{\mathrm{i}}{2}(\gamma_5(\gamma_\mu\gamma_\nu - \gamma_\nu\gamma_\mu) - (\gamma_\mu\gamma_\nu - \gamma_\nu\gamma_\mu)\gamma_5) = 0 . \tag{2.149}$$

We know from (2.115) that spinors transform under proper Lorentz transformations with the operator

$$\hat{S}(\hat{a}) = \exp\left(-\frac{\mathrm{i}}{4}\hat{\sigma}_{\mu\nu}\omega_n^{\mu\nu}\right) = \exp\left[-\frac{\mathrm{i}}{4}\omega\sigma_{\mu\nu}(\hat{I}_n)^{\mu\nu}\right] . \tag{2.150}$$

Because of (2.149) it follows that

$$[\hat{S}(\hat{a}), \gamma_5]_- = 0 . \tag{2.151}$$

The spinors are transformed under space inversion with the operator [see (2.133)]

$$\hat{P} = \mathrm{e}^{\mathrm{i}\varphi}\gamma^0 ,$$

and it is immediately obvious that

$$[\hat{P}, \gamma^5]_- = 0 \tag{2.152}$$

holds. It can easily be computed that

$$\begin{aligned} \bar{\psi}'(x')\psi'(x') &= \psi'^\dagger(x')\gamma^0\psi'(x') = \psi^\dagger(x)\hat{S}^\dagger\gamma^0\hat{S}\psi(x) \\ &= \psi^\dagger(x)\gamma^0\hat{S}^{-1}\hat{S}\psi(x) = \bar{\psi}(x)\psi(x) , \end{aligned} \tag{2.153}$$

showing that in all Lorentz systems

$$\bar{\psi}(x)\psi(x) = \bar{\psi}(x)\hat{\Gamma}^S\psi(x) \tag{2.154}$$

has the same value. It is an *invariant* or *scalar*. Similarly, it can be checked [Gr 81b] that

$$\bar{\psi}'(x')\gamma_5\psi'(x') = \bar{\psi}(x)\hat{S}^{-1}\gamma_5\hat{S}\psi(x) = \det(a^\mu_\nu)\,\bar{\psi}(x)\gamma_5\psi(x)$$

is a pseudoscalar,

$$\bar{\psi}'(x')\gamma^\nu\psi'(x') = a^\nu_\mu\bar{\psi}(x)\gamma^\mu\psi(x)$$

is a vector;

$$\bar{\psi}'(x')\gamma^5\gamma^\nu\psi'(x') = \det(a^\mu_\nu)\,a^\nu_\mu\bar{\psi}(x)\gamma^5\gamma^\mu\psi(x) \tag{2.155}$$

is a pseudovector;

$$\bar{\psi}'(x')\hat{\sigma}^{\mu\nu}\psi'(x') = a^\mu_\alpha a^\nu_\beta\bar{\psi}(x)\hat{\sigma}^{\alpha\beta}\psi(x)$$

is a tensor of second rank. "Pseudo" indicates that the corresponding quantity behaves as under proper Lorentz transformation, but changes sign under improper Lorentz transformations.

2.8 Gauge Invariant Coupling of Electromagnetic and Spinor Field

Electrodynamics is gauge invariant, meaning that the electrodynamic laws, i.e. the Maxwell equations, are not changed by a gauge transformation of the vector potential $A_\mu(x)$

$$A'_\mu(x) = A_\mu(x) + \frac{\partial\chi(x)}{\partial x^\mu} , \tag{2.156}$$

where $\chi(x)$ is an arbitrary function.

To ensure invariance of quantum mechanics under gauge transformations two more conditions are necessary.

1) The particle field (spinor field) ψ must transform under gauge transformations according to

$$\psi'(x) = \exp\left[-\frac{\mathrm{i}e}{\hbar c}\chi(x)\right]\psi(x)\,, \tag{2.157}$$

i.e. $\psi(x)$ obtains a space-time dependent phase $\chi(x)$.

2) Then it is necessary to couple the ψ and A_μ fields minimally, meaning that everywhere in the equation of motion terms of the form $\hat{p}_\mu\psi(x)$ are replaced by

$$\hat{p}_\mu\psi(x) \to \left(\hat{p}_\mu - \frac{e}{c}A_\mu(x)\right)\psi(x)\,. \tag{2.158}$$

This then introduces (minimal) coupling of the two fields. For the Dirac equation, where $\hat{p}_\mu$ appears in the form $\gamma^\mu\hat{p}_\mu$, the minimal coupling term is given by

$$\hat{H}_{\text{int}} = -\frac{e\gamma^\mu}{c}A_\mu(x)\,. \tag{2.159}$$

One immediately recognizes that these requirements of gauge transformations and couplings indeed leave quantum mechanics gauge invariant, because

$$\begin{aligned}\left(\hat{p}_\mu - \frac{e}{c}A'_\mu\right)\psi'(x) &= \left(\mathrm{i}\hbar\frac{\partial}{\partial x^\mu} - \frac{e}{c}\left(A_\mu + \frac{\partial\chi}{\partial x^\mu}\right)\right)\psi(x)\exp\left[-\frac{\mathrm{i}e}{\hbar c}\chi(x)\right]\\ &= \left[\left(\mathrm{i}\hbar\frac{\partial}{\partial x^\mu} - \frac{e}{c}A_\mu\right)\psi(x)\right]\exp\left[-\frac{\mathrm{i}e}{\hbar c}\chi(x)\right]\\ &= \left[\left(\hat{p}_\mu - \frac{e}{c}A_\mu\right)\psi(x)\right]\exp\left[-\frac{\mathrm{i}e}{\hbar c}\chi(x)\right].\end{aligned} \tag{2.160}$$

Since the common phase factor $\exp[-(\mathrm{i}e/\hbar c)\chi(x)]$ appears now in all terms of a wave equation, it drops out. In particular, for the Dirac equation (2.79)

$$\gamma^\mu\left(\hat{p}_\mu - \frac{e}{c}A'_\mu\right)\psi'(x) - m_0c\psi'(x) = 0\,, \tag{2.161}$$

which is, because of (2.160), equivalent to

$$\gamma^\mu\left(\hat{p}_\mu - \frac{e}{c}A_\mu\right)\psi(x) - m_0c\psi(x) = 0\,. \tag{2.162}$$

This then proves that $\psi'(x)$ together with $A'_\mu(x)$ obeys the same wave equation as $\psi(x)$ together with $A_\mu(x)$. The theory is thus gauge invariant. This fact shall be used often below.

Bibliographical Notes

In Chap. 2 we have summarized essential facts of elementary relativistic quantum mechanics. More depth is given by the standard text books [Bj 64, Sc 79, Ba 80b, Be 71, Gr 81b].

3. Dirac Particles in External Potentials

Chapter 3 treats the spectra and wave functions for a Dirac particle in various external potentials. It deepens our understanding and widens the range of applicability of the Dirac equation.

3.1 A Dirac Particle in a One-Dimensional Square Well Potential

A potential of depth V_0 and extension $-a/2 \leq x \leq +a/2$ is sketched in Figure 3.1. The Dirac equation in the three regions I, II, III is readily established by (2.56):

$$\text{I, III:} \quad (\hat{\alpha} \cdot \hat{p} + \hat{\beta} m_0)\, \psi(x) = E\, \psi(x)\,, \tag{3.1a}$$

$$\text{II:} \quad (\hat{\alpha} \cdot \hat{p} + \hat{\beta} m_0)\, \psi(x) = (E - V_0)\, \psi(x)\,. \tag{3.1b}$$

We have set $\hbar = c = 1$, so simplifying the following calculations. The energy E can be positive or negative, i.e. we consider both the particle and antiparticle solutions. The precise distinction between them is discussed below. The spinor depends only on x, which is the coordinate of the one dimension considered. Therefore $(\hat{\alpha}_2 \hat{p}_2 + \hat{\alpha}_3 \hat{p}_3)\, \psi(x) = 0$. The corresponding solutions are the free solutions (2.50) of the Dirac equation, in which for (3.1b) E is replaced by $(E - V_0)$. We consider particle solutions and antiparticle solutions with spin up only. The spinors with spin down are degenerate with the spin-up spinors, since no spin-flip occurs at the walls because the potential is spin-independent, and therefore both types are completely decoupled. The energy of these spinors may have any value between $-\infty < E < +\infty$ so that both particle and antiparticle solutions will be obtained. We can make the ansatz

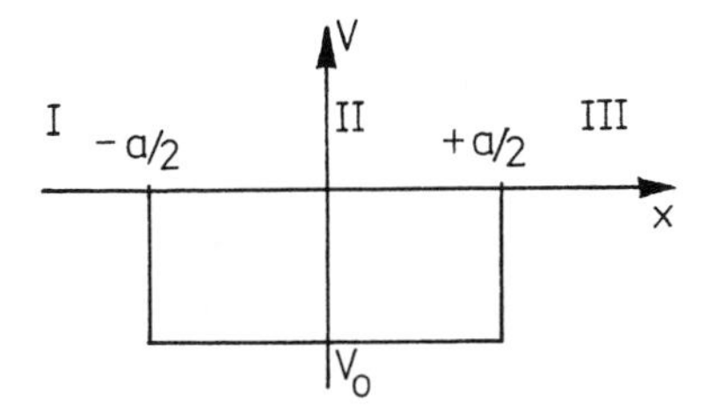

Fig. 3.1. One-dimensional square well potential with depth $V = -V_0$ and extension $-a/2 \leq x \leq a/2$

$$\text{I)}\quad \psi_{\mathrm{I}} = A\begin{pmatrix}1\\0\\\dfrac{k_1}{E+m_0}\\0\end{pmatrix}\mathrm{e}^{\mathrm{i}k_1x} + A'\begin{pmatrix}1\\0\\\dfrac{-k_1}{E+m_0}\\0\end{pmatrix}\mathrm{e}^{-\mathrm{i}k_1x}$$

$$\text{with}\quad k_1^2 = E^2 - m_0^2\,,\qquad x \le -\frac{a}{2}\,. \tag{3.2a}$$

$$\text{II)}\quad \psi_{\mathrm{II}} = B\begin{pmatrix}1\\0\\\dfrac{k_2}{(E-V_0)+m_0}\\0\end{pmatrix}\mathrm{e}^{\mathrm{i}k_2x} + B'\begin{pmatrix}1\\0\\\dfrac{-k_2}{(E-V_0)+m_0}\\0\end{pmatrix}\mathrm{e}^{-\mathrm{i}k_2x}$$

$$\text{with}\quad k_2^2 = (E-V_0)^2 - m_0^2\,,\qquad -\frac{a}{2} \le x \le \frac{a}{2}\,. \tag{3.2b}$$

$$\text{III)}\quad \psi_{\mathrm{III}} = C\begin{pmatrix}1\\0\\\dfrac{k_1}{E+m_0}\\0\end{pmatrix}\mathrm{e}^{\mathrm{i}k_1x} + C'\begin{pmatrix}1\\0\\\dfrac{-k_1}{E+m_0}\\0\end{pmatrix}\mathrm{e}^{-\mathrm{i}k_1x}$$

$$\text{with}\quad k_1^2 = E^2 - m_0^2\,,\qquad x \ge \frac{a}{2}\,. \tag{3.2c}$$

Here, $\hbar = c = 1$, but at the end $\hbar$ and c will be inserted again so that the final results are in the usual units. The wave function has to be continuous at the borders of the three regions, following from the validity of the continuity equation $\partial j^\mu/\partial x^\mu = 0$.

Hence at

$x = -a/2$:

$$A\,\mathrm{e}^{-\mathrm{i}k_1a/2} + A'\,\mathrm{e}^{\mathrm{i}k_1a/2} = B\mathrm{e}^{-\mathrm{i}k_2a/2} + B'\,\mathrm{e}^{+\mathrm{i}k_2a/2}\,, \tag{3.3}$$

$$(A\,\mathrm{e}^{-\mathrm{i}k_1a/2} - A'\,\mathrm{e}^{\mathrm{i}k_1a/2})\frac{k_1}{E+m_0} = (B\mathrm{e}^{-\mathrm{i}k_2a/2} - B'\,\mathrm{e}^{+\mathrm{i}k_2a/2})\frac{k_2}{(E-V_0)+m_0}\,. \tag{3.4}$$

$x = +a/2$:

$$B\mathrm{e}^{\mathrm{i}k_2a/2} + B'\,\mathrm{e}^{-\mathrm{i}k_2a/2} = C\mathrm{e}^{\mathrm{i}k_1a/2} + C'\,\mathrm{e}^{-\mathrm{i}k_1a/2}\,, \tag{3.5}$$

$$(B\mathrm{e}^{\mathrm{i}k_2a/2} - B'\,\mathrm{e}^{-\mathrm{i}k_2a/2})\frac{k_2}{(E-V_0)+m_0} = (C\mathrm{e}^{\mathrm{i}k_1a/2} - C'\,\mathrm{e}^{-\mathrm{i}k_1a/2})\frac{k_1}{E+m_0}\,. \tag{3.6}$$

Defining

$$\gamma = \frac{k_1}{E+m_0}\frac{(E-V_0)+m_0}{k_2} = \sqrt{\frac{(E-m_0)(E-V_0+m_0)}{(E+m_0)(E-V_0-m_0)}}\,, \tag{3.7}$$

(3.3 – 6) can be written in the following matrix notation:

$$\begin{pmatrix} A \\ A' \end{pmatrix} = \frac{1}{2}\begin{bmatrix} \dfrac{\gamma+1}{\gamma}\,\mathrm{e}^{\mathrm{i}(k_1-k_2)a/2} & \dfrac{\gamma-1}{\gamma}\,\mathrm{e}^{\mathrm{i}(k_1+k_2)a/2} \\ \dfrac{\gamma-1}{\gamma}\,\mathrm{e}^{-\mathrm{i}(k_1+k_2)a/2} & \dfrac{\gamma+1}{\gamma}\,\mathrm{e}^{\mathrm{i}(k_2-k_1)a/2} \end{bmatrix}\begin{pmatrix} B \\ B' \end{pmatrix}, \tag{3.8}$$

$$\begin{pmatrix} B \\ B' \end{pmatrix} = \frac{1}{2}\begin{pmatrix} (1+\gamma)\mathrm{e}^{\mathrm{i}(k_1-k_2)a/2} & (1-\gamma)\mathrm{e}^{-\mathrm{i}(k_1+k_2)a/2} \\ (1-\gamma)\mathrm{e}^{\mathrm{i}(k_1+k_2)a/2} & (1+\gamma)\mathrm{e}^{\mathrm{i}(k_2-k_1)a/2} \end{pmatrix}\begin{pmatrix} C \\ C' \end{pmatrix}. \tag{3.9}$$

Inserting (3.9) into (3.8) gives

$$\begin{pmatrix} A \\ A' \end{pmatrix} = \frac{1}{4\gamma}\begin{pmatrix} (1+\gamma)^2\mathrm{e}^{\mathrm{i}(k_1-k_2)a} - (1-\gamma)^2\mathrm{e}^{\mathrm{i}(k_1+k_2)a}, & -(1-\gamma^2)(\mathrm{e}^{\mathrm{i}k_2a}-\mathrm{e}^{-\mathrm{i}k_2a}) \\ -(1-\gamma^2)(\mathrm{e}^{-\mathrm{i}k_2a}-\mathrm{e}^{\mathrm{i}k_2a}), & (1+\gamma)^2\mathrm{e}^{\mathrm{i}(k_2-k_1)a} - (1-\gamma)^2\mathrm{e}^{-\mathrm{i}(k_1+k_2)a} \end{pmatrix}\begin{pmatrix} C \\ C' \end{pmatrix}. \tag{3.10}$$

These are two equations with four unknowns A, A', C, C'. A third equation is the normalization condition

$$\int \psi^\dagger \psi(x)\,dx = 1\,. \tag{3.11}$$

Thus, in general, one of the four quantities A, A', C, C' can be chosen freely, to satisfy the boundary conditions (see below). Let us now first discuss qualitatively the solutions for various energies. For that purpose it is useful to look at Figs. 3.2, 3. In Fig. 3.2 it is assumed that $|V_0| < 2m_0$ whilst in Fig. 3.3 $|V_0| > 2m_0$. In Fig. 3.2 four energy domains can be distinguished.

1) $E > m_0$. Free electrons travel from left to right and are scattered by the (attractive) potential. If the length of the potential well a is an integer multiple of the wavelength in domain II, a potential resonance appears, i.e. the probability for finding the electron within the well is especially large.

2) $m_0 + V_0 < E < m_0$. The bound states are found within this energy range, characterized by an exponentially decaying probability in domains I and III.

3) $-m_0 + V_0 < E < -m_0$. Chapter 4 shows that negative energy states are to be interpreted as wave functions of antiparticles, i.e. of positrons. Incoming positrons (continuum states with negative energy) "feel" a repulsive potential

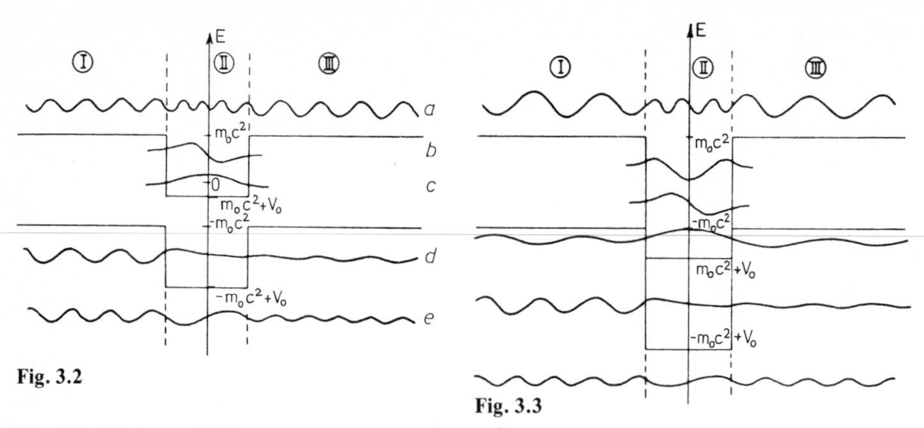

Fig. 3.2. The potential depth V_0 is smaller than $2m_0c^2$. The qualitative behaviour of the large components of the wave function is shown for a few typical energies: (*a*) shows an electron wave function of positive energy, (*b*, *c*) represent bound states. The wave functions (*d*, *e*) are typical for the negative energy continuum

Fig. 3.3. The potential depth V_0 is larger than $2m_0c^2$. The qualitative behaviour of wave functions with various energies is indicated in analogy to Fig. 3.2

(because of their opposite charge), at which they will be scattered. Since the probability of finding such a particle in domain II decreases exponentially, the reflection of the positrons is large (the transmission decreases with a and increases with $|E|$).

4) $E < -m_0 + V_0$. The positrons are scattered at the repulsive potential. Again, potential resonances might appear if the extension of well a is an integer multiple of the wavelength in domain II.

Figure 3.3 illustrates the case for $|V_0| > 2m_0$ for which an additional energy interval with $m_0 + V_0 < E < -m_0$ appears, showing a new feature of the wave function. In this energy interval electronic bound states with $E < -m_0$ are possible. Their wave functions, however, do not decrease exponentially in domains I and III, but, instead, join a continuum wave with the same energy $E < -m_0$. Hence the wave function has a finite probability outside the potential well. Later we discuss hole theory, which assumes that all states with $E < -m_0$ are occupied by electrons. A hole in this "Dirac sea" is essentially identified with a positron, the antiparticle of the electron. Therefore we can illustrate such wave functions by saying that a hole in such a state will escape the potential well with a certain probability, moving away to infinity as a positron. If such a bound state has by some means been prepared to be empty, it will, as time proceeds, be occupied by an electron from the filled Dirac sea. Hence in this case spontaneous electron-positron pair production becomes possible. An empty bound state is filled spontaneously by an electron from the occupied negative energy continuum and a positron (a hole) escapes to infinity. Because of the spin degeneracy two such pairs are created. Such a potential well is called "*overcritical with respect to* $e^+ - e^-$ *pair creation*" (Chap. 6).

Thus, while bound states in the energy domain $-m_0 \leq E \leq m_0$ can be empty without causing instability of the total system, it is not possible to keep bound states which "dived" into the negative energy continuum empty for a long time: they will be filled spontaneously. In other words, the hole in such states has a finite lifetime and the state therefore a finite decay width. It will be shown that in the energy interval $m_0 + V_0 \leq E \leq -m_0$ no energetically sharp bound states (as for states in the interval $-m_0 \leq E \leq m_0$) exist. The wave function of such states obtains a resonating structure, peaking around the expected binding energy of the bound state.

After these qualitative considerations we shall now specify the statements made above via the solution of (3.8 – 11). We distinguish two cases:

a) $|E| > m_0$, i.e. k_1 is real

b) $|E| < m_0$, i.e. k_1 is imaginary .

Solutions of type (a) are generally called *scattering states*. Let us first consider case (b), for which (3.8 – 10) simplify considerably, because A and C' must vanish, so that ψ_{I} and ψ_{III} do not grow exponentially, i.e. they should be normalizable. Hence (3.10) yields

$$0 = \frac{1}{4\gamma} \mathrm{e}^{\mathrm{i}k_1 a} C((1+\gamma)^2 \mathrm{e}^{-\mathrm{i}k_2 a} - (1-\gamma)^2 \mathrm{e}^{\mathrm{i}k_2 a}) . \tag{3.12}$$

Here C must be $\neq 0$, otherwise the total wave function would be zero. It thus follows that

$$\frac{1+\gamma}{1-\gamma} \mathrm{e}^{-\mathrm{i}k_2 a} = \frac{1-\gamma}{1+\gamma} \mathrm{e}^{\mathrm{i}k_2 a} . \tag{3.13}$$

As long as k_2 is real (which holds in Fig. 3.2 for $m_0 + V_0 < E < m_0$ and in Fig. 3.3 for the total interval $-m_0 < E < m_0$), γ is pure imaginary, because k_1 is imaginary. Therefore (3.13) simply states that

$$\frac{1+\gamma}{1-\gamma} e^{-ik_2 a} = \left[\frac{(1+\gamma)}{(1-\gamma)} \mathrm{e}^{-ik_2 a} \right]^* , \tag{3.14}$$

where (*) denotes complex conjugation. This can also be expressed by

$$\mathrm{Im} \left\{ \frac{1+\gamma}{1-\gamma} \mathrm{e}^{-ik_2 a} \right\} = 0 . \tag{3.15}$$

Setting $\gamma = \mathrm{i}\Gamma$, $(\Gamma \in \mathbb{R})$, it follows that

$$\frac{2\Gamma}{1-\Gamma^2} = \tan k_2 a . \tag{3.16}$$

Inserting now Γ from (3.7) gives

$$k_2 \cot k_2 a = f(k_2)\,, \quad \text{where}$$

$$f(k_2) = -\frac{E V_0}{\varkappa_1} - \varkappa_1 \quad \text{with} \quad k_1 = \mathrm{i}\varkappa_1\,. \tag{3.17}$$

This equation determines the energies of the bound states. Note that (3.13) has no solution, if in Fig. 3.2 $-m_0 < E < m_0 + V_0$, i.e. if k_2 is imaginary:

$$\frac{(1+\gamma)^2}{(1-\gamma)^2}\, \mathrm{e}^{2\varkappa_2 a} \neq 1\,, \quad (k_2 = \mathrm{i}\varkappa_2)\,. \tag{3.18}$$

This is because $|(1+\mathrm{i}\Gamma)^2/(1-\mathrm{i}\Gamma)^2| = 1$, $\exp(2\varkappa_2 a) > 1$. Hence, in this energy domain, the wave function vanishes identically.

Equation (3.17) can be solved approximately by a graphical method, which suffices to obtain a qualitative survey about the behaviour of the bound states. We proceed with that first; afterwards we determine the exact energy spectrum of the bound states by numerical solution of (3.17).

We know that

$$k_2 \cot k_2 a = \frac{1}{a}(k_2 a) \cot k_2 a \xrightarrow[k_2 a \to 0]{} \frac{1}{a}\,,$$

$$f(k_2) \xrightarrow[k_2 \to 0]{} \frac{m_0 V_0}{\sqrt{-2 m_0 V_0 - V_0^2}} < 0\,, \tag{3.19}$$

$$f(k_2) \xrightarrow[k_2 \to \sqrt{V_0^2 - 2 m_0 V_0}]{} +\infty \quad (\text{corresponding to } E \to m_0)\,.$$

For $V_0 < -2m_0$ the limit $f(k_2 \to 0)$ becomes imaginary. Hence for (3.17) a graph as depicted in Fig. 3.4 can be drawn. We see immediately that always at least one bound state exists, independent of the depth V_0 of the potential, agreeing with the corresponding non-relativistic problem [Me 70], but contrasting to the corresponding three-dimensional problem; because of the angular momentum barrier, not every three-dimensional well contains a bound state, but only potential wells deeper than certain depth V_0 do.

The numerical solution of (3.17) can be obtained for various sets of parameters V_0, a. A resulting energy spectrum is shown in Fig. 3.5 for $a = 10\,\lambda_e = 10\,\hbar/m_0 c$. It is seen that for increasing $|V_0|$ more and more bound states appear. At $V_0 = -2.04\, m_0 c^2$ the *potential becomes overcritical*, i.e. the lowest bound state "dives" into the lower continuum, in which it becomes a resonance, as can be seen by inspecting the transmission coefficient (see below).

Let us now consider the scattering states. Again several energy domains exist:

a) k_2 and γ are real for $E > m_0$ and for $E < -m_0 + V_0$. In the overcritical case the additional domain $m_0 + V_0 < E < -m_0$ exists where the conditions are fulfilled.

b) k_2 and γ are imaginary for $V_0 - m_0 < E < V + m_0$.

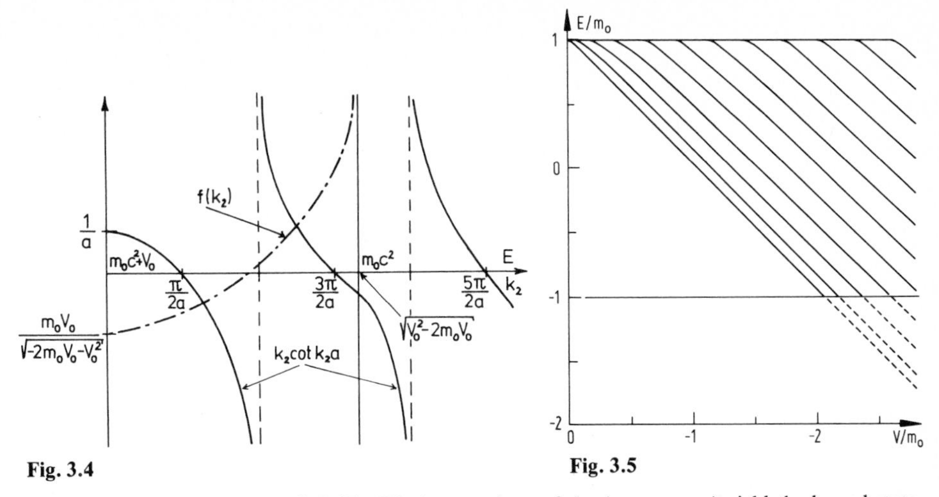

Fig. 3.4

Fig. 3.5

Fig. 3.4. Graphical solution of (3.17). The intersections of the (– · –, ——) yield the bound state energies

Fig. 3.5. Eigenvalue spectrum of electrons bound in a one-dimensional square well potential of extension 10 λ_e. (– – –) show those energies at which the "dived states" resonate, as can be deduced from the maxima of the transmission coefficient (Fig. 3.6)

At first we choose one of the constants A, A', C, C' arbitrarily assuming that there is no wave impacting on the potential from the right-hand side, i.e. $C' = 0$. Thus, C can simply be interpreted as that fraction of a wave with amplitude A moving from the left through the potential well. The term proportional to A' is the wave reflected at the potential.

It is now possible to define a transmission coefficient T and a phase shift δ by

$$\frac{C}{A} = \sqrt{T}\mathrm{e}^{-\mathrm{i}\delta}. \tag{3.20}$$

This means that the outgoing wave is reduced in its amplitude by the factor $\sqrt{T}$ and shifted in its phase relative to the incoming wave by δ.

From (3.10) for real k_2

$$\begin{aligned} T = \left|\frac{C}{A}\right|^2 &= \left[\cos^2 k_2 a + \left(\frac{1+\gamma^2}{2\gamma}\right)^2 \sin^2 k_2 a\right]^{-1} \\ &= \left[1 + \left(\frac{1-\gamma^2}{2\gamma}\right)^2 \sin^2 k_2 a\right]^{-1} \leq 1\,. \end{aligned} \tag{3.21}$$

The phase shift δ follows from [see (3.10)]

$$\frac{1}{\sqrt{T}}\mathrm{e}^{\mathrm{i}\delta}\mathrm{e}^{-\mathrm{i}k_1 a} = \cos k_2 a - \mathrm{i}\left(\frac{1+\gamma^2}{2\gamma}\right)\sin k_2 a\,, \tag{3.22}$$

which can be written as

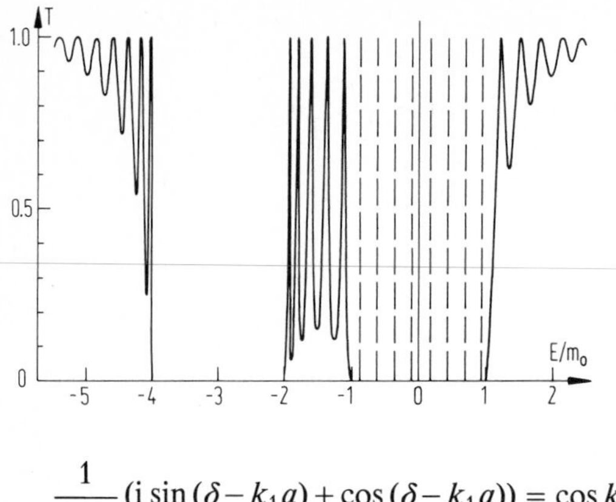

Fig. 3.6. The transmission coefficient for scattering states of a one-dimensional square well potential of depth $V = -3\,m_0c^2$ and extension $a = 10\,\lambda_e$. The energies of the bound states are indicated by (– – –)

$$\frac{1}{\sqrt{T}}(\mathrm{i}\sin(\delta-k_1a)+\cos(\delta-k_1a)) = \cos k_2a - \mathrm{i}\frac{(1+\gamma^2)}{2\gamma}\sin k_2a\,. \tag{3.23}$$

Separating real and imaginary parts and eliminating $\sqrt{T}$ yields

$$-\tan(k_1a-\delta) = -\frac{1+\gamma^2}{2\gamma}\tan k_2a\,, \tag{3.24}$$

or, respectively,

$$\delta = k_1a - \arctan\left(\frac{1+\gamma^2}{2\gamma}\tan k_2a\right). \tag{3.25}$$

If, on the other hand, k_2 is imaginary, (3.10) gives instead of (3.21) the following transmission coefficient

$$T = \left[1+\left(\frac{1+\Gamma^2}{2\Gamma}\right)^2\sinh^2\varkappa_2a\right]^{-1}\leq 1\,, \tag{3.26}$$

where $\gamma = \mathrm{i}\Gamma$ and $k_2 = \mathrm{i}\varkappa_2$. Instead of (3.25), the phase shift now reads

$$\delta = k_1a - \arctan\left(\frac{1-\Gamma^2}{2\Gamma}\tanh\varkappa_2a\right). \tag{3.27}$$

The transmission coefficient for a potential depth $V_0 = -3\,m_0$ and an extension $a = 10\,\lambda_e = 10\,\hbar/m_0c$ is plotted in Fig. 3.6. We have immediately chosen an overcritical potential, because the undercritical case will be the same as before except for the non-appearance of states in the domain $m_0 + V_0 < E < -m_0$ in Fig. 3.6. Pronouced resonance structures in the electron continuum $E > m_0$ and in the negative energy continuum for energies above the potential barrier $|E| > |V_0 - m_0|$ $(E < -m_0 + V_0)$ appear. Positrons (electrons of negative energy) with small kinetic energy $(-m_0 + V_0 < E < m_0 + V_0)$ penetrate the barrier only with a probability which decreases exponentially with the extension of the potential well a. Therefore T is practically zero in this case. However, in the energy domain $m_0 + V_0 < E < -m_0$ the possibility exists that the incoming wave coin-

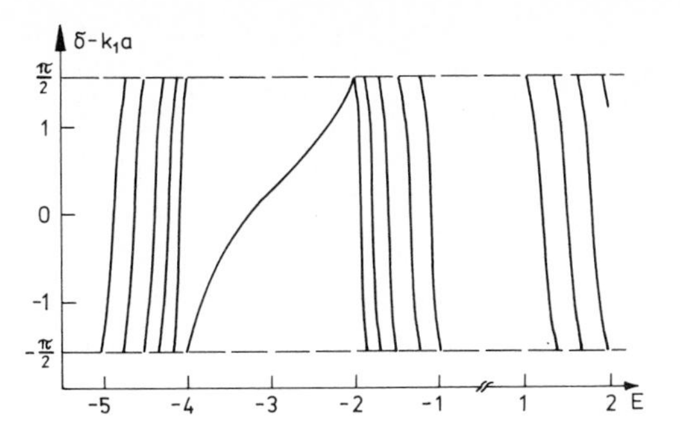

Fig. 3.7. The phase shift δ for scattering states of a one-dimensional square well potential of depth $V_0 = 3\,m_0c^2$ and extension $a = 10\,\lambda_e$ as a function of energy E in units of m_0c^2

cides with the energy of an overcritical, quasi-bound state and, consequently, can more or less penetrate the potential well undisturbed. Indeed, at that energy (Fig. 3.6) where one would expect the quasi-bound state (by extrapolation from the spectrum of the bound states) $T = 1$ arises. The bound state, which has dived into the negative energy continuum, obviously becomes a resonance below $E = -m_0$. Such resonances do not exist for undercritical potentials. They are responsible for spontaneous pair creation.

Figure 3.7 depicts graphically the scattering phases δ. As can be seen from (3.21, 25–27) the transmission coefficient T becomes 1 if $\delta - k_1 a = 0$ (mod π), and T becomes minimal for $\delta - k_1 a = \pi/2$ (mod π). In the latter case, i.e. for minimal T, the reflection coefficient is maximal, agreeing with the well-known fact from scattering theory that scattering cross-sections become maximal if the scattering phase goes through $\pi/2$.

Thus, in the energy domain $m_0 + V_0 < E < -m_0$, one can identify the energies of the dived bound states through $\delta - k_1 a = 0$ (mod π). These energies appear exactly where extrapolation from the bound state spectrum (Fig. 3.5) predicts.

3.2 A Dirac Particle in a Scalar, One-Dimensional Square Well Potential

The potential depth is again V_0, its linear extension a. The potential, however, is not considered as the time-like component of a four-vector, as in the last section, but as a scalar potential, meaning that it is considered as an r-dependent mass and therefore coupled into the Dirac equation together with the mass. Scalar couplings play a role in many areas of modern physics, e.g. for spin-0 mesons like the σ meson, on which the so-called σ model of *Gell-Mann* and *Lévy* [Ge 60] and of *Lee* and *Wick* [Le 74] is based. Also the soliton-bag model of quarks [Le 81, Wi 81] relies on a scalar coupling between the quarks (3.28) and a σ field.

As in the last section, we define three domains I ($x < -a/2$), II ($-a/2 \le x \le a/2$) and III ($x > a/2$). The Dirac equation then reads in the various domains

$$\text{I, III:} \quad (\hat{\alpha}\cdot\hat{p}+\hat{\beta}m_0)\,\psi = E\psi\,,$$

$$\text{II:} \quad (\hat{\alpha}\cdot\hat{p}+\hat{\beta}(m_0+V_0))\,\psi = E\psi\,, \qquad (V_0<0)\,. \tag{3.28}$$

In contrast to vector coupling, which yielded in domain II the replacement of E by $(E-V_0)$, we now have to replace m_0 by (m_0+V_0) in this domain. While the vector coupling of the potential affected the electron $(E>0)$ and positron $(E<0)$ wave functions differently (if electrons were attracted by the potential, the positrons were repelled by the same potential and vice versa) and hence yielded an asymmetric energy spectrum (bound states existed for only one kind of particle, either for electrons or for positrons), the scalar coupling acts on particles and antiparticles alike. For vector coupling of a potential the particle couples to the charge (which is different for particles and antiparticles) while for scalar coupling the particle couples to the mass (which is the same for particles and antiparticles). In the last case we thus expect a symmetric energy spectrum, i.e. for electrons as well positrons bound states exist in the same potential. Therefore we expect an overcritical behaviour already for $V_0 \leq -m_0$; even in this relatively shallow potential electronic and positronic states may cross in principle. What happens afterwards is discussed in detail further below.

Let us now proceed as for vector coupling. Again the momentum in domains I and III is given by

$$k_1^2 = E^2-(m_0)^2\,, \tag{3.29}$$

while in domain II

$$k_2^2 = E^2-(m_0+V_0)^2\,. \tag{3.30}$$

We can also again denote the conditions for continuity of the wave function at $x=-a/2$ and $x=a/2$, respectively. In analogy to vector coupling we define

$$\gamma = \frac{k_1}{(E+m_0)}\,\frac{(E+m_0+V_0)}{k_2} = \sqrt{\frac{(E-m_0)(E+m_0+V_0)}{(E+m_0)(E-m_0-V_0)}}\,. \tag{3.31}$$

Note that $\gamma\to 1/\gamma$ if $E\to -E$. With the same notation as for vector coupling (Sect. 3.1), (3.10) can be used, realizing that γ is now given by (3.31). Thus,

$$\begin{pmatrix} A \\ A' \end{pmatrix} = \frac{1}{4\gamma}\begin{pmatrix} (1+\gamma)^2 e^{i(k_1-k_2)a}-(1-\gamma)^2 e^{i(k_1+k_2)a}\,, & -(1-\gamma^2)(e^{ik_2a}-e^{-ik_2a}) \\ -(1-\gamma^2)(e^{-ik_2a}-e^{ik_2a})\,, & (1+\gamma)^2 e^{i(k_2-k_1)a}-(1-\gamma)^2 e^{-i(k_1+k_2)a} \end{pmatrix}\begin{pmatrix} C \\ C' \end{pmatrix}. \tag{3.32}$$

These equations are invariant under the transformation $E\to -E$ (i.e. $\gamma\to 1/\gamma$), which reflects the particle–antiparticle symmetry.

The following cases are discussed separately:

1) $|E| < m_0$ (bound states)

2) $|E| > m_0$ (scattering states of the electron and positron continuum).

We first consider the bound states, for which the following eigenvalue equation holds

$$\frac{1+\gamma}{1-\gamma}\,\mathrm{e}^{-\mathrm{i}k_2 a} = \frac{1-\gamma}{1+\gamma}\,\mathrm{e}^{\mathrm{i}k_2 a}. \tag{3.33}$$

This equation has solutions only as long as k_2 is real, i.e. $|E| > |m_0 + V_0|$ or, equivalently $E > m_0 + V_0$, for electrons and $E < -(m_0 + V_0)$ for positrons. Under these circumstances (3.33) again yields

$$\tan k_2 a = \frac{2\Gamma}{1-\Gamma^2}, \tag{3.34}$$

where $\gamma = \mathrm{i}\Gamma\,(\Gamma \in \mathbb{R})$ and

$$k_2 \cot k_2 a = -\left(\frac{m_0 V_0}{\varkappa_1} + \varkappa_1\right) \equiv f(k_2), \tag{3.35}$$

where $k_1 = \mathrm{i}\varkappa_1$, respectively. Also this equation is, of course, symmetric in $\pm E$. One has

$$f(k_2) \xrightarrow[k_2 \to 0]{} \frac{m_0 V_0 + V_0^2}{\sqrt{-2m_0 V_0 - V_0^2}} \begin{cases} <0 & \text{for } |V_0| < m_0, \\ >0 & \text{for } |V_0| > m_0, \\ \to\infty & \text{for } V_0 \to -2m_0, \end{cases} \tag{3.36}$$

$$f(k_2) \xrightarrow[|E| \to m_0]{} +\infty, \quad \text{(corresponds to } k_2 \to \sqrt{-2m_0 c^2 V_0 - V_0^2}\text{)}, \tag{3.37}$$

$$f(k_2) = \frac{m_0 V_0 + V_0^2 + k_2^2}{\sqrt{-2m_0 V_0 - V_0^2 - k_2^2}}, \quad (0 \le k_2 < \sqrt{-2m_0 c^2 V_0 - V_0^2}). \tag{3.38}$$

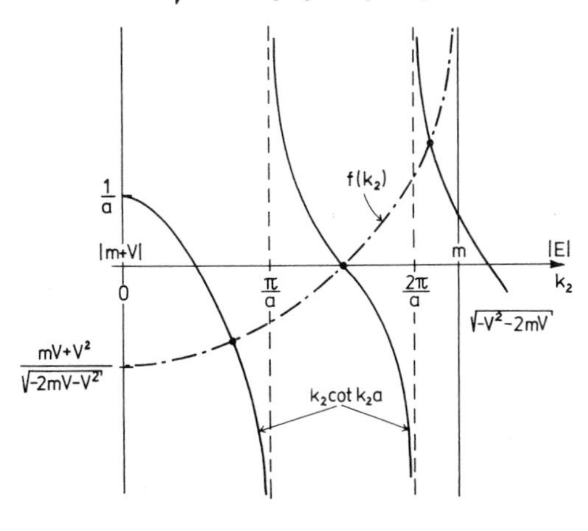

Fig. 3.8. The graphical solution of (3.36)

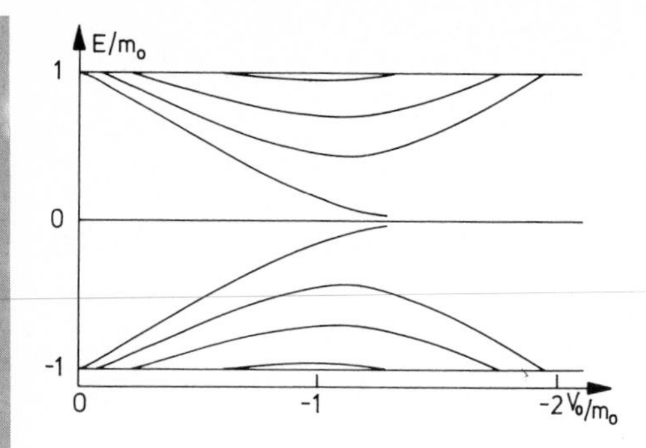

Fig. 3.9. Energy spectrum of the Dirac equation with a scalarly coupled one-dimensional square well potential of extension $a = 10\,\lambda_e$ as a function of the potential depth V

As long as $|V_0| < m_0$, the graphical solution of (3.35) is similar to that for the vector-coupling case (Fig. 3.4), except that it must be continued symmetrically to negative energies, Fig. 3.8.

Now, if $|V_0|$ increases, the intersection of $f(k_2)$ with the ordinate axis in Fig. 3.8 shifts upwards. Correspondingly the lowest eigenvalues shifts to smaller values of k_2. For $(m_0 V_0 + V_0^2)/\sqrt{-2 m_0 V_0 - V_0^2} > 1/a$ the lowest eigenvalue disappears at $k_2 = 0$, i.e. at $|E| = |m_0 + V_0| > 0$. Hence the lowest bound state never reaches the energy $E = 0$. This can also be expressed in the following way: the deepest bound electron state will never cross a corresponding positron state, i.e. one which can be traced back to lower continuum.

If $|V_0|$ is further increased, the ordinate $f(k_2)$ is also increased and, therefore, the higher lying states approach the eigenvalues $k_2 = n\pi/a$ with the eigenenergies $E^2 = (n\pi/a)^2 + (m_0 + V_0)^2$. For $|V_0| \to 2m_0$ $f(k_2 = 0)$ diverges, and the energies approach

$$E^2 = \left(\frac{n\pi}{a}\right)^2 + m_0^2 > m_0^2 .$$

This means that for $|V_0| \to 2m_0$ the bound states disappear. With increasing potential depth all bound states disappear, one after another. This property is graphically depicted in Fig. 3.9, which was obtained by numerically solving (3.35). This energy diagram is typical for a square well potential. One can investigate this problem also in three dimensions and include a scalar Coulomb-type potential, Sect. 3.4 and [So 73a].

In this case all bound states approach the energy $|E| = 0$ asymptotically ($\alpha' \to \infty$), so that again no crossings between electron and positron states appear (see Fig. 3.13).

We thus find – in contrast to vector-coupled potentials – that for scalar coupled potentials spontaneous pair creation never appears, independent of the depth of the potential. This property can be traced back to the absolute positivity of the Dirac-Hamiltonian for scalar-coupled potentials [So 82c].

The qualitatively different behaviour of bound states for differently coupled α/r potentials can easily be understood by observing that due to scalar coupling the particles obtain an effective mass

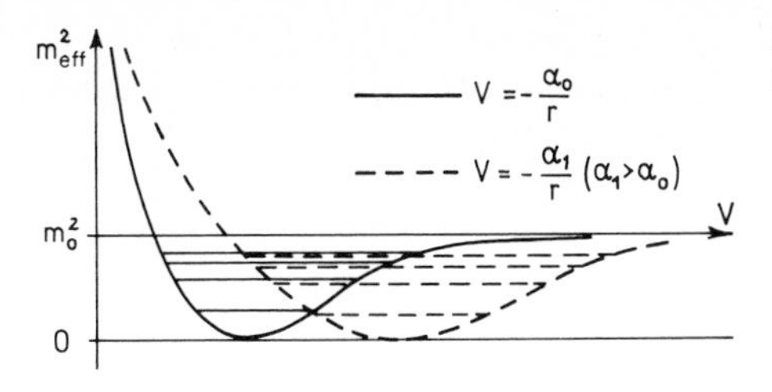

Fig. 3.10. The effective, r-dependent mass for two scalarly coupled α/r potentials

$$m_{\text{eff}}c^2 = |m_0c^2 + V(r)|. \tag{3.39}$$

For the α/r potential this effective mass is schematized in Fig. 3.10, together with some bound states.

A region with $m_{\text{eff}} < m_0$, in which bound states are possible, always exists. For increasing coupling strength α the bound state wave functions shift towards larger r values; at the same time the effective potential minimum becomes broader, so that the energy values $|E|$ are lowered and for $\alpha \to \infty$ approach $|E| = 0$. Because $m_{\text{eff}} \geq 0$ always, then $|E| \geq 0$ always, so that electron and positron states can never cross. It is true that for $\alpha \to \infty$ the energy gap between electron and positron states decreases, but it never becomes zero. Hence energy-less spontaneous pair production is never possible.

Let us shortly qualitatively discuss the scattering solutions with $|E| > m_0$, again considering $C' = 0$ [Sect. 3.1, after (3.19)], investigating the transmission coefficient $T = |C/A|^2$. In the domain $|E| > m_0$ naturally $|E| > m_0 + V_0$, but generally not $|E|^2 > (m_0 + V_0)^2$, if $V_0 < -2m_0$. This means that for $(V_0 < -2m_0)$ the electrons "feel" a potential barrier if they have energies in the energy interval

$$m_0 < E < |m_0 + V_0|.$$

A similar result holds also for positrons. Thus, while for $|V_0| < 2m_0$ resonances appear in both continua, similar to those in Fig. 3.6 for energies $E > m_0$, for $|V_0| < 2m_0$ an additional energy domain arises, where electrons and positrons "feel" a potential barrier, thus lowering the transmission coefficient significantly. This corresponds to the situation of Fig. 3.5 in the energy interval $-m_0 + V_0 < E < m_0 + V_0$.

3.3 A Dirac Particle in a Spherical Well

Let us now search for the solutions of the stationary Dirac equation

$$(\hat{\alpha} \cdot \hat{p}c + \hat{\beta}m_0c^2)\,\psi(r) = [E - V(r)]\,\psi(r) \tag{3.40}$$

with a spherical potential

$$V(r) = V(r) = \begin{cases} -V_0 & \text{for} \quad r \leq R_0 \\ 0 & \text{for} \quad r > R_0 \end{cases}. \tag{3.41}$$

In this section we shall at first keep c and $\hbar$ explicitly in the equations, later $\hbar = c = 1$ will be introduced again. First the kinetic energy operator $\hat{\alpha} \cdot \hat{p}$ is transformed into spherical polar coordinates using the identity

$$\begin{aligned} \nabla &= e_r(e_r \cdot \nabla) - e_r \times (e_r \times \nabla) \\ &= e_r \frac{\partial}{\partial r} - \mathrm{i} \frac{e_r}{r} \times \hat{L} \end{aligned} \tag{3.42}$$

with the orbital angular momentum operator (in units of $\hbar$)

$$\hat{L} = -\mathrm{i}(r \times \nabla) = -\mathrm{i}(r e_r \times \nabla) \,.$$

This yields

$$\hat{\alpha} \cdot \hat{p} = -\mathrm{i}\, \hat{\alpha}_r \frac{\partial}{\partial r} - \frac{1}{r} \hat{\alpha} \cdot (e_r \times \hat{L}) \,. \tag{3.43}$$

Using the identity

$$(\hat{\alpha} \cdot A)(\hat{\alpha} \cdot B) = A \cdot B + \mathrm{i} \hat{\Sigma} \cdot (A \times B) \tag{3.44}$$

for $A = e_r$, $B = \hat{L}$ leads to

$$\begin{aligned} \hat{\alpha}_r(\hat{\alpha} \cdot \hat{L}) &= e_r \cdot \hat{L} + \mathrm{i} \hat{\Sigma} \cdot (e_r \times \hat{L}) \\ &= \mathrm{i}\, \hat{\sigma} \cdot (e_r \times \hat{L}) \end{aligned} \tag{3.45}$$

and, by multiplying from the right by $\gamma_5 = \begin{pmatrix} 0 & 1\!\!1 \\ 1\!\!1 & 0 \end{pmatrix}$ to

$$\hat{\alpha}_r(\hat{\sigma} \cdot \hat{L}) = \mathrm{i}\, \hat{\alpha} \cdot (e_r \times \hat{L}) \,. \tag{3.46}$$

Therefore (3.43) becomes

$$\begin{aligned} \hat{\alpha} \cdot \hat{p} &= -\mathrm{i}\, \hat{\alpha}_r \frac{\partial}{\partial r} + \mathrm{i} \frac{\hat{\alpha}_r}{r} (\hat{\sigma} \cdot \hat{L}) \\ &= -\mathrm{i}\, \hat{\alpha}_r \left(\frac{\partial}{\partial r} + \frac{1}{r} - \frac{\hat{\beta}}{r} \hat{\varkappa} \right), \end{aligned} \tag{3.47}$$

where

$$\hat{\varkappa} = \hat{\beta}(\hat{\sigma} \cdot \hat{L} + 1) \,. \tag{3.48}$$

For the eigenfunctions of (3.40) we make the following ansatz:

$$\psi(r) = \begin{pmatrix} g(r) & \chi^{\mu}_{\varkappa}(\vartheta, \varphi) \\ \mathrm{i} f(r) & \chi^{\mu}_{-\varkappa}(\vartheta, \varphi) \end{pmatrix}, \tag{3.49}$$

thus separating the r and ϑ-φ motions. The 2-spinors $\chi^{\mu}_{\varkappa}$ fulfil [Ro 61, Gr 81b]:

$$\begin{aligned} \hat{\beta}\hat{\varkappa}\chi^{\mu}_{\varkappa} &= -\varkappa\chi^{\mu}_{\varkappa}\,, \\ \hat{\beta}\hat{\varkappa}\chi^{\mu}_{-\varkappa} &= \varkappa\chi^{\mu}_{-\varkappa}\,, \\ \hat{j}_z\chi^{\mu}_{\varkappa} &= \mu\chi^{\mu}_{\varkappa}\,, \qquad \varkappa = (-1, 1, -2, 2, -3, 3\ldots)\,. \end{aligned} \tag{3.50}$$

Hence the two coupled differential equations for $g(r)$ and $f(r)$ follow

$$\begin{aligned} [E - V(r) - m_0c^2]\, g(r) &= \left[-\left(\frac{d}{dr} + \frac{1}{r}\right) + \frac{\varkappa}{r}\right] f(r)\,, \\ [E - V(r) + m_0c^2]\, f(r) &= \left(\frac{d}{dr} + \frac{1}{r} + \frac{\varkappa}{r}\right) g(r)\,. \end{aligned} \tag{3.51}$$

Instead of using $f(r)$ and $g(r)$, it is often convenient to deal with

$$\begin{aligned} u_1(r) &= r g(r)\,, \\ u_2(r) &= r f(r)\,, \end{aligned} \tag{3.52}$$

for which the differential equations read

$$\frac{d}{dr}\begin{pmatrix} u_1(r) \\ u_2(r) \end{pmatrix} = \begin{pmatrix} -\dfrac{\varkappa}{r} & E + m_0c^2 - V(r) \\ -(E - m_0c^2 - V(r)) & \dfrac{\varkappa}{r} \end{pmatrix} \begin{pmatrix} u_1(r) \\ u_2(r) \end{pmatrix}. \tag{3.53}$$

For constant potentials [$V(r) = -V_0$] (3.53) has the following solutions:

a) If $\hbar^2k^2c^2 \equiv (E+V_0)^2 - m_0^2c^4 > 0$:

$$\begin{aligned} u_1(r) &= r[a_1 j_{l_\varkappa}(kr) + a_2 n_{l_\varkappa}(kr)]\,, \\ u_2(r) &= \frac{\varkappa}{|\varkappa|}\,\frac{kr}{E + V_0 + m_0c^2}[a_1 j_{l_{-\varkappa}}(kr) + a_2 n_{l_{-\varkappa}}(kr)]\,, \end{aligned} \tag{3.54}$$

where

$$l_\varkappa = \begin{cases} \varkappa & \text{for} \quad \varkappa > 0 \\ -\varkappa - 1 & \text{for} \quad \varkappa < 0 \end{cases}, \tag{3.55}$$

$$l_{-\varkappa} = \begin{cases} -\varkappa & \text{for} \quad -\varkappa > 0 \\ \varkappa - 1 & \text{for} \quad -\varkappa < 0 \end{cases}. \tag{3.56}$$

b) If $\hbar^2K^2c^2 \equiv m_0^2c^4 - (E+V_0)^2 > 0$:

$$\begin{aligned} u_1(r) &= r\sqrt{\frac{2Kr}{\pi}}\,[b_1 K_{l_\varkappa+1/2}(Kr) + b_2 I_{l_\varkappa+1/2}(Kr)]\,, \\ u_2(r) &= \frac{Kr}{E + V_0 + m_0c^2}\sqrt{\frac{2Kr}{\pi}}\,[-b_1 K_{l_{-\varkappa}+1/2}(Kr) + b_2 I_{l_{-\varkappa}+1/2}(Kr)]\,. \end{aligned} \tag{3.57}$$

The j_l, n_l are the regular and irregular spherical Bessel functions, respectively; $K_{l+1/2}$, $I_{l+1/2}$ are the modified spherical Bessel functions. Their asymptotic behaviour is given by

$$\begin{aligned} j_l(z) &\xrightarrow[z\to 0]{} \frac{1}{(2l+1)!!} z^l, \\ n_l(z) &\xrightarrow[z\to 0]{} -(2l-1)!!\frac{1}{z^{l+1}}, \end{aligned} \tag{3.58}$$

$$\begin{aligned} \sqrt{\frac{\pi}{2z}} I_{n+1/2}(z) &\xrightarrow[z\to 0]{} \frac{1}{(2n+1)!!} z^n, \\ \sqrt{\frac{\pi}{2z}} K_{n+1/2}(z) &\xrightarrow[z\to 0]{} (-)^n (2n-1)!! z^{-n-1}, \end{aligned} \tag{3.59}$$

$$\begin{aligned} \sqrt{\frac{\pi}{2z}} I_{n+1/2}(z) &\xrightarrow[z\to \infty]{} \sim \frac{e^z}{z}, \\ \sqrt{\frac{\pi}{2z}} K_{n+1/2}(z) &\xrightarrow[z\to \infty]{} \sim \frac{e^{-z}}{z}. \end{aligned} \tag{3.60}$$

Before continuing, let us denote the explicit form of the $\chi^\mu_\varkappa(\vartheta, \varphi)$ of (3.49, 50):

$$\chi^\mu_\varkappa(\vartheta, \varphi) = \sum_{m=-1/2,+1/2} (l_k \tfrac{1}{2} j; \mu - m, m, \mu)\, Y_{l_k, \mu-m}(\vartheta, \varphi)\, \tilde{\chi}^m_{1/2}, \tag{3.61}$$

where

$$\tilde{\chi}^{1/2}_{1/2} = \begin{pmatrix} 1 \\ 0 \end{pmatrix} \quad \text{and} \quad \tilde{\chi}^{-1/2}_{1/2} = \begin{pmatrix} 0 \\ 1 \end{pmatrix} \tag{3.62}$$

are the standard spinors.

Our aim now is to determine the bound states [Pi 69]. For those $E > V_0 + m_0$, $-m_0 < E < m_0$, where $\hbar = c = 1$. In the interior of the potential well we have to take the solution (3.54) with $a_2 = 0$, so that the wave function remains normalizable at the origin. In the exterior the solutions (3.57) with $b_2 = 0$ can be normalized even at infinity, (3.60). Both solutions will be smoothly connected at $r = R_0$. The normalization constants a_1 and b_1 can be eliminated by connecting smoothly at $r = R_0$ the ratio $u_1(r)/u_2(r)$. This yields

$$\frac{\varkappa}{|\varkappa|} \frac{R_0 j_{l_\varkappa}(kR_0)}{kR_0 j_{l_{-\varkappa}}(kR_0)} (E + V_0 + m_0) = -\frac{R_0 K_{l_\varkappa+1/2}(KR_0)}{KR_0 K_{l_{-\varkappa}+1/2}(KR_0)} (E + m_0), \tag{3.63}$$

or

$$\frac{j_{l_\varkappa}(kR_0)}{j_{l_{-\varkappa}}(kR_0)} = -\frac{\varkappa}{|\varkappa|} \frac{k}{K} \frac{K_{l_\varkappa+1/2}(KR_0)}{K_{l_{-\varkappa}+1/2}(KR_0)} \frac{(E+m_0)}{(E+V_0+m_0)}, \tag{3.64}$$

with

$$k = \sqrt{(E+V_0)^2 - m_0^2}\,; \quad K = \sqrt{m_0^2 - E^2}\,. \tag{3.65}$$

For $|\varkappa|=1$ these relations can be solved (nearly) analytically. For s states ($\varkappa=-1$, $l_\varkappa=0$, $l_{-\varkappa}=1$)

$$\frac{kR_0\sin(kR_0)}{\sin kR_0-kR_0\cos kR_0}=+\frac{k}{K}\,\frac{\mathrm{e}^{-KR_0}}{\mathrm{e}^{-KR_0}\left(1+\dfrac{1}{KR_0}\right)}\,\frac{E+m_0}{E+V_0+m_0}\,, \tag{3.66}$$

which after some manipulations becomes

$$\tan\left[R_0\sqrt{(E+V_0)^2-m_0^2}\right]\cdot\sqrt{\frac{E+V_0+m_0}{E+V_0-m_0}}\left[\frac{1}{R_0}\left(\frac{1}{E+V_0+m_0}-\frac{1}{E+m_0}\right)-\sqrt{\frac{m_0-E}{m_0+E}}\right]=1\,. \tag{3.67}$$

Analogously, for the $p_{1/2}$ states ($\varkappa=1$, $l_\varkappa=1$, $l_{-\varkappa}=0$)

$$\frac{kR_0\sin(kR_0)}{\sin kR_0-kR_0\cos kR_0}=-\frac{K}{k}\,\frac{1}{1+\dfrac{1}{KR_0}}\,\frac{E+V_0+m_0}{E+m_0} \tag{3.68}$$

and

$$\tan\left[R_0\sqrt{(E+V_0)^2-m_0^2}\right]\cdot\sqrt{\frac{E+V_0-m_0}{E+V_0+m_0}}\left[\frac{1}{R_0}\left(\frac{1}{E+V_0+m_0}+\frac{1}{m_0-E}\right)+\sqrt{\frac{m_0+E}{m_0-E}}\right]=1\,. \tag{3.69}$$

Another form of (3.66) is

$$kR_0\cot(kR_0)-1=\frac{m_0R_0\sqrt{1-E^2/m_0^2}+1}{1+E/m_0}\left[1+\sqrt{1+\left(\frac{kR_0}{m_0R_0}\right)^2}\right], \tag{3.70}$$

and of (3.68)

$$1-kR_0\cot kR_0=\frac{m_0R_0\sqrt{1-E^2/m_0^2}+1}{1-E/m_0}\left[1+\sqrt{1+\left(\frac{kR_0}{m_0R_0}\right)^2}\right], \tag{3.71}$$

Equations (3.67, 69) determine the eigenvalues of the energy for the $s_{1/2}$ and $p_{1/2}$ states, respectively. If R_0 is assumed to be small, i.e. $m_0R_0\ll 1$, these equations can be solved approximately by expanding in terms of m_0R_0. A short calculation leads to Table 3.1.

In the opposite limit ($m_0R_0\gg 1$) approximate solutions can be obtained from (3.70, 71). They are analogous to the case above and are listed in Table 3.2.

The $p_{1/2}$ states are always less bound than the $s_{1/2}$ states, which is intuitively clear because of the $l=1$ angular momentum of the $p_{1/2}$ states. However, even for the $s_{1/2}$ states which have $l=0$ a given radius R_0 needs a certain potential depth V_0 to get at least one bound state. This is in contrast to the one-dimensional potential well (Sect. 3.1), where one bound state always existed. The reason for that is that a Dirac particle in three dimensions has an angular momentum barrier

Table 3.1. $n = 1, 2, 3\ldots$ numbers the states. Depth V_0 of the potential well for various typical energies ($E = m_0, 0, -m_0$) and for $s_{1/2}$ and $p_{1/2}$ states ($\varkappa = -1, 1$) for $m_0 R_0 \ll 1$

E	$V_0\ (\varkappa = -1)$	$V_0\ (\varkappa = +1)$
m_0	$\frac{n\pi}{R_0} - 3m_0$	$\frac{n\pi}{R_0} - m_0$
0	$\frac{n\pi}{R_0} - m_0$	$\frac{n\pi}{R_0} + m_0$
$-m_0$	$\frac{n\pi}{R_0} + m_0$	$\frac{n\pi}{R_0} + 3m_0$

Table 3.2. Depth of the potential well for various typical energies and for $s_{1/2}$ and $p_{1/2}$ states for $m_0 R_0 \gg 1$

E	$V_0\ (\varkappa = -1)$	$V_0\ (\varkappa = +1)$
m_0	$\frac{(n-\frac{1}{2})^2\pi^2}{2m_0 R_0^2}$	$\frac{n^2\pi^2}{2m_0 R_0^2}$
0	$m_0\left(1 + \frac{n^2\pi^2}{2m_0^2 R_0^2}\right)$	$m_0\left(1 + \frac{(n+\frac{1}{2})^2\pi^2}{2m_0^2 R_0^2}\right)$
$-m_0$	$2m_0\left(1 + \frac{n^2\pi^2}{4m_0^2 R_0^2}\right)$	$2m_0\left(1 + \frac{(n+\frac{1}{2})^2\pi^2}{4m_0^2 R_0^2}\right)$

even for s states, because of the spin. This can be seen if (3.54) are decoupled. In a second differentiation an insertion for constant V_0 yields

$$
\begin{aligned}
g'' - \left\{[(E+V_0)^2 - m_0^2] - \frac{\varkappa(\varkappa+1)}{r^2}\right\} g(r) &= 0\,, \\
f'' + \left\{[(E+V_0)^2 - m_0^2] - \frac{\varkappa(\varkappa-1)}{r^2}\right\} f(r) &= 0\,.
\end{aligned}
\tag{3.72}
$$

Even though for s states ($\varkappa = -1$) the angular momentum barrier for the large component $g(r)$ vanishes, the equation for the small component $f(r)$ contains an angular momentum term which must first be balanced by the potential V_0 before a bound state – even for s waves – can appear.

Figure 3.11 shows the $1s_{1/2}$ eigenvalues obtained numerically for various potential depths and different radii R_0. For both $m_0R_0 \ll 1$ and $m_0R_0 \gg 1$, $E(V_0)$ has nearly linear dependence on the potential depth V_0.

As in the one-dimensional case we can also determine scattering phases for continuum states. For the s states this can be done relatively easily, while for the higher states the continuity conditions at $r = R_0$ are quite involved. We therefore consider the scattering phases for s waves only.

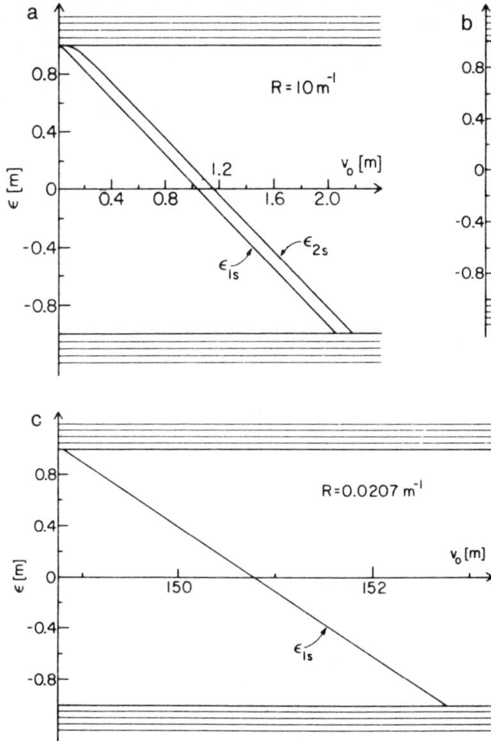

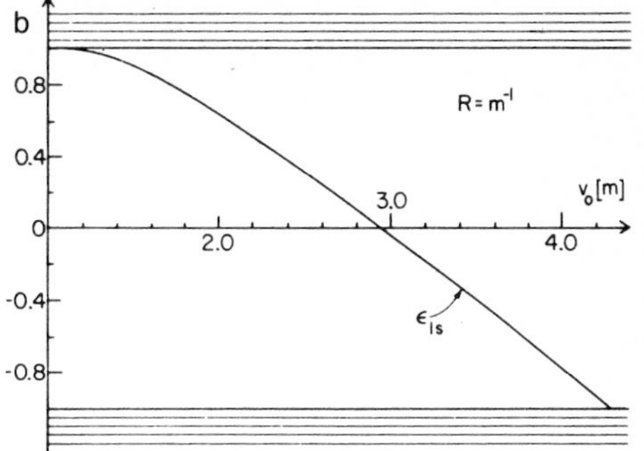

Fig. 3.11 a–c. 1*s* eigenvalues as functions of the potential strength for a square well potential. Results for the 2*s* level are also shown in **(a)**

First the solution inside the well, i.e.

$$
\begin{aligned}
&u_1^{\mathrm{i}}(r) = a_1 \sin k_{\mathrm{i}} r\,, \\
&u_2^{\mathrm{i}}(r) = -a_1 \sqrt{\frac{E+V_0-m_0}{E+V_0+m_0}} \left(\frac{\sin k_{\mathrm{i}} r}{k_{\mathrm{i}} r} - \cos k_{\mathrm{i}} r \right), \qquad k_{\mathrm{i}}^2 = (E+V_0)^2 - m_0^2\,,
\end{aligned}
\tag{3.73}
$$

has to be joined at $r = R_0$ with the exterior solution

$$
\begin{aligned}
&u_1^{\mathrm{e}}(r) = b_1 \sin k_{\mathrm{e}} r - b_2 \cos k_{\mathrm{e}} r\,, \\
&u_2^{\mathrm{e}}(r) = -\sqrt{\frac{E-m_0}{E+m_0}} \left[b_1 \left(\frac{\sin k_{\mathrm{e}} r}{k_{\mathrm{e}} r} - \cos k_{\mathrm{e}} r \right) - b_2 \left(\frac{\cos k_{\mathrm{e}} r}{k_{\mathrm{e}} r} + \sin k_{\mathrm{e}} r \right) \right], \\
&\qquad k_{\mathrm{e}}^2 = E^2 - m_0^2\,.
\end{aligned}
\tag{3.74}
$$

Introducing

$$
\begin{aligned}
&\gamma \equiv \sqrt{\frac{(E+V_0-m_0)(E+m_0)}{(E+V_0+m_0)(E-m_0)}}\,, \\
&\Delta_i \equiv k_{\mathrm{i}} R_0\,, \\
&\Delta_e \equiv k_{\mathrm{e}} R_0\,,
\end{aligned}
\tag{3.75}
$$

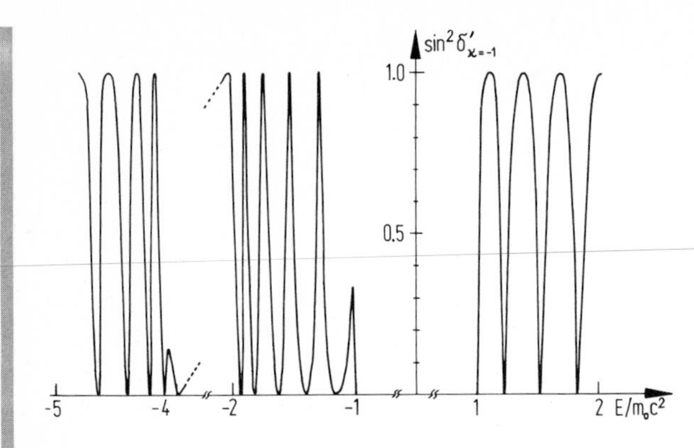

Fig. 3.12. The $s_{1/2}$ phase shift as a function of energy. The potential well is $V_0 = 3\,m_0 c^2$ deep and extends to $R_0 = 10\,\lambda_e$

the continuity of the wave function at $r = R_0$ yields

$$b_1 = a_1\left[-\gamma\frac{\sin\Delta_i}{\Delta_i}\cos\Delta_e + \gamma\cos\Delta_e\cos\Delta_i + \sin\Delta_i\frac{\cos\Delta_e}{\Delta_e} + \sin\Delta_i\sin\Delta_e\right]$$

$$b_2 = a_1\left[-\gamma\frac{\sin\Delta_i}{\Delta_i}\sin\Delta_e + \gamma\sin\Delta_e\cos\Delta_i + \sin\Delta_i\frac{\sin\Delta_e}{\Delta_e} - \sin\Delta_i\cos\Delta_e\right]. \tag{3.76}$$

The phase shift is defined via the asymptotical behaviour of the exterior solution

$$\begin{aligned} u_1^e(r) &\underset{r\to\infty}{\longrightarrow} A\,a_1\sin(k_e r + \delta)\,, \\ u_2^e(r) &\underset{r\to\infty}{\longrightarrow} A\sqrt{\frac{E-m_0}{E+m_0}}\,a_1\cos(k_e r + \delta)\,. \end{aligned} \tag{3.77}$$

Inserting (3.76) into (3.74) and comparing the result with (3.77) yields

$$\cos\delta = \frac{b_1}{A\,a_1}\,;\qquad \sin\delta = -\frac{b_2}{A\,a_1}\,, \tag{3.78}$$

and after some manipulations

$$\delta = -\Delta_e + \operatorname{arcot}\left(\frac{1}{\Delta_e} - \frac{\gamma}{\Delta_i} + \gamma\cot\Delta_i\right) = -k_e R_0 + \operatorname{arcot}\left[\gamma\cot(k_i R_0) + \frac{1}{R_0}\sqrt{\frac{E+m_0}{E-m_0}}\left(\frac{1}{E+m_0} - \frac{1}{(E+V_0+m_0)}\right)\right]. \tag{3.79}$$

As in the one-dimensional case $k_e R_0$ is the same for all states. Therefore, it is usually included in the definition of the phase shift:

$$\delta'_{\varkappa=-1} = \delta + k_e R_0 = \operatorname{arccot}\left[\gamma \cot(k_i R_0) + \frac{1}{R_0}\sqrt{\frac{E+m_0}{E-m_0}} \times \left(\frac{1}{E+m_0} - \frac{1}{E+V_0+m_0}\right)\right]. \tag{3.80}$$

This phase shift is illustrated in Fig. 3.12 for the s states in a potential of depth $V_0 = 3\,m_0 c^2$ and radius $R_0 = 10\,\lambda\!\!\!^-_e$. As in the one-dimensional case, the maxima of $\sin^2\delta'_{\varkappa=-1}$ [corresponding to $\delta'_{\varkappa=-1} = (2n+1)\,\pi/2$] signal resonances and, as before, here resonances also appear in the energy domain $m_0 - V_0 < E < m_0$. They correspond to dived s states in the overcritical potential. Comparison with Fig. 3.11 reveals that these resonances appear precisely where expected from a simple extrapolation of the binding energies $E(V_0)$.

3.4 Solutions of the Dirac Equation for a Coulomb and a Scalar 1/r Potential

Let us now solve the Dirac equation for a mixed external potential with $1/r$-radial dependence, consisting of a scalar part and Coulomb potential. The Coulomb potential is coupled to the Dirac equation via minimal coupling, while the scalar potential is added to the mass term [Do 71, So 73a]. The latter can therefore be understood as an effective, coordinate dependent mass. In the same way as one can visualize the Coulomb potential to originate from the exchange of massless photons between the nucleus and the lepton, the scalar potential of the form $-\alpha'/r$ could be due to the exchange of massless scalar mesons. The σ meson is such a scalar meson, but is has a large mass and, consequently, the corresponding potential is of very short range. Our investigations of the scalar potential in the Dirac equation thus have to be understood as an academically interesting model. We denote the Coulomb potential by

$$V_1(r) = -\frac{\hbar c \alpha'}{r}, \tag{3.81}$$

and the scalar potential by

$$V_2(r) = -\frac{\hbar c \alpha''}{r}. \tag{3.82}$$

Here α' and α'' are the electromagnetic and scalar coupling constants, respectively. The stationary Dirac equation reads

$$[c\hat{\alpha}\cdot\hat{p} + \hat{\beta}(m_0c^2 + V_2(r)) - (E - V_1(r))]\,\psi(r) = 0\,. \tag{3.83}$$

(In this section we keep $\hbar$ and c explicitly.)

Since the potentials are spherically symmetric, we can proceed directly as in Sect. 3.3. The differential equations for the "large" and "small" spinor radial functions $u_1(r)$ and $u_2(r)$ now read

$$\begin{aligned}\frac{du_1(r)}{dr} &= -\frac{\varkappa}{r}u_1(r) + \frac{1}{\hbar c}[E+m_0c^2+V_2(r)-V_1(r)]\,u_2(r)\,,\\ \frac{du_2(r)}{dr} &= +\frac{\varkappa}{r}u_2(r) - \frac{1}{\hbar c}[E-m_0c^2-V_2(r)-V_1(r)]\,u_1(r)\,.\end{aligned} \tag{3.84}$$

They are the analogues of (3.53). Inserting $V_1(r)$ and $V_2(r)$ according to (3.81, 82) yields

$$\begin{aligned}\frac{du_1(r)}{dr} &= -\frac{\varkappa}{r}u_1(r) + \left[\frac{E+m_0c^2}{\hbar c}+\frac{1}{r}(\alpha'-\alpha'')\right]u_2(r)\,,\\ \frac{du_2(r)}{dr} &= +\frac{\varkappa}{r}u_2(r) - \left[\frac{E-m_0c^2}{\hbar c}+\frac{1}{r}(\alpha'+\alpha'')\right]u_1(r)\,,\end{aligned} \tag{3.85}$$

which for $r\to 0$ have the form

$$\begin{aligned}\frac{du_1(r)}{dr} &= -\frac{\varkappa}{r}u_1(r) + \frac{1}{r}(\alpha'-\alpha'')\,u_2(r)\,,\\ \frac{du_2(r)}{dr} &= +\frac{\varkappa}{r}u_2(r) - \frac{1}{r}(\alpha'+\alpha'')\,u_1(r)\,.\end{aligned} \tag{3.86}$$

With the ansatz

$$u_1(r) = ar^{\gamma}; \qquad u_2(r) = br^{\gamma} \tag{3.87}$$

the two linear equations for a and b arise

$$\begin{aligned}a(\gamma+\varkappa)-b(\alpha'-\alpha'') &= 0\,,\\ a(\alpha'+\alpha'')+b(\gamma-\varkappa) &= 0\,,\end{aligned} \tag{3.88}$$

whose determinant vanishes if

$$\gamma = \pm\sqrt{\varkappa^2-\alpha'^2+\alpha''^2}\,. \tag{3.89}$$

For the wave functions to be normalizable, γ must be positive. For $Z\alpha > \sqrt{3/2}$, i.e. $|\gamma| < 1/2$, both solutions (corresponding to both signs for γ) are normalizable:

$$\int_0^\infty (u_1^2+u_2^2)\,dr \sim \int_0^\infty dr\, r^{\pm 2|\gamma|}\,.$$

However, the negative sign must be ruled out, because subparts of the Hamiltonian, e.g. the potential, must be finite. With the substitution

$$\varrho = 2\lambda r\,, \qquad \text{where} \quad \lambda = \frac{\sqrt{m_0^2 c^4 - E^2}}{\hbar c}\,, \tag{3.90}$$

the original differential equations (3.86) transform into

$$\begin{aligned}
\frac{du_1(\varrho)}{d\varrho} &= -\frac{\varkappa}{\varrho}u_1(\varrho) + \left(\frac{E+m_0c^2}{\hbar c 2\lambda} + \frac{(\alpha'-\alpha'')}{\varrho}\right]u_2(\varrho)\,,\\
\frac{du_2(\varrho)}{d\varrho} &= +\frac{\varkappa}{\varrho}u_2(\varrho) - \left(\frac{E-m_0c^2}{\hbar c 2\lambda} + \frac{(\alpha'+\alpha'')}{\varrho}\right)u_1(\varrho)\,,
\end{aligned} \tag{3.91}$$

and with the ansatz

$$\begin{aligned}
u_1(\varrho) &= \sqrt{(m_0c^2)+E}\,e^{-\varrho/2}(\Phi_1+\Phi_2)\,,\\
u_2(\varrho) &= \sqrt{(m_0c^2)-E}\,e^{-\varrho/2}(\Phi_1-\Phi_2)\,,
\end{aligned} \tag{3.92}$$

one gets

$$\begin{aligned}
&-\frac{1}{2}(\Phi_1+\Phi_2) + \frac{d\Phi_1}{d\varrho} + \frac{d\Phi_2}{d\varrho}\\
&= -\frac{\varkappa}{\varrho}(\Phi_1+\Phi_2) + \left(\frac{E+m_0c^2}{2\hbar c\lambda} + \frac{(\alpha'-\alpha'')}{\varrho}\right)\cdot\frac{(m_0c^2-E)}{2\hbar c\lambda}(\Phi_1-\Phi_2)\\
&-\frac{1}{2}(\Phi_1-\Phi_2) + \frac{d\Phi_1}{d\varrho} - \frac{d\Phi_2}{d\varrho}\\
&= \frac{\varkappa}{\varrho}(\Phi_1-\Phi_2) - \left(\frac{E-m_0c^2}{2\hbar c\lambda} + \frac{(\alpha'+\alpha'')}{\varrho}\right)\cdot\frac{(m_0c^2+E)}{2\hbar c\lambda}(\Phi_1+\Phi_2)\,.
\end{aligned} \tag{3.93}$$

Adding and subtracting both equations yields respectively

$$\begin{aligned}
-\Phi_1 + 2\frac{d\Phi_1}{d\varrho} &= -\frac{2\varkappa}{\varrho}\Phi_2 + \Phi_1 + \frac{(\alpha'-\alpha'')}{\varrho}\left(\frac{m_0c^2-E}{\hbar c\lambda}\right)(\Phi_1-\Phi_2)\\
&\quad - \frac{(\alpha'+\alpha'')}{\varrho}\left(\frac{m_0c^2+E}{\hbar c\lambda}\right)(\Phi_1+\Phi_2)\,,\\
-\Phi_2 + 2\frac{d\Phi_2}{d\varrho} &= -\frac{2\varkappa}{\varrho}\Phi_1 - \Phi_2 + \frac{(\alpha'-\alpha'')}{\varrho}\left(\frac{m_0c^2-E}{\hbar c\lambda}\right)(\Phi_1-\Phi_2)\\
&\quad + \frac{(\alpha'+\alpha'')}{\varrho}\left(\frac{m_0c^2+E}{\hbar c\lambda}\right)(\Phi_1+\Phi_2)\,,
\end{aligned} \tag{3.94}$$

and after ordering

$$\frac{d\Phi_1}{d\varrho} = \left(1 - \frac{\alpha' E}{\hbar c \lambda \varrho} - \frac{\alpha'' m_0 c^2}{\hbar c \lambda \varrho}\right) \Phi_1 - \left(\frac{\varkappa}{\varrho} + \frac{\alpha' m_0 c^2}{\hbar c \lambda \varrho} + \frac{\alpha'' E}{\hbar c \lambda \varrho}\right) \Phi_2 ,$$

$$\frac{d\Phi_2}{d\varrho} = \left(-\frac{\varkappa}{\varrho} + \frac{\alpha' m_0 c^2}{\hbar c \lambda \varrho} + \frac{\alpha'' E}{\hbar c \lambda \varrho}\right) \Phi_1 + \left(\frac{\alpha' E}{\hbar c \lambda \varrho} + \frac{\alpha'' m_0 c^2}{\hbar c \lambda \varrho}\right) \Phi_2 . \quad (3.95)$$

It is now sensible to split-off the asymptotic behaviour at the origin and to make an ansatz for Φ_1 and Φ_2 in terms of a potential series:

$$\Phi_1(\varrho) = \varrho^{\gamma} \sum_{m=0}^{\infty} \alpha_m \varrho^m ,$$

$$\Phi_2(\varrho) = \varrho^{\gamma} \sum_{m=0}^{\infty} \beta_m \varrho^m . \quad (3.96)$$

Inserting this into (3.95) and comparing equal powers in ϱ gives

$$\alpha_m(m+\gamma) = \alpha_{m-1} - \left(\frac{\alpha' E}{\hbar c \lambda} + \frac{\alpha'' m_0 c^2}{\hbar c \lambda}\right) \alpha_m - \left(\varkappa + \frac{\alpha' m_0 c^2}{\hbar c \lambda} + \frac{\alpha'' E}{\hbar c \lambda}\right) \beta_m ,$$

$$\beta_m(m+\gamma) = \left(-\varkappa + \frac{\alpha' m_0 c^2}{\hbar c \lambda} + \frac{\alpha'' E}{\hbar c \lambda}\right) \alpha_m + \left(\frac{\alpha' E}{\hbar c \lambda} + \frac{\alpha'' m_0 c^2}{\hbar c \lambda}\right) \beta_m . \quad (3.97)$$

The second of these equations yields

$$\frac{\beta_m}{\alpha_m} = \frac{\varkappa - \dfrac{\alpha' m_0 c^2}{\hbar c \lambda} - \dfrac{\alpha'' E}{\hbar c \lambda}}{n' - m} , \quad \text{where}$$

$$n' = \frac{\alpha' E}{\hbar c \lambda} + \frac{\alpha'' m_0 c^2}{\hbar c \lambda} - \gamma . \quad (3.98)$$

Obviously, only for $n' = 0, 1, 2, \ldots,$ the power series (3.96) become polynomials and thus ensure normalizability. Defining the principal quantum number n by

$$n = n' + j + \tfrac{1}{2} = 1, 2, 3 \ldots , \quad \text{then} \quad (3.99)$$

$$n' + \gamma = \frac{\alpha' E + \alpha'' m_0 c^2}{\sqrt{(m_0 c^2)^2 - E^2}} = n - j - \tfrac{1}{2} + \gamma , \quad (3.100)$$

from which follows

$$E^2 + \frac{2\alpha' \alpha'' m_0 c^2}{\alpha'^2 + (n - j - \frac{1}{2} + \gamma)^2} E + (m_0 c^2)^2 \frac{\alpha''^2 - (n - j - \frac{1}{2} + \gamma)^2}{\alpha'^2 + (n - j - \frac{1}{2} + \gamma)^2} = 0 . \quad (3.101)$$

Its solution yields the energy eigenvalues

$$E = m_0 c^2 \left\{ \frac{-\alpha' \alpha''}{\alpha'^2 + (n-j-\frac{1}{2}+\gamma)^2} \pm \sqrt{\left[\left(\frac{\alpha' \alpha''}{\alpha'^2 + (n-j-\frac{1}{2}+\gamma)^2} \right)^2 - \frac{\alpha''^2 - (n-j-\frac{1}{2}+\gamma)^2}{\alpha'^2 + (n-j-\frac{1}{2}+\gamma)^2} \right]} \right\} . \quad (3.102)$$

To understand this formula, it is convenient to consider a few special cases listed below.

3.4.1 Pure Scalar Potential

$$\begin{aligned} \alpha' &= 0 , \\ \gamma &= \sqrt{\alpha''^2 + \varkappa^2} , \\ E &= \pm m_0 c^2 \left(1 - \frac{\alpha''^2}{(n-j-\frac{1}{2}+\gamma)^2} \right)^{1/2} . \end{aligned} \quad (3.103)$$

Obviously two branches of solutions exist within the gap $-m_0c^2 \le E \le m_0c^2$ reflecting that the scalar potential does not distinguish between particles and antiparticles. Attention should also be paid to ordering of the levels: $1s_{1/2}$, $2p_{3/2}$, $2s_{1/2}$ and $2p_{1/2}$, $3s_{1/2}$, etc. For $\alpha'' \to \infty$ the energy levels E approach zero, Fig. 3.13.

The states with negative energy correspond to antiparticles. With growing coupling constant α'', the particle and antiparticle states approach each other, but never cross. Critical behaviour as for the Coulomb case (see below) does not occur in this case. The Dirac vacuum stays stable for scalar potentials.

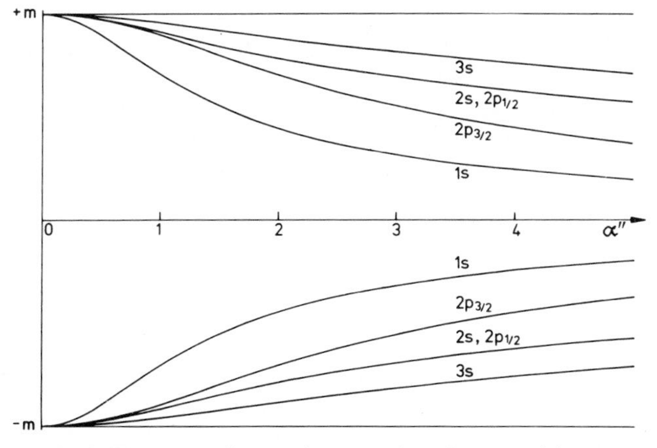

Fig. 3.13. The energy diagram for a purely scalar potential $V_2(r) = -\hbar c\alpha''/r$ as a function of the coupling strength α''

3.4.2 Pure Coulomb Potential

$$\alpha'' = 0\,,$$
$$\gamma = \sqrt{\varkappa^2 - \alpha'^2} = \sqrt{(j+\tfrac{1}{2})^2 - \alpha'^2}\,, \tag{3.104}$$
$$E = m_0c^2\left(1 + \frac{\alpha'^2}{(n-j-\frac{1}{2}+\gamma)^2}\right)^{-1/2}\,.$$

The positive root has been chosen to ensure that the electron states enter the gap $-m_0c^2 \leq E \leq m_0c^2$ from above if the coupling constant α' increases from zero. In other words, $\lim_{\alpha' \to 0} E\,(\alpha') = +m_0c^2$ for all electron states. The positive sign for the energy in (3.104) can also be inferred from (3.100) for $\alpha'' = 0$. Its left-hand side is always positive; hence E must also be always positive. The result (3.104) is *Sommerfeld's fine-structure formula*. The energies depend only on the principal quantum number n, on α' and on the central charge Z which enters $\alpha' = Ze^2/\hbar c = Z\alpha$. The ionization energy of an electron is thus

$$\begin{aligned} E_{\text{ionization}} &= m_0c^2\left[1 - \frac{1}{\sqrt{1 + \dfrac{(Z\alpha)^2}{(n - |\varkappa| + \sqrt{\varkappa^2 - Z^2\alpha^2})^2}}}\right] \\ &\approx m_0c^2(Z\alpha)^2\left[\frac{1}{2n^2} + \frac{(Z\alpha)^2}{2n^3}\left(\frac{1}{|\varkappa|} - \frac{3}{4n}\right)\right]. \end{aligned} \tag{3.105}$$

The energies obviously do not depend on the sign of $Z\alpha$, which means that electrons in a repulsive potential seem to have the same energies as electrons in an attractive potential. However, to be consistent with (3.100) for $\alpha'' = 0$, whose left-hand side is positive, for negative α' E must also be negative, i.e. the over-all sign of the energies (3.104), chosen positive for electrons in (3.104), will then change. This paradoxical result, that electrons in a repulsive potential also exhibit bound states, is interpreted and better understandable below in connection with charge conjugation.

For the states with $j = 1/2$ the energy eigenvalues are determined for all values of Z with $Z\alpha \leq 1$, i.e. up to $Z = 137$. The $1s_{1/2}$ energy is, for example, given by

$$E_{1s_{1/2}} = m_0c^2\sqrt{1 - Z^2\alpha^2}\,, \tag{3.106}$$

which becomes zero at $Z = 137$. The slope of the curve $E_{1s_{1/2}}(Z)$, i.e.

$$\frac{dE_{1s_{1/2}}(Z)}{dZ} = -m_0c^2\frac{Z\alpha}{\sqrt{1 - Z^2\alpha^2}} \xrightarrow[Z \to 137]{} \infty\,.$$

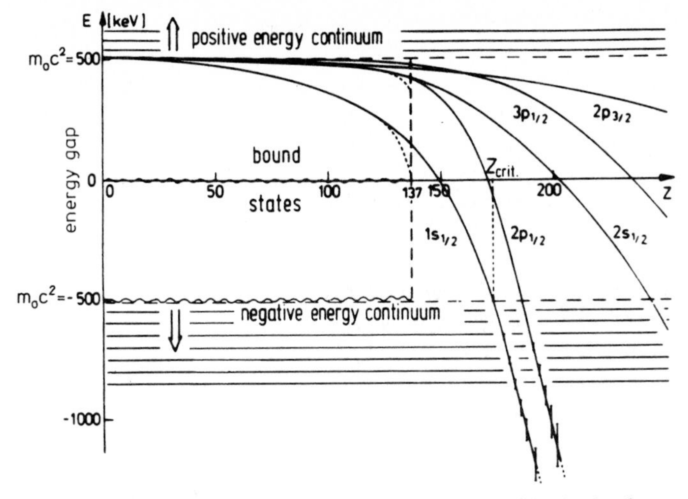

Fig. 3.14. Energies of the Dirac equation for an electron in a Coulomb-central field. The dots indicate the energies for point nuclei (fine-structure formula), which exist for $s_{1/2}$ and $p_{1/2}$ states only up to $Z = 137$. (———) represent the energies for extended nuclei. The overcritical case ($Z > Z_{\mathrm{crit}} \approx 172$) is discussed in Chap. 6 [Mü 72b]

For $Z\alpha > |\varkappa|$ the energies (3.104) become imaginary. It thus seems as if the point nucleus with $Z > 1/\alpha$ no longer can support $s_{1/2}$ and $p_{1/2}$ states. This strange result will become more transparent in Chap. 6 in connection with the overcritical vacuum.

A few values for the binding energy of $1s_{1/2}$ electrons according to the fine structure formula (3.106) are listed in Table 3.3. The energy values are compared with results from a calculation taking into account the finite size of the nucleus. Figure 3.14 graphically illustrates the function $E_n(Z)$.

Because the Coulomb potential is so important for later discussion, we also present the wave functions explicitly, in Sect. 3.5, so that the general discussion here is not interrupted.

Table 3.3. The binding energies of $1s_{1/2}$ electrons in keV as a function of the central charge Z. The energies assuming a point nucleus (Sommerfeld's fine structure formula) are compared with results for extended nuclei. For this calculation the potential of a homogeneously charged sphere with radius $R = 1.2A^{1/3}$ fm and $A = 2.5Z$ was used

Z	E_b^{point}	E_b^{ext}	Z	E_b^{point}	E_b^{ext}
10	− 1.362	− 1.362	100	−161.615	− 161.166
20	− 5.472	− 5.472	110	−206.256	− 204.890
30	− 12.396	− 12.395	120	−264.246	− 259.693
40	− 22.254	− 22.253	130	−349.368	− 330.749
50	− 35.229	− 35.227	137	−499.288	− 394.741
60	− 51.585	− 51.578	140	–	− 427.012
70	− 71.699	− 71.679	150	–	− 563.062
80	− 96.117	− 96.062	160	–	− 758.490
90	−125.657	−125.502	169	–	−1001.154

3.4.3 Coulomb and Scalar Potential of Equal Strength ($\alpha' = \alpha''$)

$$\alpha' = \alpha'' ,$$
$$\gamma = \sqrt{\varkappa^2 - \alpha'^2 + \alpha''^2} = |\varkappa| , \qquad (3.107)$$
$$E = m_0 c^2 \left[-\frac{\alpha'^2}{\alpha'^2 + n^2} \pm \frac{n^2}{\alpha'^2 + n^2} \right] .$$

The minus sign seems to yield the solution $E = -m_0 c^2$ which, however, contradicts (3.100). This possibility has therefore no physical realization. The plus sign in (3.107) gives the energy levels as

$$E = m_0 c^2 \left(1 - \frac{2\alpha'^2}{\alpha'^2 + n^2} \right) , \qquad (3.108)$$

which for $\alpha' \to \infty$ reach the lower continuum $\lim_{\alpha' \to \infty} E = -m_0 c^2$. However, the levels never dive, Fig. 3.15.

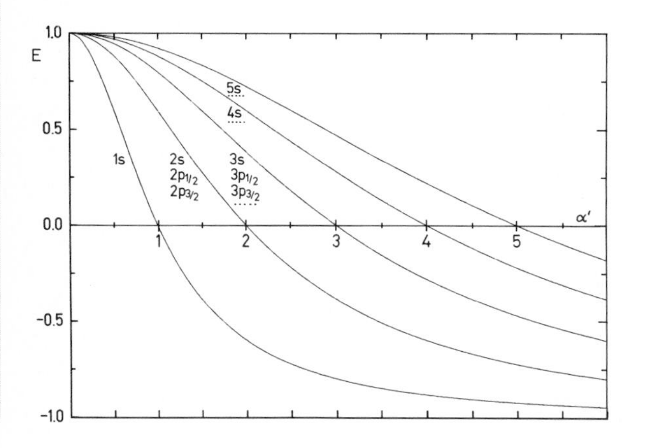

Fig. 3.15. Energy level diagram for Dirac particles with Coulomb and scalar potential of equal strength as functions of the coupling constant α'

3.5 Stationary Continuum Waves for a Dirac Particle in a Coulomb Potential

The continuum waves of Dirac particles in an external potential is very important in calculating most QED effects. In particular, the decay of the neutral vacuum into a charged one and the various processes in overcritical fields, like ionization of superheavy quasimolecules, induced decay of the vacuum, shake-off of the vacuum polarization, require knowing the continuum structure for extended superheavy nuclei. As a first step we shall solve here the Dirac equation for continuum states in a pure $1/r$-Coulomb potential.

The starting point is again the set of coupled differential equations (3.85) for the radial functions $u_1(r)$ and $u_2(r)$, where now $\alpha'' = 0$:

$$
\begin{aligned}
\frac{du_1(r)}{dr} &= -\frac{\varkappa}{r}u_1(r) + \left(\frac{E+m_0c^2}{\hbar c} + \frac{\alpha'}{r}\right)u_2(r) \\
\frac{du_2(r)}{dr} &= \frac{\varkappa}{r}u_2(r) - \left(\frac{E-m_0c^2}{\hbar c} + \frac{\alpha'}{r}\right)u_1(r) \,,
\end{aligned}
\tag{3.109}
$$

where $\alpha' = Z\alpha = Ze^2/\hbar c = Z/137$. It is convenient to introduce the new coordinate

$$
\begin{aligned}
x &= 2\mathrm{i}pr \\
p &= \frac{\sqrt{E^2-(m_0c^2)^2}}{\hbar c} = \mathrm{i}\lambda \,.
\end{aligned}
\tag{3.110}
$$

With $dx/dr = 2\mathrm{i}p$ and dividing by $2\mathrm{i}p$, (3.109) transform into

$$
\begin{aligned}
\frac{du_1(x)}{dx} &= -\frac{\varkappa}{x}u_1(x) + \left(\frac{E+m_0c^2}{2\mathrm{i}p\hbar c} + \frac{Z\alpha}{x}\right)u_2(x) \\
\frac{du_2(x)}{dx} &= \frac{\varkappa}{x}u_2(x) - \left(\frac{E-m_0c^2}{2\mathrm{i}p\hbar c} + \frac{Z\alpha}{x}\right)u_1(x) \,.
\end{aligned}
\tag{3.111}
$$

In analogy to (3.92) the following ansatz is useful:

a) for positive energies ($E > m_0c^2$):

$$
\begin{aligned}
u_1(x) &= \sqrt{E+m_0c^2}\,[\Phi_1(x) + \Phi_2(x)] \,, \\
u_2(x) &= \mathrm{i}\sqrt{E-m_0c^2}\,[\Phi_1(x) - \Phi_2(x)] \,;
\end{aligned}
\tag{3.112}
$$

b) for negative energies ($E < -m_0c^2$):

$$
\begin{aligned}
u_1(x) &= \sqrt{-E-m_0c^2}\,[\Phi_1(x) + \Phi_2(x)] \,, \\
u_2(x) &= -\mathrm{i}\sqrt{-E+m_0c^2}\,[\Phi_1(x) - \Phi_2(x)] \,.
\end{aligned}
\tag{3.113}
$$

Inserting (3.112) into (3.111), then dividing by $\sqrt{E+m_0c^2}$ and $\mathrm{i}\sqrt{E-m_0c^2}$, respectively, yields

$$
\begin{aligned}
\text{a)}\quad \frac{d\Phi_1}{dx} + \frac{d\Phi_2}{dx} &= -\frac{\varkappa}{x}(\Phi_1 + \Phi_2) + \left(\frac{E+m_0c^2}{2\mathrm{i}p\hbar c} + \frac{Z\alpha}{x}\right)\frac{\mathrm{i}\sqrt{E-m_0c^2}}{\sqrt{E+m_0c^2}}(\Phi_1 - \Phi_2) \,, \\
\frac{d\Phi_1}{dx} - \frac{d\Phi_2}{dx} &= \frac{\varkappa}{x}(\Phi_1 - \Phi_2) - \left(\frac{E-m_0c^2}{2\mathrm{i}p\hbar c} + \frac{Z\alpha}{x}\right)\frac{\sqrt{E+m_0c^2}}{\mathrm{i}\sqrt{E-m_0c^2}}(\Phi_1 + \Phi_2) \,,
\end{aligned}
\tag{3.114}
$$

and inserting (3.113) into (3.111) then dividing by $\sqrt{-E-m_0c^2}$ and $-\mathrm{i}\sqrt{-E+m_0c^2}$, respectively, yields similarly

$$\text{b)}\quad \frac{d\Phi_1}{dx}+\frac{d\Phi_2}{dx} = -\frac{\varkappa}{x}(\Phi_1+\Phi_2)+\left(\frac{E+m_0c^2}{2\mathrm{i}p\hbar c}+\frac{Z\alpha}{x}\right) \times\frac{(-\mathrm{i})\sqrt{-E+m_0c^2}}{\sqrt{-E-m_0c^2}}(\Phi_1-\Phi_2)\,,$$

$$\frac{d\Phi_1}{dx}-\frac{d\Phi_2}{dx} = \frac{\varkappa}{x}(\Phi_1-\Phi_2)-\left(\frac{E-m_0c^2}{2\mathrm{i}p\hbar c}+\frac{Z\alpha}{x}\right) \times\frac{\sqrt{-E-m_0c^2}}{(-\mathrm{i})\sqrt{-E+m_0c^2}}(\Phi_1+\Phi_2)\,. \tag{3.115}$$

Adding and subtracting these equations leads after simplification to sets of coupled equations for Φ_1 and Φ_2:

$$\text{a)}\quad \frac{d\Phi_1(x)}{dx} = \left(\frac{1}{2}+\frac{\mathrm{i}Z\alpha E}{\hbar cpx}\right)\Phi_1+\left(-\frac{\varkappa}{x}+\frac{\mathrm{i}Z\alpha m_0c^2}{\hbar cpx}\right)\Phi_2\,,$$

$$\frac{d\Phi_2(x)}{dx} = \left(-\frac{1}{2}-\frac{\mathrm{i}Z\alpha E}{\hbar cpx}\right)\Phi_2+\left(-\frac{\varkappa}{x}-\frac{\mathrm{i}Z\alpha m_0c^2}{\hbar cpx}\right)\Phi_1\,. \tag{3.116}$$

$$\text{b)}\quad \frac{d\Phi_1(x)}{dx} = \left(\frac{1}{2}+\frac{\mathrm{i}Z\alpha E}{\hbar cpx}\right)\Phi_1+\left(-\frac{\varkappa}{x}+\frac{\mathrm{i}Z\alpha m_0c^2}{\hbar cpx}\right)\Phi_2\,,$$

$$\frac{d\Phi_2(x)}{dx} = \left(-\frac{1}{2}-\frac{\mathrm{i}Z\alpha E}{\hbar cpx}\right)\Phi_2+\left(-\frac{\varkappa}{x}-\frac{\mathrm{i}Z\alpha m_0c^2}{\hbar cpx}\right)\Phi_1\,. \tag{3.117}$$

Obviously both systems of differential equations (3.116, 117) are equal. In other words, the positive and negative energy continuum states obey the same radial differential equations. There is still another interesting property of the functions Φ_1 and Φ_2. It can be discovered if the complex conjugate of (3.116) is taken, which is

$$\frac{d\Phi_1^*(x)}{dx} = -\left(\frac{1}{2}+\frac{\mathrm{i}Z\alpha E}{\hbar cpx}\right)\Phi_1^*-\left(\frac{\varkappa}{x}+\frac{\mathrm{i}Z\alpha E}{\hbar cpx}\right)\Phi_2^*\,,$$

$$\frac{d\Phi_2^*(x)}{dx} = +\left(\frac{1}{2}+\frac{\mathrm{i}Z\alpha E}{\hbar cpx}\right)\Phi_2^*+\left(-\frac{\varkappa}{x}+\frac{\mathrm{i}Z\alpha m_0c^2}{\hbar cpx}\right)\Phi_1^*\,. \tag{3.118}$$

Here $x^* = -x$, because of (3.110). The coupled equations (3.118) are the same as the coupled equations (3.114) if

$$\Phi_1 = \Phi_2^* \quad \text{and} \quad \Phi_2 = \Phi_1^*\,. \tag{3.119}$$

This is the necessary condition for the functions $u_1(x)$ and $u_2(x)$ to be real, (3.112). The u_1-, u_2 radial functions thus describe standing waves at infinity. Eliminating Φ_2 from (3.116) gives a second-order differential equation for Φ_1

$$\frac{d^2\Phi_1}{dx^2}+\frac{1}{x}\frac{d\Phi_1}{dx}-\left[\frac{1}{4}+\left(\frac{1}{2}+\mathrm{i}\frac{Z\alpha E}{\hbar cp}\right)\frac{1}{x}+\frac{\gamma^2}{x^2}\right]\Phi_1(x)=0\,, \tag{3.120}$$

with $\gamma^2=\varkappa^2-(Z\alpha)^2$. If $\Phi_1(x)$ is known, $\Phi_2(x)$ can be calculated from it with (3.119). Searching now for a solution of the differential equation (3.120), it is helpful to substitute

$$M(x)=\sqrt{x}\,\Phi_1(x) \tag{3.121}$$

which obeys the new differential equation

$$\frac{d^2M(x)}{dx^2}-\left[\frac{1}{4}+\left(\frac{1}{2}+\frac{\mathrm{i}Z\alpha E}{\hbar cp}\right)\frac{1}{x}+\frac{\gamma^2-\frac{1}{4}}{x^2}\right]M(x)=0\,. \tag{3.122}$$

As is well known, two fundamental independent solutions exist, a regular and an irregular one. The irregularity occurs at the origin. The solution of this second-order differential equation, which is regular at $x=0$, is the Whittaker function [Ab 65]

$$M_{-(\mathrm{i}y+1/2),\,+\gamma}(x)=x^{\gamma+1/2}\mathrm{e}^{-x/2}{}_1F_1(\gamma+1+\mathrm{i}y,2\gamma+1;x) \tag{3.123}$$

where

$$y=\frac{Z\alpha E}{\hbar cp}\,. \tag{3.124}$$

The irregular solution is obtained by replacing $\gamma\rightarrow-\gamma$ in (3.15). With that, the function Φ_1 can be calculated, according to (3.121) as

$$\begin{aligned}\Phi_1(r)&=N(\gamma+\mathrm{i}y)\,\mathrm{e}^{\mathrm{i}\eta}(2pr)^\gamma\mathrm{e}^{-\mathrm{i}pr}{}_1F_1(\gamma+1+\mathrm{i}y,2\gamma+1;2\mathrm{i}pr)\\&\equiv N(\gamma+\mathrm{i}y)\mathrm{e}^{\mathrm{i}\eta}(2p)^\gamma\Phi(r)\,.\end{aligned} \tag{3.125}$$

Here the phase η has been introduced. It must be determined so that indeed $\Phi_2=\Phi_1^*$, as it should be according to (3.119). With $x=\mathrm{i}2pr$ and $d/dr=2\mathrm{i}p\,d/dx$, following the first part of (3.116),

$$\frac{d\Phi_1(r)}{dr}=\left(\mathrm{i}p+\frac{\mathrm{i}Z\alpha E}{\hbar cpr}\right)\Phi_1(r)+\left(\frac{-\varkappa}{r}+\frac{\mathrm{i}Z\alpha m_0c^2}{\hbar cpr}\right)\Phi_2(r) \tag{3.126}$$

and, after inserting (3.125),

$$\begin{aligned}N(\gamma+\mathrm{i}y)\mathrm{e}^{\mathrm{i}\eta}(2p)^\gamma\frac{d\Phi(r)}{dr}&=\left(\mathrm{i}p+\frac{\mathrm{i}Z\alpha E}{\hbar cpr}\right)N(\gamma+\mathrm{i}y)\mathrm{e}^{\mathrm{i}\eta}(2p)^\gamma\Phi(r)\\&\quad+\left(\frac{-\varkappa}{r}+\frac{\mathrm{i}Z\alpha m_0c^2}{\hbar cpr}\right)N(\gamma-\mathrm{i}y)\mathrm{e}^{-\mathrm{i}\eta}(2p)^\gamma\Phi^*(r)\,,\end{aligned} \tag{3.127}$$

which determines η to be

$$e^{-2i\eta} = -\frac{\gamma + iy}{\gamma - iy} \times \left(\frac{-r\hbar cp}{-\varkappa\hbar cp + iZ\alpha m_0c^2} \frac{\frac{d\Phi(r)}{dr}}{\Phi^*(r)} + \frac{i\hbar cp^2 r + iZ\alpha E}{-\varkappa\hbar cp + iZ\alpha m_0c^2} \frac{\Phi}{\Phi^*} \right) . \tag{3.128}$$

A longer, but straightforward calculation then yields [Gr 81b]

$$e^{-2i\eta} = \frac{\varkappa - iE\frac{m_0c^2}{E}}{\gamma + iy} . \tag{3.129}$$

With that and (3.125) the radial functions $u_1(r)$ and $u_2(r)$ can now be determined from (3.112, 113):

$$\begin{aligned} u_1(r) &= C_1 N (2pr)^\gamma [(\gamma + iy) e^{-ipr + i\eta} {}_1F_1(\gamma + 1 + iy, 2\gamma + 1; 2ipr) + \text{c.c.}] , \\ u_2(r) &= iC_2 N (2pr)^\gamma [(\gamma + iy) e^{-ipr + i\eta} {}_1F_1(\gamma + 1 + iy, 2\gamma + 1; 2ipr) - \text{c.c.}] , \end{aligned} \tag{3.130}$$

where c.c. means complex conjugate and

$$\begin{aligned} C_1 &= \sqrt{E + m_0c^2} && \text{for} \quad E > m_0c^2 , \\ &= \sqrt{-E - m_0c^2} && \text{for} \quad E < -m_0c^2 ; \\ C_2 &= \sqrt{E - m_0c^2} && \text{for} \quad E > m_0c^2 , \\ &= \sqrt{-E + m_0c^2} && \text{for} \quad E < -m_0c^2 . \end{aligned} \tag{3.131}$$

Here N is a normalization factor. If it is chosen so that

$$N = \frac{|\Gamma(\gamma + iy)| e^{\pi y/2}}{2\sqrt{\pi p}\,\Gamma(2\gamma + 1)} , \tag{3.132}$$

the spinors

$$\psi_{E,\varkappa m} = \begin{pmatrix} \frac{u_1(r)}{r} \chi_{\varkappa m} \\ i\frac{u_2(r)}{r} \chi_{-\varkappa m} \end{pmatrix}$$

are normalized to energy δ functions,

$$\int d^3r\, \psi^\dagger_{E',\varkappa' m'} \psi_{E,\varkappa m} = \delta_{\varkappa\varkappa'} \delta_{mm'} \delta(E' - E) . \tag{3.133}$$

The radial functions $u_1(r)$ and $u_2(r)$ furthermore have the following asymptotic form:

$$u_1(r) \xrightarrow[r\to\infty]{} \frac{1}{\sqrt{\pi p}} C_1 \cos(pr+\delta)\,,$$
$$u_2(r) \xrightarrow[r\to\infty]{} -\frac{1}{\sqrt{\pi p}} C_2 \sin(pr+\delta)\,, \tag{3.134}$$

where

$$\delta = y \ln 2pr - \arg\Gamma(\gamma + \mathrm{i}y) - \frac{\pi\gamma}{2} + \eta \tag{3.135}$$

is the so-called *Coulomb phase.* Its peculiarity is the r-dependent logarithmic term. It is also derived again in a more general context in Sect. 6.4 (6.135). It is useful to denote finally the complete radial wave functions for Coulomb potentials:

$$u_1(r) = \frac{C_1(2pr)^\gamma \mathrm{e}^{(\pi/2)y}|\Gamma(\gamma+\mathrm{i}y)|}{2\sqrt{\pi p}\,\Gamma(2\gamma+1)} \times [\mathrm{e}^{-\mathrm{i}pr+\mathrm{i}\eta}(\gamma+\mathrm{i}y)\,{}_1F_1(\gamma+1+\mathrm{i}y, 2\gamma+1; 2\mathrm{i}pr) + \mathrm{c.c.}]\,,$$
$$u_2(r) = \frac{C_2(2pr)^\gamma \mathrm{e}^{(\pi/2)y}|\Gamma(\gamma+\mathrm{i}y)|}{2\sqrt{\pi p}\,\Gamma(2\gamma+1)} \times [\mathrm{e}^{-\mathrm{i}pr+\mathrm{i}\eta}(\gamma+\mathrm{i}y)\,{}_1F_1(\gamma+1+\mathrm{i}y, 2\gamma+1; 2\mathrm{i}pr) - \mathrm{c.c.}]\,. \tag{3.136}$$

Determining the continuum spinors is, for the calculation, a rather involved problem. One can also construct running waves, besides the standing waves, discussed here [Ro 61]. It should be remarked, however, that calculating scattering processes of relativistic electrons on nuclei and quantum electrodynamical processes requires knowing the continuum spinors. For calculating the new QED processes in strong fields that appear in heavy-ion collisions, the continuum spinors must even be known for extended nuclei (and not just for point nuclei), Sect. 6.5. In this connection, to determine continuum waves for two Coulomb centres at a distance R is a difficult, as yet unsolved, but every essential mathematical problem. Recently significant progress has been reported in [Schl 84b].

Bibliographical Notes

The various problems discussed in Chap. 3 are partly original and partly discussed in special literature [Do 71, So 73a, 82c]. The solution of the Dirac particle in the Coulomb potential of an extended nucleus is discussed in great detail in [Pi 69], which started the development of QED of strong fields. Valuable technical points can be found in [Ro 61].

4. The Hole Theory

To describe a world which is stable against electromagnetic decay the negative energy solutions of the Dirac equation have to be reinterpreted. In Dirac's hole picture all states in the lower continuum are assumed to be occupied originally. Empty levels are identified with positrons having positive charge and positive energy. We discuss the physical successes and shortcomings of this model. The ensuing new kind of symmetry mediating between particles and antiparticles, charge conjugation symmetry, is studied for both free and bound states. In addition the action of parity and time reversal transformations is treated.

4.1 The "Dirac Sea"

Up to this point, the solutions of the Dirac equation with negative energy remain a riddle. The very existence of such negative energy solutions leads to great difficulties in the single-particle interpretation of Dirac states. Consider, for example, an electron in an atom whose spectrum is displayed in Fig. 4.1.

The bound states with $E < m_0c^2$, just below the positive energy continuum starting at m_0c^2, do in general agree well with experiment. Therefore one has no doubt that these states are the bound states of the one-electron atom. An electron in the most deeply bound $1s$ state can now make successive transitions to the even lower states of the negative energy continuum by emitting x-rays. This process will never stop. The atom (and hence the whole world) would therefore be unstable. In fact, this decay would proceed in less than 10^{-9} s [Gr 81a, p. 465]. Our world would therefore not exist at all, if this decay were possible. The

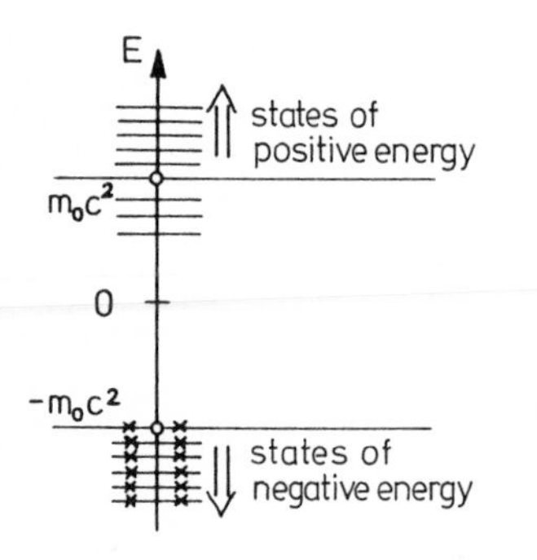

Fig. 4.1. Scheme of the radiation catastrophe of a radiating electron in an atom. By emission of x-rays, the electron can jump to energetically deeper and deeper states, thus making the atom unstable

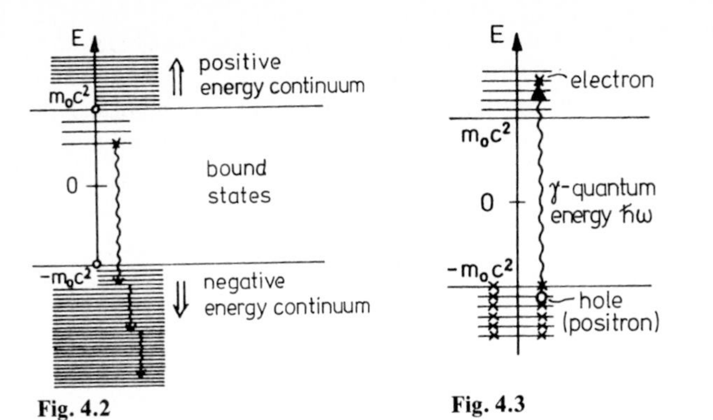

Fig. 4.2

Fig. 4.3

Fig. 4.2. In hole theory negative energy states occupied by electrons (×). Each state contains two electrons, one with spin up and one with spin down. This "Dirac sea" represents the vacuum

Fig. 4.3. A photon of energy $\hbar\omega > 2m_0c^2$ creates an electron – electron-hole configuration ($e^- - e^+$ pair creation). The hole is interpreted as a positron

problem of how the electrons are saved from "falling into the negative energy continuum" is therefore of principal and practical nature. If coupling to the radiation field did not exist, the atoms would be stable. If it is turned on, the transition probability to increasingly, lower states in the negative energy continuum grows rapidly. This is, of course, nonsense.

Dirac solved this dilemma [Di 30a, b, Op 30] by assuming that all negative energy continuum states are occupied by electrons, Fig. 4.2. The *vacuum state* is then characterized by having no real electrons present. Only the states below $-m_0c^2$, i.e. the states with $E \lesssim -m_0c^2$, are all occupied by electrons; the states with $E > m_0c^2$ are all empty.

In other words, the vacuum is characterized by the Fermi surface at $E_F = -m_0c^2$. This guarantees the stability of the atoms, since transitions to the negative energy continuum are now blocked because of the Pauli principle. On the other hand, the new picture of the vacuum also implies new processes. If, for example, a photon of energy $\hbar\omega > 2m_0c^2$ is absorbed by an electron of the negative energy continuum (which is also sometimes called "Dirac sea"), the electron can be lifted to a positive energy continuum state, leaving a hole behind in the Dirac sea, Fig. 4.3.

Thus a photon creates pairs consisting of the electron and its hole.

The hole has the properties of a particle of the same mass as the electron, but of opposite charge. The latter can be recognized if the electron of charge e recombines with the hole of charge e_+. Then, because of charge conservation,

$$e + e_+ = 0 , \tag{4.1}$$

where it is assumed that the total charge of the vacuum is zero. Thus the charge of the hole is

$$e_+ = -e , \tag{4.2}$$

i.e. opposite to that of the electron. The hole is the *antiparticle of the electron*; it is called *positron*. The creation of an electron – electron-hole configuration is thus *electron – positron pair creation*, for which the threshold energy obviously is $2m_0c^2$. Consequently, the process whereby the electron recombines with the hole under emission of a photon is called *pair annihilation*. The energy balance for pair creation reads

$$\begin{aligned}\hbar\omega &= E_{\text{electron with positive energy}} - E_{\text{electron with negative energy}} \\ &= \sqrt{(c\boldsymbol{p})^2+(m_0c^2)^2} - \left[-\sqrt{(c\boldsymbol{p}')^2+(m_0c^2)^2}\right] \\ &= E_{\text{electron}} + E_{\text{positron}}\,.\end{aligned} \tag{4.3}$$

Accordingly the electron has the positive energy

$$E_{\text{electron}} = +\sqrt{(c\boldsymbol{p})^2+(m_0c^2)^2}\,. \tag{4.4}$$

This is nothing new. However, according to (4.3) the positron (electron with negative energy) can also have positive energy

$$E_{\text{positron}} = -(E_{\text{electron of negative energy}}) = +\sqrt{(c\boldsymbol{p}')^2+(m_0c^2)^2}\,, \tag{4.5}$$

which is now a new result. In the special case of vanishing positron momentum $\boldsymbol{p}'$, (4.5) yields the rest energy of the positron

$$(E_{\text{positron}})_{\text{rest}} = m_0c^2\,, \tag{4.6}$$

which is the same as that of the electron. We have thus proven the earlier statement that the hole (positron) can be considered as a particle of the same mass but opposite charge, (4.2), as the electron.

Let us now consider the momentum balance. The photon has a momentum $\hbar\boldsymbol{k}$, shared by the electron and the positron. (In reality such a process is possible only in the vicinity of a nucleus. For simplicity, this additional complication is dismissed here.) Thus

$$\hbar\boldsymbol{k} + (\boldsymbol{p}')_{\text{electron of negative energy}} = (\boldsymbol{p})_{\text{electron of positive energy}} \quad \text{or}$$

$$\begin{aligned}\hbar\boldsymbol{k} &= (\boldsymbol{p})_{\text{electron of positive energy}} - (\boldsymbol{p}')_{\text{electron of negative energy}} \\ &= (\boldsymbol{p})_{\text{electron}} + (\boldsymbol{p}')_{\text{positron}}\,.\end{aligned} \tag{4.7}$$

Accordingly the positron has opposite momentum to the electron of negative energy:

$$\boldsymbol{p}'_{\text{positron}} = -\boldsymbol{p}'_{\text{electron of negative energy}}\,. \tag{4.8}$$

Indeed, a missing electron of negative energy with momentum $\boldsymbol{p}'$ should behave like a positively charged particle of the same mass but with opposite momentum.

The most important result of the hole theory is that it gives for the first time a quantum mechanical *model of the vacuum*, i.e. in the naive sense a model of particle-free space. It is the state consisting of all the negative energy continuum single-particle states occupied by electrons, i.e. the "Dirac sea". As a vacuum it should have energy (mass) zero and should contain no charge. It is clear, however, that the model of the vacuum in its present form does not fulfil these requirements: the negative energy continuum states filled with electrons contain a total energy of $-\infty$ and also an infinite negative charge. Therefore both are renormalized to zero, i.e. the negative total charge and energy of the occupied free negative energy continuum states are dropped and only their changes are considered if, for example, an external field is applied. In other words, one redefines the charge and mass zero points so that the *Dirac* sea has zero mass and zero charge. This procedure is neither aesthetic nor satisfactory, but possible.

A somewhat improved model is possible, but first let us discuss the charge conjugation symmetry. It is important to notice, however, that the vacuum can be modified, e.g. through external fields. Such fields can distort the wave functions of the electrons in the Dirac sea (which in the vacuum state are plane waves) and thus lead to a displacement charge relative to the field-free situation. This is called *vacuum polarization.* Still another point should be stressed: Hole-theory is a many-body theory describing particles (electrons) and antiparticles (positrons), i.e. particles of both charges. The simple probability interpretation of the square of the Dirac wave function, as it was originally conceived in the single-particle theory, is now no longer possible, because the creation and annihilation of electron – positron pairs must now be considered in the total wave function.

4.1.1 Historical Context

Let us consider for a moment our present achievements. The Klein-Gordon equation was abandoned during the first years of relativistic quantum mechanics, because it did not seem to allow for a proper single-particle interpretation. Also at the beginning the appearance of negative energy continuum states was difficult to explain. This led Dirac, who wanted to create a more satisfactory relativistic single-particle theory, to the equation named after him, i.e. to the Dirac equation. As we now see, the difficulties associated with negative energy continuum states necessarily lead to a many-body theory, i.e. hole theory. It then seems logical to abandon the Dirac equation.

On the other hand, one recognizes the great success of the Dirac equation on many occasions (explanation of the spin of the electron, the spin-orbit interaction, the g factor, the atomic fine structure, etc.). More than that, generalizing the single-particle Dirac theory to hole theory offers new and impressive successes. The prediction of the positron as the antiparticle of the electron has been experimentally fully confirmed, including the correct threshold behaviour; also the above-mentioned vacuum polarization has been verified as a part of the Lamb shift. The physics behind the hole theory has become the underlying model for quantum electrodynamics, which is today the best confirmed theory not only in physics, but also in other sciences.

Dirac's original historic motivation led to the Dirac equation. The relativistic theory for particles with spin thus created had to be reinterpreted and generalized (hole theory) to be freed of contradictions, so enabling even better success. Even though Dirac's original motivation is invalid from our present stand point, it has led physicists on the right track. It often happens in the history of science in general that a great breakthrough is reached via larger or minor traps.

We shall keep the Dirac equation and its generalization and reinterpretation as hole theory and shall continue in the following to improve it further as the underlying model for quantum electrodynamics.

4.2 Charge Conjugation Symmetry

The hole theory leads to a new fundamental theory: the electrons have antiparticles, the positrons. In general, any particle has its antiparticle. Sometimes particle and antiparticle are identical, in which case the particle must be neutral, of course. We now have to formulate this symmetry mathematically more precisely, whereby a wave function for electrons of negative energy gives rise to a wave function for the positron and vice versa.

According to the hole theory a hole in the Dirac sea represents a positron, which has the same mass as the electron (4.6), positive energy (4.5), opposite charge (4.2) and opposite momentum (4.8). A unique correspondence between negative energy solutions of the Dirac equation for electrons and the eigenfunction of the positrons exists. The electrons obey the Dirac equation

$$\left(\mathrm{i}\hbar \nabla\!\!\!\!/ - \frac{e}{c} A\!\!\!/ - m_0 c\right) \psi(x) = 0 \,. \tag{4.9}$$

Here e denotes the electron charge. To find this correspondence the positrons must have all the properties of positively charged electrons. Denoting the positron wave function by $\psi_c(x)$, then

$$\left(\mathrm{i}\hbar \nabla\!\!\!\!/ + \frac{e}{c} A\!\!\!/ - m_0 c\right) \psi_c(x) = 0 \,. \tag{4.10}$$

Note particularly that this positron wave function $\psi_c(x)$ is a *positive energy solution* of (4.10). At this point note that it is unessential which of the two equations (4.9, 10) is considered to be "original" and "charge conjugated". Historically the electrons, i.e. (4.9), have been considered as particles and thus as a starting point of the theory; in principle it could have been the other way around. The sign of the charge e in the "original" equation is unessential. Had (4.10) been the original starting equation, the free positrons would have had the same spectrum as the free electrons of (4.9). Its negative energy states would have been filled with positrons, just as hole theory requires. The electrons would have then appeared as positron holes with wave functions of negative energy as solutions of (4.10), Fig. 4.4.

Our task is now to find an operator connecting the solutions $\psi(x)$ of (4.9) with the solutions $\psi_c(x)$ of (4.10). Inspecting both (4.9, 10) shows that this

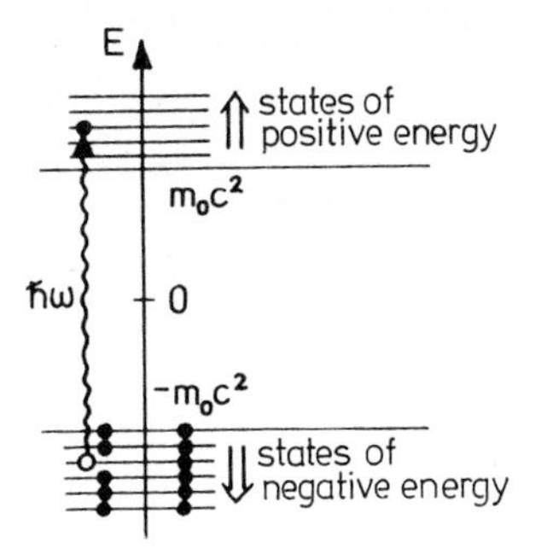

Fig. 4.4. The vacuum in hole theory, when positrons are considered as "original" particles. The continuum states of negative energy are occupied by positrons (●). A positron hole then corresponds to an electron

operator has to change the *relative* sign of $\mathrm{i}\hbar\nabla$ and $e/c\boldsymbol{A}$, most easily achievable by complex conjugation:

$$\left(\mathrm{i}\hbar\frac{\partial}{\partial x^\mu}\right)^* = -\mathrm{i}\hbar\frac{\partial}{\partial x^\mu}$$
$$A_\mu^* = A_\mu . \tag{4.11}$$

The electromagnetic potential A_μ is always real. Thus, applying complex conjugation to (4.9) yields

$$\left[\left(\mathrm{i}\hbar\frac{\partial}{\partial x^\mu}+\frac{e}{c}A_\mu\right)\gamma^{\mu *}+m_0c\right]\psi^*(x)=0 . \tag{4.12}$$

(One should recognize here that the sign change of the mass term introduced in the Dirac equation (4.12) is immaterial. In other words, Dirac equations of the form $(\not{p}-e/c\not{A}\pm m_0c)\psi(x)=0$ are equivalent; they have the same spectrum. This is easily shown by multiplying the Dirac equation by γ_5. The spinor $\gamma_5\psi(x)$ then obeys the Dirac equation with opposite sign to the mass term.)

Now it is possible, to find a non-singular matrix

$$\hat{U}=\hat{C}\gamma^0 , \tag{4.13}$$

obeying the following relation

$$\hat{U}\gamma^{\mu *}\hat{U}^{-1}=-\gamma^\mu . \tag{4.14}$$

We shall determine $\hat{U}$ below, but for the moment we assume that it exists and continue. Multiplying (4.12) by $\hat{U}$ gives

$$\left[\left(\mathrm{i}\hbar\frac{\partial}{\partial x^\mu}+\frac{e}{c}A_\mu\right)\hat{U}\gamma^{\mu *}\hat{U}^{-1}+m_0c\right]\hat{U}\psi^*(x)=0 , \tag{4.15}$$

which can finally be written as

$$\left[\left(\mathrm{i}\hbar\frac{\partial}{\partial x^\mu}+\frac{e}{c}A_\mu\right)\gamma^\mu-m_0c\right]\hat{U}\psi^*(x)=0 , \quad \text{or} \tag{4.16a}$$

$$\left[\left(\mathrm{i}\hbar\nabla\!\!\!/ + \frac{e}{c}A\!\!\!/\right) - m_0 c\right]\hat{U}\psi^*(x) = 0\,. \tag{4.16b}$$

Comparison with (4.10) reveals that

$$\psi_c = \hat{U}\psi^*(x) \equiv \hat{C}\gamma^0\psi^*(x) = \hat{C}\bar{\psi}^{\mathrm{T}}\,. \tag{4.17}$$

The superscript T indicates "transposition", i.e.

$$\bar{\psi}^{\mathrm{T}} = (\psi^\dagger\gamma^0)^{\mathrm{T}} = \gamma^{0\mathrm{T}}\psi^{\dagger\mathrm{T}} = \gamma^0(\psi^{*\mathrm{T}})^{\mathrm{T}} = \gamma^0\psi^* \tag{4.18}$$

because $\gamma^{0\mathrm{T}} = \gamma^0$ and $\psi^\dagger(x) = \psi^*(x)^{\mathrm{T}}$. Equation (4.16b) has the same form as (4.10): in particular, the sign of the electromagnetic coupling has changed relative to the original equation (4.9). This completes the correspondence between (4.9) and (4.10).

To determine $\hat{U}$ explicitly, (4.14) is rewritten as

$$\hat{C}\gamma^0\gamma^{\mu *}(\hat{C}\gamma^0)^{-1} = \hat{C}\gamma^0\gamma^{\mu *}\gamma^0\hat{C}^{-1} = -\gamma^\mu\,. \tag{4.19}$$

With the explicit representation for the γ matrices

$$\gamma^i = \begin{pmatrix} 0 & \sigma_i \\ -\sigma_i & 0 \end{pmatrix}, \quad \gamma^0 = \begin{pmatrix} \mathbb{1} & 0 \\ 0 & -\mathbb{1} \end{pmatrix}, \tag{4.20}$$

the following relation can be checked:

$$\gamma^0\gamma^{\mu *}\gamma^0 = \gamma^{\mu\mathrm{T}}\,. \tag{4.21}$$

With that, (4.19) becomes

$$(\hat{C}^{-1})^{\mathrm{T}}\gamma^\mu(\hat{C})^{\mathrm{T}} = -\gamma^{\mu\mathrm{T}}\,. \tag{4.22}$$

Now, because

$$\gamma^{1\mathrm{T}} = -\gamma^1\,, \quad \gamma^{2\mathrm{T}} = +\gamma^2\,, \quad \gamma^{3\mathrm{T}} = -\gamma^3\,, \quad \gamma^{0\mathrm{T}} = \gamma_0 \tag{4.23}$$

one concludes that $\hat{C}$ has to commute with γ^1 and γ^3 and to anticommute with γ^2 and γ^0. Therefore

$$\hat{C} = \mathrm{i}\gamma^2\gamma^0 \tag{4.24}$$

is an appropriate choice for the operator $\hat{C}$. It obeys the relations

$$\hat{C} = \mathrm{i}\gamma^2\gamma^0 = -\hat{C}^{-1} = -\hat{C}^\dagger = -\hat{C}^T\,. \tag{4.25}$$

This shows directly that $\hat{C}$ is non-singular, because the inverse $\hat{C}^{-1}$ has been explicitly constructed. Even though $\hat{C}$ in (4.24) has been constructed in a special representation (4.20), $\hat{C}$ can be transformed into any other representation via a unitary transformation. The operator $\hat{C}$ in (4.24) contains with the factor i an arbitrary phase, which is unimportant in any of the following investigations. The charge-conjugated state $\psi_c(x)$ of $\psi(x)$ is thus given by

$$\begin{aligned}\psi_c(x) &= \hat{C}'\,\psi(x) = \hat{C}\gamma^0\hat{K}\,\psi(x) = \hat{C}\gamma^0\,\psi^*(x)\\ &= \hat{C}\bar{\psi}^{\mathrm{T}}(x) = \mathrm{i}\,\gamma^2\,\psi^*(x)\,.\end{aligned} \tag{4.26}$$

The operator $\hat{K}$ denotes complex conjugation. The wave equation for $\psi_c(x)$ differs according to (4.10) from the wave equation (4.9) for $\psi(x)$ only by the sign of the charge e. Therefore one can state: *if $\psi(x)$ describes the motion of a Dirac particle of mass m_0 and charge e in a potential $A_\mu(x)$, then $\psi_c(x)$ describes the motion of a Dirac particle of the same mass m_0 and opposite charge $(-e)$ in the same potential $A_\mu(x)$.*

The spinors $\psi(x)$ and $\psi_c(x)$ are charge conjugated with respect to each other. Using (4.21, 26) and $(\gamma^2)^{\mathrm{T}} = \gamma^2$ one finds easily that

$$(\psi_c(x))_c = \psi(x)\,. \tag{4.27}$$

Thus the correspondence between $\psi_c(x)$ and $\psi(x)$ is reciprocal. Furthermore, one can straightforwardly verify the following interesting relationships concerning expectation values of various operators. If

$$\langle|\hat{Q}|\rangle = \langle\psi(x)|\hat{Q}|\psi(x)\rangle \tag{4.28}$$

is the expectation value of an operator $\hat{Q}$ in the state $\psi(x)$, then the expectation value of the same operator $\hat{Q}$ in the charge-conjugated state $\psi_c(x)$ is given by

$$\begin{aligned}\langle|\hat{Q}|\rangle_c &= \langle\psi_c(x)|\hat{Q}|\psi_c(x)\rangle = \int(\mathrm{i}\,\gamma^2\psi^*(x))^\dagger\hat{Q}(\mathrm{i}\,\gamma^2\psi^*(x))\,d^3x\\ &= \int\psi^{*\dagger}(x)(\gamma^2)^\dagger\hat{Q}\gamma^2\psi^*(x)\,d^3x = \int\psi^{*\dagger}(x)\,\gamma^0\gamma^2\gamma^0\hat{Q}\gamma^2\psi^*(x)\,d^3x\\ &= -\int\psi^{*\dagger}(x)\,\gamma^2\gamma^0\gamma^0\hat{Q}\gamma^2\psi^*(x)\,d^3x\\ &= -\int\psi^{*\dagger}(x)\,\gamma^2\hat{Q}\gamma^2\psi^*(x)\,d^3x\\ &= -\left(\int\psi^\dagger(x)(\gamma^2\hat{Q}\gamma^2)^*\,\psi(x)\,d^3x\right)^*\\ &= -\langle\psi(x)|\gamma^{2*}\hat{Q}^*\gamma^{2*}|\psi(x)\rangle^*\\ &= -\langle\psi(x)|\gamma^2\hat{Q}^*\gamma^2|\psi(x)\rangle^*\,.\end{aligned} \tag{4.29}$$

With that the following relations can easily be verified [Gr 81a]:

$$\langle|\hat{\beta}|\rangle_c = -\langle|\hat{\beta}|\rangle\,, \tag{4.30a}$$

$$\langle|\boldsymbol{x}|\rangle_c = \langle|\boldsymbol{x}|\rangle\,, \tag{4.30b}$$

$$\langle|\hat{\alpha}_i|\rangle_c = \langle|\hat{\alpha}_i|\rangle\,, \tag{4.30c}$$

$$\langle|\hat{\boldsymbol{p}}|\rangle_c = -\langle|\hat{\boldsymbol{p}}|\rangle\,, \tag{4.30d}$$

$$\psi_c^\dagger \psi_c(x) = \psi^\dagger \psi(x)\,, \tag{4.30e}$$

$$\psi_c^\dagger \hat{\boldsymbol{\alpha}}\, \psi_c(x) = \psi^\dagger(x)\, \hat{\boldsymbol{\alpha}}\, \psi(x)\,, \tag{4.30f}$$

$$\langle|\hat{\boldsymbol{\sigma}}|\rangle_c = -\langle|\hat{\boldsymbol{\sigma}}|\rangle\,, \tag{4.30g}$$

$$\langle|\hat{\boldsymbol{L}}|\rangle_c = -\langle|\hat{\boldsymbol{L}}|\rangle\,, \quad \hat{\boldsymbol{L}} = \boldsymbol{r}\times\hat{\boldsymbol{p}}\,, \tag{4.30h}$$

$$\langle|\hat{\boldsymbol{J}}|\rangle_c = -\langle|\hat{\boldsymbol{J}}|\rangle\,, \quad \hat{\boldsymbol{J}} = \boldsymbol{r}\times\hat{\boldsymbol{p}} + \tfrac{1}{2}\hat{\boldsymbol{\sigma}}\,. \tag{4.30i}$$

Since the Hamiltonians for the two Dirac equations (4.9, 10) are

$$\begin{aligned} \hat{H}(e) &= c\hat{\alpha}^i\left(\hat{p}_i + \frac{e}{c}A_i\right) + eA_0 + \hat{\beta}m_0c^2 \\ &= c\hat{\boldsymbol{\alpha}}\cdot\left(\hat{\boldsymbol{p}} - \frac{e}{c}\boldsymbol{A}\right) + eA_0 + \hat{\beta}m_0c^2\,, \\ \hat{H}(-e) &= c\hat{\alpha}^i\left(\hat{p}_i - \frac{e}{c}A_i\right) - eA_0 + \hat{\beta}m_0c^2 \\ &= c\hat{\boldsymbol{\alpha}}\cdot\left(\hat{\boldsymbol{p}} + \frac{e}{c}\boldsymbol{A}\right) - eA_0 + \hat{\beta}m_0c^2\,, \end{aligned} \tag{4.31}$$

then

$$\langle|\hat{H}(-e)|\rangle_c = -\langle|\hat{H}(e)|\rangle\,. \tag{4.32}$$

Results (4.30, 32) are most interesting. Obviously the charge-conjugated wave functions $\psi_c(x)$ lead to the same probability density distribution and probability current distribution in all space-time points, (4.30e, f) as the wave functions $\psi(x)$. Therefore the electric charge density and electric current density have opposite signs for $\psi(x)$ and $\psi_c(x)$. Equation (4.30i, 32) express the important result that the charge-conjugated state $\psi_c(x)$ has opposite momentum, angular momentum and energy to the original state $\psi(x)$. These are the properties (4.5, 8) of the hole theory in most concise form: *charge conjugation changes the sign of momentum, angular momentum and energy.* These properties allow the solutions of negative energy, charge e, spin $\boldsymbol{s}$, momentum $\boldsymbol{p}$ to be identified with the wave function of a particle of positive energy, charge $-e$, spin $-\boldsymbol{s}$ and momentum $-\boldsymbol{p}$.

Let us illustrate charge conjugation by applying it to the wave function of a resting electron of negative energy with spin ↓. This wave function is

$$\psi^4 = \frac{1}{\sqrt{2\pi\hbar}^3}\begin{pmatrix}0\\0\\0\\1\end{pmatrix}\exp\left(+\mathrm{i}\frac{m_0c^2}{\hbar}t\right)\,.$$

The corresponding charge-conjugated solution (positron solution) is

$$(\psi^4)_c = \mathrm{i}\gamma^2\psi^{4*} = \mathrm{i}\begin{pmatrix} 0 & 0 & 0 & -\mathrm{i} \\ 0 & 0 & \mathrm{i} & 0 \\ 0 & \mathrm{i} & 0 & 0 \\ -\mathrm{i} & 0 & 0 & 0 \end{pmatrix}\begin{pmatrix} 0 \\ 0 \\ 0 \\ 1 \end{pmatrix}\left(\frac{1}{\sqrt{2\pi\hbar}^3}\exp\left(+\mathrm{i}\frac{m_0c^2}{\hbar}t\right)\right)^*$$

$$= \frac{1}{\sqrt{2\pi\hbar}^3}\begin{pmatrix} 1 \\ 0 \\ 0 \\ 0 \end{pmatrix}\exp\left(-\mathrm{i}\frac{m_0c^2}{\hbar}t\right) = \psi^1 .$$

A similar result is obtained for the wave function of a resting electron of negative energy with spin $\uparrow$,

$$\psi^3 = \frac{1}{\sqrt{2\pi\hbar}^3}\begin{pmatrix} 0 \\ 0 \\ 1 \\ 0 \end{pmatrix}\exp\left(+\mathrm{i}\frac{m_0c^2}{\hbar}t\right),$$

and charge conjugation yields

$$(\psi^3)_c = \frac{(-1)}{\sqrt{2\pi\hbar}^3}\begin{pmatrix} 0 \\ 1 \\ 0 \\ 0 \end{pmatrix}\exp\left(-\mathrm{i}\frac{m_0c^2}{\hbar}t\right).$$

Here an unessential phase factor (-1) appears. But the example shows explicitly the operation of charge conjugation: the absence of a resting electron of negative energy and spin $\uparrow$ or $\downarrow$ is equivalent to the presence of a resting positron with positive energy and spin $\downarrow$ or $\uparrow$, respectively. In the field-free case there is no difference between electrons and positrons. The example clearly shows that the transformation in the field-free case simply leads to other electron solutions. This example also elucidates the redefinition of the spinors $\omega^3(x)$, $\omega^4(x)$, which is often convenient:

$$\begin{aligned} \omega^1(p) &= u(p, u_z) , \\ \omega^2(p) &= u(p, -u_z) , \\ \omega^4(p) &= v(p, u_z) , \\ \omega^3(p) &= v(p, -u_z) , \end{aligned} \tag{4.33}$$

where $\omega^r(p)$ are the free spinors given by

$$\omega^1(p) = \sqrt{\frac{E+m_0c^2}{2m_0c^2}}\begin{pmatrix} 1 \\ 0 \\ \dfrac{p_zc}{E+m_0c^2} \\ \dfrac{p_+c}{E+m_0c^2} \end{pmatrix},$$

$$\omega^2(p) = \sqrt{\frac{E+m_0c^2}{2m_0c^2}} \begin{pmatrix} 0 \\ 1 \\ \dfrac{p_-c}{E+m_0c^2} \\ \dfrac{-p_zc}{E+m_0c^2} \end{pmatrix},$$

$$\omega^3(p) = \sqrt{\frac{E+m_0c^2}{2m_0c^2}} \begin{pmatrix} \dfrac{p_zc}{E+m_0c^2} \\ \dfrac{p_+c}{E+m_0c^2} \\ 1 \\ 0 \end{pmatrix}, \tag{4.34}$$

$$\omega^4(p) = \sqrt{\frac{E+m_0c^2}{2m_0c^2}} \begin{pmatrix} \dfrac{p_-c}{E+m_0c^2} \\ \dfrac{-p_zc}{E+m_0c^2} \\ 0 \\ 1 \end{pmatrix}.$$

An electron of negative energy with spin ↑ or ↓ corresponds to a positron with spin ↓ or ↑, respectively. The spinors $v(p,s)$ of (4.33) describe positrons with spin projection s.

4.3 Charge Conjugation of States in External Potential

Consider now an electron in an attractive Coulomb potential

$$eA_0(r) = -\frac{Ze^2}{r}. \tag{4.35}$$

Chapter 3 shows that in this case the spectrum consists of the positive energy continuum $m_0c^2 \leq E \leq \infty$, a number of discrete bound states with $-m_0c^2 \leq E \leq m_0c^2$ (this number is infinite for a pure Coulomb potential) and a negative energy continuum with $-\infty \leq E \leq -m_0c^2$ (Fig. 4.5).

If charge conjugation is applied, the negative energy continuum transforms into a positive energy continuum and vice versa. Let us be very clear on that: an electron continuum state of negative energy corresponds to a continuum state of positive energy of a particle of the same mass m_0, but opposite charge, i.e. to a positive continuum state of a positron in the *same potential*, or, which leads to the same, to a positive continuum state of an electron in a repulsive Coulomb

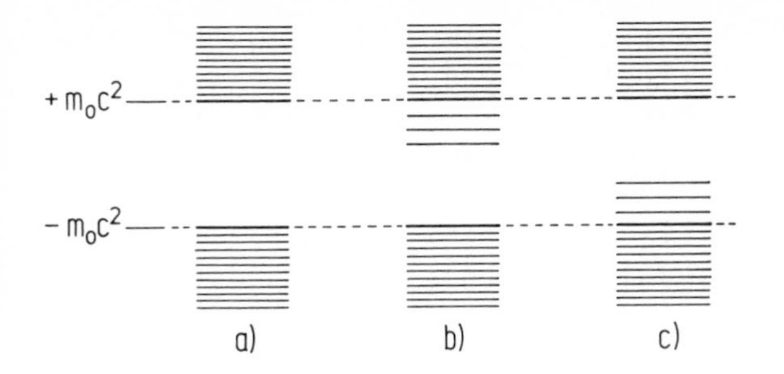

Fig. 4.5a – c. Energy spectrum of a Dirac electron: **(a)** free electron without potential; **(b)** electron in an attractive potential; **(c)** electron in a repulsive potential

potential. In this correspondence the energy changes its sign and because of γ^2 in the operator $\hat{C}\gamma^0 = \mathrm{i}\gamma^2$ the large (upper) components of the spinor are exchanged with the small (lower) ones. However, *probability density and current density are unchanged.*

The spectrum of an electron in a repulsive potential is shown in (c) of Fig. 4.5. It contains a number of bound states (precisely the same number as the electron in the attractive Coulomb potential of the same strength), which are located near the lower continuum. If the repulsive potential is adiabatically "turned on", these bound positron states emerge out of the lower (negative energy) continuum in the same way as the bound electron states emerge out of the positive energy continuum if the attractive Coulomb potential for the electron is "turned on" (Fig. 4.5b). The positive energy continuum of the repulsive potential corresponds exactly to the negative energy continuum in the attractive potential. This symmetry has important consequences for constructing the vacuum configuration in both hole and field theories.

Earlier we defined the vacuum as a configuration with all negative energy states occupied by electrons. We can now redefine it in the following way: because of the complete symmetry between electrons and positrons due to charge conjugation, the *vacuum is the symmetrized sea between electrons and positrons*, Fig. 4.6.

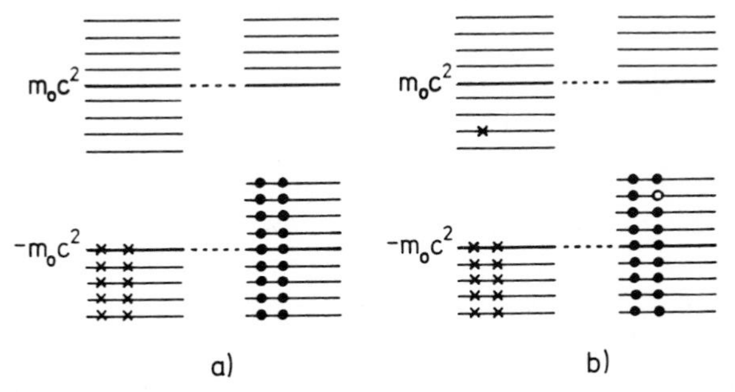

Fig. 4.6a, b. (a) Vacuum state consisting of electrons (×) and positrons (●) in single-particle states of negative energy. **(b)** Configuration representing one electron in the second strongest bound single-particle state. A positron hole in the second strongest bound positron level cannot be distinguished from the electron configuration. Both are equivalent. This is the physical content of charge-conjugation symmetry

In Fig. 4.6b an electron is in the second strongest bound single electron level. It cannot be distinguished from a positron hole in the second strongest bound (and occupied) positron level. The bound positron levels have emerged (due to the repulsive potential for the positrons) from the negative energy continuum. Both configurations exist always equally probable side by side. If they could be distinguished, the particle-antiparticle symmetry (charge conjugation symmetry) would be broken. Then two kinds of electrons and positrons would exist. On the other hand, the presence of the symmetry also enables one of the two sides, usually the positron side, to be ignored. Thus we are back at the vacuum picture of hole theory.

We remark that in the field-free case the thus defined *symmetrical vacuum state* has obviously a zero total charge, but still infinite energy. The latter has to be renormalized to zero, as done in field theory.

4.3.1 Historical Note

It is amusing to read early papers on the quantum theory of radiation, e.g. *Fermi's* article [Fe 32], which appeared between the invention of the Dirac equation in 1928 and the discovery of the positron by C. D. Anderson in 1933. In these papers one tried, for example, to identify the electron hole with the proton. A whole variety of arguments was presented to argue that the mass of the hole had to be larger than that of the electron, predicted by hole theory to be equal. On the other hand, according to hole theory it should also be possible that an electron and an electron hole annihilate under *emission of two photons*. The probability for this process to occur was calculated by *Oppenheimer, Dirac* and by *Tamm* [Di 30b, Op 30, Ta 30], with the result that matter would be annihilated in very short time. However, when the positron was discovered, all the problems which before seemed to cause serious difficulties for the theory now emerged as one of its greatest triumphs. The prediction of the existence of this previously unknown particle has to be considered as one of the most outstanding successes of theoretical physics.

4.4 Parity and Time-Reversal Symmetry

In this section we shall investigate the effects of time-reversal symmetry on Dirac states and their connection with parity and charge conjugation. Like space inversion (parity), time inversion is an *improper Lorentz transformation*.

Another important symmetry of the coupled Dirac $[\psi(x)]$ and photon fields $[A_\mu(x)]$ is gauge invariance. As known, it is assured by minimal coupling $(\hat{p}_\mu \rightarrow \hat{p}_\mu - (eA_\mu)/c)$. However, this aspect is not pursued further because it is less important for our discussion here. In the renormalization procedure of QED, gauge invariance plays a very important role, see below.

4.4.1 Parity Invariance

The parity transformation or space inversion is known from (2.138) to be

$$\psi'(\boldsymbol{x}', t) = \hat{P}\psi(\boldsymbol{x}, t) = \psi'(-\boldsymbol{x}, t) = \mathrm{e}^{\mathrm{i}\varphi}\gamma^0\psi(\boldsymbol{x}, t)\,, \tag{4.36}$$

where $\boldsymbol{x}' = -\boldsymbol{x}$. The spinor $\psi'(\boldsymbol{x}', t)$ is usually called the *space-inverted spinor* or the space-mirrored wave function. For plane waves space inversion changes the momentum sign, but not that of spin, just as expected for the behaviour of the corresponding classical quantities. The parity transformation changes various operators in the following way:

$$\begin{aligned}
\hat{P}\boldsymbol{x}\hat{P}^{-1} &= \boldsymbol{x}' = -\boldsymbol{x}\,,\\
\hat{P}x_0\hat{P}^{-1} &= x_0' = x_0\,,\\
\hat{P}\hat{\boldsymbol{p}}\hat{P}^{-1} &= \hat{\boldsymbol{p}}' = -\hat{\boldsymbol{p}}\,,\\
\hat{P}\hat{p}_0\hat{P}^{-1} &= \hat{p}_0' = \hat{p}_0\,,\\
\hat{P}A_0(\boldsymbol{x}, t)\hat{P}^{-1} &= A_0'(\boldsymbol{x}', t) = A_0(\boldsymbol{x}, t)\,,\\
\hat{P}\boldsymbol{A}(\boldsymbol{x}, t)\hat{P}^{-1} &= \boldsymbol{A}'(\boldsymbol{x}', t) = -\boldsymbol{A}(\boldsymbol{x}, t)\,,
\end{aligned} \tag{4.37}$$

where again $\boldsymbol{x}' = -\boldsymbol{x}$. The first four relations are immediately obvious, while the last two express the scalar and vectorial nature of the potential and vector potential, respectively. If $\hat{P}$ is applied to the Dirac equation, then

$$\begin{aligned}
&\hat{P}\left(\hat{p\!\!/} - \frac{e}{c}A\!\!\!/ - m_0c\right)\psi(\boldsymbol{x}, t) = 0\,;\\
&\hat{P}\left[\hat{p}_0\gamma^0 + \hat{p}_i\gamma^i - \frac{e}{c}A_0(\boldsymbol{x}, t)\gamma^0 - \frac{e}{c}A_i(\boldsymbol{x}, t)\gamma^i - m_0c\right]\hat{P}^{-1}\hat{P}\psi(\boldsymbol{x}, t) = 0\,;\\
&\left\{\hat{p}_0\gamma^0 + (-\hat{p}_i)(-\gamma^i) - \frac{e}{c}A_0(\boldsymbol{x}, t)\gamma^0 - \frac{e}{c}[-A_i(\boldsymbol{x}, t)](-\gamma^i) - m_0c\right\}\\
&\qquad\times\hat{P}\psi(\boldsymbol{x}, t) = 0\,;\\
&\left(\hat{p\!\!/} - \frac{e}{c}A\!\!\!/ - m_0c\right)\hat{P}\psi(\boldsymbol{x}, t) = 0\,.
\end{aligned} \tag{4.38}$$

This means that the parity-transformed wave function $\psi'(\boldsymbol{x}', t) = \hat{P}\psi(\boldsymbol{x}, t) = \psi'(-\boldsymbol{x}, t)$ obeys the same Dirac equation as the original wave function $\psi(\boldsymbol{x}, t)$. The parity transformation $\hat{P}$ leaves the Dirac equation and all physical observables unchanged.

What are now the physical consequences of parity invariance? To answer this question clearly consider observations of the state described by the wave function $\psi(\boldsymbol{x}, t)$. These observations are filmed. The film, however, is exposed via a mirror. In other words, the camera photographs the state $\psi(\boldsymbol{x}, t)$ in a mirror (Fig. 4.7). The dynamics of the state $\psi(\boldsymbol{x}, t)$ are called parity invariant, because the events on the film (which are mirror images of the real events) could be direct,

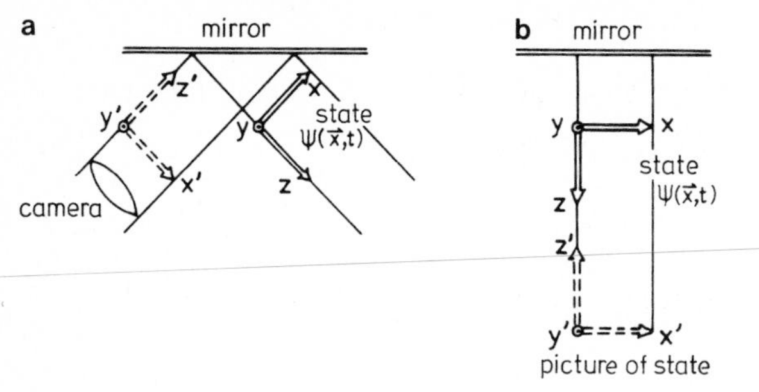

Fig. 4.7 a, b. Parity invariance: The state $\psi(x, t)$ is represented by the Dreibein with x, y, z axes, filmed by the camera via the mirror. Therefore, the Dreibein appears on the film with an inverted z axis (because of the mirror image). If the camera were situated directly beside (before or under) the state, (**b**) results, where the picture (Dreibein x', y', z') again has an inverted z axis

i.e. real events themselves. Reviewing the mirror images on the film, we are unable to say whether we see the real state or its mirror image. In parity-invariant dynamics both the image and the mirror image are equally possible observations. Note that the mirror image is *not* identical with a full space inversion, because, as can also be seen from Fig. 4.7b, a mirror inverts only the z axis, i.e. the axis perpendicular to the mirror. To obtain a complete space inversion, there must also be a rotation through π around the z axis, which is, however, a proper Lorentz transformation. The above-described mirror reflection is thus equivalent to a space inversion plus a proper Lorentz transformation (rotation). Since the theory is invariant under proper Lorentz transformations, the mirror film contains exactly the cogent information on parity invariance.

4.4.2 Time-Reversal Symmetry

Let us now discuss time-reversal symmetry. Here it is useful to illustrate the physical content of this symmetry by the film example already used above to describe parity. The film takes pictures of observations of state $\psi(\boldsymbol{x}, t)$ in a certain time order. The mirror is now unnecessary. However, contrary to the parity example above, we now move the film backwards. One then calls the *dynamics time-reversal invariant* if the set of observations on the film backwards moving could have been seen equally well on the forwards moving film. One must be unable to recognize by looking at the observations whether the film is running forwards or backwards. In other words, both the observations on the forwards and backwards moving film should be physically realizable in state $\psi(\boldsymbol{x}, t)$.

In our case of Dirac theory the dynamics are called time-reversal invariant, if under the transformation

$$\begin{aligned} t' &= -t\,, \\ \boldsymbol{x}' &= \boldsymbol{x} \end{aligned} \tag{4.39}$$

the form of the Dirac equation is unchanged; also the prescriptions for the interpretation should be unaltered. We return to this latter point below. The transformed wave function

$$\psi_{n'}(\boldsymbol{x}',t') = \hat{T}\psi_n(\boldsymbol{x},t) = \psi_{n'}(\boldsymbol{x},-t) = \hat{T}\psi_n(\boldsymbol{x}',-t') \tag{4.40}$$

describes a Dirac particle moving backwards in time. This state (wave function) is physically realizable if $\psi_{n'}(\boldsymbol{x}',t')$ also obeys the Dirac equation. Equation (4.40) is a special case of the general definition (2.89); the special case consists in the special improper Lorentz transformation of time inversion (4.39). Here $\hat{T}$ *is an operator acting on the spinor, but not the space* ($\boldsymbol{x}$) *and time* (t) coordinates. This point should be emphasized, especially in connection with the general symmetry consideration following (2.89).

Now we turn to the explicit construction of time-reversal symmetry. To that purpose the Dirac equation in Schrödinger form is then

$$\begin{aligned} \mathrm{i}\hbar\frac{\partial\psi_n(\boldsymbol{x},t)}{\partial t} &= \hat{H}(\boldsymbol{x},t)\,\psi_n(\boldsymbol{x},t) \\ &= \left\{c\hat{\boldsymbol{\alpha}}\left[-\mathrm{i}\hbar\nabla-\frac{e}{c}\boldsymbol{A}(\boldsymbol{x},t)\right]+\hat{\beta}m_0c^2+eA_0(\boldsymbol{x},t)\right\}\psi_n(\boldsymbol{x},t)\,. \end{aligned} \tag{4.41}$$

All the quantum numbers of the spinor are characterized by n. Time inversion is applied to (4.41) by multiplying it from the left by $\hat{T}$:

$$\hat{T}\mathrm{i}\hbar\hat{T}^{-1}\hat{T}\frac{\partial}{\partial t}\hat{T}^{-1}\hat{T}\psi_n(\boldsymbol{x},t) = \hat{T}\hat{H}(\boldsymbol{x},t)\hat{T}^{-1}\hat{T}\psi_n(\boldsymbol{x},t)\,. \tag{4.42}$$

Here, $\hat{T}\mathrm{i}\hat{T}^{-1}$ leaves open whether the $\hat{T}$ operator eventually contains also a complex conjugation. From the special Lorentz transformation (4.39), the definition (4.40) for time inversion and the fact that $\hat{T}(\partial/\partial t)\hat{T}^{-1} = \partial/\partial t = -\partial/\partial t'$ one gets

$$-\hat{T}\mathrm{i}\hbar\hat{T}^{-1}\frac{\partial}{\partial t'}\psi_{n'}(\boldsymbol{x}',t') = \hat{T}\hat{H}(\boldsymbol{x},t)\hat{T}^{-1}\psi_{n'}(\boldsymbol{x}',t')\,. \tag{4.43}$$

Now, to have time-reversal symmetry, $\psi_{n'}(\boldsymbol{x}',t') = \hat{T}\psi_n(\boldsymbol{x},t)$ should obey the same Schrödinger-Dirac equation as $\psi_n(\boldsymbol{x},t)$, (4.41). Hence

$$\mathrm{i}\hbar\frac{\partial}{\partial t'}\psi_{n'}(\boldsymbol{x}',t') = \hat{H}'(\boldsymbol{x}',t')\,\psi_{n'}(\boldsymbol{x}',t')\,, \tag{4.44}$$

where

$$\hat{H}'(\boldsymbol{x}',t') = \hat{H}(\boldsymbol{x},-t) = \hat{H}(\boldsymbol{x},t) \tag{4.45}$$

has to be valid. This is the only way the observer of the backwards moving film is unable to distinguish whether he sees the forwards or the backwards moving film. Comparison of (4.44) with (4.43) yields two possibilities. For both equations to be the same, we can require either

$$\hat{T}\mathrm{i}\hat{T}^{-1} = \mathrm{i} \quad \text{and} \quad \hat{T}\hat{H}(\boldsymbol{x},t)\hat{T}^{-1} \equiv \hat{H}'(\boldsymbol{x}',t') \stackrel{!}{=} -\hat{H}(\boldsymbol{x},t) \tag{4.46}$$

or

$$\hat{T}\mathrm{i}\hat{T}^{-1} = -\mathrm{i} \quad \text{and} \quad \hat{T}\hat{H}(\boldsymbol{x},t)\hat{T}^{-1} \equiv \hat{H}'(\boldsymbol{x}',t') = \hat{H}(\boldsymbol{x},-t) \stackrel{!}{=} H(\boldsymbol{x},t)\,. \tag{4.47}$$

The last possibility is better because the requirement

$$\hat{H}'(\boldsymbol{x}',t') = \hat{T}\hat{H}(\boldsymbol{x},t)\hat{T}^{-1} = \hat{H}(\boldsymbol{x},t) \tag{4.48}$$

naturally fits into the general scheme of symmetry transformations. It agrees with what we have assumed to be valid in (4.45) already. Furthermore, condition (4.46) implies that in the special case of time-independent Hamiltonians, the time-reversal operation changes the spectrum of the Hamiltonian, which is physically unacceptable. This can also be seen by considering the explicit form of $\hat{H}(\boldsymbol{x},t)$:

$$\hat{H}(\boldsymbol{x},t) = c\hat{\boldsymbol{\alpha}}\cdot\left[-\mathrm{i}\hbar\nabla - \frac{e}{c}\boldsymbol{A}(\boldsymbol{x},t)\right] + \hat{\beta}m_0c^2 + eA_0(\boldsymbol{x},t)\,. \tag{4.49}$$

From this follows that the conditions (4.46) cannot be fulfilled. The vector potential $\boldsymbol{A}(\boldsymbol{x},t)$ that appears is generated by electric currents $\boldsymbol{j}(\boldsymbol{x},t)$, which change sign under the transformation $t' = -t$. Thus

$$\hat{T}\boldsymbol{A}(\boldsymbol{x},t)\hat{T}^{-1} = \boldsymbol{A}(\boldsymbol{x}',t') = \boldsymbol{A}(\boldsymbol{x},-t) = -\boldsymbol{A}(\boldsymbol{x},t)\,. \tag{4.50}$$

The Coulomb potential $A_0(\boldsymbol{x},t)$ on the other hand, is generated by the charge density $\varrho(\boldsymbol{x},t)$ which is not changed under time inversion. Therefore

$$\hat{T}A_0(\boldsymbol{x},t)\hat{T}^{-1} = A_0(\boldsymbol{x}',t') = A_0(\boldsymbol{x},-t) = A_0(\boldsymbol{x},t)\,. \tag{4.51}$$

Also the following relations are true:

$$\begin{aligned} \hat{T}\nabla\hat{T}^{-1} &= \nabla\,, \\ \hat{T}\boldsymbol{x}\hat{T}^{-1} &= \boldsymbol{x}\,, \end{aligned} \tag{4.52}$$

because the $\hat{T}$ operation does not affect the space coordinates. Thus, due to the term $\hat{\boldsymbol{\alpha}}\cdot\boldsymbol{A}(\boldsymbol{x},t)$, condition (4.46) yields

$$\hat{T}\hat{\boldsymbol{\alpha}}\hat{T}^{-1} = \hat{\boldsymbol{\alpha}}\,,$$

and, due to the term $\hat{\boldsymbol{\alpha}}\cdot\mathrm{i}\nabla$, it yields at the same time

$$\hat{T}\hat{\boldsymbol{\alpha}}\hat{T}^{-1} = -\hat{\boldsymbol{\alpha}}\,,$$

which are incompatible. Therefore only possibility (4.47) is left, and this is indeed consistently fulfillable. Because of $\hat{T}\mathrm{i}\hat{T}^{-1} = -\mathrm{i}$,

$$\begin{aligned} \hat{H}'(\boldsymbol{x}',t') = \hat{T}\hat{H}(\boldsymbol{x},t)\hat{T}^{-1} = \hat{T}c\hat{\boldsymbol{\alpha}}\hat{T}^{-1}\Big[&(\hat{T}-\mathrm{i}\hbar\hat{T}^{-1})(\hat{T}\nabla\hat{T}^{-1}) \\ &- \frac{e}{c}\hat{T}\boldsymbol{A}(\boldsymbol{x},t)\hat{T}^{-1}\Big] + m_0c^2\hat{T}\hat{\beta}\hat{T}^{-1} + e\hat{T}A_0(\boldsymbol{x},t)\hat{T}^{-1} \end{aligned}$$

$$= -\hat{T}(c\hat{\alpha})\hat{T}^{-1}\left[-\mathrm{i}\hbar\nabla - \frac{e}{c}A(x,t)\right] + m_0c^2\hat{T}\hat{\beta}\hat{T}^{-1}$$

$$= \quad \hat{H}(x,t)\,. \tag{4.53}$$

Here (4.50 – 52) have been used in the penultimate step and

$$\begin{aligned}\hat{T}\hat{\alpha}\hat{T}^{-1} &= -\hat{\alpha}\,,\\ \hat{T}\hat{\beta}\hat{T}^{-1} &= \hat{\beta}\,,\end{aligned} \tag{4.54}$$

in the last step.

These latter conditions (4.54) must be investigated further. Because of $\hat{T}\mathrm{i}\hat{T}^{-1} = -\mathrm{i}$ the operation $\hat{T}$ must contain a complex conjugation $\hat{K}$. Therefore

$$\hat{T} = \hat{T}_0\hat{K}\,, \tag{4.55}$$

where $\hat{T}_0$ is a matrix, still to be determined. Inserting (4.55) into (4.54) yields

$$\hat{T}_0\hat{\alpha}^*\hat{T}_0^{-1} = -\hat{\alpha}\,, \qquad \hat{T}_0\gamma^{i*}\hat{T}_0^{-1} = -\gamma^i\,,$$

or

$$\hat{T}_0\hat{\beta}\hat{T}_0^{-1} = \hat{\beta}\,, \qquad \hat{T}_0\gamma^0\hat{T}_0^{-1} = \gamma^0\,. \tag{4.56}$$

Since in our representation (1.13) only $\hat{\alpha}_2$ is purely imaginary and all other matrices are real, conditions (4.56) read explicitly

$$\begin{aligned}\hat{T}_0\hat{\alpha}_1\hat{T}_0^{-1} &= -\hat{\alpha}_1\,,\\ \hat{T}_0\hat{\alpha}_2\hat{T}_0^{-1} &= \hat{\alpha}_2\,,\\ \hat{T}_0\hat{\alpha}_3\hat{T}_0^{-1} &= -\hat{\alpha}_3\,,\\ \hat{T}_0\hat{\beta}\hat{T}_0^{-1} &= \hat{\beta}\,.\end{aligned} \tag{4.57}$$

Remembering the commutation relations (1.10), one realizes that

$$\hat{T}_0 = -\mathrm{i}\hat{\alpha}_1\hat{\alpha}_3\,, \qquad \hat{T}_0^{-1} = \mathrm{i}\hat{\alpha}_3\hat{\alpha}_1 \tag{4.58}$$

fulfils all the conditions (4.57). The factor i ensures the Hermiticity of $\hat{T}_0$. With $\gamma^0 = \hat{\beta}$ and $\gamma^i = \hat{\beta}\hat{\alpha}_i$ [cf. (1.70)] the total time-reversal operator (4.55) becomes

$$\hat{T} = -\mathrm{i}\hat{\alpha}_1\hat{\alpha}_3\hat{K} = \mathrm{i}\gamma^1\gamma^3\hat{K} = \hat{T}_0\hat{K}\,. \tag{4.59}$$

This completes the construction of the time-reversal operator $\hat{T}$. One more remark is appropriate. The time-reversed equation (4.44) is also form invariant, because we have constructed it in this way. Comparing (4.44, 41) makes this obvious.

Now we shall show that the time inversion $\hat{T}$ manifests also classical intuitive ideas on time reversal. Hence we investigate the effect of $\hat{T}$ on certain observables (expectation values of certain operators) in the state $\psi_n(x,t)$ and in its time-reversed state $\psi_{n'}(\boldsymbol{x}',t')$. We have stated earlier that the rules to interpret the wave function

$$\begin{aligned} &\psi_{n'}(\boldsymbol{x}',t') = \hat{T}\psi_n(\boldsymbol{x},t)\,; \\ &\boldsymbol{x}' = \boldsymbol{x}\,, \\ &t' = -t \end{aligned} \tag{4.60}$$

are unchanged, meaning that an observable formed bilinearly from $\psi_{n'}(\boldsymbol{x}',t')$ and $\psi^\dagger_{n'}(\boldsymbol{x}',t')$ has to be interpreted (i.e. it has to have the same physical meaning) as the observable from $\psi_n(\boldsymbol{x},t)$ and $\psi^\dagger_n(\boldsymbol{x},t)$. The observables should then show the intuitively expected special properties under time inversion. The following examples illustrate this. We shall prove that

$$j'_\mu(x') = j^\mu(x) \tag{4.61}$$

and

$$\begin{aligned} \langle|\boldsymbol{x}|\rangle' &= \langle|\boldsymbol{x}|\rangle\,, \\ \langle|\hat{\boldsymbol{p}}'|\rangle' &= -\langle|\hat{\boldsymbol{p}}|\rangle\,. \end{aligned} \tag{4.62}$$

Note that in (4.61) there are lower indices on the lhs and upper indices on the rhs. Clearly, (4.61, 62) express an intuitive idea about the behaviour of current, density, position and momentum under time inversion.

To prove these relations we start with

$$\begin{aligned} &\psi_{n'}(\boldsymbol{x}',t') = \hat{T}_0\psi_n^*(\boldsymbol{x},t)\,; \quad \boldsymbol{x}' = \boldsymbol{x}\,, \quad t' = -t\,, \quad \text{and} \\ &\psi^\dagger_{n'}(\boldsymbol{x}',t') = \tilde{\psi}_n(\boldsymbol{x},t)\,\hat{T}_0^\dagger\,. \end{aligned} \tag{4.63}$$

The tilde ($\tilde{\ }$) denotes transposition. Since $\hat{T}_0 = \mathrm{i}\gamma^1\gamma^3$ [cf. (4.59)], and since

$$\gamma^{\mu\dagger} = \gamma^0\gamma^\mu\gamma^0\,, \tag{4.64}$$

one verifies easily that

$$\hat{T}_0^\dagger = \mathrm{i}\gamma^{3\dagger}\gamma^{1\dagger} = -\mathrm{i}\gamma^0\gamma^3\gamma^1\gamma^0 = -\mathrm{i}\gamma^0\gamma^0\gamma^3\gamma^1 = -\mathrm{i}(-\gamma^1\gamma^3) = \hat{T}_0\,, \tag{4.65}$$

and

$$\hat{T}_0\hat{T}_0 = \mathrm{i}\gamma^1\gamma^3 \quad \mathrm{i}\gamma^1\gamma^3 = -\gamma^1\gamma^3\gamma^1\gamma^3 = \gamma^1\gamma^3\gamma^3\gamma^1 = \mathbb{1}\,. \tag{4.66}$$

This proves the Hermiticity and unitarity of $\hat{T}_0$. With $\hat{T}_0 = \mathrm{i}\gamma^1\gamma^3$ one calculates directly

$$\begin{aligned} &\hat{T}_0\gamma^1\hat{T}_0^{-1} = \hat{T}_0\gamma^1\hat{T}_0 = -\gamma^1 = -\gamma^{1*}\,, \quad \hat{T}_0\gamma^2\hat{T}_0 = \gamma^2 = -\gamma^{2*}\,, \\ &\hat{T}_0\gamma^3\hat{T}_0^{-1} = \hat{T}_0\gamma^3\hat{T}_0 = -\gamma^3 = -\gamma^{3*}\,, \quad \hat{T}_0\gamma^0\hat{T}_0 = \gamma^0 = \gamma^{0*}\,, \end{aligned} \tag{4.67}$$

because γ^1, γ^3, γ^0 are real and γ^2 is purely imaginary. These four relations can be combined into one, namely

$$\hat{T}\gamma^{\mu}\hat{T}^{-1} = \gamma_{\mu}^{*} . \tag{4.68}$$

With these remarks we calculate the time-reversed current density

$$j_{n'}^{\mu}(\boldsymbol{x}',t') = \bar{\psi}_{n'}(\boldsymbol{x}',t')\,\gamma^{\mu}\psi_{n'}(\boldsymbol{x}',t') \tag{4.69}$$

and obtain

$$\begin{aligned}
j_{n'}^{\mu}(\boldsymbol{x}',t') &= (\psi_{n'}^{\dagger}(\boldsymbol{x}',t')\,\gamma_0)\,\gamma^{\mu}\psi_{n'}(\boldsymbol{x}',t') \\
&= (\hat{T}\psi_n(\boldsymbol{x},t))^{\dagger}\gamma_0\gamma^{\mu}\hat{T}\psi_n(\boldsymbol{x},t) \\
&= (\hat{T}_0\hat{K}\psi_n(\boldsymbol{x},t))^{\dagger}\gamma_0\gamma^{\mu}\hat{T}_0\hat{K}\psi_n(\boldsymbol{x},t) \\
&= (\hat{K}\psi_n(\boldsymbol{x},t))^{\dagger}\hat{T}_0\gamma_0\gamma^{\mu}\hat{T}_0\hat{K}\psi_n(\boldsymbol{x},t) \\
&= \tilde{\psi}_n(\boldsymbol{x},t)\,\gamma^0\gamma_{\mu}^{*}\psi_n^{*}(\boldsymbol{x},t) \\
&= (\psi_n(\boldsymbol{x},t))_{\alpha}(\gamma^0\gamma_{\mu}^{*})_{\alpha\beta}(\psi_n^{*}(\boldsymbol{x},t))_{\beta} \\
&= (\psi_n^{*}(\boldsymbol{x},t))_{\beta}\widetilde{(\gamma^0\gamma_{\mu}^{*})}_{\beta\alpha}(\psi_n(\boldsymbol{x},t))_{\alpha} \\
&= \tilde{\psi}_n^{*}(\boldsymbol{x},t)\,\tilde{\gamma}_{\mu}^{*}\gamma_0\psi_n(\boldsymbol{x},t) \\
&= \tilde{\psi}_n^{*}(\boldsymbol{x},t)\,\gamma_{\mu}^{\dagger}\gamma_0\psi_n(\boldsymbol{x},t) \\
&= \psi_n^{\dagger}(\boldsymbol{x},t)\,\gamma_0\gamma_{\mu}\gamma_0\gamma_0\psi_n(\boldsymbol{x},t) \\
&= \bar{\psi}_n(\boldsymbol{x},t)\,\gamma_{\mu}\psi_n(\boldsymbol{x},t) \\
&= (j_n)_{\mu}(\boldsymbol{x},t) ,
\end{aligned} \tag{4.70}$$

where ($\sim\!\sim\!\sim$) denotes matrix transposition.

Relations (4.62) are proven similarly and hence we have shown that the observables indeed obey the intuitively expected transformation under time reversal, i.e. currents and momenta reverse their sign, whereas densities do not.

Bibliographical Notes

As additional literature to Chap. 4 [Me 66, Vol. 2, Chap. 20] is recommended. Also the corresponding sections on parity and time-reversal symmetries are useful in this reference.

General aspects of symmetries in quantum mechanics are discussed extensively by *Greiner* in [Gr 79a].

5. The Klein Paradox

The Klein paradox is connected with (but not identical to) the phenomena of vacuum change in supercritical fields. Since the latter are one of the major objectives of this book, we shall discuss the Klein paradox extensively. It is named after *O. Klein*, who in 1929 studied the scattering of electrons from a potential barrier of height V_0 [Kl 29]. We first study this paradox purely from the single-particle stand point to interpret the Dirac equation, i.e. without use of hole theory, as Klein did. Later we introduce the hole-theory interpretation, as was originally done by *Hund* [Hu 54].

5.1 The Klein Paradox in the Single-Particle Interpretation of the Dirac Equation

Consider a case when electrons have energy E and momentum p_z, a potential step appears at $z = 0$ and the potential is infinitely extended along the positive z axis (Fig. 5.1).

For free electrons

$$\left(\frac{E}{c}\right)^2 = p^2 + m_0^2 c^2 , \tag{5.1a}$$

while for electrons in the constant potential

$$\left(\frac{E - V_0}{c}\right)^2 = \bar{p}^2 + m_0^2 c^2 \tag{5.1b}$$

holds. The Dirac equation and its adjoint equation read

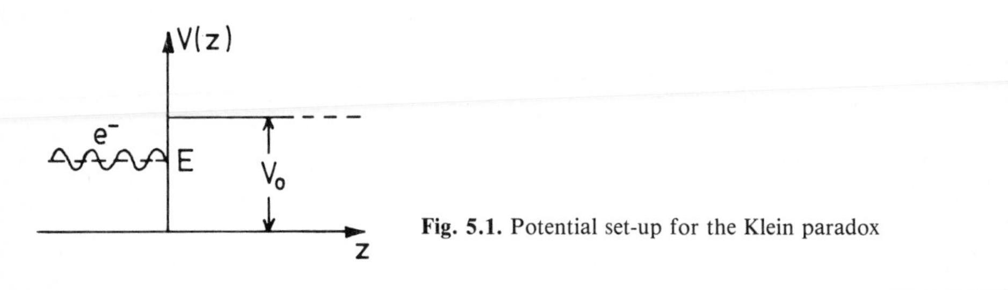

Fig. 5.1. Potential set-up for the Klein paradox

$$\left[\frac{E-eV}{c}-m_0c\hat{\beta}\right]\psi+\mathrm{i}\hbar\sum_{k=1}^{3}\hat{\alpha}_k\frac{\partial\psi}{\partial x_k}=0\,,$$
$$\bar{\psi}\left[\frac{E-eV}{c}-m_0c\hat{\beta}\right]+\mathrm{i}\hbar\sum_{k=1}^{3}\frac{\partial\bar{\psi}}{\partial x_k}\hat{\alpha}_k=0\,. \tag{5.2}$$

The potential barrier lies at $z=0$; hence we set

$$\begin{aligned} eV&=V_0 \quad \text{for} \quad z\geq 0\,,\\ eV&=0 \quad \text{for} \quad z<0\,, \end{aligned} \tag{5.3}$$

and for the incoming plane wave obtain

$$\psi_i=u_i\exp[-\mathrm{i}(Et-pz)/\hbar]\,, \tag{5.4}$$

where u_i obeys ($\hat{\alpha}\equiv\hat{\alpha}_3$)

$$\left[\frac{E}{c}-\hat{\alpha}p-\hat{\beta}m_0c\right]u_i=0\,. \tag{5.5}$$

Since we require $u_i\neq 0$, we conclude because of $\hat{\alpha}\hat{\beta}+\hat{\beta}\hat{\alpha}=0$ that

$$E^2/c^2=p^2+m_0^2c^2\,,$$

which is (5.1a). We chose $E>0$, because we are interested in an incoming electron. The reflected wave has to have momentum $-p$, while the momentum $\bar{p}$ of the penetrating wave is given by (5.1b). For small V_0, $\bar{p}$ is real and positive, so that

$$\psi_r=u_r\exp[-\mathrm{i}(Et+pz)/\hbar]\,,\qquad \psi_p=u_p\exp[-\mathrm{i}(Et-\bar{p}z)/\hbar]\,, \tag{5.6}$$

where according to (5.2) u_r and u_p obey

$$\begin{aligned} &\left[\frac{E}{c}+\hat{\alpha}p-m_0c\hat{\beta}\right]u_r=0 \quad \text{and}\\ &\left[\frac{E-V_0}{c}-\hat{\alpha}\bar{p}-m_0c\hat{\beta}\right]u_p=0 \end{aligned} \tag{5.7}$$

respectively. The total wave function has to be smooth at the barrier ($z=0$):

$$u_i+u_r=u_p\,. \tag{5.8}$$

From (5.4, 7) it follows that

$$\left[\frac{E}{c}-m_0c\hat{\beta}\right](u_i+u_r)=+\hat{\alpha}p(u_i-u_r)\,, \tag{5.9}$$

and from (5.7, 8)

$$\left[\frac{E}{c} - m_0 c\hat{\beta}\right](u_i + u_r) = \left(\frac{V_0}{c} + \hat{\alpha}\bar{p}\right)(u_i + u_r) \tag{5.10}$$

and hence

$$\left[\frac{V_0}{c} + \hat{\alpha}\bar{p}\right](u_i + u_r) = +\hat{\alpha}p(u_i - u_r) \quad \text{or} \tag{5.11}$$

$$\left[\frac{V_0}{c} + \hat{\alpha}(p+\bar{p})\right]u_r = -\left[\frac{V_0}{c} - \hat{\alpha}(p-\bar{p})\right]u_i\,. \tag{5.12}$$

Multiplying both left sides by $[V_0 - \hat{\alpha}(p+\bar{p})/c]$ and using $\hat{\alpha}^2 = 1$ and (5.1, 5) yields

$$u_r = \frac{2\,V_0/c(-E/c + \hat{\alpha}p)}{(V_0/c)^2 - (p+\bar{p})^2}\,u_i \equiv \hat{r}u_i\,. \tag{5.13}$$

Analogously

$$u_r^\dagger = u_i^\dagger \hat{r}\,, \qquad \text{i.e.} \tag{5.14}$$

$$u_r^\dagger u_r = \left(\frac{2\,V_0/c}{(V_0/c)^2 - (p+\bar{p})^2}\right)^2 u_i^\dagger \cdot \left(-\frac{E}{c} + \hat{\alpha}p\right)^2 u_i\,. \tag{5.15}$$

Using the identity

$$+cu_i^\dagger \hat{\alpha} u_i = \frac{pc^2}{E}\,u_i^\dagger u_i\,, \tag{5.16}$$

which follows easily from the equations of motion (5.2, 5) for $u_i^\dagger$ and u_i, (5.15) can be further simplified to

$$\begin{aligned} u_r^\dagger u_r &= \left(\frac{2\,V_0/c}{(V_0/c)^2 - (p+\bar{p})^2}\right)^2 \left[\left(\frac{E^2}{c^2} + p^2\right)u_i^\dagger u_i - \frac{2E_p}{c}\,u_i^\dagger \hat{\alpha} u_i\right] \\ &= \left(\frac{2\,V_0/c\,m_0 c}{(V_0/c)^2 - (p+\bar{p})^2}\right)^2 u_i^\dagger u_i \equiv R\,u_i^\dagger u_i\,. \end{aligned} \tag{5.17}$$

Clearly R gives the fraction of electrons reflected at $z = 0$. For $V_0 = 0$ one gets $R = 0$; while for $V_0 \to (E - m_0 c^2)$, i.e. $\bar{p} = 0$ according to (5.1b), R approaches 1, which means all electrons are then reflected. If the potential barrier V_0 grows even further, i.e. $V_0 > E - m_0 c^2$, $\bar{p}$ becomes purely imaginary. Then for the penetrating wave

$$\psi_p = u_p \exp\left(-\mathrm{i}\frac{E}{\hbar}t - \mu z\right), \tag{5.18}$$

where μ is now real and must be positive, because otherwise the density of the wave inside the potential would grow to infinity. Because of (5.6) $\bar{p} = +\mathrm{i}\hbar\mu$ so that (5.13) now becomes

$$u_r = -\frac{(2V_0/c)(E/c - \hat{\alpha}p)}{(V_0/c)^2 - (p + \mathrm{i}\hbar\mu)^2}u_i\,,$$
$$u_r^\dagger = -u_i^\dagger\frac{(2V_0/c)(E/c - \hat{\alpha}p)}{(V_0/c)^2 - (p - \mathrm{i}\hbar\mu)^2}\,, \tag{5.19}$$

and hence

$$u_r^\dagger u_r = \frac{\left(2\dfrac{V_0}{c}\right)^2\left(\dfrac{E^2}{c^2} - p^2\right)}{\left[\left(\dfrac{V_0}{c} + p\right)^2 + \mu^2\hbar^2\right]\left[\left(\dfrac{V_0}{c} - p\right)^2 + \mu^2\hbar^2\right]}u_i^\dagger u_i\,. \tag{5.20}$$

With (5.1) it then follows that

$$\bar{p}^2 = p^2 - \frac{V_0(2E - V_0)}{c^2}\,, \tag{5.21}$$

and since $\bar{p}^2 = -\mu^2\hbar^2$:

$$\left(\frac{V_0}{c} \pm p\right)^2 + \mu^2\hbar^2 = 2\frac{V_0}{c}\left(\frac{E}{c} \pm p\right)\,. \tag{5.22}$$

Thus, comparing (5.22) with (5.20) we conclude that in this case

$$u_r^\dagger u_r = u_i^\dagger u_i\,. \tag{5.23}$$

The reflected current equals the incoming current. Behind the surface of the barrier there is an exponentially decaying wave solution (5.18). According to (5.21) the condition for this case is

$$p^2 < \frac{V_0(2E - V_0)}{c^2}\,.$$

For growing V_0 this condition is initially fulfilled, i.e. as soon as V_0 becomes larger than

$$E - c\sqrt{\frac{E^2}{c^2} - p^2} = E - m_0c^2\,.$$

If V_0 becomes even larger, μ also grows initially, as can be seen from (5.21). For $V_0 = E$, μ becomes maximal and then decreases again to become zero at $V_0 = E + m_0c^2$. For even larger values, i.e. $V_0 > E + m_0c^2$, $\bar{p}$ becomes real again, so that (5.13, 17) are again solutions of the problem. However, then the kinetic energy $E - V_0$ becomes negative, i.e. this is a classically forbidden situation. The group velocity, given according to (5.16, 1) by

$$v_{\mathrm{gr}} = \frac{c^2}{E - V_0} \bar{p} , \tag{5.24}$$

in this range has the height of the potential barrier opposite to the momentum $\bar{p}$, if $\bar{p}$ is chosen positive ($\bar{p} > 0$). Since the group velocity characterizes the velocity of a wave packet, it seems as if the penetrating wave packet would come in from $z = +\infty$. This, however, contradicts the condition which allows an incoming wave packet only from $z = -\infty$. We thus must allow also the negative root in (5.1b), i.e. negative values for $\bar{p}$. This condition does not, however, follow from the Dirac equation, but is dictated by the boundary conditions [Do 71]. This is not considered in the discussion of the Klein paradox in some text books [Bj 64], where the reflection coefficient

$$R = \left(\frac{1 - r}{1 + r} \right)^2 , \tag{5.25}$$

with

$$r = \frac{\bar{p}}{p} \cdot \frac{E + m_0 c^2}{E - V_0 + m_0 c^2} .$$

For $V_0 > E + m_0 c^2$ the denominator becomes negative. However, since $\bar{p}$ must also be chosen negative because of the boundary condition r is always positive. Therefore the reflection coefficient R is always smaller than 1: $R < 1$, and not, as stated in some text books, $R > 1$.

We thus see that for $V_0 > E + m_0 c^2$ a fraction of the electrons penetrates the potential barrier by changing their kinetic energy from an originally positive to a negative value. Because of (5.1) the group velocity of the penetrating electrons is

$$\frac{c^2}{V_0 - E} |\bar{p}| = c \sqrt{1 - \left(\frac{m_0 c^2}{V_0 - E} \right)^2} , \tag{5.26}$$

which vanishes just for $V_0 = E + m_0 c^2$ and approaches the velocity of light for $V_0 \to \infty$. For $V_0 = E + m_0 c^2$ the reflection coefficient of (5.17) approaches just 1 ($R \to 1$), which means total reflection. For increasing V_0 the reflection coefficient R decreases to

$$\alpha = R_{\min} = \lim_{V_0 \to \infty} R(V_0) = \frac{(E/c) - p}{(E/c) + p} . \tag{5.27}$$

The corresponding fraction of electrons penetrating the wall is therefore

$$\beta = \frac{2p}{(E/c) + p} . \tag{5.28}$$

Here β is the *penetration factor*. Obviously $\alpha+\beta=1$. For $p=m_0 c$ (i.e. for electrons with a velocity amounting to about 80% of the velocity of light) we obtain with (5.5)

$$\beta = \frac{2}{\sqrt{2}+1} \approx 0.83 \, .$$

This means that 83% of the incoming electrons penetrate the potential barrier. This large penetration factor exists also if V_0 is not infinite, but amounts to only a few rest masses ($V_0 \gtrsim 3 m_0 c^2$).

Calculations by *Sauter* [Sa 31], who used a smooth barrier, indicated that this classically completely unexpected large penetration exists only if the width d of the potential barrier (d is the distance between $V=0$ and $V=V_0$) is less than the Compton wavelength of the electron

$$d \lesssim \frac{\hbar}{m_0 c} \, . \tag{5.29}$$

The unexpectedly large size of this penetration factor is known as *Klein's paradox*.

The interpretation of Klein's paradox given here is purely from the standpoint of single-particle interpretation of the Dirac equation. In this framework it is not necessary to assume pair production, as *Bjorken* and *Drell* (loc. cit.) did already at this stage. In the next section it is shown how the boundary condition (no incoming wave from $z=+\infty$) changes in hole theory, where $\bar{p}<0$ need not pertain.

5.2 Klein's Paradox and Hole Theory

The hole theory is essential to understand the paradoxical result of the last section in a better, more satisfying way. Let us therefore again consider an electron wave impacting on a potential step of height $V_0 > m_0 c^2 + E$, Fig. 5.2.

The Dirac equation for a plane wave in z direction and spin up reads

a) for region I:

$$(c\hat{\alpha}_3 \hat{p}_z + m_0 c^2 \hat{\beta})\,\psi_{\mathrm{I}} = E\,\psi_{\mathrm{I}} \, , \tag{5.30}$$

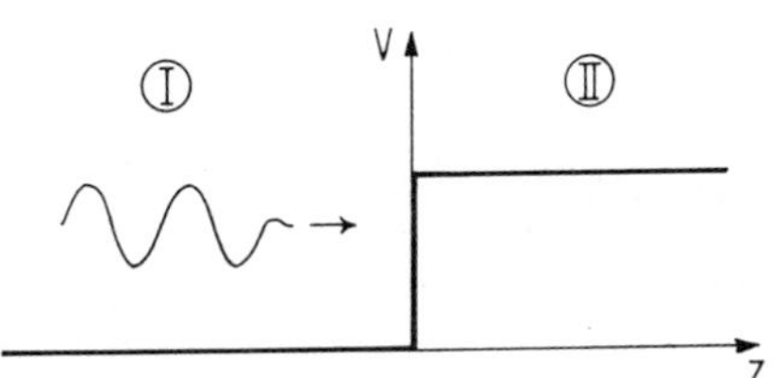

Fig. 5.2. Electron wave impacting on a potential step of height V_0

b) for region II:

$$(c\hat{\alpha}_3\hat{p}_z + m_0c^2\hat{\beta})\,\psi_{\mathrm{II}} = (E - V_0)\,\psi_{\mathrm{II}}\,. \tag{5.31}$$

The solutions are

$$\psi_{\mathrm{I}} = a\begin{pmatrix} 1 \\ 0 \\ \dfrac{p_1c}{E+m_0c^2} \\ 0 \end{pmatrix} \exp\left(\mathrm{i}\frac{p_1z}{\hbar}\right); \quad p_1c = \sqrt{E^2 - m_0^2c^4}\,, \tag{5.32}$$

$$\psi_{\mathrm{II}} = b\begin{pmatrix} 1 \\ 0 \\ \dfrac{-p_2z}{V_0-E-m_0c^2} \\ 0 \end{pmatrix} \exp\left(\mathrm{i}\frac{p_2z}{\hbar}\right); \quad p_2c = \sqrt{(V_0-E)^2 - m_0^2c^4}\,. \tag{5.33}$$

Essential for understanding Klein's paradox is the fact that for $V_0 > E + m_0c^2$ the momentum p_2 becomes real again, which means that in region II plane running waves exist. This is understandable only if a second energy continuum exists corresponding to the negative energy continuum of the free Dirac equation. The impacting wave (5.32) is partly reflected. It is denoted by

$$\psi_{\mathrm{I}}^{\mathrm{refl}} = c\begin{pmatrix} 1 \\ 0 \\ \dfrac{-p_1c}{E+m_0c^2} \\ 0 \end{pmatrix} \exp\left(-\mathrm{i}\frac{p_1z}{\hbar}\right) \tag{5.34}$$

and some fraction of the impacting wave continues into the potential. At the barrier ($z = 0$) the wave functions from inside and outside should be the same

$$\psi_{\mathrm{I}}(z=0) + \psi_{\mathrm{I}}^{\mathrm{refl}}(z=0) = \psi_{\mathrm{II}}(z=0)\,. \tag{5.35}$$

This directly yields the following two equations for a, b, c:

$$\begin{aligned} a + c &= b \\ a - c &= -b\frac{p_2}{p_1}\,\frac{E+m_0c^2}{V_0-E-m_0c^2} = -b\sqrt{\frac{(V_0-E+m_0c^2)(E+m_0c^2)}{(V_0-E-m_0c^2)(E-m_0c^2)}} \\ &=: -b\gamma\,, \end{aligned} \tag{5.36}$$

from which follow

$$
\begin{aligned}
a &= \frac{b}{2}(1-\gamma)\,, \\
c &= \frac{b}{2}(1+\gamma)\,,
\end{aligned} \tag{5.37}
$$

and thus

$$
\begin{aligned}
\frac{c}{a} &= \frac{1+\gamma}{1-\gamma}\,, \\
\frac{b}{a} &= \frac{2}{1-\gamma}\,.
\end{aligned} \tag{5.38}
$$

From the expression for the particle current

$$
\boldsymbol{j}(x) = c\,\psi^\dagger(x)\,\hat{\boldsymbol{\alpha}}\,\psi(x) \tag{5.39}
$$

and

$$
\begin{aligned}
\psi_1^\dagger\hat{\alpha}_1 &= a^*\left(0, \frac{p_1 c}{E+m_0c^2}, 0, 1\right)\exp\left(-\mathrm{i}\frac{p_1 z}{\hbar}\right), \\
\psi_1^\dagger\hat{\alpha}_2 &= a^*\left(0, \frac{-\mathrm{i}p_1 c}{E+m_0c^2}, 0, -\mathrm{i}\right)\exp\left(-\mathrm{i}\frac{p_1 z}{\hbar}\right), \\
\psi_1^\dagger\hat{\alpha}_3 &= a^*\left(\frac{p_1 c}{E+m_0c^2}, 0, 1, 0\right)\exp\left(-\mathrm{i}\frac{p_1 z}{\hbar}\right),
\end{aligned} \tag{5.40}
$$

follows

$$
\begin{aligned}
\boldsymbol{j}_\mathrm{I} &= a^*a\frac{2p_1c^2}{E+m_0c^2}\boldsymbol{e}_z\,, \\
\boldsymbol{j}_\mathrm{I}^\mathrm{refl} &= -c^*c\frac{2p_1c^2}{E+m_0c^2}\boldsymbol{e}_z\,, \\
\boldsymbol{j}_\mathrm{II} &= -b^*b\frac{2p_2c^2}{V_0-E-m_0c^2}\boldsymbol{e}_z\,.
\end{aligned} \tag{5.41}
$$

Hence for the ratios of the currents [note that γ of (5.38) is real] it follows that

$$
\frac{|\boldsymbol{j}_\mathrm{I}^\mathrm{refl}|}{|\boldsymbol{j}_\mathrm{I}|} = \frac{(1+\gamma)^2}{(1-\gamma)^2}\,; \qquad \frac{|\boldsymbol{j}_\mathrm{II}|}{|\boldsymbol{j}_\mathrm{I}|} = \frac{4\,|-\gamma\,|}{(1-\gamma)^2} = \frac{4\gamma}{(1-\gamma)^2}\,. \tag{5.42}
$$

Equation (5.36) indicates that $\gamma > 1$ for $V_0 > E + m_0c^2$, and thus

$$
|\boldsymbol{j}_\mathrm{I}^\mathrm{refl}| > |\boldsymbol{j}_\mathrm{I}|\,. \tag{5.43}
$$

This result agrees with the observation that $\boldsymbol{j}_\mathrm{II}$ is directed in the negative z direction, (5.41), which indicates the emission of electrons from region II into

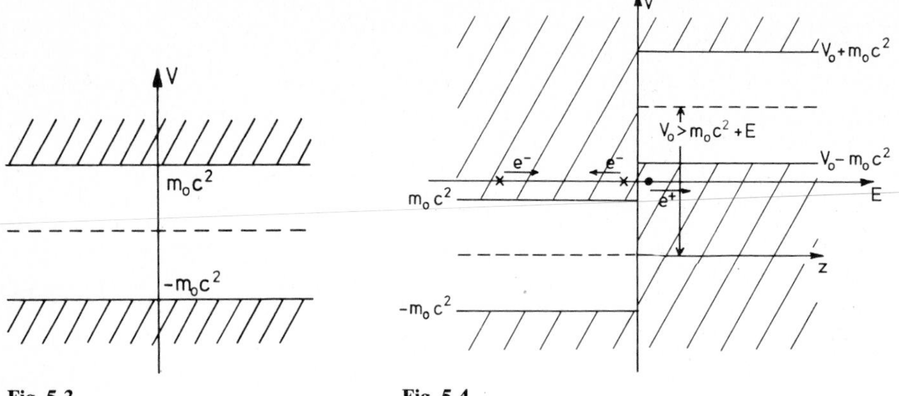

Fig. 5.3 **Fig. 5.4**

Fig. 5.3. Positive and negative energy continua of the free Dirac equation

Fig. 5.4. Upper and lower continua in the regions with and without potential. For $V_0 > m_0c^2+E$, the electron impacting from the left is confronted with electrons from the lower, occupied continuum at the right

region I. However, according to our assumptions there are no electrons in region II. Hence the results must be reinterpreted. This is achieved by interpreting the Dirac field as a many-body problem, i.e. by means of the hole theory. In hole theory the formally obtained solutions to negative energy are taken seriously and thus two electron continua exist (Fig. 5.3).

As outlined in Chap. 4, the negative energy states must be occupied by electrons to stabilize the vacuum. This hypothesis now allows the following explanation for the Klein paradox. If the potential $V_0 > m_0c^2+E$, where E is the energy of the electron in region I, then the energies of the level spectrum in region II are lifted by V_0. As seen in Fig. 5.4, then a part of the positive energy spectrum of region I overlaps with a fraction of the lower energy continuum of region II. Therefore, electrons impacting on the potential barrier from the left can knock out electrons from the occupied lower continuum states at the right. This explains that the reflected electron current is larger than the incoming electron current, (5.43). In the domain of the potential (region II) a positron current (i.e. hole current) is produced.

We can now understand within this picture, why according to (5.33) plane waves may exist in region II. These are *positron waves*. Furthermore, one can also understand the sign of the current $j_{\rm II}$ of (5.41). It is a positron current in $+\boldsymbol{e}_z$ direction, which is equivalent to an electron current in $-\boldsymbol{e}_z$ direction. Indeed from (5.41)

$$\begin{aligned} j_{\rm I}+j_{\rm I}^{\rm refl} &= j_{\rm I}\left(1-\frac{|c^2|}{|a^2|}\right) \\ &= j_{\rm I}\left(\frac{-4\gamma}{(1-\gamma)^2}\right) = j_{\rm II}\,. \end{aligned} \tag{5.44}$$

Since the holes created in region II are interpreted as positrons, one can simply explain the Klein paradox in the following way: the incoming electrons induce creation of electron-positron pairs at the potential barrier. This effects is connected with the induced decay of the vacuum in overcritical fields, as proposed and discussed by the Frankfurt school [Sm 74, Re 81b], see Sect. 12.2. We return to the Klein paradox in Sect. 10.7, where we discuss spontaneous pair creation by the supercritical potential step in the context of quantum field theory.

Bibliographical Notes

The original papers by *Klein* [Kl 29] are recommended for additional reading. Also recommended are [Sa 31] and Chap. 32 in [Hu 54]. Some aspects are also discussed in [Gr 81a].

6. Resonant States in Supercritical Fields

If the strength of an external electromagnetic potential exceeds a critical value, spontaneous positron creation sets in and the neutral vacuum of QED becomes unstable. In this situation the spectrum of the Dirac equation for a particle in the external field exhibits one or more resonances in the lower continuum. The energies, widths and wavefunctions of these resonances will be discussed using an analytic description. Also an exact treatment of the Dirac continuum solutions for the Coulomb potential generated by an extended nucleus is presented and the vacuum charge distribution is studied.

6.1 Resonances in the Negative Energy Continuum

Ordinary QED of weak fields rests on perturbation theory, as indicated in Chap. 7, where Feynman's perturbation theory based on hole theory is outlined [Gr 84c]. Clearly, perturbation theory has limited applicability; in particular the convergence of such a theory naturally depends on the strength of the perturbation potential. If the perturbation potential is so strong that the binding energy of electrons exceeds $2m_0c^2$, perturbation theory in $Z\alpha$ is impossible. A potential of such strength is called *overcritical* [Mü 72a – c]. In such a potential an empty electron state can be filled spontaneously by an electron under simultaneous emission of a positron, which can be simply guessed from the energy balance, since the energy for pair creation is smaller than the binding energy of an electron.

The basic processes in overcritical fields are outlined within the framework of hole theory. As shown, one is – as we think – led to the most fundamental process of a field theory (here quantum electrodynamics), which is the *change of its vacuum structure* [Pi 69, Mü 72, Ra 74].

First consider that in quantum mechanics an unstable state may be described as a resonance. Consider, for example, the two states with energy E_0 and E_1 in the potential in Fig. 6.1. The state at E_0 is stable.

The state at E_1, however, is quasi-stable, i.e. a particle in that state is originally localized within the potential, but creeps out after some time. If particles are scattered with energies in the vicinity of E_1, a resonance appears in the scattering cross-section at $E = E_1$. The property of tunnelling through the barrier is contained in the eigenfunction of the full Hamiltonian, because besides the bound state part of that wave function it also contains the oscillating part

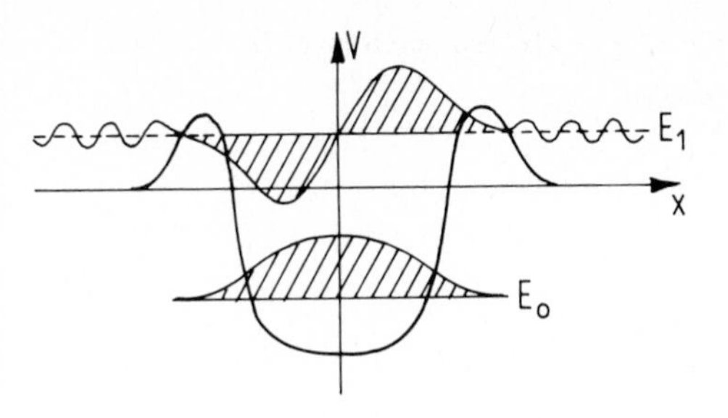

Fig. 6.1. Bound and resonance states in quantum mechanics

outside the potential well. In many cases it is useful to split up the potential V to $V = V_{cr} + V'$, so that V_{cr} stabilizes the state at E_1. Its instability (decay) can afterwards be treated (often as a perturbation) by considering the effects of V'.

We shall now consider the concrete case of strong (deep) external potentials in the Dirac equation. Let

$$V(r) = -Ze^2 U(r) \tag{6.1}$$

be the Coulomb potential; $U(r)$ is simply the radial dependence of that potential, which is asymptotically proportional to $1/r$. The factor Z has been removed to demonstrate the Z dependence of the resonance parameters below. For increasing Z, the binding energy of the electron increases. The *critical charge* Z_{cr} is defined such that the $1s$ state ψ_{1s}^{cr} is still bound with energy E_{1s} within the gap $-m_0c^2 \leq E_{1s} < m_0c^2$, but close to $-m_0c^2$, i.e.

$$E_{1s} \approx -m_0c^2 . \tag{6.2}$$

If the charge Z is increased by one unit, i.e. for $Z = Z_{cr} + 1$, the $1s$ state joins the lower (negative energy) continuum. For even higher charges Z other bound states may follow. We shall first discuss the "diving" of one bound state and later outline the more complex situation of several overcritical states (see Sect. 6.3).

6.2 One Bound State Diving into One Continuum

For overcritical potentials $Z > Z_{cr}$

$$V(r) = V_{cr}(r) + V'(r) , \tag{6.3}$$

where

$$V'(r) = -(Z - Z_{cr})e^2 U(r) \equiv -Z'e^2 U(r) . \tag{6.4}$$

Then Z' is called the *overcritical central charge.* The Dirac Hamiltonian reads

$$\begin{aligned} \hat{H} &= \hat{\alpha} \cdot \hat{p} + \hat{\beta} m_0 c^2 + V(r) \\ &= \hat{H}_{cr} + V'(r) \end{aligned} \tag{6.5}$$

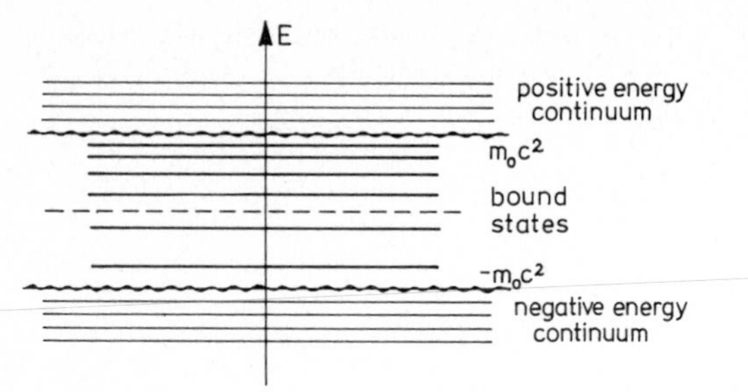

Fig. 6.2. Spectrum of $\hat{H}_{\text{cr}}$

and obviously splits into $\hat{H}_{\text{cr}}$ and the overcritical potential $V'(r)$. Figure 6.2 illustrates the spectrum of $\hat{H}_{\text{cr}}$. Its characteristic feature is that the $1s$ bound state is just slightly above the negative energy continuum threshold.

Let us focus on the ψ_{1s}^{cr} bound state, just at the edge of the negative energy continuum, and the states ψ_E of the negative energy continuum themselves. They obey

$$\begin{aligned} &\hat{H}_{\text{cr}}\,\psi_{1s}^{\text{cr}} = E_{1s}\,\psi_{1s}^{\text{cr}} \\ &\hat{H}_{\text{cr}}\,\psi_E^{\text{cr}} = E\,\psi_E^{\text{cr}}\,, \qquad E < -m_0c^2\,, \end{aligned} \tag{6.6}$$

where

$$\begin{aligned} &\langle \psi_{1s}^{\text{cr}} | \psi_{1s}^{\text{cr}} \rangle = 1 \qquad \text{and} \\ &\langle \psi_E^{\text{cr}} | \psi_{E'}^{\text{cr}} \rangle = \delta(E-E')\,. \end{aligned} \tag{6.7}$$

The equivalent set of solutions for the overcritical operator $\hat{H}$, (6.5), is

$$\hat{H}\,\Psi_E = E\,\Psi_E\,, \qquad E < -m_0c^2\,. \tag{6.8}$$

They can also be normalized to δ functions:

$$\langle \Psi_E | \Psi_{E'} \rangle = \delta(E-E')\,. \tag{6.9}$$

We assume that the reduced set of basis states (6.6) of the undercritical Hamiltonian sufficies to determine the eigenstates (6.8) of the overcritical Hamiltonian (6.5) and their resonance properties. This is indeed a reasonable approximation for illustrating the principal aspect of the problem most clearly. It is shown below how the exact continuum solutions (6.8) are obtained numerically, then compared with the present results. The simplification of the calculation by using the restricted basis set ψ_{1s}^{cr}, ψ_E^{cr} of (6.6) rests upon the assumption that only the most deeply bound state close to the negative energy continuum threshold interacts strongly with that continuum, if the overcritical potential $V'(r)$ of (6.4) is turned on. Hence we make the following ansatz to solve (6.8):

$$\Psi_E(\boldsymbol{x}) = a(E)\,\psi_{1s}^{\text{cr}}(\boldsymbol{x}) + \int_{-\infty}^{-m_0c^2} h_{E'}(E)\,\psi_{E'}^{\text{cr}}(\boldsymbol{x})\,dE'\,, \tag{6.10}$$

where the expansion coefficients $a(E)$ and $h_{E'}(E)$ are exactly defined by the projections

$$\begin{aligned} a(E) &= \int d^3x\, \psi_{1s}^{\mathrm{cr}\dagger}(\boldsymbol{x})\, \Psi_E(\boldsymbol{x}) , \\ h_{E'}(E) &= \int d^3x\, \psi_{E'}^{\mathrm{cr}\dagger}(\boldsymbol{x})\, \Psi_E(\boldsymbol{x}) . \end{aligned} \tag{6.11}$$

This is, of course, useful only if $\Psi_E(\boldsymbol{x})$ is known, which is not the case, however. We therefore insert ansatz (6.10) into (6.8):

$$(\hat{H}_{\mathrm{cr}} + V')\, \Psi_E = E\, \Psi_E . \tag{6.12}$$

After the projections $\int d^3x\, \psi_{1s}^{\mathrm{cr}\dagger} \ldots$ and $\int d^3x\, \psi_{E'}^{\mathrm{cr}\dagger} \ldots$ then

$$\begin{aligned} [E-(E_{1s}+\Delta E_{1s})]\, a(E) &= \int dE'\, V_{E'}^{*}\, h_{E'}(E) , \\ (E-E')\, h_{E'}(E) &= V_{E'}\, a(E) + \int dE''\, U_{E'E''}\, h_{E''}(E) , \end{aligned} \tag{6.13}$$

where

$$\begin{aligned} V_E &= \int d^3x\, [\psi_E^{\mathrm{cr}\dagger}(\boldsymbol{x})\, V'(\boldsymbol{x})\, \psi_{1s}^{\mathrm{cr}}(\boldsymbol{x})] \quad \text{and} \\ U_{E'E} &= \int d^3x\, [\psi_{E'}^{\mathrm{cr}\dagger}(\boldsymbol{x})\, V'(\boldsymbol{x})\, \psi_E^{\mathrm{cr}}(\boldsymbol{x})] . \end{aligned} \tag{6.14}$$

Obviously V_E describes the mixing between the $1s$ bound state $\psi_{1s}^{\mathrm{cr}}(\boldsymbol{x})$ and the negative energy continuum $\psi_E^{\mathrm{cr}}(\boldsymbol{x})$ due to the overcritical potential $V'(\boldsymbol{x})$; similarly $U_{E'E}$ is the continuum-continuum interaction due to $V'(\boldsymbol{x})$. The quantity ΔE_{1s} is given by

$$\Delta E_{1s} = \int d^3x\, [\psi_{1s}^{\mathrm{cr}\dagger}(\boldsymbol{x})\, V'(\boldsymbol{x})\, \psi_{1s}^{\mathrm{cr}}(\boldsymbol{x})] \tag{6.15}$$

and represents the lowest order energy shift of the $1s$ state.

With the expansion (6.10) the normalization condition (6.9) becomes

$$\begin{aligned} \delta(E-E') &= \int d^3x\, \Psi_E^{\dagger} \Psi_{E'}(\boldsymbol{x}) \\ &= a^*(E)\, a(E') + \int_{-\infty}^{-m_0c^2} dE''\, h_{E''}^{*}(E)\, h_{E''}(E') . \end{aligned} \tag{6.16}$$

Thus there are three equations (6.13, 16) for the unknown functions $a(E)$ and $h_{E'}(E)$. They are analytically solvable if one assumes

$$\int d^3E''\, U_{E'E''}\, h_{E''}(E) = 0 . \tag{6.17}$$

This can, in principle, be achieved by a prediagonalization of the continuum states. Suppose the new continuum states are χ_E^{cr}, and are expressed in terms of the continuum states ψ_E^{cr} through

$$\chi_E^{\mathrm{cr}} = \int dE'\, M_{EE'}\, \psi_{E'}^{\mathrm{cr}} . \tag{6.18}$$

The matrix $M_{EE'}$ is obviously connected with the overcritical potential $V'(\boldsymbol{x})$ and is chosen such that the new continuum states $\chi_E^{\mathrm{cr}}(\boldsymbol{x})$ are diagonal in the overcritical potential $V'(x)$. Quite often such a prediagonalization is rather cumbersome. More precisely, the eigenstates of the projected Dirac Hamiltonian $\hat{P}\hat{H}\hat{P}$ with $\hat{P} = \mathbb{1} - |\varphi_{1s}^{\rm cr}\rangle\langle\varphi_{1s}^{\rm cr}|$ have to be constructed, Sect. 10.3. If we simply neglect the second term of the second equation of (6.13), i.e. simply setting

$$\int dE''\, U_{E'E''} h_{E''}(E) \approx 0\,, \tag{6.19}$$

the error here can be expected to be small, because the rearrangement of the continuum from the states $\psi_E^{\rm cr}$ to $\chi_E^{\rm cr}$ should not contain any essential effect.

The total set of equations to be solved is

$$(E - E_{1s} - \Delta E_{1s})\, a(E) = \int_{-\infty}^{-m_0c^2} dE'\, V_{E'}^* h_{E'}(E)\,, \tag{6.20a}$$

$$(E - E')\, h_{E'}(E) = V_{E'} a(E)\,, \tag{6.20b}$$

$$a^*(E)\, a(E') + \int_{-\infty}^{-m_0c^2} dE''\, h_{E''}^*(E)\, h_{E''}(E') = \delta(E - E')\,. \tag{6.20c}$$

Assume now that

$$h_{E'}(E) = C_{E'}(E)\, a(E)\,. \tag{6.21}$$

Such a factorization is always possible, because of the unknown function $C_{E'}(E)$. Then obviously $a(E)$ factors out of (6.20a, b) and can be determined by the normalization equation (6.20c). Equation (6.20b) yields

$$(E - E')\, C_{E'}(E) = V_{E'}\,, \tag{6.22}$$

which suggests that $C_{E'}(E)$ must be of the form

$$C_{E'}(E) = \frac{V_{E'}}{E - E'} = P\frac{V_{E'}}{E - E'} + g(E)\, V_E\, \delta(E - E')\,. \tag{6.23}$$

One should now recognize an important step. Normally scattering problems involve conditions implying $g(E) = \pm \mathrm{i}\pi$, corresponding to ensuring that only outgoing or incoming solutions are present. (In Sect. 6.5, it is shown that the different conditions imply only a different choice of phase.) Here we want to describe the mixture of stationary states, whereby *Fano's* procedure is used [Fa 62]. With ansatz (6.23) the function $g(E)$ is still unknown. We determine it later and find that it is a real function. The symbol P indicates that the principal value shall be taken. Inserting (6.21, 23) into (6.20a) factors out $a(E)$ and yields an equation for $g(E)$:

$$(E - E_{1s} - \Delta E_{1s}) = \int V_{E'}^* \left[P\frac{V_{E'}}{E - E'} + g(E)\, V_E\, \delta(E - E') \right] dE'\,,$$

from which follows

$$g(E)=\frac{1}{|V_E|^2}\left[(E-E_{1s}-\Delta E_{1s})-P\int dE'\,\frac{|V_{E'}|^2}{E-E'}\right]$$
$$=\frac{1}{|V_E|^2}\left[(E-E_{1s}-\Delta E_{1s})-F(E)\right], \tag{6.24}$$

where

$$F(E)=P\int\frac{|V_{E'}|^2}{E-E'}\,dE'\ . \tag{6.25}$$

This determines $g(E)$, which is indeed real.

For purposes which become clear later, it is convenient to introduce the phase shift $\Delta(E)$ through

$$\tan\Delta(E)=-\frac{\pi}{g(E)}\quad\text{or}\quad\Delta(E)=-\arctan\frac{\pi}{g(E)}\ . \tag{6.26}$$

Let us now understand why $\Delta(E)$ is a phase shift. Clearly we are dealing with a continuum problem. If the continuum states $\psi_{E'}^{\mathrm{cr}}$ are represented by stationary wave functions with the asymptotic behaviour

$$\psi_{E'}^{\mathrm{cr}}(r)\underset{r\to\infty}{\longrightarrow}\sin[k(E')r+\delta_{E'}]\ ,$$

their superposition with the coefficients $h_{E'}(E)$ [cf. (6.10, 21, 23)] yields

$$\begin{aligned}
&\int dE'\,h_{E'}(E)\,\psi_{E'}^{\mathrm{cr}}(r)\\
&=\int dE'\,a(E)\,C_{E'}(E)\,\psi_{E'}^{\mathrm{cr}}(r)\\
&\underset{r\to\infty}{\longrightarrow}a(E)\int dE'\left[P\frac{V_{E'}}{E-E'}+g(E)\,V_E\,\delta(E-E')\right]\sin[k(E')r+\delta]\\
&=a(E)\left\{P\int\frac{V_{E'}\sin[k(E')r+\delta_{E'}]}{E-E'}\,dE'+g(E)\,V_E\sin[k(E)r+\delta_{E'}]\right\}.
\end{aligned}$$

The principal value integral can be carried out in the following way:

$$\begin{aligned}
P\int dE'\,\frac{V_{E'}\sin[k(E')r+\delta_{E'}]}{E-E'}&=\frac{1}{2}\int\frac{V_{E'}\sin[k(E')r+\delta_{E'}]}{E-E'+\mathrm{i}\varepsilon}\,dE'\\
&\quad+\frac{1}{2}\int\frac{V_{E'}\sin[k(E')r+\delta_{E'}]}{E-E'-\mathrm{i}\varepsilon}\,dE'\ .
\end{aligned}$$

Now, since

$$k(E')=\frac{1}{\hbar c}\sqrt{E'^2-(m_0c^2)^2}$$

× E+iε

× E−iε

E'

Poles appearing in the evaluation of the principal value integral

one concludes

$$\mathrm{Im}\{k(E')\} \gtrless 0 \quad \text{if} \quad \mathrm{Im}\{E'\} \lessgtr 0$$

since $\mathrm{Re}\{E'\} < 0$, and therefore

$$\begin{aligned} P\int \frac{V_{E'} \sin[k(E')r+\delta]}{E-E'} dE' = &\frac{1}{4\mathrm{i}} \int \frac{V_{E'} \exp\{\mathrm{i}[k(E')r+\delta_{E'}]\}}{E-E'+\mathrm{i}\varepsilon} dE' \\ &- \frac{1}{4\mathrm{i}} \int \frac{V_{E'} \exp\{-\mathrm{i}[k(E')r+\delta_{E'}]\}}{E-E'+\mathrm{i}\varepsilon} dE' \\ &+ \frac{1}{4\mathrm{i}} \int \frac{V_{E'} \exp\{+\mathrm{i}[k(E')r+\delta_{E'}]\}}{E-E'-\mathrm{i}\varepsilon} dE' \\ &- \frac{1}{4\mathrm{i}} \int \frac{V_{E'} \exp\{-\mathrm{i}[k(E')r+\delta_{E'}]\}}{E-E'-\mathrm{i}\varepsilon} dE' \,. \end{aligned} \tag{6.27}$$

Obviously, for $\mathrm{Im}\{k(E')\} > 0$, the first and third integrals have to be closed in the upper plane, while for $\mathrm{Im}\{k(E')\} < 0$, the second and fourth integrals must be closed in the lower plane. Hence (Fig. 6.3)

$$\begin{aligned} P\int \frac{V_{E'} \sin[k(E')r+\delta]}{E-E'} dE' = &\frac{1}{4\mathrm{i}} \cdot 0 \\ &- \frac{1}{4\mathrm{i}} (-2\pi\mathrm{i})\, V_E \exp\{-\mathrm{i}[k(E)r+\delta_E]\} \\ &+ \frac{1}{4\mathrm{i}} (2\pi\mathrm{i})\, V_E \exp\{\mathrm{i}[k(E)r+\delta_E]\} \\ &- \frac{1}{4\mathrm{i}} \cdot 0 \\ = &+\pi V_E \cos[k(E)r+\delta_E] \,, \end{aligned}$$

and therefore

$$\begin{aligned} \int dE' h_{E'}(E) \psi_E^{\mathrm{cr}}(r) &\underset{r\to\infty}{=} a(E) V_E\{+\pi \cos[k(E)R+\delta_E] + g(E) \sin[k(E)r+\delta_E]\} \\ &= a(E) V_E \sqrt{\pi^2+g^2(E)} \sin[k(E)r+\delta_E - \Delta(E)] \,. \end{aligned} \tag{6.27a}$$

The Δ represents the phase shift (6.26), which the continuum obtains due to the configuration interaction of $\psi_{E'}$ with the bound state ψ_{1s}^{cr}. This completes the argument to prove that $\Delta(E)$ is a phase shift.

We now come back to our original task of solving (6.20) and determining the coefficient $a(E)$. Since (6.20a, b) have already been used, we focus on the orthonormalization condition (6.20b). Substituting (6.21, 23) yields

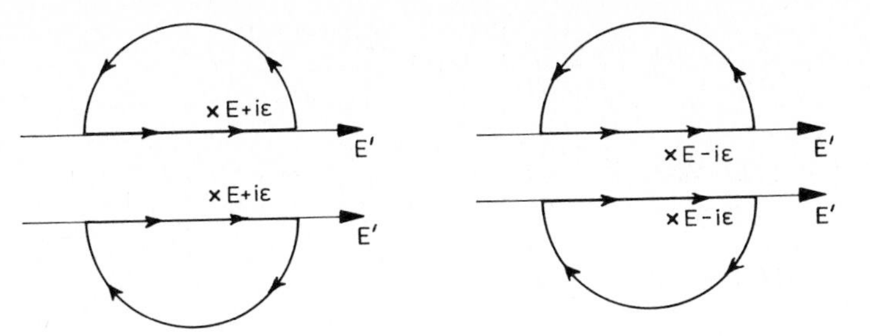

Fig. 6.3. Integration path and location of poles for the four integrals appearing in (6.27)

$$
a^*(E)\left\{1+\int dE''\,V^*_{E''}\left[P\frac{1}{E-E''}+g(E)\,\delta(E-E'')\right]\right.
$$
$$
\left.\times\left[P\frac{1}{E'-E''}+g(E')\,\delta(E'-E'')\right]V_{E''}\right\}a(E')=\delta(E-E')\,,
$$

which can be transformed into

$$
a^*(E)\,a(E')\left[1+\int dE''\left(\frac{P}{E-E''}\,\frac{P}{E'-E''}\,|V_{E''}|^2\right)+P\frac{|V_{E'}|^2 g(E')}{E-E'}\right.
$$
$$
\left.+P\frac{|V_E|^2 g(E)}{E'-E}+g^2(E)\,|V_E|^2\,\delta(E-E')\right]=\delta(E-E')\,. \tag{6.28}
$$

The P symbols in front of, e.g., $P(|V_{E'}|^2/E-E')$ indicate that the principal value should be taken when integrating either over E or E'. Particular attention must be paid to the integral over E'' in (6.28) at the point of double singularity $E=E'$. We show in Sect. 6.5 that this factor can be written as

$$
\frac{P}{E-E''}\,\frac{P}{E'-E''}=-\frac{P}{E-E'}\left(\frac{P}{E-E''}-\frac{P}{E'-E''}\right)
$$
$$
+\pi^2\delta(E-E')\,\delta(E''-\tfrac{1}{2}(E'+E))\,. \tag{6.29a}
$$

It thus can be properly resolved into partial fractions plus a singular term. In (6.28) the well-known relations have been used

$$
\delta(E-E'')\,\delta(E'-E'')=\delta(E-E')\,\delta(E-E'')=\delta(E-E')\,\delta(E'-E'')
$$
$$
=\delta(E-E')\,\delta[E''-\tfrac{1}{2}(E+E')]
$$

and

$$
\delta(E-E'')f(E'')=\delta(E-E'')f(E)\,. \tag{6.29b}
$$

Substituting (6.29a) into (6.28) yields

$$a^*(E)a(E')\left[1-\frac{P}{E-E'}\left(P\int\frac{dE''|V_{E''}|^2}{E-E''}-P\int\frac{dE''|V_{E''}|^2}{E'-E''}\right)\right.$$
$$+\pi^2|V_{E'}|^2\delta(E-E')+P\frac{|V_{E'}|^2g(E')}{E-E'}-P\frac{|V_E|^2g(E)}{E-E'}$$
$$\left.+g^2(E)|V_E|^2\delta(E-E')\right]=\delta(E-E')\,, \tag{6.30}$$

and by rearranging the terms and using definition (6.25)

$$a^*(E)a(E)[\pi^2|V_E|^2+g^2(E)|V_E|^2]\,\delta(E-E')$$
$$+a^*(E)\left\{1-\frac{P}{E-E'}[F(E)-F(E')+g(E')|V_{E'}|^2-g(E)|V_E|^2]\right\}a(E')$$
$$=\delta(E-E')\,. \tag{6.31}$$

The expression in braces { } vanishes, because of (6.24):

$$1-\frac{P}{E-E'}[F(E)-F(E')+g(E)|V_E|^2-g(E')|V_{E'}|^2]$$
$$=1-\frac{P}{E-E'}\{F(E)-F(E')+[E-E_{1s}-\Delta E_{1s}-F(E)]$$
$$-[E'-E_{1s}-\Delta E_{1s}-F(E')]\}$$
$$=1-\frac{P}{E-E'}(E-E')=1-1=0\,.$$

Hence

$$|a(E)|^2[\pi^2|V_E|^2+g^2(E)|V_E|^2]=1$$

results, from which follows

$$|a(E)|^2=\frac{1}{|V_E|^2[\pi^2+g^2(E)]}=\frac{|V_E|^2}{[E-E_{1s}-\Delta E_{1s}-F(E)]^2+\pi^2|V_E|^4} \tag{6.32a}$$

or

$$a(E)=+\sqrt{\frac{|V_E|^2}{[E-E_{1s}-\Delta E_{1s}-F(E)]^2+\pi^2|V_E|^4}}=\frac{\sin\Delta(E)}{\pi V_E}\,, \tag{6.32b}$$

if the phase is chosen such that $a(E)$ is real. In the last step $a(E)$ was expressed in terms of the phase shift $\Delta(E)$ introduced in (6.26).

The function $h_{E'}(E)$ defined in (6.10) and further specified in (6.21, 23) can now also be denoted in the convenient form

$$h_{E'}(E)=C_{E'}(E)a(E)=P\frac{V_{E'}a(E)}{E-E'}+\frac{[E-E_{1s}-\Delta E_{1s}-F(E)]}{\sqrt{\pi^2|V_E|^4+[E-E_{1s}-\Delta E_s-F(E)]^2}}$$

$$\times \sqrt{\frac{V_{E'}}{V_E}}\,\delta(E-E') \tag{6.33a}$$

$$= P\frac{V_{E'}\sin\Delta(E)}{\pi V_E(E-E')} + g(E)\,V_E a(E)\,\delta(E-E')$$

$$= P\frac{V_{E'}\sin\Delta(E)}{\pi V_E(E-E')} - \cos\Delta(E)\,\delta(E-E')\,, \tag{6.33b}$$

where for the second term we have used (6.29, 26). Note that in the limit of vanishing supercriticality, i.e. $V_E \to 0$, $h_{E'}(E)$ reduces to the $\delta(E-E')$ function, as (6.33) reveals. Furthermore the following useful properties of $a(E)$ are obtained:

$$|a(E)|^2 = \frac{|V_E|^2}{[E-E_{1s}-\Delta E_{1s}-F(E)]^2+\pi^2|V_E|^4} \tag{6.34a}$$

$$= \frac{1}{2\pi\mathrm{i}}\left[\frac{1}{[E-E_{1s}-\Delta E_{1s}-F(E)]-\mathrm{i}\pi|V_E|^2} - \frac{1}{[E-E_{1s}-\Delta E_{1s}-F(E)]+\mathrm{i}\pi|V_E|^2}\right] \tag{6.34b}$$

and

$$\int_{-\infty}^{-m_0c^2} |a(E)|^2 dE = 1\,. \tag{6.35}$$

To interpret the results (6.34, 35) it is helpful to introduce the width Γ_E defined as

$$\Gamma_E = 2\pi|V_E|^2 \tag{6.36}$$

which is in general energy (E) dependent. With that, (6.34a) reads

$$|a(E)|^2 = \frac{1}{2\pi}\,\frac{\Gamma_E}{\{E-[E_{1s}+\Delta E_{1s}+F(E)]\}^2+\Gamma_E^2/4}\,. \tag{6.37}$$

Since $|a(E)|^2$ is the probability for finding the former bound state ψ_{1s}^{cr} [cf. (6.10)], (6.37) shows that this former bound state is imbedded in the continuum in a resonance-type fashion. The resonance lies at the energy

$$E_{\mathrm{res}} = E_{1s}+\Delta E_{1s}+F(E) \tag{6.38}$$

and has the width (6.36):

$$\Gamma_E = 2\pi|V_E|^2\,.$$

Since, because of (6.4), the overcritical potential $V'(r)$ is proportional to the overcritical charge Z', i.e. $V'(r) \sim Z'$, it follows from (6.14, 15) that

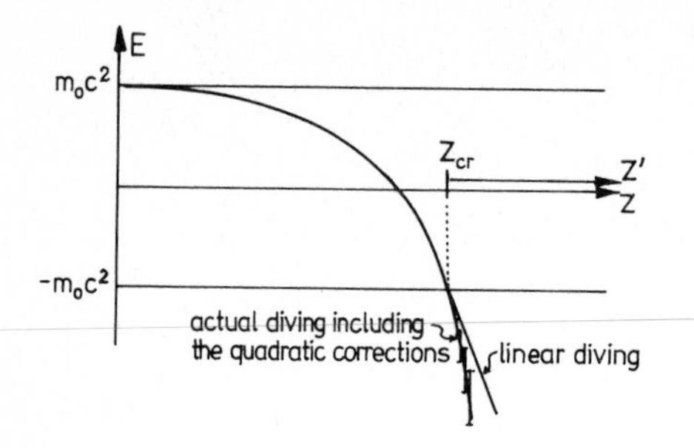

Fig. 6.4. Diving of a bound state in overcritical potentials (here: Coulomb potential). Z' is the overcritical central charge. The vertical bars indicate the width Γ_E of the dived bound state which grows quadratically with Z' [Mü 72a]

$$\Delta E_{1s} \sim Z' , \qquad \Gamma_E \sim Z'^2 , \tag{6.39}$$

and

$$F(E) \sim Z'^2 .$$

Embedding the former bound state ψ_{1s}^{cr} into the negative energy continuum, where it becomes a resonance, is called *"diving" of the bound state.* Following (6.38, 39) the diving is dominantly linear in the overcritical charge Z' with some (smaller) deviations quadratic in Z'. The linearity is due to ΔE_{1s}, the quadratic deviations stem from $F(E)$. While the bound state dives, it obtains a width $\Gamma_E \sim Z'^2$, i.e. the width grows essentially quadratically with the diving depth (which is approximately given by ΔE_{1s} as long as the energy dependence of V_E and the term $F(E)$ is neglected), Fig. 6.4.

Before continuing with the mathematical discussion, let us briefly interpret this new phenomenon physically. We do this in the framework of hole theory, which is completely equivalent to the field theoretical formulation of QED. The field theoretical interpretation of these results is given in Sects. 10.3 – 5.

We shall distinguish two cases, namely (a) filled K shell, and (b) empty K shell.

6.2.1 Filled *K* Shell

The filled K shell can also be characterized by a Fermi surface just above the $1s$ state. If the proton number of an undercritical nucleus (i.e. with $Z < Z_{\mathrm{cr}}$) is steadily increased, the energy of the K shell electrons decreases until it reaches $-m_0c^2$ at $Z = Z_{\mathrm{cr}}$. During this process the spatial extension of the K shell electron charge distribution also decreases, i.e. the bound state wave function becomes more and more localized. This is illustrated in Fig. 6.5, where the K shell charge distribution is shown for various elements. When Z grows beyond Z_{cr} the bound $1s$ state ceases to exist, but this does not mean that the K electron cloud becomes delocalized. Indeed, according to (6.10), the bound state ψ_{1s} is shared by the negative energy continuum states in a typical resonance-type manner. It is *spread out* over an energy range of the order Γ_E, (6.37), and due to the bound ψ_{1s}^{cr} state admixture the negative energy continuum wave functions strongly distort

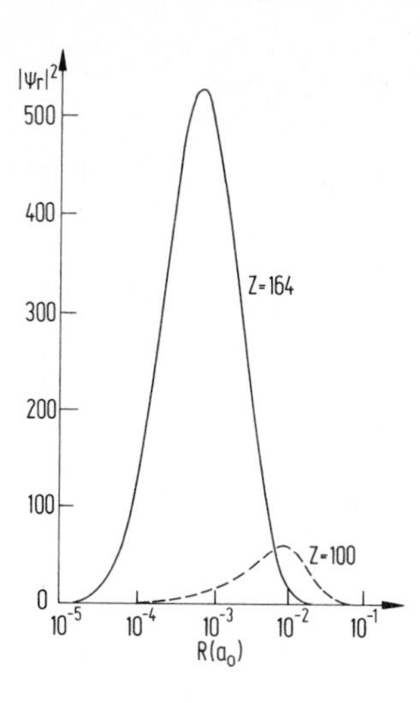

Fig. 6.5. The charge distribution $r^2\psi^\dagger\psi(r)$ of K electrons for heavy, and giant nuclei. The maximum of this distribution (Bohr radius) shifts towards smaller radii as Z increases while the tail of the distribution increases proportionally to $\exp\{-2r\sqrt{(m_e c^2)^2-[E-V(r)]^2}\}$ where E is the energy and $V(r)$ the potential

around the nucleus. Such spreading widths are quite well known in nuclear physics, where a property (wave function) of a certain state is spread out over many densely lying background states [Da 65]. This additional distortion of the negative energy continuum due to the bound state can be called *real vacuum polarization*, because it is caused by a real electron state, which joined the "vacuum states", i.e. the negative energy continuum states occupied by electrons (in the framework of hole theory). Indeed, as shown in Chap. 9, the vacuum polarization charge is given by

$$\varrho^{(u)}_{\text{vac pol}}(\boldsymbol{x}) = \frac{e}{2}\left[\sum_{E_n<-m_0c^2}\psi_n^\dagger\psi_n(\boldsymbol{x}) - \sum_{E_p>-m_0c^2}\psi_p^\dagger\psi_p(\boldsymbol{x})\right], \tag{6.40}$$

i.e. by the difference of the density of the electron-occupied negative energy states and that of the states above the Fermi surface (which, besides the bound states in the gap, are essentially the positive energy continuum states) multiplied by half the electron charge e. In principle $\sum$ denotes both a sum over bound states and an integral over continuum states. For simplicity we assume in the following that the continua are discretized, e.g. by solving the Dirac equation in a box, so that only sums appear. The superscript u indicates that $\varrho^{(u)}_{\text{vac pol}}(\boldsymbol{x})$ represents the vacuum polarization charge in an undercritical situation. Further, $\varrho^{(u)}_{\text{vac pol}}(\boldsymbol{x})$ is a *displacement charge*. Namely, in the absence of external fields, ψ_n and ψ_p are simply plane waves and cancel each other in (6.40). If an external field is turned on, the waves ψ_n, ψ_p are distorted, thus (6.40) represents a displacement charge with

$$Q^{(u)}_{\text{vac pol}} = \int \varrho^{(u)}_{\text{vac pol}}(\boldsymbol{x})\,d^3x = 0\,. \tag{6.41}$$

In the undercritical case the K shell, i.e. ψ_{1s}, belongs to the group of the ψ_p states in (6.40). This fact is clarified by rewriting (6.40):

$$\varrho_{\text{vac pol}}^{(u)}(\boldsymbol{x}) = \frac{e}{2}\left[\sum_{E_n<-m_0c^2} \psi_n^\dagger \psi_n(\boldsymbol{x}) - \psi_{1s\uparrow}^\dagger \psi_{1s\uparrow}(\boldsymbol{x}) - \psi_{1s\downarrow}^\dagger \psi_{1s\downarrow}(\boldsymbol{x}) - \sum_{\substack{E_p>-m_0c^2\\ E_p\neq 1s\uparrow,1s\downarrow}} \psi_p^\dagger \psi_p(\boldsymbol{x})\right], \tag{6.42}$$

where we have explicitly removed the contribution of the $1s$ state from the sum $\sum_{E_p}$.

Suppose now there is an overcritical situation; the Fermi surface is still at $E_f = -m_0c^2$. However the K shell ψ_{1s} now belongs to the group of ψ_n states. Therefore for the overcritical vacuum polarization density

$$\begin{aligned}
\varrho_{\text{vac pol}}^{(o)}(\boldsymbol{x}) &= \frac{e}{2}\left[\psi_{1s\uparrow}^\dagger \psi_{1s\uparrow}(\boldsymbol{x}) + \psi_{1s\downarrow}^\dagger \psi_{1s\downarrow}(\boldsymbol{x}) + \sum_{\substack{E_n<-m_0c^2\\ E_n\neq 1s\uparrow,1s\downarrow}} \psi_n^\dagger \psi_n(\boldsymbol{x}) - \sum_{E_p>-m_0c^2} \psi_p^\dagger \psi_p(\boldsymbol{x})\right]\\
&= \frac{e}{2}\left[2\psi_{1s\uparrow}^\dagger \psi_{1s\uparrow}(\boldsymbol{x}) + 2\psi_{1s\downarrow}^\dagger \psi_{1s\downarrow}(\boldsymbol{x}) + \sum_{\substack{E_n<-m_0c^2\\ E_n\neq 1s\uparrow,1s\downarrow}} \psi_n^\dagger \psi_n(\boldsymbol{x}) \right.\\
&\quad \left. - \psi_{1s\uparrow}^\dagger \psi_{1s\uparrow}(\boldsymbol{x}) - \psi_{1s\downarrow}^\dagger \psi_{1s\downarrow}(\boldsymbol{x}) - \sum_{E_p>-m_0c^2} \psi_p^\dagger \psi_p(\boldsymbol{x})\right]\\
&= e\,\psi_{1s\uparrow}^\dagger \psi_{1s\uparrow}(\boldsymbol{x}) + e\,\psi_{1s\downarrow}^\dagger \psi_{1s\downarrow}(\boldsymbol{x}) + \varrho_{\text{vac pol}}^{(u)}(\boldsymbol{x})\,.
\end{aligned} \tag{6.43}$$

Clearly, because of (6.41),

$$\begin{aligned}
\int \varrho_{\text{vac pol}}^{(o)}(\boldsymbol{x})\,d^3x &= e\int[\psi_{1s\uparrow}^\dagger \psi_{1s\uparrow}(\boldsymbol{x}) + \psi_{1s\downarrow}^\dagger \psi_{1s\downarrow}(\boldsymbol{x})]\,d^3x\\
&= 2e \neq 0\,,
\end{aligned} \tag{6.44}$$

the vacuum now contains the real charge $2e$, namely that of the K shell, which is imbedded the vacuum in the overcritical case. This is mathematically expressed in (6.10) and particularly (6.35). Thus, the K shell electron cloud remains localized in $\boldsymbol{x}$ space. The surprising result that the K shell wave function ψ_{1s} spreads out in energy by a width Γ_E (6.37), though the electron cloud does not decay, is resolved by the fact that it is distributed in energy over many states. In other words, the dived bound state obtains a *spreading width* in the negative energy continuum.

This can be further illustrated in the following way. Consider the *gedanken-experiment* of solving the Dirac equation with the cut-off Coulomb potential inside a finite sphere of radius a. Certain boundary conditions on the sphere have to be fulfilled. Thus the continuum is discretized (Figs. 6.6, 7). Figure 6.6a illustrates the situation just before diving, i.e. at $Z \approx Z_{\text{cr}}$, while (b) shows the overcritical case where the $1s$ bound state joined the lower continuum (which is discretized) and spread over it. The diving process itself is illustrated in Fig. 6.7. If all levels (the $1s$ level and the lower continuum) are occupied, the K shell still exists, but is spread out energetically. One might ask how this spreading mani-

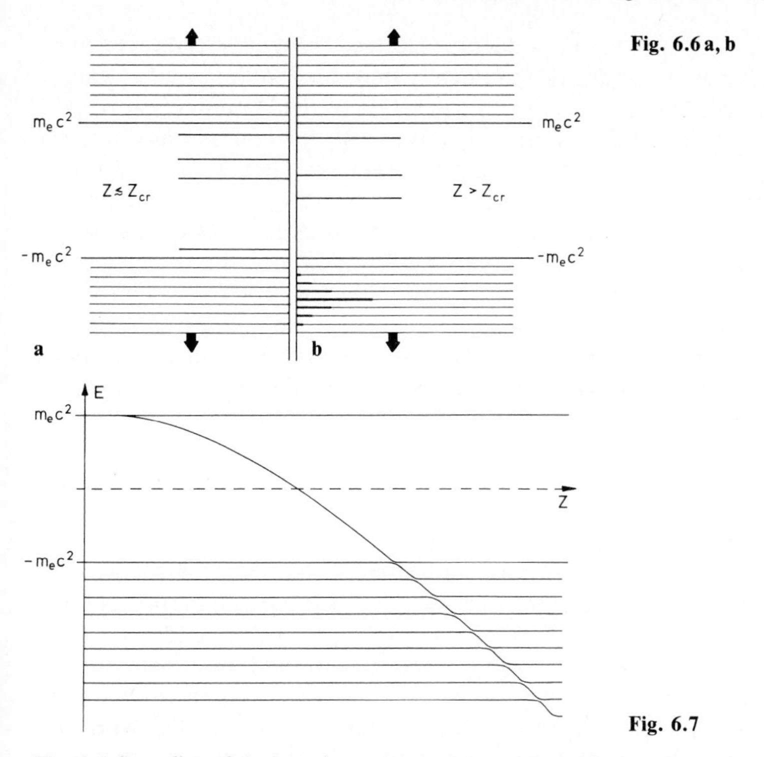

Fig. 6.6. Spreading of the bound state (▬▬) over the negative energy continuum states (——). After diving (*b*), i.e. for $Z > Z_{cr}$, the negative energy continuum obtains one more state, the $1s$ bound state. It is now spread, however, over the negative energy continuum states. (*a*) Spectrum before and (*b*) after diving [Mü 72b]

Fig. 6.7. Qualitative behaviour of the "diving" bound state in the negative energy continuum states. No level crossing occurs. Only the property ψ_{1s}, i.e. the wave function of the bound state, "passes" through the negative continuum [Mü 72c]

fest itself physically. The answer is simple. Suppose one shines x-rays on a supercritical nucleus with empty L shell. The x-ray absorption line from the $1s \rightarrow 2p$ transitions then acquire an additional width, the spreading width, identical with the positron escape width discussed next. If one interprets this behaviour as *real vacuum polarization*, as in (6.40–44), it should be recognized that the new vacuum in overcritical fields is doubly charged. Those electrons which in the undercritical case represented the K electrons have now become part of the vacuum. We are thus led to the concept of a *charged vacuum in overcritical fields*. The neutral vacuum, i.e. the ground state of ordinary (undercritical field) QED becomes unstable in overcritical fields, as shown below.

6.2.2 Empty *K* Shell

Let us now discuss the physical phenomena which occur during the diving process if the bound state $|\psi_{1s}\rangle$ is unoccupied. The situation now differs from that above. If the central charge is suddenly increased from undercritical to overcritical values, the empty state $\psi_{1s}^{\rm cr}$ is suddenly imbedded in the negative energy continuum. This configuration is not an eigenstate of the Hamiltonian. After a short time the imbedded bound state $\psi_{1s}^{\rm cr}$ mixes with the continuum states $\psi_E^{\rm cr}$ so forming the new, overcritical continuum $\Psi_E(\boldsymbol{x})$ (6.10). Here $\Psi_E(\boldsymbol{x})$ is an eigenstate of the overcritical Hamiltonian. Due to this mixture an electron from the occupied lower continuum levels $\psi_{E\uparrow}^{\rm cr}$ can, without energy transfer, i.e. spontaneously, fill the empty bound state $\psi_{1s\uparrow}$ (and similarly to the spin $\downarrow$ levels), leaving a hole in a continuum state $\psi_{E\uparrow}^{\rm cr}$. The latter has to be interpreted as a positron with spin $\downarrow$, which escapes the overcritical field space by moving to infinity (the positron is repelled from the central nucleus). The kinetic energy of the escaping positron is not sharp, but has a Breit-Wigner type distribution as given by (6.37). The width Γ_E is now the *positron escape width* and the time

$$\tau_{E_r} = \frac{\hbar}{\Gamma_{E_r}} \tag{6.45}$$

is the *decay time for the empty K shell* imbedded in the negative energy continuum. Precisely put, *the neutral electron-positron vacuum of quantum electrodynamics is unstable in overcritical electric fields.* It decays under positron emission into a *charged vacuum*, which is stable due to the Pauli principle. The last statement means that after two positrons ($\uparrow$ and $\downarrow$) have been emitted and, consequently, the previously empty, dived K shell has been filled with two electrons, the Pauli principle prevents further decay due to positron emission.

The physics of this *fundamental change of the ground state of QED* is illustrated in Fig. 6.8. The empty space in the box (the space free of real particles) represents the vacuum.

The central nucleus therein serves only as a source (generator) of the strong electric field. Otherwise it is a spectator. If the charge of the central nucleus is smaller than the critical charge, i.e. for $Z < Z_{\rm cr}$, the empty space (vacuum) stays empty for ever. This is the *neutral vacuum*. In the overcritical situation, however, there are always two electrons present (Fig. 6.8b), respresenting the *charged*

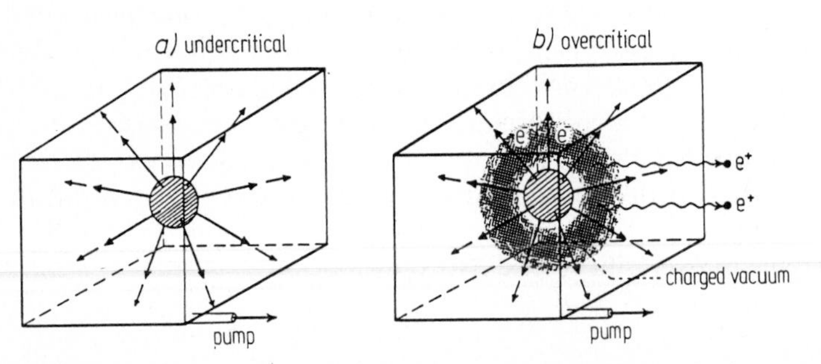

Fig. 6.8a, b. The vacuum is represented by a box containing only the central nucleus as a source (generator) for the strong electric field (indicated by arrows). It is stable and neutral in the undercritical case (**a**). In the overcritical case (**b**) the vacuum is no longer empty, but contains two electrons

vacuum. If they were pumped away by an "elementary particle pump" (this would need much energy since the electrons are bound by more than $2m_0c^2$), a neutral vacuum will be created. However, within the time

$$\tau_E = \frac{\hbar}{\Gamma_E} = \frac{\hbar}{2\pi} \frac{1}{|V_E|^2},$$

(6.45, 36), the electron cloud reestablishes itself and two positrons in continuum states would float around in the box. They are easily pumped out, so that the old situation is reestablished. Only the charged vacuum is stable in overcritical fields; the space without particles, i.e. the neutral vacuum, decays by positron emission into the charged vacuum.

6.3 Two and More Bound States Imbedded in One Continuum

As can be guessed, the $2p_{1/2}$ bound state is the next one to reach the negative energy continuum if the central charge is further increased to $\bar{Z}_{cr} \approx 185$. Again this number depends on the details of the model for the nuclear charge distribution. It seems at first that for nuclei with $Z > \bar{Z}_{cr}$ two bound states are imbedded in the continuum. However, the $1s$ state is imbedded into the s negative energy continuum and the $2p_{1/2}$ bound state is imbedded into the $p_{1/2}$ negative energy continuum. The sets

$$\begin{aligned} &\{|1s_{1/2}\rangle, \quad |s_{1/2}E\rangle\} \\ &\{|2p_{1/2}\rangle, \quad |p_{1/2}E\rangle\} \end{aligned} \tag{6.46}$$

consisting of bound states and continuum wave functions are pairwise orthogonal and the overcritical monopole interaction $V'(r)$ of (6.5) does not generate any matrix element between states belonging to different sets. (The dipole and quadrupole potentials occurring in a heavy-ion collision $(Z_1 \neq Z_2)$ would generate transitions between various states, because angular momentum and parity $(Z_1 \neq Z_2)$ are not conserved in the two-centre problem.) Hence interaction of bound states with the continua can be treated separately for the $1s_{1/2}$, $2p_{1/2}, \ldots,$ levels.

The problem of two (or more) bound states interacting with one continuum does not occur until $Z \gtrsim 245$. At $\bar{\bar{Z}}_{cr} \approx 245$ the $2s_{1/2}$ bound state dives into the negative energy continuum. Though it is unlikely that nuclei with $Z > 245$ can be created or even simulated during heavy-ion collisions, we now present the solution of this problem. (If sufficiently intense heavy-ion beams were available, it is plausible that simultaneous collisions among three heavy nuclei, e.g. $U + U + U$, occur with sufficient probability to make experiments possible.) This adds to our understanding of overcritical systems and charged vacua and is important for the study of overcritical point nuclei and shielding of central charge, Sect. 16.5. Furthermore, the solution presented can be immediately applied also to the case where only the bound $1s$ state dives, but the influence of a finite number of bound ns level is considered.

We start with a set of N discrete states

$$\psi_1^{\mathrm{cr}}, \psi_2^{\mathrm{cr}}, \ldots \psi_N^{\mathrm{cr}} \tag{6.47}$$

interacting with the continuum

$$\psi_E^{\mathrm{cr}} . \tag{6.48}$$

Both the bound states ψ_n^{cr} (which can be thought of being, e.g., the $ns_{1/2}$ bound states) and the continuum ψ_E^{cr} (which can be the $s_{1/2}$ continuum states) are solutions of the just undercritical Dirac Hamiltonian $\hat{H}_{\mathrm{cr}}$ introduced in (6.5). Thus

$$\begin{aligned} \hat{H}_{\mathrm{cr}} \psi_n^{\mathrm{cr}} &= E_n \psi_n^{\mathrm{cr}} , \\ \hat{H}_{\mathrm{cr}} \psi_E^{\mathrm{cr}} &= E \psi_E^{\mathrm{cr}} \end{aligned} \tag{6.49}$$

with the normalization

$$\begin{aligned} \langle \psi_n^{\mathrm{cr}} | \psi_{n'}^{\mathrm{cr}} \rangle &= \delta_{nn'} , \\ \langle \psi_n^{\mathrm{cr}} | \psi_E^{\mathrm{cr}} \rangle &= 0 , \\ \langle \psi_E^{\mathrm{cr}} | \psi_{E'}^{\mathrm{cr}} \rangle &= \delta(E-E') . \end{aligned} \tag{6.50}$$

Now we construct a solution for the overcritical Hamiltonian $\hat{H} = \hat{H}_{\mathrm{cr}} + V'(\boldsymbol{x})$ by diagonalizing it on the basis of (6.47, 48). The required matrix elements are

$$\begin{aligned} \langle \psi_m^{\mathrm{cr}} | \hat{H} | \psi_n^{\mathrm{cr}} \rangle &= \langle \psi_m^{\mathrm{cr}} | \hat{H}_{\mathrm{cr}} | \psi_n^{\mathrm{cr}} \rangle + \langle \psi_m^{\mathrm{cr}} | V' | \psi_n^{\mathrm{cr}} \rangle \\ &= E_n \delta_{nm} + \Delta E_{mn} , \end{aligned} \tag{6.51}$$

$$\langle \psi_{E'}^{\mathrm{cr}} | \hat{H} | \psi_n^{\mathrm{cr}} \rangle = \langle \psi_{E'}^{\mathrm{cr}} | V' | \psi_n^{\mathrm{cr}} \rangle \equiv V_{E'n} , \tag{6.52}$$

$$\begin{aligned} \langle \psi_{E''}^{\mathrm{cr}} | \hat{H} | \psi_{E'}^{\mathrm{cr}} \rangle &= \langle \psi_{E''}^{\mathrm{cr}} | \hat{H}_{\mathrm{cr}} | \psi_{E'}^{\mathrm{cr}} \rangle + \langle \psi_{E''}^{\mathrm{cr}} | V' | \psi_{E'}^{\mathrm{cr}} \rangle \\ &= E' \delta(E'-E'') + U_{E''E'} . \end{aligned} \tag{6.53}$$

As before, the matrix $U_{E''E'}$ describes the rearrangement of the continuum states due to the overcritical potential $V'(\boldsymbol{x})$. Assume again that this matrix can be prediagonalized (or neglected) and proceed as if the continuum is already chosen such that $U_{E''E'}$ vanishes. The eigenvectors we seek now have the form

$$\Psi_E(\boldsymbol{x}) = \sum_n a_n(E) \psi_n^{\mathrm{cr}} + \int_{|E'| \geq m_0 c^2} dE' h_{E'}(E) \psi_{E'}^{\mathrm{cr}} , \tag{6.54}$$

and there must be one such eigenvector for each value of E in the energy range considered. Note that the continuum integral in (6.54) now extends over the positive and negative energy continua, and Ψ_E has to obey

$$\hat{H} \Psi_E(\boldsymbol{x}) = E \Psi_E(\boldsymbol{x}) . \tag{6.55}$$

Taking matrix elements we arrive at the system of linear equations

$$E_n a_n + \sum_m \Delta E_{nm} a_m + \int dE' \, V_{nE'} h_{E'}(E) = E a_n \,, \tag{6.56a}$$

$$\sum_n V_{E'n} a_n(E) + E' h_{E'}(E) = E h_{E'}(E) \,, \tag{6.56b}$$

$$\sum_n a_n^*(E) a_n(E') + \int dE'' \, h^*_{E''}(E) h_{E''}(E') = \delta(E-E') \,. \tag{6.56c}$$

The last equation expresses the normalization of $\Psi_E(x)$ (6.54). The set of coupled equations (6.56) corresponds to the former set (6.20). Similarly to (6.21) we make the ansatz

$$h_{E'}(E) = C_{E'}(E) \sum_n V_{E'n} a_n(E) \tag{6.57}$$

and insert it into (6.56b), thus obtaining

$$(E-E') C_{E'}(E) = 1 \,. \tag{6.58}$$

This suggests a solution of the form

$$C_{E'}(E) = \frac{P}{E-E'} + g(E) \, \delta(E-E') \,, \tag{6.59}$$

where again $g(E)$ is an unknown function which can be determined similarly as in (6.23 – 25). Inserting (6.57, 59) into (6.56a) yields

$$(E_n - E) a_n(E) + \sum_m \Delta E_{nm} a_m(E) + \int dE' \, C_{E'}(E) V_{nE'} \sum_m V_{E'm} a_m(E) = 0$$

or

$$(E_n - E) a_n(E) + \sum_m [\Delta E_{nm} + F_{nm}(E)] a_m(E) + g(E) V_{nE} \sum_m V_{Em} a_m(E) = 0 \,, \tag{6.60}$$

where

$$F_{nm}(E) = P \int dE' \, \frac{V_{nE'} V_{E'm}}{E-E'} \tag{6.61}$$

describes the second-order interaction of the bound states via the continuum. In the next stage of solution the matrix

$$E_n \delta_{nm} + \Delta E_{nm} + F_{nm} \tag{6.62}$$

can be diagonalized, that is, we consider the effect of the interaction matrix $\Delta E_{nm} + F_{nm}$ upon the discrete states. This effect perturbs the states ψ_n^{cr} and their energies E_n and replaces them by new states

$$\bar{\psi}_n^{\text{cr}} = \sum_m A_{mn} \psi_m^{\text{cr}} \tag{6.63}$$

with energies $\bar{E}_n$, obtained by solving the system

$$E_n A_{nm} + \sum_\nu (\Delta E_{n\nu} + F_{n\nu}) A_{\nu m} = A_{nm} \bar{E}_m \,. \tag{6.64}$$

When off-diagonal terms $(\Delta E_{nm}+F_{nm})$ are small compared to diagonal ones, E_m are approximately given in first-order perturbation theory by

$$\bar{E}_n \approx E_n + \Delta E_{nn} + F_{nn} \,. \tag{6.65}$$

This approximation certainly holds for the $1s$ and higher ns states, which are separated by more than 500 keV. Replacing the energies E_n by $\bar{E}_n$ corresponds to replacing E_{1s} by $E_{1s}+\Delta E_{1s}+F(E)$ in the earlier case (6.24, 25, 37). Since the matrix $F_{n\nu}(E)$ depends on E, (6.61), though only weakly, the solutions of (6.64) are also functions of E, i.e., $A_{mn}(E)$, $\bar{E}_n(E)$.

Assuming now that the coefficients A_{mn} and energies $\bar{E}_n$ have been obtained, the coefficients $a_n(E)$ in (6.60) are replaced by new coefficients $\bar{a}_n$, setting

$$a_n(E) = \sum_\nu A_{n\nu}(E)\bar{a}_\nu(E) \,. \tag{6.66}$$

The matrix $E_{nm}+F_{nm}$ can now be eliminated from (6.60) by means of (6.64), and (6.60) becomes

$$\sum_\nu A_{n\nu}\bar{E}_\nu\bar{a}_\nu + g(E)\, V_{nE} \sum_{m,\nu} V_{Em}A_{m\nu}\bar{a}_\nu = E \sum_\nu A_{n\nu}\bar{a}_\nu \,. \tag{6.67}$$

Multiplication by $(A^{-1})_{\mu n}$, summation over n and application of the orthonormality

$$\sum_n (A^{-1})_{\mu n}A_{n\nu} = \delta_{\mu\nu}$$

yields finally

$$\bar{E}_\nu\bar{a}_\nu + g(E)\, \bar{V}_{\nu E} \sum_\mu \bar{V}_{E\mu}\bar{a}_\mu = E\bar{a}_\nu \,, \tag{6.68}$$

where

$$\bar{V}_{E\mu} = \sum_m V_{Em}A_{m\mu} \,. \tag{6.69}$$

The next step is to solve the system (6.68). It is a matrix equation for $\bar{a}_\nu$, which has a matrix of the form

$$M_{\nu\mu} = \sqrt{g(E)}\, \bar{V}_{\nu E} \sqrt{g(E)}\, \bar{V}_{E\mu} \tag{6.70}$$

and thus belongs to the class whose solutions are expressed in terms of a *polarizability function*. Multiplication of (6.68) by $\bar{V}_{E\nu}/E-\bar{E}_\nu$ and summation over ν yields

$$\sum_\nu \frac{\bar{V}_{E\nu}\bar{a}_\nu\bar{E}_\nu}{E-\bar{E}_\nu} + g(E) \sum_\nu \frac{\bar{V}_{E\nu}\bar{V}_{\nu E}}{E-\bar{E}_\nu} \sum_\mu \bar{V}_{E\mu}\bar{a}_\mu = E \sum_\nu \frac{\bar{V}_{E\nu}\bar{a}_\nu}{E-\bar{E}_\nu}$$

or

$$-\sum_\nu \bar{V}_{E\nu}\bar{a}_\nu \frac{(E-\bar{E}_\nu)}{(E-\bar{E}_\nu)} + g(E) \sum_\nu \frac{\bar{V}_{E\nu}\bar{V}_{\nu E}}{E-\bar{E}_\nu} \sum_\mu \bar{V}_{E\mu}\bar{a}_\mu = 0 \,,$$

which with (6.69) leads to the consistency requirement

$$g(E)\sum_\nu \frac{|\bar{V}_{\nu E}|^2}{E-\bar{E}_\nu} = 1\,. \tag{6.71}$$

It plays the role of a secular equation and determines the eigenvalue $g(E)$. Under this condition, (6.68) is solved by

$$\bar{a}_\nu(E) = g(E)\frac{\bar{V}_{\nu E}}{E-\bar{E}_\nu}\sum_\mu \bar{V}_{E\mu}\bar{a}_\mu(E) \tag{6.72}$$

in terms of the quantity

$$\begin{aligned}\sum_\mu \bar{V}_{E\mu}\bar{a}_\mu(E) &= \sum_{\substack{m\\ \mu,n}} V_{Em}A_{m\mu}(A^{-1})_{\mu n}a_n(E)\\ &= \sum_m V_{Em}a_m(E)\end{aligned} \tag{6.73}$$

where (6.66, 69) were used. Since the same term $\sum_n V_{E'n}a_n(E)$ appears in (6.57) for the continuum amplitudes $h_{E'}(E)$, it can be treated as a normalization constant $N(E',E)$. Thus for (6.72, 57)

$$\begin{aligned}\bar{a}_\nu(E) &= g(E)\frac{\bar{V}_{\nu E}}{E-\bar{E}_\nu}N(E,E)\\ h_{E'}(E) &= C_{E'}(E)N(E'E) = \left[\frac{P}{E-E'} + g(E)\,\delta(E-E')\right]N(E',E)\,,\end{aligned} \tag{6.74}$$

where

$$N(E'E) = \sum_m V_{E'm}a_m(E)\,. \tag{6.75}$$

The "eigenvalue" function $g(E)$ follows from (6.71)

$$g(E) = \frac{1}{\displaystyle\sum_\nu \frac{|\bar{V}_{\nu E}|^2}{E-\bar{E}_\nu}} \tag{6.76}$$

and is obviously real. For later use it is appropriate to write this in the form

$$\frac{\pi}{g(E)} = \sum_\nu \frac{\pi|\bar{V}_{E\nu}|^2}{E-\bar{E}_\nu} = -\tan\Delta \equiv -\sum_\nu \tan\Delta_\nu\,. \tag{6.77}$$

Here Δ and Δ_ν turn out to be phase shifts, whose meaning is discussed below.

The remaining determination of the normalization constant $N(E',E)$ in (6.74, 75) is analogous to determining $a(E)$ from (6.28). We start with the normalization condition (6.56c) and follow the calculation in detail. From

$$\sum_n a_n^*(E)a_n(E') + \int dE''\,h_{E''}(E)\,h_{E''}(E') = \delta(E-E') \tag{6.56c}$$

after utilizing (6.57, 59) and the reality of $g(E)$ in (6.76) follows

$$\sum_n a_n^*(E)a_n(E') + \int dE'' \left\{\left[\frac{P}{E-E''} + g(E)\,\delta(E-E'')\right] \sum_n V_{E''n}^* a_n^*(E)\right\}$$
$$\times \left\{\left[\frac{P}{E'-E''} + g(E')\,\delta(E'-E'')\right] \sum_m V_{E''m} a_m(E')\right\} = \delta(E-E') . \quad (6.78)$$

The integral breaks up into

$$\begin{aligned}
&\text{a)}\quad \int dE'' \frac{P}{(E-E'')}\,\frac{P}{(E'-E'')} \sum_{n,m} V_{E''n}^* a_n^*(E)\, V_{E''m} a_m(E') ,\\
&\text{b)}\quad \int dE'' \frac{P}{E-E''}\, g(E')\,\delta(E'-E'') \sum_{n,m} V_{E''n}^* a_n^*(E)\, V_{E''m} a_m(E') ,\\
&\text{c)}\quad \int dE'' \frac{P}{E'-E''}\, g(E)\,\delta(E-E'') \sum_{n,m} V_{E''n}^* a_n^*(E)\, V_{E''m} a_m(E') ,\\
&\text{d)}\quad \int dE''\, g(E)\, g(E')\,\delta(E-E'')\,\delta(E'-E'') \sum_{n,m} V_{E''n}^* a_n^*(E)\, V_{E''m} a_m(E') ,
\end{aligned} \quad (6.79)$$

so that the total integral of (6.78) is given by the sum of these terms

$$\int dE'' \ldots = (a) + (b) + (c) + (d) . \quad (6.80)$$

They are investigated separately.

a) Utilizing the identity (6.29a), this integral becomes

$$\sum_{n,m} a_n^*(E) \left[-\frac{P}{E-E'} \left(P\int dE'' \frac{V_{E''n}^* V_{E''m}}{E-E''} - P\int dE'' \frac{V_{E''n}^* V_{E''m}}{E'-E''} \right) \right.$$
$$\left. + \pi^2 \delta(E-E')\, V_{En}^* V_{Em} \right] a_m(E')$$
$$= \sum_{n,m} a_n^*(E) \left\{ -\frac{P}{E-E'} [F_{nm}(E) - F_{nm}(E')] + \pi^2 \delta(E-E')\, V_{En}^* V_{Em} \right\} a_m(E') . \quad (6.81)$$

b) This integral can immediately be written down as

$$\frac{P}{E-E'}\, g(E') \sum_{n,m} a_n^*(E)\, V_{E'n}^* V_{E'm} a_m(E') . \quad (6.82)$$

c) This integral is similar to (b) and is

$$-\frac{P}{E-E'}\, g(E) \sum_{n,m} a_n^*(E)\, V_{En}^* V_{Em} a_m(E') . \quad (6.83)$$

d) Direct evaluation yields

$$g^2(E)\,\delta(E'-E)\sum_{n,m} a_n^*(E)\,V_{En}^*\,V_{Em}\,a_m(E')\,. \tag{6.84}$$

The total integral can now be added and inserted into (6.78) and regrouped:

$$\begin{aligned}\sum_n a_n^*(E)\,a_n(E') + \sum_{n,m} a_n^*(E)\Bigg\{ -\frac{P}{E-E'}\,[F_{nm}(E)+g(E)\,V_{En}^*\,V_{Em}\\ -F_{nm}(E')-g(E')\,V_{E'n}^*\,V_{E'm}]\Bigg\}\,a_m(E')\\ +[\pi^2+g^2(E)]\,\delta(E-E')\sum_{n,m} a_n^*(E)\,V_{En}^*\,V_{Em}\,a_m(E) = \delta(E-E')\,,\end{aligned} \tag{6.85}$$

where E replaces E' in the last term, which is possible because of the $\delta(E-E')$ function in front. A further simplification is possible via (6.60) which states that

$$\sum_m [F_{nm}(E')+g(E')\,V_{E'm}]\,a_m(E') = (E'-E_n)\,a_n(E') - \sum_m \Delta E_{nm}\,a_m(E') \tag{6.86a}$$

and

$$\sum_n a_n^*(E)\,[F_{nm}(E)+g(E)\,V_{En}^*\,V_{Em}] = (E-E_m)\,a_m^*(E) - \sum_n \Delta E_{mn}\,a_n^*(E)\,. \tag{6.86b}$$

Then last relation follows from the first one by taking the complex conjugate, recognizing that

$$F_{nm}^*(E) = F_{mn}(E)\,, \tag{6.87}$$

and by finally interchanging n and m.

Now, with (6.86), (6.85) becomes

$$\begin{aligned}\sum_n a_n^*(E)\,a_n(E') - \frac{P}{E-E'}\Bigg[\sum_m (E-E_m)\,a_m^*(E)\,a_m(E')\\ -\sum_n (E'-E_n)\,a_n^*(E)\,a_n(E')\Bigg]\\ +[\pi^2+g^2(E)]\,\delta(E'-E)\sum_{n,m} a_n^*(E)\,V_{En}^*\,V_{Em}\,a_m(E) = \delta(E-E')\,.\end{aligned} \tag{6.88}$$

Changing the summation index of the first term in the bracket [ ] from m to n yields

$$\begin{aligned}\sum_n a_n^*(E)\,a_n(E') - \frac{P}{E-E'}\left[(E-E')\sum_n a_n^*(E)\,a_n(E')\right]\\ +[\pi^2+g^2(E)]\,\delta(E-E')\left|\sum_m V_{Em}\,a_m(E)\right|^2 = \delta(E-E')\,.\end{aligned} \tag{6.89}$$

Obviously the first two terms cancel, so that

$$\left|\sum_m V_{Em}\,a_m(E)\right|^2 = \frac{1}{\pi^2+g^2(E)} \tag{6.90}$$

finally follows. This determines the normalization factor (6.73) and also $\bar{a}_\nu(E)$ of (6.72). After using (6.77) the final normalized solution of this problem may now be given in the form

$$\begin{aligned}\bar{a}_\nu(E) &= \frac{g(E)}{\sqrt{\pi^2+g^2(E)}}\,\frac{\bar{V}_{\nu E}}{E-\bar{E}_\nu} = -\frac{\pi}{\tan\Delta}\,\frac{1}{\sqrt{\pi^2+(\pi^2/\tan^2\Delta)}}\left(-\frac{\tan\Delta_\nu}{\pi\bar{V}^*_{\nu E}}\right)\\ &= \frac{1}{\pi\bar{V}^*_{\nu E}}\,\frac{1}{\sqrt{1+\tan^2\Delta}}\tan\Delta_\nu\\ &= \frac{1}{\pi\bar{V}^*_{\nu E}}\cos\Delta\tan\Delta_\nu\\ &= \frac{1}{\pi\bar{V}_{E\nu}}\cos\Delta\tan\Delta_\nu\end{aligned} \tag{6.91a}$$

$$\begin{aligned}h_{E'}(E) &= C_{E'}(E)\sum_n V_{E'n}a_n(E) = \left[\frac{P}{E-E'}+g(E)\,\delta(E-E')\right]\sum_n V_{E'n}a_n(E)\\ &= \frac{P}{E-E'}\sum_n V_{E'n}a_n(E)+g(E)\sum_n V_{En}a_n(E)\,\delta(E-E')\\ &\approx \frac{P}{E-E'}\sum_n \bar{V}_{E'n}\bar{a}_n(E)+\frac{g(E)}{\sqrt{\pi^2+g^2(E)}}\,\delta(E-E')\\ &= \cos\Delta\left[\sum_n P\frac{\bar{V}_{E'n}}{E-E'}\,\frac{\tan\Delta_n}{\pi\bar{V}_{En}}-\delta(E'-E)\right].\end{aligned} \tag{6.91b}$$

In the penultimate step we have assumed that

$$\sum_n V_{E'n}a_n(E) \approx \sum_n \bar{V}_{E'n}\bar{a}_n(E)\,, \tag{6.92}$$

i.e. that the transformation matrix $A_{n\nu}(E)$ in (6.66) does not depend on E, which is only approximately true. It is obvious that (6.91) reduces for a single level [only one particular value of the $\sum_n$ in (6.91b) contributes] to the result (6.33).

Let us now discuss the implications of these results for overcritical fields. As mentioned earlier (6.65), the off-diagonal terms of E_{nm} and F_{nm} are expected to be small compared to the diagonal ones, so that (6.65) holds and, moreover,

$$\bar{a}_n \approx a_n\,. \tag{6.93}$$

With these approximations one obtains from (6.91a)

$$|\bar{a}_n(E)|^2 = \frac{|V_{En}|^2}{(E-\bar{E}_n)^2+\pi^2\left(\sum_m |V_{Em}|^2\frac{E-\bar{E}_n}{E-\bar{E}_m}\right)^2}\,. \tag{6.94}$$

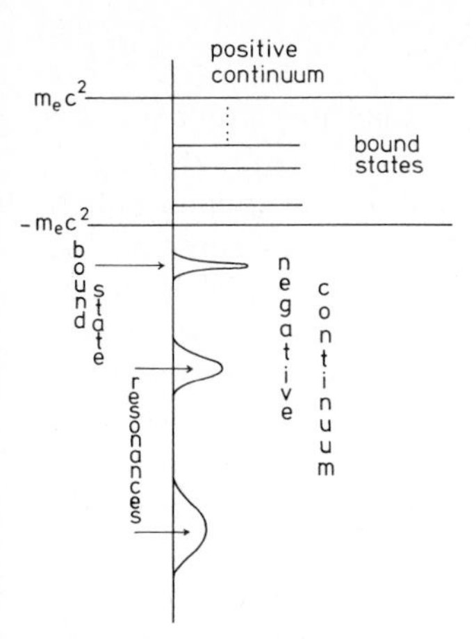

Fig. 6.9. Scheme of the distortion of the negative energy continuum when several bound states have binding energies larger than $2m_ec^2$. The bound states have become resonances. The deeper bound states have generally larger spreading width than the weaker bound states [Mü 72b]

If the discrete levels $\bar{\psi}_n^{\text{cr}}$ with energy $\bar{E}_n$ are well apart in energy, one can perform for $E \approx \bar{E}_n$ the further approximation

$$|\bar{a}_n(E \approx \bar{E}_n)|^2 \approx \frac{|V_{En}|^2}{(E-\bar{E}_n)^2 + \pi^2 |V_{En}|^4} = \frac{1}{2\pi} \frac{\Gamma_n(E)}{(E-\bar{E}_n)^2 + \Gamma_n^2(E)/4} . \tag{6.95}$$

This is equal to the earlier result (6.37), if (6.65) is taken into account. In general, however, (6.94) represents a *multilevel Breit-Wigner distribution*:

$$|\bar{a}_n(E)|^2 = \frac{\frac{1}{2\pi} \Gamma_n(E)}{(E-E_n-\Delta E_{nn}-F_{nn})^2 + \frac{1}{4}\left[\sum_m \Gamma_m(E) \frac{E-\bar{E}_n}{E-\bar{E}_m}\right]^2} . \tag{6.96}$$

If only the lowest s level is imbedded in the negative energy continuum, and the correct (diagonalized) values for $\bar{E}_n$ are used, this formula modifies (6.37) due to interaction with higher bound s orbitals. This modification mainly shifts levels of the resonance from $E_0 + \Delta E_0$ to $\bar{E}_0$, where $\bar{E}_0$ is the energy resulting from the diagonalization of (6.64). One expects, of course, that $\bar{E}_0 < E_0 + \Delta E_0$. When more bound states occur with binding energies larger than $2m_0c^2$, i.e. for the overcritical charges $Z' = Z - Z_{\text{cr}} = 50 - 100$, the negative energy continuum becomes more distorted, Fig. 6.9.

6.4 One Bound State Imbedded in Several Continua

We generalize now the treatment in Sect. 6.2 to the case where the overcritical interaction couples the bound state φ to several different continua

$$\psi_{E'}, \chi_{E'}, \ldots \quad (6.97)$$

These continua may be distinguished by suitable quantum numbers. For example, they may represent the $s_\uparrow$, $s_\downarrow$, $p_{1/2\uparrow}$, $p_{1/2\downarrow}$, $p_{3/2\mu}$, $d_{3/2\mu}$, $d_{5/2\nu}$, etc., continua. For a spherical supercritical atom the $1s$ bound state couples only to the s continuum, because of the spherical symmetry. An example where this symmetry is broken is given for an overcritical giant molecule, consisting of two heavy nuclei, equal or unequal, like U+U or U+Cm at a distance R. In such *giant molecules* the interaction of the two central charges contains many multipoles (monopole, dipole, quadrupole, etc.), and therefore also the overcritical potential V' (6.3) contains various multipoles. Consequently, the matrix elements between the bound state φ, which also has no good angular momentum, and the various multipole continua generally exist. We do not specify the continua here, but only assume that they may be distinguished by suitable quantum numbers (e.g. s, p, d states). For simplicity we shall refer to two continua only, unless some purpose is served by explicit reference to n continua.

The eigenvectors to be determined have the form

$$\Psi_{hE} = a^h(E)\,\varphi^{\mathrm{cr}} + \int dE'\,[b^h_{E'}(E)\,\psi^{\mathrm{cr}}_{E'} + c^h_{E'}(E)\,\chi^{\mathrm{cr}}_{E'}]\,. \quad (6.98a)$$

The index h indicates a set of n parameters required to specify Ψ, since at each energy an n-fold degeneracy exists. Then, clearly, the coefficients $a^h(E)$, $b^h_{E'}(E)$, $c^h_{E'}(E)$ are functions of E and h. As shown, determining $(n-1)$ orthogonal continua is almost trivial, because they can be determined so that for these $n-1$ continua the interaction with the bound state φ^{cr} vanishes, i.e. $a^h(E)=0$ for $h=1,2\ldots(n-1)$. The single remaining linearly independent eigenvector is then determined by the methods presented before, Sect. 6.1, 6.2.

The following matrix elements of the overcritical Dirac Hamiltonian

$$\hat{H} = \hat{H}_{\mathrm{cr}} + V' \quad (6.99)$$

in the basis of subcritical states φ^{cr}, $\psi^{\mathrm{cr}}_{E'}$, $\chi^{\mathrm{cr}}_{E'}$ are relevant for later discussion:

$$\langle\varphi^{\mathrm{cr}}|\hat{H}|\varphi^{\mathrm{cr}}\rangle = E_0 + \Delta E_0\,, \qquad \Delta E_0 = \langle\varphi^{\mathrm{cr}}|V'|\varphi^{\mathrm{cr}}\rangle, \quad (6.100a)$$

$$\langle\psi_{E'}|\hat{H}|\varphi^{\mathrm{cr}}\rangle = V_{E'}\,, \quad (6.100b)$$

$$\langle\chi_{E'}|\hat{H}|\varphi^{\mathrm{cr}}\rangle = W_{E'}\,, \quad (6.100c)$$

$$\langle\psi_{E''}|\hat{H}|\psi_{E'}\rangle = \langle\chi_{E''}|\hat{H}|\chi_{E'}\rangle = E'\,\delta(E'-E'')\,, \quad (6.100d)$$

$$\langle\psi_{E''}|\hat{H}|\chi_{E'}\rangle = 0\,. \quad (6.100e)$$

The last two formulas imply that the continuum-continuum interaction due to the overcritical potential V' has already been diagonalized. The prediagonalized continuum states are $\psi_{E'}$, $\chi_{E'}$, i.e. without the superscript "critical" as in (6.98). This is, of course, not easy to achieve practically (Sect. 10.4 gives a method of practically achieving prediagonalization). Nevertheless, the assumption is made in order to focus on the interaction of the continua with the discrete state. From now on the superscript "critical" on $\psi_{E'}^{\text{cr}}$ and $\chi_{E'}^{\text{cr}}$ of (6.98) is dropped, i.e. we start with the expansion

$$\Psi_{hE} = a^h(E)\,\varphi^{\text{cr}} + \int dE'\,[b_{E'}^h(E)\,\psi_{E'} + c_{E'}^h(E)\,\chi_{E'}] \,. \tag{6.98b}$$

The system of equations analogous to (6.20) is then

$$(E_0 + \Delta E_0)\,a(E) + \int dE'\,[V_{E'}^* b_{E'}(E) + W_{E'}^* c_{E'}(E)] = E\,a(E)\,, \tag{6.101a}$$

$$V_{E'}\,a(E) + E'\,b_{E'}(E) = E\,b_{E'}(E)\,, \tag{6.101b}$$

$$W_{E'}\,a(E) + E'\,c_{E'}(E) = E\,c_{E'}(E)\,. \tag{6.101c}$$

Here superscript h is deleted for clarity. After multiplying by $(V_{E'}^*, W_{E'}^*)$ and $(W_{E'}, -V_{E'})$ respectively, (6.101b, c) can be combined to form

$$\begin{aligned}[]&[|V_{E'}|^2 + |W_{E'}|^2]\,a(E) + E'\,[V_{E'}^* b_{E'}(E) + W_{E'}^* c_{E'}(E)] \\ &\quad = E[V_{E'}^* b_{E'}(E) + W_{E'}^* c_{E'}(E)]\,,\end{aligned} \tag{6.102}$$

$$E'\,[W_{E'} b_{E'}(E) - V_{E'} c_{E'}(E)] = E\,[W_{E'} b_{E'}(E) - V_{E'} c_{E'}(E)]\,. \tag{6.103}$$

Equation (6.102) contains the same combination of $b_{E'}(E)$ and $c_{E'}(E)$ as (6.101a) and with (6.101a) forms a system equivalent to (6.20a, b). On the other hand, (6.103) is completely decoupled from $a(E)$. Equations (6.101a, 102) are solved in analogy to (6.20a, b) ff. for the dependent variables $a(E)$ and $V_{E'}^* b_{E'} + W_{E'}^* c_{E'}$. In fact, solution (6.23, 24, 32) can be applied directly provided $|V_{E'}|^2$ is replaced by $|V_{E'}|^2 + |W_{E'}|^2$. The solution of (6.103) for $W_{E'} b_{E'}(E) - V_{E'} c_{E'}(E)$ is simply $\delta(E' - E)$ to within a normalization factor. Hence we set

$$\begin{aligned} a^{(1)}(E) &= \frac{\sin\bar{\Delta}}{\pi(|V_E|^2 + |W_E|^2)^{1/2}} \\ &= \sqrt{\frac{|V_E|^2 + |W_E|^2}{[E - E_0 - \Delta E_0 - G(E)]^2 + \pi^2(|V_E|^2 + |W_E|^2)}} \end{aligned} \tag{6.104}$$

$$\begin{aligned} &V_{E'}^* b_{E'}^{(1)}(E) + W_{E'}^* c_{E'}^{(1)}(E) \\ &= a^{(1)}(E)\left[P\frac{|V_{E'}|^2 + |W_{E'}|^2}{E - E'} + g(E)(|V_{E'}|^2 + |W_{E'}|^2)\,\delta(E - E')\right] \\ &= a^{(1)}(E)\left\{P\frac{|V_{E'}|^2 + |W_{E'}|^2}{E - E'} + [E - E_0 - \Delta E_0 G(E)]\,\delta(E' - E)\right\} \\ &= \frac{|V_{E'}|^2 + |W_{E'}|^2}{\pi\sqrt{|V_{E'}|^2 + |W_{E'}|^2}}\left[P\frac{\sin\bar{\Delta}}{E - E'} - \cos\bar{\Delta}\,\delta(E - E')\right], \end{aligned} \tag{6.105}$$

where

$$\bar{\Delta} = -\arctan \frac{\pi(|V_E|^2 + |W_E|^2)}{E - E_0 - \Delta E_0 - G(E)} \tag{6.106}$$

and

$$G(E) = P\int dE' \frac{|V_{E'}|^2 + |W_{E'}|^2}{E - E'} . \tag{6.107}$$

Equations (6.104 – 107) are the straightforward analogues of (6.32b, 23, 26, 24a) respectively. The corresponding equations are in fact identical if the above-mentioned replacements are executed. Note, however, the bracket in the last form of (6.105), which seems to be slightly different from the analogous equation (6.23b), but is, in fact, the same! Since $b_{E'}(E)$ and $c_{E'}(E)$ must have symmetric structure with respect to $V_{E'}$ and $W_{E'}$, the following solutions for $b_{E'}(E)$ and $c_{E'}(E)$ themselves emerge from (6.105):

$$\begin{aligned} b_{E'}^{(1)}(E) &= \frac{V_{E'}}{\sqrt{|V_E|^2 + |W_E|^2}} \left[\frac{1}{\pi} P \frac{\sin\bar{\Delta}}{E - E'} - \cos\bar{\Delta}\, \delta(E - E') \right] , \\ c_{E'}^{(1)}(E) &= \frac{W_{E'}}{\sqrt{|V_E|^2 + |W_E|^2}} \left[\frac{1}{\pi} P \frac{\sin\bar{\Delta}}{E - E'} - \cos\bar{\Delta}\, \delta(E - E') \right] . \end{aligned} \tag{6.108}$$

They satisfy also (6.103), as is easily checked. The second orthogonal continuum is the following solution of (6.101):

$$a^{(2)}(E) = 0 , \tag{6.109a}$$

$$b_{E'}^{(2)}(E) = \frac{W_E^*}{\sqrt{|V_E|^2 + |W_E|^2}} \delta(E - E') , \tag{6.109b}$$

$$c_{E'}^{(2)}(E) = -\frac{V_E^*}{\sqrt{|V_E|^2 + |W_E|^2}} \delta(E - E') , \tag{6.109c}$$

as can immediately be verified.

Clearly, if there are $n > 2$ continua, the extension of the formulas is straightforward; additional terms have to be added to $|V_E|^2 + |W_E|^2$ wherever the expression appears in (6.104 – 109), and additional dependent variables $d_{E'}^{(1)}(E) \ldots$, etc., which are analogously constructed as, e.g., (6.108) have to be considered. The second solution (6.109) is extended to a set of $(n-1)$ orthogonal solutions of the n variable homogeneous equation

$$V_{E'}^* b_{E'}(E) + W_{E'}^* c_{E'}(E) + \ldots = 0 , \tag{6.110}$$

which results from the analogous solution of (6.102, 103) for $a(E) = 0$.

The overcritical vacuum is then represented by the two (or more) continuum states

$$\Psi_{1E},\ \Psi_{2E}, \ldots , \tag{6.111}$$

all occupied by electrons. The dived bound state is contained in Ψ_{1E} and the overcritical vacuum is again charged by two units, if the diving state is a $s_{1/2}$ or $p_{1/2}$ state. The other occupied continua Ψ_{2E}, Ψ_{3E}, ... etc. are modified also, but do not contain additional real charge. They contribute solely to vacuum polarization.

6.5 Overcritical Continuum States

In Sect. 3.5 we considered standing Coulomb waves for an undercritical $1/r$ potential. Let us summarize the essential results.

In the coordinates

$$x = 2\mathrm{i}pr$$
$$p = \frac{\sqrt{E^2-(m_0c^2)}}{\hbar c} \equiv \mathrm{i}\lambda \tag{6.112}$$

and with the transformation of the radial functions $u_1(r)\ u_2(r)$

$$\begin{pmatrix} u_1(r) \\ u_2(r) \end{pmatrix} = \begin{pmatrix} \sqrt{E+m_0c^2} & \sqrt{E+m_0c^2} \\ \mathrm{i}\sqrt{E-m_0c^2} & -\mathrm{i}\sqrt{E-m_0c^2} \end{pmatrix} \begin{pmatrix} \Phi_1 \\ \Phi_2 \end{pmatrix}, \tag{6.113}$$

we obtained the second-order differential equation for $\Phi_1(x)$:

$$\frac{d^2\Phi_1(x)}{dx^2} + \frac{1}{x}\frac{d\Phi_1(x)}{dx} - \left[\frac{1}{4} + \left(\frac{1}{2} + \mathrm{i}y\right)\frac{1}{x} + \frac{\gamma^2}{x^2}\right]\Phi_1 = 0\,. \tag{6.114}$$

Quantity $\Phi_2(x)$ can be expressed through $\Phi_1(x)$ rather simply, namely

$$\Phi_2(x) = \Phi_1^*(x)\,. \tag{6.115}$$

The following abbreviations were used

$$y = \frac{Z\alpha E}{\hbar c p}, \qquad \gamma = +\sqrt{\varkappa^2-(Z\alpha)^2}\,. \tag{6.116}$$

For $Z\alpha < |\varkappa|$ the quantity γ is real. For $Z\alpha > |\varkappa|$, γ is imaginary and the root with $\mathrm{Im}\{\gamma\} > 0$ is chosen. It is well known that the fundamental system of independent solutions of (6.114) is given by [Sl 60]

$$\begin{aligned} \Phi_1^{(\pm)} &= x^{\pm\gamma}\mathrm{e}^{-1/2x}\,{}_1F_1(1\pm\gamma+\mathrm{i}y, \pm 2\mathrm{i}\gamma+1; x) \\ &= x^{-1/2}M_{-(\mathrm{i}y+1/2),\,\pm\gamma}(x)\,. \end{aligned} \tag{6.117}$$

Obviously $\Phi_1^{(+)}(x)$ is regular and $\Phi_1^{(-)}(x)$ irregular at the origin for $Z\alpha < |\varkappa|$. We return to this point later when discussing the point nucleus problem (Sect. 6.4.2). For $Z\alpha > |\varkappa|$ both solutions oscillate rapidly

$$\Phi_1^{(\pm)} \propto x^{\pm\gamma} = \mathrm{e}^{\pm\mathrm{i}|y|\ln x} \tag{6.118}$$

as $x \to 0$, but $|\Phi_1^{(\pm)}|$ remains limited and therefore integrable.

If Φ_1 is determined, Φ_2 can be obtained either through (6.115) (as in Sect. 3.5) or from the coupled system of linear differential equations obeyed by Φ_1 and Φ_2 (cf. (3.116) of Sect. 3.5). Then Φ_2 is calculated to be

$$\Phi_2(x) = -\frac{x}{\varkappa - \mathrm{i}y\dfrac{m_0c^2}{E}}\left[\frac{d}{dx} - \frac{1}{2} - \frac{\mathrm{i}y}{x}\right]\Phi_1(x)\,. \tag{6.119}$$

6.5.1 Continuum Solutions for Extended Nuclei

To demonstrate the occurrence of resonances in the positron spectrum for overcritical fields from a different point of view to that in Sects. 6.2–4, we now treat the continuum problem for extended nuclei, following [Mü 73b]. The main task is to determine phase shifts, as compared to the pure Coulomb phases described in Sect. 3.5. This then allows us to make the central, extended charge overcritical and to follow the appearance of resonances in the negative energy continuum. In fact, the investigations here culminate in a rather exact method to determine the resonances due to dived bound states. This complements the discussions in Sect. 6.2–4 and gives more insight with higher numerical accuracy into the various processes related to the overcriticality phenomenon.

We first outline the solutions in the conventional region of nuclear charge.

$Z\alpha < |\varkappa|$. In view of (6.117), the most general ansatz for Φ_1 is

$$\Phi_1(x) = x^{-1/2}(a_+ M_{-(\mathrm{i}y+1/2),\gamma}(x) + a_- M_{-(\mathrm{i}y+1/2),-\gamma}(x))\,. \tag{6.120}$$

The coefficients a_+ and a_- must still be determined. By means of (6.119) we then find

$$\begin{aligned}\Phi_2^*(x) = {} & [a_+^*(-\gamma-\mathrm{i}y)\,\mathrm{e}^{-\mathrm{i}\pi\gamma}M_{-(\mathrm{i}y+1/2),\gamma}(x) + a_-^*(\gamma-\mathrm{i}y)\,\mathrm{e}^{\mathrm{i}\pi\gamma}M_{-(\mathrm{i}y+1/2),\gamma}(x)] \\ & \times \frac{x^{-1/2}}{\varkappa+\mathrm{i}y\dfrac{m_0c^2}{E}}\,.\end{aligned} \tag{6.121}$$

Taking (6.115) into account leads to the following requirement for the coefficients a_+, a_-:

$$\begin{aligned} a_+ &= a_+^*\frac{(-\gamma-\mathrm{i}y)\,\mathrm{e}^{-\mathrm{i}\pi\gamma}}{\varkappa+\mathrm{i}y\dfrac{m_0c^2}{E}}\,, \\ a_- &= a_-^*\frac{(\gamma-\mathrm{i}y)\,\mathrm{e}^{+\mathrm{i}\pi\gamma}}{\varkappa+\mathrm{i}y\dfrac{m_0c^2}{E}}\,. \end{aligned} \tag{6.122}$$

Writing now

$$a_+ = N e^{i\alpha_+} \cos\eta$$
$$a_- = N e^{-i\alpha_-} \sin\eta \qquad (6.123)$$

conditions (6.122) can be rewritten in terms of the phases α_+, α_- as

$$e^{2i\alpha_+} = -\frac{(\gamma+iy)\,e^{-i\pi\gamma}}{\varkappa+iy\dfrac{m_0c^2}{E}},$$
$$e^{2i\alpha_-} = \frac{(\gamma-iy)\,e^{+i\pi\gamma}}{\varkappa+iy\dfrac{m_0c^2}{E}}. \qquad (6.124)$$

Hence the phases are in fact fixed and only one parameter, namely η of (6.123), is still free. The superposition (6.120) can now be written as

$$\Phi_1(x) = Nx^{-1/2}[\cos\eta\, e^{i\alpha_+} M_{-(iy+1/2),\gamma}(x) + \sin\eta\, e^{i\alpha_-} M_{-(iy+1/2),-\gamma}(x)]\,. \qquad (6.125)$$

With the help of the asymptotic expressions for the Whittaker functions [Sl 60] the asymptotic expansion of $\Phi_1(x)$ can be calculated to be

$$\Phi_1(x) \xrightarrow[x\to\infty]{} N\left[\cos\eta\, e^{i\alpha_+}\frac{\Gamma(2\gamma+1)}{\Gamma(\gamma+1+iy)} + \sin\eta\, e^{i\alpha_-}\frac{\Gamma(-2\gamma+1)}{\Gamma(-\gamma+1+iy)}\right]$$
$$\times \exp\left[-\frac{\pi}{2}y + i(pr + y\ln 2pr)\right]. \qquad (6.126)$$

Since the total continuum spinor $\Psi_E(r)$ given by [see Sect. 3.5, (3.130, 131) and also Sect. 3.3, (3.49)]

$$\psi_{E,\varkappa m}(r) = \frac{1}{r}\begin{pmatrix} u_1(r)\,\chi_{\varkappa m} \\ iu_2(r)\,\chi_{-\varkappa m}\end{pmatrix} \qquad (6.127)$$

should be normalized to delta functions, i.e.

$$\int d^3x\, \psi^\dagger_{E,\varkappa,m}(x)\,\psi_{E',\varkappa',m'}(x) = \delta_{\varkappa\varkappa'}\,\delta_{mm'}\,\delta(E-E')\,, \qquad (6.128)$$

one verifies with the help of (6.113, 115) that the Φ_1 function should have the following asymptotic form

$$\Phi_1 \xrightarrow[r\to\infty]{} \frac{1}{2\sqrt{\pi p}}\, e^{i(pr+\Delta)}\,. \qquad (6.129)$$

Comparing this with (6.126) allows N and Δ to be determined.

$$N = \frac{e^{\pi/2y}}{2\sqrt{\pi p}\left|\cos\eta\, e^{i\alpha_+}\frac{\Gamma(2\gamma+1)}{\Gamma(\gamma+1+iy)} + \sin\eta\, e^{i\alpha_-}\frac{\Gamma(-2\gamma+1)}{\Gamma(-\gamma+1+iy)}\right|} \tag{6.130}$$

and

$$\Delta = y\ln 2pr + \arg\left[e^{i\alpha_+}\frac{\Gamma(2\gamma+1)}{\Gamma(\gamma+1+iy)} + \tan\eta\, e^{i\alpha_-}\frac{\Gamma(-2\gamma+1)}{\Gamma(-\gamma+1+iy)}\right]. \tag{6.131}$$

As mentioned above, α_+ and α_- are determined by (6.124); the only free parameter which still has to be fixed is η. It is determined from the matching condition

$$\frac{u_1(R)}{u_2(R)} = \frac{u_1^{(i)}(R)}{u_2^{(i)}(R)} \tag{6.132}$$

at the finite nuclear radius R [cf. the discussion of the Dirac particle in a spherical well, Sect. 3.3, (3.63)]. The radial functions inside the spherical nucleus are $u_{1,2}^{(i)}$. Inserting (6.123) into (6.120), then this together with (6.115) into (6.113), one can finally calculate from (6.132) the desired value for η, which depends, of course, on the nuclear radius R. The result of this straightforward but lengthy calculation is

$$\tan\eta = -\frac{u_2^{(i)}(R)\,\mathrm{Re}\{B_+(2ipR)\} \pm \sqrt{\dfrac{E-m_0c^2}{E+m_0c^2}}\,u_1^{(i)}(R)\,\mathrm{Im}\{B_+(2ipR)\}}{u_2^{(i)}(R)\,\mathrm{Re}\{B_-(2ipR)\} \pm \sqrt{\dfrac{E-m_0c^2}{E+m_0c^2}}\,u_1^{(i)}(R)\,\mathrm{Im}\{B_-(2ipR)\}}, \tag{6.133}$$

where

$$B_\pm(x) = \frac{1}{\sqrt{x}}\,e^{i\alpha_\pm}M_{-(iy+1/2),\,\pm\gamma}(x)\,. \tag{6.134}$$

The $(+)$ sign in (6.133) is valid for $E > m_0c^2$ and the $(-)$ sign for $E < -m_0c^2$. When the extended nucleus shrinks to a point $(R\to 0)$, $\eta\to 0$ because $M_{-(iy+1/2),\,-\gamma}(x)$ diverges for $x\to 0$; hence the denominator of (6.133) also diverges. Then the *Coulomb phase shift* results from (6.131), namely

$$\begin{aligned}
\Delta &= y\ln 2pr + \alpha_+ + \arg\Gamma(2\gamma+1) - \arg\Gamma(\gamma+1+iy)\\
&= y\ln 2pr + \frac{1}{2}\left[\arg(-1) + \arg(\gamma+iy) - \arg\left(\varkappa + iy\frac{m_0c^2}{E}\right) - \pi\gamma\right]\\
&\qquad + \arg\Gamma(2\gamma+1) - \arg[(\gamma+iy)\,\Gamma(\gamma+iy)]\\
&= y\ln 2pr - \arg\Gamma(\gamma+iy) - \frac{\pi\gamma}{2}\\
&\qquad + \frac{1}{2}\left[\arg(-1) - \arg(\gamma+iy) - \arg\left(\varkappa + iy\frac{m_0c^2}{E}\right)\right]\\
&= y\ln 2pr - \arg\Gamma(\gamma+iy) - \frac{\pi\gamma}{2} + \frac{1}{2}\arg\left[-\frac{\varkappa - iy\,m_0c^2/E}{(\gamma+iy)}\right].
\end{aligned} \tag{6.135}$$

Here we used the well-known properties $\Gamma(Z+1) = Z\Gamma(Z)$, $\arg Z^* = \arg 1/Z$ and $\arg \Gamma(2\gamma+1) = 0$, because $\Gamma(2\gamma+1)$ is real. Note that (6.135) is identical with (3.135), Sect. 3.5.

$Z\alpha > |\varkappa|$. Since this situation is most important for the overcriticality problem, the characteristic features are more explicitly treated. Again, as in (6.120), the most general ansatz for Φ_1 is

$$\Phi_1(x) = \frac{1}{\sqrt{x}} [a_+ M_{-(\mathrm{i}y+1/2),\mathrm{i}\gamma}(x) + a_- M_{-(\mathrm{i}y+1/2),\,-\mathrm{i}\gamma}(x)] \,, \tag{6.136}$$

where, because of (6.116), γ is now imaginary. For lucidity, this feature is made explicit by replacing

$$\gamma \to \mathrm{i}\gamma = \mathrm{i}\sqrt{(Z\alpha)^2 - \varkappa^2} \,. \tag{6.137}$$

From (6.119) the analogous expression to (6.121) is calculated as

$$\begin{aligned}\Phi_2^*(x) = \frac{x^{-1/2}}{\left(\varkappa + \mathrm{i}y \dfrac{m_0 c^2}{E}\right)} & [a_+^* (\mathrm{i}\gamma - \mathrm{i}y) \mathrm{e}^{-\pi\gamma} M_{-(\mathrm{i}y+1/2),\,-\mathrm{i}\gamma}(x) \\ & + a_-^* (-\mathrm{i}\gamma - \mathrm{i}y) \mathrm{e}^{\pi\gamma} M_{-(\mathrm{i}y+1/2),\,+\mathrm{i}\gamma}(x)] \,.\end{aligned} \tag{6.138}$$

As before, this must be identical to (6.136), which yields relations analogous to (6.122):

$$\begin{aligned} a_- &= \frac{a_+^* (\mathrm{i}\gamma - \mathrm{i}y) \mathrm{e}^{-\pi\gamma}}{\varkappa + \mathrm{i}y \dfrac{m_0 c^2}{E}} \,, \\ a_+ &= \frac{a_-^* (\mathrm{i}\gamma + \mathrm{i}y) \mathrm{e}^{\pi\gamma}}{\varkappa + \mathrm{i}y \dfrac{m_0 c^2}{E}} \,. \end{aligned} \tag{6.139}$$

From these a_+ and a_- can easily be determined up to an arbitrary phase η and a normalization factor N:

$$\begin{aligned} a_+ &= N \mathrm{e}^{\mathrm{i}\eta} \,, \\ a_- &= N \mathrm{e}^{-\mathrm{i}\eta} \mathrm{e}^{-\pi\gamma} \frac{\mathrm{i}\gamma - \mathrm{i}y}{\varkappa + \mathrm{i}y (m_0 c^2/E)} \,. \end{aligned} \tag{6.140}$$

Inserting this into (6.136) gives

$$\begin{aligned}\Phi_1(x) = \frac{N}{\sqrt{x}} \Bigg[& \mathrm{e}^{\mathrm{i}\eta} M_{-(\mathrm{i}y+1/2),\mathrm{i}\gamma}(x) \\ & + \mathrm{e}^{(-\mathrm{i}\eta - \pi\gamma)} \frac{\mathrm{i}\gamma - \mathrm{i}y}{\varkappa + \mathrm{i}y (m_0 c^2)/E} M_{-(\mathrm{i}y+1/2),\,-\mathrm{i}\gamma}(x) \Bigg] \,.\end{aligned} \tag{6.141}$$

As in (6.126), the asymptotic form of $\Phi_1(x)$ can now easily be determined.

$$\Phi_1(x) \xrightarrow[x\to\infty]{} N\left[\mathrm{e}^{\mathrm{i}\eta}\frac{\Gamma(2\mathrm{i}\gamma+1)}{\Gamma(\mathrm{i}\gamma+1+\mathrm{i}y)}+\mathrm{e}^{(-\mathrm{i}\eta-\pi y)}\frac{\mathrm{i}\gamma-\mathrm{i}y}{\varkappa+\mathrm{i}y(m_0c^2)/E}\right.$$
$$\left.\times\frac{\Gamma(-2\mathrm{i}\gamma+1)}{\Gamma(-\mathrm{i}\gamma+1+\mathrm{i}y)}\right]\exp[-\tfrac{1}{2}\pi y+\mathrm{i}(pr+y\ln 2pr)]\,. \tag{6.142}$$

Normalizing the total spinors (6.127) to δ functions (6.128) requires the asymptotic form (6.129). Hence, comparing (6.142) with (6.129) yields the normalization factor

$$N=\frac{\mathrm{e}^{1/2\pi y}/(2\sqrt{\pi p})}{\left|\mathrm{e}^{\mathrm{i}\eta}\frac{\Gamma(2\mathrm{i}\gamma+1)}{\Gamma(\mathrm{i}\gamma+1+\mathrm{i}y)}+\mathrm{e}^{(-\mathrm{i}\eta-\pi y)}\frac{\mathrm{i}\gamma-\mathrm{i}y}{\varkappa+\mathrm{i}y(m_0c^2)/E}\frac{\Gamma(-2\mathrm{i}\gamma+1)}{\Gamma(-\mathrm{i}\gamma+1+\mathrm{i}y)}\right|} \tag{6.143}$$

and the scattering phase

$$\Delta=y\ln 2pr+\arg\left[\mathrm{e}^{\mathrm{i}\eta}\frac{\Gamma(2\mathrm{i}\gamma+1)}{\Gamma(\mathrm{i}\gamma+1+\mathrm{i}y)}+\mathrm{e}^{(-\mathrm{i}\eta-\pi y)}\right.$$
$$\left.\times\frac{\mathrm{i}\gamma-\mathrm{i}y}{\varkappa+\mathrm{i}y(m_0c^2)/E}\frac{\Gamma(-2\mathrm{i}\gamma+1)}{\Gamma(-\mathrm{i}\gamma+1+\mathrm{i}y)}\right]. \tag{6.144}$$

The phase η is still unknown, but can be determined through the matching condition (6.132). The procedure is the same as in the undercritical case $Z\alpha<|\varkappa|$. Introducing the abbreviations

$$B_{\pm}(x)=\frac{1}{\sqrt{x}}\left[M_{-(\mathrm{i}y+1/2),\mathrm{i}\gamma}(x)\pm\mathrm{e}^{-\pi y}\frac{\mathrm{i}\gamma-\mathrm{i}y}{\varkappa+\mathrm{i}y(m_0c^2)/E}M_{-(\mathrm{i}y+1/2),-\mathrm{i}\gamma}(x)\right] \tag{6.145}$$

yields

$$\tan\eta=\frac{u_2^{(i)}(R)\,\mathrm{Re}\{B_+(2\mathrm{i}pR)\}\pm\sqrt{\frac{E-m_0c^2}{E+m_0c^2}}\,u_1^{(i)}(R)\,\mathrm{Im}\{B_+(2\mathrm{i}pR)\}}{u_2^{(i)}(R)\,\mathrm{Im}\{B_-(2\mathrm{i}pR)\}\mp\sqrt{\frac{(E-m_0c^2)}{(E+m_0c^2)}}\,u_1^{(i)}(R)\,\mathrm{Re}\{B_-(2\mathrm{i}pR)\}}\,, \tag{6.146}$$

where the upper sign is valid for $E>m_0c^2$ and the lower sign for $E<-m_0c^2$. Note both the similarity and differences between (6.146, 133).

6.5.2 Comments on the Point Nucleus Problem for $Z\alpha>|\varkappa|$

The overcritical point nucleus problem is pertinent in understanding the physical significance of perpendicular diving of the fine-structure levels as $Z\to 1/\alpha$ (Sect. 3.4). This issue is difficult and is treated in greater detail in Sect. 16.4, where is shown that the vacuum shields itself in the case of point nuclei so that $Z\alpha$

never exceeds 1. In other words, the highest possible charge for a point nucleus allowed by QED is

$$(Z_{\max})_{\text{point nucleus}} = \frac{1}{\alpha} = 137 \,. \tag{6.147}$$

Here follows a preliminary consideration of the behaviour of the phase shifts in the point nucleus limit.

The usual procedure to obtain solutions of the Dirac equation for a point-like Coulomb charge with $Z\alpha > |\varkappa|$ fails, because all solutions of the wave equation are square integrable for $r \to 0$. This is not crucial, however, since any point charge (mass point) is a mathematical and not a physical object. It has to be understood (defined) as the limit (idealization) of the more realistic, extended charges.

The realistic situation is accounted for by cutting off the $(1/r)$ behaviour of the Coulomb potential at some small distance R, Fig. 6.10.

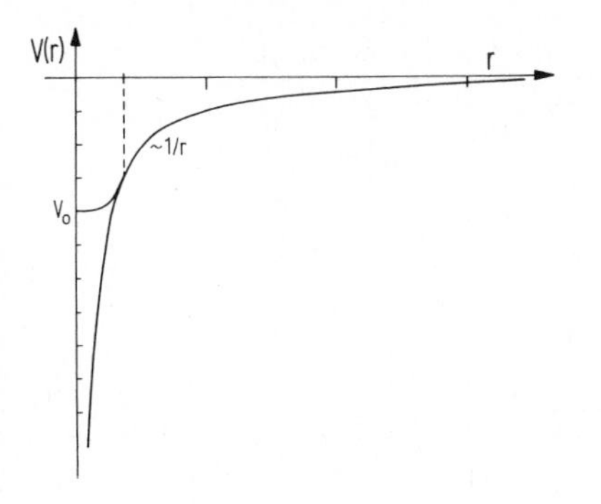

Fig. 6.10. Coulomb potential for a point charge is Ze^2/r everywhere; for a finite charge distribution the potential is cut off at some short distance R, which measures the extent of the source. V_0 is the depth of the potential

Only if an unambiguous limit of the relevant physical quantities (energies of bound states, phase shifts of continuum states) exists for $R \to 0$ may we justly speak of the corresponding limiting solution as the physical solution of the idealized "point" problem. For $Z\alpha < |\varkappa|$ this procedure leads directly to the well-known point nucleus solutions described in Sect. 3.4, 5, particularly with respect to the phase shifts in (6.135).

For $Z\alpha > |\varkappa|$ such a point limit for the phase shifts cannot be defined [Mü 72c, 73b]. Assume that the central charge distribution has a radius R small enough so

$$pR = \sqrt{E^2 - m_0 c^2} R \ll 1 \,. \tag{6.148}$$

This condition is fulfilled for real nuclei and low or intermediate electron (or positron) energies ($|E| \gtrsim m_0 c^2$). Then we may investigate the quantities $B_\pm(2\mathrm{i}pR)$ of (6.145), which are crucial in the phase shift formula (6.146). In the limit $R \to 0$

$$\begin{aligned} B_\pm(2\mathrm{i}pR) &\xrightarrow[R\to 0]{} (2\mathrm{i}pR)^{\mathrm{i}\gamma} \pm \mathrm{e}^{-\pi\gamma} \frac{\mathrm{i}\gamma - \mathrm{i}y}{\varkappa + \mathrm{i}y(m_0c^2)/E} (2\mathrm{i}pR)^{-\mathrm{i}\gamma} \\ &= \mathrm{e}^{-(\pi/2)\gamma}\left(\mathrm{e}^{\mathrm{i}\gamma \ln 2pR} \pm \frac{\mathrm{i}\gamma - \mathrm{i}y}{\varkappa + \mathrm{i}y(m_0c^2)/E}\, \mathrm{e}^{-\mathrm{i}\gamma \ln 2pR}\right) . \end{aligned} \tag{6.149}$$

Inserting this expression into the phase shift formula (6.146), dividing numerator and denominator by the common factor

$$\cos(\gamma \ln 2pR)\left[u_2^{(i)}(R)\,\mathrm{Im}\left\{\frac{\mathrm{i}\gamma-\mathrm{i}y}{\varkappa+\mathrm{i}y(m_0c^2)/E}\right\}\right.$$
$$\left.\pm\sqrt{\frac{E-m_0c^2}{E+m_0c^2}}\,u_1^{(i)}(R)\left(1-\mathrm{Re}\left\{\frac{\mathrm{i}\gamma-\mathrm{i}y}{\varkappa+\mathrm{i}y(m_0c^2)/E}\right\}\right)\right], \tag{6.150}$$

and applying the addition theorem for the tangent gives

$$\tan\eta = -\tan[\gamma \ln 2pR + \varepsilon(R)]\,. \tag{6.151}$$

Here a phase $\varepsilon(R)$ has been introduced, given by

$$\tan\varepsilon(R) = \frac{u_2^{(i)}(R)\left[1+\mathrm{Re}\left\{\frac{\mathrm{i}\gamma-\mathrm{i}y}{\varkappa+\mathrm{i}y(m_0c^2)/E}\right\}\right] \pm \sqrt{\frac{E-m_0c^2}{E+m_0c^2}}}{u_2^{(i)}(R)\,\mathrm{Im}\left\{\frac{\mathrm{i}\gamma-\mathrm{i}y}{\varkappa+\mathrm{i}y(m_0c^2)/E}\right\} \mp \sqrt{\frac{E-m_0c^2}{E+m_0c^2}}}$$
$$\frac{\cdot\, u_1^{(i)}(R)\,\mathrm{Im}\left\{\frac{\mathrm{i}\gamma-\mathrm{i}y}{\varkappa+\mathrm{i}y(m_0c^2)/E}\right\}}{\cdot\, u_1^{(i)}(R)\left(\mathrm{Re}\left\{\frac{\mathrm{i}\gamma-\mathrm{i}y}{\varkappa+\mathrm{i}y(m_0c^2)/E}\right\}-1\right)}\,. \tag{6.152}$$

This phase $\varepsilon(R)$ depends on R via the interior solutions $u_1^{(i)}(R)$ and $u_2^{(i)}(R)$. To show that $\varepsilon(R)$ tends to a finite limit ε_0 for $R\to 0$, i.e. that

$$\lim_{R\to 0}\varepsilon(R) = \varepsilon_0\,, \tag{6.153}$$

the asymptotic form $(r\to 0)$ of the radial Dirac equation must be examined:

$$\frac{d}{dr}\begin{pmatrix} u_1^{(i)}(r) \\ u_2^{(i)}(r)\end{pmatrix} = \begin{pmatrix} -\varkappa/r & [E+m_0c^2-V(r)] \\ -[E-m_0c^2-V(r)] & \varkappa/r\end{pmatrix}\begin{pmatrix} u_1^{(i)}(r) \\ u_2^{(i)}(r)\end{pmatrix}. \tag{6.154}$$

Since $V(r)$ is assumed to have a finite cut off for $r\to 0$, i.e.

$$V(r) \xrightarrow[r\to 0]{} V_0\,, \tag{6.155}$$

one may replace $V(r)$ by V_0 on the rhs of (6.154) and then easily obtain the following solutions:

$$\begin{aligned} u_1^{(i)}(r) &\xrightarrow[r\to 0]{} A\,\frac{m_0c^2+(E-V_0)}{2|\varkappa|+1}\,r^{|\varkappa|+1}\,, \\ u_2^{(i)}(r) &\xrightarrow[r\to 0]{} A\,r^{|\varkappa|}\,, \quad (\varkappa>0) \end{aligned} \tag{6.156}$$

and

$$\begin{aligned} u_1^{(i)}(r) &\xrightarrow[r\to 0]{} A r^{|\varkappa|}\,, \\ u_2^{(i)}(r) &\xrightarrow[r\to 0]{} A\frac{m_0c^2-(E-V_0)}{2|\varkappa|+1}r^{|\varkappa|+1}\,. \end{aligned} \qquad (\varkappa<0) \tag{6.157}$$

Without loss of generality we can restrict ourselves to $\varkappa<0$ (e.g. the $S_{1/2}$ solutions), where obviously

$$\frac{u_2^{(i)}(r)}{u_1^{(i)}(r)} \xrightarrow[r\to 0]{} \frac{m_0c^2-(E-V_0)}{2|\varkappa|+1}r\,. \tag{6.158}$$

Choosing for simplicity a straight cut off, i.e. $V_0=-Z\alpha/R$, we obtain

$$\frac{u_2^{(i)}(R)}{u_1^{(i)}(R)} \xrightarrow[R\to 0]{} \frac{(m_0c^2-E-Z\alpha/R)R}{2|\varkappa|+1} \xrightarrow[R\to 0]{} \frac{-Z\alpha}{2|\varkappa|+1}\,, \tag{6.159}$$

because $pR\ll 1$ is required above and therefore the terms $(E-m_0c^2)R$ can be neglected. Inserting this result into (6.152), the limit of $\varepsilon(R)$ for $R\to 0$ can be calculated:

$$\tan\varepsilon(R) \xrightarrow[R\to 0]{} \tan\varepsilon_0$$

$$= \frac{-\dfrac{Z\alpha}{2|\varkappa|+1}\left(\mathrm{Re}\left\{\dfrac{\mathrm{i}\gamma-\mathrm{i}y}{\varkappa+\mathrm{i}y(m_0c^2)/E}\right\}+1\right)}{\begin{array}{c}-\dfrac{Z\alpha}{2|\varkappa|+1}\mathrm{Im}\left\{\dfrac{\mathrm{i}\gamma-\mathrm{i}y}{\varkappa+\mathrm{i}y(m_0c^2)/E}\right\} \\ \pm\sqrt{\dfrac{E-m_0c^2}{E+m_0c^2}}\mathrm{Im}\left\{\dfrac{\mathrm{i}\gamma-\mathrm{i}y}{\varkappa+\mathrm{i}y(m_0c^2)/E}\right\}\end{array}} \Bigg/ \ldots$$

$$\tan\varepsilon_0 = \frac{\begin{array}{c}-\dfrac{Z\alpha}{2|\varkappa|+1}\left(\mathrm{Re}\left\{\dfrac{\mathrm{i}\gamma-\mathrm{i}y}{\varkappa+\mathrm{i}y(m_0c^2)/E}\right\}+1\right) \\ -\dfrac{Z\alpha}{2|\varkappa|+1}\mathrm{Im}\left\{\dfrac{\mathrm{i}\gamma-\mathrm{i}y}{\varkappa+\mathrm{i}y(m_0c^2)/E}\right\} \\ \pm\sqrt{\dfrac{E-m_0c^2}{E+m_0c^2}}\mathrm{Im}\left\{\dfrac{\mathrm{i}\gamma-\mathrm{i}y}{\varkappa+\mathrm{i}y(m_0c^2)/E}\right\}\end{array}}{\mp\sqrt{\dfrac{E-m_0c^2}{E+m_0c^2}}\left(\mathrm{Re}\left\{\dfrac{\mathrm{i}\gamma-\mathrm{i}y}{\varkappa+\mathrm{i}y(m_0c^2)/E}\right\}-1\right)}\,, \tag{6.160}$$

which is independent of R. Hence, from (6.151) follows

$$\eta \xrightarrow[R\to 0]{} = -(\gamma\ln 2pR+\varepsilon_0)\,, \tag{6.161}$$

which diverges when the central charge shrinks to a point. Since R enters the total scattering phase Δ of (6.144) solely via $\eta(R)$, no point limit for Δ can be defined. The same holds for $\varkappa>0$ if the limit

$$\frac{u_2^{(i)}(R)}{u_1^{(i)}(R)} \xrightarrow[R\to 0]{} -\frac{2|\varkappa|+1}{Z\alpha} \tag{6.162}$$

is used in (6.152) to derive an expression similar to (6.160).

As mentioned earlier, the overcritical point charge problem is treated in Sect. 16.4. The central charge is highly shielded by the charge of the vacuum. The most interesting result is that the effective point charge consisting of the central charge and the vacuum charge will always be undercritical, i.e. $Z_{\text{eff}} = Z - Z_{\text{vac}} \lesssim 137$, so that the ill-defined situation just encountered disappears. Nature, somehow, prevents divergent and ill-defined situations!

6.5.3 The Physical Phase Shifts

We defined and calculated the phase shifts Δ in (6.129, 131, 133, 135, 144, 146). This Δ can be split up into a $\delta_{\log}$ part, which contains the logarithmic phase characteristic for the long-range Coulomb potential, (6.135), and other parts which are *independent of the multipolarity* of the scattering wave. In other words, $\delta_{\log}$ is common to all partial waves. The other part δ contained in Δ is physically more relevant. It depends on the multipolarity of the scattering wave and can be experimentally deduced from the measured differential cross-sections. It is therefore plausible that it vanishes for low energies

$$\lim_{p \to 0} \delta = 0 . \tag{6.163}$$

Thus

$$\Delta = \delta_{\log} + \delta . \tag{6.164}$$

Requirement (6.163) is not fulfilled for Δ; only the "physical" phase δ vanishes as the energy becomes small. We shall show now how the physically meaningful scattering phase δ can be extracted from Δ so that (6.163) is satisfied.

Let us first consider the most transparent case, namely scattering off a point nucleus with $Z\alpha < |\varkappa|$, for which Δ is given by (6.135). According to (6.116) $y \to \infty$ for $p \to 0$. Hence we need the asymptotic behaviour ($|y| \to \infty$) of the complex gamma function [Ab 65]:

$$\arg \Gamma(\gamma - \mathrm{i}y) \xrightarrow[|y| \to \infty]{} \frac{\pi}{2}\gamma - y(\ln|y| - 1) + \frac{1}{4}\pi \operatorname{sign} y . \tag{6.165}$$

Furthermore,

$$\lim_{|y| \to \infty} \arg \left(-\frac{\varkappa - \mathrm{i}y(m_0c^2)/E}{\gamma + \mathrm{i}y} \right) = \begin{cases} 0 & \text{for} \quad y > 0 \\ \pi & \text{for} \quad y < 0 \end{cases} . \tag{6.166}$$

From these relations it is seen that the phase shift Δ approaches the limit

$$\Delta \xrightarrow[|y| \to \infty]{} y\left(\ln \frac{2pr}{|y|} + 1\right) + \frac{\pi}{4} \equiv \delta_{\log} . \tag{6.167}$$

As stated above, this *logarithmic phase shift* is independent of the angular momentum $\varkappa$. Hence, the physical phase shift δ can be extracted according to (6.164) as

$$\delta \equiv \Delta - \delta_{\log} \xrightarrow[|E| \to m_0 c^2]{} 0 \; ; \tag{6.168}$$

and it is immediately evident that it indeed vanishes for $p \to 0$. So far the procedure for Coulomb scattering was illustrated for $1/r$ potentials.

For extended nuclei, the phase shift can be handled in exactly the same way, but the calculations become much more involved. Therefore, we give the final results only. For $Z\alpha > |\varkappa|$ and $|y| \to \infty$ the phase shift for extended nuclear charges (6.144) tends to precisely the same limit $\delta_{\log}$ defined in (6.167). Therefore

$$\delta \equiv \Delta - \delta_{\log} \tag{6.169}$$

can be called the *physical scattering phase shift* for all $Z\alpha$, for extended nuclei as well as for point nuclei.

One other property of phase shifts is worthy of mention, namely their behaviour under extremely large energies of the incident particle. We again restrict our analysis to scattering on a point charge potential with $Z\alpha < |\varkappa|$. For $E \to \pm\infty$, from (6.116)

$$y = \frac{Z\alpha E}{\hbar c p} \xrightarrow[E \to \pm\infty]{} \pm Z\alpha \tag{6.170}$$

and hence from (6.135)

$$\Delta \xrightarrow[E \to \pm\infty]{} \pm Z\alpha \ln 2pr - \arg\Gamma(\gamma \pm \mathrm{i} Z\alpha) \mp \arctan\frac{Z\alpha}{\gamma} - \frac{\pi}{2}[\gamma \mp (1 + \operatorname{sign}\varkappa)] \; . \tag{6.171}$$

Also the physical scattering phase δ defined in (6.169) has a non-zero value in this limit ($E \to \pm\infty$)

$$\delta \xrightarrow[E \to \pm\infty]{} \arg\Gamma(\gamma + 1 \mp \mathrm{i} Z\alpha) \pm Z\alpha \ln(Z\alpha - 1) - \frac{\pi}{2}[\gamma \mp (1 + \operatorname{sign}\varkappa)] - \frac{\pi}{4} \; . \tag{6.172}$$

This contrasts to the behaviour of non-relativistic scattering phases, where $y = Z\alpha/p \to 0$ for $E \to +\infty$ and hence $\delta \to 0$. The difference is easily understood by the fact that in the non-relativistic treatment the velocity of the incident particle increases beyond bound if $E \to \infty$. The interaction time, and hence the magnitude of interaction, approach a zero limit and the scattering phase vanishes, whereas in the relativistic theory the interaction time goes to a non-zero limit, because of the limiting velocity of light.

6.5.4 Resonances in the Lower Continuum for $Z > Z_{\mathrm{cr}}$

We are now able to return to the question of what happens to the $1s$ bound state if the nuclear charge Z exceeds the critical charge Z_{cr}. It was shown in Sect. 3.4 that for $Z \to Z_{\mathrm{cr}}$ (~ 172 for the $1s$ state) the bound state approaches the negative energy continuum. From the physical point of view it is clear that for $Z > Z_{\mathrm{cr}}$ the previously bound state interacts with the continuum wave functions and becomes a resonance. This was extensively discussed in Sects. 6.1 – 3 and is illustrated in a

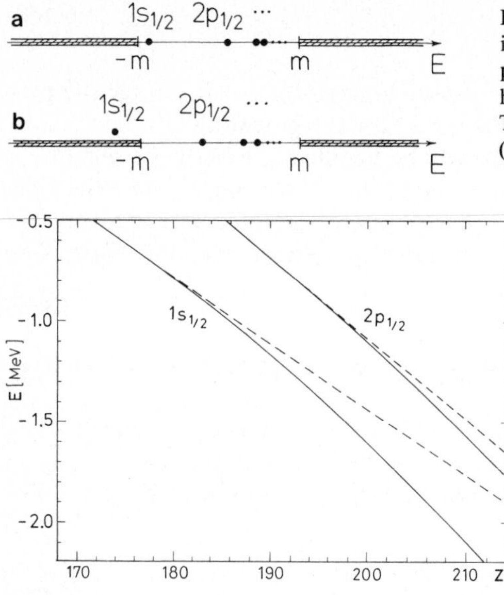

Fig. 6.11a, b. Single-particle electron spectrum in undercritical (**a**) and overcritical (**b**) central potentials. In (**b**) the $1s$ state (indicated by $1s$) has joined the negative energy continuum. The two continua are shown as dashed areas: (**a**) $Z < Z_{cr}$, (**b**) $Z > Z_{cr}$ [Mü 73b]

Fig. 6.12. The diving of the $1s_{1/2}$ and $2p_{1/2}$ levels in overcritical central Coulomb potentials. (– – –) indicate the "linear" diving; (———) give the exact results obtained from phase shift analysis [Mü 73b]

different way in Fig. 6.11. This resonance should show up in the scattering phase shifts for the $s_{1/2}$ continuum wave functions ($E < -m_0c^2$) of overcritical atoms ($Z > Z_{cr}$). In fact, the numerical evaluation of the resonance condition

$$\delta(E = E_{res}) = \frac{\pi}{2} \tag{6.173}$$

with the physical phase δ given by (6.169, 144) yields for extended nuclei a resonance energy with the characteristic linear Z dependence (Fig. 6.12). Such a quasi-linear dependence is expected for the diving bound state according to our previous discussions, (6.38, 39). The same holds again when the central charge Z is increased beyond Z_{cr} $(2p_{1/2}) \approx 185$, so that the $2p_{1/2}$ bound state penetrates into the lower continuum. Of course, this resonance is found then in the $p_{1/2}$ continuum phase shifts.

For energy $Z > Z_{cr}$ the resonance is characterized by its resonance energy E_{res} and width Γ. By virtue of its simple Z dependence, the numerically obtained resonance energy can be parametrized by

$$E_{res} = -(\delta Z' + \tau Z'^2), \tag{6.174}$$

Z' being the overcritical charge $Z' = Z - Z_{cr}$. Similarly the numerically obtained width Γ can be well described by

$$\Gamma = \gamma Z'^2 \tag{6.175}$$

as long as $Z' \gtrsim 3$, i.e. a little bit beyond the diving threshold. Precise numerical values for δ, τ and γ are listed in Table 6.1. Especially the values of Z_{cr} are

Table 6.1. Numerically obtained parameters for the critical charge, the resonance energy and the width of the dived $s_{1/2}$ and $p_{1/2}$ bound states

	$1s_{1/2}$	$2p_{1/2}$
Z_{cr}	171.5	185.5
δ [keV]	29.0	37.8
τ [keV]	0.33	0.22
γ [keV]	0.04	0.08

somewhat dependent on the model of nuclear charge distribution used, causing an uncertainty within one or two units of charge.

These numerical results can be understood by interpreting them in the light of the autoionization formalism described in Sects. 6.2–4. As shown in (6.10), the s continuum states $\Psi_E(x)$ can be expanded with $E < -m_0c^2$ for $Z > Z_{cr}$ into the undercritical continuum $\psi_E^{cr}(x)$ $(E < -m_0c^2)$ and the $1s$ bound state $\psi_{1s}^{cr}(x)$ defined in (6.6). Remember that $\psi_{1s}^{cr}(x)$ and $\psi_E^{cr}(x)$ are solutions of the total Dirac equation for $Z = Z_{cr}$, i.e. just before diving. It is useful for the following discussion to summarize the essential results of Sect. 6.1:

$$\Psi_E(x) = a(E)\,\psi_{1s}^{cr}(x) + \int_{-\infty}^{\infty} h_{E'}(E)\,\psi_{E'}^{cr}(x)\,dE' \, , \tag{6.10}$$

$$|a(E)|^2 = \frac{1}{2\pi}\,\frac{\Gamma_E}{[E-E_{1s}-\Delta E_{1s}-F(E)]^2+\Gamma_E^2/4} = \frac{2}{\pi\Gamma_E}\sin^2\Delta(E)\, , \tag{6.37, 32b}$$

$$\Gamma(E) = 2\pi|V_E|^2 = 2\pi|\langle\psi_E^{cr}|V'(x)|\psi_{1s}^{cr}\rangle|^2 \equiv \gamma Z'^2\, , \tag{6.36, 39}$$

$$\Delta E_{1s} = \langle\psi_{1s}^{cr}|V'(x)|\psi_{1s}^{cr}\rangle \equiv -\delta Z'\, , \tag{6.15, 39}$$

$$F(E) = +P\int\frac{|V_{E'}|^2}{E-E'}\,dE' \equiv -\tau Z'^2\, . \tag{6.25, 39}$$

Obviously for the resonance energy

$$E_{res} = \Delta E_{1s} + F(E) = -(\delta Z' + \tau Z'^2) \tag{6.176}$$

and also the width

$$\Gamma_E = \gamma Z'^2 \tag{6.177}$$

completely agree with the numerical results (6.174, 175) with respect to the Z' dependence. In fact, numerical estimates of the parameters δ, γ within the framework of autoionization formalism are practically identical with the values listed for the phase shift analysis in Table 6.1. The connection between the autoionization formalism and the exact continuum wave functions derived in this section

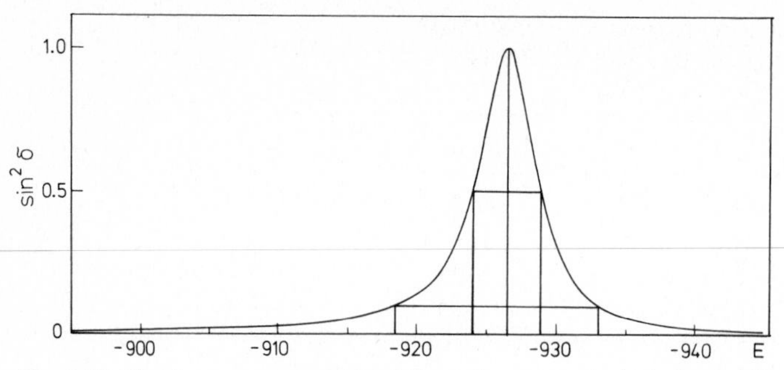

Fig. 6.13. The $1s$ resonance obtained from the exact physical phase shift δ calculated with (6.169, 144) for a giant atom with $Z = 184$ (with respect to $Z = 181$). The maximum of the resonance appears at $E_0 = -926.4$ keV, the width $\Gamma = 4.8$ keV [Mü 72c]. The K shell imbedded in the negative energy continuum has resonance character

is completed by remarking that the scattering phase $\Delta(E)$ due to the admixed bound state is given by [cf. (6.37, 32b)]

$$|a(E)|^2 = \frac{2}{\pi \Gamma_E} \sin^2 \Delta(E) . \tag{6.32b}$$

This is illustrated for the $1s$ resonance for $Z = 184$ in Fig. 6.13.

6.5.5 The Vacuum Charge Distribution

This discussion is completed by determining the charge distribution of the overcritical vacuum. The question to be answered is: how can we extract from the exact overcritical single-particle solutions given in (6.141) and by the phase shifts (6.144) the vacuum charge distribution $\varrho^{(o)}_{\text{vac pol}}(\boldsymbol{x})$ and, in particular, the bound state components of the wave functions characterizing the overcritical vacuum?

The vacuum polarization density in the overcritical case is given by [cf. (6.40, 43)]

$$\varrho^{(o)}_{\text{vac pol}}(\boldsymbol{x}) = \frac{e}{2}\left[\int\limits_{-\infty}^{-m_0c^2} \Psi_E^* \Psi_E(\boldsymbol{x})\, dE - \sum\!\!\!\!\!\!\int\limits_{-m_0c^2}^{+\infty} \Psi_E^\dagger \Psi_E(\boldsymbol{x})\, dE\right], \tag{6.178}$$

and in the undercritical case by

$$\varrho^{(u)}_{\text{vac pol}}(\boldsymbol{x}) = \frac{e}{2}\left[\int\limits_{-\infty}^{-m_0c^2} \psi_E^{\text{cr}\dagger} \psi_E^{\text{cr}}(\boldsymbol{x})\, dE - \int\limits_{-m_0c^2}^{+\infty} \psi_E^{\text{cr}\dagger} \psi_E^{\text{cr}}(\boldsymbol{x})\, dE\right]. \tag{6.179}$$

Here $\Psi_e(\boldsymbol{x})$ are the continuum wave functions already calculated in this chapter, e.g. (6.141, 143, 144), which have to be merged with (6.113, 115) and finally inserted into (6.127). Let us denote them in detail:

$$\Psi_{E,\varkappa m}(\boldsymbol{x}) = \frac{1}{r}\begin{pmatrix} \sqrt{E + m_0c^2}\,(\Phi_1 + \Phi_2)\,\chi_{\varkappa m} \\ -\sqrt{E - m_0c^2}\,(\Phi_1 - \Phi_2)\,\chi_{-\varkappa m} \end{pmatrix}$$

$$= \frac{1}{r}\begin{pmatrix} \sqrt{E+m_0c^2}(\Phi_1+\Phi_1^*)\,\chi_{\varkappa m} \\ -\sqrt{E-m_0c^2}(\Phi_1-\Phi_1^*)\,\chi_{-\varkappa m} \end{pmatrix}$$

$$= \frac{2}{r}\begin{pmatrix} \sqrt{E+m_0c^2}\,\mathrm{Re}\{\Phi_1\}\,\chi_{\varkappa m} \\ -\mathrm{i}\sqrt{E-m_0c^2}\,\mathrm{Im}\{\Phi_1\}\,\chi_{\varkappa m} \end{pmatrix} = \frac{1}{r}\begin{pmatrix} u_1(r)\,\chi_{\varkappa m} \\ -\mathrm{i}u_2(r)\,\chi_{-\varkappa m} \end{pmatrix}. \tag{6.180}$$

Here Φ_1 is given by (6.141, 143, 144).

The $\sum\!\!\!\!\!\!\int$ part of (6.178, 179) also contains bound states from within the gap. We are interested in calculating the change of the vacuum polarization density in the overcritical case with respect to the undercritical situation, i.e.

$$\Delta\varrho_{\mathrm{vac\,pol}}(\boldsymbol{x}) = \varrho^{(\mathrm{o})}_{\mathrm{vac\,pol}}(\boldsymbol{x}) - \varrho^{(\mathrm{u})}_{\mathrm{vac\,pol}}(\boldsymbol{x})\,. \tag{6.181}$$

Following the discussions leading to (6.43, 44) we expect that $\Delta\varrho_{\mathrm{vac\,pol}}$ essentially yields the charge distribution of the dived shells, i.e. the real vacuum polarization charge; perhaps modified by some very tiny effects stemming from the change of the vacuum displacement charge. To see how this occurs, let us analyse the various arguments by using the expansion (6.10) of the overcritical waves instead of the numerical solution (6.180), i.e.

$$\Psi_E(\boldsymbol{x}) = a(E)\,\psi^{\mathrm{cr}}_{1s}(\boldsymbol{x}) + \sum\!\!\!\!\!\!\int_{E'} h_{E'}(E)\,\psi^{\mathrm{cr}}_{E'}(\boldsymbol{x})\,dE'\,. \tag{6.182}$$

The second term has been modified compared to (6.10) in that $\sum\!\!\!\!\!\!\int_{E'}$ also includes the bound ψ^{cr}_{ns} states with $n>1$ and also the positive energy s continuum. The coefficient $h_{E'}(E)$ is then slightly modified from (6.37), but for all practical purposes is the same (see the discussions in Sect. 6.2). Using the orthogonality and completeness of the sets $\{\Psi_E\}$ and $\{\psi^{\mathrm{cr}}_{1s},\psi^{\mathrm{cr}}_{E'}\}$ we can derive

$$\begin{aligned} &\int a^*(E)\,a(E)\,dE \\ &\quad = \int\langle\Psi_E(\boldsymbol{x})\,|\,\psi^{\mathrm{cr}}_{1s}(\boldsymbol{x})\rangle^\dagger\langle\Psi_E(\boldsymbol{x}')\,|\,\psi^{\mathrm{cr}}_{1s}(\boldsymbol{x}')\rangle\,dE \\ &\quad = \int d^3x\int d^3x'\int dE\;\Psi_E(\boldsymbol{x})\,\psi^{\mathrm{cr}\dagger}_{1s}(\boldsymbol{x})\,\Psi^\dagger_E(\boldsymbol{x}')\,\psi^{\mathrm{cr}}_{1s}(\boldsymbol{x}') \\ &\quad = \int d^3x\int d^3x'\;\psi^{\mathrm{cr}\dagger}_{1s}(\boldsymbol{x})\,\psi^{\mathrm{cr}}_{1s}(\boldsymbol{x}')\,\delta(\boldsymbol{x}'-\boldsymbol{x})\,\mathbb{1} = 1\,, \end{aligned} \tag{6.183a}$$

$$\begin{aligned} &\int a(E)^\dagger h_{E'}(E)\,dE \\ &\quad = \int\langle\Psi_E(\boldsymbol{x})\,|\,\psi^{\mathrm{cr}}_{1s}(\boldsymbol{x})\rangle^\dagger\langle\Psi_E(\boldsymbol{x}')\,|\,\psi^{\mathrm{cr}}_{E'}(\boldsymbol{x}')\rangle \\ &\quad = \int d^3x\int d^3x'\int dE\;\Psi_E(\boldsymbol{x})\,\psi^{\mathrm{cr}\dagger}_{1s}(\boldsymbol{x})\,\Psi_E(\boldsymbol{x}')\,\psi^{\mathrm{cr}}_{E'}(\boldsymbol{x}') \\ &\quad = \int d^3x\int d^3x'\;\psi^{\mathrm{cr}\dagger}_{1s}(\boldsymbol{x})\,\psi^{\mathrm{cr}}_{E'}(\boldsymbol{x}')\,\delta(\boldsymbol{x}'-\boldsymbol{x})\,\mathbb{1} = 0\,, \end{aligned} \tag{6.183b}$$

and analogously

$$\int h^\dagger_{E'}(E)\,h_{E''}(E)\,dE = \delta(E'-E'')\,. \tag{6.183c}$$

We then get, e.g., for the first integral appearing in (6.178)

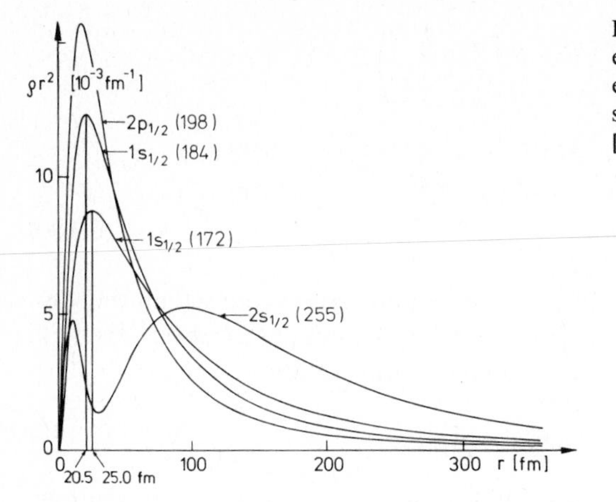

Fig. 6.14. Density distribution of several electronic shells imbedded in the negative energy continuum (and the precritical K shell for reference) obtained with (6.181) [Mü 72c, 73b]

$$
\begin{aligned}
\int dE\, \Psi_E^\dagger(\boldsymbol{x})\, \Psi_E(\boldsymbol{x}) &= \int a(E)^\dagger a(E)\, \psi_{1s}^{\mathrm{cr}\dagger}(\boldsymbol{x})\, \psi_{1s}^{\mathrm{cr}}(\boldsymbol{x})\, dE \\
&\quad + 2\int\int a(E)^\dagger h_{E'}(E)\, \psi_{1s}^{\mathrm{cr}\dagger}(\boldsymbol{x})\, \psi_{E'}^{\mathrm{cr}}(\boldsymbol{x})\, dE'\, dE \\
&\quad + \int\int\int h_{E'}^\dagger(E)\, h_{E''}(E)\, \psi_{E'}^{\mathrm{cr}\dagger}(\boldsymbol{x})\, \psi_{E''}^{\mathrm{cr}}(\boldsymbol{x})\, dE'\, dE'' \\
&= \psi_{1s}^{\mathrm{cr}}(\boldsymbol{x})\, \psi_{1s}^{\mathrm{cr}}(\boldsymbol{x}) + \int \psi_{E'}^{\mathrm{cr}\dagger}(\boldsymbol{x})\, \psi_{E'}^{\mathrm{cr}}(\boldsymbol{x})\, dE' \; .
\end{aligned}
\tag{6.184}
$$

Hence, as expected, the density of the dived $1s$ shell $\psi_{1s}^{\mathrm{cr}\dagger}(\boldsymbol{x})\, \psi_{1s}^{\mathrm{cr}}(\boldsymbol{x})$ is contained in this integral. It is weighted in (6.178) by the factor $\frac{1}{2}$; the other factor $\frac{1}{2}$ comes from the second term in (6.178) absolutely similarly to (6.43). This second half of the $1s$ shell density comes, so to speak, from the fact that the virtual vacuum polarization density $\varrho_{\mathrm{vac\,pol}}^{(\mathrm{u})}(\boldsymbol{x})$ is extracted from the overcritical vacuum polarization density $\varrho_{\mathrm{vac\,pol}}^{(\mathrm{o})}(\boldsymbol{x})$. Hence $\Delta\varrho_{\mathrm{vac\,pol}}(\boldsymbol{x})$ of (6.181), indeed, essentially contains the density distribution of the dived shells. If the exact wave functions (6.180) are used, shrinking of the shell after diving can be exhibited. This is demonstrated in Fig. 6.14, which shows the $1s$ electron distribution shortly before diving (i.e. at $Z_{\mathrm{cr}} \approx 172$) and the *real vacuum polarization charge* for various degrees of overcriticality ($Z = 184$, 255). For $Z = 198$ the charged vacuum also contains the $2p_{1/2}$ shell, which dives at $\bar{Z}_{\mathrm{cr}} \approx 183$, Fig. 6.14.

At this point the overcritical point charge problem can be further elucidated. Equation (6.161) showed that for a shrinking nuclear charge radius ($R \to 0$) the phase shift η diverges as

$$
-\gamma \ln 2pR \, . \tag{6.185}
$$

If we require that the resonance condition (6.173) is satisfied while $R \to 0$, then

$$
p_{\mathrm{res}} R = \mathrm{const} \, , \tag{6.186}
$$

which through (6.112) implies

$$E_{\text{res}} = -\sqrt{(m_0c^2)^2 + \frac{\text{const}^2}{R^2}} \to -\frac{\text{const}}{R}\,. \tag{6.187}$$

Thus, for a given Z all bound states with $Z\alpha > |\varkappa|$ obtain an infinite binding energy when $R \to 0$. This behaviour demonstrates again that the physical situation of a point nucleus with $Z\alpha \to 1$ is highly overcritical, i.e. the vacuum becomes highly charged. Because of the high vacuum charge a self-consistent treatment of the vacuum is required, Chap. 16.

Nevertheless, already the peculiar asymptotic behaviour of the Sommerfeld fine-structure formula at $Z = 137$ can be understood (Sect. 3.4). The perpendicular tangent of, e.g., the $E_{1s}(Z)$ binding energy curve as a function of nuclear charge Z for $Z \to 137$ simply indicates "perpendicular diving" of the overcritical electronic shells in pure $1/r$ potentials, Figs. 6.15, 16. Figure 6.15 shows the overcritical charge $Z_{\text{cr}}(R)$ for a given extension R of the homogeneous central charge. As $R \to 0$, $Z_{\text{cr}}(R)$ approaches 137. Figure 6.16 shows the energy of the overcritical K shell as a function of R for a given central charge $Z = 184$. As indicated in (6.187), the K shell dives like $1/R$ for a shrinking nuclear radius.

Finally, a remark concerning the threshold behaviour of the overcritical continuum wave functions at $E = -m_0c^2$ is appropriate. Our interest focusses on the probability of a particle in a negative energy continuum wave function near the field creating centre, i.e.

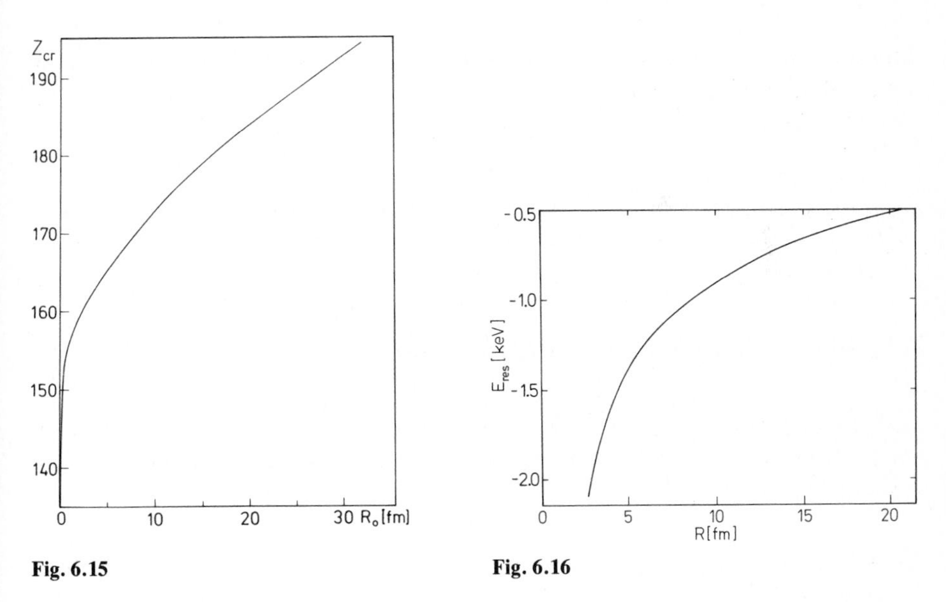

Fig. 6.15

Fig. 6.16

Fig. 6.15. The extension R of the central nuclear charge determines the critical charge Z_{cr}. As $R \to 0$ (limiting point nucleus), Z_{cr} approaches 137 (α^{-1})

Fig. 6.16. The K shell energy in the overcritical field of a $Z = 184$ nucleus as a function of the nuclear charge radius R. Diving starts at about 21 fm and continues for decreasing R. The curve roughly corresponds to the internuclear distance in a collision of two uranium nuclei

$$(r^2|\psi_E(r)|^2)_{\substack{r\to 0\\ E\to -m_0c^2}} \sim e^{+2\pi y} = \exp\left(\frac{-2\pi Z\alpha m_0 c^2}{p}\right). \tag{6.188}$$

This result can be inferred from (6.141, 143) and by using (6.117). For $p\to 0$ the magnitude of the wave function at a small value of r, say the nuclear radius, is determined by the normalization constant N, (6.143), since the other factors remain finite. The gamma functions in the denominator of (6.143) behave like

$$|\Gamma(1+\mathrm{i}y)| \sim \sqrt{2\pi|y|}\exp\left(-\frac{\pi}{2}|y|\right)$$

for $y\to\pm\infty$. Therefore the normalization factor N behaves as

$$N\sim\exp\left(\frac{\pi}{2}y\right)\exp\left(-\frac{\pi}{2}|y|\right).$$

For the lower continuum $y\to-\infty$, this leads to the exponential suppression factor in (6.188). Thus for $E\to -m_0c^2$, $p\to 0$, the wave function is totally expelled from the vicinity of the nucleus. This behaviour implies that the bound state-continuum interaction characterized by the width (6.36), i.e. by

$$\Gamma_E = 2\pi|\langle\psi_E^{\mathrm{cr}}|V'|\psi_{1s}^{\mathrm{cr}}\rangle|^2,$$

vanishes exponentially as $E\to -m_0c^2$. This accounts for the deviations from the simple Z' dependence of

$$\Gamma = \gamma Z'^2 \tag{6.175}$$

for $Z'<3$. It also ensures the existence of the principal value integral $F(E)$ of (6.25) at the threshold $E=-m_0c^2$.

6.6 Some Useful Mathematical Relations

On several occasions above some mathematical relations have been used which might be not commonly known and so deserve to be properly derived. We shall state and prove these relations here.

a) $$P\left(\frac{1}{x}\right) = \frac{\mathrm{i}}{2}\int_{-\infty}^{\infty}\operatorname{sign}\xi\, e^{-\mathrm{i}\xi x}\,d\xi \tag{6.189a}$$

or equivalently

$$\operatorname{sign}\xi = -\frac{\mathrm{i}}{\pi}\int_{-\infty}^{\infty}P\left(\frac{1}{x}\right)e^{+\mathrm{i}\xi x}\,dx. \tag{6.189b}$$

Version (6.189b) can be proven by writing

$$I(\xi) \equiv -\frac{\mathrm{i}}{\pi}\int_{-\infty}^{\infty} P\left(\frac{1}{x}\right) \mathrm{e}^{\mathrm{i}\xi x} dx = \frac{1}{\pi}\int_{-\infty}^{\infty} P\left(\frac{1}{x}\right) \sin \xi x \, dx$$

$$= \frac{2}{\pi} \lim_{\varepsilon \to 0} \int_{\varepsilon}^{\infty} \frac{\sin \xi x}{x} dx . \tag{6.190}$$

The limes $\varepsilon \to 0$ can be carried out by defining

$$\sin(z)/z|_{z=0} = 1 ,$$

i.e. by a holomorphic continuation of $\sin(z)/z$ into the origin. Since $\sin(z)/z$ also has no pole, the integration path of (6.190) can be carried into the complex plane in the following way:

$$I(\xi) = \frac{2}{\pi}\int_{0}^{\infty} \frac{\sin \xi x}{\xi x} d(\xi x) = \frac{1}{\pi}\int_{-\infty}^{\infty} \frac{\sin \xi x}{\xi x} d(\xi x)$$

$$= \frac{1}{\pi} \Theta(\xi) \int_{-\infty}^{\infty} \frac{\sin z}{z} dz - \frac{1}{\pi} \Theta(-\xi) \int_{-\infty}^{\infty} \frac{\sin z}{z} dz$$

$$= \frac{\mathrm{sign}(\xi)}{\pi} \int_{-\infty}^{\infty} \frac{\sin z}{z} dz = \frac{\mathrm{sign}(\xi)}{\pi} \int_{\mathscr{C}} \frac{\sin z}{z} dz$$

$$= \frac{\mathrm{sign}(\xi)}{2\pi \mathrm{i}} \left[\int_{\mathscr{C}} \frac{\mathrm{e}^{\mathrm{i}z}}{z} dz - \int_{\mathscr{C}} \frac{\mathrm{e}^{-\mathrm{i}z}}{z} dz \right] . \tag{6.191}$$

Here we used

$$\mathrm{sign}\, \xi = \Theta(\xi) - \Theta(-\xi) \tag{6.192}$$

and introduced the integration path shown in Fig. 6.17.

The last two integrals along $\mathscr{C}$ in (6.191) can easily be calculated by closing $\mathscr{C}$ in the upper and lower halplane respectively. Note that $\exp(\mathrm{i}z)/z$ and $\exp(-\mathrm{i}z)/z$ have a pole at $z = 0$ while $\sin(z)/z$ does not (Fig. 6.18):

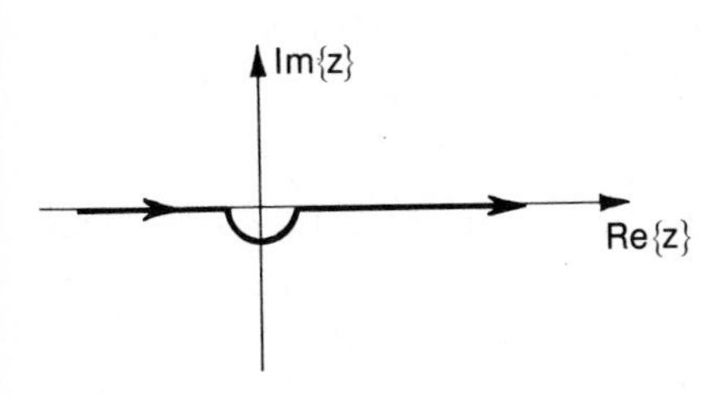

Fig. 6.17. The integration path $\mathscr{C}$

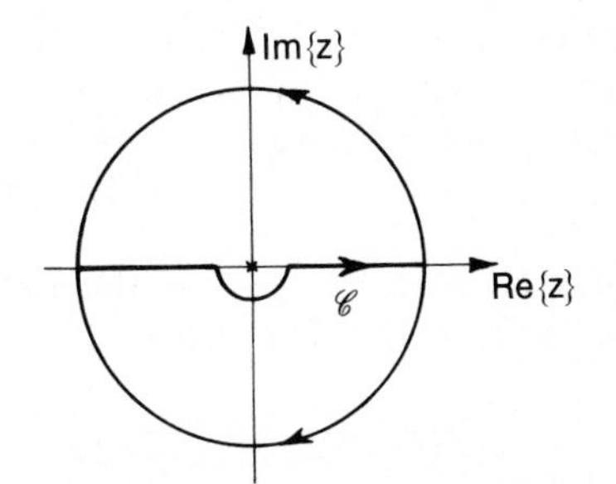

Fig. 6.18. The two integration paths used in calculating the integrals (6.193a, b)

$$\frac{1}{2\pi i}\int_{\mathscr{C}}\frac{e^{iz}}{z}dz=\frac{1}{2\pi i}2\pi i+\lim_{R\to\infty}\int_{+\pi}^{0}\frac{\exp(i\cos\varphi R-R\sin\varphi)}{Re^{i\varphi}2\pi i}d\varphi=1\,, \quad (6.193a)$$

and

$$\frac{1}{2\pi i}\int_{\mathscr{C}}\frac{e^{-iz}}{z}dz=\lim_{R\to 0}\int_{\pi}^{2\pi}\frac{\exp(-i\cos\varphi R+R\sin\varphi)}{Re^{-i\varphi}2\pi i}d\varphi=0\,. \quad (6.193b)$$

Hence

$$\operatorname{sign}\xi=-\frac{i}{\pi}\int_{-\infty}^{\infty}P\left(\frac{1}{x}\right)e^{i\xi x}dx\,, \quad (6.189b)$$

which proves the conjecture.

b) $$\frac{P}{(\bar{E}-E')}\,\frac{P}{(E-E')}=\frac{P}{(\bar{E}-E)}\left(\frac{P}{E-E'}-\frac{P}{\bar{E}-E'}\right)+\pi^2\delta[E'-\tfrac{1}{2}(E+\bar{E})]\,\delta(E-\bar{E})\,. \quad (6.194)$$

This is the Poincaré-Bertrand formula [Po 10, Be 21]. It is proven by starting from (6.189a), yielding

$$\frac{P}{\bar{E}-E'}=\frac{i}{2}\int_{-\infty}^{\infty}\operatorname{sign}(\xi)\,e^{-i\xi(\bar{E}-E')}d\xi\,,$$

and with $\xi=-2\pi k$, $\operatorname{sign}\xi=-\operatorname{sign}k$,

$$\int_{-\infty}^{\infty}\ldots d\xi=-\int_{+\infty}^{-\infty}2\pi\ldots dk:$$

$$\frac{P}{\bar{E}-E'}=i\pi\int_{-\infty}^{\infty}\operatorname{sign}(k)\exp[2\pi ik(\bar{E}-E')]\,dk\,. \quad (6.195)$$

Another way to obtain this result is the following. One has

$$P\left(\frac{1}{x}\right)=\frac{1}{2}\lim_{\varepsilon\to 0}\left(\frac{1}{x+i\varepsilon}+\frac{1}{x-i\varepsilon}\right)=\lim_{\varepsilon\to 0}\frac{x}{x^2+\varepsilon^2}\,,$$

and therefore

$$-\frac{i}{\pi}\int_{-\infty}^{\infty}P\left(\frac{1}{x}\right)e^{i\xi x}dx=-\frac{i}{2\pi}\lim_{\varepsilon\to 0}\int_{-\infty}^{\infty}\frac{e^{i\xi x}}{x+i\varepsilon}dx-\frac{i}{2\pi}\lim_{\varepsilon\to 0}\int_{-\infty}^{\infty}\frac{e^{i\xi x}}{x-i\varepsilon}dx\,.$$

The integrals must be closed in the upper plane for $\xi>0$ and in the lower plane for $\xi<0$ (Fig. 6.19). This yields

$$-\frac{i}{2\pi}\lim_{\varepsilon\to 0}\theta(\xi)\,e^{i\xi i\varepsilon}2\pi i-\frac{i}{2\pi}\lim_{\varepsilon\to 0}\theta(-\xi)\,e^{i\xi(-i\varepsilon)}(-2\pi i)$$
$$=\theta(\xi)-\theta(-\xi)=\operatorname{sign}\xi\,,$$

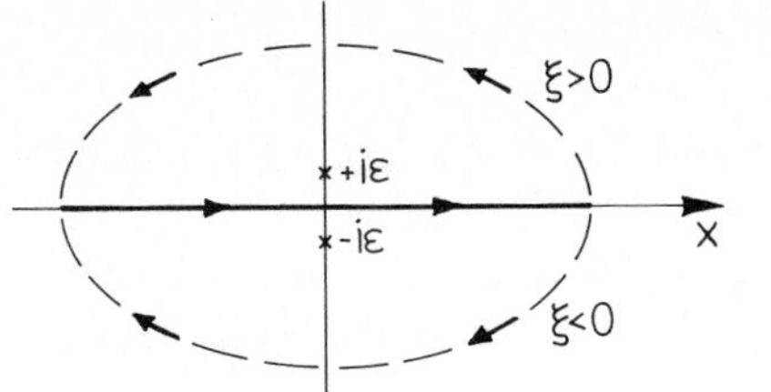

Fig. 6.19. Integration path for the integrals $\int_{-\infty}^{\infty} \frac{e^{i\xi x}}{x \pm i\varepsilon} dx$ for $\xi > 0$ and $\xi < 0$

which agrees with (6.189b).

Hence

$$\frac{P}{(\bar{E}-E')} \frac{P}{(E-E')} = -\pi^2 \int_{-\infty}^{\infty} dk \int_{-\infty}^{\infty} dk' \operatorname{sign}(k) \operatorname{sign}(k') \times \exp\{2\pi i [k(\bar{E}-E') + k'(E-E')]\} . \tag{6.196}$$

Substituting

$$u = k + k' , \quad v = \tfrac{1}{2}(k - k') ,$$

or

$$k = \tfrac{1}{2}(u + 2v) , \quad k' = \tfrac{1}{2}(u - 2v) \tag{6.197}$$

gives

$$\left| \frac{\partial(k, k')}{\partial(u, v)} \right| = \begin{vmatrix} \frac{1}{2} & 1 \\ \frac{1}{2} & -1 \end{vmatrix} = 1$$

and

$$\frac{P}{(\bar{E}-E')} \frac{P}{(E-E')} = -\pi^2 \int_{-\infty}^{\infty} du \int_{-\infty}^{\infty} dv \operatorname{sign}(u^2 - 4v^2) \times \exp\{2\pi i [u \tfrac{1}{2}(\bar{E} + E - 2E') + v(\bar{E} - E)]\} . \tag{6.198}$$

Here we used

$$\operatorname{sign}(k) \operatorname{sign}(k') = \operatorname{sign}(kk') = \operatorname{sign}(\tfrac{1}{4}(u^2 - 4v^2)) = \operatorname{sign}(u^2 - 4v^2) . \tag{6.199}$$

In the following we shall also use the identity

$$\operatorname{sign}(x) = -1 + 2\Theta(x) , \tag{6.200}$$

where $\Theta(x)$ is the Heaviside step function. Hence (2.198) can also be written as

$$\frac{P}{(\bar{E}-E')} \frac{P}{(E-E')} = \pi^2 \int_{-\infty}^{\infty} du \exp\left\{2\pi i u \left[\frac{(\bar{E}+E)}{2} - E'\right]\right\} \cdot \int_{-\infty}^{\infty} dv (1 - 2\Theta(u^2 - 4v^2)) \exp[2\pi i v(\bar{E} - E)] . \tag{6.201}$$

The last integral splits into two calculatable parts:

$$\int_{-\infty}^{\infty} dv \exp[2\pi i v(\bar{E}-E)] = \frac{1}{2\pi}\int_{-\infty}^{\infty} d\xi \exp[-i\xi(E-\bar{E})] = \delta(E-\bar{E}), \quad (6.202)$$

$$-2\int_{-\infty}^{\infty} dv\, \Theta(u^2-4v^2)\exp[2\pi i v(\bar{E}-E)] = -2\int_{-1/2|u|}^{1/2|u|} dv \exp[2\pi i v(\bar{E}-E)]$$

$$= -2\left\{\frac{1}{2\pi i(\bar{E}-E)}\exp[2\pi i v(\bar{E}-E)]\right\}_{-1/2|u|}^{+1/2|u|} = -\frac{4i\sin(\pi(\bar{E}-E)|u|)}{2\pi i(\bar{E}-E)}$$

$$= -\frac{2}{\pi}\frac{\sin[\pi|u|(\bar{E}-E)]}{(\bar{E}-E)}$$

$$= -\frac{2}{\pi}\sin(u)\frac{1}{(\bar{E}-E)}\sin(\pi u(\bar{E}-E)). \quad (6.203)$$

Hence (6.201) becomes

$$\frac{P}{(\bar{E}-E')}\frac{P}{(E-E')} = \pi^2\int_{-\infty}^{\infty} du \exp\{2\pi i u[\tfrac{1}{2}(\bar{E}+E)-E']\}$$

$$\times\left\{\delta(E-\bar{E}) - \frac{2}{\pi}\mathrm{sign}(u)\frac{1}{\bar{E}-E}\frac{1}{2i}\right.$$

$$\left.\times\left[\exp\left[2\pi i u\frac{(\bar{E}-E)}{2}\right] - \exp\left[-2\pi i u\frac{(\bar{E}-E)}{2}\right]\right]\right\}$$

$$= \pi^2\delta(E-\bar{E})\,\delta[\tfrac{1}{2}(\bar{E}+E)-E')]$$

$$+\frac{i\pi^2}{\pi(\bar{E}-E)}\int_{-\infty}^{\infty} du\,\mathrm{sign}(u)\{\exp[2\pi i u(\bar{E}-E')]$$

$$-\exp[+2\pi i u(E-E')\}. \quad (6.204)$$

Using (6.195) finally gives

$$\frac{P}{(\bar{E}-E')}\frac{P}{(E-E')} = \pi^2\delta[\tfrac{1}{2}(\bar{E}+E)-E']\,\delta(\bar{E}-E)$$

$$+\frac{i\pi}{(\bar{E}-E)}\left[\frac{i}{\pi}\frac{P}{(\bar{E}-E')} - \frac{i}{\pi}\frac{P}{(E-E')}\right]$$

$$= \frac{1}{\bar{E}-E}\left(\frac{P}{(E-E')} - \frac{P}{(\bar{E}-E')}\right)$$

$$+\pi^2\delta(E'-\tfrac{1}{2}(\bar{E}+E))\,\delta(\bar{E}-E), \quad (6.205)$$

which completes the proof.

For completeness we shall add a second proof of (6.194). We start from the well-known relation

$$\frac{1}{x \pm \mathrm{i}\varepsilon} = P\left(\frac{1}{x}\right) \mp \mathrm{i}\pi\delta(x) \tag{6.206}$$

and conclude

$$\begin{aligned} P\frac{1}{(x-y)}&\left[P\left(\frac{1}{y}\right) - P\left(\frac{1}{x}\right)\right] \\ &= \frac{1}{(x-y)+\mathrm{i}\varepsilon}\left[P\left(\frac{1}{x}\right) - P\left(\frac{1}{y}\right)\right] + O \\ &= \frac{1}{(x-y)+\mathrm{i}\varepsilon}\left(\frac{1}{y-\mathrm{i}\eta} - \frac{1}{x+\mathrm{i}\sigma}\right) + \frac{\mathrm{i}\pi}{(x-y)+\mathrm{i}\varepsilon}(\delta(y)+\delta(x)) \\ &= \frac{1}{(y-\mathrm{i}\eta)(x+\mathrm{i}\sigma)} + \mathrm{i}\pi\,\delta(y)\frac{1}{x+\mathrm{i}\varepsilon} - \mathrm{i}\pi\,\delta(x)\frac{1}{y-\mathrm{i}\varepsilon}\,. \end{aligned} \tag{6.207}$$

We can also conclude from (6.206), that

$$\begin{aligned} P\left(\frac{1}{x}\right)P\left(\frac{1}{y}\right) = &\frac{1}{(x+\mathrm{i}\varepsilon)}\,\frac{1}{(y-\mathrm{i}\eta)} - \mathrm{i}\pi\,\delta(x)\frac{1}{y-\mathrm{i}\eta} + \mathrm{i}\pi\,\delta(y)\frac{1}{x+\mathrm{i}\varepsilon} \\ &+ \pi^2\delta(x)\,\delta(y)\,. \end{aligned} \tag{6.208}$$

The small quantities ε, η, σ determine solely the integration path; their (small) magnitude is irrelevant. Mathematically this means that (6.206) is true only in the limes $\varepsilon \to 0$. Therefore from (6.207, 208) follows

$$P\left(\frac{1}{x}\right)P\left(\frac{1}{y}\right) = P\left(\frac{1}{(x-y)}\right)\left[P\left(\frac{1}{y}\right) - P\left(\frac{1}{x}\right)\right] + \pi^2\delta(x)\,\delta(y) \tag{6.209}$$

or, setting $x = \bar{E} - E'$, $y = E - E'$:

$$\begin{aligned} P\left(\frac{1}{\bar{E}-E'}\right)P\left(\frac{1}{E-E'}\right) = P\left(\frac{1}{(\bar{E}-E)}\right)&\left[P\left(\frac{1}{E-E'}\right) - P\left(\frac{1}{(\bar{E}-E')}\right)\right] \\ &+ \pi^2\delta(\bar{E}-E')\,\delta(E-E')\,. \end{aligned} \tag{6.210}$$

We have thus again derived (6.194), but in a much faster and more elegant way than before.

6.6.1 A Different Choice of Phases

The supercritical states are expanded in (sub)critical basis as

$$\Psi_E(\boldsymbol{x}) = a(E)\,\psi_{1s}^{\mathrm{cr}}(\boldsymbol{x}) + \int_{-\infty}^{-m} h_{E'}(E)\,\psi_{E'}^{\mathrm{cr}}(\boldsymbol{x})\,dE' \,. \tag{6.211}$$

Hence

$$\begin{aligned} a(E) &= \langle \psi_{1s}^{\mathrm{cr}} | \Psi_E \rangle \\ h_{E'}(E) &= \langle \psi_{E'}^{\mathrm{cr}} | \Psi_E \rangle \,. \end{aligned} \tag{6.212}$$

In Sect. 6.1 [(6.36) ff.] it was shown that

$$a(E) = \frac{|V_E|}{\sqrt{(E-E_R)^2 + \pi^2 |V_E|^4}} \tag{6.213}$$

$$h_{E'}(E) = P\frac{V_{E'}\,a(E)}{E-E'} + \frac{(E-E_R)}{|V_E|^2}\,V_E'\,a(E)\,\delta(E-E')\,, \quad \text{with} \tag{6.214}$$

$$\begin{aligned} V_E &= \langle \psi_E^{\mathrm{cr}} | V' | \psi_{1s}^{\mathrm{cr}} \rangle \\ E_R &= E_{1s}^{\mathrm{cr}} + \langle \psi_{1s}^{\mathrm{cr}} | V' | \psi_{1s}^{\mathrm{cr}} \rangle + P\int_{-\infty}^{-m} \frac{|V_{E'}|^2}{E-E'}\,dE' \,. \end{aligned} \tag{6.215}$$

Then (6.214) can be rewritten as

$$h_{E'}(E) = \frac{1}{\sqrt{(E-E_R)^2 + \pi^2 |V_E|^4}} \Bigg[P\frac{V_{E'}|V_E|}{E-E'} + \underbrace{(E-E_R)\frac{V_{E'}|V_E|}{|V_E|^2}\,\delta(E-E')}_{=\frac{V_E}{|V_E|}\,\delta(E-E')} \Bigg] . \tag{6.214a}$$

A different choice of phases was used by *Rafelski* et al. [Ra 78b], namely

$$\tilde{a}(E) = \frac{V_E^*}{E - E_R + \mathrm{i}\pi |V_E|^2} \tag{6.216}$$

$$\tilde{h}_{E'}(E) = \lim_{\eta\to 0} \frac{V_{E'}\,\tilde{a}(E)}{E - E' + \mathrm{i}\eta} + \delta(E-E') \,. \tag{6.217}$$

Using

$$\lim_{\xi\to 0} \frac{1}{E-E'+\mathrm{i}\xi} = P\frac{1}{E-E'} - \mathrm{i}\pi\,\delta(E-E') \tag{6.218}$$

allows (6.217) to be rewritten as

$$\begin{aligned} \tilde{h}_{E'}(E) = \frac{1}{E-E_R+\mathrm{i}\pi|V_E|^2} \Bigg\{ & P\frac{V_{E'}V_E^*}{E-E'} - \mathrm{i}\pi\, \underbrace{V_{E'}V_E^*\,\delta(E-E')}_{=|V_E|^2\delta(E-E')} \\ & + (E-E_R+\mathrm{i}\pi|V_E|^2)\,\delta(E-E') \Bigg\} \end{aligned}$$

$$= \frac{1}{E - E_R + \mathrm{i}\pi |V_E|^2} \left\{ P \frac{V_{E'} V_E^*}{E - E'} + (E - E_R)\, \delta(E - E') \right\}. \qquad (6.217')$$

It is now convenient to introduce the phases $\delta(E)$ and $\Delta(E)$ through

$$\frac{1}{E - E_R + \mathrm{i}\pi |V_E|^2} = \frac{1}{\sqrt{(E - E_R)^2 + \pi^2 |V_E|^4}}\, \mathrm{e}^{\mathrm{i}\delta(E)} \qquad (6.219)$$

$$V_E^* = |V_E| \cdot \mathrm{e}^{-\mathrm{i}\Delta(E)}. \qquad (6.220)$$

Let Ψ_E and $\tilde{\Psi}_E$ be defined by (6.211), but with the amplitudes replaced by the choice (6.216, 217), so that

$$\Psi_E(\boldsymbol{x}) = \frac{1}{\sqrt{(E - E_R)^2 + \pi^2 |V_E|^4}} \left\{ |V_E| \left[\psi_{1s}^{\mathrm{cr}}(\boldsymbol{x}) + P \int_{-\infty}^{-m} \frac{V_{E'} \psi_{E'}^{\mathrm{cr}}(\boldsymbol{x})}{E - E'} dE' \right] + \mathrm{e}^{\mathrm{i}\Delta(E)} (E - E_R)\, \psi_E^{\mathrm{cr}}(\boldsymbol{x}) \right\} \qquad (6.221)$$

and

$$\tilde{\Psi}_E(\boldsymbol{x}) = \frac{\mathrm{e}^{\mathrm{i}\delta(E)}}{\sqrt{(E - E_R)^2 + \pi^2 |V_E|^4}} \left\{ \mathrm{e}^{-\mathrm{i}\Delta(E)} |V_E| \left[\psi_{1s}^{\mathrm{cr}}(\boldsymbol{x}) + P \int_{-\infty}^{-m} \frac{V_{E'} \psi_{E'}^{\mathrm{cr}}(\boldsymbol{x})}{E - E'} dE \right] + (E - E_R)\, \psi_E^{\mathrm{cr}}(\boldsymbol{x}) \right\}. \qquad (6.222)$$

Hence it is immediately evident that both wave functions are connected by

$$\tilde{\Psi}_E(\boldsymbol{x}) = \exp\{\mathrm{i}[\delta(E) - \Delta(E)]\}\, \Psi_E(\boldsymbol{x}), \qquad (6.223)$$

i.e. the two conventions differ just by an additional E-dependent, but spatially constant phase $(\delta(E) - \Delta(E))$ in the definition of the supercritical states:

$$\exp\{\mathrm{i}[\delta(E) - \Delta(E)]\} = \sqrt{\frac{E - E_R - \mathrm{i}\pi |V_E|^2}{E - E_R + \mathrm{i}\pi |V_E|^2}} \cdot \frac{V_E^*}{|V_E|}. \qquad (6.224)$$

Bibliographical Notes

Chapter 6 is crucial for understanding overcriticality. The autoionization formalism used to explain the essential physical steps can be found in *Fano's* original paper [Fa 61]. The change of the vacuum structure and most of the underlying physical ideas can be found in our original papers [Pi 69, Mü 72a – c, 73b, Ra 74].

7. Quantum Electrodynamics of Weak Fields

Two ways of presenting ordinary quantum electrodynamics of weak fields exist. The first, more formal in nature, starts with the quantization of wave fields; the second, more easily dealt with, stems from *Stückelberg* and *Feynman* [St 41, Fe 48] and uses propagator techniques. We shall review here in condensed form this second approach and apply it to scattering processes. It is our goal to derive formulas for transition probabilities and scattering processes within the quantized version of Dirac's hole theory for electrons and positrons. These calculations are in principle exact, but in practice use perturbation theory. Since creation and annihilation processes of electron – positron pairs have to be dealt with in the formalism, it must necessarily be relativistic.

In Feynman's propagator approach the scattering processes are described by integral equations. The leading idea is that, in the spirit of the hole theory, positrons are electrons of negative energy moving backward in time. However, to be as self-contained and lucid as possible, we first recapitulate the propagator concept in non-relativistic quantum mechanics. A more detailed, recent presentation of perturbative QED can be found in [Gr 84b]. Propagator methods, which are exact in the external field but perturbative in α, are discussed in Chap. 15. [For simplicity we mostly put $\hbar = c = 1$ in this chapter.]

7.1 The Non-Relativistic Propagator

Let us consider a quantum mechanical scattering process in three dimensions, in which a particle collides with a fixed centre of force or with another particle. Such a scattering process proceeds according to the scheme in Fig. 7.1. The collimators K focus the incoming beam so there is no interference between the incoming and scattered waves at observation point P. Note that such an incoming collimated beam is generally not a plane wave $\exp(\mathrm{i}kz)$, extended over all space, but a superposition of many plane waves with wave vectors centring around $\boldsymbol{k}$, i.e. a wave packet. Nevertheless, for simplicity, the wave packet shall be represented by a plane wave. Bear in mind that, e.g., in point P of Fig. 7.1, there is no interference between incoming and scattered waves. In other words, if one uses plane waves, such interferences have to be explicitly excluded.

A typical question of a scattering problem is then what happens, as a function of time, with a wave packet which represents in the far past a particle approaching a scattering centre, and into what does it develop in the future? The gener-

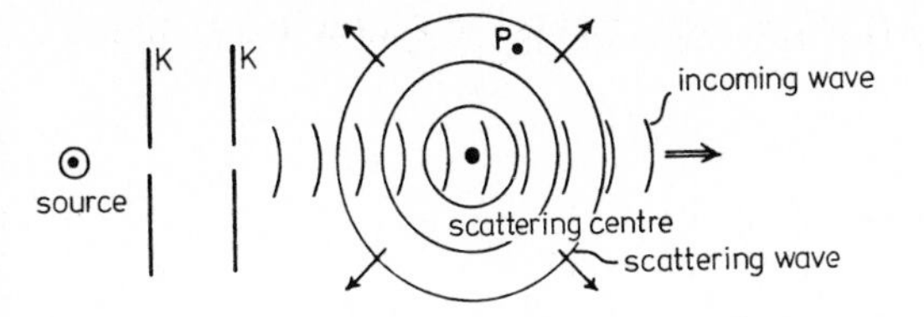

Fig. 7.1. Measurement in a scattering process. The collimators K ensure that there is no interference of incoming and scattering waves at, e.g., the observation point P

alized Huygens' principle helps clarify and formulate this problem. Assume the wave function $\psi(\boldsymbol{x}, t)$ is known at time t. Then the wave at any later time t' at space point $\boldsymbol{x}'$ can be found as a superposition of spherical waves emerging from all points $\boldsymbol{x}$:

$$\psi(\boldsymbol{x}', t') = \mathrm{i} \int d^3x \, G(\boldsymbol{x}' t' | \boldsymbol{x} t) \, \psi(\boldsymbol{x}, t) . \tag{7.1}$$

Here $\psi(\boldsymbol{x}', t')$ is the wave at $\boldsymbol{x}'$ at time t'. The integration extends over all space. The quantity $G(\boldsymbol{x}' t' | \boldsymbol{x} t)$ is called *Green's function* or *propagator*, which is important in that if it is known, according to (7.1) one can calculate the *final wave* $\psi(\boldsymbol{x}', t')$ from a given *initial wave* $\psi(\boldsymbol{x}, t)$. Knowing $G(\boldsymbol{x}' t' | \boldsymbol{x} t)$ solves the whole scattering problem.

Even though we have argued that t' is later than $t (t' > t)$ when (7.1) was established intuitively, this is not necessary, since t' can also be earlier than t $(t' < t)$. However, to solve many physical problems it is desirable to distinguish clearly between these two cases. To extend the wave forward in time one defines the *retarded Green's function*

$$G^+(\boldsymbol{x}' t' | \boldsymbol{x} t) = \begin{cases} G(\boldsymbol{x}' t' | \boldsymbol{x} t) & \text{for} \quad t' > t \\ 0 & \text{for} \quad t' < t \end{cases} . \tag{7.2}$$

With that and the step function $\Theta(\tau) = \begin{cases} 1 \text{ for } \tau > 0 \\ 0 \text{ for } \tau < 0 \end{cases}$, (7.1) then yields the forward extension of the wave:

$$\psi^+(x') \equiv \Theta(t' - t) \, \psi(\boldsymbol{x}', t') = \mathrm{i} \int G^+(\boldsymbol{x}' t' | \boldsymbol{x} t) \, \psi(\boldsymbol{x}, t) \, d^3x . \tag{7.3}$$

Similarly the *advanced Green's function* G^- is defined as

$$G^-(\boldsymbol{x}' t' | \boldsymbol{x} t) = \begin{cases} -G(\boldsymbol{x}' t' | \boldsymbol{x} t) & \text{for} \quad t' < t \\ 0 & \text{for} \quad t' > t \end{cases} , \tag{7.4}$$

so that the development of the wave backwards in time is

$$\psi^-(x') \equiv \Theta(t - t') \, \psi(\boldsymbol{x}', t') = -\mathrm{i} \int G^-(\boldsymbol{x}' t' | \boldsymbol{x} t) \, \psi(\boldsymbol{x}, t) \, d^3x . \tag{7.5}$$

Green's functions obey a *differential equation*, obtained by applying to (7.3 or 5) the Schrödinger operator

$$\left(\mathrm{i} \frac{\partial}{\partial t'} - \hat{H}(x') \right) \Theta(t' - t) \, \psi(x') = \mathrm{i} \int d^3x \left[\mathrm{i} \frac{\partial}{\partial t'} - H(x') \right] G^+(x' | x) \, \psi(x) . \tag{7.6}$$

Recognizing that $d\Theta(\tau)/d\tau = \delta(\tau)$ and that $\psi(x')$ obeys the Schrödinger equation

$$\mathrm{i}\,\partial\psi(x')/\partial t = \hat{H}(x')\,\psi(x')\,, \tag{7.7}$$

a straightforward calculation yields

$$\left(\mathrm{i}\frac{\partial}{\partial t'}\Theta(t'-t)\right)\psi(x') + \Theta(t'-t)\left[\mathrm{i}\frac{\partial}{\partial t'} - \hat{H}(x')\right]\psi(x') = \mathrm{i}\,\delta(t'-t)\,\psi(x')\;, \tag{7.8}$$

and therefore

$$\mathrm{i}\int d^3x\left\{\left[\mathrm{i}\frac{\partial}{\partial t'} - \hat{H}(x')\right]G^+(x'|x) - \delta^3(\boldsymbol{x}'-\boldsymbol{x})\,\delta(t'-t)\right\}\psi(x) = 0\,. \tag{7.9}$$

Consequently one deduces

$$\left(\mathrm{i}\frac{\partial}{\partial t'} - \hat{H}(x')\right)G^+(x'|x) = \delta^4(x'-x)\,. \tag{7.10}$$

The same differential equation is obtained for $G^-(x'|x)$; only the boundary conditions are different. The corresponding *free Green's functions* $G_0^\pm(x'|x)$ obey a similar differential equation to the total Green's function (7.10), the only difference being that $\hat{H}$ is replaced by the free Hamiltonian $\hat{H}_0$, i.e.

$$\left(\mathrm{i}\frac{\partial}{\partial t'} - \hat{H}_0(x')\right)G_0^\pm(x'|x) = \delta^4(x'-x)\,. \tag{7.11}$$

Splitting the total Hamiltonian $\hat{H}$ into the free Hamiltonian $\hat{H}_0$ and an interaction $V(x)$

$$\hat{H} = \hat{H}_0 + V(x) \tag{7.12}$$

enables (7.10) to be rewritten as

$$\begin{aligned}\left[\mathrm{i}\frac{\partial}{\partial t'} - \hat{H}_0(x')\right]G^+(x'|x) &= \delta^4(x'-x) + \hat{V}(x')\,G^+(x'|x)\\ &= \int d^4x''\,\delta^4(x'-x'')\,[\delta^4(x''-x) + V(x'')G^+(x''|x)]\,.\end{aligned} \tag{7.13}$$

The integral form of the rhs of this equation is useful, because due to the linearity of the differential equation we can immediately write the formal solution for the total Green's function $G^+(x'|x)$ expressed in terms of the free Green's function $G_0^+(x'|x)$, namely

$$\begin{aligned}G^+(x'|x) &= \int d^4x''\,G_0^+(x'|x'')\,[\delta^4(x''-x) + V(x'')\,G^+(x''|x)]\\ &= G_0^+(x'|x) + \int d^4x''\,G_0^+(x'|x'')\,V(x'')\,G^+(x''|x)\,.\end{aligned} \tag{7.14}$$

This is the *integral equation* for the total Green's function $G(x'|x)$ in terms of the free Green's function $G_0(x'|x)$, which is supposed to be known. It can be solved by iteration, starting with $G_0(x'|x)$. One then immediately deduces from (7.14) the following multiple-scattering series for $G(x'|x)$ [or for $G^+(x'|x)$, depending on the boundary conditions]:

$$\begin{aligned} G^+(x'|x) = G_0^+(x'|x) &+ \int d^4x_1\, G_0^+(x'|x_1)\, V(x_1)\, G_0^+(x_1|x) \\ &+ \int d^4x_1\, d^4x_2\, G_0^+(x'|x_1)\, V(x_1)\, G_0^+(x_1|x_2)\, V(x_2)\, G_0^+(x_2|x) \\ &+ \cdots\, . \end{aligned} \tag{7.15}$$

This series is assumed to be convergent. In all practical cases this assumption has to be checked. Also the possibility of bound states in the potential $V(x)$ is ignored. Bound states may occur as poles of $G^+(x)$ after summation of the infinite series. Explicit examples for e.g., $V(x) = V_0\,\delta(x)$ are possible.

Inserting (7.14) into (7.3), for the time development into the future of an initial state $\varphi(x) = \varphi(\boldsymbol{x}, t)$

$$\begin{aligned} \psi^+(x') &= \mathrm{i}\int d^3x\, G^+(x'|x)\, \varphi(x) \\ &= \mathrm{i}\int d^3x\, [G_0^+(x'|x) + \int d^4x_1\, G_0^+(x'|x_1)\, V(x_1)\, G^+(x_1|x)]\, \varphi(x) \\ &= \varphi(x') + \int d^4x_1\, G_0^+(x'|x_1)\, V(x_1)\, \mathrm{i}\int d^3x\, G^+(x_1|x)\, \varphi(x) \\ &= \varphi(x') + \underbrace{\int d^4x_1\, G_0^+(x'|x_1)\, V(x_1)\, \psi^+(x_1)}_{\text{scattered wave}}\, , \end{aligned} \tag{7.16}$$

where the last term represents the *scattered wave*. Equation (7.16) is an integral equation for $\psi^+(x)$, which corresponds to the integral equation (7.14) for the Green's function. It is clear that (7.16) does not solve the scattering problem, because the unknown wave packet $\psi^+(x)$ also appears on the rhs under the integral. Nevertheless, formal progress has been achieved, because (7.16) can be solved for small scattering potentials $V(x)$ by iteration.

7.2 The S Matrix

Let us consider now a scattering problem in which the impacting wave packet $\varphi(\boldsymbol{x}, t)$ describes a particle in the distant past, and $\varphi(x)$ is known. We search for that wave which develops out of $\varphi(x)$ by interactions with the potential $V(x)$. The scattering problem is idealized so that initially no scattering occurs; the initial wave $\varphi(x)$ is a solution of the free wave equation, obeying certain initial conditions. This can be achieved mathematically, if the interaction $V(\boldsymbol{x}, t)$ is localized in time,

$$V(\boldsymbol{x}, t) \xrightarrow[\text{for } t \to -\infty]{} 0\, . \tag{7.17}$$

This means that the interaction is *turned on adiabatically.* The adiabatic switching on prevents initial disturbances due to the switching. In the limit $t \to -\infty$ the exact wave $\psi(\boldsymbol{x}, t)$ approaches the impacting initial wave:

$$\lim_{t \to -\infty} \psi(\boldsymbol{x}, t) = \varphi(\boldsymbol{x}, t) . \tag{7.18}$$

In the distant future the exact wave $\psi^+(\boldsymbol{x}', t')$ is then, according to (7.16), given by

$$\begin{aligned} \psi^+(\boldsymbol{x}', t') &= \lim_{t \to -\infty} \mathrm{i} \int d^3x \, G^+(\boldsymbol{x}' t' | \boldsymbol{x} t) \, \varphi(\boldsymbol{x}, t) \\ &= \varphi(x') + \underbrace{\int d^4x_1 \, G_0^+(x' | x_1) \, V(x_1) \, \psi^+(x_1)}_{\text{scattered wave}} . \end{aligned} \tag{7.19}$$

The $\psi^+(x_1)$ is that exact wave which develops out of the initial wave packet (7.18). As stated earlier, the second term of (7.19)

$$\int d^4x_1 \, G_0^+(x' | x_1) \, V(x_1) \, \psi^+(x_1) \tag{7.20}$$

represents the scattering wave, which contains all single and multiple scatter events. We now also assume that the potential $V(\boldsymbol{x}', t')$ is also adiabatically switched off in the future

$$\lim_{t' \to \infty} V(\boldsymbol{x}', t') = 0 . \tag{7.21}$$

Hence the exact wave $\psi^+(\boldsymbol{x}', t')$ is finally free, i.e., it no longer changes and is fully developed [of course $\psi^+(\boldsymbol{x}', t')$ still changes, but in a trivial way according to $\hat{H}_0$]. If the interaction is over, the wave field has finalized. All information on the scattered wave is then contained in the probability amplitudes with which a particle is scattered for $t' \to \infty$ from an initial state φ_{i} into the final states $\varphi_{\mathrm{f}}(\boldsymbol{x}', t')$ which can be taken as plane waves, for example:

$$\varphi_{\mathrm{f}}(\boldsymbol{x}', t') = \frac{1}{\sqrt{2\pi\hbar}^{\,3}} \exp[\mathrm{i}(\boldsymbol{k}_{\mathrm{f}} \cdot \boldsymbol{x}' - \omega_{\mathrm{f}} t')] . \tag{7.22}$$

Plane waves can be used for the initial (φ_{i}) and final (φ_{f}) states, because the potential is assumed to vanish in the distant past and far future (adiabatic turning on and off). The probability amplitudes are the elements of *Heisenberg's scattering or S matrix* [He 46, see also Mø 45, Wh 37].

$$\begin{aligned} S_{\mathrm{fi}} &= \lim_{t' \to \infty} \langle \varphi_{\mathrm{f}}(\boldsymbol{x}', t') | \psi_{\mathrm{i}}^+(\boldsymbol{x}', t') \rangle \\ &= \lim_{t' \to \infty} \int \varphi_{\mathrm{f}}^*(\boldsymbol{x}', t') \, \psi_{\mathrm{i}}^+(\boldsymbol{x}', t') \, d^3x' \\ &= \lim_{\substack{t' \to \infty \\ t \to -\infty}} \mathrm{i} \int\!\!\int \varphi_{\mathrm{f}}^*(\boldsymbol{x}', t') \, G^+(\boldsymbol{x}' t' | \boldsymbol{x} t) \, \varphi_{\mathrm{i}}(\boldsymbol{x}, t) \, d^3x' \, d^3x \end{aligned}$$

$$= \lim_{t' \to \infty} \int d^3x' \, \varphi_f^* (\boldsymbol{x}', t') [\varphi_i(\boldsymbol{x}', t') + \int d^4x \, G_0^+ (\boldsymbol{x}' t' | \boldsymbol{x} t) \, V(\boldsymbol{x}, t) \, \psi_i^+ (\boldsymbol{x}, t)]$$

$$= \delta^3(\boldsymbol{k}_f - \boldsymbol{k}_i) + \lim_{t' \to \infty} \int d^3x' \, d^4x \, \varphi_f^* (\boldsymbol{x}', t') \, G_0^+ (\boldsymbol{x}' t' | \boldsymbol{x} t) \, V(\boldsymbol{x}, t) \, \psi_i^+ (\boldsymbol{x}, t) \, . \tag{7.23}$$

It goes without saying that $\psi_i^+ (\boldsymbol{x}, t)$ is that solution (7.19) of the wave equation which develops due to the scattering process out of the incoming, initial $(t \to -\infty)$ plane wave $\varphi_i(\boldsymbol{x}, t)$ with momentum $\boldsymbol{k}$. The limits $t \to \pm\infty$ always mean

$t \to$ very long, finite time T

at which the particle wave packet does not interact with the potential $V(\boldsymbol{x}, t)$ any more. Typical times are, for example, the scattering time or the production time or the detection time of a particle (wave).

Calculating the iterative solution obtained from (7.16), namely

$$\begin{aligned} \psi_i^+ (x') = \varphi_i(x') &+ \int d^4x_1 \, G_0^+ (x' | x_1) \, V(x_1) \, \varphi_i(x_1) \\ &+ \int d^4x_1 d^4x_2 \, G_0^+ (x' | x_1) \, V(x_1) \, G_0^+ (x_1 | x_2) \, V(x_2) \, \varphi_i(x_2) \\ &+ \cdots \, , \end{aligned} \tag{7.24}$$

and inserting it into (7.23) expands the S matrix into multiple-scattering processes. The result is the so-called Born series:

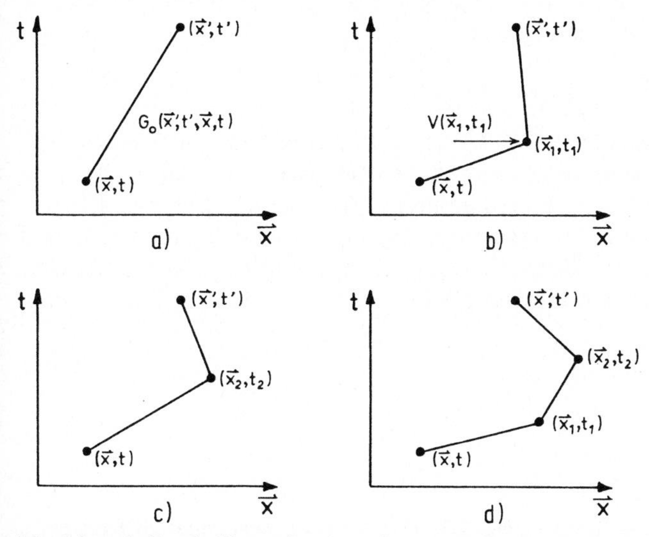

Fig. 7.2 a – d. Graphs illustrating the various scattering processes; **(a)** describes the free propagation of a particle from space-time point x to space-time point x'. In (b) the particle propagates also from x to x', but is scattered once by the potential at space-time point x_1; **(c)** is similar to **(b)**, except the scattering occurs at x_2 instead of x_1; **(d)** represents a double-scattering process

$$
\begin{aligned}
S_{\mathrm{fi}} = \delta^3(\boldsymbol{k}_\mathrm{f}-\boldsymbol{k}_\mathrm{i}) &+ \lim_{t'\to\infty} \iint d^3x'\, d^4x_1\, \varphi_\mathrm{f}^*(\boldsymbol{x}',t')\, G_0^+(\boldsymbol{x}'t'|\boldsymbol{x}_1 t_1)\, V(x_1)\, \varphi_\mathrm{i}(x_1) \\
&+ \lim_{t'\to\infty} \iiint d^3x'\, d^4x_1\, d^4x_2\, \varphi_\mathrm{f}^*(\boldsymbol{x}',t')\, G_0^+(x'|x_1)\, V(x_1) \\
&\times G_0^+(x_1|x_2)\, V(x_2)\, \varphi_\mathrm{i}(x_2) \\
&+ \cdots \; . \qquad (7.25)
\end{aligned}
$$

The first term $\delta^3(\boldsymbol{k}_\mathrm{f}-\boldsymbol{k}_\mathrm{i})$ does not describe any scattering; it characterizes the flux of particles without any interaction (scattering). The second term describes single, the third term, double scattering, etc. Some of these contributions are illustrated by graphs in Fig. 7.2. They add coherently to the total S-matrix element.

7.3 Propagator for Electrons and Positrons

We shall now generalize the non-relativistic propagator technique to the relativistic theory of electrons and positrons. Important, then, is the non-relativistic idea according to which the propagator $G^+(x'|x)$ is the probability amplitude for the propagation of a wave from space-time point x to space-time point x'. According to (7.15) this amplitude can be decomposed into a sum of *multiple-scattering amplitudes*; the nth order scattering amplitude can be illustrated by a graph as in Fig. 7.3. As can be seen from (7.15), the nth order scattering amplitude consists of n factors describing the propagation of the particle between the individual scattering processes. Each single line of the graph represents a Green's function. For example, the line

x_{i-1} ●————● x_i

represents the Green's function $G_0^+(x_i|x_{i-1})$, i.e. the amplitude for free propagation of a wave known at space-time point x_{i-1} to the space-time point x_i. At x_i the wave is scattered with the probability amplitude $V(x_i)$ per unit volume of space-time. The scattering wave thus generated then propagates again freely and forwards in time with the amplitude $G_0^+(x_{i+1}|x_i)$ from x_i to x_{i+1}. At x_{i+1} the next scattering takes place, etc. The total amplitude is then the sum over all space-time

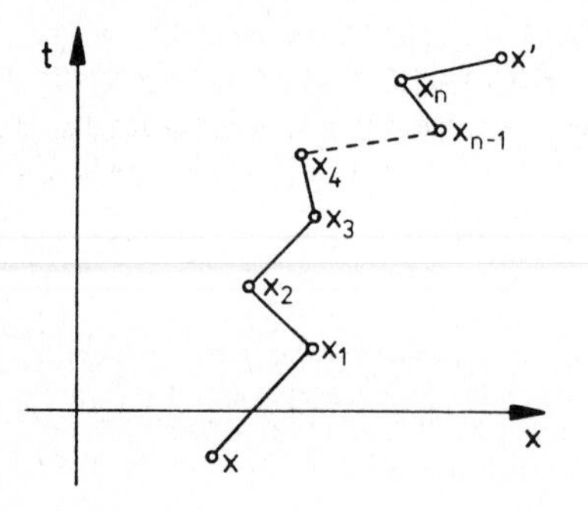

Fig. 7.3. The nth order contribution to the Green's function $G^+(x'|x)$, with multiple scattering of the particle

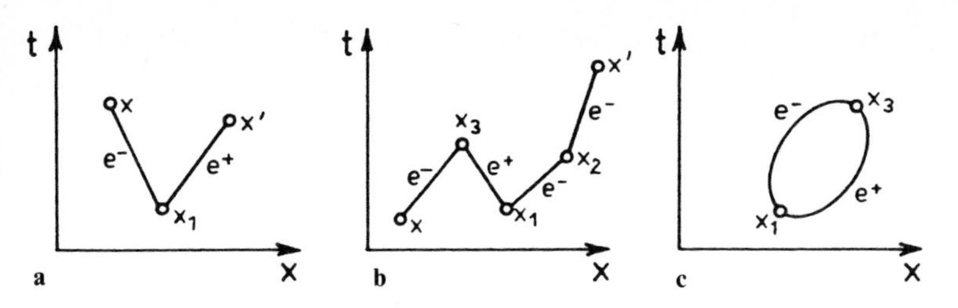

Fig. 7.4a – c. Simple examples of processes in electron – positron field theory: (**a**) electron – positron pair creation; (**b**) electron scattering with intermediate pair creation; (**c**) closed loop, to be identified with vacuum polarization

points at which an interaction appears. The interaction points are called *vertices*. The scattering process may also be described by saying the interaction annihilates at the ith vertex a particle which has moved freely to x_i and creates a new particle which moves from there freely to x_{i+1}. Here $t_{i+1} > t_i$ always.

This interpretation of a scattering process is suitable for generalization to relativistic hole theory, because the full space-time structure of the interaction is considered. Hamilton's formalism, on the other hand, would have the disadvantage of exhibiting the time particularly and thus equality of space and time would not be manifest.

It is our goal to develop the calculational methods for scattering processes in Dirac's hole theory in analogy to the non-relativistic propagator theory. Special attention has to be paid to the pair creation and annihilation processes, unfamiliar to non-relativistic theory. Many of the rules are obtained intuitively with the dynamics inherent in the Dirac equation. A mathematically rigorous derivation of the rules can be found in [Gr 84b].

Let us first survey the typical processes to be described in a relativistic theory, summarized in Fig. 7.4 as graphs, whose meaning must be discussed in detail. The first process (Fig. 7.4a) is *electron – positron pair creation*. The pair is created by the interaction (potential) at space-time point x_1. Both the electron and positron then move forward in time; the electron to space-time point x, the positron to x'.

Figure 7.4b represents *scattering of an incoming electron* (from space-time point x), propagating forward in time, undergoing complicated interactions on its way and finally appearing at x'. At space-time point x_1 on the way from x to x' an electron – positron pair is created by the interaction (potential), and both particles of this pair move forward in time. At x_3 the positron of the pair meets the original (incoming) electron and both annihilate. The electron of the pair is annihilated by the interaction at space-time point x_2 and simultaneously a new electron is created propagating freely to x'. The last step can also be expressed by saying that the electron of the pair generated at x_1 is scattered by the interaction (potential) at x_2 and finally propagates to x'.

Figure 7.4c shows pair creation at space-time point x_1. The pair propagates to x_3, where it annihilates again. It was present only for a short time; it is a *virtual pair*. This process is typical of *vacuum polarization*. The interaction (potential) generates virtual e^+e^- pairs and thus polarizes the vacuum. These simple con-

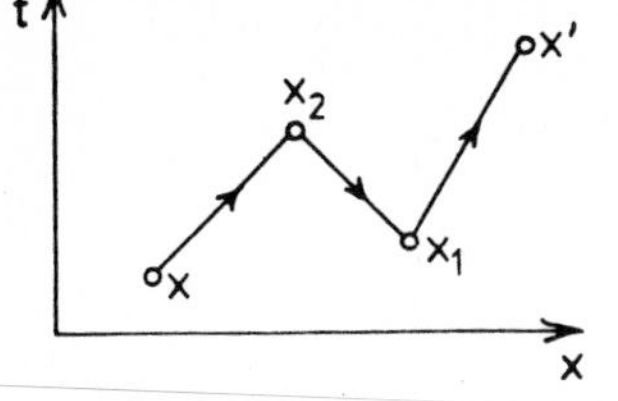

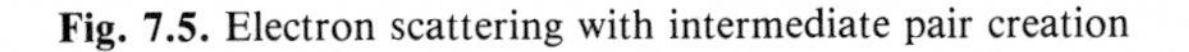
Fig. 7.5. Electron scattering with intermediate pair creation

siderations already indicate that a relativistic theory contains more than a non-relativistic theory: not only is the propagator for the motion of a particle (electron) necessary, which, for example, propagates from x_1 to x_2, but so is the amplitude for creating an electron positron pair, which develops in time and is partly or completely annihilated afterwards. This e^+e^- amplitude must first be set up so that the various processes of Fig. 7.4 can be calculated, and the relevant total amplitudes can be constructed. Of course, as in non-relativistic propagator theory, one has to sum (integrate) over all interaction points between initial and final space-time. In general, the total scattering amplitude contains both electron and positron amplitudes.

From hole theory (Chap. 4) it is known that a positron corresponds to a hole in the sea of negative energy continuum states filled with electrons. The annihilation of a positron at a space-time point is therefore equivalent to the creation of an electron with negative energy at this space-time point. Hence it is plausible that for Fig. 7.4b, the amplitude for creating a positron at x_1 and its annihilation at x_3 is identical with the amplitude for creating an electron of negative energy at x_3 and its annihilation at x_1. The space-time point x_1 is (time wise) before x_3. Therefore, a pair creation process as in Figs. 7.4a, b automatically involves the following definition of positrons: *Positrons with positive energy, moving forward in space-time, are in propagator language identical with electrons of negative energy, moving backward in space-time.*

This definition stems from Stückelberg and Feynman. A process as shown in Fig. 7.5 can therefore be interpreted as an electron propagating forward in time from x to x_2, where it is scattered by $V(x_2)$ into a state of negative energy, in which it propagates backward in time to x_1. At x_1 it is again scattered by $V(x_1)$ into a state of positive energy, moving in that state forward in time to x'. Alternatively, we could also describe the process by saying that the electron (of positive energy) moves forward in time from x to x_2, where it annihilates due to the interaction $V(x_2)$ with a positron from an e^+e^- pair created at x_1. The electron of the pair created by $V(x_1)$ at x_1 propagates forward in time to x'.

Similarly, there are two ways to express the physical process described by closed-loop graphs as in Fig. 7.4c. The interaction $V(x_1)$ generates at x_1 and e^+e^- pair, which propagates forward in time and is annihilated by $V(x_2)$ at x_2. Alternatively expressed, the potential $V(x_1)$ scatters an electron at x_1 (from the Dirac sea) into a positive energy state, moving this state forward in time to be rescattered at x_2 by $V(x_2)$ into a negative energy state, in which it moves backward in space-time to the original point x_1, i.e. the electron generated at x_1 is scattered backward in time at x_2, so that it disappears again at x_1.

Let us now describe these processes formally. Hence we construct the Green's function for electrons and positrons and call the relativistic propagator

$$S_{\mathrm{SF}}(x'|x)\,, \tag{7.26}$$

where the index "SF" refers to Stückelberg and Feynman. Very often this propagator is simply called the Feynman propagator. It obeys the differential equation

$$\sum_{\lambda=1}^{4}\left[\gamma_\mu\left(\mathrm{i}\hbar\frac{\partial}{\partial x'_\mu}-\frac{e}{c}A^\mu(x')\right)-m_0c\right]_{\alpha\lambda}(S_{\mathrm{SF}})_{\lambda\beta}(x'|x)$$
$$=\hbar\delta_{\alpha\beta}\delta^4(x'-x)\,. \tag{7.27}$$

Here $S_{\mathrm{SF}}(x'|x)$ is obviously a 4×4 matrix, because the γ matrices are of this dimension. Setting, as usual, $\hbar=c=1$ and writing $e\to e/\hbar c$, $m_0\to m_0c/\hbar$, (7.27) can be denoted in matrix form

$$(\mathrm{i}\not{\nabla}'-e\not{A}'-m_0)S_{\mathrm{SF}}(x'|x)=\mathbb{1}\,\delta^4(x'-x)\,. \tag{7.28}$$

Compared to the non-relativistic definition of the propagator, (7.10), is a minor modification. In (7.10) there was an operator

$$\mathrm{i}\frac{\partial}{\partial t'}-\hat{H}(x')$$

on the lhs, in (7.28) this form has been multiplied by γ^0, so that the covariant operator $(\mathrm{i}\not{\nabla}'-e\not{A}'-m_0)$ appears. The unit matrix $\mathbb{1}$ on the rhs of (7.28) is often deleted, giving

$$(\mathrm{i}\not{\nabla}'-e\not{A}'-m_0)S_{\mathrm{SF}}(x'|x)=\delta^4(x'-x)\,. \tag{7.29}$$

The *free propagator* $S_0^{\mathrm{SF}}(x'|x)$ then obviously obeys

$$(\mathrm{i}\not{\nabla}'-m_0)S_0(x'|x)=\delta^4(x'-x) \tag{7.30}$$

(the superscript "SF" on the free propagator is henceforth dropped). It can be calculated in momentum space. Therefore we realize, as in non-relativistic theory, that it has to depend on the space-time distance $x-x'$, i.e.

$$S_0(x'|x)=S_0(x'-x)=\int\frac{d^4p}{(2\pi)^4}\,\mathrm{e}^{-\mathrm{i}p\cdot(x'-x)}S_0(p)\,. \tag{7.31}$$

Then after substitution into (7.30)

$$\int\frac{d^4p}{(2\pi)^4}(\not{p}-m_0)S_0(p)\,\mathrm{e}^{-\mathrm{i}p\cdot(x'-x)}=\int\frac{d^4p}{(2\pi)^4}\,\mathrm{e}^{-\mathrm{i}p\cdot(x'-x)} \tag{7.32}$$

and hence

$$(\not{p} - m_0) S_0(p) = \mathbb{1} , \tag{7.33}$$

or, denoted in detail

$$\sum_{\lambda=1}^{4} (\not{p} - m_0)_{\alpha\lambda} (S_0(p))_{\lambda\beta} = \delta_{\alpha\beta} . \tag{7.34}$$

Multiplying (7.33) from the left by $(\not{p} + m_0)$ yields

$$(\not{p} + m_0)(\not{p} - m_0) S_0(p) = (\not{p} + m_0)$$

and, because of $\not{p}\not{p} = p^2$,

$$(p^2 - m_0^2) S_0(p) = (\not{p} + m_0) . \tag{7.35}$$

Therefore, for $p^2 \neq m_0$,

$$S_0(p) = \frac{\not{p} + m_0}{p^2 - m_0^2} ; \quad p^2 \neq m_0^2 . \tag{7.36}$$

To determine $S_0(p)$ completely, one has to treat the singularity at

$$p^2 = m_0^2 , \quad \text{i.e.} \quad p_0^2 - \boldsymbol{p}^2 = m_0^2 \quad \text{or}$$

$$p_0 = \pm \sqrt{\boldsymbol{p}^2 + m_0^2} . \tag{7.37}$$

This is achieved by incorporating the boundary conditions of $S_0(x' - x)$.

Now the interpretation according to which the positrons are electrons of negative energy moving backward in time is considered within the Fourier representation (7.31) and the Fourier amplitude (7.36) by a special choice of the integration path over energy (p_0 integration):

$$\begin{aligned} S_0(x' - x) &= \int \frac{d^4p}{(2\pi)^4} S_0(p) \exp[-\mathrm{i} p \cdot (x' - x)] \\ &= \int \frac{d^4p}{(2\pi)^4} S_0(p) \exp\{-\mathrm{i}[p_0(t' - t) - \boldsymbol{p} \cdot (\boldsymbol{x}' - \boldsymbol{x})]\} \\ &= \int \frac{d^3p}{(2\pi)^3} \exp[\mathrm{i}\boldsymbol{p} \cdot (\boldsymbol{x}' - \boldsymbol{x})] \int_{\mathscr{C}} \frac{dp_0}{2\pi} \frac{\exp[-\mathrm{i} p_0(t' - t)]}{p^2 - m_0^2} (\not{p} + m_0) . \end{aligned} \tag{7.38}$$

For $t' > t$ the integration path $\mathscr{C}$ is closed in the lower half-plane ($\mathscr{C}_1$), because then the integral over the lower semicircle does not contribute if $p_0 = \varrho \exp(\mathrm{i}\varphi)$ is inserted and the limit $\varrho \to \infty$ is taken. According to the residue theorem the pole at

$$p_0 = +E = +\sqrt{\boldsymbol{p}^2 + m_0^2}$$

now contributes (Fig. 7.6), giving

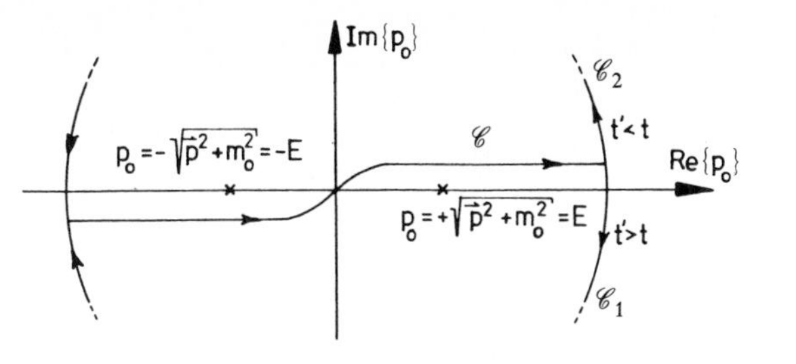

Fig. 7.6. Integration path $\mathscr{C}$ leading to the Stückelberg-Feynman propagator. The singularities at $p_0 = -E$ and $p_0 = +E$ are indicated

$$
\begin{aligned}
&\int\limits_{\mathscr{C}+\mathscr{C}_1} \frac{dp_0}{2\pi} \frac{\exp[-\mathrm{i}p_0(t'-t)]}{p_0^2 - \boldsymbol{p}^2 - m_0^2}(\not{p} + m_0) \\
&= \int\limits_{\mathscr{C}+\mathscr{C}_1} \frac{dp_0}{2\pi} \frac{\exp[-\mathrm{i}p_0(t'-t)]}{(p_0 - E)(p_0 + E)}(p_0\gamma^0 + p_i\gamma^i + m_0) \\
&= -2\pi\mathrm{i}\frac{\exp[-\mathrm{i}E(t'-t)]}{2\pi 2E}(E\gamma^0 - \boldsymbol{p}\cdot\boldsymbol{\gamma} + m_0)\,.
\end{aligned}
\tag{7.39}
$$

Hence (7.38) becomes

$$
S_0(x'-x) = -\mathrm{i}\int \frac{d^3p}{(2\pi)^3} \exp[\mathrm{i}\boldsymbol{p}(\boldsymbol{x}'-\boldsymbol{x})] \exp[-\mathrm{i}E(t'-t)] \frac{(E\gamma^0 - \boldsymbol{p}\cdot\boldsymbol{\gamma} + m_0)}{2E}\,,
\tag{7.40}
$$

where $t' > t$.

The minus sign in (7.39) stems from the integration in the mathematically negative sense. The thus obtained propagator (7.40) describes a motion from x to x' forwards in time (t' is always $> t$) and at $x' = (\boldsymbol{x}', t')$ it contains only components with positive energy, because the factor $E = +\sqrt{\boldsymbol{p}^2 + m_0^2}$ in the exponent of $\exp[-\mathrm{i}E(t'-t)]$ is positive.

Let us now consider the propagation backwards in time, when $t' - t$ is negative and the p_0 integration in (7.38) must be closed in the upper halfplane, so that the contribution of the infinitely extended semicircle vanishes. The integral in (7.39) goes over $\mathscr{C} + \mathscr{C}_2$, to which only the pole at

$$
p_0 = -E = -\sqrt{\boldsymbol{p}^2 + m_0^2}
$$

contributes. The result in this case is

$$
\begin{aligned}
&\int\limits_{\mathscr{C}+\mathscr{C}_2} \frac{dp_0}{2\pi} \frac{\exp[-\mathrm{i}p_0(t'-t)]}{(p_0 - E)(p_0 + E)}(p_0\gamma^0 + p_i\gamma^i + m_0) \\
&= 2\pi\mathrm{i}\frac{\exp[-\mathrm{i}(-E)(t'-t)]}{2\pi(-2E)}(-E\gamma^0 - \boldsymbol{p}\cdot\boldsymbol{\gamma} + m_0)\,,
\end{aligned}
\tag{7.41}
$$

so that the propagator (7.38) becomes

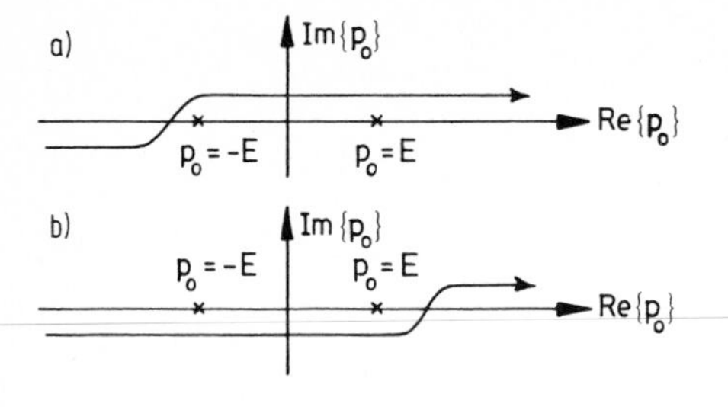

Fig. 7.7. Two different integration paths, which would have led to contradictions with hole theory, in particular with the position of the Fermi surface (which states have to be counted as particles and which as antiparticles)

$$S_0(x'-x) = -\mathrm{i}\int \frac{d^3p}{(2\pi)^3}\exp[\mathrm{i}\boldsymbol{p}\cdot(\boldsymbol{x}'-\boldsymbol{x})]$$

$$\times\exp[+\mathrm{i}E(t'-t)]\,\frac{(-E\gamma^0-\boldsymbol{p}\cdot\boldsymbol{\gamma}+m_0)}{2E}\,. \tag{7.42}$$

where $t' < t$.

This propagator describes the motion backwards in time ($t' < t$) and it consists of waves with negative energy, as can be seen from the factors

$$\exp[-\mathrm{i}(-E)(t'-t)]$$

appearing in (7.42). Such waves do not exist in a non-relativistic theory, because no pole at negative energy $p_0 = -E = -\sqrt{\boldsymbol{p}^2+m_0^2}$ exists. In a relativistic theory, on the other hand, such waves cannot be prevented since they are inherent to the problem.

A different integral path $\mathscr{C}$, e.g. as in Fig. 7.7, gives rise to waves of negative energy in the propagator, which would propagate forwards in time (Fig. 7.7a) or waves of positive energy which would propagate backwards in time (Fig. 7.7b). The path chosen for the Stückelberg-Feynman propagator in Fig. 7.6 ensures that all waves of positive energy propagate forwards in time and all waves with negative energy, backwards in time. *These waves of negative energy propagating backwards in time are identified with positrons; the waves of positive energy propagating forwards in time are identified as electrons, following hole theory.* Which states are interpreted as electrons and which as positrons is determined by the vacuum (characterized by the Fermi energy E_{F}), i.e. by the assumption that states with $E < E_{\mathrm{F}}$ are occupied by particles (electrons) and holes therein are identified as antiparticles (positrons). This choice of the vacuum enters the propagator essentially as the change in the integration path of Fig. 7.6 from the lower to the upper half of the p_0 plane (E plane). In an atom the Fermi surface E_{f} usually lies at a bound state. Then the propagator has to be determined so that the switch in the p_0 integration from the lower to the upper half of the p_0 plane occurs at the particular energy E_{F} of the bound state, representing the Fermi surface. Another illustrative example is the Fermi-gas model [Gr 84b, Fe 71].

One can characterize the integration path $\mathscr{C}$ belonging to the Stückelberg-Feynman propagator $S_0(x'-x)$ (Fig. 7.6) by simply adding $+\mathrm{i}\varepsilon$ to the denominator of (7.38) with the understanding that the limes $\varepsilon\to 0$ has to be taken:

$$S_0(x'-x) = \int \frac{d^4p}{(2\pi)^4} \frac{\exp[-\mathrm{i}p\cdot(x'-x)]}{p^2-m_0^2+\mathrm{i}\varepsilon} (\not{p}+m_0)\,. \tag{7.43}$$

Then the poles of positive energy appear at

$$p_0 = +\sqrt{\boldsymbol{p}^2+m_0^2-\mathrm{i}\varepsilon} = \sqrt{\boldsymbol{p}^2+m_0^2} - \mathrm{i}\eta(\varepsilon)\,, \tag{7.44a}$$

i.e. below the p_0 axis, and the poles of negative energy at

$$p_0 = -\sqrt{\boldsymbol{p}^2+m_0^2-\mathrm{i}\varepsilon} = -\sqrt{\boldsymbol{p}^2+m_0^2} + \mathrm{i}\eta(\varepsilon)\,, \tag{7.44b}$$

i.e. above the real p_0 axis, exactly as the integral path $\mathscr{C}$ of Fig. 7.6 specifies.

The two propagators (7.40, 42) for waves of positive and negative energy, respectively, can be amalgamated into one expression using the projection operators

$$\hat{\Lambda}_\pm(p) = \frac{\pm\not{p}+m_0}{2m_0}$$

or

$$\hat{\Lambda}_r(p) = \frac{\varepsilon_r\not{p}+m_0}{2m_0}\,. \tag{7.45}$$

Here

$$\varepsilon_r = \begin{cases} +1 \text{ for waves of positive energy } (r=1,2) \\ -1 \text{ for waves of negative energy } (r=3,4)\,. \end{cases}$$

It is easily verified that $\hat{\Lambda}_+$ projects from a wave packet those components with positive energy and $\hat{\Lambda}_-$ those components with negative energy. If in addition the three-momentum $\boldsymbol{p}$ is replaced by $-\boldsymbol{p}$ in (7.42) (which does not matter because of the integration over all the momentum space $\int d^3p$), then for the total Stückelberg-Feynman propagator:

$$\begin{aligned} S_0(x'-x) = &-\mathrm{i}\int \frac{d^3p}{(2\pi)^3} \Bigg[\exp[-\mathrm{i}(+E)(t'-t)] \exp[+\mathrm{i}\boldsymbol{p}\cdot(\boldsymbol{x}'-\boldsymbol{x})] \\ &\times \frac{(+E\gamma^0-\boldsymbol{p}\cdot\boldsymbol{\gamma}+m_0)}{2E} \theta(t'-t) + \exp[-\mathrm{i}(-E)(t'-t)] \\ &\times \exp[-\mathrm{i}\boldsymbol{p}\cdot(\boldsymbol{x}'-\boldsymbol{x})] \frac{(-E\gamma^0+\boldsymbol{p}\cdot\boldsymbol{\gamma}+m_0)}{2E} \theta(t-t') \Bigg] \\ = &-\mathrm{i}\int \frac{d^3p}{(2\pi)^3} \frac{m_0}{E} \Bigg[\frac{p_0\gamma^0+p_i\gamma^i+m_0}{2m_0} \exp\{-\mathrm{i}[p_0(t'-t)-\boldsymbol{p}(\boldsymbol{x}'-\boldsymbol{x})]\} \\ &\times \theta(t'-t) + \frac{-p_0\gamma^0-p_i\gamma^i+m_0}{2m_0} \\ &\times \exp\{+\mathrm{i}[p_0(t'-t)-\boldsymbol{p}\cdot(\boldsymbol{x}'-\boldsymbol{x})]\}\theta(t-t') \Bigg] \\ = &-\mathrm{i}\int \frac{d^3p}{(2\pi)^3} \left(\frac{m_0}{E}\right) \Bigg[\frac{\not{p}+m_0}{2m_0} \exp[-\mathrm{i}p\cdot(x'-x)]\,\theta(t'-t) \end{aligned}$$

$$\left. + \frac{-\not{p}+m_0}{2m_0} \exp[+\mathrm{i}p\cdot(x'-x)]\,\theta(t-t') \right]$$

$$= -\mathrm{i}\int \frac{d^3p}{(2\pi)^3}\left(\frac{m_0}{E}\right)[\hat{\Lambda}_+(p)\exp[-\mathrm{i}p\cdot(x'-x)]\,\theta(t'-t)$$

$$+\hat{\Lambda}_-(p)\exp[+\mathrm{i}p\cdot(x'-x)]\,\theta(t-t')] \,. \tag{7.46}$$

Furthermore, we now use the well-known free Dirac waves denoted by [Gr 81a]

$$\psi_p^r = \sqrt{\frac{m_0}{E}}\frac{1}{\sqrt{2\pi}^3}\,\omega^r(p)\exp(-\mathrm{i}\varepsilon_r p\cdot x) \qquad (r=1,2,3,4)\,. \tag{7.47}$$

Equation (7.46) can be brought to the form

$$S_0(x'-x) = -\mathrm{i}\,\theta(t'-t)\int d^3p \sum_{r=1}^{2} \psi_p^r(x')\,\bar{\psi}_p^r(x)$$

$$+\mathrm{i}\,\theta(t-t')\int d^3p \sum_{r=3}^{4} \psi_p^r(x')\,\bar{\psi}_p^r(x)\,, \tag{7.48}$$

which is quite compact and useful. With it we can easily determine the following relations for wave packets of positive ($\psi^{(+E)}$) and negative energy ($\psi^{(-E)}$):

$$\begin{aligned}\theta(t'-t)\,\psi^{(+E)}(x') &= \mathrm{i}\int S_0(x'-x)\,\gamma_0\,\psi^{(+E)}(x)\,d^3x\\ \theta(t-t')\,\psi^{(-E)}(x') &= -\mathrm{i}\int S_0(x'-x)\,\gamma_0\,\psi^{(-E)}(x)\,d^3x\,.\end{aligned} \tag{7.49}$$

They exhibit explicitly the fact that wave packets of positive energy (electrons) propagate forwards in time and wave packets of negative energy (positrons) propagate backwards in time. The propagator $S_0(x'-x)$ is called the free *Stückelberg-Feynman propagator*, because it was introduced by Stückelberg in 1941 into the electron-positron theory and used extensively in 1948 by Feynman for calculating physical processes [St 41, Fe 48].

7.4 Relativistic Scattering Theory

With (7.46, 48) the free propagator for the electron-positron field is known. Similarly to (7.14, 23, 25), we can now (at least formally) determine the complete Green's function and the S matrix for the $\mathrm{e}^+\mathrm{e}^-$ field with interaction. Indeed, below we identify and calculate some electron and positron scattering processes. For that purpose we rewrite the differential equation (7.28) for the exact propagator $S_{\mathrm{SF}}(x'-x)$ in a form analogous to (7.13) for the non-relativistic propagator:

$$(\mathrm{i}\nabla\!\!\!\!/_{x'} - m_0)\,S_{\mathrm{SF}}(x'-x) = \int d^4y\,\delta^4(x'-y)\,[\delta^4(y-x)+e\not{A}(y)\,S_{\mathrm{SF}}(y-x)]\,. \tag{7.50}$$

The source term on the rhs consists of a superposition of point sources in space-time with the strength function $[\delta^4(y-x)+e\not{A}(y)S_{SF}(y-x)]$.

As in the non-relativistic case (7.13), (7.50) is also linear. Hence we can immediately obtain by integration the following *integral equation for the exact Feynman-Stückelberg* propagator:

$$\begin{aligned} S_{SF}(x'-x) &= \int d^4y\, S_0(x'-y)[\delta^4(y-x)+e\not{A}(y)S_{SF}(y-x)] \\ &= S_0(x'-x)+e\int d^4y\, S_0(x'-y)\not{A}(y)S_{SF}(y-x)\,, \end{aligned} \tag{7.51}$$

which corresponds to (7.14) of the non-relativistic theory. This is an integral equation for the full propagator $S_{SF}(x'-x)$ in terms of the free propagator $S_0(x'-x)$, which is supposed to be known. Iteration of (7.51) leads to the expansion

$$\begin{aligned} S_{SF}(x'-x) = S_0(x'-x) &+ e\int d^4x_1 S_0(x'-x_1)\not{A}(x_1)S_0(x_1-x) \\ &+ e^2\int d^4x_1 d^4x_2 S_0(x'-x_1)\not{A}(x_1)S_0(x_1-x_2) \\ &\qquad \not{A}(x_2)S_0(x_2-x) \\ &+ \cdots\,, \end{aligned} \tag{7.52}$$

which, again, has to be seen in analogy to the non-relativistic equation (7.15). We are now in a position to denote (formally) the exact solution of the Dirac equation

$$(\mathrm{i}\not{\nabla}_x - m_0)\,\Psi(x) = e\not{A}(x)\,\Psi(x) \tag{7.53}$$

if in addition the Feynman and Stückelberg boundary conditions are requested. In complete analogy to (7.16),

$$\Psi(x) = \psi(x) + \underbrace{\int d^4y\, S_0(x-y)\,e\not{A}(y)\,\Psi(y)}_{\text{scattered wave}}\,. \tag{7.54}$$

The potential $V(x)$ in (7.16) is now replaced by $e\not{A}(y)$. The second term on the rhs is the scattered wave. Because of the special property (7.48) of the Stückelberg-Feynman propagator, it contains only positive energy solutions in the future and only negative energy solutions in the distant past. Indeed,

$$\Psi(x)-\psi(x) \;\Rightarrow\; \int d^3p \sum_{r=1}^{2} \psi_p^r(x)\left[-\mathrm{i}e\int d^4y\, \bar{\psi}_p^r(y)\not{A}(y)\,\Psi(y)\right] \tag{7.55a}$$

for $t\to+\infty$ and

$$\Psi(x)-\psi(x) \;\Rightarrow\; \int d^3p \sum_{r=3}^{4} \psi_p^r(x)\left[+\mathrm{i}e\int d^4y\, \bar{\psi}_p^r(y)\not{A}(y)\,\Psi(y)\right] \tag{7.55b}$$

for $t\to-\infty$. One should recognize that here x corresponds to x' of (7.48) and y to x of (7.48). Similarly, t of (7.55) corresponds to t' of (7.48).

Equation (7.54) formulates the relativistic scattering problem, which fulfills the requirements of hole theory contained in the Stückelberg-Feynman propagator, which – due to the special choice of the integration path (Fig. 7.6) – inherently contains the Fermi energy, i.e. which states are occupied and which not. More than that, according to (7.55a) an electron, after scattering at an external potential $A(y)$, cannot drop into the (occupied) sea of negative energy states, but can scatter only into the (unoccupied) states of positive energy. Positrons, on the other hand, which are described as electrons of negative energy moving backward in time, only scatter backwards to earlier times, into other states of negative energy.

Let us now construct the S-matrix elements. They are defined as earlier in (7.23). If $\psi_f(x)$ is the plane wave registered at the end of the scattering process with the quantum numbers f ($\varepsilon_f = 1$ for waves of positive energy in the future and $\varepsilon_f = -1$ for waves of negative energy in the past), with (7.54, 55, 48) then

$$\begin{aligned} S_{fi} &= \langle \psi_f(x) | \Psi_i(x) \rangle = \langle \psi_f(x) | \psi_i(x) + \int d^4y\, S_0(x-y) e A(y)\, \Psi_i(y) \rangle \\ &= \delta_{fi} - \mathrm{i} e\, \theta(t_x - t_y) \left\langle \psi_f(x) \Big| \int d^3p \sum_{r=1}^{2} \psi_p^r(x) \int d^4y\, \bar{\psi}_p^r(y) A(y)\, \Psi_i(y) \right\rangle \\ &\quad + \mathrm{i} e\, \theta(t_y - t_x) \left\langle \psi_f(x) \Big| \int d^3p \sum_{r=3}^{4} \psi_p^r(x) \int d^3y\, \bar{\psi}_p^r(y) A(y)\, \Psi_i(y) \right\rangle \\ &= \delta_{fi} - \mathrm{i} e \varepsilon_f \int d^4y\, \bar{\psi}_f(y) A(y)\, \Psi_i(y) \,. \end{aligned} \tag{7.56}$$

The final step summarizes the two last term of the penultimate line: depending on whether ψ_f is an electron or a positron, the first or the second of these two terms survives. In (7.56) $\Psi_i(x)$ represents that incoming wave which, according to the Stückelberg-Feynman boundary conditions, is either an incoming $\psi_i(y)$ wave of positive energy for $y_0 \to -\infty$ with the quantum numbers i ($\varepsilon_i = +1$) *or* an incoming wave of negative energy for $y_0 \to +\infty$, which propagates backward in time ($\varepsilon_i = -1$).

We shall now discuss in some detail how various scattering processes are contained in (7.56). At first let us consider the "normal" scattering process of electron scattering in which case

$$\Psi_i(y) \xrightarrow[y_0 \to -\infty]{} \psi_i^{(+E)}(y) \,, \tag{7.57}$$

where $\psi_i^{(+E)}(y)$ represents an incoming electron wave of positive energy. The nth order contribution to the S-matrix element (7.56) is then given by

$$\begin{aligned} &-\mathrm{i} e^n \int d^4y_1 \ldots d^4y_n\, \bar{\psi}_f^{(+E)}(y_n) A(y_n) S_0(y_n - y_{n-1}) A(y_{n-1}) \ldots \\ &\cdot S_0(y_2 - y_1) A(y_1)\, \psi_i^{(+E)}(y_1) \,. \end{aligned} \tag{7.58}$$

This expression also contains graphs of type

and ,

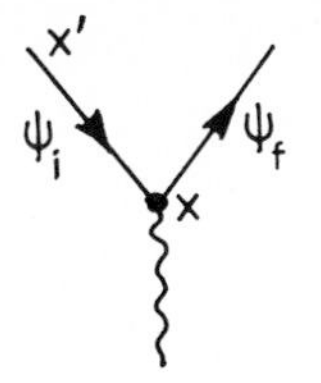

Fig. 7.8. Element of the graph for pair creation

i.e. both electron scattering and also intermediate pair creation can occur, because the various integrations $\int d^4y_n \ldots$ also contain time orders $y^0_{n+1} < y^0_n$. Hence contributions from the second part of the propagator (7.48) are possible.

Let us shortly describe how pair creation arises in this formalism. The initial wave function $\Psi_i(y)$ must approach a plane wave of negative energy for $y_0 \to +\infty$. This state, propagating backward in time, represents a positron.

Let

p_-, s_- and

p_+, s_+

denote momentum and spin for electron and positron, respectively. For $t \to +\infty$ the positron wave function must be represented by a plane wave of negative energy with the quantum numbers $(-p_+, -s_+, \varepsilon = -1)$. This wave, propagating backward in time, is "incoming" into the vertex, i.e.

$$\Psi_i(y) \xrightarrow[y_0 \to \infty]{} \psi_i^{(-E)}(y) = \sqrt{\frac{m_0}{E_+}} \frac{1}{\sqrt{2\pi}^3} v(p_+, s) \, \mathrm{e}^{+\mathrm{i}p_+x} . \tag{7.59}$$

The final wave function $\bar{\psi}_f$ in the pair creation process (Fig. 7.8) is a solution of positive energy with the quantum numbers $(p_-, s_-, \varepsilon = +1)$, which describes an electron.

Let us once more repeat the essential ideas: from hole theory it is known that a missing electron of negative energy, four-momentum $-p_+$, spin $-s_+$ has to be interpreted as a positron with four-momentum $+p_+$ and polarization $+s_+$. In the propagator formalism discussed here we calculate the amplitude for creating a positron at x and its propagation forward in time into the free state (p_+, s_+) at x' by computing the amplitude of an electron of negative energy (with four-momentum $-p_+$, spin $-s_+$) propagating backward in time from x' to x, where it is scattered into a state of positive energy, propagating from thereon forward in time. Figure 7.9 summarizes pair creation in first and second order. Note especially that two diagrams exist for the second-order process, one with an additional electron scattering and one with an additional positron scattering. These two diagrams are different in the time ordering of the two scattering processes.

Let us now turn to *pair annihilation*. It is in lowest order represented by the graph in Fig. 7.10. Here $\Psi_i(y)$ must be a solution of (7.55) which for $t \to -\infty$ approaches $\psi_i^{(+E)}(y)$. This represents an electron of positive energy, propagating forward in time, as the arrow in Fig. 7.10 indicates. At the interaction point

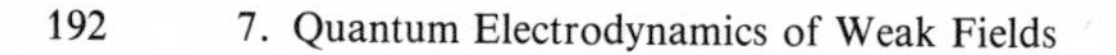

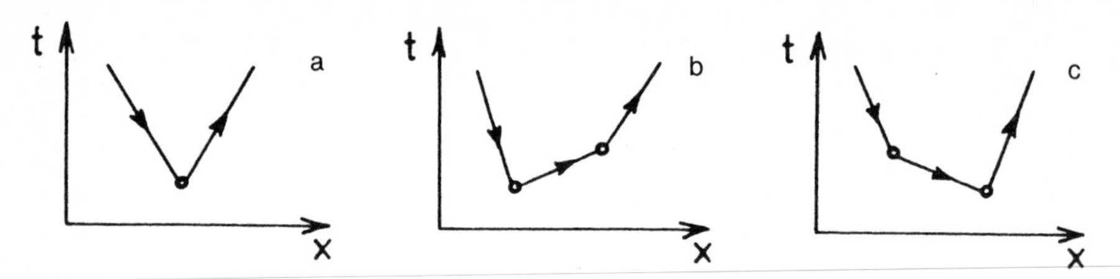

Fig. 7.9 a – c. Feynman diagrams in space and time for pair creation: (**a**) pair creation in lowest order; (**b**) and (**c**) are second-order diagrams with different time ordering. In (**b**) the electron undergoes an additional scattering; in (**c**) the positron scatters a second time

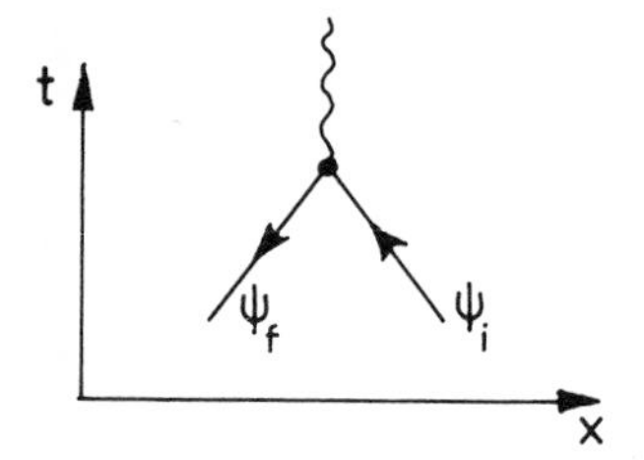

Fig. 7.10. Pair annihilation in lowest order

(vertex) it is scattered into a state of negative energy, propagating backwards in time. Because of (7.56), the pair annihilation amplitude in nth order for scattering into the given final state $\psi_f^{(-E)}$ (with quantum numbers $\boldsymbol{p}_+$, s_+, $\varepsilon_f = -1$) is given by

$$\mathrm{i}e^n \int d^4y_1 \ldots d^4y_n \, \bar{\psi}_f^{(-E)}(y_n) \not{A}(y_n) \, S_0(y_n - y_{n-1}) \ldots \not{A}(y_1) \, \psi_i^{(+E)}(y_1) \, . \tag{7.60}$$

In hole theory this amplitude of nth order describes the scattering of an electron of positive energy into a state of negative energy with momentum $-\boldsymbol{p}_+$ and spin $-s_+$. Clearly, this state has to be empty for $t \to -\infty$. This simply means that a hole (i.e. a positron) with four-momentum p_+ and polarization s_+ has to be available for $t \to -\infty$. The difference in sign of the total amplitude (7.60) relative to the amplitude (7.58) originates in the different signs of the electron and positron parts of the Stückelberg-Feynman propagator (7.48) [cf. (7.55, 56)].

Finally we consider *positron scattering*, which is in lowest order represented by Fig. 7.11. The incoming wave is a positron with quantum numbers ($\boldsymbol{p}'_+$, s'_+, $\varepsilon_i = -1$) represented by an electron of negative energy with quantum numbers ($\boldsymbol{p}'_+$, $-s'_+$, $\varepsilon_i = -1$) and the final state (outcoming wave) is also represented by an electron of negative energy with quantum numbers ($-\boldsymbol{p}_+$, $-s_+$, $\varepsilon_f = -1$). Recognize that the outgoing positron (Fig. 7.11 b) is to be identified with the incoming electron of negative energy, etc.

In Fig. 7.11 b a positron propagates forward in time in the incoming state $\psi_i(\boldsymbol{p}'_+, s'_+)$ and is scattered into the outgoing state $\psi_f(\boldsymbol{p}_+, s_+)$. In hole theory the outgoing positron $\psi_f(\boldsymbol{p}_+, s_+)$ is interpreted as an incoming electron of negative energy $\psi_i^{(-E)}(-\boldsymbol{p}_+, -s_+)$ (propagating backward in time), and scattered into an outgoing electron of negative energy $\psi_f^{(-E)}(-\boldsymbol{p}'_+, -s'_+)$ (also propagating backward in time).

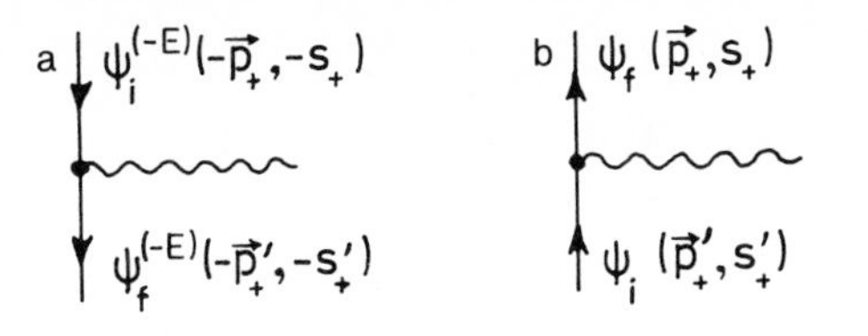

Fig. 7.11 a, b. Positron scattering in lowest order, representable in two ways: (**a**) hole theory interpretation, (**b**) positron interpretation

Figure 7.11 a describes the process from the calculational point of view, while (b) describes the real physics, which deals with positrons and not with electrons of negative energy.

For clarity the wave functions entering the calculation (Fig. 7.11 a) are denoted by

$$\psi_i^{(-E)}(-\boldsymbol{p}_+,-s_+)=\sqrt{\frac{m_0}{E}}\frac{1}{\sqrt{2\pi}^3}v(p_+s_+)\mathrm{e}^{+\mathrm{i}p_+\cdot x}$$

$$\psi_f^{(-E)}(-\boldsymbol{p}'_+,-s'_+)=\sqrt{\frac{m_0}{E}}\frac{1}{\sqrt{2\pi}^3}v(p'_+s'_+)\mathrm{e}^{+\mathrm{i}p'_+\cdot x}.$$

The negative energy $(-E)$, negative momentum $(-\boldsymbol{p}_+)$ and negative spin $(-s_+)$ are taken care of by the negative exponent in the plane waves $\{\exp[-\mathrm{i}(-p_+)\cdot x] = \exp(+\mathrm{i}p_+\cdot x)\}$ and in the construction of $v(p_+s_+)=\omega^4(p_+s_+)$, $v(p_+-s_+) = \omega^3(p_+s_+)$.

Bibliographical Notes

Standard QED of weak fields is dealt with in many books [Bj 64, Be 71, Sc 79, Ba 80b]. The original papers by *Stückelberg* [St 41] and *Feynman* [Fe 48] are recommended for additional reading. A more detailed, recent presentation of perturbative QED can be found in [Gr 84b]. Propagator methods, which are exact in the external fields but perturbative in α, are discussed in Chap. 15.

8. The Classical Dirac Field Interacting with a Classical Electromagnetic Field – Formal Properties

Before giving the second quantization of the field of a Dirac particle, we must deepen our understanding of the formal properties of the Dirac equation. This is the subject of Chap. 8 which consequently contains a disproportionate amount of formal developments compared to other chapters. We start by casting the formalism of the Dirac equation into Hamiltonian form, because this is necessary for canonical quantization. The canonical formalism leads to the conservation laws of the classical Dirac theory, such as charge and energy conservation. The discussion continues with how the general solution of the Dirac equation can be represented by a superposition of eigenfunctions of the Dirac Hamiltonian. The extension to time-dependent external potentials leads to the time-evolution operator and the single-particle scattering matrix. Finally, we discuss the various elementary functions or propagators of the Dirac equation, which are crucial in the theory of the quantized Dirac field.

Concerning our notation, the circumflex ceases to denote the Hilbert space operators in the remainder of the book, since the circumflex will now indicate Fock space, i.e. second quantization, operators starting from Chap. 9. The concurrent use of two operator symbols would make the equations too complicated. We henceforth use natural units $\hbar = c = 1$.

8.1 Field Equations in Hamiltonian Form

The Dirac equation for a relativistic electron in an external electromagnetic field $A_\mu(x)$ can be derived from the Lagrangian density

$$\mathscr{L}_{\mathrm{D}} = \frac{\mathrm{i}}{2}\left[\bar{\psi}(x)\,\gamma^\mu \frac{\partial \psi}{\partial x^\mu} - \frac{\partial \bar{\psi}}{\partial x^\mu}\,\gamma^\mu \psi(x)\right] - \mathrm{e}\,\bar{\psi}(x)\,\gamma^\mu A_\mu(x)\,\psi(x) - m_e \bar{\psi}(x)\,\psi(x) \tag{8.1}$$

by invoking the action principle

$$\delta \int d^4x\, \mathscr{L}_{\mathrm{D}}(\bar{\psi}, \psi; x) = 0\,. \tag{8.2}$$

The γ matrices are those defined in Sect. 2.1, and $\bar{\psi}$ is introduced as

$$\bar{\psi}(x) = \psi^\dagger(x)\,\gamma^0 \tag{8.3}$$

to make $\int d^4x\, \mathscr{L}_D$ self-adjoint. This choice is necessary because the γ matrices are not Hermitian, but obey

$$\gamma^{\mu\dagger} = \gamma^0\gamma^\mu\gamma^0, \quad \text{i.e.} \quad \gamma^{0\dagger} = \gamma^0, \quad \boldsymbol{\gamma}^\dagger = -\boldsymbol{\gamma}, \tag{8.4}$$

which is easily verified via (2.75, 76). Equation (8.3) gives

$$[\bar\psi\gamma^\mu\psi]^\dagger = \psi^\dagger\gamma^{\mu\dagger}\bar\psi^\dagger = \psi^\dagger\gamma^0\gamma^\mu\gamma^0\gamma^0\psi = \bar\psi\gamma^\mu\psi\,. \tag{8.5}$$

Since $\psi(x)$ is a complex field, the variation is understood to be carried out independently with respect to ψ and $\bar\psi$. Variation with respect to $\bar\psi$ gives the Dirac equation

$$\gamma^\mu\left(\mathrm{i}\frac{\partial}{\partial x^\mu} - eA_\mu\right)\psi - m_e\psi = 0\,, \tag{8.6}$$

and variation with respect to ψ gives the conjugate equation

$$\left(\mathrm{i}\frac{\partial}{\partial x^\mu} + eA_\mu\right)\bar\psi\gamma^\mu + m_e\bar\psi = 0\,. \tag{8.7}$$

The part of $\mathscr{L}_D$ describing the interaction of the Dirac field with the electromagnetic field can be written as $(-j^\mu A_\mu)$ with the Dirac current four-vector

$$j^\mu = e\bar\psi\gamma^\mu\psi\,. \tag{8.8}$$

It is easy to show that this current satisfies a continuity equation. To do this, (8.6) is multiplied by $\bar\psi$ from the left, (8.7) by ψ from the right and both equations are added:

$$\bar\psi\gamma^\mu\mathrm{i}\left(\frac{\partial\psi}{\partial x^\mu}\right) + \mathrm{i}\left(\frac{\partial\bar\psi}{\partial x^\mu}\right)\gamma^\mu\psi = \mathrm{i}\frac{\partial}{\partial x^\mu}(\bar\psi\gamma^\mu\psi) = 0\,. \tag{8.9}$$

It is clear from these considerations that $j^0 = e\bar\psi\gamma^0\psi = e\psi^\dagger\psi$, the electronic charge density, is not an invariant but is the time-like component of a vector. On the other hand, $\bar\psi\psi$ *is* a relativistic invariant, but is not directly related to an observable physical quantity.

To go over to a Hamiltonian formalism, the momenta conjugate to ψ and $\bar\psi$ are defined as usual:

$$\Pi_\psi(\boldsymbol{x},t) = \frac{\delta}{\delta\dot\psi(\boldsymbol{x},t)}\int d^3y\,\mathscr{L}_D(\bar\psi,\psi;\boldsymbol{y},t) = \frac{\mathrm{i}}{2}\bar\psi(\boldsymbol{x},t)\gamma^0 = \frac{\mathrm{i}}{2}\psi^\dagger(\boldsymbol{x},t) \tag{8.10a}$$

$$\Pi_{\bar\psi}(\boldsymbol{x},t) = \frac{\delta}{\delta\dot{\bar\psi}(\boldsymbol{x},t)}\int d^3y\,\mathscr{L}_D(\bar\psi,\psi;\boldsymbol{y},t) = -\frac{\mathrm{i}}{2}\gamma^0\psi(\boldsymbol{x},t)\,. \tag{8.10b}$$

As the canonical momenta do not contain the time derivatives of ψ and $\bar{\psi}$, it is impossible to express $\dot{\psi}$ and $\dot{\bar{\psi}}$ in terms of Π_ψ or $\Pi_{\bar{\psi}}$. Despite this, the Hamiltonian density turns out to be independent of $\dot{\psi}$ and $\dot{\bar{\psi}}$, as straightforward calculation shows:

$$\begin{aligned}\mathscr{H} &= \Pi_\psi \dot{\psi} + \Pi_{\bar{\psi}} \dot{\bar{\psi}} - \mathscr{L}_\mathrm{D} \\ &= \frac{\mathrm{i}}{2} \bar{\psi} \gamma^0 \dot{\psi} - \frac{\mathrm{i}}{2} \dot{\bar{\psi}} \gamma^0 \psi - \frac{\mathrm{i}}{2} \bar{\psi} \gamma^\mu \frac{\partial \psi}{\partial x^\mu} + \frac{\mathrm{i}}{2} \frac{\partial \bar{\psi}}{\partial x^\mu} \gamma^\mu \psi + e \bar{\psi} \gamma^\mu A_\mu \psi + m_e \bar{\psi} \psi \\ &= \tfrac{1}{2} [\bar{\psi} \boldsymbol{\gamma} \cdot (-\mathrm{i} \nabla \psi) + (\mathrm{i} \nabla \bar{\psi}) \cdot \boldsymbol{\gamma} \psi] + e \bar{\psi} \gamma^\mu A_\mu \psi + m_e \bar{\psi} \psi \,. \end{aligned} \tag{8.11}$$

We may even utilize (8.10) to eliminate $\bar{\psi}$ and $\Pi_{\bar{\psi}}$ from the Hamiltonian density, such that only one independent field function $\psi(\boldsymbol{x}, t)$ (i.e. one for every space and spin variable) and its conjugate momentum $\Pi \equiv 2\Pi_\psi = \mathrm{i}\,\psi^\dagger$ remain:

$$\begin{aligned}\mathscr{H}(x) &= \tfrac{1}{2} [-\Pi(x) \gamma^0 \boldsymbol{\gamma} \cdot \nabla \psi(x) + (\nabla \Pi(x)) \gamma^0 \cdot \boldsymbol{\gamma} \psi(x)] \\ &\quad - \mathrm{i} \Pi(x) \gamma^0 (e \gamma^\mu A_\mu(x) + m_e) \psi(x) \,. \end{aligned} \tag{8.12}$$

For practical purposes it is often useful to separate the term describing the interaction with the electromagnetic field into a term involving the Coulomb potential V and one containing the vector potential $\boldsymbol{A}$, where $eA_\mu = (V, -e\boldsymbol{A})$. We further introduce the non-covariant Dirac matrices $\beta = \gamma^0$ and $\boldsymbol{\alpha} = \gamma^0 \boldsymbol{\gamma}$, enabling the Hamiltonian function in its final form to be written:

$$\begin{aligned}H = \int d^3x\, \mathscr{H}(x) &= \int d^3x \{\tfrac{1}{2} (\nabla + \mathrm{i} e \boldsymbol{A}(x)) \Pi(x) \cdot \boldsymbol{\alpha} \psi(x) \\ &\quad - \tfrac{1}{2} \Pi(x) \boldsymbol{\alpha} \cdot (\nabla - \mathrm{i} e \boldsymbol{A}(x)) \psi(x) - \mathrm{i} \Pi(x) (\beta m_e + V(x)) \psi(x)\} \,. \end{aligned} \tag{8.13}$$

This form of the Dirac Hamiltonian is explicitly Hermitian, because (in contrast to the γ matrices) the $\boldsymbol{\alpha}$ matrices are Hermitian (4×4) matrices as shown in (2.11). The equations of motion follow as usual by varying the Hamiltonian with respect to the Dirac field ψ and its conjugate momentum field Π. For the Dirac equation then

$$\begin{aligned}\mathrm{i} \dot{\psi}(\boldsymbol{x}, t) &= \mathrm{i} \frac{\delta}{\delta \Pi(\boldsymbol{x}, t)} \int d^3y\, \mathscr{H}(\psi(y), \Pi(y); y, t) \\ &= \mathrm{i} \int d^3y \{\tfrac{1}{2} (\nabla_y + \mathrm{i} e \boldsymbol{A}(y)) \delta^3(x - y) \cdot \boldsymbol{\alpha} \psi(y) \\ &\quad - \tfrac{1}{2} \delta^3(x - y) \boldsymbol{\alpha} \cdot (\nabla_y - \mathrm{i} e \boldsymbol{A}(y)) \psi(y) - \mathrm{i} \delta^3(x - y) (\beta m_e + V(y)) \psi(y)\} \\ &= [\boldsymbol{\alpha} \cdot (-\mathrm{i} \nabla_x - e \boldsymbol{A}(x)) + V(x) + \beta m_e] \psi(\boldsymbol{x}, t) \\ &\equiv H_\mathrm{D} \psi(\boldsymbol{x}, t) \,, \end{aligned} \tag{8.14}$$

while the adjoint equation follows from

$$\dot{\Pi}(\boldsymbol{x},t) = -\frac{\delta}{\delta\psi(\boldsymbol{x},t)}\int d^3y\,\mathscr{H}(\psi(y),\Pi(y);y,t)$$
$$= -\mathrm{i}[(-\mathrm{i}\nabla+e\boldsymbol{A})\Pi(\boldsymbol{x},t)\cdot\boldsymbol{\alpha}-\Pi(\boldsymbol{x},t)(V+\beta m_e)] \tag{8.15}$$

on substituting $\Pi=\mathrm{i}\,\psi^\dagger$.

We have thus succeeded in casting the Dirac theory into a Hamiltonian form involving canonical field variables, which is particularly suited for quantization of the field theory. Before proceeding to the quantized theory, however, we shall explore the properties of the classical theory further, especially its conservation laws and propagation of solutions of classical field equations.

8.2 Conservation Laws

In a dynamical theory that can be derived from an action principle, every invariance under a symmetry operation signals the existence of a conservation law. This fact is known as *Noether's* theorem [No 18, Hi 51]. To account fully for the effects of a symmetry operation, such as translation, rotation, etc., the Dirac Lagrangian (8.1) must be supplemented by the Lagrangians for the free electromagnetic field and the external sources which produce the field:

$$\mathscr{L}_{\mathrm{tot}} = \mathscr{L}_{\mathrm{D}} - \tfrac{1}{4}F_{\mu\nu}F^{\mu\nu} - J^{\mu}_{\mathrm{ext}}A_\mu + \mathscr{L}_{\mathrm{ext}}\,, \tag{8.16}$$

where

$$F_{\mu\nu} = \frac{\partial}{\partial x^\mu}A_\nu - \frac{\partial}{\partial x^\nu}A_\mu \tag{8.17}$$

are the electromagnetic field strengths. In our notation, $F_{\mu\nu}F^{\mu\nu}/2=\boldsymbol{B}^2-\boldsymbol{E}^2$. By assuming that the total action $\int d^4x\,\mathscr{L}_{\mathrm{tot}}$ is invariant under arbitrary operations of some kind, one can show that a conserved quantity exists [Bj 65]. The advantage of this Lagrangian approach is that it provides a constructive derivation of the conserved quantities, but the drawback is that it does not form a suitable basis for canonical quantization. Relativistic quantum field theory can also be derived from a quantum action principle [Schw 53], but this approach is less intuitive. The Lagrangian also plays a central role in quantization through functional integrals [Fe 65, It 80, Ra 81].

As we are more interested in the quantized theory than in the classical one, at least as concerns the Dirac field, the Hamiltonian derivation of a conservation law is discussed already on the classical level. The example is the conservation of electrical charge or, being equivalent at this level of theory, of fermion number.

Assume that $C(x)$ is some physical, space- and time-dependent quantity built up of bilinear expressions in the Dirac field. Classical Hamiltonian theory then states that the time variation of $C(x,t)$ is given by the classical Poisson bracket of C with the Dirac Hamiltonian function (8.13) plus a possible explicit time dependence caused by the external electromagnetic field:

$$\frac{d}{dt}C(\boldsymbol{x},t) = \frac{\partial}{\partial t}C(\boldsymbol{x},t) + \{H, C(\boldsymbol{x},t)\}_{\text{Poisson}}$$

$$\equiv \frac{\partial C(\boldsymbol{x},t)}{\partial t} + \int d^3z \left(\frac{\delta H(t)}{\delta \psi(z,t)} \frac{\delta C(\boldsymbol{x},t)}{\delta \Pi(z,t)} - \frac{\delta C(\boldsymbol{x},t)}{\delta \psi(z,t)} \frac{\delta H(t)}{\delta \Pi(z,t)} \right) . \tag{8.18}$$

The variational derivatives of H have already been calculated in (8.14, 15):

$$\frac{\delta H}{\delta \psi(z,t)} = \mathrm{i}[(-\mathrm{i}\nabla + e\boldsymbol{A})\,\Pi(z,t)\cdot\boldsymbol{\alpha} - \Pi(z,t)(V+\beta m_e)] ,$$

$$\frac{\delta H}{\delta \Pi(z,t)} = \frac{1}{\mathrm{i}}[\boldsymbol{\alpha}\cdot(-\mathrm{i}\nabla - e\boldsymbol{A}) + V + \beta m_e]\,\psi(z,t) .$$

If the resulting expression is the divergence of a vector that can be rightfully assumed to vanish at the boundary of the system under consideration (i.e. usually at "spatial infinity"), then $\int d^3x\, C(\boldsymbol{x},t)$ does not change with time.

To illustrate this we take for C the electric charge density

$$j^0(\boldsymbol{x},t) = e\bar{\psi}\gamma^0\psi = \mathrm{i}e\Pi(\boldsymbol{x},t)\,\psi(\boldsymbol{x},t) . \tag{8.19}$$

Evidently, j^0 does not contain an explicit time dependence.

The variations with respect to ψ and Π yield:

$$\frac{\delta j^0(\boldsymbol{x},t)}{\delta \psi(z,t)} = \mathrm{i}e\Pi(z,t)\,\delta^3(x-z)$$

$$\frac{\delta j^0(\boldsymbol{x},t)}{\delta \Pi(z,t)} = \mathrm{i}e\psi(z,t)\,\delta^3(x-z) .$$

With these results, the Poisson bracket is

$$\begin{aligned}\{H, j^0(\boldsymbol{x},t)\}_{\text{Poisson}} &= \int d^3z\,[\mathrm{i}(-\mathrm{i}\nabla + e\boldsymbol{A})\,\Pi(z,t)\cdot\boldsymbol{\alpha} - \Pi(z,t)(V+\beta m_e)] \\ &\quad \cdot \mathrm{i}e\psi(z,t)\,\delta(\boldsymbol{x}-z) - \int d^3z\,\mathrm{i}e\Pi(z,t)\,\delta(\boldsymbol{x}-z) \\ &\quad \cdot \frac{1}{\mathrm{i}}[\boldsymbol{\alpha}\cdot(-\mathrm{i}\nabla - e\boldsymbol{A}) + V + \beta m_e]\,\psi(z,t) \\ &= \mathrm{i}e\nabla\cdot[\Pi(\boldsymbol{x},t)\,\boldsymbol{\alpha}\,\psi(\boldsymbol{x},t)] = -e\nabla\cdot(\psi^\dagger\boldsymbol{\alpha}\psi) .\end{aligned} \tag{8.20}$$

The total charge is defined as

$$Q(t) = \int d^3x\, j^0(\boldsymbol{x},t) . \tag{8.21}$$

Therefore $dQ/dt = 0$ if no current flows through the boundary of the integration volume. That this conservation law can be derived from the Hamiltonian of the Dirac field alone is connected with the fact that the electromagnetic field does not carry charge. In more general cases, the full Hamiltonian as derived from the total Lagrangian $\mathcal{L}_{\text{tot}}$ (8.16) must be substituted into the Poisson bracket. An example is the Dirac field interaction with a Yang-Mills field [It 80].

Charge conservation is a consequence of the gauge invariance of the theory. Other invariance properties are those with respect to translations and Lorentz transformations (including spatial rotations). Invariance of the total Lagrangian with respect to translations conserves energy and momentum:

$$\partial_\nu T^\nu_\mu = 0 \quad \text{with} \tag{8.22}$$

$$T^\nu_\mu = T^\nu_\mu(\text{Dirac}) + T^\nu_\mu(\text{EM}) + T^\nu_\mu(\text{ext}) \quad \text{and} \tag{8.23a}$$

$$\begin{aligned} T_{\mu\nu}(\text{Dirac}) = {} & \frac{\mathrm{i}}{4}[\bar\psi\gamma_\mu\partial_\nu\psi + \bar\psi\gamma_\nu\partial_\mu\psi - (\partial_\nu\bar\psi)\gamma_\mu\psi - (\partial_\mu\bar\psi)\gamma_\nu\psi] \\ & - \frac{e}{2}(\bar\psi\gamma_\mu A_\nu\psi + \bar\psi\gamma_\nu A_\mu\psi) - g_{\mu\nu}\mathcal{L}_{\text{D}}, \end{aligned} \tag{8.23b}$$

$$T_{\mu\nu}(\text{EM}) = F_{\mu\lambda}F^{\nu\lambda} - \tfrac{1}{4}g_{\mu\nu}F_{\varkappa\lambda}F^{\varkappa\lambda}. \tag{8.23c}$$

Here $T^\nu_\mu(\text{ext})$ stands for the energy-momentum tensor of the source of the external field. As with every local quantity satisfying a continuity equation, the definition of T^ν_μ is not unique. Indeed, an arbitrary quantity whose four-divergence vanishes identically can be added. Further conditions must be imposed for a unique definition. Our choice has been motivated by the wish that T^ν_μ as well as its three contributions be gauge invariant, a necessary requirement for T^ν_μ to be an observable quantity.

The energy-momentum tensor is in fact an observable, i.e. measurable, quantity because it is the source of graviation. It enters into the Einstein equations for the potential field [Ad 65]

$$R_{\mu\nu} - \tfrac{1}{2}g_{\mu\nu}R = 8\pi\varkappa T_{\mu\nu}. \tag{8.24}$$

Since the space-time metric does not depend on the electromagnetic gauge, the tensor $T_{\mu\nu}$ must also be gauge invariant.

Equation (8.24) may be obtained from the graviational action principle $\int d^4x\sqrt{-g}(R - 16\pi\varkappa\mathcal{L})$, where $\mathcal{L}$ is the matter Lagrangian and R the curvature scalar, by variation with respect to the metric tensor $g_{\mu\nu}$. Thus a gauge-invariant definition of $T_{\mu\nu}$ arises:

$$T^{\mu\nu} = \frac{\partial\mathcal{L}}{\partial g_{\mu\nu}} - g^{\mu\nu}\mathcal{L}, \tag{8.25}$$

which gives the result (8.23) for the coupled Dirac-Maxwell fields [Schm 68].

For practical purposes, it is useful to separate from T^ν_μ a part depending solely on the external source, and to assume that the source is not affected by the presence of the Dirac particle (this is a rather good approximation for an electron in the field of a nucleus, but already worse for a muon which can cause considerable polarization of the nucleus). Now $F^{\mu\nu} = F^{\mu\nu}_{\text{ext}} + f_{\mu\nu}$ and $F^{\mu\nu}_{\text{ext}}$ is defined to be the "bare" external field produced by the external source alone in the absence of coupling to the Dirac field ψ. Due to the linearity of the Maxwell equations, the two contributions satisfy independent equations:

$$\frac{\partial}{\partial x^\nu} F^{\mu\nu}_{\text{ext}} = J^\mu_{\text{ext}} , \qquad \frac{\partial}{\partial x^\nu} f^{\mu\nu} = e\bar{\psi}\gamma^\mu\psi . \tag{8.26}$$

We can decompose $T^\nu_\mu(\text{EM})$ into three terms:

$$T^\nu_\mu(\text{EM}) = T^\nu_\mu(\text{EM, ext}) + T^\nu_\mu(\text{EM, self}) + T^\nu_\mu(\text{EM, int}) . \tag{8.27}$$

The first term contains only $F^{\mu\nu}_{\text{ext}}$, the second only $f_{\mu\nu}$, while the last term stands for all expressions mixed in $F^{\mu\nu}_{\text{ext}}$ and $f^{\mu\nu}$.

In the external field approach one assumes that the source, at least in first approximation, is not affected by the presence of the Dirac particle, and therefore $T^\nu_\mu(\text{ext})$ and $T^\nu_\mu(\text{EM, ext})$ remain unchanged. In the physical system we have in mind, i.e. electrons in the Coulomb field of heavy nuclei, this condition is satisfied to a high degree. The electromagnetic energy density of the source is of the order $Z^2\alpha/R^4$, where R is the nuclear radius, while the interaction energy density of an electronic K-shell wave function is approximately $Z\alpha(Z\alpha m_e)^4$. Even for $Z\alpha \sim 1$ the electronic energy density is smaller by many orders of magnitude because the nuclear radius R is much smaller than the electron Compton wavelength m_e^{-1}. We shall assume, moreover, that the external field is much stronger than the field generated by the particle itself, allowing us to neglect $T^\nu_\mu(\text{EM, self})$. For an atom this condition requires that $Z \gg 1$. As the only component of T^ν_μ that will be of further interest to us is the energy density, we now discuss the case $\mu = \nu = 0$ in the following. In this case we have:

$$\begin{aligned} T^0_0(\text{Dirac}) &= \frac{\mathrm{i}}{2}(\bar{\psi}\gamma^0\dot{\psi} - \dot{\bar{\psi}}\gamma^0\psi) - e\bar{\psi}\gamma^0 A^{\text{ext}}_0\psi - \mathcal{L}_{\text{D}} \\ &= -\frac{\mathrm{i}}{2}[\bar{\psi}\boldsymbol{\gamma}\cdot\nabla\psi - (\nabla\bar{\psi})\cdot\boldsymbol{\gamma}\psi] - e\bar{\psi}\boldsymbol{\gamma}\cdot\boldsymbol{A}_{\text{ext}}\psi + m_e\bar{\psi}\psi \end{aligned} \tag{8.28a}$$

$$T^0_0(\text{EM, int}) = \boldsymbol{E}_{\text{ext}}\cdot\boldsymbol{e} + \boldsymbol{B}_{\text{ext}}\cdot\boldsymbol{b} , \tag{8.28b}$$

where small letters denote the fields generated by the Dirac particle.

Using the equations connecting the field strengths with the potential

$$\boldsymbol{E}_{\text{ext}} = -\nabla A^{\text{ext}}_0 - \dot{\boldsymbol{A}}_{\text{ext}} , \qquad \boldsymbol{B}_{\text{ext}} = \nabla\times\boldsymbol{A}_{\text{ext}} , \tag{8.29a}$$

and the second equation (8.26)

$$\nabla\cdot\boldsymbol{e} = e\bar{\psi}\gamma^0\psi , \tag{8.29b}$$

it is seen that the total energy density of the Dirac field is related to the Hamiltonian density (8.11) by

$$T^0_0(\text{Dirac}) + T^0_0(\text{EM, int}) = \mathcal{H} - \nabla\cdot(A^{\text{ext}}_0\boldsymbol{e}) - \dot{\boldsymbol{A}}_{\text{ext}}\cdot\boldsymbol{e} + (\nabla\times\boldsymbol{A}_{\text{ext}})\cdot\boldsymbol{b} . \tag{8.30}$$

The difference between T_0^0 and $\mathscr{H}$ is discussed in [We 49].

If the external field is purely electrostatic, as for an electron in the field of a heavy nucleus, then the total energy of the Dirac particle differs from the Hamiltonian function H, (8.13), only by a surface term:

$$W_{\mathrm{D}} = \int d^3x\,[T_0^0(\mathrm{Dirac}) + T_0^0(\mathrm{EM, int})] = H - \oint d\sigma\, A_0^{\mathrm{ext}}\boldsymbol{n}\cdot\boldsymbol{e}\,. \tag{8.31}$$

The integral is over a remote surface "at infinity" and $\boldsymbol{n}$ is the unit normal vector on the surface. The most important case is where $eA_0^{\mathrm{ext}} = V$ is generated by a localized source such as a nucleus. Then V becomes constant at infinity and by virtue of Gauss' theorem together with (8.29b)

$$W_{\mathrm{D}} = H - V(\infty)\,\mathscr{N}\,, \tag{8.32}$$

where $\mathscr{N}$ is the norm of the wave function $\int d^3x\,\psi^\dagger\psi$. The constant $V(\infty)$ may, of course, be gauged to zero, but it is reassuring to see that this arbitrary gauge dependence is taken care of in the definition of the total energy W_{D}. The Hamiltonian function, on the other hand, is explicitly gauge dependent! One may finally employ the operator H_{D} from (8.14) to write the total energy as the expectation value of the operator $H_{\mathrm{D}} - V(\infty)$ valid for localized wave functions:

$$W_{\mathrm{D}} = \int d^3x\,\psi^\dagger[H_{\mathrm{D}} - V(\infty)]\,\psi\,. \tag{8.33}$$

This expression is most useful if ψ is an eigenfunction of H_{D}.

We conclude with a comment on (8.26). In quantum field theory, the interaction between electrons and the electromagnetic field leads to the existence of a "vacuum polarization" four-current density $(e\bar{\psi}\gamma^\mu\psi)_{\mathrm{vac}}$ in the absence of any real electron or positron. This polarization current (or charge) produces a contribution $f_{\mathrm{vac}}^{\mu\nu}$ to the electromagnetic field around the source that cannot be eliminated even if all Dirac particles were removed from the vicinity of the source. The first part of (8.26) therefore defines, strictly speaking, the *bare* external field $F_{\mathrm{ext,\,bare}}^{\mu\nu}$ which must be distinguished from the *effective* external field $F_{\mathrm{ext,\,eff}}^{\mu\nu} = F_{\mathrm{ext,\,bare}}^{\mu\nu} + f_{\mathrm{vac}}^{\mu\nu}$. While the effective external field is experimentally observable, the bare external field is a fictitious, albeit formally useful, construction. With the bare field the discussion following (8.26) completely carries over to the second quantized Dirac theory, if the rhs (8.26) is replaced by the expectation value of the Dirac current operator in the true state of the system, including the vacuum expectation value.

8.3 Representation by Energy Eigenmodes

It is obviously impossible to give the explicit general solution of the Dirac equation for arbitrary potentials $V(\boldsymbol{x}, t)$, $\boldsymbol{A}(\boldsymbol{x}, t)$. For many purposes, however, it is important to have a representation of the general solution in terms of a suitably chosen, complete set of basis wave functions.

8.3.1 Time-Independent Potentials

If the external field is time independent, we may seek stationary solutions of (8.14) in the form

$$\psi_n(\boldsymbol{x},t)=\varphi_n(\boldsymbol{x})\exp(-\mathrm{i}E_n t)\,, \tag{8.34}$$

where the functions $\varphi_n(\boldsymbol{x})$ are eigenmodes of the operator

$$H_{\mathrm{D}}=\boldsymbol{\alpha}\cdot(-\mathrm{i}\nabla-e\boldsymbol{A})+V+\beta m_e\,. \tag{8.35}$$

For weak, i.e. non-singular and subcritical, potentials $V(\boldsymbol{x})$ this operator is self-adjoint and has only real "frequency" eigenvalues E_n. The eigenvalues can be either positive or negative, as amply discussed above. Since H_{D} is self-adjoint, the eigenmodes form a complete orthonormal system of solutions $\{\varphi_n\}$.

For a normal hydrogen-like atom, a discrete set of states φ_n exists, representing the bound states, and two continua φ_E, with $E>m_e c^2$ and $E<-m_e c^2$, representing the scattering states for electrons and positrons, respectively. For every value of the continuum energy E, there is an infinite number of degenerate solutions that here may be classified according to angular momentum quantum numbers. The degeneracy can be resolved at least partially by introducing boundary conditions at a finite surface (box or sphere quantization). The orthogonality and completeness relations for the eigenfunctions are:

$$\int d^3x\,\varphi_n^\dagger(\boldsymbol{x})\,\varphi_m(\boldsymbol{x})=\delta_{nm} \tag{8.36}$$

$$\sum_n \varphi_n^{(\alpha)}(\boldsymbol{x})\,\varphi_n^{\dagger(\beta)}(\boldsymbol{x}')=\delta_{\alpha\beta}\delta(\boldsymbol{x}-\boldsymbol{x}')\,. \tag{8.37}$$

The spinor indices α and β in (8.37) have been explicitly denoted, while in (8.36) summation over spinor indices is simplicitly understood. Every physical wave function can be expanded in terms of these states φ_n, so that the general solution of the Dirac equation for a time-independent potential reads:

$$\psi(\boldsymbol{x},t)=\sum_n a_n\varphi_n(\boldsymbol{x})\exp(-\mathrm{i}E_n t) \tag{8.38}$$

with constant coefficients a_n.

The simplest case is that of vanishing potential, i.e. the free electron or positron, in which there are no bound states, and the scattering states can be classified according to the particle momentum and spin orientation. These plane wave solutions are given in (2.50). In many cases, however, it is more convenient to expand in other complete sets of wave functions, in particular when strong external fields are present.

8.3.2 Explicitly Time-Dependent Potentials

In the typical situation realizable in the laboratory, the potentials $V,\boldsymbol{A}$ vary only for an intermediate period of time, but had constant values in the distant past ($t\to-\infty$) and become constant again in the distant future ($t\to+\infty$), Fig. 8.1.

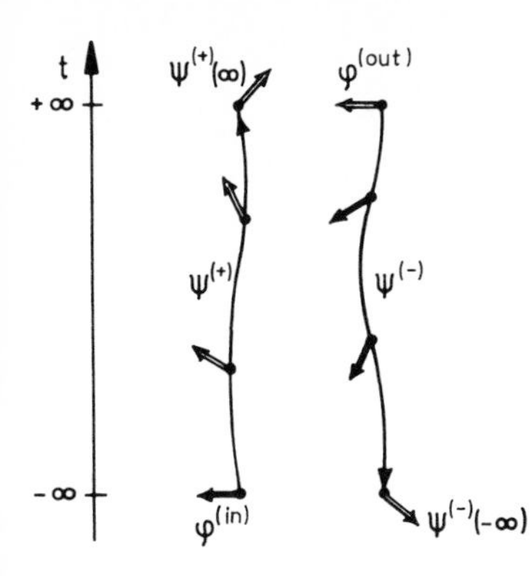

Fig. 8.1. Forward and backward time evolution of the solutions of the time-dependent Dirac equation

The word "distant" applies here to time intervals on the atomic scale. The typical experiment we have in mind is a scattering experiment, or a situation where a new source of electromagnetic field is switched on and the reaction of electrons or positrons to the field change is studied.

In these cases, it makes sense to start from a complete set of eigenmodes $\varphi_n^{(\text{in})}(\boldsymbol{x})$ of the Dirac Hamiltonian before the time change occurs, i.e. $H_\text{D}(\boldsymbol{x}, t=-\infty)$. Integrating the Dirac equation forward in time gives a set of solutions $\psi_n^{(+)}(\boldsymbol{x}, t)$ of the full time-dependent Dirac equation which satisfies the boundary condition

$$\psi_n^{(+)}(\boldsymbol{x}, t) \xrightarrow{t\to-\infty} \varphi_n^{(\text{in})}(\boldsymbol{x}) \exp(-\mathrm{i}E_n^{(\text{in})} t) . \tag{8.39}$$

The wave function $\psi_n^{(+)}$ describes the fate of an electron (or positron) that has been initially prepared to occupy the stationary state $\varphi_n^{(\text{in})}$.

Since H_D is Hermitian, the $\psi_n^{(+)}(\boldsymbol{x}, t)$ form a complete, orthonormal basis set at every instant of time:

$$\int d^3x\, \psi_n^{(+)\dagger}(\boldsymbol{x}, t)\, \psi_m^{(+)}(\boldsymbol{x}, t) = \delta_{nm} \tag{8.40}$$

$$\sum_n \psi_{n,\alpha}^{(+)}(\boldsymbol{x}, t)\, \psi_{n,\beta}^{(+)\dagger}(\boldsymbol{x}', t) = \delta_{\alpha\beta}\delta(\boldsymbol{x}-\boldsymbol{x}') , \tag{8.41}$$

where α, β in the second equation again denote the spinor indices. The orthogonality relation is easily proven by showing that the scalar product does not change with time:

$$\begin{aligned}\frac{d}{dt}\int d^3x\, \psi_n^\dagger \psi_m &= \int d^3x\,(\dot\psi_n^\dagger \psi_m + \psi_n^\dagger \dot\psi_m)\\ &= \int d^3x\,[(-\mathrm{i}H_\text{D}\psi_n)^\dagger \psi_m + \psi_n^\dagger(-\mathrm{i}H_\text{D}\psi_m)]\\ &= \mathrm{i}\int d^3x\,[\psi_n^\dagger H_\text{D}^\dagger \psi_m - \psi_n^\dagger H_\text{D}\psi_m] = 0 .\end{aligned} \tag{8.42}$$

A similar proof can be worked out for the completeness relation.

Of course, one can also think backwards: an electron occupying an eigenstate $\psi_n^{(\text{out})}$ of the Dirac Hamiltonian $H_\text{D}(\boldsymbol{x}, t=+\infty)$ after the potentials have changed can be traced backward in time by means of the Dirac equation. This gives rise to another complete, orthonormal set of basis functions $\psi_n^{(-)}(\boldsymbol{x}, t)$ satisfying the boundary condition

$$\psi_n^{(-)}(\boldsymbol{x},t) \xrightarrow{t\to+\infty} \varphi_n^{(\mathrm{out})}(\boldsymbol{x})\exp(-\mathrm{i}E_n^{(\mathrm{out})}t)\,. \tag{8.43}$$

The properties of and the connection between these solutions can be most elegantly formulated with the help of the time-development operator $U(t,t')$, defined to effect the evolution of the solutions of the time-dependent Dirac equation:

$$|\psi(t)\rangle = U(t,t')\,|\psi(t')\rangle\,. \tag{8.44}$$

This operator must obviously satisfy

$$\mathrm{i}\frac{\partial}{\partial t}U(t,t') = H_\mathrm{D}(t)\,U(t,t') \tag{8.45}$$

with the initial condition $U(t,t) = 1$.

The Hermiticity of H_D implies that U is a unitary operator

$$U^\dagger(t,t') = U^{-1}(t,t') = U(t',t)\,. \tag{8.46}$$

For time-independent potentials, (8.45) is solved by

$$U(t,t') = \exp[-\mathrm{i}H_\mathrm{D}(t-t')]\,, \tag{8.47}$$

and the unitarity is obvious in this case. If $H_\mathrm{D}(t)$ is time dependent, only a formal solution of the differential equation (8.45) can be found by successive approximations. Writing $U^{(n)}$ for the nth approximation,

$$\mathrm{i}\frac{\partial}{\partial t}U^{(n)}(t,t') = H_\mathrm{D}(t)\,U^{(n-1)}(t,t')\,.$$

Starting with the initial condition $U^{(n)}(t,t) = 1$, then successively

$$U^{(n)}(t,t') = 1 - \mathrm{i}\int_{t'}^{t} dt_1 H_\mathrm{D}(t_1)\,U^{(n-1)}(t_1,t')\,,$$

$$U^{(1)}(t,t') = 1 - \mathrm{i}\int_{t'}^{t} dt_1 H_\mathrm{D}(t_1)$$

$$\begin{aligned} U^{(2)}(t,t') &= U^{(1)}(t,t') + (-\mathrm{i})^2\int_{t'}^{t} dt_1 H_\mathrm{D}(t_1)\int_{t'}^{t_1} dt_2 H_\mathrm{D}(t_2) \\ &= U^{(1)}(t,t') + \tfrac{1}{2}(-\mathrm{i})^2\int_{t'}^{t} dt_1\int_{t'}^{t} dt_2\, T[H_\mathrm{D}(t_1)H_\mathrm{D}(t_2)]\,, \qquad \text{etc.} \end{aligned}$$

Here "T" denotes time ordering of the bracketed operators:

$$T[H_\mathrm{D}(t_1)H_\mathrm{D}(t_2)] = \begin{cases} H_\mathrm{D}(t_1)H_\mathrm{D}(t_2) & \text{for} \quad t_1 \geq t_2 \\ H_\mathrm{D}(t_2)H_\mathrm{D}(t_1) & \text{for} \quad t_2 \geq t_1 \end{cases}.$$

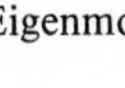

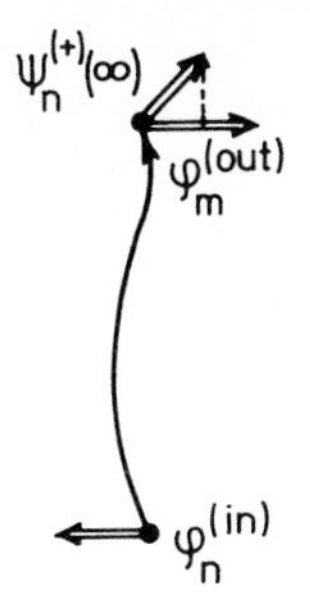

Fig. 8.2. The S-matrix element is defined as overlap between a stationary out state and that state resulting from the time evolution of a stationary in state

The solution [Dy 49, Schw 61] then is

$$U(t,t') = 1 + \sum_{n=1}^{\infty} \frac{(-\mathrm{i})^n}{n!} \int_{t'}^{t} dt_1 \ldots \int_{t'}^{t} dt_n \, T[H_\mathrm{D}(t_1)\ldots H_\mathrm{D}(t_n)] \,,$$

which is often written symbolically as

$$U(t,t') = T\left[\exp\left(-\mathrm{i}\int_{t'}^{t} d\tau\, H_\mathrm{D}(\tau)\right)\right] .$$

Because unitary transformations always conserve the scalar product, it is also clear that the orthogonality and completeness relations between wave functions $\psi_n^{(+)}$ or $\psi_n^{(-)}$ are unchanged by the time evolution. Of particular interest is the amplitude S_{mn} of an electron starting in state $\varphi_n^{(\mathrm{in})}$ and ending in state $\varphi_m^{(\mathrm{out})}$. This amplitude is obtained as the overlap between the solutions $\psi_n^{(+)}$ and $\psi_m^{(-)}$ of the time-dependent Dirac equation [cf. (7.23)]:

$$S_{mn} = \int d^3x \, \psi_m^{(-)\dagger}(\boldsymbol{x},t)\, \psi_n^{(+)}(\boldsymbol{x},t) \,. \tag{8.48}$$

Due to the unitarity of the operator U, the value of S_{mn} does not depend on the value of t where (8.48) is evaluated, so assume $t \to +\infty$. Then $\psi_m^{(-)}$ can be represented by its asymptotic form (8.43) and $\psi_n^{(+)}$ can be generated from its asymptotic form (8.39) by means of the time evolution operator. Hence

$$S_{mn} = \lim_{\substack{t\to\infty \\ t'\to -\infty}} \langle \varphi_m^{(\mathrm{out})} | U(t,t') | \varphi_n^{(\mathrm{in})} \rangle \exp[\mathrm{i}E_m^{(\mathrm{out})} t - \mathrm{i}E_n^{(\mathrm{in})} t'] \,. \tag{8.49}$$

The limit certainly exists if the potentials vary only during a finite interval of time. One may then define S_{mn} to be the matrix element of the scattering operator $\hat{S}$ (Fig. 8.2):

$$\hat{S} = \lim_{\substack{t\to\infty \\ t'\to -\infty}} \exp(\mathrm{i}H_\mathrm{D}(t)\,t)\, U(t,t')\, \exp(-\mathrm{i}H_\mathrm{D}(t')\,t') \tag{8.50}$$

and

$$S_{mn} = \int d^3x \, \varphi_m^{(\mathrm{out})}(\boldsymbol{x})\, \hat{S}\, \varphi_n^{(\mathrm{in})}(\boldsymbol{x}) \,. \tag{8.51}$$

If the operator $\hat{S}$ or its matrix elements S_{mn} are known, all observables at the time the potentials have become constant again can be calculated. Below is shown that S_{mn} also carry all the information about pair creation and the development of many-particle states in the independent particle approximation.

8.4 The Elementary Field Functions

In field theory, the equations of motion determine the propagation of a field from one instant of time to another; if these could be solved exactly, the result of every experiment could be predicted. For interacting fields, however, an explicit solution is generally not obtainable. Instead, one has to find approximate solutions. Many of these approximation schemes start from non-interacting fields, for which explicit solutions can be readily found, and treat the interaction as a (not necessarily small) perturbation. It is then necessary to know the propagation of the perturbation due to an inhomogeneous term in the wave equation.

This was the point of view taken in Chap. 7, where perturbation theory based on free propagators and Feynman diagrams was developed. In the presence of a strong external field and deeply bound states, however, an expansion of the exact propagator into powers of the external field is not meaningful. When the external field is that of a nucleus, diagrams differing only in the power of the coupling constant $(Z\alpha)$ contribute similar magnitudes, if $Z\alpha \gtrsim 1$. Examples of such diagrams are shown in Fig. 8.3. On the other hand, diagrams involving virtual photon lines describing the interaction among the electrons (see Fig. 8.4) contain powers of the coupling constant α in addtion to the powers of $(Z\alpha)$. Each factor α reduces the size of the contribution from a diagram by about two orders of magnitude. It is therefore reasonable to treat the interactions due to the virtual radiation field in perturbation theory along the lines of Chap. 7, but to work with

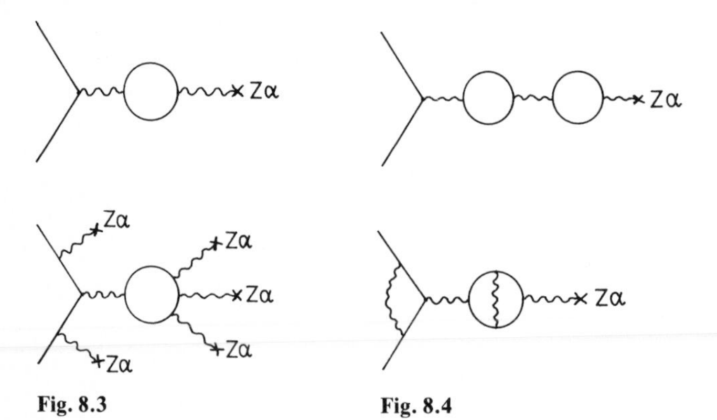

Fig. 8.3. Feynman diagrams differing only by powers of $(Z\alpha) \sim 1$ give comparable contributions

Fig. 8.4. Diagrams differing in powers of α do *not* give contributions of similar size

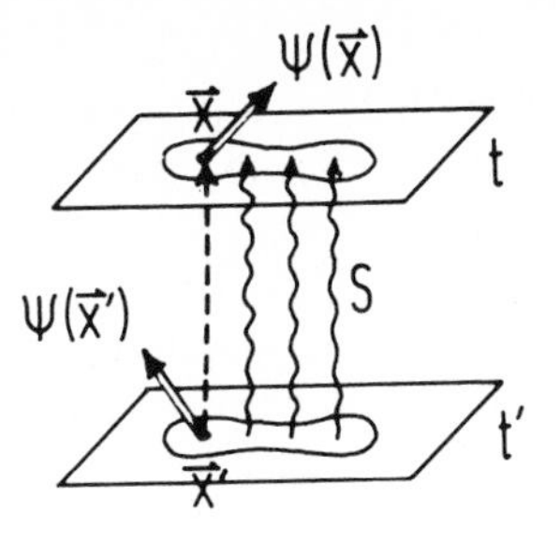

Fig. 8.5. The homogeneous propagator or Schwinger function $S(x,x')$ describes the propagation of a solution of the Dirac equation from one space-like hypersurface (time t') to another (time t)

the *exact* propagator in the strong external field. We elaborate on this approach, called the Furry picture [Fu 51], in Chap. 15. Here we develop only the formalism of exact propagators in an external field, needed to describe the decay of the vacuum in second quantization.

Propagation of a wave function or a disturbance can be effected by an integral transformation, having as kernel a solution of the homogeneous or the elementary inhomogeneous field equation

$$\left[\gamma^\mu\left(\mathrm{i}\frac{\partial}{\partial x^\mu}-eA_\mu\right)-m_e\right]S(x,x')=0 \tag{8.52}$$

$$\left[\gamma^\mu\left(\mathrm{i}\frac{\partial}{\partial x^\mu}-eA_\mu\right)-m_e\right]S_{\mathrm{inh}}(x,x')=\delta(x-x')\,. \tag{8.53}$$

Either equation must be supplemented by a boundary condition in order to define a unique function $S(x,x')$. Note that S is actually a (4×4) spin matrix function.

The most widely used initial condition for the homogeneous equation is

$$S(\boldsymbol{x},t;\boldsymbol{x}',t)=-\mathrm{i}\gamma^0\delta(\boldsymbol{x}-\boldsymbol{x}') \quad \text{for} \quad t=t'\,. \tag{8.54}$$

This function is called the *Schwinger function* for the Dirac equation and it enables a solution of the Dirac equation to be propagated from one space-like hypersurface to another (see also Fig. 8.5):

$$\psi(\boldsymbol{x},t)=\mathrm{i}\int d^3x'\,S(\boldsymbol{x},t;\boldsymbol{x}',t')\,\gamma^0\psi(\boldsymbol{x}',t') \tag{8.55}$$

$$\bar{\psi}(\boldsymbol{x},t)=\mathrm{i}\int d^3x\,\bar{\psi}(\boldsymbol{x}',t')\,\gamma^0S(\boldsymbol{x}',t';\boldsymbol{x},t)\,. \tag{8.56}$$

From (8.55) it is obvious that $(\mathrm{i}S\gamma^0)$ is the integration kernel of the time development operator $U(t,t')$ introduced in (8.44).

Although S can be determined explicitly only for the free Dirac field and for a few simple potentials (constant electric or magnetic field, Coulomb potential, etc.), it is possible to give a representation in terms of a complete set of solutions of the Dirac equation introduced above:

$$S_{\alpha\beta}(\boldsymbol{x},t;\boldsymbol{x}',t') = -\mathrm{i}\sum_n \psi^{(+)}_{n,\alpha}(\boldsymbol{x},t)\,\bar{\psi}^{(+)}_{n,\beta}(\boldsymbol{x}',t')$$

$$= -\mathrm{i}\sum_n \varphi_{n,\alpha}(\boldsymbol{x})\,\bar{\varphi}_{n,\beta}(\boldsymbol{x})\exp[-\mathrm{i}\varepsilon_n(t-t')]\,. \tag{8.57}$$

The last form holds only for time-independent potentials. When the Dirac operator of (8.52) is applied to this expression for the homogeneous propagator, it acts only on the function $\psi^{(+)}_n(\boldsymbol{x},t)$, giving zero because $\psi^{(+)}_n$ was chosen as the solution of the Dirac equation. The representation of S in terms of a set of solutions of the time-dependent Dirac equation relies heavily on the completeness of these solutions. This is brought out clearly in the limit $t\to t'$, where (8.54, 57) are combined to yield the completeness relation

$$\delta_{\alpha\beta}\delta(\boldsymbol{x}-\boldsymbol{x}') = \sum_n \psi^{(+)}_{n,\alpha}(\boldsymbol{x},t)\,\psi^{\dagger(+)}_{n,\beta}(\boldsymbol{x}',t)\,.$$

For the free Dirac field, S is a function of the coordinate differences $\boldsymbol{x}-\boldsymbol{x}',t-t'$, and may be most conveniently given in the four-dimensional Fourier representation similar to (7.38)

$$S^0(x,x') = \oint_C \frac{d^4k}{(2\pi)^4}\exp[-\mathrm{i}k_\mu(x-x')^\mu]\cdot\frac{\gamma^\mu k_\mu + m_e}{k^2-m_e^2}\,, \tag{8.58}$$

where the integration path in the complex k_0 plane is chosen according to Fig. 8.6. This relation is easily derived either by substituting the plane wave solutions (2.50) into the general formula for S, or directly by transforming the free Dirac equation into momentum space [Bj 65].

It is now easy to obtain solutions of the inhomogenous equation (8.53), the Green's functions or propagators in the proper sense. As these functions denote the propagation by the Dirac field of a disturbance, such as an additional

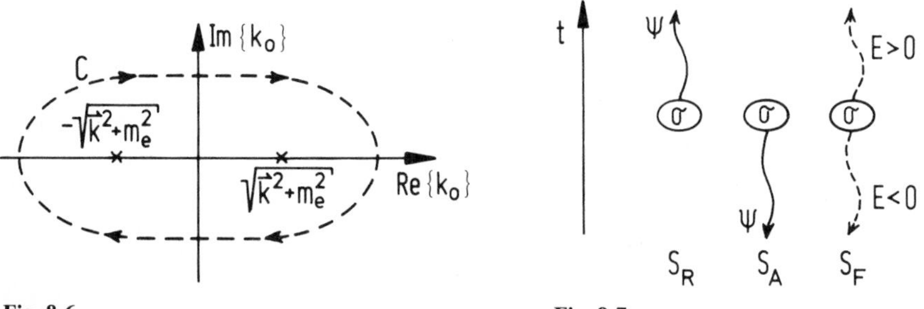

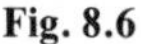
Fig. 8.6 **Fig. 8.7**

Fig. 8.6. Integration contour in the complex energy plane for the Schwinger function of the free Dirac field

Fig. 8.7. Action of the various propagators: the retarded propagator S_R propagates a perturbation into the future, while the advanced propagator S_A describes which modification of the past wave function would have resulted in the perturbation. The Feynman propagator is half retarded (for positive frequencies) and half advanced (for negative frequencies) to describe the propagation of particles and antiparticles correctly

potential, it is important to impose the appropriate boundary conditions. We shall study three cases, the retarded propagator S_R, the advanced propagator S_A and the Feynman propagator S_F (see Fig. 8.7). All three enable a particular solution of the Dirac equation to be found with an inhomogeneous spinor term

$$\left[\gamma^\mu\left(\mathrm{i}\frac{\partial}{\partial x^\mu}-eA_\mu\right)-m_e\right]\psi(x)=\sigma(x) \tag{8.59}$$

in the form

$$\psi(x)=\int d^4x'\,S_{\mathrm{inh}}(x,x')\,\sigma(x')\,, \tag{8.60}$$

where, of course, any solution of the homogeneous equation may be added. A typical case is an additional perturbing electromagnetic potential A'_μ. Then (8.60) does not permit an explicit solution but it gives an integral representation for $\psi(x)$ with the "source" term $\sigma(x')=e\gamma^\mu A'_\mu(x')\,\psi(x')$. This representation serves as the starting point for a perturbation expansion and the familiar diagrammatical representation of the scattering problem [see Chap. 7].

The retarded propagator is constructed so as to propagate only into the future, whereas the advanced propagates backwards in time:

$$S_R(\boldsymbol{x},t;\boldsymbol{x}',t')=\Theta(t-t')\,S(\boldsymbol{x},t;\boldsymbol{x}',t') \tag{8.61a}$$

$$S_A(\boldsymbol{x},t;\boldsymbol{x}',t')=-\Theta(t'-t)\,S(\boldsymbol{x},t;\boldsymbol{x}',t')\,. \tag{8.61b}$$

Remember that S was a solution of the homogeneous Dirac equation, one easily verifies that S_R and S_A satisfy the imhomogeneous equation (8.53):

$$\begin{aligned}\left[\gamma^\mu\left(\mathrm{i}\frac{\partial}{\partial x^\mu}-eA_\mu\right)-m_e\right]S_R&=S(\boldsymbol{x},t;\boldsymbol{x}',t')\,\mathrm{i}\gamma^0\frac{\partial}{\partial t}\,\Theta(t-t')\\&=\delta(x-x')\,,\end{aligned} \tag{8.62}$$

where we used the boundary condition for S at equal times (8.54). Similarly, S_A is also shown to satisfy (8.53). The main importance of the advanced propagator lies in the fact that it propagates perturbations of the *adjoint* wave function *forward* in time.

As discussed in Chap. 7, the Stückelberg-Feynman interpretation of the negative energy solutions of the Dirac equation is that they represent electrons moving backwards in time, corresponding to positrons moving forward in time. In this sense the physical propagator of Dirac particles is that which is retarded for positive energy solutions and advanced for negative energy solutions. This propagator is called the Feynman propagator S_F, and it is formally defined as

$$\begin{aligned}\mathrm{i}S_F(\boldsymbol{x},t;\boldsymbol{x}',t')&=\Theta(t-t')\sum_n\Theta(E_n)\,\varphi_n(\boldsymbol{x})\,\bar{\varphi}_n(\boldsymbol{x}')\exp[-\mathrm{i}E_n(t-t')]\\&\quad-\Theta(t'-t)\sum_n\Theta(-E_n)\,\varphi_n(\boldsymbol{x})\,\bar{\varphi}_n(\boldsymbol{x}')\exp[-\mathrm{i}E_n(t-t')]\\&\equiv\sum_n\mathrm{sgn}(E_n)\,\Theta(E_n(t-t'))\,\varphi_n(\boldsymbol{x})\,\bar{\varphi}_n(\boldsymbol{x}')\exp[-\mathrm{i}E_n(t-t')]\,.\end{aligned} \tag{8.63}$$

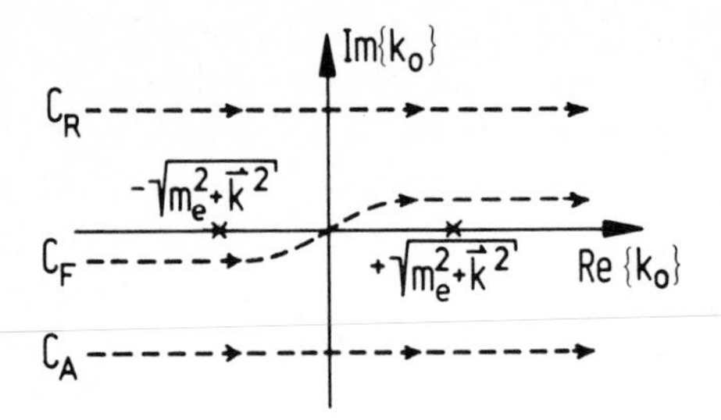

Fig. 8.8. Integration contours in the complex energy plane for the propagators S_{R}, S_{A} and S_{F}. The mixed behaviour of S_{F} for positive and negative frequencies is clearly exhibited

This definition is the generalization of (7.48) for the Feynman-Stückelberg propagator. Because it contains the time step functions, it also satisfies the inhomogeneous Dirac equation (8.53).

For free electrons, the various propagators are given by integral representations analogous to (8.58), but with different integration paths (Fig. 8.8)

$$S_i^0(x,x') = \int_{C_i} \frac{d^4k}{(2\pi)^4} \exp[-\mathrm{i}k_\mu(x-x')^\mu] \frac{\gamma^\mu k_\mu + m}{k^2-m^2} \quad (i=\mathrm{R},\mathrm{A},\mathrm{F}) . \tag{8.64}$$

According to (7.43), the free Feynman propagator can also be written as

$$S_{\mathrm{F}}^0(x,x') = \lim_{\varepsilon\to 0} \int \frac{d^4k}{(2\pi)^4} \mathrm{e}^{-\mathrm{i}k(x-x')} \frac{\gamma k + m}{k^2-m^2+\mathrm{i}\varepsilon} , \tag{8.65}$$

where the k_0 integration runs along the real axis.

As the separation into positive and negative frequencies makes sense only in time-independent potentials, our present definition of S_{F} is valid only for this case. For variable potentials that become asymptotically constant (in the distant past and future), definition (8.63) can be generalized by dividing the complete set of wave functions according to their asymptotic frequencies:

$$\mathrm{i}S^{(\mathrm{in/out})}_{\mathrm{F}(xt|x't')} = \sum_n \Theta(E_n(\mp\infty)(t-t'))\, \psi_n^{(\pm)}(x,t)\, \bar{\psi}_n^{(\pm)}(x',t') , \tag{8.66}$$

where the functions $\psi_n^{(\pm)}$ are those defined in (8.39, 43). As for the functions $\psi_n^{(\pm)}$ themselves, the propagators $S_{\mathrm{F}}^{(\mathrm{in})}$, $S_{\mathrm{F}}^{(\mathrm{out})}$ have little physical meaning outside their respective asymptotic regions. Because the Feynman propagators, however, implicitly contain the definition of electron versus positron wave functions, the difference between $S_{\mathrm{F}}^{(\mathrm{in})}$ and $S_{\mathrm{F}}^{(\mathrm{out})}$ contains all the information about pair production by the time-dependent external potential. Since this can be treated consistently only in a second quantized theory, the discussion is postponed to Sects. 9.6, 7, and 10.5.

Recently, *Rumpf* defined particle and antiparticle states applicable to certain situations where the external potential does not approach a constant value asymptotically. His technique uses the properties of the wave functions $\psi_n^{(\pm)}$ for complex values of the rest mass m. *Rumpf's* definition is manifestly covariant, and it is most usefully applied to situations where the external field is not localized but extended over all space, in particular to particles in gravitational fields (curved space-time). For details, see [Ru 78, 79].

Bibliographical Notes

The subject of Chaps. 8, 9 is treated in many excellent textbooks on quantum field theory. Some personal preferences are [Bj 65], from which we take the conventions used throughout this book and [Ja 55], probably the most complete book on QED.

Other recommendations are [It 80, Kä 58, Ni 69a, We 49, Ro 60, 69, Bo 59].

9. Second Quantization of the Dirac Field and Definition of the Vacuum

Chapter 6 treated at length the instability of the neutral Dirac vacuum in a sufficiently strong, i.e. supercritical, external field and its spontaneous decay into a charged vacuum. The discussion then was based on Dirac's intuitive hole picture of the electron – positron vacuum combined with the theory of spontaneous autoionization processes. While such an amalgamation of intuitive, pictorial ideas and relativistic quantum mechanics may serve as a useful guide to the underlying physical phenomena, it is equally clear that it cannot rigorously prove our assertions. This proof should more properly be sought within the framework of relativistic quantum field theory which enables a consistent description of processes involving virtual and real particle creation.

The aim of this chapter is thus to show by quantum field theory that the vacuum in supercritical fields indeed carries charge, and that it has less energy than the normal, neutral vacuum of subcritical fields. We begin with the canonical quantization of the Dirac field and introduce the normal vacuum and the Fock space built on it, also deriving the formal expressions for the charge and energy of the Dirac vacuum. Then the invariance properties of the quantum field theory are treated, both for continuous and discrete symmetries, particularly emphasizing the operation of charge conjugation. We then resume our analysis of the properties of the vacuum state, showing how the charged vacuum appears naturally as the stable equilibrium state (or ground state) of an open system which contains an electric field of supercritical strength.

After a short discussion of the role of the Feynman propagator in relativistic quantum field theory, we are ready for the ultimate goal of this chapter, i.e. to calculate the charge and energy of the vacuum state. First the formal expressions for these quantities are transcribed into a more easily manageable form involving the Feynman propagator, then they are analyzed with regard to the contributions from virtual and real vacuum polarization. It is shown that the charge of the vacuum for subcritical potentials becomes non-zero only when one or several bound states have joined the antiparticle continuum. If the number of these states is denoted by n_{vac}, the charge of the vacuum is $q_{\text{vac}} = e n_{\text{vac}}$. Similarly, the supercritical bound states always contribute negatively to the vacuum energy, causing a sudden drop of the energy at the critical field strength. This underscores that the transition from neutral to charged vacuum can be viewed as a spontaneous discharge process connected with a decrease in the energy of the electric field.

9.1 Canonical Quantization of the Dirac Field

The transition from the classical theory of the Dirac field to the associated quantum theory is most easily made in the framework of Hamiltonian formalism, in which the independent variables are the field coordinates $\psi_\alpha(\boldsymbol{x},t)$ and the conjugate momenta $\Pi_\alpha(\boldsymbol{x},t) = \mathrm{i}\,\psi_\alpha^\dagger(\boldsymbol{x},t)$, $(\alpha = 1,\ldots,4)$. The quantization rule is formally the same as in the ordinary quantum mechanics of a point particle: ψ_α and Π_α are to be replaced by operators $\hat{\psi}_\alpha(\boldsymbol{x},t)$ and $\hat{\Pi}_\alpha(\boldsymbol{x},t)$, one each for every spin index α and every space point $\boldsymbol{x}$. In the Heisenberg picture these operators are functions of time. The basic meaning of the operators $\hat{\psi}$ and $\hat{\Pi}$ is that $\hat{\psi}(\boldsymbol{x},t)$ annihilates an electron (or, as will be shown, creates a positron) at point x and time t. As $\hat{\Pi} = \mathrm{i}\,\hat{\psi}^\dagger$, the field momentum operator does precisely the opposite, i.e. it creates an electron or destroys a positron. An important point always to be kept in mind is that the four-spinor $\psi(\boldsymbol{x},t)$ is *not* an observable; all observables are built from bilinear expressions in the Dirac wave function. This was evident in the discussion of observables like charge density, energy, momentum, etc.

The action of these operators on a state $|\Omega\rangle$ of the Dirac field is prescribed by two independent requirements. The first one defines how field operators *commute* with each other, while the second defines the normal ground state of the field, i.e. the *vacuum*.

Let us begin with the commutation relations. As the Dirac field describes particles of spin-1/2, like electrons or muons, which are Fermions, they must obey the Pauli principle. This is ensured by requiring that field operators at different spatial points, but taken at the same time t, *anti*commute rather than commute

$$\{\hat{\psi}_\alpha(\boldsymbol{x},t),\,\hat{\psi}_\beta(\boldsymbol{x}',t)\} = 0 \tag{9.1a}$$

$$\{\hat{\Pi}_\alpha(\boldsymbol{x},t),\hat{\Pi}_\beta(\boldsymbol{x}',t)\} = 0\,, \tag{9.1b}$$

where we denote $\{A,B\} = AB + BA$. The effect of this special choice becomes evident when $\boldsymbol{x} = \boldsymbol{x}'$, $\alpha = \beta$,

$$\hat{\psi}_\alpha(\boldsymbol{x},t)\,\hat{\psi}_\alpha(\boldsymbol{x},t) = \hat{\Pi}_\alpha(\boldsymbol{x},t)\,\hat{\Pi}_\alpha(\boldsymbol{x},t) = 0\,,$$

i.e. it is impossible to create or annihilate two particles at the same space-time point. This is the essence of the Pauli principle.

The Pauli principle is, of course, a well-founded experimental fact. However, there are also grave theoretical reasons demanding the quantization with anti-commutators instead of commutators. Probably the most obvious argument is that from studying the ordinary, unquantized Dirac equation it is clear that infinitely many single-particle states of negative energy exist. When these are occupied in the framework of a many-body theory, energy is released. Quantization with commutators would permit an arbitrarily large number of particles to occupy each state, so that the energy of any given state could be lowered further by adding particles to the negative energy states. Thus the spectrum of the quantized Hamiltonian would not be bounded from below. This problem is obviously resolved by quantizing with anticommutators, as discussed above. Other theoretical arguments include causality of the propagators of the quantized theory, the requirement that (anti-)commutators at unequal times remain c numbers, etc. [St 64, Bj 65, Ni 69b, Ro 69].

The anticommutator between the field operator and the conjugate momentum operator requires

$$\{\hat{\psi}_\alpha(\boldsymbol{x},t),\hat{\Pi}_\beta(\boldsymbol{x}',t)\} = \mathrm{i}\{\hat{\psi}_\alpha(\boldsymbol{x},t),\hat{\psi}^\dagger_\beta(\boldsymbol{x},t)\} = \mathrm{i}\,\delta_{\alpha\beta}\delta(\boldsymbol{x}-\boldsymbol{x}') \tag{9.2a}$$

or, after multiplying with γ^0, the equivalent form

$$\{\hat{\psi}_\alpha(\boldsymbol{x},t),\hat{\bar{\psi}}_\beta(\boldsymbol{x}',t)\} = \gamma^0_{\alpha\beta}\delta(\boldsymbol{x}-\boldsymbol{x}')\,. \tag{9.2b}$$

To exhibit the full relativistic covariance of the theory, it is best to work with the Heisenberg model. In this model the state vector $|\Omega\rangle$ of any system is time independent and the full time evolution is described by the operators, here $\hat{\psi}(\boldsymbol{x},t)$ and $\hat{\psi}^\dagger(\boldsymbol{x},t)$. The equation of motion for the field operator is Heisenberg's equation

$$\frac{\partial}{\partial t}\hat{\psi}(\boldsymbol{x},t) = \mathrm{i}[\hat{H}(t),\hat{\psi}(\boldsymbol{x},t)]\,, \tag{9.3}$$

where the quantized Hamiltonian $\hat{H}$ is obtained from the classical Hamilton function $H(t)$, (8.13), by replacing the field functions ψ and Π by the corresponding field operators $\hat{\psi}$ and $\hat{\Pi}$. In doing so, we must consider the order of the field operators because they do not commute as the c-number functions did. Various choices are possible; we shall take the form that is symmetric under any possible exchange of the operators:

$$\begin{aligned}\hat{H} &= \int d^3x\,\mathscr{H}(\hat{\psi},\hat{\Pi}=\mathrm{i}\hat{\psi}^\dagger;\boldsymbol{x},t)_{\mathrm{sym}}\\ &= \tfrac{1}{2}\int d^3x\{\tfrac{1}{2}[\hat{\psi}^\dagger(\boldsymbol{x},t),\boldsymbol{\alpha}\cdot(-\mathrm{i}\nabla-e\boldsymbol{A})\,\hat{\psi}(\boldsymbol{x},t)]\\ &\quad+\tfrac{1}{2}[(\mathrm{i}\nabla-e\boldsymbol{A})\,\hat{\psi}^\dagger(\boldsymbol{x},t)\cdot\boldsymbol{\alpha},\hat{\psi}(\boldsymbol{x},t)]\\ &\quad+[\hat{\psi}^\dagger(\boldsymbol{x},t),(\beta m_e+V)\,\hat{\psi}(\boldsymbol{x},t)]\}\,.\end{aligned} \tag{9.4}$$

For simplicity, $\hat{\Pi}$ is henceforth identified with $\mathrm{i}\hat{\psi}^\dagger$. For every order of $\hat{\psi}^\dagger$ and $\hat{\psi}$ the summation over spin indices is implicitly understood, e.g.

$$[\hat{\psi}^\dagger,\beta\hat{\psi}] \equiv \psi^\dagger\beta\hat{\psi}-(\beta\hat{\psi})^{\mathrm{T}}(\hat{\psi}^\dagger)^{\mathrm{T}} = \psi^\dagger_\alpha\beta_{\alpha\beta}\hat{\psi}_\beta-\beta_{\alpha\beta}\hat{\psi}_\beta\hat{\psi}^\dagger_\alpha\,. \tag{9.5}$$

Later it is shown that this particular choice of $\hat{H}$ is actually forced by the requirement of charge symmetry of the theory (Sect. 9.5).

It is possible to simplify (9.4) somewhat by partial integration applied to the second term, thereby shifting the action of the gradient from $\hat{\psi}^\dagger$ to $\hat{\psi}$:

$$\begin{aligned}\hat{H} &= \tfrac{1}{2}\int d^3x[\hat{\psi}^\dagger(\boldsymbol{x},t),(\boldsymbol{\alpha}\cdot(-\mathrm{i}\nabla-e\boldsymbol{A})+\beta m_e+V)\,\hat{\psi}(\boldsymbol{x},t)]\\ &\quad+\tfrac{\mathrm{i}}{2}\int d^3x\,\nabla\cdot[\hat{\psi}^\dagger(\boldsymbol{x},t),\boldsymbol{\alpha}\hat{\psi}(\boldsymbol{x},t)]\\ &\equiv \tfrac{1}{2}\int d^3x[\hat{\psi}^\dagger(\boldsymbol{x},t),H_{\mathrm{D}}\hat{\psi}(\boldsymbol{x},t)]+\tfrac{\mathrm{i}}{2}\oint d\boldsymbol{\Sigma}\cdot[\hat{\psi}^\dagger,\boldsymbol{\alpha}\hat{\psi}]\,.\end{aligned} \tag{9.6}$$

Here the abbreviation (8.17) for the Hamiltonian of the Dirac equations has been used and $\oint d\Sigma$ denotes an integration over the boundary of the volume of our system (usually taken at "infinity").

The operator in the surface integral is the spatial part of the current operator as obtained from the classical expression (8.8):

$$\hat{j}^\mu = \frac{e}{2}[\hat{\bar{\psi}}, \gamma^\mu \hat{\psi}] \,. \tag{9.7}$$

Thus, the surface integral vanishes if the operator $\hat{H}$ is taken between localized states, giving

$$\langle \mathrm{loc}' | \hat{H} | \mathrm{loc} \rangle = \tfrac{1}{2} \int d^3x \, \langle \mathrm{loc}' | [\hat{\psi}^\dagger, H_\mathrm{D} \hat{\psi}] \, | \mathrm{loc} \rangle \,, \tag{9.8}$$

where $|\mathrm{loc}\rangle$ and $|\mathrm{loc}'\rangle$ denote spatially localized Fock space states.

In the remainder of this chapter we always assume that the localization condition is satisfied and replace the full form of $\hat{H}$ by the much shorter (9.8). In doing so we have to keep in mind that all states involving particles in continuum wave functions should, in principle, be properly constructed from wave packets.

Substituting the Hamilton operator into the Heisenberg equation of motion (9.3), and employing the equal-time commutation relations (9.1, 2), in agreement with the correspondence principle

$$\begin{aligned}
\frac{\partial}{\partial t} \hat{\psi}(\boldsymbol{x}, t) &= \frac{\mathrm{i}}{2} \int d^3x' \, [[\hat{\psi}^\dagger(\boldsymbol{x}', t), H'_\mathrm{D} \hat{\psi}(\boldsymbol{x}', t)], \hat{\psi}(\boldsymbol{x}, t)] \\
&= \frac{\mathrm{i}}{2} \int d^3x' \, \{\hat{\psi}^\dagger(\boldsymbol{x}', t), \{H'_\mathrm{D} \hat{\psi}(\boldsymbol{x}', t), \hat{\psi}(\boldsymbol{x}, t)\}\} \\
&\quad - \frac{\mathrm{i}}{2} \int d^3x' \, \{\{\hat{\psi}^\dagger(\boldsymbol{x}', t), \hat{\psi}(\boldsymbol{x}, t)\}, H'_\mathrm{D} \hat{\psi}(\boldsymbol{x}', t)\} \\
&= -\frac{\mathrm{i}}{2} \int d^3x' \, \{\delta^3(\boldsymbol{x} - \boldsymbol{x}'), H_\mathrm{D} \hat{\psi}(\boldsymbol{x}', t)\} = -\mathrm{i} H_\mathrm{D} \hat{\psi}(\boldsymbol{x}, t) \,,
\end{aligned} \tag{9.9}$$

where the operator identity $[[\hat{A}, \hat{B}], \hat{C}] = \{\hat{A}, \{\hat{B}, \hat{C}\}\} - \{\{\hat{A}, \hat{C}\}, \hat{B}\}$ was used. This means that the field operator in the Heisenberg picture satisfies the ordinary Dirac equation, but is now taken as an operator equation among operators.

Knowing the time evolution of $\hat{\psi}(\boldsymbol{x}, t)$, one can, at least in principle, also evaluate the anticommutators at unequal times. Equation (9.9) states that the operators $\hat{\psi}$ and $\hat{\psi}^\dagger$ do not become mixed with time, and therefore

$$\{\hat{\psi}_\alpha(\boldsymbol{x}, t), \hat{\psi}_\beta(\boldsymbol{x}', t')\} = \{\hat{\psi}^\dagger_\alpha(\boldsymbol{x}, t), \hat{\psi}^\dagger_\beta(\boldsymbol{x}', t')\} = 0 \,, \tag{9.10}$$

stating that particles remain particles forever, and antiparticles remain antiparticles. The anticommutator between $\hat{\psi}$ and $\hat{\bar{\psi}}$ evolves into a complicated function of the two space-time points, but remains a c number [Ni 69b]

$$\{\hat{\psi}_\alpha(\boldsymbol{x}, t), \hat{\bar{\psi}}_\beta(\boldsymbol{x}', t')\} = \mathrm{i} S_{\alpha\beta}(\boldsymbol{x}, t; \boldsymbol{x}', t') \,. \tag{9.11}$$

On applying the operator $\mathrm{i}(\partial/\partial t) - H_\mathrm{D}(\boldsymbol{x}, t)$ to this equation, it follows from (9.9) that the matrix function S satisfies the homogeneous Dirac equation

$$\left[\mathrm{i}\frac{\partial}{\partial t} - H_\mathrm{D}(\boldsymbol{x}, t)\right] S(\boldsymbol{x}, t; \boldsymbol{x}', t') = 0 . \tag{9.12}$$

Moreover, on evaluating the lhs of (9.11) at $t' = t$, the equal-time commutation relation (9.2′) supplies the initial condition

$$S(\boldsymbol{x}, t; \boldsymbol{x}', t) = -\mathrm{i}\gamma^0 \delta(\boldsymbol{x} - \boldsymbol{x}') . \tag{9.13}$$

This means that S is precisely the Schwinger-Green function as defined in Sect. 8.4.

To make contact with the operators for physical particles, we must now leave the coordinate space representation of the field operators and turn to a representation of a complete set of solutions of the Dirac equation discussed in Sect. 8.3. Indeed, with a set of functions $\psi_n(\boldsymbol{x}, t)$ forming a complete basis at every instant t, the field operator may be decomposed into operators related to each of these wave functions:

$$\hat{b}_n(t) = \int d^3x\, \psi_n^\dagger(\boldsymbol{x}, t)\, \hat{\psi}(\boldsymbol{x}, t) . \tag{9.14}$$

The functions ψ_n being complete in 3-space, the definition may be inverted with the help of (8.41), yielding

$$\hat{\psi}(\boldsymbol{x}, t) = \sum_n \hat{b}_n(t)\, \psi_n(\boldsymbol{x}, t) . \tag{9.14′}$$

The operators $\hat{b}_n(t)$ could, in principle, vary with time. However, by inserting this expansion into the equation of motion for $\hat{\psi}$ (9.9), and remembering that the ψ_n were chosen as solutions of the classical Dirac equation, then

$$\begin{aligned} 0 = \left(\frac{\partial}{\partial t} + \mathrm{i}H_\mathrm{D}\right) \hat{\psi}(\boldsymbol{x}, t) &= \sum_n \dot{\hat{b}}_n(t)\, \psi_n(\boldsymbol{x}, t) - \mathrm{i} \sum_n \hat{b}_n(t) \left(\mathrm{i}\frac{\partial}{\partial t} - H_\mathrm{D}\right) \psi_n(\boldsymbol{x}, t) \\ &= \sum_n \dot{\hat{b}}_n(t)\, \psi_n(\boldsymbol{x}, t) . \end{aligned} \tag{9.15}$$

Projection with a particular function $\psi_m(t)$ shows that the operators $\hat{b}_n$ are actually time independent

$$\hat{\psi}(\boldsymbol{x}, t) = \sum_n \hat{b}_n\, \psi_n(\boldsymbol{x}, t) . \tag{9.16}$$

This result enables the operator $\hat{b}_n$ to be identified with the annihilation operator for an electron in the state ψ_n.

A word of caution is necessary here. The "particles" thus defined are physical particles only when the external electromagnetic potential is constant in time. In general, however, the operators describing the result of a laboratory experiment are time dependent in the Heisenberg picture, and a particle in the state ψ_n will not be recognized as a "physical" particle after some time has elapsed. This is discussed in more detail in Chap. 10, when treating the time evolution of the vacuum in supercritical fields.

Here we assume the potentials to be constant in time, i.e. we assume that the basis of stationary wave functions

$$\psi_n(\boldsymbol{x}, t) = \Phi_n(\boldsymbol{x})\, \mathrm{e}^{-\mathrm{i}E_n t} \tag{9.17}$$

is complete and orthonormal (this holds true if the Hamiltonian H_D is self-adjoint).

Projecting out wave functions ϕ_n and ϕ_m gives the following commutation relations for the operators $\hat{b}_n, \hat{b}_m$:

$$\{\hat{b}_n, \hat{b}_m\} = \{\hat{b}_n^\dagger, \hat{b}_m^\dagger\} = 0 \tag{9.18}$$

$$\{\hat{b}_n, \hat{b}_m^\dagger\} = \delta_{nm}\,. \tag{9.19}$$

These are the standard anticommutation relations for the creation and annihilation operators which refer to a certain stationary state. We emphasize once more that although these relations are valid for time-dependent external potentials as well, they can be taken as statements about *physical*, i.e. measurable, particles only in static potentials.

Finally, the Hamiltonian $\hat{H}$ can be expressed in terms of the particle operators for the stationary case:

$$\begin{aligned}\hat{H} &= \tfrac{1}{2}\int d^3x\,[\hat{\psi}^\dagger, H_\mathrm{D}\hat{\psi}] = \tfrac{1}{2}\sum_n E_n[\hat{b}_n^\dagger, \hat{b}_n] \\ &= \sum_n E_n(\hat{b}_n^\dagger\hat{b}_n - \tfrac{1}{2}) = \sum_n E_n(\hat{N}_n - \tfrac{1}{2})\,.\end{aligned} \tag{9.20}$$

The operator

$$\hat{N}_n = \hat{b}_n^\dagger \hat{b}_n \tag{9.21}$$

is called the number operator for particles with the quantum numbers n. Using (9.18, 19) immediately gives

$$\hat{N}_n^2 = \hat{b}_n^\dagger\hat{b}_n\hat{b}_n^\dagger\hat{b}_n = \hat{b}_n^\dagger(1 - \hat{b}_n^\dagger b_n)\hat{b}_n = \hat{N}_n - \hat{b}_n^\dagger\hat{b}_n^\dagger\hat{b}_n\hat{b}_n = \hat{N}_n\,. \tag{9.22}$$

This shows that the only eigenvalues of the number operator are 0 and 1, i.e. that at most one electron can occupy any given state. Again the Pauli principle is a direct consequence of the quantization with anticommutators.

However, we are faced with a different problem already mentioned: the energy eigenvalues E_n are not positive definite, as discussed in detail in Chap. 2. Thus, the expectation value of $\hat{H}$ can take arbitrarily large negative values. The way out of this dilemma is provided by the observation that for every n the contribution to $\hat{H}$ is bounded below, even if E_n is negative, for any state $|\Omega\rangle$:

$$\frac{\langle\Omega|E_n(\hat{N}_n - \frac{1}{2})|\Omega\rangle}{\langle\Omega|\Omega\rangle} \geqslant \left\{\begin{matrix} -\frac{1}{2}E_n & (E_n \geqslant 0) \\ -\frac{1}{2}(-E_n) & (E_n < 0)\end{matrix}\right\} = -\tfrac{1}{2}|E_n|\,.$$

Commuting $\hat{b}_n$ and $\hat{b}_n^\dagger$, the Hamiltonian can also be written as

$$\hat{H} = \sum_n (-E_n)(\hat{b}_n \hat{b}_n^\dagger - \tfrac{1}{2}) . \tag{9.23}$$

The operator $\hat{b}_n \hat{b}_n^\dagger = 1 - \hat{N}_n$ has the same properties as $\hat{N}_n$, it is a number operator for holes, i.e. for the absence of particles. This procedure amounts to interchanging the roles of the operators $\hat{b}_n$ and $\hat{b}_n^\dagger$. Instead of calling $\hat{b}_n$ a destruction operator for an electron in state ϕ_n, it can be called a creation operator for a vacancy in state ϕ_n, and vice versa with $\hat{b}_n^\dagger$. Obviously this must be done for all states Φ_n with negative energy eigenvalue in order to reverse the sign of the contributions from these states. In the next sections we discuss how this is related to the definition of the vacuum state.

9.2 Fock Space and the Vacuum State (I)

So far we have deliberately refrained from discussing the states on which the field operators act. The problem of indefiniteness of the Hamilton operator encountered at the end of the last section indicates that we have reached a point where this can no longer be ignored.

Let us start from the state $|0, \text{bare}\rangle$, defined to be an eigenstate of the number operator $\hat{N}_n$ with eigenvalue zero for every n:

$$\hat{N}_n |0, \text{bare}\rangle = \hat{b}_n^\dagger \hat{b}_n |0, \text{bare}\rangle = 0 . \tag{9.24}$$

Obviously, this definition requires that every particle annihilation operator annihilates the state:

$$\hat{b}_n |0, \text{bare}\rangle = 0 \qquad \text{for all} \quad n . \tag{9.25}$$

All other states are built by acting upon $|0, \text{bare}\rangle$ with an arbitrary number of operators of the type $\hat{b}_n^\dagger$. A finite number of states with negative contribution to $\hat{H}$ would not harm, because then the spectrum of $\hat{H}$ would still have a finite lower bound. The minimum value would be assumed when all those states are occupied. Because of the relations

$$[\hat{N}_n, \hat{b}_n^\dagger] = \hat{b}_n^\dagger ; \qquad [\hat{N}_n, \hat{N}_m] = 0 , \tag{9.26}$$

which are an immediate consequence of the commutation relations (9.18, 19), the state

$$|n_1, \ldots, n_k, \text{bare}\rangle \equiv \hat{b}_{n_1}^\dagger \ldots \hat{b}_{n_k}^\dagger |0, \text{bare}\rangle \qquad (n_i \text{ all different}) \tag{9.27}$$

is an eigenstate of the number operators $\hat{N}_{n_1}, \ldots, \hat{N}_{n_k}$ with eigenvalue one, i.e. it contains precisely k particles, one in every state $\phi_{n_1}, \ldots, \phi_{n_k}$. Defining the norm of $|0, \text{bare}\rangle$ to be

$$\langle 0, \text{bare} | 0, \text{bare} \rangle = 1 \, , \tag{9.28}$$

then every scalar product of the type

$$\langle n'_1, \ldots, n'_{\bar{k}}, \text{bare} | n_1, \ldots, n_k, \text{bare} \rangle \tag{9.29}$$

can be readily evaluated by the commutation relations and definition (9.25). The scalar product vanishes, except where $k = \bar{k}$, and the numbers $(n'_1, \ldots, n'_{\bar{k}})$ are simply a permutation of the numbers $(n_1, \ldots, n_k)$. In this case the result is ± 1, depending on whether the permutation is even or odd. If an index n_i occurs twice, the state vanishes because of the Pauli principle. With these definitions, the space spanned by all normalizable linear combinations of states of the type (9.27) forms a Hilbert space, called the *Fock* space, on which the field operators act.

We are now ready to return to our Hamiltonian (9.20):

$$\hat{H} = \sum_n E_n (\hat{N}_n - \tfrac{1}{2}) \, .$$

According to the discussion at the end of the last section, the state $|0, \text{bare}\rangle$ is not the state with the lowest expectation value of $\hat{H}$. Rather, energy is gained by filling all states with negative energy eigenvalue E_n. This state is called the true ground state or vacuum state

$$|0\rangle = \prod_{n<0} \hat{b}_n^\dagger |0, \text{bare}\rangle \, , \tag{9.30}$$

where the symbol "$n<0$" means product of all single particle states with negative energy eigenvalue:

$$\text{“}n < 0\text{”} = \{n \,|\, E_n < 0\} \, . \tag{9.31}$$

The state $|0\rangle$ is precisely the Fock space equivalent of the Dirac picture of a filled sea of negative energy electrons introduced in Chap. 4 and Sect. 9.2 with heuristic arguments (Fig. 9.1). Now

$$E_{\text{vac}} = \langle 0 | \hat{H} | 0 \rangle = -\frac{1}{2} \left[\sum_{n<0} (-E_n) + \sum_{n>0} E_n \right] = -\frac{1}{2} \sum_n |E_n| \tag{9.32}$$

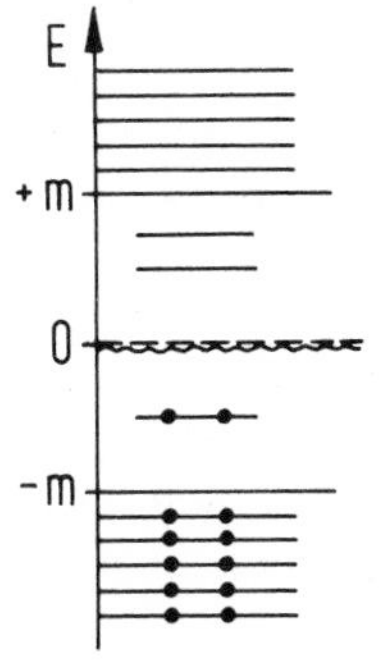

Fig. 9.1. Definition of the normal vacuum state: all levels with negative energy eigenvalue are occupied, all positive energy levels are empty. Holes in the negative energy levels are identified as antiparticles

and any other state $|\Omega\rangle$ has higher energy

$$\langle\Omega|\hat{H}|\Omega\rangle \geqslant \langle 0|\hat{H}|0\rangle . \tag{9.33}$$

With the definition (9.30) of the ground state $|0\rangle$, all "creation" operators $\hat{b}_n^\dagger$ associated with negative energy wave functions (i.e. with $n < F_0$, F_0 being the n-value of the single particle state characterizing the Fermi surface) annihilate $|0\rangle$, and not the "annihilation" operators $\hat{b}_n$, as in the state $|0, \text{bare}\rangle$. It is therefore natural to utilize the mentioned symmetry between the operators $\hat{b}_n$ and $\hat{b}_n^\dagger$ to define new creation and destruction operators $\hat{d}_n^\dagger$ and $\hat{d}_n$ for the negative energy states by:

$$\hat{d}_n^\dagger = \hat{b}_n , \quad \hat{d}_n = \hat{b}_n^\dagger \quad \text{for} \quad E_n < 0 . \tag{9.34}$$

The new destruction operators have the property of annihilating the ground state:

$$\left.\begin{array}{lll} \hat{b}_n|0\rangle = 0 & \text{for} & E_n > 0 \\ \hat{d}_n|0\rangle = 0 & \text{for} & E_n < 0 \end{array}\right\} . \tag{9.35}$$

The new particles thus defined (they are actually the holes in the old definition) are called *antiparticles*, or positrons. Their properties are the same as those of the electrons, with the sole exception of the charge, which is one positive unit ($-e$, e being the electron charge). With this new definition, the Hamilton operator $\hat{H}$ takes the form

$$\begin{aligned} \hat{H} &= \tfrac{1}{2} \sum_{n<0} E_n [\hat{d}_n, \hat{d}_n^\dagger] + \tfrac{1}{2} \sum_{n>0} E_n [\hat{b}_n^\dagger, \hat{b}_n] \\ &= \sum_{n<0} (-E_n)\hat{\bar{N}}_n + \sum_{n>0} E_n \hat{N}_n + E_{\text{vac}} , \end{aligned} \tag{9.36}$$

where the number operator for antiparticles was introduced

$$\hat{\bar{N}}_n = \hat{d}_n^\dagger \hat{d}_n \quad (E_n < 0) \tag{9.37}$$

and the vacuum energy separated (9.32).

It is customary to renormalize the energy operator by discarding the infinite contribution of the vacuum energy with the claim that only *differences* in energy are measurable. However, the energy eigenvalues E_n, and accordingly the vacuum energy E_{vac}, are functionals of the external electromagnetic potential. Therefore, the vacuum energy cannot be disregarded when states belonging to two different external potentials are compared. It is shown in Sect. 9.6 that E_{vac} does, indeed, depend on the strength of the potential, such that in the general case the vacuum energy must be included to obtain the correct energy balance.

After the vacuum energy has been separated, all operators in $\hat{H}$ are so ordered that the destruction operators are always to the right of the creation operators. This (particular) form of an operator is called *normal ordered*, indicated by enclosing the symbol of the operator in colons:

$$:\hat{H}: = \sum_{n<0} (-E_n)\hat{\bar{N}}_n + \sum_{n>0} E_n \hat{N}_n \,. \tag{9.38}$$

This form of the Hamiltonian can always be used to calculate the energy difference between Fock space states with different electron – positron configurations. Care must be exercised when energies are compared which pertain to different external potentials.

Let us now consider the conservation laws in the quantum field theory. These, in turn, and especially the conservation law of electric charge, shed new light on the meaning of various possible choices of the physical vacuum state.

9.3 Poincaré Invariance of the Quantum Theory

Since the space-time dependence of the field operator $\hat{\psi}(\boldsymbol{x}, t) \equiv \hat{\psi}(x)$ is governed by the same differential equation that holds for the classical Dirac wave function $\psi(x)$, the transformation properties of the field operator under space-time transformations must be the same. But the field operator $\hat{\psi}(x)$ taken at some space-time point x is an operator acting in a Hilbert space of state vectors. If the quantum theory is to be invariant under a certain symmetry operation, such as a Lorentz transformation, then the transformed field operator $\hat{\psi}'(x)$ and the original operator $\hat{\psi}(x)$ must be related by a unitary transformation. Only then will they have the same eigenvalues and transition matrix elements. Thus

$$\hat{\psi}'(x) = \hat{U}\hat{\psi}(x)\hat{U}^\dagger \tag{9.39}$$

with some unitary transformation $\hat{U}$ acting in the Hilbert space of state vectors $|\Omega\rangle$. For $\hat{\psi}'$ to be in the Heisenberg picture when $\hat{\psi}$ is (only then will ψ' satisfy the covariant Dirac equation), the transformation $\hat{U}$ must not depend on time, i.e. $\hat{U}$ must commute with the Hamiltonian.

As the external potential $A_\mu(x)$ in general destroys the invariance under Lorentz transformations and translations, we shall discuss only the *free* Dirac field. Again, it suffices to treat infinitesimal transformations, because the Lorentz transformations and translations taken together form the so-called Poincaré group. According to (2.98) the coordinates in the original and the transformed systems are related by

$$x^{\mu\prime} = x^\mu + \Delta\omega^\mu_\nu x^\nu + \varepsilon^\mu \,. \tag{9.40}$$

The transformation is denoted by the symbol L, i.e. $x' = L(x)$. Applying the correspondence principle to the considerations of Sect. 2.5, the connection between $\hat{\psi}'$ and $\hat{\psi}$ must be given by

$$\hat{\psi}'(x') = S\hat{\psi}(x) = \hat{\psi}(x) - \frac{\mathrm{i}}{4}\sigma_{\mu\nu}\Delta\omega^{\mu\nu}\hat{\psi}(x) \,. \tag{9.41}$$

From Taylor's theorem the lhs is

$$\hat{\psi}'(x') = \hat{\psi}'(x) + \frac{\partial \hat{\psi}'(x)}{\partial x^\mu}(\Delta\omega^\mu_{\ \nu} x^\nu + \varepsilon^\mu) \tag{9.42}$$

if we retain only terms linear in $\Delta\omega^\mu_{\ \nu}$ and ε^μ. Combining the last two equations gives

$$\hat{\psi}'(x) = \hat{\psi}(x) + \frac{1}{2}\left[\left(x_\mu \frac{\partial}{\partial x^\nu} - x_\nu \frac{\partial}{\partial x^\mu}\right)\hat{\psi}(x) - \frac{\mathrm{i}}{2}\sigma_{\mu\nu}\hat{\psi}(x)\right]\Delta\omega^{\mu\nu} - \varepsilon^\mu \frac{\partial \hat{\psi}(x)}{\partial x^\mu}, \tag{9.43}$$

where we used the antisymmetry of $\Delta\omega^{\mu\nu}$ and replaced $\hat{\psi}'$ by $\hat{\psi}$ in terms that are linear in the infinitesimals $\Delta\omega$ and ε.

On the other hand, for an infinitesimal transformation L the unitary operator $\hat{U}$ from (9.39) differs only slightly from the identity, so that

$$\hat{U}(L) = \mathbb{1} + \mathrm{i}\hat{K}(L) \tag{9.44}$$

with an infinitesimal Hermitian operator $\hat{K}$. Equation (9.39) then takes the form:

$$\hat{\psi}'(x) = \hat{\psi}(x) + \mathrm{i}[\hat{K}, \hat{\psi}(x)]\,. \tag{9.45}$$

Equating this condition with the infinitesimal transformation law (9.43) we obtain an equation for the operator $\hat{K}(L)$. Since $\hat{K}$ must depend linearly on $\Delta\omega$ and ε, it may be decomposed to

$$\hat{K}(L) = \tfrac{1}{2}\hat{M}_{\mu\nu}\Delta\omega^{\mu\nu} - \hat{P}_\mu \varepsilon^\mu \tag{9.46}$$

with yet unknown operators $\hat{M}_{\mu\nu}$ and $\hat{P}_\mu$.

The corresponding operator conditions for $\hat{M}_{\mu\nu}$ and $\hat{P}_\mu$ are obtained from (9.43) as:

$$\mathrm{i}[\hat{P}_\mu, \hat{\psi}(x)] = \frac{\partial \hat{\psi}(x)}{\partial x^\mu} \tag{9.47}$$

$$\mathrm{i}[\hat{M}_{\mu\nu}, \hat{\psi}(x)] = \left(x_\mu \frac{\partial}{\partial x^\nu} - x_\nu \frac{\partial}{\partial x^\mu}\right)\hat{\psi}(x) - \frac{\mathrm{i}}{2}\sigma_{\mu\nu}\hat{\psi}(x)\,. \tag{9.48}$$

If, and only if, we can find time-independent Hermitian operators $\hat{P}_\mu$ and $\hat{M}_{\mu\nu}$ satisfying these relations, then the quantum theory is invariant under Lorentz transformations and translations. Invariance of the classical theory does not automatically imply the existence of an invariance of the quantized theory. However, it is a strong indication that such an invariance should exist. But the matter is even more complicated: even if the quantized Hamiltonian is invariant, the vacuum state and the Fock space built on it need not exhibit symmetry.

In this case one speaks of "spontaneous" symmetry breaking, a typical example being the breaking of rotational symmetry by a spontaneously magnetized ferromagnet (see [Be 74, Le 81] for an introduction).

For the Dirac field it is simple to construct the operators $\hat{P}_\mu$ and $\hat{M}_{\mu\nu}$. It turns out that it is sufficient to take the expressions for the energy momentum four-vector and the angular momentum tensor of a classical Dirac particle, respectively, and replace the wave function by the field operator wherever it occurs. As in the definition of the Hamilton operator $\hat{H}$, the question as to the order of field operators must be solved by explicit symmetrization:

$$\hat{P}_\mu = \frac{\mathrm{i}}{2}\int d^3x\,[\hat{\psi}^\dagger(\boldsymbol{x},t), \frac{\partial}{\partial x^\mu}\,\hat{\psi}(\boldsymbol{x},t)]\,, \tag{9.49}$$

$$\hat{M}_{\mu\nu} = \frac{\mathrm{i}}{2}\int d^3x\left[\hat{\psi}^\dagger(\boldsymbol{x},t), \left(x_\mu\frac{\partial}{\partial x^\nu} - x_\nu\frac{\partial}{\partial x^\mu} - \frac{\mathrm{i}}{2}\sigma_{\mu\nu}\right)\hat{\psi}(\boldsymbol{x},t)\right]. \tag{9.50}$$

As an example we verify explicitly that the ansatz (9.49) satisfies (9.47). One must first show that the operators commute with the Hamiltonian. This is trivial for $\hat{P}_0$, because with the equation of motion (9.9)

$$\hat{P}_0 = \frac{\mathrm{i}}{2}\int d^3x\left[\hat{\psi}^\dagger(\boldsymbol{x},t), \frac{\partial}{\partial t}\,\hat{\psi}(\boldsymbol{x},t)\right] = \frac{1}{2}\int d^3x\,[\hat{\psi}^\dagger(\boldsymbol{x},t), H_{\mathrm{D}}\hat{\psi}(\boldsymbol{x},t)] \equiv \hat{H} \tag{9.51}$$

is just the Hamiltonian, hence it commutes with $\hat{H}$. For the vector components $\hat{\boldsymbol{P}} = (\hat{P}^1, \hat{P}^2, \hat{P}^3)$ the calculation is lengthy but straightforward. Remembering that our convention for the metric tensor implies

$$\left(\frac{\partial}{\partial x_1}, \frac{\partial}{\partial x_2}, \frac{\partial}{\partial x_3}\right) = \left(-\frac{\partial}{\partial x^1}, -\frac{\partial}{\partial x^2}, -\frac{\partial}{\partial x^3}\right) = -\nabla\,,$$

the commutator between $\hat{H}$ and $\hat{\boldsymbol{P}}$ is explicitly

$$[\hat{H}, \hat{\boldsymbol{P}}] = -\frac{\mathrm{i}}{4}\int d^3x\,d^3x'\,[[\hat{\psi}^\dagger(\boldsymbol{x},t), H_{\mathrm{D}}^{(x)}\,\hat{\psi}(\boldsymbol{x},t)], [\hat{\psi}^\dagger(\boldsymbol{x}',t), \nabla'\,\hat{\psi}(\boldsymbol{x}',t)]]\,. \tag{9.52}$$

The commutator under the integral is symbolically $[[A,B],[C,D]]$. Because of the anticommutation relations (9.1, 2) all the anticommutators $\{A,C\}$, $\{A,D\}$, $\{B,C\}$, $\{B,D\}$ are C numbers in Fock space and hence commute with Fock space operators. Under this condition the following operator identity holds:

$$\begin{aligned}
[[A,B],[C,D]] &= A\{B,C\}D + \{B,C\}AD - DA\{B,C\} - D\{B,C\}A \\
&\quad - B\{A,C\}D - \{A,C\}BD + DB\{A,C\} + D\{A,C\}B \\
&\quad - A\{B,D\}C - \{B,D\}AC + CA\{B,D\} + C\{B,D\}A \\
&\quad + B\{A,D\}C + \{A,D\}BC - CB\{A,D\} - C\{A,D\}B \\
&= 2[A,D]\{B,C\} - 2[B,D]\{A,C\} - 2[A,C]\{B,D\} \\
&\quad + 2[B,C]\{A,D\}\,.
\end{aligned}$$

In our special case, the anticommutators $\{A, C\}$ and $\{B, D\}$ actually vanish by virtue of (9.1). Therefore

$$
\begin{aligned}
[\hat{H}, \hat{\boldsymbol{P}}] = & -\frac{\mathrm{i}}{2} \int d^3x\, d^3x' \left([\hat{\psi}^\dagger(\boldsymbol{x}, t), \nabla' \hat{\psi}(\boldsymbol{x}', t)]\, H_{\mathrm{D}}^{(x)} \delta(\boldsymbol{x}-\boldsymbol{x}') \right. \\
& \left. + [H_{\mathrm{D}}^{(x)} \hat{\psi}(\boldsymbol{x}, t), \hat{\psi}^\dagger(\boldsymbol{x}', t)]\, \nabla' \delta(\boldsymbol{x}-\boldsymbol{x}') \right) \\
= & -\frac{\mathrm{i} m_e}{2} \int d^3x\, \nabla [\hat{\psi}^\dagger(\boldsymbol{x}, t), \beta \hat{\psi}(\boldsymbol{x}, t)]
\end{aligned}
\tag{9.53}
$$

where we have integrated by parts. For localized states this gives a vanishing surface term:

$$
\left\langle \mathrm{loc}' \left| \frac{d}{dt} \hat{\boldsymbol{P}} \right| \mathrm{loc} \right\rangle = 0 \, . \tag{9.54}
$$

In deriving (9.52) it must be remembered that $\hat{H}$ is time independent, allowing t to be chosen the same in the expressions for $\hat{H}$ and $\hat{\boldsymbol{P}}$. That $\hat{P}_\mu$ satisfies (9.47) is easily verified by direct calculation:

$$
\begin{aligned}
\mathrm{i}[\hat{P}_\mu, \hat{\psi}(\boldsymbol{x}, t)] &= -\frac{1}{2} \int d^3x' \left[\left[\hat{\psi}^\dagger(\boldsymbol{x}', t), \frac{\partial}{\partial x'^\mu} \hat{\psi}(\boldsymbol{x}', t) \right], \hat{\psi}(\boldsymbol{x}, t) \right] \\
&= -\frac{1}{2} \int d^3x' \left(\left\{ \hat{\psi}^\dagger(\boldsymbol{x}', t), \left\{ \frac{\partial}{\partial x'^\mu} \hat{\psi}(\boldsymbol{x}', t), \hat{\psi}(\boldsymbol{x}, t) \right\} \right\} \right. \\
&\quad \left. - \left\{ \{\hat{\psi}^\dagger(\boldsymbol{x}', t), \hat{\psi}(\boldsymbol{x}, t)\}, \frac{\partial}{\partial x'^\mu} \hat{\psi}(\boldsymbol{x}', t) \right\} \right) \\
&= \int d^3x' \left\{ \delta(\boldsymbol{x}-\boldsymbol{x}'), \frac{\partial}{\partial x'^\mu} \hat{\psi}(\boldsymbol{x}', t) \right\} = \frac{\partial}{\partial x^\mu} \hat{\psi}(\boldsymbol{x}, t) \, .
\end{aligned}
\tag{9.55}
$$

Here we utilized the identity

$$
[[\hat{A}, \hat{B}], \hat{C}] = \{\hat{A}, \{\hat{B}, \hat{C}\}\} - \{\{\hat{A}, \hat{C}\}, \hat{B}\} \tag{9.56}
$$

which expresses the commutators in terms of anticommutators. The second set of equations for $\hat{M}_{\mu\nu}$ can be verified in the same way.

We have thus established two important results.

1) The equation for the quantized free Dirac field is covariant under Poincaré transformations of the coordinate system. The Dirac field transforms according to (9.39), where the unitary operator $\hat{U}$ is given by

$$
\hat{U} = \exp\left(\frac{\mathrm{i}}{2} \hat{M}_{\mu\nu} \Delta\omega^{\mu\nu} - \mathrm{i} \hat{P}_\mu \varepsilon^\mu \right) . \tag{9.57}
$$

2) The operators $\hat{P}_\mu$ and $\hat{M}_{\mu\nu}$ are the quantized versions of the expressions for energy, momentum and angular momentum of the classical Dirac field. They

commute with the Hamiltonian (in the Fock space of localized states) and, being Hermitian operators, correspond to observable constants of motion.

In the presence of an external potential, the energy momentum operator $\hat{P}_\mu$ must be replaced by the gauge-invariant expression

$$\hat{\Pi}_\mu = \frac{1}{2}\int d^3x \left[\hat{\psi}^\dagger(\boldsymbol{x},t), \left(\mathrm{i}\frac{\partial}{\partial x^\mu} - eA_\mu \right) \hat{\psi}(\boldsymbol{x},t) \right] . \tag{9.58}$$

Equations (9.47) are then replaced by

$$\mathrm{i}[\hat{\Pi}_\mu, \hat{\psi}(x)] = \left(\frac{\partial}{\partial x^\mu} + \mathrm{i}eA_\mu \right) \hat{\psi}(\boldsymbol{x},t) , \tag{9.59}$$

which does not express anything like the Poincaré invariance of the theory. It is also seen that $\hat{\Pi}_0$ is not equal to the Hamiltonian, but instead the quantum analogue of the gauge invariant expression for the energy of the Dirac field, (8.27a), which must be supplemented by the energy residing in the electromagnetic field.

Calculating the time derivative of the gauge invariant momentum for localized states along the lines of (9.53), using (9.7) for the Dirac current operator, gives

$$\begin{aligned} \langle \mathrm{loc}' \left| \frac{d}{dt}\hat{\boldsymbol{\Pi}} \right| \mathrm{loc} \rangle &= \langle \mathrm{loc}' \left| \frac{\partial}{\partial t}\hat{\boldsymbol{\Pi}} + \mathrm{i}[\hat{H},\hat{\boldsymbol{\Pi}}] \right| \mathrm{loc} \rangle \\ &= \langle \mathrm{loc}' | \hat{j}^0\boldsymbol{E} + \hat{\boldsymbol{j}} \times \boldsymbol{B} | \mathrm{loc} \rangle . \end{aligned} \tag{9.60}$$

This is the expression for the Lorentz force in the quantized theory.

9.4 Gauge Invariance and Discrete Symmetries

As known from the classical Dirac theory, there are further symmetry transformations related either to internal degrees of freedom or to discrete space-time transformations. The most notable internal symmetry is known as gauge invariance. The classical fields are invariant under the transformation

$$\psi'(x) = \mathrm{e}^{-\mathrm{i}e\chi(x)}\psi(x) , \qquad A'_\mu(x) = A_\mu(x) + \frac{\partial \chi}{\partial x^\mu} . \tag{9.61}$$

It was shown that this invariance gives rise to the continuity equation

$$\frac{\partial}{\partial x^\mu} j^\mu(x) = \frac{\partial}{\partial x^\mu}(e\bar{\psi}\gamma^\mu\psi) = 0 \tag{9.62}$$

which leads to the conserved charge

$$Q = \int d^3x \, j^0(\boldsymbol{x}, t) = e \int d^3x \, \psi^\dagger(x) \, \psi(x) \, . \tag{9.63}$$

Reasoning as in the last section, we can argue now that the same transformation applies to the operator of the quantized Dirac field

$$\hat{\psi}'(x) = \mathrm{e}^{-\mathrm{i}e\chi(x)} \, \hat{\psi}(x) \, , \tag{9.61'}$$

whereas $A(x)$ remains a classical, unquantized field in this context. Since $\hat{\psi}'$ and $\hat{\psi}$ represent the same physical field, they must be related by a unitary transformation,

$$\hat{\psi}'(x) = \hat{U}[\chi] \, \hat{\psi}(x) \, \hat{U}[\chi]^{-1} \, , \qquad \hat{U} = \exp(\mathrm{i} \, \hat{W}) \, . \tag{9.64}$$

For infinitesimal transformations $\delta\chi(x)$, expanding in first order,

$$\hat{\psi}(x) + \mathrm{i}[\hat{W}[\delta\chi], \, \hat{\psi}(x)] = \hat{\psi}'(x) = \hat{\psi}(x) - \mathrm{i} e \, \delta\chi(x) \, \hat{\psi}(x) \, , \tag{9.65}$$

and by comparison we conclude that $\hat{W}$ must satisfy

$$[\hat{W}, \hat{\psi}(x)] = -e \delta\chi(x) \, \hat{\psi}(x) \, . \tag{9.66}$$

This relation is certainly satisfied by the explicitly symmetrized expression

$$\hat{W}[\delta\chi] = \frac{e}{2} \int d^3x \, [\hat{\psi}^\dagger(\boldsymbol{x}, t), \, \hat{\psi}(\boldsymbol{x}, t)] \, \delta\chi(x) = \int d^3x \, \hat{j}^0(x) \, \delta\chi(x) \, . \tag{9.67}$$

This is again easily shown if we use of operator identity (9.56) and the equal-time anticommutation relations (9.1, 2):

$$\begin{aligned}
[\hat{W}, \hat{\psi}(\boldsymbol{x}, t)] &= \frac{e}{2} \int d^3x' \, \delta\chi(\boldsymbol{x}', t) (\{\hat{\psi}^\dagger(\boldsymbol{x}', t), \{\hat{\psi}(\boldsymbol{x}', t), \, \hat{\psi}(\boldsymbol{x}, t)\}\} \\
&\quad - \{\{\hat{\psi}^\dagger(\boldsymbol{x}', t), \, \hat{\psi}(\boldsymbol{x}, t)\}, \, \hat{\psi}(\boldsymbol{x}', t)\}) \\
&= -e \int d^3x' \, \delta\chi(\boldsymbol{x}', t) \, \hat{\psi}(\boldsymbol{x}', t) \, \delta^3(\boldsymbol{x} - \boldsymbol{x}') = -e \, \delta\chi(\boldsymbol{x}, t) \, \hat{\psi}(\boldsymbol{x}, t) \, .
\end{aligned}$$

Thus this is the unitary operator which implements gauge transformations in the quantum theory.

Taking the special case of space-time independent gauge transformations χ gives the quantum charge operator

$$\hat{Q} = \frac{1}{\chi} \hat{W}[\chi] = \frac{e}{2} \int d^3x \, [\hat{\psi}^\dagger(\boldsymbol{x}, t), \, \hat{\psi}(\boldsymbol{x}, t)] \, . \tag{9.68}$$

For localized states, $\hat{Q}$ commutes with the Hamiltonian. To prove this we apply the same method as used to calculate the time derivative of the momentum operator in (9.51 – 54). The operator identity (9.53) gives:

$$[\hat{H},\hat{Q}] = \frac{e}{4}\int d^3x\, d^3x' [[\hat{\psi}^\dagger(\boldsymbol{x},t), H_D\hat{\psi}(\boldsymbol{x},t)], [\hat{\psi}^\dagger(\boldsymbol{x}',t), \hat{\psi}(\boldsymbol{x}',t)]]$$

$$= \frac{e}{2}\int d^3x\, d^3x' ([\hat{\psi}^\dagger(\boldsymbol{x},t), \hat{\psi}(\boldsymbol{x}',t)] H_D(x)\, \delta^3(\boldsymbol{x}-\boldsymbol{x}')$$

$$+ [H_D\hat{\psi}(\boldsymbol{x},t), \hat{\psi}^\dagger(\boldsymbol{x}',t)]\, \delta^3(\boldsymbol{x}-\boldsymbol{x}'))$$

$$= \frac{e}{2}\int d^3x ([(\boldsymbol{\alpha}\cdot(-\boldsymbol{p}-e\boldsymbol{A}) + \beta m_e + V)\, \hat{\psi}^\dagger(\boldsymbol{x},t), \hat{\psi}(\boldsymbol{x},t)]$$

$$- [\hat{\psi}^\dagger(\boldsymbol{x},t), (\boldsymbol{\alpha}\cdot(\boldsymbol{p}-e\boldsymbol{A}) + \beta m_e + U)\, \hat{\psi}(\boldsymbol{x},t)])$$

$$= \frac{\mathrm{i}e}{2}\int d^3x\, \nabla\cdot [\hat{\psi}^\dagger(\boldsymbol{x},t), \boldsymbol{\alpha}\hat{\psi}(\boldsymbol{x},t)]\,.$$

For localized states the integral vanishes by virtue of Gauss' theorem:

$$\langle \mathrm{loc}' | [\hat{H},\hat{Q}] | \mathrm{loc}\rangle = \frac{\mathrm{i}e}{2}\langle \mathrm{loc}' | \oint d\boldsymbol{\Sigma}\cdot [\hat{\psi}^\dagger, \boldsymbol{\alpha}\hat{\psi}] | \mathrm{loc}\rangle = 0\,, \tag{9.69}$$

which proves that the electric charge is a constant of motion also in the quantum theory. Whereas the conservation of energy, momentum and angular momentum of the Dirac field is violated when a generic external field is present, the charge of the Dirac field is still conserved because the electromagnetic field does not carry charge. This is intimately connected with the fact that the group of electromagnetic gauge transformations $U(1)$ is Abelian, whereas the Lorentz and Poincaré groups are not. We remark in passing that in gauge theories based on more complicated gauge transformations (so-called Yang-Mills theories, e.g. quantum chromodynamics) the gauge field carries a kind of charge, too, and (9.69) does not hold [It 80, Ai 82].

The discrete symmetry transformations are more difficult to treat, because here we cannot begin with the infinitesimal transformations, but have to treat the finite transformation law at once. Since this involves slightly different techniques than the derivation of the continuous transformation laws, we shall deal with the charge conjugation law in detail. Here a unitary Fock space operator $\hat{\mathscr{C}}$ that accomplishes the classical charge conjugation transformation on the field operator is necessary [cf. (4.17)]

$$\hat{\mathscr{C}}\,\hat{\psi}(x)\hat{\mathscr{C}}^\dagger = C\hat{\bar{\psi}}^{\mathrm{T}}(x) = C\gamma^{0\mathrm{T}}\,\hat{\psi}^*(x) \equiv \hat{\psi}^{\mathrm{c}}(x) \tag{9.70}$$

and

$$\hat{\mathscr{C}}\,\hat{\bar{\psi}}(x)\hat{\mathscr{C}}^\dagger = -\,\hat{\psi}^{\mathrm{T}}(x)\, C^\dagger = \hat{\bar{\psi}}^{\mathrm{c}}(x)\,. \tag{9.71}$$

The symbol "T" denotes spin matrix transposition, "*" denotes Hermitian conjugate except in spin space, and "†" is the full Hermitian conjugate, i.e. the product of the * and T operations. Further, C is a spin matrix satisfying $C\gamma_\mu = -\gamma_\mu^{\mathrm{T}} C$. The choice $C = \mathrm{i}\gamma^2\gamma^0$ in the standard representation of the γ matrices gives $C = -C^{\mathrm{T}} = -C^\dagger$ [cf. (4.24, 25)].

We shall now prove that

$$\hat{\mathscr{C}} = \exp\left[\pm\frac{\mathrm{i}\pi}{4}\int d^3x([\hat{\bar{\psi}},\hat{\psi}]-(\hat{\bar{\psi}}\hat{\psi}^{\mathrm{c}}+\hat{\bar{\psi}}^{\mathrm{c}}\hat{\psi}))\right] \tag{9.72}$$

has the desired properties. The sign in the exponent can be selected freely, because $\hat{\mathscr{C}}$ is idempotent and hence equal to its own inverse. The second term in (9.72) is constructed to be Hermitian and invariant under charge conjugation:

$$\hat{\mathscr{C}}(\hat{\bar{\psi}}\hat{\psi}^{\mathrm{c}})\hat{\mathscr{C}}^{\dagger} = (\hat{\mathscr{C}}\hat{\bar{\psi}}\hat{\mathscr{C}}^{\dagger})(\hat{\mathscr{C}}\hat{\psi}^{\mathrm{c}}\hat{\mathscr{C}}^{\dagger}) = \hat{\bar{\psi}}^{\mathrm{c}}\hat{\psi}\,, \tag{9.73a}$$

$$(\hat{\bar{\psi}}\hat{\psi}^{\mathrm{c}})^{\dagger} = \hat{\psi}^{\mathrm{c}\dagger}\hat{\bar{\psi}}^{\dagger} = \hat{\psi}^{\mathrm{c}}\gamma^0\hat{\psi} = \hat{\bar{\psi}}^{\mathrm{c}}\hat{\psi}\,. \tag{9.73b}$$

Note that the explicit nature of $\hat{\mathscr{C}}$ has not been used, but only its existence which, of course, still has to be demonstrated. This demonstration proceeds best by writing

$$\hat{\mathscr{C}} = \exp(\pm\mathrm{i}\hat{\Gamma}),\, \hat{\Gamma}^{\dagger} = \hat{\Gamma} = \frac{\pi}{4}\int d^3x([\hat{\bar{\psi}},\hat{\psi}]-\hat{\bar{\psi}}\hat{\psi}^{\mathrm{c}}-\hat{\bar{\psi}}^{\mathrm{c}}\hat{\psi})\,, \tag{9.74}$$

whereupon (9.70) takes the form

$$\hat{\mathscr{C}}\hat{\psi}\hat{\mathscr{C}}^{\dagger} = \mathrm{e}^{\pm\mathrm{i}\hat{\Gamma}}\hat{\psi}\mathrm{e}^{\mp\mathrm{i}\hat{\Gamma}} = \hat{\psi} + \sum_{n=1}^{\infty}\frac{(\pm\mathrm{i})^n}{n!}[\hat{\Gamma},\hat{\psi}]^{(n)}\,. \tag{9.75}$$

Here we have introduced the notation

$$[A,B]^{(1)} = [A,B]\,, \qquad [A,B]^{(n)} = [A,[A,B]^{(n-1)}] \tag{9.76}$$

for the n-fold commutator.

Explicit calculation gives

$$\begin{aligned}
[\hat{\Gamma},\hat{\psi}(x)] &= \frac{\pi}{4}\int d^3x'\,[[\hat{\bar{\psi}}(x'),\hat{\psi}(x')]-(\hat{\bar{\psi}}(x')\hat{\psi}^{\mathrm{c}}(x')+\hat{\bar{\psi}}^{\mathrm{c}}(x')\hat{\psi}(x')),\hat{\psi}(x)]\\
&= -\frac{\pi}{2}\gamma^0(\hat{\psi}(x)-\hat{\psi}^{\mathrm{c}}(x))\,,\\
[\hat{\Gamma},[\hat{\Gamma},\hat{\psi}(x)]] &= -\frac{\pi}{2}\gamma^0[\hat{\Gamma},\hat{\psi}(x)-\hat{\psi}^{\mathrm{c}}(x)] = \frac{\pi^2}{2}(\hat{\psi}(x)-\hat{\psi}^{\mathrm{c}}(x))\,,\\
&\;\;\vdots\\
[\hat{\Gamma},\hat{\psi}(x)]^{(n)} &= \frac{(-\pi)^n}{2}(\gamma^0)^n(\hat{\psi}(x)-\hat{\psi}^{\mathrm{c}}(x))\,.
\end{aligned} \tag{9.77}$$

To evaluate the commutators we used $C^{\mathrm{T}} = -C$, which makes it very important to keep careful track of the spin indices, in order to arrive at the correct result. Let us show in detail for the first (9.77) how the calculation goes, using (9.56):

$$\begin{aligned}
[[\hat{\bar{\psi}}_\alpha(x'),\hat{\psi}_\alpha(x')],\hat{\psi}_\beta(x)] &= \{\hat{\bar{\varphi}}_\alpha(x'),\{\hat{\psi}_\alpha(x'),\hat{\psi}_\beta(x)\}\}-\{\{\hat{\bar{\psi}}_\alpha(x'),\hat{\psi}_\beta(x)\},\hat{\psi}_\alpha(x')\}\\
&= -2\,\delta(x-x')\,\gamma^0_{\beta\alpha}\hat{\psi}_\alpha(x') = -2\,\delta(x-x')\,(\gamma^0\hat{\psi}(x))_\beta\,,
\end{aligned}$$

$$
\begin{aligned}
[\hat{\bar{\psi}}_\alpha(x')\,\hat{\psi}^{\rm c}_\alpha(x'),\,\hat{\psi}_\beta(x)] &= [\hat{\bar{\psi}}_\alpha(x')\,C_{\alpha\gamma}\hat{\bar{\psi}}_\gamma(x'),\,\hat{\psi}_\beta(x)] \\
&= \hat{\bar{\psi}}_\alpha(x')\,C_{\alpha\gamma}\{\hat{\bar{\psi}}_\gamma(x'),\,\hat{\psi}_\beta(x)\} - \{\hat{\bar{\psi}}_\alpha(x'),\,\hat{\psi}_\beta(x)\}\,C_{\alpha\gamma}\bar{\psi}_\gamma(x') \\
&= (\hat{\bar{\psi}}_\alpha(x')\,C_{\alpha\gamma}\gamma^0_{\beta\gamma} - \gamma^0_{\beta\alpha}C_{\alpha\gamma}\hat{\bar{\psi}}_\gamma(x'))\,\delta(x-x') \\
&= \delta(x-x')\,(\gamma^0(C^{\rm T}-C)\,\hat{\bar{\psi}}(x))_\beta \\
&= -2\,\delta(x-x')\,(\gamma^0\,\hat{\psi}^{\rm c}(x))_\beta\,,
\end{aligned}
$$

because $C^{\rm T} = -C$ and $[AB, C] = A\{B, C\} - \{A, C\}B$.

Finally

$$
[\hat{\psi}^{\rm c}_\alpha(x')\,\hat{\psi}_\alpha(x'),\,\hat{\psi}_\beta(x)] = [-\,\hat{\psi}_\gamma(x')\,C^\dagger_{\gamma\alpha}\hat{\psi}_\alpha(x'),\,\hat{\psi}_\beta(x)] = 0\,,
$$

since $\hat{\psi}(x)$ anticommutes with $\hat{\psi}(x')$. Combining these three results, the first equation in (9.77) is proven. The rest of the proof is straightforward, as (9.75) gives

$$
\begin{aligned}
\mathscr{C}\,\hat{\psi}\,\mathscr{C}^\dagger &= \hat{\psi} + \left(\sum_{n=1}^{\infty}\frac{(\mp\mathrm{i}\,\pi\,\gamma^0)^n}{n!}\right)\cdot\frac{1}{2}(\hat{\psi}-\hat{\psi}^{\rm c}) \\
&= \hat{\psi} + \frac{1}{2}(\exp(\mp\mathrm{i}\,\pi\,\gamma^0)-1)\cdot(\hat{\psi}-\hat{\psi}^{\rm c}) = \hat{\psi}^{\rm c}\,.
\end{aligned}
\tag{9.78}
$$

In the last step one uses the fact that γ^0 has the eigenvalues (± 1), therefore $\exp(\mp\mathrm{i}\,\pi\,\gamma^0)$ has only the eigenvalue (-1) and must be equal to the negative unit matrix. A similar calculation shows that (9.72) gives as desired

$$
\hat{\mathscr{C}}\,\hat{\psi}^\dagger\hat{\mathscr{C}}^\dagger = \hat{\psi}^{\rm c\dagger}\,.
\tag{9.79}
$$

We now turn to the effect of charge conjugation on bilinear field operators, taking the current operator as a typical example. Since $\hat{\Gamma}$ and hence also $\mathscr{C}$ are scalars in spin space,

$$
\mathscr{C}\,(\hat{\bar{\psi}}\gamma^\mu\hat{\psi})\,\mathscr{C}^\dagger = (\mathscr{C}\,\hat{\bar{\psi}}\,\mathscr{C}^\dagger)\,\gamma^\mu(\mathscr{C}\,\hat{\psi}\,\mathscr{C}^\dagger) = \hat{\bar{\psi}}^{\rm c}\gamma^\mu\hat{\psi}^{\rm c} = \hat{\psi}^{\rm T}\gamma^{\mu{\rm T}}\,\hat{\bar{\psi}}^{\rm T}\,,
\tag{9.80}
$$

where heavy use was made of the properties of the matrix C. Because $\hat{\psi}$ and $\hat{\bar{\psi}}$ do not commute, $\hat{\bar{\psi}}\gamma^\mu\hat{\psi}$ do not transform properly, so the correct choice of the electric current operator must be

$$
j^\mu(x) = \frac{e}{2}(\hat{\bar{\psi}}\gamma^\mu\hat{\psi} - \hat{\bar{\psi}}^{\rm c}\gamma^\mu\hat{\psi}^{\rm c}) = \frac{e}{2}(\hat{\bar{\psi}}\gamma^\mu\hat{\psi} - (\gamma^\mu\hat{\psi})^{\rm T}\hat{\bar{\psi}}^{\rm T})\,.
\tag{9.81}
$$

In our somewhat free notation adopted in (9.5) this expression is just the commutator

$$
j^\mu(x) = \frac{e}{2}[\hat{\bar{\psi}},\,\gamma^\mu\hat{\psi}]\,.
\tag{9.81$'$}
$$

We now see why bilinear field operators have to be antisymmetrized, viz., to transform correctly under charge conjugation:

$$\hat{\mathscr{C}}\hat{j}^\mu\hat{\mathscr{C}}^\dagger = -\hat{j}^\mu\,; \qquad \hat{\mathscr{C}}\hat{H}_{A_\mu=0}\hat{\mathscr{C}}^\dagger = \hat{H}_{A_\mu=0}\,. \tag{9.82}$$

The latter relation also shows that the unitary operator $\hat{\mathscr{C}}$ is time independent. Multiplying by $\hat{\mathscr{C}}$ from the right gives

$$[\hat{\mathscr{C}}, \hat{H}_{A_\mu=0}] = 0\,. \tag{9.83}$$

Charge conjugation invariance of the Hamiltonian is of course destroyed if an external field is present. In this general case one obtains

$$\hat{\mathscr{C}}\,\hat{H}[A_\mu]\,\hat{\mathscr{C}}^\dagger = \hat{H}[-A_\mu]\,, \tag{9.84}$$

which corresponds to the "classical" relation obtained in (4.31, 32). The relation for the electric current operator remains unaffected, because it does not contain the external potential explicitly.

We finally note without proof the operators that implement the parity transformation and time reversal, denoted by $\hat{\mathscr{P}}$ and $\hat{\mathscr{T}}$, respectively. By the same kind of reasoning as above,

$$\begin{aligned} &\hat{\mathscr{P}}\,\hat{\psi}(\boldsymbol{x},t)\hat{\mathscr{P}}^\dagger = \gamma^0\hat{\psi}(-\boldsymbol{x},t) \qquad \text{with} \\ &\hat{\mathscr{P}} = \exp\left\{\pm\frac{\pi}{4}\int d^3x\,[\hat{\bar{\psi}}(\boldsymbol{x},t), \hat{\psi}(\boldsymbol{x},t) - \gamma^0\hat{\psi}(-\boldsymbol{x},t)]\right\} \end{aligned} \tag{9.85}$$

for the parity transformation and

$$\begin{aligned} &\hat{\mathscr{T}}\,\hat{\psi}(\boldsymbol{x},t)\,\hat{\mathscr{T}}^\dagger = T\hat{\psi}(\boldsymbol{x},-t) \qquad \text{with} \\ &T\hat{\psi}(\boldsymbol{x},-t) = \mathrm{i}\,\gamma^1\gamma^3\,\hat{\psi}^*(\boldsymbol{x},t) \end{aligned} \tag{9.86}$$

for the operation of time reversal [Chap. 4, (4.59)].

9.5 The Vacuum State (II)

We can now resume the discussion on what should be taken as the vacuum state. For the free Dirac field, (9.83), the charge conjugation operator commutes with the Hamiltonian. Since symmetries should not unnecessarily be broken by the dynamics of the theory, it is natural to choose the vacuum state $|0\rangle$ as an eigenstate of $\hat{\mathscr{C}}$. On the other hand, the charge operator $\hat{Q}$ is also known to commute with the Hamiltonian (9.69), so the vacuum could be an eigenstate of $\hat{Q}$ as well. However, (9.82) states that $\hat{\mathscr{C}}$ and $\hat{Q}$ do not commute, rather

$$\hat{\mathscr{C}}\,\hat{Q} = -\hat{Q}\,\hat{\mathscr{C}}\,. \tag{9.87}$$

As $\hat{\mathscr{C}}^2 = 1$, the eigenvalues of $\hat{\mathscr{C}}$ are ± 1, and the only possible choice is

$$\hat{Q}\,|0\rangle = 0 \tag{9.88}$$

that satisfies our requirements and does not contradict (9.87). The Fock space expression for the charge operator is

$$\hat{Q} = \frac{e}{2}\int d^3x\,[\hat{\psi}^\dagger(x), \hat{\psi}(x)] = \frac{e}{2}\sum_{n>F_0}[\hat{b}_n^\dagger, \hat{b}_n] + \frac{e}{2}\sum_{n<F_0}[\hat{d}_n, \hat{d}_n^\dagger]\,, \tag{9.89}$$

where F_0 defines the state $|0\rangle$ according to the sign of energy, as in (9.31). Remembering that there is an equal number of positive and negative energy eigenstates in the absence of an external potential,

$$\begin{aligned}\hat{Q} &= e\left(\sum_{n>F_0}\hat{b}_n^\dagger\hat{b}_n - \sum_{n<F_0}\hat{d}_n^\dagger\hat{d}_n\right) - \frac{e}{2}\left(\sum_{n>F_0} - \sum_{n<F_0}\right)\\ &= e\left(\sum_{n>F_0}\hat{b}_n^\dagger\hat{b}_n - \sum_{n<F_0}\hat{d}_n^\dagger\hat{d}_n\right).\end{aligned} \tag{9.90}$$

Since $\hat{b}_n|0\rangle = \hat{d}_n|0\rangle = 0$ for the state $|0\rangle$ defined by (9.30), $|0\rangle$ is indeed annihilated by the charge operator. Thus for the free Dirac field the state $|0\rangle$ is not only the state with lowest energy, but it is also singled out as being the state of highest possible symmetry. Indeed, calculation reveals that $|0\rangle$ satiesfies [Bj 65]

$$\hat{\boldsymbol{P}}|0\rangle = \hat{M}_{ik}|0) = \hat{Q}|0\rangle = 0\,; \qquad \hat{\mathscr{C}}|0\rangle = \hat{\mathscr{P}}|0\rangle = \hat{\mathscr{T}}|0\rangle = |0\rangle\,. \tag{9.91}$$

These symmetries are obviously lost in the presence of an external field. Only $\hat{Q}$ still commutes with the Hamiltonian according to (9.69), thus being a constant of motion. This means that even in an external potential, the full Fock space $\mathscr{V}$ can be divided into subspaces $\mathscr{V}_q$, each pertaining to a different integer value q of charge, Fig. 9.2. In principle, in a *static* potential, we can seek the lowest energy state in each of these subspaces. In the present context $|0, q\rangle$ denotes these states, where the integer q implies that they are eigenstates of the charge operator with eigenvalue qe. One of these states has lower energy than all the others (accidental degeneracies not counted); this state is called the true ground state $|0\rangle$ in

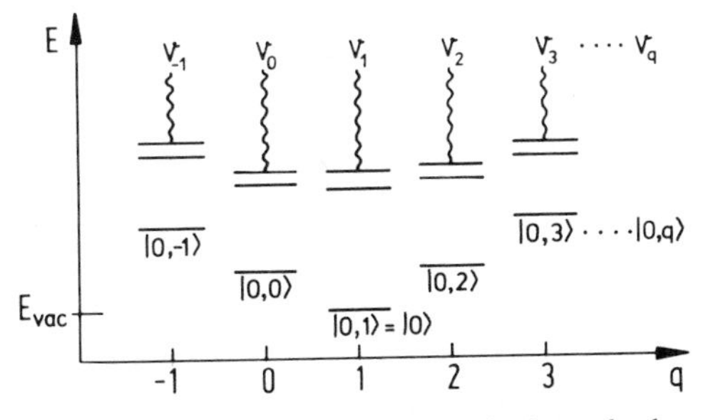

Fig. 9.2. Vacuum states for differently charged subspaces $\mathscr{V}_q$

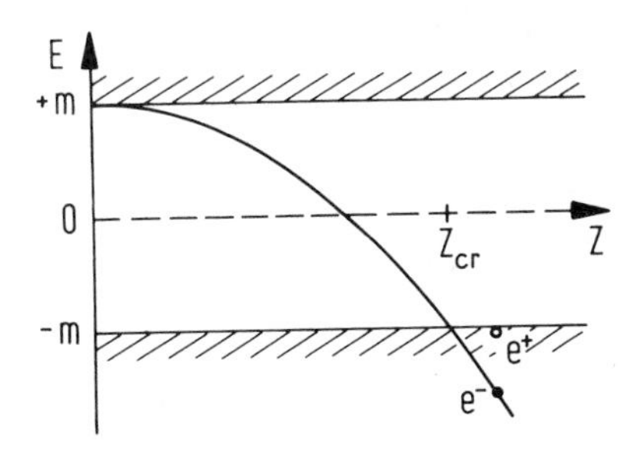

Fig. 9.3. The state $|0, q=0\rangle$ for a supercritical potential

Sect. 9.2. However, correct as this assignation may be, it is of little practical value because a closed system cannot jump between states belonging to different subspaces $\mathscr{V}_q$.

If any external potential A_μ can be adiabatically switched on, like $\lambda(t)A_\mu$ with a function $\lambda(t)$ that rises infinitely slowly from 0 to 1, and supposing the (closed) system to start out in the fully symmetric vacuum state of the free Dirac equation, it will remain forever in the subspace $\mathscr{V}_0$ of neutral states of the electron – positron field. We therefore expect to find it in the state $|0, q=0\rangle$ after the potential has acquired its full strength. If the external potential is supercritical, this state turns out to be rather awkward: it consists of all bound states (to be precise: bound state resonances) of more than $2m_e$ binding energy occupied by particles and the same number of antiparticles in scattering states of virtually zero kinetic energy (Fig. 9.3).

It is important to note that this situation already involves a change in the symmetry of the state $|0, q=0\rangle$, i.e. a phase transition, at the point where the potential becomes supercritical. Up to this moment, the diving bound state n_0 was a particle state; from this point on it is counted as an antiparticle state, and an initial antiparticle state of vanishing kinetic energy is transferred to the set of particle states:

$$\begin{aligned} \hat{b}_{n_0}|0, q=0\rangle &= 0\,, \quad \text{if} \quad A_\mu \quad \text{is subcritical}\,, \\ \hat{d}_{n_0}|0, q=0\rangle &= 0\,, \quad \text{if} \quad A_\mu \quad \text{is supercritical}\,. \end{aligned} \tag{9.92}$$

This effect can be described in a simpler way, when we recall that the charge of a microscopic system is conserved only when it is absolutely closed [Mü 83a]. Such a situation is never realized in practice. If the system contains particles in free scattering states, these particles will, in general, eventually escape (Fig. 9.4). If they are charged particles, they will carry charge out of the system, as expressed in (9.69), which states that charge is conserved only when there is no electric flux through the boundary:

$$\frac{d}{dt}\hat{Q} = \mathrm{i}[\hat{H}, \hat{Q}] = -\frac{e}{2}\int d^3x\, \nabla\cdot[\hat{\psi}^\dagger, \alpha\hat{\psi}] = -\oint d\Sigma(\hat{j}\cdot n)\,. \tag{9.93}$$

The same conclusion is contained in (9.6), which states that the usual expression for the Hamiltonian applicable to localized states $\hat{H}_{\text{loc}}$ has an imaginary part when there is a flux through the boundary:

$$\hat{H}_{\text{loc}} = \frac{1}{2}\int d^3x\,[\hat{\psi}^\dagger, H_{\text{D}}\hat{\psi}] = \hat{H} - \frac{\mathrm{i}}{e}\oint d\Sigma(\hat{j}\cdot n) = \hat{H} + \frac{\mathrm{i}}{e}\frac{d}{dt}\hat{Q}\,. \tag{9.94}$$

This means that in practice the microscopic system can make a transition from one charge subspace $\mathscr{V}_q$ to another $\mathscr{V}_{q'}$ by emitting a particle or antiparticle, i.e. a change of the total charge of the system is inevitably connected with an emission process.

Consequently, the chemical potential μ of the emitted particle species must be considered. If the system is imbedded in empty space, then $\mu^\pm = -(m_e \pm V(\infty))$,

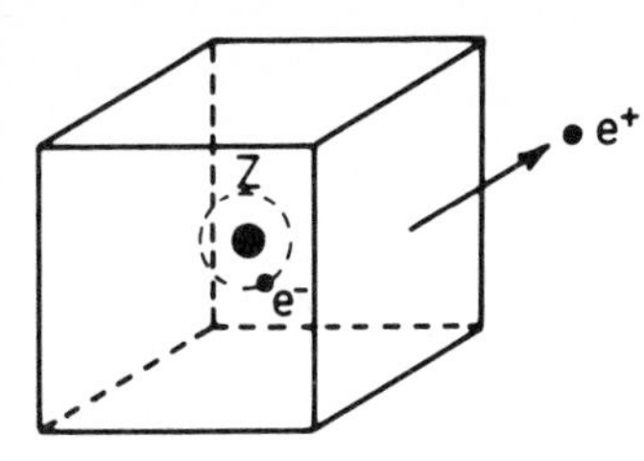

Fig. 9.4. The antiparticle, which is spontaneously created in a free scattering state of the supercritical field, escapes through the boundary of the system. It carries away charge, leaving a charged vacuum behind. The charged vacuum is characterized by a Fermi energy $E_F = -[m_e \pm V(\infty)]$

where the two signs are valid for particles and antiparticles, respectively. This means, e.g., that an emitted antiparticle (positron) carries away an amount of energy equal to $[m_e - V(\infty)]$, or that a particle can be supplied to the system from the outside only by creating an additional antiparticle at the expense of energy. Thermodynamics indicates that the equilibrium state of the system minimizes the expectation value of $(\hat{H} - \mu\hat{N} - \bar{\mu}\hat{\bar{N}})$, where $\hat{N}, \hat{\bar{N}}$ are the operators that count particles or antiparticles. These operators can be re-expressed in terms of the charge operator, and one finally obtains the condition that the equilibrium state $|0, \text{equil}\rangle$ is determined by

$$\langle 0, \text{equil}|\hat{H} + \left[m_e \operatorname{sgn}\left(\frac{\hat{Q}}{e}\right) - V(\infty)\right]\frac{\hat{Q}}{e}|0, \text{equil}\rangle = \min . \tag{9.95}$$

Considering only potentials attractive to particles ($V < 0$) and the gauge where $V(\infty)$ vanishes, the charge of the equilibrium state, $q_0 e$, satisfies

$$\langle 0, q_0|\hat{H} + \frac{m_e}{e}\hat{Q}|0, q_0\rangle = \langle 0, q_0|\hat{H}|0, q_0\rangle + m_e q_0 = \min . \tag{9.96}$$

To see what this implies we go back for a moment to the representation of the field operator (9.14) where we had not yet distinguished between particle and antiparticle states. Taking (9.20) and the analogous expression for the charge operator gives

$$\begin{aligned}\hat{H} + \frac{m_e}{e}\hat{Q} &= \sum_n E_n\left(\hat{N}_n - \frac{1}{2}\right) + \frac{m_e}{e}\, e\sum_n\left(\hat{N}_n - \frac{1}{2}\right)\\ &= \sum_n (E_n + m_e)\left(\hat{N}_n - \frac{1}{2}\right).\end{aligned} \tag{9.97}$$

Repeating the discussion in Sect. 9.2, the Fock space state minimizing the expectation value of $\hat{H} + (m_e Q/e)$ must be that in which all levels with $E_n + m_e < 0$ are occupied. These are all positron scattering states combined with all electronic bound states of binding energy higher than $2m_e$. Thus the equilibrium state $|0, q_0\rangle$ is characterized by the Fermi energy $E_F = -m_e$:

$$\text{“}n < m_e\text{”} = \{n|E_n < -m_e\} . \tag{9.98}$$

The field operator is quantized according to

$$\hat{\psi}(\boldsymbol{x}, t) = \sum_{E_n > -m_e} \hat{b}_n \varphi_n(\boldsymbol{x}) \,\mathrm{e}^{-\mathrm{i}E_n t} + \sum_{E_n < -m_e} \hat{d}_n^\dagger \varphi_n(\boldsymbol{x}) \,\mathrm{e}^{-\mathrm{i}E_n t} \tag{9.99}$$

with the specifications

$$\begin{aligned} \hat{b}_n |0, q_0\rangle &= 0\,, \quad (E_n \geqslant -m_e)\,, \\ \hat{d}_n |0, q_0\rangle &= 0\,, \quad (E_n < -m_e)\,. \end{aligned} \tag{9.100}$$

This state is called the "*charged vacuum*" state [Mü 72b, Ra 74, Ze 72, Fu 73a]. Whenever a microscopic system is prepared as neutral in a weak external field, it will be found in this state after the potential has been increased to full strength and sufficient time has elapsed for the equilibrium to be established. Here q_0 equals the number of bound states that have become supercritical during this process.

Let us summarize our results. There are three different possible definitions of the vacuum state.

1) The state of absolutely lowest energy $|0\rangle$, characterized by the Fermi energy $E_{\mathrm{F}} = 0$. Particle and antiparticle states are divided according to the sign of the energy eigenvalue. Due to conservation of electric charge, a microscopic system can not often reach this state.

2) If the system is surrounded by an impenetrable boundary, its charge is exactly conserved. It will then settle down in the state $|0, q\rangle$, where q is the initial charge of the system. The state consists of all levels below $E = -m_e$ filled except for the q levels closest to the top.

3) In practice, the system will change its charge by antiparticle (or particle) emission until it reaches the "charged vacuum" equilibrium state $|0, q_0\rangle$, characterized by the Fermi energy $E_{\mathrm{F}} = -m_e$. All states below $E = -m_e$ are antiparticle states. Whenever sufficient time is available, the system will *spontaneously* occupy this state.

Of the three definitions, the charged vacuum is therefore the most important. In Chap. 10 we investigate how the charged vacuum state is approached when the external potential becomes supercritical.

We finally make a few comments on time-dependent potentials. Then the Hamiltonian $\hat{H}$ (9.6) is explicitly time dependent and energy is not conserved. Although it is still possible to find states $|\Omega, t_0\rangle$ that are eigenstates of the Hamiltonian at some arbitrarily chosen instant of time t_0, they will not in general remain eigenstates of $H(t)$ at different times $t \neq t_0$. In particular, the Heisenberg state $|0, t_0\rangle$ giving the lowest possible expectation value of $\hat{H}(t_0)$ will not do so for $t \neq t_0$. We conclude that there is no analogue of the vacuum state, as defined above, in the generic case of time-dependent external potentials.

However, if the potential $A_\mu(\boldsymbol{x}, t)$ approaches constant values at $t = -\infty$ (the "in region") and $t = +\infty$ (the "out region"), then vacuum states in these two regions can be defined separately, the in vacuum $|0, \mathrm{in}\rangle$ and the out vacuum $|0, \mathrm{out}\rangle$, respectively. In general, then $|0, \mathrm{in}\rangle \neq |0, \mathrm{out}\rangle$ and the vacuum-vacuum amplitude $W_0[A_\mu]$ defined by

$$\exp(\mathrm{i}\,W_0[A_\mu]) = \langle 0, \mathrm{out}\,|0, \mathrm{in}\rangle$$

describes the amplitude for the persistence of the vacuum state. Knowing $W_0[A_\mu]$ for all possible external field functions enables the vacuum polarization current to be calculated. (This method, originally due to *Schwinger* [Schw 51] has been developed into a powerful tool in quantum field theory by *De Witt* [DeW 67].) We shall pursue this approach further in Sect. 10.5, when we treat the properties of the vacuum in a constant external field.

9.6 The Feynman Propagator

As discussed at length in previous chapters, the vacuum state must be defined by specifying which of the single-particle states $\psi_n(x)$ are counted as particle states and which as antiparticles, or – in Dirac's hole picture – which states are empty and which are filled. The same information is contained in the Feynman propagator $S_\mathrm{F}(x,x')$, (8.63), which we introduced even before quantizing the Dirac field. It is for precisely this reason that all physical properties of the vacuum state, the definition of particles and antiparticles and their wave functions, can be deduced from the Feynman propagator. Chapter 10 describes how this can be achieved in practice for the examples of vacuum charge and vacuum energy, but it is clear that this must be possible in principle from our general argument. This property makes the Feynman propagator an extremely useful tool of quantum field theory: $S_\mathrm{F}(x,x')$ is a "normal" c-number function, albeit a highly singular one, and still it enables statements to be expressed about the quantized theory.

It must be shown precisely how the Feynman propagator is related to the field operator and the vacuum state. We start by generalizing the definition (8.63) to the general case where the vacuum state may be charged, characterized by a Fermi energy E_F and the set of "occupied" states

$$\text{“}n<F\text{”} = \{n\,|E_n<E_\mathrm{F}\}\,. \tag{9.101}$$

In analogy to (7.48) for the free propagator the definition then becomes

$$\begin{aligned}
\mathrm{i}\,S_\mathrm{F}(x,x') &= \Theta(t-t')\sum_n \Theta(E_n-E_\mathrm{F})\,\psi_n(x)\,\bar{\psi}_n(x') \\
&\quad - \Theta(t'-t)\sum_n \Theta(E_\mathrm{F}-E_n)\,\psi_n(x)\,\bar{\psi}_n(x') \\
&= \Theta(t-t')\sum_{n>F}\psi_n(x)\,\bar{\psi}_n(x') - \Theta(t'-t)\sum_{n<F}\psi_n(x)\,\bar{\psi}_n(x')\,.
\end{aligned} \tag{9.102}$$

Using the definitions (9.99, 100) to rewrite the individual terms as vacuum expectation values of appropriate number operators gives

$$\begin{aligned}
\sum_{n>F}\psi_n(x)\,\bar{\psi}_n(x') &= \sum_{n>F}\psi_n(x)\,\bar{\psi}_n(x')\,\langle 0,F|\hat{b}_n\hat{b}_n^\dagger|0,F\rangle \\
&= \langle 0,F|\,\hat{\psi}(x)\hat{\bar{\psi}}(x')\,|0,F\rangle\,,
\end{aligned} \tag{9.103a}$$

$$\sum_{n<F} \psi_n(x)\,\bar{\psi}_n(x') = \sum_{n<F} \psi_n(x)\,\bar{\psi}_n(x')\langle 0,F|\hat{d}_n\hat{d}_n^\dagger|0,F\rangle$$
$$= \langle 0,F|\hat{\bar{\psi}}(x')\,\hat{\psi}(x)\,|0,F\rangle\,. \tag{9.103b}$$

Combining the two expressions, the Feynman propagator can be expressed in the form

$$\mathrm{i}\,S_\mathrm{F}(x,x') = \langle 0,F|\Theta(t-t')\,\hat{\psi}(x)\hat{\bar{\psi}}(x') - \Theta(t'-t)\,\hat{\bar{\psi}}(x')\,\hat{\psi}(x)\,|0,F\rangle\,, \tag{9.104}$$

i.e. as the vacuum expectation value of the operator

$$T[\hat{\psi}(x)\hat{\bar{\psi}}(x')] = \begin{cases} \hat{\psi}(x)\hat{\bar{\psi}}(x') & \text{for} \quad t>t'\,, \\ -\hat{\bar{\psi}}(x')\,\hat{\psi}(x) & \text{for} \quad t<t'\,. \end{cases} \tag{9.105}$$

The step functions ensure that the time argument of the field operator to the right is always smaller than that to the left, hence "T" is called the *operator of time ordering.* (This symbol occurs in the framework of the classical Dirac theory in the general solution for the time-evolution operator $U(t,t')$ in Sect. 8.3. The sole difference with respect to the old definition is the minus sign, derived from the anticommutativity of the Dirac field operators.)

The final result

$$\mathrm{i}\,S_\mathrm{F}(x,x') = \langle 0,F|T(\hat{\psi}(x)\hat{\bar{\psi}}(x'))\,|0,F\rangle \tag{9.106}$$

expresses neatly the connection between the Feynman propagator, the field operator and the vacuum state. *It is obviously not sufficient to know the field operator* $\hat{\psi}(x)$ *alone, which is a solution of the equation of motion* (9.9). *To determine* S_F *uniquely, one has to know the vacuum state as well.* That means that the Feynman propagator is generally different for the three vacuum states discussed in Chap. 8: it is more proper to write $S_\mathrm{F}(x,x'|F)$ and distinguish $S_\mathrm{F}(x,x'|E_\mathrm{F}=0)$, $S_\mathrm{F}(x,x'|E_\mathrm{F}=-m_e)$ and so on. In the following, if not otherwise stated, we implicitly assume that S_F is defined with respect to $E_\mathrm{F}=-m_e$, i.e. to the neutral vacuum for subcritical potential and to the charged vacuum for supercritical potentials.

The Feynman propagator for explicitly time-dependent fields is discussed in more details in the Sect. 9.9.

9.7 Charge and Energy of the Vacuum (I)

After the meaning and nature of the vacuum state have been clarified, we are in a position to discuss the charge and energy associated with it. According to (9.90, 32), they are given by

$$q_\mathrm{vac} = \frac{e}{2}\left(\sum_{n<F} 1 - \sum_{n>F} 1\right)\,, \tag{9.107}$$

$$E_\mathrm{vac} = \frac{1}{2}\left(\sum_{n<F} E_n - \sum_{n>F} E_n\right)\,, \tag{9.108}$$

where F characterizes the boundary between particle and antiparticle states with the possible choices F_0 or F_{-m}. Clearly, the above expressions for q_{vac} and E_{vac} make little sense as they stand, because they involve divergent sums.

To obtain well-defined expressions, we demand that in the absence of an external potential the vacuum must carry no charge and no energy, and subtract these formal expressions. Even then (9.107, 108) remain infinite. Therefore, we expand in powers of the external potential and apply appropriate renormalization prescriptions to the lowest order terms [Bj 65, It 80, Gr 84b]. Resumming the series expansion gives definite expressions for q_{vac} and E_{vac} that are correct in all orders of the external field.

We shall discuss two different ways to carry this programme through. In this chapter, we express the charge and energy of the vacuum in terms of the Feynman propagator and show how they are related to each other; in Sect. 9.8 we proceed more intuitively and express the charge and energy of the vacuum state as integrals over the phase shift of the wave functions for electrons and positrons.

Let us start with the propagator method. To incorporate the field theoretic definition of the Feynman propagator (9.106), we artificially split the arguments of the field operators entering into the current operator (9.81). The correct time ordering is

$$\begin{aligned}\langle 0|\hat{j}^{\mu}(x)|0\rangle &= \frac{e}{2}\langle 0|\hat{\bar{\psi}}_{\alpha}(x)\,\gamma^{\mu}_{\alpha\beta}\,\hat{\psi}_{\beta}(x)-\gamma^{\mu}_{\alpha\beta}\hat{\psi}_{\beta}(x)\,\hat{\bar{\psi}}_{\alpha}(x)\,|0\rangle \\ &= \frac{e}{2}\,\gamma^{\mu}_{\alpha\beta}(\langle 0|\hat{\bar{\psi}}_{\alpha}(x')\,\hat{\psi}_{\beta}(x)\,|0\rangle_{x'\to x+0} \\ &\quad -\langle 0|\hat{\psi}_{\beta}(x)\,\hat{\bar{\psi}}_{\alpha}(x')\,|0\rangle_{x'\to x-0})\,,\end{aligned} \tag{9.109}$$

where $x'\to x\pm 0$ means the limit

$$\lim_{x'\to x}\ \lim_{t'\to t\pm 0}\ .$$

Comparing this with (9.106) shows that in each limit there is just that part of the Feynman propagator which contributes:

$$\begin{aligned}\langle 0|\hat{j}^{\mu}(x)|0\rangle &= -\mathrm{i}\frac{e}{2}\,\gamma^{\mu}_{\alpha\beta}[S_{\mathrm{F}}^{\beta\alpha}(x,x'\to x+0)+S_{\mathrm{F}}^{\beta\alpha}(x,x'\to x-0)] \\ &\equiv -\mathrm{i}e\,\gamma^{\mu}_{\alpha\beta}[S_{\mathrm{F}}^{\beta\alpha}(x,x')]_{x'\to x} \\ &= -\mathrm{i}e\,\mathrm{tr}[\gamma^{\mu}S_{\mathrm{F}}(x,x')]_{x'\to x}\,.\end{aligned} \tag{9.110}$$

Brackets [ ] mean the symmetric limit

$$[\;]_{x'\to x}\equiv\frac{1}{2}\lim_{x'\to x}\left(\lim_{t'\to t+0}+\lim_{t'\to t-0}\right). \tag{9.111}$$

The total vacuum charge is the space integral over (9.110):

$$q_{\text{vac}} = \int d^3x \langle 0|\hat{j}^0(x)|0\rangle = -\mathrm{i}e\int d^3x \,\mathrm{tr}\,[\gamma^0 S_\mathrm{F}(x,x')]_{x'\to x}\,. \tag{9.112}$$

The energy of the vacuum can be expressed similarly by starting from the Hamiltonian (9.6), dropping the surface term,

$$\begin{aligned} E_{\text{vac}} &= \langle 0|\hat{H}|0\rangle = \frac{1}{2}\int d^3x \langle 0|[\hat{\bar{\psi}}(x)\gamma^0, H_\mathrm{D}\hat{\psi}(x)]|0\rangle \\ &= \frac{\mathrm{i}}{2}\int d^3x \langle 0|\left[\hat{\bar{\psi}}(x)\gamma^0, \frac{\partial}{\partial t}\hat{\psi}(x)\right]|0\rangle\,. \end{aligned} \tag{9.113}$$

The last identity results from the field equation (9.9). Splitting the space-time arguments as before,

$$\begin{aligned} &\frac{1}{2}\langle 0|\left[\hat{\bar{\psi}}(x)\gamma^0, \frac{\partial}{\partial t}\hat{\psi}(x)\right]|0\rangle \\ &= \frac{1}{2}\gamma^0_{\alpha\beta}\frac{\partial}{\partial t}(\langle 0|\hat{\bar{\psi}}_\alpha(x')\hat{\psi}_\beta(x)|0\rangle_{x'\to x+0} - \langle 0|\hat{\psi}_\beta(x)\hat{\bar{\psi}}_\alpha(x')|0\rangle_{x'\to x-0}) \\ &= \frac{1}{2}\gamma^0_{\alpha\beta}\frac{\partial}{\partial t}(-\mathrm{i}S_\mathrm{F}^{\beta\alpha}(x,x'\to x+0) - \mathrm{i}S_\mathrm{F}^{\beta\alpha}(x,x'\to x-0)) \\ &= -\mathrm{i}\gamma^0_{\alpha\beta}\left[\frac{\partial}{\partial t}S_\mathrm{F}^{\beta\alpha}(x,x')\right]_{x'\to x}\,, \end{aligned}$$

an analogous expression for the vacuum energy in terms of the Feynman propagator is obtained:

$$\begin{aligned} E_{\text{vac}} &= \int d^3x \,\mathrm{tr}\left[\gamma^0\frac{\partial}{\partial t}S_\mathrm{F}(x,x')\right]_{x'\to x} \\ &= \int d^3x \,\mathrm{tr}\,[\gamma^0 H_\mathrm{D}(x)S_\mathrm{F}(x,x')]_{x'\to x}\,. \end{aligned} \tag{9.114}$$

Charge and energy of the vacuum depend on the strength of the external potential $A_\mu(x)$ because the Feynman propagator does. How S_F depends on A_μ is determined by the inhomogeneous Dirac equation (8.53):

$$\left[\gamma^\mu\left(\mathrm{i}\frac{\partial}{\partial x^\mu} - eA_\mu\right) - m\right]S_\mathrm{F}(x,x'|A_\mu) = \delta(x-x')\,. \tag{9.115}$$

We will now argue that it is not possible to measure the absolute value of the energy and charge of the vacuum, but only their change with variations of the potential. For microscopic systems, this statement is certainly correct. On a large scale, however, the absolute value of the vacuum energy is, in principle, a measurable quantity, because energy gravitates. The energy density of the vacuum must therefore be expected to contribute to the energy density in the universe. To what extent this is really the case is still an unsolved problem. A proper resolution of it must probably wait until a consistent theory of quantum gravity has been developed (see [DeW 75] and references therein).

We are therefore free to subtract from q_{vac} and E_{vac} the expressions for vanishing potential A_μ, i.e. for the free Dirac field,

$$q'_{\text{vac}}[A_\mu] = q_{\text{vac}}[A_\mu] - q_{\text{vac}}[A_\mu = 0] = q_{\text{vac}}[A_\mu] \,, \tag{9.116a}$$

$$E'_{\text{vac}}[A_\mu] = E_{\text{vac}}[A_\mu] - E_{\text{vac}}[A_\mu = 0] \,. \tag{9.116b}$$

Thus (9.114) is replaced by

$$E'_{\text{vac}}[A_\mu] = \int d^3x \, \text{tr}\left[\gamma^0 \frac{\partial}{\partial t}\left(S_{\text{F}}(x, x' \,|\, A_\mu) - S_{\text{F}}(x, x' \,|\, 0)\right)\right]_{x' \to x} \,, \tag{9.117}$$

the change in the vacuum energy being due to the presence of an interaction. (In the special case when the interaction can be expressed as a boundary condition, E'_{vac} is commonly known as *Casimir energy* [Ca 48].) Treating only the *change* of the vacuum energy, we can go one step further and ask how $E_{\text{vac}}[A_\mu]$ changes under an infinitesimal change in the interaction. Therefore the strength of the external potential is scaled by a continuous parameter λ, and the derivative

$$\frac{d}{d\lambda} E_{\text{vac}}[\lambda A_\mu] = \int d^3x \, \text{tr}\left[\gamma^0 \frac{\partial}{\partial t} \frac{\partial}{\partial \lambda} S_{\text{F}}(x, x' \,|\, \lambda A_\mu)\right]_{x' \to x} \tag{9.118}$$

is investigated.

The total vacuum energy (9.117) can be written with its help as

$$E'_{\text{vac}}[A_\mu] = \int_0^1 d\lambda \, \frac{d}{d\lambda} E_{\text{vac}}[\lambda A_\mu] \,. \tag{9.119}$$

The great advantage of the differential expression (9.118) is that it can be related to the charge density of the vacuum (9.110). To study this, write the Dirac equation (9.115) for $S_{\text{F}}(x, x' \,|\, \lambda A_\mu)$ and differentiate with respect to λ:

$$\left[\gamma^\mu\left(\mathrm{i}\frac{\partial}{\partial x^\mu} - e\lambda A_\mu\right) - m_e\right] \frac{\partial}{\partial \lambda} S_{\text{F}}(x, x' \,|\, \lambda A_\mu) - e\gamma^\mu A_\mu S_{\text{F}}(x, x' \,|\, \lambda A_\mu) = 0 \,. \tag{9.120}$$

Explicitly resolving for the integrand of (9.118) yields

$$\gamma^0 \frac{\partial}{\partial t} \frac{\partial S_{\text{F}}}{\partial \lambda} = -\mathrm{i} e \gamma^\mu A_\mu S_{\text{F}} - \mathrm{i}(\boldsymbol{\gamma} \cdot (-\mathrm{i}\nabla - e\lambda \boldsymbol{A}) + m_e + e\lambda\gamma^0 A_0) \frac{\partial S_{\text{F}}}{\partial \lambda} \,. \tag{9.121}$$

Taking the Dirac trace and the symmetric limit $x' \to x$, the first term on the rhs contains the vacuum current density (9.110), while the operator in front of the second term is identified as the single-particle Hamiltonian:

$$\operatorname{tr}\left[\gamma^0 \frac{\partial}{\partial t} \frac{\partial}{\partial \lambda} S_{\mathrm{F}}(x, x' \,|\, \lambda A_\mu)\right]_{x' \to x}$$
$$= A_\mu \langle 0 | \hat{j}^\mu(x) | 0 \rangle - \mathrm{i} \operatorname{tr}\left[\gamma^0 H_{\mathrm{D}}[\lambda A_\mu] \frac{\partial}{\partial \lambda} S_{\mathrm{F}}(x, x' \,|\, \lambda A_\mu)\right]_{x' \to x} . \tag{9.122}$$

The interpretation of the first term is obvious: it describes the interaction of the charge current of the vacuum with the variation of the external field:

$$\frac{d}{d\lambda}(\lambda A_\mu) \langle 0 | \hat{j}^\mu(x) | 0 \rangle .$$

To find the meaning of the second term one must consider the representation of S_{F} in terms of eigenmodes given in (9.102) [Schw 54, Pl 83]. A change in the strength parameter λ of the potential results in a change of the energy eigenvalues $E_n(\lambda)$ and also modifies the eigenfunctions $\psi_n(x|\lambda)$. The eigenfunction change, however, is severely restricted by the requirement that their scalar product must remain unaffected by a variation of the potential:

$$\frac{\partial}{\partial \lambda} \langle \psi_n | \psi_m \rangle \equiv \frac{\partial}{\partial \lambda} \int d^3x \, \psi_n^\dagger(x|\lambda) \, \psi_m(x|\lambda) = 0 \qquad \text{for all} \quad n, m . \tag{9.123}$$

With this in mind, we differentiate (9.102) with respect to λ, obtaining in the symmetric limit (9.111)

$$\mathrm{i}\left[\frac{\partial}{\partial \lambda} S_{\mathrm{F}}(x, x' \,|\, \lambda A_\mu)\right]_{x' \to x}$$
$$= \frac{1}{2} \sum_n \left[\Theta(E_n - E_{\mathrm{F}}) \frac{\partial}{\partial \lambda} (\psi_n(x) \, \bar{\psi}_n(x')) + \frac{\partial E_n}{\partial \lambda} \delta(E_n - E_{\mathrm{F}}) \, \psi_n(x) \, \bar{\psi}_n(x')\right]_{x' \to x - 0}$$
$$- \frac{1}{2} \sum_n \left[\Theta(E_{\mathrm{F}} - E_n) \frac{\partial}{\partial \lambda} (\psi_n(x) \, \bar{\psi}_n(x')) - \frac{\partial E_n}{\partial \lambda} \delta(E_{\mathrm{F}} - E_n) \, \psi_n(x) \, \bar{\psi}_n(x')\right]_{x' \to x + 0}$$
$$= \frac{1}{2} \lim_{x' \to x} \sum_n \left[\operatorname{sgn}(E_n - E_{\mathrm{F}}) \frac{\partial}{\partial \lambda} (\psi_n(x) \, \bar{\psi}_n(x')) + 2 \frac{\partial E_n}{\partial \lambda} \delta(E_n - E_{\mathrm{F}}) \, \psi_n(x) \, \bar{\psi}_n(x')\right] . \tag{9.124}$$

Inserting this result into (9.122) gives the following contribution of the second term to the variation of the vacuum energy (9.118):

$$-\mathrm{i} \int d^3x \operatorname{tr}\left[\gamma^0 H_{\mathrm{D}}(x) \frac{\partial}{\partial \lambda} S_{\mathrm{F}}\right]_{x' \to x}$$
$$= -\frac{1}{2} \sum_n \left[\operatorname{sgn}(E_n - E_{\mathrm{F}}) \left(\left\langle \psi_n \middle| H_{\mathrm{D}} \middle| \frac{\partial \psi_n}{\partial \lambda} \right\rangle + \left\langle \frac{\partial \psi_n}{\partial \lambda} \middle| H_{\mathrm{D}} \middle| \psi_n \right\rangle \right) \right.$$
$$\left. + 2 \frac{\partial E_n}{\partial \lambda} \delta(E_n - E_{\mathrm{F}}) \langle \psi_n | H_{\mathrm{D}} | \psi_n \rangle \right]$$

$$= -\frac{1}{2}\sum_n \left[E_n \operatorname{sgn}(E_n - E_F) \frac{\partial}{\partial\lambda} \langle \psi_n | \psi_n \rangle + 2E_n \frac{\partial E_n}{\partial\lambda} \delta(E_n - E_F) \right]$$

$$= -\sum_n E_F \frac{\partial E_n}{\partial\lambda} \delta(E_n - E_F)$$

$$= E_F \frac{\partial}{\partial\lambda} \sum_n \Theta(E_F - E_n(\lambda)) \, . \tag{9.125}$$

In the last expression, the sum counts all states below the Fermi surface. It changes as a function of potential strength λ only when a single-particle state n becomes supercritical and penetrates the Fermi surface at E_F. Therefore

$$\int_0^1 d\lambda \frac{\partial}{\partial\lambda} \sum_n \Theta(E_F - E_n(\lambda)) = \sum_n [\Theta(E_F - E_n(\lambda = 1)) - \Theta(E_F - E_n(0))]$$

$$= n_{\mathrm{vac}} \, , \tag{9.126}$$

where n_{vac} is the number of supercritical states. This statement can be proven more formally by treating the variation of the vacuum charge in exactly the same way. With (9.124)

$$\frac{\partial q_{\mathrm{vac}}}{\partial\lambda} = -\mathrm{i}e \int d^3x \, \mathrm{tr} \left[\gamma^0 \frac{\partial}{\partial\lambda} S_F(x, x' | \lambda A_\mu) \right]_{x' \to x}$$

$$= -\frac{e}{2} \sum_n \left[\operatorname{sgn}(E_n - E_F) \frac{\partial}{\partial\lambda} \langle \psi_n | \psi_n \rangle + 2 \frac{\partial E_n}{\partial\lambda} \delta(E_n - E_F) \right]$$

$$= -e \sum_n \frac{\partial E_n}{\partial\lambda} \delta(E_n - E_F)$$

$$= e \frac{\partial}{\partial\lambda} \sum_n \Theta(E_F - E_n(\lambda)) \, . \tag{9.127}$$

Combined with (9.126) this result formally proves the conclusion intuitively reached in the framework of Dirac's hole picture: the charge of the vacuum, in units of the electron charge, is equal to the number of supercritical electron states (or minus the number of supercritical positron states):

$$q_{\mathrm{vac}} = e n_{\mathrm{vac}} \, . \tag{9.128}$$

Collecting all the terms, (9.118, 122, 125, 127) then give

$$\frac{\partial E_{\mathrm{vac}}}{\partial\lambda} = \int d^3x \, A_\mu(x) \langle 0 | \hat{j}^\mu(x) | 0 \rangle_{\lambda A_\mu} + E_F \frac{\partial}{\partial\lambda} \left(\frac{q_{\mathrm{vac}}}{e} \right) , \tag{9.129}$$

which may be integrated to give the final result:

$$E'_{\rm vac}[A_\mu] = \int_0^1 d\lambda \frac{\partial E_{\rm vac}}{\partial \lambda}$$
$$= \int d^3x\, A_\mu(x) \int_0^1 d\lambda \langle 0|\hat{j}^\mu(x)|0\rangle_{\lambda A_\mu} + E_{\rm F}\frac{q_{\rm vac}[A_\mu]}{e} . \tag{9.130}$$

Observe that there are two mechanisms contributing to the energy of the vacuum. Firstly, the external potential A_μ has to work against the charge density $\langle 0|\hat{j}^\mu|0\rangle$ of the vacuum. Thus the first contribution is simply the electromagnetic self-energy of the vacuum charge. In addition, an energy quantum $E_{\rm F} = -m_e$ is released whenever the charge of the vacuum changes by one unit of the elementary charge, i.e. when a single-particle state of the Dirac equation becomes supercritical, the "sparking" of the vacuum lowers the vacuum energy by $-m_e$ (this is exactly the rest energy carried away by the escaping antiparticle!).

Often the differential relation (9.129) is written in the more general but less precise form [Schw 51, Pl 83]

$$\delta E_{\rm vac} = \int d^3x\, \delta A_\mu(x) \langle 0|\hat{j}^\mu(x)|0\rangle + E_{\rm F}\frac{1}{e}\delta q_{\rm vac} , \tag{9.131}$$

where $\delta A_\mu = A_\mu \delta\lambda$.

To calculate the vacuum energy, (9.110) must be evaluated for the vacuum polarization charge density, Chap. 15. At this stage we only mention that for not too supercritical fields, i.e. for Z not very much greater than $Z_{\rm cr}$, the vacuum charge density can be approximately written as the sum of the first-order virtual vacuum polarization (the so-called Uehling term) and the contribution from the individual supercritical bound states. After renormalization the Uehling charge density is

$$\langle 0|\hat{j}^\mu_{(1)}(x)|0\rangle_{\rm ren} = \int d^4x'\, \Pi^{(1)}_{\rm ren}(x-x')\, j^\mu_{\rm ext}(x') \tag{9.132}$$

with the so-called polarization function (Fig. 9.5)

$$\Pi^{(1)}_{\rm ren}(x-x') = \int \frac{d^4p}{(2\pi)^4} \exp[-{\rm i}p_\mu(x^\mu - x'^\mu)]\, \Pi^{(1)}_{\rm ren}(p) ,$$
$$\Pi^{(1)}_{\rm ren}(p) = -\frac{e^2}{12\pi^2}\left[\frac{5}{3} + \frac{4m_e^2}{p^2} - \left(1+\frac{2m_e^2}{p^2}\right)\sqrt{1-\frac{4m_e^2}{p^2}}\,\ln\frac{\sqrt{1-\frac{4m_e^2}{p^2}}+1}{\sqrt{1-\frac{4m_e^2}{p^2}}-1}\right] . \tag{9.133}$$

(For the derivation of these equations, see Sect. 14.4.)

The Uehling charge density has the property of a displacement charge, i.e. its integral over all space vanishes, so that it does not contribute to the vacuum charge $q_{\rm vac}$. In the vicinity of the field-generating external charge it has the same sign, but far away the opposite sign. (This is contrary to the intuitive picture of a polarization charge, which leads one to expect that oppositely charged particles

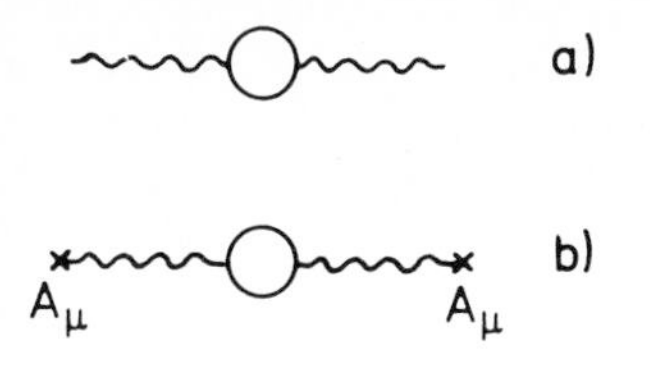

Fig. 9.5. (a) Feynman diagram for the lowest order vacuum polarization function $\Pi^{(1)}$; **(b)** graphical representation of the vacuum energy $E^{(2)}_{\rm vac}$ in second-order perturbation theory

are attracted. However, the result is consistent with the expression for the vacuum charge in hole theory: namely, the continuum states of negative energy are *repelled* by an *attractive* potential, because they represent *positron* wave functions, but they are considered as filled with *electrons*. Therefore the negative vacuum charges are repelled by the positive central charge of a nucleus.) Consequently, the first term in (9.130),

$$\int d^3x\, A_\mu(x) \int_0^1 d\lambda\, \langle 0|\hat{j}^\mu_{(1)}(x)|0\rangle_{\lambda A_\mu},$$

is always positive, growing as the square of the strength of the external field. Since $\langle 0|\hat{j}^\mu_{(1)}(x)|0\rangle_{\lambda A_\mu}$ is linear in the strength parameter λ, integration over λ can be carried out, yielding the contribution

$$\tfrac{1}{2}\int d^3x \int d^4x'\, A_\mu(x)\, \Pi^{(1)}_{\rm ren}(x-x')\, j^\mu_{\rm ext}(x') \tag{9.134}$$

to the energy of the vacuum.

The contribution from the supercritical bound states can be estimated when we cheat somewhat, pretending that their charge distributions could be represented by eigenstates of the Dirac equation. If $\psi(x|\lambda)$ is an eigenstate of the Hamiltonian $H^{(\lambda)}_{\rm D}$ to the potential λA_μ, it satisfies [He 37, Fe 39]

$$\begin{aligned}\left\langle \psi(\lambda)\left|\frac{\partial H_{\rm D}}{\partial\lambda}\right|\psi(\lambda)\right\rangle &= \frac{\partial}{\partial\lambda}\langle\psi(\lambda)|H^{(\lambda)}_{\rm D}|\psi(\lambda)\rangle \\ &\quad - \left\langle\frac{\partial\psi(\lambda)}{\partial\lambda}\left|H^{(\lambda)}_{\rm D}\right|\psi(\lambda)\right\rangle - \left\langle\psi(\lambda)\left|H^{(\lambda)}_{\rm D}\right|\frac{\partial\psi}{\partial\lambda}\right\rangle \\ &= \frac{\partial}{\partial\lambda}E(\lambda) - E(\lambda)\frac{\partial}{\partial\lambda}\langle\psi(\lambda)|\psi(\lambda)\rangle = \frac{\partial E}{\partial\lambda},\end{aligned} \tag{9.135}$$

where the eigenvalue equation $E(\lambda)\,\psi(\lambda) = H^{(\lambda)}_{\rm D}\,\psi(\lambda)$ and (9.123) have been repeatedly used. With this relation it is seen that an eigenstate $\psi(x|\lambda)$ contributes to the first part of $\partial E_{\rm vac}/\partial\lambda$ in (9.129) according to

$$\begin{aligned} e\int d^3x\, A_\mu(x)\,\bar\psi(\lambda)\,\gamma^\mu\psi(\lambda) &= \langle\psi(\lambda)|e\gamma^0\gamma^\mu A_\mu|\psi(\lambda)\rangle \\ &= \left\langle\psi(\lambda)\left|\frac{\partial}{\partial\lambda}(-\mathrm{i}\gamma^0\boldsymbol{\gamma}\cdot\nabla + m_e\gamma^0 + e\gamma^0\gamma^\mu\lambda A_\mu)\right|\psi(\lambda)\right\rangle \\ &= \left\langle\psi(\lambda)\left|\frac{\partial H_{\rm D}}{\partial\lambda}\right|\psi(\lambda)\right\rangle = \frac{\partial E(\lambda)}{\partial\lambda}.\end{aligned} \tag{9.136}$$

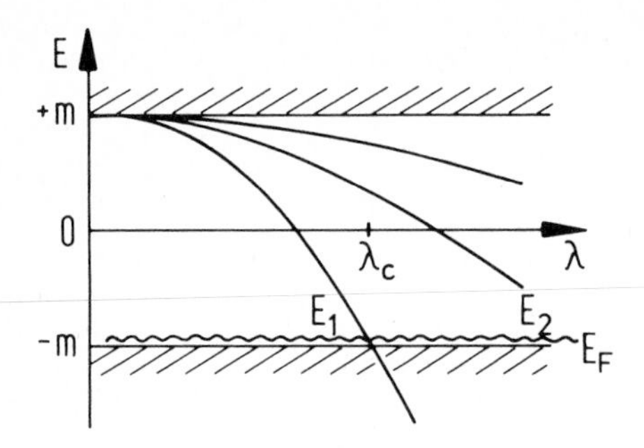

Fig. 9.6. Enumeration of the bound states which become successively supercritical with growing strength parameter λ of the external potential A_μ

Of course, a given bound state contributes to the energy of the vacuum only when it has become supercritical. If $\lambda_c A_\mu$ is the critical potential, the contribution to the vacuum energy is

$$\int_0^1 d\lambda \frac{\partial E}{\partial \lambda} \Theta(\lambda-\lambda_c) = \int_{\lambda_c}^1 d\lambda \frac{\partial E}{\partial \lambda} = E(\lambda=1) - E(\lambda_c) = E(\lambda=1) - E_F . \tag{9.137}$$

For several supercritical states labelled by $n = 1, 2, \ldots, n_{vac}$ [cf. (9.126) and Fig. 9.6], the total contribution is

$$\sum_{n=1}^{n_{vac}} (E_n[A_\mu] - E_F) = \sum_{n=1}^{n_{vac}} E_n[A_\mu] - n_{vac} E_F . \tag{9.138}$$

According to (9.128) the second part $n_{vac}E_F$ just cancels the second contribution to the vacuum energy in (9.130). Replacing the virtual vacuum polarization charge density by the Uehling term, as discussed above, gives the approximate relation

$$\begin{aligned} E'_{vac}[A_\mu] &\approx \frac{1}{2} \int d^3x \int d^4x' A_\mu(x)\, \Pi^{(1)}_{ren}(x-x')\, j^\mu_{ext}(x') + \sum_{n=1}^{n_{vac}} E_n[A_\mu] \\ &\equiv E^{(2)}_{vac}[A_\mu] + \sum_{n=1}^{n_{vac}} E_n[A_\mu] . \end{aligned} \tag{9.139}$$

Here $E^{(2)}_{vac}$ stands for the vacuum energy as calculated in second-order perturbation theory.

The dependence of the vacuum energy on the strength of an external potential and the influence of supercritical bound states are illustrated by two examples. In Fig. 9.7 the vacuum energy is shown for a single nucleus of charge Z as a function of Z. Up to $Z_{cr} \approx 170$ the vacuum energy increases like Z^2, but at Z_{cr} a sudden drop by $2m_e$ occurs due to the interaction with the nucleus of the two $1s$ electrons contained in the supercritical vacuum. At $Z'_{cr} \sim 185$ the two $2p_{1/2}$ states join the Dirac sea and the vacuum energy drops by another $2m_e$. When Z is increased further, the growth in the binding energy of the four electrons of the real vacuum polarization overcompensates the further increase of the contribution $E^{(2)}_{vac}$ from the virtual vacuum polarization, i.e. the energy of the electron – positron vacuum actually decreases with increasing nuclear charge. (Of course, the nuclear Coulomb energy grows much more rapidly, further increasing the total energy.

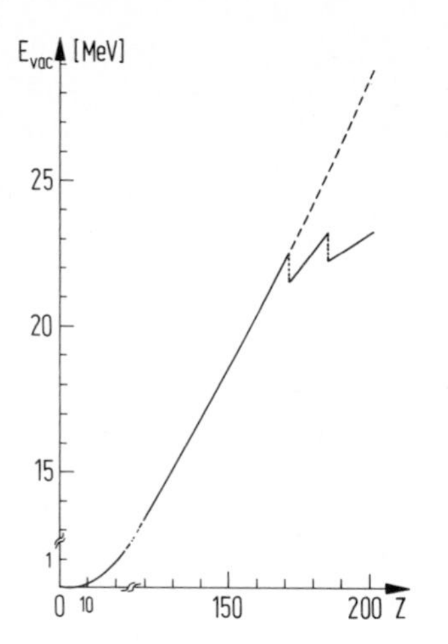

Fig. 9.7. Energy of the electron – positron vacuum in the Coulomb field of a nucleus. At the critical charges $Z_{cr} = 172$, $Z'_{cr} = 185$, etc., the vacuum becomes charged and reduces its energy by an amount of $2m_ec^2 \sim 1$ MeV by emitting two positrons (from [Pl 83])

Only in the extreme supercritical case $q_{vac} \sim Ze$ is the vacuum energy a dominant contribution to the total energy; see Sect. 16.4.)

The second example is the energy of the vacuum in the electric field of two nuclei of charge Z_1 and Z_2 separated by a distance R, which exists during a close collision of two atoms (Fig. 9.8). We discuss the solutions of the Dirac equation for two Coulomb centres in Chap. 11, but here we anticipate the results. For supercritical total charge $Z_1 + Z_2 > Z_{cr}$, the $1s$ state becomes supercritical at some internuclear separation R_{cr}, and for distances $R < R_{cr}$ the stable vacuum is charged. For the symmetric system U + U ($Z_1 = Z_2 = 92$), which has been subject to extensive experimental investigations during the past years, the critical distance is $R_{cr} \sim 36$ fm for two bare nuclei.

When one calculates the vacuum energy for the U + U system as a function of R, it is found that initially E'_{vac} grows with shrinking separation of the nuclei (Fig. 9.9). Finally, at R_{cr}, the vacuum energy suddenly drops by $2m_e$ due to the

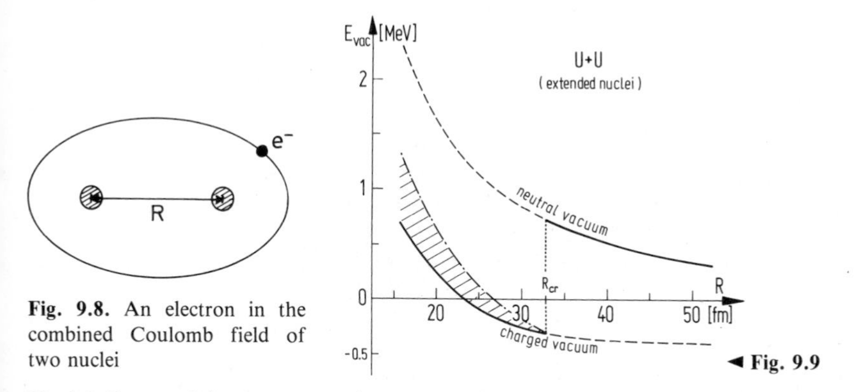

Fig. 9.8. An electron in the combined Coulomb field of two nuclei

◄ **Fig. 9.9**

Fig. 9.9. Energy of the electron – positron vacuum in the Coulomb field of two bare uranium nuclei ($Z = 92$) separated by a distance R. At the critical distance $R_{cr} = 36$ fm the vacuum becomes charged, and its energy is reduced by $2m_e$ due to spontaneous emission of two positrons [Pl 83]. For distances below R_{cr} the vacuum energy grows less rapidly because the vacuum charge screenes the nuclear Coulomb repulsion

emission of two positrons. Upon further approach of the nuclei the energy of the vacuum continues to grow, but less rapidly than before R_{cr} as a result of the binding effect of the two electrons in the supercritical $1s$ state.

9.8 Charge and Energy of the Vacuum (II)

In the previous section we derived exact expressions for the charge and energy of the vacuum, (9.128, 130). To evaluate the vacuum energy and to exhibit clearly the contribution of the real vacuum polarization we had to make severe approximations, which centred around the *ad hoc* assumption that the vacuum polarization charge density can be written as a sum of the first-order Uehling term and the contributions from the supercritical bound states. To describe these, in turn, we had to pretend that they could be represented by eigenstates of the Dirac equation. However, the detailed discussion in Chap. 6 within the framework of autoionization formalism, indicates that this cannot be true. In the supercritical external field these bound states are dissolved into resonances in the negative energy continuum of the Dirac equation.

A proper description of their contribution to the charge density of the vacuum, therefore, requires carefully treating the influence of the external potential on the continuum states. Since the presence of a resonance shows up most clearly in the energy dependence of the phase shift, we base (following Schwinger's treatment of subcritical potentials [Schw 54]) our improved treatment on a phase-shift analysis of the scattering states of the Dirac equation, starting directly from (9.107, 108) for the charge and energy of the vacuum state [Mü 83a].

We shall make the simplifying assumption that there is no external vector potential $A = 0$, and that the electrostatic potential is spherically symmetric, $eA_0 = V(r)$. To be able to count states, we impose a suitable boundary condition on the Dirac wave functions at a finite, but large radius R. For good angular momentum waves

$$\psi_{\varkappa,\mu}(\boldsymbol{x}) = \begin{pmatrix} f(r) & \chi^{\mu}_{\varkappa}(\vartheta,\varphi) \\ \mathrm{i}\,g(r) & \chi^{\mu}_{-\varkappa}(\vartheta,\varphi) \end{pmatrix} \tag{9.140}$$

appropriate boundary conditions are $f(R) = 0$, or $g(R) = 0$. Let us first suppose that the external potential is weak and there is a well-defined gap separating positive energy and negative energy solutions of the Dirac equation as indicated in Fig. 9.10. For every value of the angular momentum quantum numbers $\varkappa, \mu$, the states with energy eigenvalues $E_n > m_e$ can then be counted according to growing energy $(0 < E_1 < E_2 < \ldots)$, while the states with eigenvalues $\bar{E}_n < m_e$ can be ordered by decreasing energy $(0 > \bar{E}_1 > \bar{E}_2 > \ldots)$.

With a vanishing external potential, the spectrum is symmetric around $E = 0$, i.e. $E_n = -\bar{E}_n$. Then it is clear that $q_{\mathrm{vac}}[V\equiv 0] = 0$, but

$$E_{\mathrm{vac}}[V\equiv 0] = -\sum_{\varkappa,\mu}\sum_{n=1}^{\infty} E_n^{\varkappa,\mu}[V\equiv 0] \tag{9.141}$$

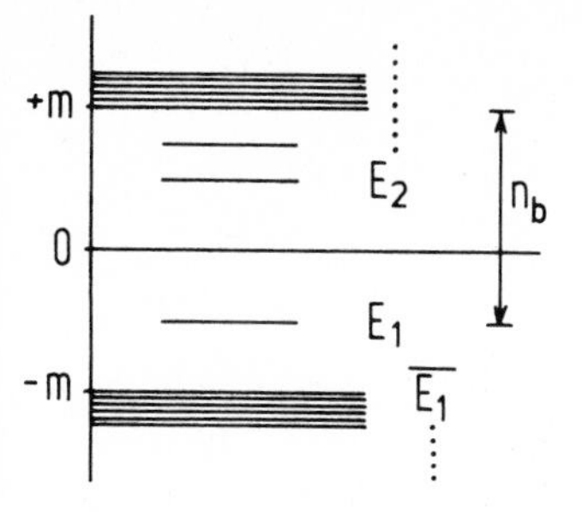

Fig. 9.10. Spectrum of the Dirac Hamiltonian: E_n count the particle states ($\bar{E}_n$ the antiparticle states) ordered by increasing (decreasing) energy eigenvalue

does not vanish. Since we have to define *some* reference point from which to measure energies, it seems reasonable to take the (infinite) energy of the Dirac vacuum in a vanishing potential as the standard. In this spirit we shall henceforth be interested only in the difference, (9.116b),

$$E'_{\text{vac}}[V] = E_{\text{vac}}[V] - E_{\text{vac}}[V \equiv 0]$$
$$= -\frac{1}{2} \sum_{\varkappa,\mu} \sum_{n=1}^{\infty} [E_n^{\varkappa\mu}[V] - E_n^{\varkappa\mu}[0] - (\bar{E}_n^{\varkappa\mu}[V] - \bar{E}_n^{\varkappa\mu}[0])] . \tag{9.142}$$

Now assume that the potential $V(r)$ is of finite range and that $V(\infty) = 0$. Then states can be distinguished for which the energy eigenvalue becomes independent of the boundary condition as R is increased further and further, i.e. the bound states, with $-m_e \leq \lim_{R\to\infty} E_n(R) < +m_e$, and those states that form the continuum in the limit $R \to \infty$. At distances larger than the range of the potential, these continuum states have wave functions behaving like

$$f_n(R) \sim \sin(k_n r + \Delta_n[V]) \tag{9.143}$$

with $k_n^2 = E_n^2 - m_e^2$, and an analogous expression for the antiparticle states. For large R, the boundary condition $f_n(R) = 0$ therefore gives

$$n\pi = k_n R + \Delta_n[V] . \tag{9.144}$$

The phase shift Δ_n is a functional of the potential $V(r)$ and is independent of n for vanishing potential (the constant value $\Delta[0]$ is different for various values of angular momentum $\varkappa$). When the boundary at R is made to recede to infinity, $E_n[V]$ and $E_n[0]$ get infinitesimally close, such that

$$E_n[V] - E_n[0] \to$$
$$\to \left.\frac{dE}{dk}\right|_{k_n} (k_n[V] - k_n[0]) = -\frac{1}{R} \left.\frac{dE}{dk}\right|_{k_n} (\Delta_n[V] - \Delta_n[0]) . \tag{9.145}$$

While the separation between the energy eigenvalues approaches zero, the number of states in a given momentum interval dk grows with R according to (9.144):

$$dn = \frac{R}{\pi} dk + \frac{1}{\pi}(\Delta_{n+dn} - \Delta_n) . \tag{9.146}$$

In the limit $R \to \infty$ the discrete sum in (9.142) becomes an integral:

$$\sum_{n=n_b+1}^{\infty} (E_n[V] - E_n[0]) \to$$

$$\to -\int_0^{\infty} dk \frac{dE}{dk} \cdot \frac{1}{\pi} \delta_k[V] = -\frac{1}{\pi} \int_{m_e}^{\infty} dE\, \delta_E[V] . \tag{9.147}$$

Here n_b counts the number of positive energy bound particle states and $\delta[V] = \Delta[V] - \Delta[0]$ accounts for the phase shift caused by the potential. The full expression for the vacuum energy (9.142) then reads

$$E'_{\mathrm{vac}}[V] = \sum_{\varkappa,\mu} \left\{ \frac{1}{2} \sum_{n=1}^{n_b} (m_e - E_n^{\varkappa\mu}[V]) + \frac{1}{2} \sum_{n=1}^{\bar{n}_b} (m_e + \bar{E}_n^{\varkappa\mu}[V]) \right\}$$

$$+ \frac{1}{2\pi} \int_{m_e}^{\infty} dE (\delta_E[V] + \delta_{-E}[V]) \tag{9.148}$$

since the bound states correspond to continuum states of infinitesimal kinetic energy in the case of vanishing potential. Here $\bar{n}_b$ stands for the number of bound antiparticle states, which may be present if the potential has different signs in different regions of space. This is not usual in QED (an academic example would be a diatomic molecule formed by a nucleus and an antinucleus), but such situations occur normally in quantum chromodynamics [Va 83].

For the vacuum charge q_{vac} the second term in (9.146) is rewritten as

$$\Delta_{n+dn} - \Delta_n \to \frac{d\Delta}{dk} dk \tag{9.149}$$

and substituted into (9.107), while going to the continuum limit:

$$q_{\mathrm{vac}} = -\frac{e}{2} \sum_{\varkappa,\mu} \left\{ n_b^{(\varkappa\mu)} + \frac{1}{\pi} \int_0^{\infty} dk \left(R + \frac{d\Delta}{dk} \right) - \bar{n}_b^{(\varkappa\mu)} - \frac{1}{\pi} \int_0^{\infty} dk \left(R + \frac{d\bar{\Delta}}{dk} \right) \right\}$$

$$= -\frac{e}{2} \sum_{\varkappa,\mu} \left\{ n_b - \bar{n}_b + \frac{1}{\pi} \int_{m_e}^{\infty} dE \frac{d}{dE} (\delta_E[V] - \delta_{-E}[V]) \right\} . \tag{9.150}$$

As they stand, (9.148, 150) still involve divergent expressions. To see this, we have to investigate the high-energy behaviour of the phase shifts $\delta_E^{(\varkappa\mu)}$. If the potential $V(r)$ has no singularities, the high-energy limit is exactly described by the WKB approximation which gives the following closed expression for the phase shift:

$$\delta_{\mathrm{WKB}} = \int_{r_0[V]}^{\infty} dr \sqrt{(E-V)^2 - \frac{\varkappa^2}{r^2} - m_e^2} - \int_{r_0}^{\infty} dr \sqrt{E^2 - \frac{\varkappa^2}{r^2} - m_e^2} , \tag{9.151}$$

where r_0, $r_0[V]$ are the radii where the square roots vanish, respectively (the classical turning points). The square root expressions occurring in (9.151) are just the classical radial momenta with and without potential, abbreviated by $p_r[V]$ and p_r. For large values of energy E, it is illustrative to expand $p_r[V]$ into a power series in the potential:

$$p_r[V] = p_r - \frac{E}{p_r} V - \frac{m_e^2 + \varkappa^2/r^2}{2p_r^3} V^2 - \frac{m_e^2 + \varkappa^2/r^2}{2p_r^5} E V^3 - \frac{(m_e^2 + \varkappa^2/r^2)(4E^2 + m_e^2 + \varkappa^2/r^2)}{8p_r^7} V^4 \pm \cdots . \tag{9.152}$$

Taking the limit $E \to \infty$,

$$\lim_{E\to\infty} \delta_E^{(\varkappa\mu)} = -\operatorname{sgn}(E) \int_0^\infty V(r)\, dr \tag{9.153}$$

independent of the angular momentum quantum numbers. Consequently, the sum over $\varkappa$ and μ in (9.150) diverges:

$$\sum_{\varkappa,\mu} \int_{m_e}^{\infty} dE \frac{d}{dE}(\delta_E - \delta_{-E}) = \sum_{\varkappa,\mu} \left[-2\int_0^\infty V(r)\, dr - \delta_{m_e}^{(\varkappa\mu)} + \delta_{-m_e}^{(\varkappa\mu)} \right] . \tag{9.154}$$

Since the term linear in V cancels in the phase shift integral occurring in (9.148), the integral is finite for every angular momentum channel. Still, the sum over $\varkappa$, μ diverges, as can be seen in the following way. Owing to the spherical symmetry of $V(r)$, the phase shift actually does not depend on the magnetic quantum number μ, and thus

$$\sum_{\varkappa,\mu} = \sum_{\varkappa} 2|\varkappa| .$$

Now p_r involves $\varkappa$ linearly, which makes the term in the expansion of $p_r[V]$ that is quadratic in V behave like $|\varkappa|^{-1}$. As a result, the $\varkappa$ summation is divergent. One easily checks that the V^3 term cancels and that the quartic term behaves as $|\varkappa|^{-3}$, and thus gives a finite contribution, etc.

We conclude that if the phase shift is expanded in powers of the external potential, the divergences arise from the lowest order terms contributing to the vacuum charge and energy, respectively. That the divergences are concentrated in the lowest order terms of a series expansion is typical of renormalizable field theory. This could have been expected since QED is renormalizable. In addition, it turns out that the divergent terms can be absorbed in a redefinition (the "renormalization") of the electric coupling constant e, Chap. 14.3. Accepting that this programme can be carried through, we can give exact expressions for the vacuum charge and energy, simply by subtracting the first-order (Born) phase shift $\delta_E^{1(\varkappa\mu)}$ from the exact phase shift $\delta_E^{(\varkappa\mu)}$ for the vacuum charge and the first- and second-order phase shifts $\delta_E^{2(\varkappa\mu)}$ for the vacuum energy:

$$\delta_E'^{(\varkappa\mu)} = \delta_E^{(\varkappa\mu)} - \delta_E^{1(\varkappa\mu)} ; \qquad \delta_E''^{(\varkappa\mu)} = \delta_E'^{(\varkappa\mu)} - \delta_E^{2(\varkappa\mu)} . \tag{9.155}$$

Denoting the part of the vacuum charge that is linear in the potential by $q_{\text{vac}}^{(1)}$ and the part of the vacuum energy quadratic in V by $E_{\text{vac}}^{(2)}$, then

$$q_{\text{vac}} = q_{\text{vac}}^{(1)} - \frac{e}{2} \sum_{\varkappa,\mu} \left\{ n_{\text{b}} - \bar{n}_{\text{b}} - \frac{1}{\pi} \delta'_{m_e} + \frac{1}{\pi} \delta'_{-m_e} \right\} \tag{9.156}$$

$$E_{\text{vac}} = E_{\text{vac}}^{(2)} + \frac{1}{2} \sum_{\varkappa,\mu} \left\{ \sum_{n=1}^{n_{\text{b}}} (m_e - E_n) + \sum_{n=1}^{\bar{n}_{\text{b}}} (m_e + \bar{E}_n) + \frac{1}{\pi} \int_{m_e}^{\infty} dE (\delta''_E + \delta''_{-E}) \right\} . \tag{9.157}$$

We can draw a number of conclusions about the vacuum charge on purely physical grounds. Because of the quantization of electric charge, q_{vac} can take values only of multiples of e. Since q_{vac} was defined to be zero for vanishing potential, it cannot acquire a non-zero value in any order of perturbation theory for the external potential. To be discontinuous, $q_{\text{vac}}[V]$ cannot be of finite order in V. Thus $q_{\text{vac}}^{(1)}$ must vanish. However, q_{vac} remains zero even beyond perturbation theory, as long as the potential is not strong enough to mix positive and negative energy states. In this region of "weak" external potentials, *Levinson's* phase shift theorem is valid, which states that the phase shift at threshold is determined by the number of bound states [Le 49, Ma 85]

$$n_{\text{b}} = \pi \delta_{m_e} = \pi \delta'_{m_e} , \qquad \bar{n}_{\text{b}} = \pi \delta_{-m_e} = \pi \delta'_{-m_e} \tag{9.158}$$

for non-singular potentials of finite range. The equality of $\delta_{\pm m_e}$ and $\delta'_{\pm m_e}$ expresses that the occurrence of bound states is a non-perturbative phenomenon. Equations (9.158) immediately yield the result that the vacuum does not carry a charge for weak potentials. In general [Mü 76a]

$$q_{\text{vac}} = -\frac{e}{2} \sum_{\varkappa,\mu} \left\{ n_{\text{b}} - \frac{1}{\pi} \delta_{m_e} - \bar{n}_{\text{b}} + \frac{1}{\pi} \delta_{-m_e} \right\} , \tag{9.159}$$

where $\delta_{\pm m_e}$ replaces $\delta'_{\pm m_e}$.

For strong fields q_{vac} acquires a non-zero value in the following way. Assume that V is attractive for particles ($V \leq 0$). Consider the vacuum state with Fermi energy $E_{\text{F}} = -m_e$. At the critical strength of the potential V_{cr}, the lowest bound state crosses the line $E = -m_e$, thereby diminishing n_{b} by one, and causing a jump of π in the phase shift δ_{-m_e}. This is illustrated in Fig. 9.11, where curve 4 corresponds to a potential that has just become supercritical. Upon further increase of the potential, the bound state resonance moves through the antiparticle continuum, but does not change δ_{-m_e} further until the next bound state dives in.

The expression for the vacuum energy may be used to find the contribution of the supercritical bound state resonance to the energy of the vacuum state. In the approximation of the autoionization model, the negative energy phase shift can be written as [cf. Chap. 6, (6.26, 27a)]

$$\delta_E[V] = \delta_E[V_{\text{cr}}] - \arctan \frac{\Gamma}{2(E - E_{\text{res}})} , \tag{9.160}$$

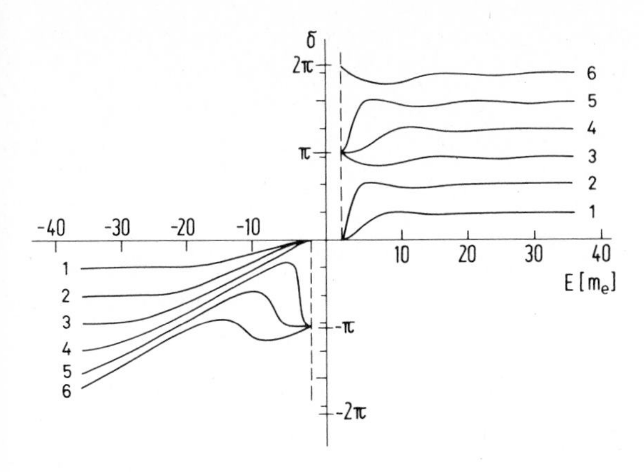

Fig. 9.11. The *s*-wave scattering phase shifts for an attractive square well potential of various strengths. The depth of the well increases proportionally to the curve label (1, 2, ..., 6)

where E_{res} is the energy of the centre of the resonance. Integrating in (9.157) gives

$$\int_{m_e}^{\infty} dE\, \delta''_{-E}[V] \simeq \int_{m_e}^{\infty} dE\, \delta''_{-E}[V_{\text{cr}}] - (m_e + E_{\text{res}})\,\pi\,, \tag{9.161}$$

if we assume the width Γ to be much smaller than the diving depth $|E_{\text{res}} + m_e|$. Keeping in mind that the resonance comes from that bound state which is now missing from the first sum in (9.157), we obtain the following approximate result for the vacuum energy in a slightly supercritical potential:

$$\begin{aligned} E_{\text{vac}}[V > V_{\text{cr}}] &= E_{\text{vac}}[V_{\text{cr}}] + \tfrac{1}{2}\{-(m_e - E_{\text{res}}) + m_e + E_{\text{res}}\} \\ &= E_{\text{vac}}[V_{\text{cr}}] + E_{\text{res}}\,. \end{aligned} \tag{9.162}$$

This result corresponds to the picture that the supercritical bound state resonance is occupied by a particle whose energy must contribute an amount E_{res} to the energy of the charged vacuum. Therefore the energy of the vacuum state with $E_{\text{F}} = -m_e$ exhibits a discontinuity of size $(-m_e)$ at the critical strength of the external potential, confirming the results obtained in Sect. 9.7 [cf. discussion following (9.130)].

Equation (9.162) is the rigorous version of (9.139) for a single supercritical state (in the narrow resonance approximation). The result of our improved treatment of the supercritical bound states can, therefore, be summarized in the statement that the energy eigenvalue of the bound state be properly replaced by the centroid of the supercritical bound state resonance.

9.9 Appendix: Feynman Propagator for Time-Dependent Fields

When the external potential is explicitly time dependent, the definition of the vacuum state depends on the instant of time to which it refers. In particular, there will be an in-vacuum state $|0, \text{in}\rangle$ defined according to the energies of stationary states at $t = -\infty$, and an out vacuum, defined according to stationary states at $t = +\infty$. This is reflected in different possible definitions of the Feynman propagator, as already pointed out in Sect. 8.4. For example, (8.66) acquires the following form in the second quantized formalism

$$\begin{aligned} \mathrm{i}S_{\mathrm{F}}^{(\mathrm{in})}(x,x') &= \langle 0, \mathrm{in} | T(\hat{\psi}(x)\hat{\bar{\psi}}(x')) | 0, \mathrm{in} \rangle \,, \\ \mathrm{i}S_{\mathrm{F}}^{(\mathrm{out})}(x,x') &= \langle 0, \mathrm{out} | T(\hat{\psi}(x)\hat{\bar{\psi}}(x')) | 0, \mathrm{out} \rangle \,. \end{aligned} \tag{9.163}$$

As discussed in Sect. 10.1 in more detail, the in and out vacua are defined by decomposition of the field operator according to the forward or backward propagating solutions of the Dirac equation, defined in (8.39, 43):

$$\begin{aligned} \hat{\psi}(x) &= \sum_{n>F} \hat{b}_n^{(\mathrm{in})} \psi_n^{(+)}(x) + \sum_{n<F} \hat{d}_n^{(\mathrm{in})\dagger} \psi_n^{(+)}(x) \,, \\ \hat{\psi}(x) &= \sum_{n>F'} \hat{b}_n^{(\mathrm{out})} \psi_n^{(-)}(x) + \sum_{n<F'} \hat{d}_n^{(\mathrm{out})\dagger} \psi_n^{(-)}(x) \,, \end{aligned} \tag{9.164}$$

and requiring

$$\begin{aligned} \hat{b}_n^{(\mathrm{in})} |0, \mathrm{in}\rangle &= \hat{d}_n^{(\mathrm{in})} |0, \mathrm{in}\rangle = 0 \,, \\ \hat{b}_n^{(\mathrm{out})} |0, \mathrm{out}\rangle &= \hat{d}_n^{(\mathrm{out})} |0, \mathrm{out}\rangle = 0 \,. \end{aligned} \tag{9.165}$$

The functions $\psi_n^{(+)}$ and $\psi_n^{(-)}$, or the operators $\hat{b}_n^{(\mathrm{in})}$ and $\hat{b}_n^{(\mathrm{out})}$, are related through the matrix elements

$$S_{mn} = \int d^3x\, \psi_m^{(-)\dagger}(\boldsymbol{x}, t)\, \psi_n^{(+)}(\boldsymbol{x}, t) \equiv \langle \psi_m^{(-)} | \psi_n^{(+)} \rangle \,, \tag{9.166}$$

introduced in (8.48) as elements of the unitary S matrix:

$$\sum_k S_{mk} S_{nk}^* = \sum_k S_{km} S_{kn}^* = \delta_{mn} \,. \tag{9.167}$$

Inserting (9.164, 165) into the definitions of the propagators (9.163) gives

$$\mathrm{i}S_{\mathrm{F}}^{(\mathrm{in})}(x,x') = \Theta(t-t') \sum_{n>F} \psi_n^{(+)}(x)\, \bar{\psi}_n^{(+)}(x') - \Theta(t'-t) \sum_{n<F} \psi_n^{(+)}(x)\, \bar{\psi}_n^{(+)}(x') \,, \tag{9.168a}$$

$$\mathrm{i}S_{\mathrm{F}}^{(\mathrm{out})}(x,x') = \Theta(t-t') \sum_{n>F'} \psi_n^{(-)}(x)\, \bar{\psi}_n^{(-)}(x') - \Theta(t'-t) \sum_{n<F'} \psi_n^{(-)}(x)\, \bar{\psi}_n^{(-)}(x') \,, \tag{9.168b}$$

i.e. the two propagators differ in terms of the complete set of wave functions entering their definition, and in terms of the Fermi level. They can be related to each other by means of the single-particle S-matrix elements (9.166):

$$\begin{aligned} \mathrm{i}S_{\mathrm{F}}^{(\mathrm{out})}(x,x') = {} & \Theta(t-t') \sum_{k,l} \left(\sum_{n>F'} S_{nk}^* S_{nl} \right) \psi_k^{(+)}(x)\, \bar{\psi}_l^{(+)}(x') \\ & - \Theta(t'-t) \sum_{k,l} \left(\sum_{n<F'} S_{nk}^* S_{nl} \right) \psi_k^{(+)}(x)\, \bar{\psi}_l^{(+)}(x') \end{aligned} \tag{9.169}$$

$$= \mathrm{i}S_{\mathrm{F}}^{(\mathrm{in})}(x,x') + \Theta(t-t') \sum_{kl} \left[\sum_{n>F'} S_{nk}^* S_{nl} - \delta_{kl}\Theta(k>F) \right] \cdot \psi_k^{(+)}(x)\,\bar{\psi}_l^{(+)}(x') - \Theta(t'-t) \sum_{k,l} \left[\sum_{n<F'} S_{nk}^* S_{nl} - \delta_{kl}\Theta(k<F) \right] \psi_k^{(+)}(x)\,\bar{\psi}_l^{(+)}(x') \,,$$

where $\Theta(k \gtrless F)$ means that the sum over k is restricted to states k above or below the in Fermi surface F. With the help of the unitarity equations (9.167) it is possible to rewrite the bracket in the first term of the difference between $S_{\mathrm{F}}^{(\mathrm{out})}$ and $S_{\mathrm{F}}^{(\mathrm{in})}$:

$$\sum_{n>F'} S_{nk}^* S_{nl} - \delta_{kl}\Theta(k>F) = \left(\delta_{kl} - \sum_{n<F'} S_{nk}^* S_{nl} \right) - [\delta_{kl} - \delta_{kl}\Theta(k<F)] = - \left[\sum_{n<F'} S_{nk}^* S_{nl} - \delta_{kl}\Theta(k<F) \right] .$$

This is just the bracket in the second term, but with opposite sign, so that the time-step functions add up to one:

$$\mathrm{i}S_{\mathrm{F}}^{(\mathrm{out})}(x,x') - \mathrm{i}S_{\mathrm{F}}^{(\mathrm{in})}(x,x') = \sum_{k,l} \left(\delta_{kl}\Theta(k<F) - \sum_{n<F'} S_{nk}^* S_{nl} \right) \psi_k^{(+)}(x)\,\bar{\psi}_l^{(+)}(x') \,. \tag{9.170}$$

If the in-vacuum state is *not* destroyed by the time-dependent external potential, i.e. if always

$$S_{nk} = 0 \quad \text{for} \quad n<F' \,, \quad k>F \,,$$

the unitary equation (9.167) guarantees that

$$\sum_{n<F'} S_{nk}^* S_{nl} = \delta_{kl}\Theta(k<F) \,,$$

and the in and out propagators coincide. When at least one element S_{nk} with $n<F'$, $k>F$ is different from zero, the out vacuum contains real electron – positron pairs compared with the in vacuum, and the two propagators are different.

The definitions (9.163 or 168) for the Feynman propagator are quite natural in the Heisenberg picture (as here), where the states are time independent. The very critical reader will have noticed, however, that they are not really the actual generalizations of the concept of the free Stückelberg-Feynman propagator $S_{\mathrm{SF}}^0(x-x')$ whose importance was so extensively elaborated in Chap. 7. There we said that S_{SF}^0 should propagate particles forwards in time, but antiparticles backwards in time. Now, when the definition of the vacuum state differs, whether it is defined at $t=-\infty$, i.e. the in vacuum, or at $t=+\infty$, i.e. the out vacuum, the forward-propagating functions $\psi_n^{(+)}$ should define a particle at $t=-\infty$, and the backward-propagating functions $\psi_n^{(-)}$, a particle at $t=+\infty$. The most general ansatz for the Stückelberg-Feynman propagator that corresponds to these requests is

$$\begin{aligned}
\mathrm{i}\tilde{S}_{\mathrm{SF}}(x,x') = \Theta(t-t') \sum_{n>F'} \sum_{m>F} \beta_{nm}\, \psi_n^{(-)}(x)\, \bar{\psi}_m^{(+)}(x') \\
- \Theta(t'-t) \sum_{n<F} \sum_{m<F'} \bar{\beta}_{nm}\, \psi_n^{(+)}(x)\, \bar{\psi}_m^{(-)}(x') \,.
\end{aligned} \tag{9.171}$$

It makes states above the Fermi surface propagate into the future and those below into the past. It also ensures that in the limit $t' \to -\infty$ (or $t \to -\infty$) an electron (or positron) is defined as a particle with respect to the in vacuum, or as particle with respect to the out vacuum in the limit $t' \to +\infty$ (or $t \to +\infty$).

The coefficients β_{nm} and $\bar{\beta}_{nm}$ have been introduced to have sufficient freedom in the ansatz (9.171) to satisfy the Dirac equation for the propagator (7.28):

$$(\gamma^\mu p_\mu - e\gamma^\mu A_\mu(x) - m_e)\, \tilde{S}_{\mathrm{SF}}(x,x') = \delta(x-x') \,. \tag{9.172}$$

When the ansatz (9.171) is inserted into this equation, the Dirac operator annihilates the functions $\psi_n^{(\pm)}(x)$, leaving only the action of the time derivative on the step functions:

$$\begin{aligned}
\gamma^0 \delta(t-t') \sum_{n>F'} \sum_{m>F} \beta_{nm}\, \psi_n^{(-)}(x)\, \bar{\psi}_m^{(+)}(x') \\
+ \gamma^0 \delta(t'-t) \sum_{n<F} \sum_{m<F'} \bar{\beta}_{nm}\, \psi_n^{(+)}(x)\, \bar{\psi}_m^{(-)}(x') = \delta(x-x') \,.
\end{aligned} \tag{9.173}$$

To obtain equations determining the β_{nm}, we cancel the time-delta function, multiply by γ^0 from the left and right, and take the scalar product with $\psi_k^{(+)}(x)$, $k>F$, from the left and $\psi_l^{(+)}(x')$, $l>F$, from the right. This yields:

$$\sum_{n>F'} \sum_{m>F} \beta_{nm} S_{nk}^* \delta_{ml} = \int d^3x\, d^3x'\, \psi_k^{(+)}(x)^\dagger \delta(\boldsymbol{x}-\boldsymbol{x}')\, \psi_l^{(+)}(x') \,, \quad \text{i.e.}$$

$$\sum_{n>F'} \beta_{nl} S_{nk}^* = \delta_{kl} \quad (k,l>F) \,, \tag{9.174a}$$

because $\langle \psi_k^{(+)} | \psi_n^{(+)} \rangle = 0$ for $k>F$, $n<F$. Similarly, equations for $\bar{\beta}_{nm}$ arise by projecting with $\psi_k^{(-)}(x)$, $k<F'$, from the left and $\psi_l^{(-)}(x')$, $l<F'$, from the right:

$$\sum_{n<F} \bar{\beta}_{nl} S_{kn} = \delta_{kl} \quad (k,l<F') \,. \tag{9.174b}$$

Another set of equations can be derived by projecting with $\psi_k^{(-)}(x)$, $k>F'$, and $\psi_l^{(-)}(x')$, $l>F'$:

$$\sum_{m>F} \beta_{km} S_{lm}^* = \delta_{kl} \quad (k,l>F') \,, \tag{9.175a}$$

or with $\psi_k^{(+)}(x)$, $k<F$, and $\psi_l^{(+)}(x')$, $l<F$:

$$\sum_{m<F'} \bar{\beta}_{km} S_{ml} = \delta_{kl} \quad (k,l<F) \,. \tag{9.175b}$$

Each of these two sets of equations determines the coefficients β_{nm} and $\bar{\beta}_{nm}$, but they cannot be explicitly resolved. Still, they suffice to derive a relation between the Stückelberg-Feynman propagator $\tilde{S}_{\rm SF}$ and the Feynman propagator $S_{\rm F}^{\rm (in)}$ defined with respect to the in vacuum:

$$\begin{aligned} &\mathrm{i}\tilde{S}_{\rm SF}(x,x')-\mathrm{i}S_{\rm F}^{\rm (in)}(x,x') \\ &\quad =\Theta(t-t')\sum_{m>F}\left[\sum_{n>F'}\beta_{nm}\psi_n^{(-)}(x)-\psi_m^{(+)}(x)\right]\bar{\psi}_m^{(+)}(x') \\ &\qquad -\Theta(t'-t)\sum_{n<F}\psi_n^{(+)}(x)\left[\sum_{m<F'}\bar{\beta}_{nm}\bar{\psi}_m^{(-)}(x')-\bar{\psi}_n^{(+)}(x')\right]. \end{aligned} \tag{9.176}$$

Let us investigate the first term of the difference, using (9.174a):

$$\begin{aligned} &\sum_{m>F}\left[\sum_{n>F'}\beta_{nm}\psi_n^{(-)}(x)-\psi_m^{(+)}(x)\right]\bar{\psi}_m^{(+)}(x') \\ &\quad =\sum_{m>F}\left[\sum_{n>F'}\sum_{k}\beta_{nm}S_{nk}^{*}\psi_k^{(+)}(x)-\psi_m^{(+)}(x)\right]\bar{\psi}_m^{(+)}(x') \\ &\quad =\sum_{m>F}\left\{\sum_{n>F'}\left[\sum_{k<F}\beta_{nm}S_{nk}^{*}\psi_k^{(+)}(x)+\delta_{mk}\psi_k^{(+)}(x)\right]-\psi_m^{(+)}(x)\right\}\bar{\psi}_m^{(+)}(x') \\ &\quad =\sum_{m>F}\sum_{k<F}\left(\sum_{n>F'}\beta_{nm}S_{nk}^{*}\right)\psi_k^{(+)}(x)\,\bar{\psi}_m^{(+)}(x'). \end{aligned} \tag{9.177a}$$

Similarly, the second term can be reduced with (9.175b). Renaming the indices yields

$$\begin{aligned} &\sum_{n<F}\psi_n^{(+)}(x)\left[\sum_{m<F'}\bar{\beta}_{nm}\bar{\psi}_m^{(-)}(x')-\bar{\psi}_n^{(-)}(x')\right] \\ &\quad =\sum_{m>F}\sum_{k<F}\left(\sum_{n<F'}\bar{\beta}_{kn}S_{nm}\right)\psi_k^{(+)}(x)\,\bar{\psi}_m^{(+)}(x'). \end{aligned} \tag{9.177b}$$

Combining (9.176, 177), the difference between the propagators is

$$\mathrm{i}\tilde{S}_{\rm SF}(x,x')-\mathrm{i}S_{\rm F}^{\rm (in)}(x,x')=\sum_{m>F}\sum_{k<F}\alpha_{km}\psi_k^{(+)}(x)\,\bar{\psi}_m^{(+)}(x'), \tag{9.178}$$

with the coefficients

$$\alpha_{km}(t,t')=\Theta(t-t')\sum_{n>F'}\beta_{nm}S_{nk}^{*}-\Theta(t'-t)\sum_{n<F'}\bar{\beta}_{kn}S_{nm}. \tag{9.179}$$

This equation can be further simplified if we derive a relation between β_{nm} and $\bar{\beta}_{kn}$ by projecting (9.173) with $\psi_k^{(+)}(x)$, $k<F$, from the left and $\psi_l^{(+)}(x')$, $l>F$, from the right:

$$\delta(t-t') \sum_{n>F'} \sum_{m>F} \beta_{nm} S^*_{nk} \delta_{ml} + \delta(t'-t) \sum_{n<F} \sum_{m<F'} \bar{\beta}_{nm} \delta_{kn} S_{ml}$$

$$= \delta(t-t') \left(\sum_{n>F'} \beta_{nl} S^*_{nk} + \sum_{n<F'} \bar{\beta}_{kn} S_{nl} \right) = 0 \,, \qquad (9.180)$$

where we renamed m to n in the second sum. This new relation enables the coefficients $\bar{\beta}_{kn}$ to be expressed by the coefficients β_{nm} in (9.179) or vice versa:

$$\alpha_{km} = \sum_{n>F'} \beta_{nm} S^*_{nk} = - \sum_{n<F'} \bar{\beta}_{kn} S_{nm} \,, \qquad (9.181)$$

i.e. the α_{km} are time independent.

When the external fields do not produce electron – positron pairs, i.e. when the in and out vacua coincide, the S-matrix elements S_{nk} for $n > F'$, $k < F$, and S_{nm} for $n < F'$, $m > F$ vanish. Equation (9.178, 179) show that the propagators $\tilde{S}_{\mathrm{SF}}$ and $S_{\mathrm{F}}^{(\mathrm{in})}$, $S_{\mathrm{F}}^{(\mathrm{out})}$ are then the same. When the external field produces real pairs, all three propagators differ from each other [Ma 78c]. Which of the propagators is most useful depends on the question asked, i.e. on the physical problem. As discussed in detail in the next section, the propagator $S_{\mathrm{F}}^{(\mathrm{in})}$ is best suited to describe properties of the (in) vacuum state, such as vacuum polarization, vacuum energy, and so forth.

Bibliographical Notes

For general treatment of second quantization of the Dirac field the textbook references given in Chap. 8 apply. Most of the fundamental papers on QED are reprinted in [Schw 58]. Extra detail can be found in the treatise by *Schweber* [Schw 61].

The role of the charged vacuum in quantum field theory was first clarified by *Rafelski* et al. [Ra 74] and also by *Fulcher* and *Klein* [Fu 73a, Fu 74]. The treatment adopted in this chapter is largely based on [Mü 83a]. The expression for the charge and energy of the vacuum state in terms of the Feynman propagator were first derived by *Schwinger* [Schw 51, 54] for weak fields. The modifications necessary for supercritical potentials were given recently by *Plunien* [Pl 85].

The connection between charge and energy of the vacuum and the scattering phase shift was noted by *Schwinger* [Schw 54]. The non-relativistic analogue of (9.160) for the vacuum charge, i.e. the number of particles contained in the ground state of an electron gas, is known as *Friedel's* sum rule [Fr 52]. *Schwinger's* result was extended to supercritical potentials in [Mü 76a, 83a]. The phase shift method has recently found an interesting application in the problem of the fermion vacuum in the presence of a magnetic monopole [Gr 83e, Ya 83].

10. Evolution of the Vacuum State in Supercritical Potentials

Three types of experiments probing a supercritical external field can be imagined, in principle. First, experiments where a subcritical but strong field is made supercritical for a certain finite period of time; then those in which a subcritical external field is rendered supercritical and made to remain so forever; and, finally, experiments that are carried out completely in a supercritical field. The prototype of the first kind of experiments – and the only one presently feasible – is the sub-Coulomb barrier collision of very heavy ions, such as uranium on uranium, Chapts. 12, 13. The second type of experiment corresponds, for instance, to the creation of a stable supercritical nucleus, initially stripped of electrons, by a nuclear fusion process. A typical experiment of the third kind is the ionization of a (hypothetical) supercritical atom and the observation of the resulting spontaneous pair creation (γ absorption accompanied by positron line emission).

This chapter treats all three types of experimental situations from the standpoint of field theory, leading to the conclusion that the instability of the neutral ground state in supercritical fields, previously asserted on the basis of heuristic arguments, is a natural consequence in the field theoretic formulation. To fulfil this goal, we have to explain how the information about observable quantities can be extracted from field theory [Re 80b, 81b].

10.1 The In/Out Formalism

Assume that all observations on a microscopic system undergoing a temporary change are made long before or long after that change takes place, at times denoted as $t = \pm\infty$ and called the in and out regions, respectively (Fig. 10.1). Assume also that the changes occur on a microscopic time scale and that the system is stationary at $t \to \pm\infty$. We shall continue to work in the Heisenberg picture, i.e. the state vector $|\Omega\rangle$ of the system is fixed and all time dependence is carried by the operators. We can now apply to the quantum theory the formalism introduced in Sect. 8.3 for the single-particle theory. In particular, we use the two complete sets of solutions of the single-particle Dirac equation, which are eigenstates in the past or future, $\psi_n^{(\pm)}(\boldsymbol{x}, t)$, to obtain two different decompositions of the field operator:

$$\hat{\psi}(\boldsymbol{x}, t) = \sum_{n>F} \hat{b}_n^{(\mathrm{in})} \psi_n^{(+)}(\boldsymbol{x}, t) + \sum_{n<F} \hat{d}_n^{(\mathrm{in})\dagger} \psi_n^{(+)}(\boldsymbol{x}, t)\,, \qquad (10.1)$$

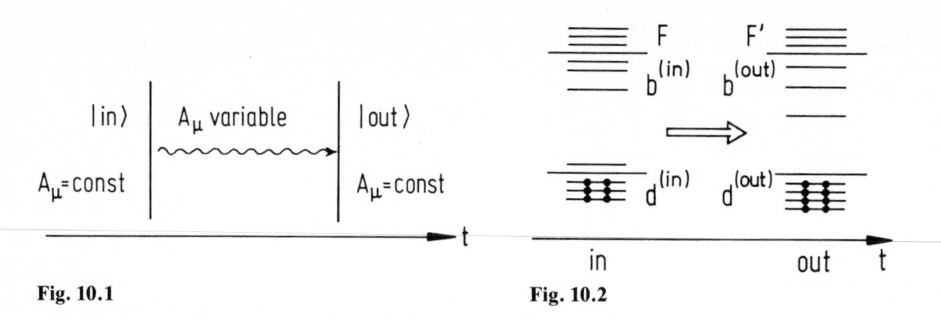

Fig. 10.1 **Fig. 10.2**

Fig. 10.1. Stationary in and out regions are connected by an intermediate region where the potential changes. Measurements are performed in the out region

Fig. 10.2. Definition of the vacuum in the in (Fermi level F) and the out regions (Fermi level F')

$$\hat{\psi}(\boldsymbol{x},t)=\sum_{n>F'}\hat{b}_n^{(\text{out})}\psi_n^{(-)}(\boldsymbol{x},t)+\sum_{n<F'}\hat{d}_n^{(\text{out})\dagger}\psi_n^{(-)}(\boldsymbol{x},t)\,. \tag{10.2}$$

The two labels F, F' denote the "filled" states (in Dirac's sense) in the in and out regions, respectively, and will be taken as the vacuum Fermi levels $E_\text{F}=E'_\text{F}=-m_e$ in the following. However, since the formalism remains equally valid for any initial state characterized by an arbitrary Fermi surface, we continue to write F and F' in this paragraph. The vacuum state (Fig. 10.2) before the experiment is then defined by the conditions

$$\left.\begin{aligned}\hat{b}_n^{(\text{in})}|0,\text{in}\rangle&=0\quad\text{for}\quad n>F\\ \hat{d}_n^{(\text{in})}|0,\text{in}\rangle&=0\quad\text{for}\quad n<F\end{aligned}\right\}, \tag{10.3}$$

whereas the vacuum state in the out region satisfies

$$\left.\begin{aligned}\hat{b}_n^{(\text{out})}|0,\text{out}\rangle&=0\quad\text{for}\quad n>F'\\ \hat{d}_n^{(\text{out})}|0,\text{out}\rangle&=0\quad\text{for}\quad n<F'\end{aligned}\right\}. \tag{10.4}$$

When the external field in the in region differs from that in the out region, $F\neq F'$ and the two vacuum states will not be the same. (Observe that the same Fermi energy $E_\text{F}=-m_e$ describes vacua of different charge for different strength of the external field, i.e. $E_\text{F}=E'_\text{F}$ does *not* imply the same division of states into particle and antiparticle states which would be denoted by $F=F'$.)

It is important to realize that the particle operators $\hat{b}_n^{(\text{in})}$, $\hat{d}_n^{(\text{in})}$, although constant in time, (9.15), differ from the particle operators $\hat{b}_n^{(\text{out})}$, $\hat{d}_n^{(\text{out})}$ defined in the out region, because $\hat{b}_n^{(\text{in})}$, $\hat{d}_n^{(\text{in})}$ *do not correspond to physical particles outside the stationary in region.* Physical particles are defined with respect to stationary wave functions, and $\psi_n^{(-)}(\boldsymbol{x},t)$ are not stationary except in the in region. In particular, they are not stationary in the out region, but in general correspond to linear superpositions of stationary wave functions there. It is possible to relate the operators for in particles to those for out particles by projecting (10.1, 2) with $\psi_m^{(-)}$. With the single-particle S-matrix elements S_{mn} defined in (8.48),

$$\hat{b}_m^{(\text{out})} = \int d^3x\, \psi_m^{(-)}(\boldsymbol{x},t)^\dagger \hat{\psi}(\boldsymbol{x},t) = \left(\sum_{n>F} \hat{b}_n^{(\text{in})} + \sum_{n<F} \hat{d}_n^{(\text{in})\dagger} \right) S_{mn}, \tag{10.5a}$$

$$\hat{d}_m^{(\text{out})} = \left(\sum_{n>F} \hat{b}_n^{(\text{in})\dagger} + \sum_{n<F} \hat{d}_n^{(\text{in})} \right) S_{mn}^* . \tag{10.5b}$$

Owing to the unitarity of the matrix S_{mn}, these relations are easily inverted to express $\hat{b}_n^{(\text{in})}$ in terms of the out operators, but this is not needed in the following.

In an experimental situation, even in a *Gedankenexperiment*, the state of the system $|\Omega\rangle$ is prepared in the in region, i.e. with respect to the in-particle operators. For example, starting with the vacuum, then $|\Omega\rangle = |0, \text{in}\rangle$; but starting with one occupied single-particle state k, $|\Omega\rangle = \hat{b}_k^{(\text{in})\dagger}|0, \text{in}\rangle$, etc. The measurements, on the other hand, are done in the out region, so the corresponding operators are defined through the out particles. Measuring the number of particles in a given state $i > F'$ corresponds to taking the expectation value of the operator

$$\hat{N}_i^{(\text{out})} = \hat{b}_i^{(\text{out})\dagger} \hat{b}_i^{(\text{out})} \quad (i > F') . \tag{10.6}$$

This matrix element is easily evaluated with the help of (10.5a, b) [Re 73, 80a]

$$\begin{aligned} N_i &= \langle 0, \text{in} | \hat{N}_i^{(\text{out})} | 0, \text{in} \rangle \\ &= \langle 0, \text{in} | \left(\sum_{n>F} \hat{b}_n^{(\text{in})\dagger} + \sum_{n<F} \hat{d}_n^{(\text{in})} \right) \left(\sum_{m>F} \hat{b}_m^{(\text{in})} + \sum_{m<F} \hat{d}_m^{(\text{in})\dagger} \right) \cdot S_{in}^* S_{im} | 0, \text{in} \rangle \\ &= \sum_{n<F} \sum_{m<F} S_{in}^* S_{im} \delta_{nm} = \sum_{n<F} |S_{in}|^2 . \end{aligned} \tag{10.7}$$

In the same way, the number of out antiparticles in a state $k < F'$ is given by

$$\bar{N}_k = \langle 0, \text{in} | \hat{d}_k^{(\text{out})\dagger} \hat{d}_k^{(\text{out})} | 0, \text{in} \rangle = \sum_{n>F} |S_{kn}|^2 = 1 - \sum_{n<F} |S_{kn}|^2 . \tag{10.8}$$

The last example is the number of pairs formed by a particle in state i and an antiparticle in state k in the out region ($i>F', k<F'$) [Re 80b]:

$$\begin{aligned} N_{ik} &= \langle 0, \text{in} | \hat{b}_i^{(\text{out})\dagger} \hat{b}_i^{(\text{out})} \hat{d}_k^{(\text{out})\dagger} \hat{d}_k^{(\text{out})} | 0, \text{in} \rangle \\ &= \sum_{n<F} \sum_{m>F} (S_{in}^* S_{in} S_{km} S_{km}^* - S_{in}^* S_{im} S_{kn} S_{km}^*) \\ &= N_i \bar{N}_k + \left| \sum_{n<F} S_{in}^* S_{kn} \right|^2 . \end{aligned} \tag{10.9}$$

Since all observables in the out region, such as energy, energy uncertainty, charge, can be expressed in terms of expectation values of operators built from the out-particle operators, every observable can be calculated from the single-particle amplitudes S_{mn}. This is a consequence of neglecting true two-body interactions between Dirac particles arising from electromagnetic interactions of electrons with other electrons or positrons. Indeed, (10.5) still holds in the Hartree-Fock approximation to QED [Re 70, 71], but becomes invalidated when correlations are taken into account [Ki 80]. In the more general case, the in and out operators become time dependent, and the operator $\hat{b}_m^{(\text{out})}$ corresponds to terms like $\hat{b}_n^{(\text{in})}$ but also to $\hat{b}_k^{(\text{in})}\hat{b}_l^{(\text{in})}\hat{b}_n^{(\text{in})}$ and other higher-order terms. In this book, however, we do not go beyond the Hartree-Fock approximation, so that (10.5) will always hold.

The Pauli principle is contained in the expressions for N_i, $\bar{N}_k$, $N_{i,k}$ via the commutation relations for the particle operators. It enters the results only if single-particle Fermi surfaces F, F' can be defined, which does not always hold for bosons, especially in supercritical fields. The Pauli principle does not directly affect the amplitudes S_{mn}, because all particles propagate independently owing to the orthogonality of the wave functions $\psi_n^{(+)}$. However, the S_{mn} follow from solutions of the Dirac equation and would be different for bosons obeying the Klein-Gordon equation.

10.2 Evolution of the Vacuum State

We are now in a position to calculate the effect of rendering the external potential supercritical on the subcritical vacuum state, Fig. 10.3. According to our considerations at the beginning of Chap. 9, two situations are treated: the potential becomes supercritical at time $t = t_0$ and (1) stays so forever; (2) the potential again becomes subcritical at a later time $t = t_0 + T$. The sub- and supercritical potentials are the same as those discussed in Chap. 6, denoted as V_{cr} and $(V_{\text{cr}} + V')$, respectively. The eigenstates in the subcritical potential are denoted by φ_m, those in the supercritical potential by ϕ_n:

$$[-\mathrm{i}\boldsymbol{\alpha}\cdot\nabla + \beta m + V_{\text{cr}}]\,\varphi_n = \varepsilon_n \varphi_n\,, \tag{10.10}$$

$$[-\mathrm{i}\boldsymbol{\alpha}\cdot\nabla + \beta m + V_{\text{cr}} + V']\,\Phi_n = E_n\,\Phi_n\,. \tag{10.11}$$

The subcritical and supercritical bases are connected by a unitary transformation

$$\Phi_n = \sum_m c_{nm}\,\varphi_m\,, \tag{10.12}$$

explicitly calculated in Chap. 6.

If V' is switched on at $t = t_0$, the time-dependent wave functions $\psi_n^{(\pm)}(\boldsymbol{x}, t)$ must solve

$$[-\mathrm{i}\boldsymbol{\alpha}\cdot\nabla + \beta m + V_{\text{cr}} + V'\,\Theta(t-t_0)]\,\psi_n(\boldsymbol{x}, t) = \mathrm{i}\,\partial\psi_n(\boldsymbol{x}, t)/\partial t\,. \tag{10.13}$$

We can make best use of the fact that the potential is static everywhere except at $t = t_0$ by evaluating the S-matrix element S_{mn} at $t = t_0$. The forward-propagating wave functions *prior* to t_0 are given by stationary subcritical functions

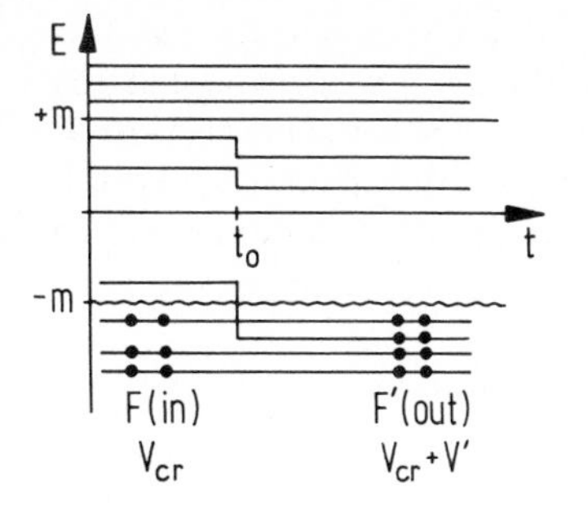

Fig. 10.3. At t_0 the subcritical potential is suddenly rendered supercritical

$$\psi_n^{(+)}(\boldsymbol{x}) = \varphi_n(\boldsymbol{x})\,\mathrm{e}^{-\mathrm{i}\varepsilon_n t} \quad (t \leqslant t_0)\,, \tag{10.14}$$

whereas the backward propagating functions *after* t_0 are stationary in the supercritical basis

$$\psi_m^{(-)}(\boldsymbol{x}, t) = \Phi_m(\boldsymbol{x})\,\mathrm{e}^{-\mathrm{i}E_m t} \quad (t \geqslant t_0)\,. \tag{10.15}$$

Thus (8.48) evaluated at $t = t_0$ yields

$$S_{mn} = \langle \Phi_m | \varphi_n \rangle \exp[\mathrm{i}(E_m - \varepsilon_n)\,t_0] = c^*_{mn} \exp[\mathrm{i}(E_m - \varepsilon_n)\,t_0]\,. \tag{10.16}$$

Except for a phase factor, the S matrix is given by the complex conjugate of the unitary transformation from the subcritical to the supercritical basis. Starting with the subcritical vacuum state F containing all states with $\varepsilon_n < -m_e$, the distribution of positrons at $t = \infty$ is given by

$$\bar{N}_E = \sum_{n>F} |c_{E,n}|^2\,, \qquad E < -m_e\,, \tag{10.17}$$

where (10.8) was used. If the Fock space is reduced to the positron continuum and the bound state that becomes supercritical, as in Sect. 6.1, the sum reduces to a single term for the bound state, $n = 0$. In the terminology of Sect. 6.1, $c_{E,0} = a(E)$, and hence from (6.32a):

$$\bar{N}_E = |a(E)|^2 = \frac{\Gamma/2\pi}{(E - E_\mathrm{r})^2 + \Gamma^2/4}\,, \tag{10.18}$$

where E_r is the resonance energy (6.38).

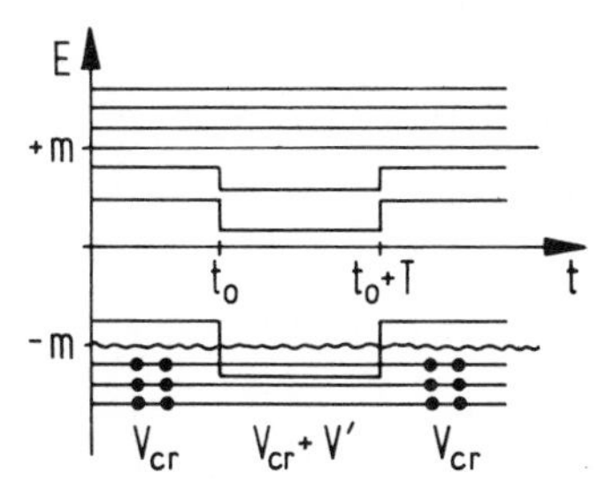

Fig. 10.4. Temporary supercritical potential: the additional potential V' is switched on at t_0 and switched off again at $t_0 + T$

We have thus shown in the framework of field theory that the neutral vacuum state decays in a supercritical external potential, producing a positron distribution centred around the supercritical quasi-bound state resonance energy E_r. To find out how the decay proceeds with time, the additional potential V' must be switched off after some time T and the extent that the neutral vacuum state has decayed during that period of time must be measured. In this situation, Fig. 10.4, it is convenient to evaluate S_{mn} at the time of the switch off, $t = t_0 + T$. For $t \geqslant t_0 + T$ the functions $\psi_m^{(-)}$ propagate in the subcritical potential, whence

$$\psi_m^{(-)}(\boldsymbol{x}, t) = \varphi_m(\boldsymbol{x})\,\mathrm{e}^{-\mathrm{i}\varepsilon_m t}, \qquad (t \geqslant t_0 + T)\,. \tag{10.19}$$

The wave functions $\psi_n^{(+)}$ are still given by (10.14) prior to t_0, but they now propagate in the supercritical potential between t_0 and $(t_0 + T)$. According to Chap. 8.4, this propagation is described by the homogeneous propagator for the supercritical potential

$$\psi_n^{(+)}(\boldsymbol{x}, t_0 + T) = \mathrm{i} \int d^3x'\, S(\boldsymbol{x}, t_0 + T; \boldsymbol{x}', t_0 \,|\, V_{\mathrm{cr}} + V')\, \gamma^0 \psi_n^{(+)}(\boldsymbol{x})\,. \tag{10.20}$$

Here S can be expressed in terms of the eigenstates of the supercritical Hamiltonian, (8.57),

$$S(\boldsymbol{x}, t; \boldsymbol{x}', t' \,|\, V_{\mathrm{cr}} + V') = -\mathrm{i} \sum_k \Phi_k(\boldsymbol{x})\, \bar{\Phi}_k(\boldsymbol{x}')\, \exp[-\mathrm{i}E_k(t - t')]\,. \tag{10.21}$$

When the last two equations are combined with (10.14) then

$$\psi_n^{(+)}(\boldsymbol{x}, t_0 + T) = \mathrm{e}^{-\mathrm{i}\varepsilon_n t_0} \sum_k \Phi_k(\boldsymbol{x})\, c_{kn}^{*}\, \mathrm{e}^{-\mathrm{i}E_k T}\,. \tag{10.22}$$

We can now calculate the single-particle S-matrix elements:

$$S_{mn} = \langle \psi_m^{(-)}(t_0 + T) \,|\, \psi_n^{(+)}(t_0 + T) \rangle = \mathrm{e}^{\mathrm{i}\varepsilon_m T} \sum_k c_{km} c_{kn}^{*}\, \mathrm{e}^{-\mathrm{i}E_k T}\,. \tag{10.23}$$

If again the Fock space is restricted to the subcritical positron states and the diving bound state ($n = 0$), the sum over intermediate supercritical states becomes an integral over the negative energy continuum:

$$|S_{mn}|^2 = \left| \int_{-\infty}^{-m_e} dE\, c_{E,m} c_{E,n}^{*}\, \mathrm{e}^{-\mathrm{i}ET} \right|^2. \tag{10.24}$$

In the terminology of the autoionization model, Sect. 6.1, $C_{E,0} = a(E)$ and $C_{E,\varepsilon} = h_\varepsilon(E)$. Neglecting the continuum-continuum interaction, the following analytic expressions describe these transformation coefficients in the narrow resonance approximation, (6.32, 33):

$$|a(E)|^2 = \frac{|V_E|^2}{(E - E_{\mathrm{r}})^2 + \frac{1}{4}\Gamma^2}\,, \qquad \Gamma = 2\pi |V_{E_{\mathrm{r}}}|^2\,, \tag{10.25}$$

$$h_\varepsilon(E) = \left[P \frac{1}{E - \varepsilon} + \frac{E - E_{\mathrm{r}}}{|V_E|^2}\, \delta(E - \varepsilon) \right] V_\varepsilon a(E)\,. \tag{10.26}$$

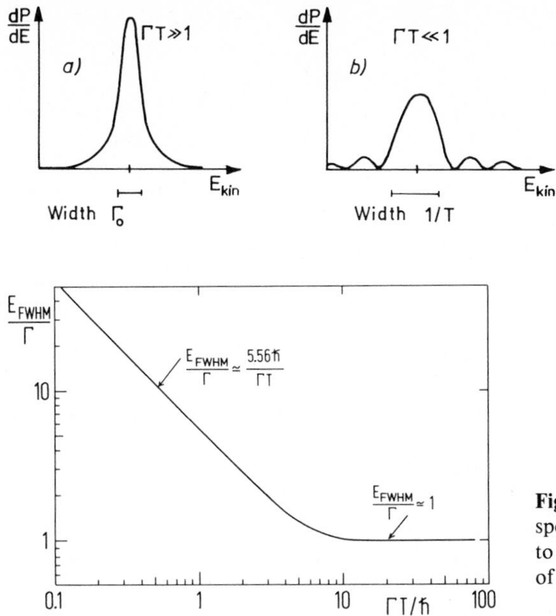

Fig. 10.5a, b. Shape of the spontaneously emitted positron spectrum for different times of duration of supercriticality

Fig. 10.6. Width of the spectrum of spontaneously emitted positrons relative to the natural linewidth Γ, as a function of the period of supercriticality

If we start with the subcritical vacuum state F_0, the positron distribution in the out region is

$$\begin{aligned}\bar{N}_\varepsilon(T) &= |S_{\varepsilon,0}|^2 = \left| \int\limits_{-\infty}^{-m_e} dE\, h_\varepsilon(E)\, a^*(E)\, \mathrm{e}^{-\mathrm{i}ET} \right|^2 \\ &= \left| V_\varepsilon \int\limits_{-\infty}^{-m_e} dE \left(\frac{1}{E-\varepsilon-\mathrm{i}\eta} + \frac{E-E_\mathrm{r}-\frac{\mathrm{i}}{2}\Gamma}{|V_E|^2}\, \delta(E-\varepsilon) \right) |a(E)|^2 \mathrm{e}^{-\mathrm{i}ET} \right|^2\end{aligned} \tag{10.27}$$

where we used $(x-\mathrm{i}\eta)^{-1} = P(1/x) + \mathrm{i}\pi\delta(x)$ with $\eta \to +0$. Closing the contour in the lower complex E plane and neglecting contributions from the finite upper boundary, only the pole of $|a(E)|^2$ at $E = E_\mathrm{r} - (\mathrm{i}\Gamma/2)$ and the delta function contribute to the integral. The result of the calculation is (see Fig. 10.5 for a graphical representation):

$$\begin{aligned}\bar{N}_\varepsilon(T) &= \left| V_\varepsilon \frac{e^{-\mathrm{i}\varepsilon T} - \exp[-\mathrm{i}(E_\mathrm{r} - \frac{\mathrm{i}}{2}\Gamma)T]}{\varepsilon - E_\mathrm{r} + \frac{\mathrm{i}}{2}\Gamma} \right|^2 \\ &= |a(\varepsilon)|^2 |1 - \exp[\mathrm{i}(\varepsilon - E_\mathrm{r} + \tfrac{\mathrm{i}}{2}\Gamma)T]|^2 .\end{aligned} \tag{10.28}$$

For times T long compared to the inverse resonance width Γ^{-1}, the positron distribution exponentially approaches that of our previous model:

$$\bar{N}_\varepsilon(T \gg \Gamma^{-1}) \sim |a(\varepsilon)|^2 (1 - 2\cos[(\varepsilon - E_\mathrm{r})T]\, \mathrm{e}^{-\frac{1}{2}\Gamma T}) , \tag{10.29}$$

but for small switch-on times T the distribution is much broader than the resonance width (Fig. 10.6):

$$\bar{N}_\varepsilon(T \ll \Gamma^{-1}) \sim |V_\varepsilon|^2 T^2 \left(\frac{\sin \frac{1}{2} T(\varepsilon - E_r)}{\frac{1}{2} T(\varepsilon - E_r)} \right)^2 . \tag{10.30}$$

The width in this case is approximately $2\pi/T$, caused by uncertainty of the energy of the dived bound state due to its short lifetime. Hence the autoionization formula (10.18) for the spectral shape of spontaneously emitted positrons is applicable only when the supercritical state lives longer than its natural lifetime $\Gamma T > 2\pi$. This result has direct bearing on the laboratory tests using heavy-ion collisions, Sect. 10.3. We can predict already that the spectrum of positrons emitted in these collisions must be broadened as compared with (10.18), the width being determined by the time T_{coll} during which the lowest bound state is supercritical.

We can also compute the number of electrons N_0 in the bound state after it has reemerged from the positron continuum, which gives the total probability that a positron has been created. On the other hand, the probability that the hole in the supercritical bound state is still present after a time T is

$$1 - N_0(T) = |S_{00}|^2 = \left| \int_{-\infty}^{-m_e} dE\, |a(E)|^2 e^{-iET} \right|^2 = e^{-\Gamma T} . \tag{10.31}$$

Again $E = +\infty$ replaces the upper limit of the integral and the contour in the lower complex E plane is closed. Only the pole at $E = E_r - (i\Gamma/2)$ contributes. Thus, the neutral vacuum state decays exponentially in a supercritical external potential, with the decay time Γ^{-1}. We conclude that the quantum field theoretic considerations fully support the results formerly obtained on the basis of intuitive physical reasoning.

10.3 Decay of a Supercritical *K* Vacancy – Projection Formalism

We now want to discuss the hypothetical situation where the K shell of a stable supercritical nucleus is ionized at time $t = t_0$. The first problem encountered when treating this case is what is the wave function of a supercritical K-shell electron? This problem has been encountered before, but we have not yet given a completely satisfactory answer, simply because there is no unique way of defining the $1s$ wave function of a supercritical atom. To do so we would have to specify an eigenvalue problem which yields the desired function as an eigenmode. We know, however, that a supercritical K vacancy must decay within about 10^{-19} s, thus it is not an eigenstate of any reasonable, physically meaningful operator.

Indeed, the apparent spatial distribution of the K vacancy could depend on the specific way it was produced. The charge distribution surrounding a supercritical nucleus contains not only the K electrons but also the displacement

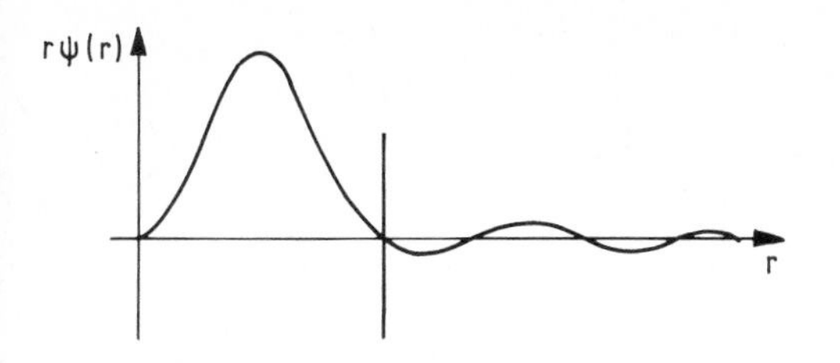

Fig. 10.7. An approximate wave function for the supercritical K shell can be obtained by cutting the oscillating tail off a negative continuum wave function in the $1s$ resonance

charge of the virtual vacuum polarization. In knocking out a total of one elementary charge from this charge cloud, it is not completely clear how much is taken from the K shell and how much from the virtual vacuum polarization density. To be sure, this argument is also valid for a subcritical atom. But then one must wait sufficiently long until the atom has settled down in a stationary state, the $1s$ state being one possibility to find out whether the electron was taken from the K shell or from the vacuum. With the supercritical atom this is not possible, because the K vacancy decays spontaneously by positron emission.

Fortunately, the dilemma is not so bad in practice, because even for a supercritical atom with charge, say, $Z \sim 175-190$, the total charge displaced in the virtual vacuum polarization amounts only to about 1/100 of an elementary charge. Crudely speaking, we could say that there is about 1% uncertainty in the definition of the wave function of a K vacancy for a supercritical atom. Actually the most primitive definition would be to take one of the negative energy continuum wave functions at the peak of the supercritical $1s$ resonance, cut it off after the resonance part before it starts its continuum oscillations, and normalize it to unity, Fig. 10.7. Other, more subtle, methods have been devised, but are of no importance here [Re 81b].

Once the K shell wave function has been chosen, denoted by Φ_R, we face the mathematically well-defined problem of finding a complete set of orthogonal wave functions containing Φ_R [Re 80b].

Such a set of wave functions helps to find a representation of the field operator $\hat{\psi}(x)$, similar to (10.1), which contains Φ_R as one of the functions. For simplicity, assume that Φ_R has been chosen to be orthogonal to the higher bound states and the positive energy continuum. Then it is sufficient to construct wave functions of the negative energy continuum $\tilde{\Phi}_E$ that are orthogonal to Φ_R, but which span together with Φ_R the same space as the continuum eigenstates:

$$\langle \Phi_R | \tilde{\Phi}_E \rangle = 0 \qquad \text{for all} \quad E < -m_e \,, \tag{10.32}$$

$$|\Phi_R\rangle\langle\Phi_R| + \int\limits_{-\infty}^{-m_e} dE\, |\tilde{\Phi}_E\rangle\langle\tilde{\Phi}_E| = \int\limits_{-\infty}^{-m_e} dE\, |\Phi_E\rangle\langle\Phi_E| \,. \tag{10.33}$$

The functions $\tilde{\Phi}_E$ can be constructed by Schmidt orthogonalization. However, the functions

$$|\Phi'_E\rangle = |\Phi_E\rangle - |\Phi_R\rangle\langle\Phi_R|\Phi_E\rangle \tag{10.34}$$

are not yet orthogonal, and one would face the problem of orthogonalizing all the functions Φ'_E among each other. Even besides the fact that there is an infinity of such functions, this is a very tricky problem, since for continuum wave functions the scalar product is not a well-defined quantity, but makes mathematical sense only as a distribution.

It is more rewarding to construct the functions $\tilde{\Phi}_E$ directly as eigenfunctions of a convenient Hermitian operator. The natural choice for this operator is the Dirac Hamiltonian H_D, but restricted to the space of functions $\tilde{\Phi}_E$. At first sight this seems tautological because we want to use the operator to construct the functions $\tilde{\Phi}_E$ in the first place. However, this choice leads to a very useful, although implicit equation for the $\tilde{\Phi}_E$ [Wa 70]. To carry this through we write (10.33) in the symbolic form [Fe 62]

$$Q+P=I\,, \quad \text{where} \tag{10.35}$$

$$Q=|\Phi_R\rangle\langle\Phi_R|\,, \tag{10.36a}$$

$$P=\int_{-\infty}^{-m_e} dE\,|\tilde{\Phi}_E\rangle\langle\tilde{\Phi}_E|\,, \tag{10.36b}$$

$$I=\int_{-\infty}^{-m_e} dE\,|\Phi_E\rangle\langle\Phi_E|\,. \tag{10.36c}$$

The operators Q, P and I are projection operators on subspaces of the total Hilbert space because one easily calculates

$$Q^2=|\Phi_R\rangle\langle\Phi_R|\Phi_R\rangle\langle\Phi_R|=|\Phi_R\rangle\langle\Phi_R|=Q\,, \tag{10.37a}$$

$$\begin{aligned} P^2 &= \int_{-\infty}^{-m_e} dE \int_{-\infty}^{-m_e} dE'\,|\tilde{\Phi}_E\rangle\langle\tilde{\Phi}_E|\tilde{\Phi}_{E'}\rangle\langle\tilde{\Phi}_{E'}| \\ &= \int_{-\infty}^{-m_e} dE \int_{-\infty}^{-m_e} dE'\,|\tilde{\Phi}_E\rangle\,\delta(E-E')\,\langle\tilde{\Phi}_{E'}| \\ &= \int_{-\infty}^{-m_e} dE\,|\tilde{\Phi}_E\rangle\langle\tilde{\Phi}_E|=P\,, \end{aligned} \tag{10.37b}$$

$$I^2=\int_{-\infty}^{-m_e} dE \int_{-\infty}^{-m_e} dE'\,|\Phi_E\rangle\langle\Phi_E|\Phi_{E'}\rangle\langle\Phi_E|=I\,, \tag{10.37c}$$

and finally

$$QP=\int_{-\infty}^{-m_e} dE\,|\Phi_R\rangle\langle\Phi_R|\tilde{\Phi}_E\rangle\langle\tilde{\Phi}_E|=0=PQ\,. \tag{10.37d}$$

Operators P and Q project on orthogonal subspaces of the space spanned by all supercritical negative continuum eigenfunctions Φ_E. Now the restriction of H_D

to the subspace projected by $P = I - Q$ is a Hermitian operator, because I, Q and therefore P are Hermitian:

$$(PH_{\mathrm{D}}P)^{\dagger} = P^{\dagger}H_{\mathrm{D}}^{\dagger}P^{\dagger} = PH_{\mathrm{D}}P\,. \tag{10.38}$$

Its eigenfunctions for $E < -m_{\mathrm{e}}$ are therefore orthogonal and form a complete basis of the P subspace. This means that the projected wave functions $\tilde{\Phi}_{\mathrm{E}}$ can be defined as solutions of the eigenvalue equation

$$E|\tilde{\Phi}_{\mathrm{E}}\rangle = PH_{\mathrm{D}}P|\tilde{\Phi}_{\mathrm{E}}\rangle \quad \text{for} \quad E < -m_{\mathrm{e}}\,. \tag{10.39}$$

Since $\tilde{\Phi}_{\mathrm{E}}$ belong to the P subspace by definition, the operator P to the right of the Hamiltonian can be dropped. If $(I - Q)$ replaces P on the left and definitions (10.36a, c) are substituted, then an implicit equation for $\tilde{\Phi}_{\mathrm{E}}(\boldsymbol{x})$ valid for $E < -m_{\mathrm{e}}$ results:

$$E\,\tilde{\Phi}_{\mathrm{E}}(\boldsymbol{x}) = H_{\mathrm{D}}(\boldsymbol{x})\,\tilde{\Phi}_{\mathrm{E}}(\boldsymbol{x}) - \langle\Phi_{\mathrm{R}}|H_{\mathrm{D}}|\tilde{\Phi}_{\mathrm{E}}\rangle\,\Phi_{\mathrm{R}}(\boldsymbol{x})\,. \tag{10.40}$$

If $\tilde{\Phi}_{\mathrm{E}}$ is a solution of this equation, its orthogonality to Φ_{R}, (10.32), is easily shown by taking the scalar product from the left with the resonance wave function Φ_{R}:

$$E\langle\Phi_{\mathrm{R}}|\tilde{\Phi}_{\mathrm{E}}\rangle = \langle\Phi_{\mathrm{R}}|H_{\mathrm{D}}|\tilde{\Phi}_{\mathrm{E}}\rangle - \langle\Phi_{\mathrm{R}}|H_{\mathrm{D}}|\tilde{\Phi}_{\mathrm{E}}\rangle\langle\Phi_{\mathrm{R}}|\Phi_{\mathrm{R}}\rangle = 0\,, \tag{10.41}$$

since Φ_{R} was assumed to be normalized to unity.

We have thus succeeded in constructing an orthogonal and complete basis of negative energy wave functions, but at the price of having to deal with an implicit eigenvalue equation, (10.40). Fortunately, a practical way of solving (10.40) exists. First one replaces the unknown constant $\langle\Phi_{\mathrm{R}}|H_{\mathrm{D}}|\tilde{\Phi}_{\mathrm{E}}\rangle$ by two arbitrarily chosen constants γ_1, γ_2 and solves the inhomogeneous Dirac equation for each of these $(i = 1, 2)$:

$$E\,\Phi_{\mathrm{E}}^{(i)}(\boldsymbol{x}) = H_{\mathrm{D}}(\boldsymbol{x})\,\Phi_{\mathrm{E}}^{(i)}(\boldsymbol{x}) + \gamma_i\,\Phi_{\mathrm{R}}(\boldsymbol{x})\,. \tag{10.42}$$

Since the general solution of the inhomogeneous equation can be written as a special solution of it plus the general solution of the homogeneous equation (which here is Φ_{E}), the solution for an arbitrary constant γ can be represented as

$$|\tilde{\Phi}_{\mathrm{E}}\rangle = a|\Phi_{\mathrm{E}}^{(1)}\rangle + b|\Phi_{\mathrm{E}}^{(2)}\rangle\,. \tag{10.43}$$

(This is most clearly seen for the choice $\gamma_1 = 0$, $\gamma_2 = 1$.)

The constants a and b can now be determined from the condition $\langle\Phi_{\mathrm{R}}|\tilde{\Phi}_{\mathrm{E}}\rangle = 0$ and the normalization of $\tilde{\Phi}_{\mathrm{E}}$. The whole projection formalism can be easily extended to the case where several supercritical bound state resonances are imbedded in the negative energy continuum, but we shall not pursue this further here.

Let us now look at the solutions of (10.40). The upper part of Fig. 10.8 shows the large and small components of an eigenfunction Φ_{E} of the Dirac Hamiltonian

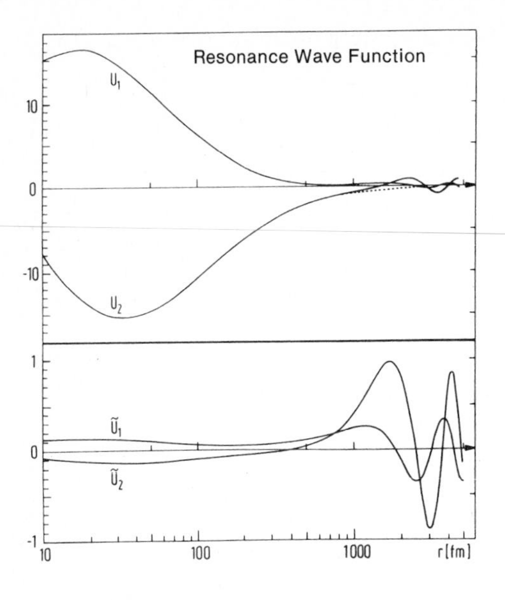

Fig. 10.8. Negative energy continuum wave functions in the resonance region for $Z = 184$. *Upper part:* continuum eigenstate Φ_E; *lower part:* projected continuum wave function $\tilde{\Phi}_E$. The large and small radial components u_1 and u_2 are shown [Re 81b]

H_D in the supercritical resonance region for the nucleus $Z = 184$. The wave function clearly has a big maximum at about $r \sim 20-30$ fm and a small oscillating tail at large distances from the nucleus. In comparison, the lower part of Fig. 10.8 shows the solution of (10.40), $\tilde{\Phi}_E$, for the same energy. The maximum at small r has completely gone, whereas the oscillating tail has remained unchanged. The success of eliminating the contribution due to Φ_R from the continuum wave function is conspicuous.

This is also evident from the energy dependence of the scattering phase shift of the modified continuum functions $\tilde{\Phi}_E$ (dashed line in Fig. 10.9) as opposed to that of the continuum eigenfunctions Φ_E (solid line). As discussed at length in Chap. 6, the jump in the phase shift $\delta(E)$ is a direct consequence of the presence of a resonance in the continuum [(6.32b) and Fig. 6.13]. The complete absence of the jump in the modified phase shift $\tilde{\delta}(E)$ proves that the projection method really works.

We now return to our discussion of the decay of a K vacancy in a supercritical atom by spontaneous positron emission. Denoting the charged supercritical vacuum state by $|0\rangle$, the state corresponding to a vacant K-shell level, ionized at time t_0, is (Fig. 10.10)

$$|R, t_0\rangle = \hat{d}^\dagger_{R,t_0}|0\rangle \,, \tag{10.44}$$

where the K-vacancy creation operator is

$$\hat{d}^\dagger_{R,t_0} = \int d^3x\, \Phi^\dagger_R(\boldsymbol{x})\, \hat{\psi}(\boldsymbol{x}, t_0) \,. \tag{10.45}$$

Why the ionization time t_0 must be specified is simply that the states do not change with time in the Heisenberg picture. As the supercritical K vacancy

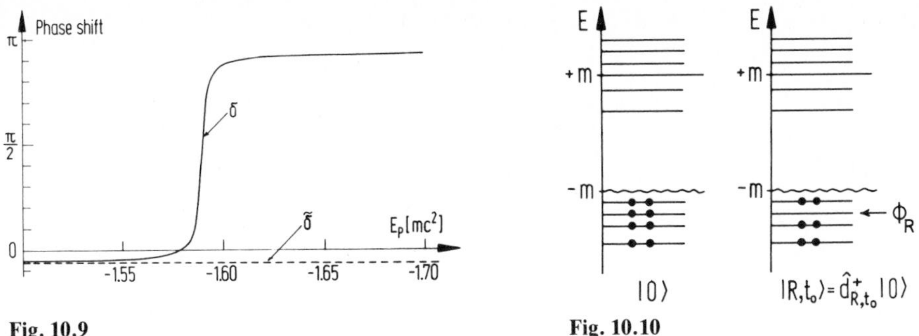

Fig. 10.9 **Fig. 10.10**

Fig. 10.9. Energy dependence of the phase shift in the negative energy continuum for $Z = 184$. (———) shows the phase shift of the eigenstates, (– – –) shows the phase shift of the projected continuum [Re 81b]

Fig. 10.10. Charged vacuum ($|0\rangle$) and neutral vacuum $|R, t_0\rangle$ in the field of a supercritical nucleus

decays, vacancies created at different times correspond to different Heisenberg states of the atom.

When examining the atom after some time T and counting the number of K vacancies, we have to use the K-vacancy number operator at the later time:

$$\bar{N}_{\mathrm{R}}(T) = \langle R, t_0 | \hat{d}^{\dagger}_{\mathrm{R}, t_0+T} \hat{d}_{\mathrm{R}, t_0+T} | R, t_0 \rangle \,. \tag{10.46}$$

Similarly, if the operator for creating a free positron at time $t_0 + T$ is

$$\hat{\tilde{d}}^{\dagger}_{\mathrm{E}, t_0+T} = \int d^3x \, \tilde{\Phi}^{\dagger}_{\mathrm{E}}(\boldsymbol{x}) \, \hat{\psi}(\boldsymbol{x}, t_0 + T) \,, \tag{10.47}$$

the number of free positrons observed after time T has elapsed is

$$\tilde{N}_{\mathrm{E}}(T) = \langle R, t_0 | \hat{\tilde{d}}^{\dagger}_{\mathrm{E}, t_0+T} \hat{\tilde{d}}_{\mathrm{E}, t_0+T} | R, t_0 \rangle \,. \tag{10.48}$$

To evaluate the expressions for $\bar{N}_{\mathrm{R}}(T)$ and $\tilde{N}_{\mathrm{E}}(T)$ the operators $\hat{d}_{\mathrm{R},t}$ and $\hat{\tilde{d}}_{\mathrm{E},t}$ for different times t must be related. (We point out once more that these operators are time independent, but if defined at different times they correspond to different physical states.) This is best achieved by using a representation of the field operator $\hat{\psi}(\boldsymbol{x}, t)$ in terms of solutions of the time-dependent Dirac equation with initial conditions specified at the ionization time t_0.

Since Φ_{R} and $\tilde{\Phi}_{\mathrm{E}}$ do not diagonalize the supercritical Hamiltonian H_{D}, stationary wave functions like $\Phi_{\mathrm{R}}(\boldsymbol{x}) \exp(-\mathrm{i}E_{\mathrm{R}}t)$ do not satisfy the time-dependent Dirac equation. However, a solution to the time-dependent Dirac equation is given by remembering that the supercritical Hamiltonian is time independent. The time evolution operator is therefore given by (8.47)

$$U(t, t_0) = \exp[-\mathrm{i}H_{\mathrm{D}}(t - t_0)] = \sum_n |\Phi_n\rangle \exp[-\mathrm{i}E_n(t - t_0)] \langle \Phi_n| \,, \tag{10.49}$$

where the sum goes over all eigenstates Φ_n of the supercritical Hamiltonian. The wave function which develops from $\Phi_{\mathrm{R}}(\boldsymbol{x})$ at $t = t_0$ is therefore

$$\psi_{\mathrm{R},t_0}^{(+)}(\boldsymbol{x},t) = \sum_n \exp[-\mathrm{i}E_n(t-t_0)] \langle \Phi_n | \Phi_\mathrm{R} \rangle \Phi_n(\boldsymbol{x})$$

$$= \int_{-\infty}^{-m_\mathrm{e}} dE' \langle \Phi_{\mathrm{E}'} | \Phi_\mathrm{R} \rangle \exp[-\mathrm{i}E'(t-t_0)] \Phi_{\mathrm{E}'}(\boldsymbol{x}) . \tag{10.50}$$

Similarly, the wave function $\tilde{\Phi}_\mathrm{E}(\boldsymbol{x})$ at $t = t_0$ evolves as

$$\tilde{\psi}_{\mathrm{E},t_0}^{(+)}(\boldsymbol{x},t) = \int_{-\infty}^{-m_\mathrm{e}} dE' \langle \Phi_{\mathrm{E}'} | \tilde{\Phi}_\mathrm{E} \rangle \exp[-\mathrm{i}E'(t-t_0)] \Phi_{\mathrm{E}'}(\boldsymbol{x}) . \tag{10.51}$$

We can now quantize the supercritical electron – positron field:

$$\hat{\psi}(\boldsymbol{x},t) = \int_{-\infty}^{-m_\mathrm{e}} dE \hat{\tilde{d}}_{\mathrm{E},t_0}^\dagger \tilde{\psi}_{\mathrm{E},t_0}^{(+)}(\boldsymbol{x},t) + \hat{d}_{\mathrm{R},t_0}^\dagger \psi_{\mathrm{R},t_0}^{(+)}(\boldsymbol{x},t)$$

$$+ \sum_{E_n > -m_\mathrm{e}} \hat{b}_n \Phi_n(\boldsymbol{x}) \exp[-\mathrm{i}E_n(t-t_0)] . \tag{10.52}$$

An identical formula holds, with t_0 replaced by $t_0 + T$, if $\hat{\psi}$ is represented by the operators at the time of observation.

It is now straightforward to relate the operators defined at times t_0 and $t_0 + T$. With the definitions (10.45) at $t_0 + T$ and (10.47), then

$$\hat{d}_{\mathrm{R},t_0+T}^\dagger = \int d^3x \, \Phi_\mathrm{R}^\dagger(\boldsymbol{x}) \hat{\psi}(\boldsymbol{x}, t_0+T)$$

$$= \hat{d}_{\mathrm{R},t_0}^\dagger \langle \Phi_\mathrm{R} | \psi_{\mathrm{R},t_0}^{(+)}(t_0+T) \rangle + \int_{-\infty}^{-m_\mathrm{e}} dE \hat{\tilde{d}}_{\mathrm{E},t_0}^\dagger \langle \Phi_\mathrm{R} | \tilde{\psi}_{\mathrm{E},t_0}^{(+)}(t_0+T) \rangle , \tag{10.53}$$

$$\hat{\tilde{d}}_{\mathrm{E},t_0+T}^\dagger = \hat{d}_{\mathrm{R},t_0}^\dagger \langle \tilde{\Phi}_\mathrm{E} | \psi_{\mathrm{R},t}^{(+)}(t_0+T) \rangle + \int_{-\infty}^{-m_\mathrm{e}} dE' \hat{\tilde{d}}_{\mathrm{E}',t_0}^\dagger \langle \tilde{\Phi}_\mathrm{E} | \tilde{\psi}_{\mathrm{E}',t_0}^{(+)}(t_0+T) \rangle , \tag{10.54}$$

because the functions Φ_R and $\tilde{\Phi}_\mathrm{E}$ are orthogonal on the higher eigenfunctions Φ_n.

In view of the definition (10.44) of the state $|R, t_0\rangle$ and the relations

$$\hat{d}_{\mathrm{R},t_0} |0\rangle = \hat{\tilde{d}}_{\mathrm{E},t_0} |0\rangle = 0 , \tag{10.55}$$

for the number of K vacancies (10.46) and free positrons (10.48), respectively

$$\bar{N}_\mathrm{R}(T) = |\langle \Phi_\mathrm{R} | \psi_{\mathrm{R},t_0}^{(+)}(t_0+T) \rangle|^2$$

$$= \left| \int_{-\infty}^{-m_\mathrm{e}} dE' |\langle \Phi_\mathrm{R} | \Phi_{\mathrm{E}'} \rangle|^2 \mathrm{e}^{-\mathrm{i}E'T} \right|^2 \quad \text{and} \tag{10.56}$$

$$\tilde{N}_\mathrm{E}(T) = |\langle \tilde{\Phi}_\mathrm{E} | \psi_{\mathrm{R},t_0}^{(+)}(t_0+T) \rangle|^2$$

$$= \left| \int_{-\infty}^{-m_\mathrm{e}} dE' \langle \tilde{\Phi}_\mathrm{E} | \Phi_{\mathrm{E}'} \rangle \mathrm{e}^{-\mathrm{i}E'T} \langle \Phi_{\mathrm{E}'} | \Phi_\mathrm{R} \rangle \right|^2 . \tag{10.57}$$

To go further, we must know the overlap matrix elements between the true eigenstates Φ_E and the projected basis functions Φ_R, $\tilde{\Phi}_E$, i.e. we must diagonalize the full Hamiltonian H_D in the modified basis. Because of the defining equation (10.39), the Hamiltonian is diagonal in the subspace of the functions $\tilde{\Phi}_E$,

$$\langle \tilde{\Phi}_E | H_D | \tilde{\Phi}_{E'} \rangle = E\,\delta(E-E') , \tag{10.58}$$

but there are off-diagonal matrix elements connecting the P and Q subspaces:

$$\langle \tilde{\Phi}_E | H_D | \Phi_R \rangle \equiv \tilde{V}_E , \tag{10.59}$$

$$\langle \Phi_R | H_D | \Phi_R \rangle \equiv E_R . \tag{10.60}$$

Comparison shows that (10.58–60) lead to an eigenvalue problem formally identical to (6.20), which formed the starting point for the solution of the supercritical eigenfunctions in the framework of autoionisation formalism. Whereas (6.20) were only approximately valid due to the neglect of coupling among the subcritical continuum states, the analogous equations derived from (10.58–60) are exact! The autoionisation method is really applicable when working with the projected basis functions.

This means that we can formally copy (6.10, 11, 32, 33) from Chap. 6, giving

$$\Phi_E(\boldsymbol{x}) = a_R(E)\,\Phi_R(\boldsymbol{x}) + \int_{-\infty}^{-m_e} dE'\,\tilde{h}_{E'}(E)\,\tilde{\Phi}_{E'}(\boldsymbol{x}) , \tag{10.61}$$

$$\langle \Phi_R | \Phi_E \rangle = a_R(E) , \tag{10.62a}$$

$$\langle \tilde{\Phi}_{E'} | \Phi_E \rangle = \tilde{h}_{E'}(E) . \tag{10.62b}$$

Here $a_R(E)$ and $\tilde{h}_{E'}(E)$ are of identical form as (10.25, 26) but with coupling $\tilde{V}_{E'}$ of the projected basis states.

When (10.62) are substituted into (10.56, 57), then

$$\bar{N}_R(T) = \left| \int_{-\infty}^{-m_e} dE'\,|a_R(E')|^2 e^{-iE'T} \right|^2 , \tag{10.63}$$

and

$$\tilde{N}_E(T) = \left| \int_{-\infty}^{-m_e} dE'\,\tilde{h}_{E'}(E)\,a_R(E')^*\,e^{-iE'T} \right|^2 . \tag{10.64}$$

It is not surprising that these equations have the same form as those describing the decay of a K vacancy when an additional potential V' is switched on (10.27, 31). Without further calculation we can therefore conclude that the K vacancy in a supercritical atom decays exponentially,

$$\bar{N}_R(T) = e^{-\tilde{\Gamma}T} , \qquad \tilde{\Gamma} = 2\pi |\tilde{V}_{E_R}|^2 , \tag{10.65}$$

emitting positrons with the spectrum

$$\tilde{N}_{\mathrm{E}}(T) = \frac{|\tilde{V}_{\mathrm{E}}|^2}{(E-E_{\mathrm{R}})^2 + \frac{1}{4}\tilde{\Gamma}^2}\,[1 - 2\cos(E-E_{\mathrm{R}})\,T\mathrm{e}^{-\frac{1}{2}\tilde{\Gamma}T} + \mathrm{e}^{-\tilde{\Gamma}T}]\,. \tag{10.66}$$

After time T much longer than the natural lifetime $\tilde{\Gamma}^{-1}$, the spectrum mirrors the shape of the bound state resonance which represents the supercritical K shell.

The similarity of the equations describing the decay of a K vacancy in a supercritical atom and the spontaneous discharge of a nucleus made supercritical by sudden increase in the nuclear charge invites the question: what is the difference? Both distributions agree in the narrow resonance approximation. Beyond this approximation, (10.17) shows that in the latter case the positrons not only originate from the bound state that becomes supercritical, but also from all higher vacant states:

$$\bar{N}_{\mathrm{E}} = |\langle \Phi_{\mathrm{E}}|\varphi_0\rangle|^2 + \sum_{E_n > E_0} |\langle \Phi_{\mathrm{E}}|\varphi_n\rangle|^2\,. \tag{10.67}$$

This expression differs from (10.57) in two respects: the contribution from higher states and the difference between φ_0 and Φ_{R}. We can understand this difference by dividing the unitary transformation between the subcritical basis $\{\varphi_n\}$ and the supercritical basis $\{\Phi_n\}$ into two steps, taking the projected resonance basis as an intermediate basis:

$$\left\{\begin{array}{l}\varphi_n,\ \varepsilon_n > \varepsilon_0\\ \varphi_0\\ \varphi_\varepsilon,\ \varepsilon < -m_{\mathrm{e}}\end{array}\right\} \to \left\{\begin{array}{l}\Phi_n,\ E_n \geqslant -m_{\mathrm{e}}\\ \Phi_{\mathrm{R}}\\ \tilde{\Phi}_{\mathrm{E}},\ E < -m_{\mathrm{e}}\end{array}\right\} \to \left\{\begin{array}{l}\Phi_n,\ E_n \geqslant -m_{\mathrm{e}}\\ \\ \Phi_{\mathrm{E}},\ E < -m_{\mathrm{e}}\end{array}\right\}. \tag{10.68}$$

Part of the created positron spectrum must be attributed to the transition from the subcritical to the projected supercritical basis, caused by the potential change. Without the sudden variation in the external potential this contribution would be missing. It is therefore appropriately called *"induced"* positron emission, in the sense that it is induced by the sudden time change of the potential $V(t) = V_{\mathrm{cr}} + V'\,\theta(t-t_0)$. In contrast, the spectrum (10.57) does not result from any variation of the external potential, so that it is called the *"spontaneous"* positron spectrum. Equation (10.67) may be formally separated into a spontaneous and a dynamically induced part:

$$\bar{N}_{\mathrm{E}} = \bar{N}_{\mathrm{E}}^{\mathrm{spont}} + \bar{N}_{\mathrm{E}}^{\mathrm{ind}}\,, \tag{10.69}$$

with $\bar{N}_{\mathrm{E}}^{\mathrm{spont}}$ given by (10.57) in the limit $T \to \infty$ and

$$\bar{N}_{\mathrm{E}}^{\mathrm{ind}} = \sum_{E_n > E_0} \int_{-\infty}^{-m_{\mathrm{e}}} dE'\,|\langle \Phi_{\mathrm{E}}|\tilde{\Phi}_{\mathrm{E'}}\rangle\langle\tilde{\Phi}_{\mathrm{E'}}|\varphi_n\rangle|^2 - \int_{-\infty}^{-m_{\mathrm{e}}} d\varepsilon\,|\langle\Phi_{\mathrm{E}}|\Phi_{\mathrm{R}}\rangle|^2\cdot|\langle\Phi_{\mathrm{R}}|\varphi_\varepsilon\rangle|^2\,. \tag{10.70}$$

This is precisely the situation continuously encountered in Chap. 12, when discussing positron creation in heavy-ion collisions. Then it is helpful to remember that the two different mechanisms, spontaneous and induced positron production, are already present in the schematic model of a sudden transition to a supercritical potential.

We finally evaluate the time evolution operator in the supercritical potential (10.49) in terms of the projected resonance basis, because this result will be needed in Chap. 12. With the help of (10.62, 63) and using the same complex integration techniques that led to (10.28), the following expression is obtained for the time evolution operator:

$$\begin{aligned} U(t_0+T, t_0) &= \sum_n |\Phi_n\rangle e^{-iE_nT}\langle \Phi_n| \\ &= \sum_{E_n>-m} |\Phi_n\rangle e^{-iE_nT}\langle \Phi_n| \\ &\quad + |\Phi_R\rangle \exp(-iE_R T - \tfrac{1}{2}\tilde{\Gamma}T)\langle \Phi_R| + \int_{-\infty}^{-m_e} dE\, |\tilde{\Phi}_E\rangle e^{-iET}\langle \tilde{\Phi}_E| \\ &\quad + \int_{-\infty}^{-m_e} dE \frac{\exp(-iET) - \exp(-iE_R T - \frac{1}{2}\tilde{\Gamma}T)}{E - E_R + \frac{i}{2}\tilde{\Gamma}} \\ &\quad \times (\tilde{V}_E|\tilde{\Phi}_E\rangle\langle \Phi_R| + \tilde{V}_E^*|\Phi_R\rangle\langle \tilde{\Phi}_E|) \\ &\quad + \int_{-\infty}^{-m_e} dE\, dE'\, \tilde{a}_{EE'}|\tilde{\Phi}_E\rangle\langle \Phi_{E'}| \end{aligned} \tag{10.71}$$

with

$$\begin{aligned} \tilde{a}_{EE'} = {} & \frac{\tilde{V}_E \tilde{V}_{E'}^*}{E - E' - i\eta}\left(\frac{e^{-iET}}{E - E_R - \frac{i}{2}\tilde{\Gamma}} - \frac{e^{-iE'T}}{E' - E_R + \frac{i}{2}\tilde{\Gamma}}\right) \\ & + \frac{\tilde{V}_E \tilde{V}_{E'}^* \exp(-iE_R T - \frac{1}{2}\tilde{\Gamma}T)}{(E - E_R + \frac{i}{2}\tilde{\Gamma})(E' - E_R + \frac{i}{2}\tilde{\Gamma})} - 2\pi i \frac{\tilde{V}_E \tilde{V}_{E'}^* |a_R(E)|^2}{E - E' - i\eta} e^{-iET}. \end{aligned} \tag{10.72}$$

This expression contains a full description of the spontaneous decay of the neutral vacuum in the supercritical potential.

10.4 Decay of the Neutral Vacuum – Schrödinger Picture

In the previous section we treated the decay of a K vacancy in a supercritical atom, i.e. the decay of the "neutral vacuum", in the Heisenberg picture where the states are time independent and all time dependence is carried by the field operator $\hat{\psi}(x, t)$. We chose the Heisenberg picture because it has a considerable formal advantage: the explicit time dependence must be found only for those operators which describe the observables of physical interest. This advantage is particularly important when discussing the decay of the neutral vacuum in collisions of two heavy ions, because then a variety of other atomic excitation processes occur, leading to complicated states of the atomic system, but which need not be investigated in detail in the Heisenberg picture.

On the other hand, the Schrödinger picture, in which all the time dependence resides in the state of the quantum system, has considerable conceptual advantages since most physicists find it much easier to think in terms of the evolution of an initially prepared state. The decay of the neutral vacuum in the static electric field of a supercritical nucleus is a sufficiently simple process to enable it to be treated in the Schrödinger picture. We shall, therefore, rederive the results of the previous section by directly solving the Schrödinger equation in Fock space [Ra 74, 78b].

In the Schrödinger picture the field operator $\hat{\psi}_S(\boldsymbol{x})$ is time independent, and the state of the system evolves according to the time-dependent Schrödinger equation

$$\mathrm{i}\frac{\partial}{\partial t}|\Omega(t)\rangle_S = \hat{H}_S|\Omega(t)\rangle_S\,, \tag{10.73}$$

where $\hat{H}_S$ is the Hamiltonian in second quantization. The transition to the Heisenberg picture is effected by the time-dependent unitary transformation $\hat{U}(t)$ which satisfies the equation of motion

$$\mathrm{i}\frac{d}{dt}\hat{U}(t) = \hat{H}\hat{U}(t)\,. \tag{10.74}$$

Obviously $\hat{U}(t)$ is just the time evolution operator in Fock space, with the explicit solution

$$\hat{U}(t) = \mathrm{e}^{-\mathrm{i}\hat{H}t}$$

for a time-independent Hamiltonian. The connection between the time-dependent Schrödinger state and the time-independent state $|\Omega\rangle_H$ of the Heisenberg picture is therefore

$$|\Omega(t)\rangle_S = \hat{U}(t)|\Omega\rangle_H = \mathrm{e}^{-\mathrm{i}\hat{H}t}|\Omega\rangle_H\,. \tag{10.75}$$

(We have chosen that the Schrödinger and Heisenberg pictures coincide at $t=0$; one can make them agree at any other time t_0 by replacing t everywhere by $t-t_0$.)

The connection between the operators is in general

$$\hat{A}_S = \hat{U}(t)\hat{A}_H\hat{U}(t)^{-1} = \mathrm{e}^{-\mathrm{i}\hat{H}t}\hat{A}_H\mathrm{e}^{\mathrm{i}\hat{H}t}\,, \tag{10.76}$$

so that the Hamiltonian remains unchanged (in the absence of an explicit time dependence) $\hat{H}_S = \hat{H}_H$. In the following assume all quantities to be in the Schrödinger picture, so subscript "S" can be dropped.

Therefore according to (9.8) the Hamiltonian is

$$\hat{H} = \tfrac{1}{2}\int d^3x\,[\hat{\psi}^\dagger(\boldsymbol{x}), H_D\hat{\psi}(\boldsymbol{x})] = \int d^3x\!: \hat{\psi}^\dagger(\boldsymbol{x})H_D\hat{\psi}(\boldsymbol{x})\!: + E_{\mathrm{vac}}\,. \tag{10.77}$$

We can represent the field operator in terms of either supercritical eigenstate basis

$$\hat{\psi}(\boldsymbol{x}) = \sum_{E_n>-m}\hat{b}_n\Phi_n(\boldsymbol{x}) + \int_{-\infty}^{-m_e} dE\,\hat{d}_E^\dagger\,\Phi_E(\boldsymbol{x})\,, \tag{10.78a}$$

or the projected basis

$$\hat{\psi}(x) = \sum_{E_n > -m} \hat{b}_n \Phi_n(x) + \hat{a}_{\mathrm{R}}^{\dagger} \Phi_{\mathrm{R}}(x) + \int_{-\infty}^{-m_e} dE \hat{\tilde{a}}_{\mathrm{E}}^{\dagger} \tilde{\Phi}_{\mathrm{E}}(x) . \tag{10.78b}$$

Taking the matrix elements in coordinate space requires to evaluate the Hamiltonian, by virtue of (10.58 – 60) the two equivalent representations of $\hat{H}$ are

$$\hat{H} = \sum_{E_n > -m_e} E_n \hat{b}_n^{\dagger} \hat{b}_n - \int_{-\infty}^{-m_e} E\, dE\, \hat{a}_{\mathrm{E}}^{\dagger} \hat{a}_{\mathrm{E}} + E_{\mathrm{vac}} , \tag{10.79a}$$

$$\begin{aligned} \hat{H} = & \sum_{E_n > -m_e} E_n \hat{b}_n^{\dagger} \hat{b}_n - E_{\mathrm{R}} \hat{a}_{\mathrm{R}}^{\dagger} \hat{a}_{\mathrm{R}} - \int_{-\infty}^{-m_e} E\, dE \hat{\tilde{a}}_{\mathrm{E}}^{\dagger} \hat{\tilde{a}}_{\mathrm{E}} \\ & - \int_{-\infty}^{-m_e} dE\, \tilde{V}_{\mathrm{E}} \hat{a}_{\mathrm{R}}^{\dagger} \hat{\tilde{a}}_{\mathrm{E}} - \int_{-\infty}^{-m_e} dE\, \tilde{V}_{\mathrm{E}}^{*} \hat{\tilde{a}}_{\mathrm{E}}^{\dagger} \hat{a}_{\mathrm{R}} + E_{\mathrm{vac}} . \end{aligned} \tag{10.79b}$$

In the eigenmode representation $\hat{H}$ is diagonal, but it contains off-diagonal terms connecting the resonance and the modified continuum wave functions in the projected basis. These off-diagonal elements describe the decay of a supercritical K vacancy, i.e. of the neutral vacuum, described by the state

$$|R\rangle = \hat{a}_{\mathrm{R}}^{\dagger} |0\rangle . \tag{10.80}$$

To find out how the neutral vacuum decays, we have to solve the Schrödinger equation (10.73) with the initial condition

$$|\Omega(0)\rangle = |R\rangle . \tag{10.81}$$

Since the Hamiltonian (10.79b) contains only one-body operators like $\hat{b}^{\dagger}\hat{b}$ or $\hat{a}^{\dagger}\hat{a}$, it cannot cause the state $|R\rangle$ to decay into more complicated, e.g. one particle – two hole, configurations. The vacancy in the resonance state can be transferred only into a state of the modified negative energy continuum. The most general ansatz for $|\Omega(t)\rangle$ is therefore

$$|\Omega(t)\rangle = y(t) \hat{a}_{\mathrm{R}}^{\dagger} |0\rangle + \int_{-\infty}^{-m_e} dE\, w_E(t) \hat{\tilde{a}}_{\mathrm{E}}^{\dagger} |0\rangle . \tag{10.82}$$

If the Hamiltonian (10.79b) acts on this state, then

$$\begin{aligned} & \mathrm{i} \frac{dy}{dt} \hat{a}_{\mathrm{R}}^{\dagger} |0\rangle + \mathrm{i} \int_{-\infty}^{-m_e} dE \frac{dw_E}{dt} \hat{\tilde{a}}_{\mathrm{E}}^{\dagger} |0\rangle \\ & \quad = -(E_{\mathrm{R}} - E_{\mathrm{vac}}) y \hat{a}_{\mathrm{R}}^{\dagger} |0\rangle - \int_{-\infty}^{-m_e} (E - E_{\mathrm{vac}})\, dE\, w_{\mathrm{E}} \hat{\tilde{a}}_{\mathrm{E}}^{\dagger} |0\rangle \\ & \qquad - \int_{-\infty}^{-m_e} dE\, \tilde{V}_E w_E \hat{a}_{\mathrm{R}}^{\dagger} |0\rangle - \int_{-\infty}^{-m_e} dE\, \tilde{V}_{\mathrm{E}}^{*} y \hat{\tilde{a}}_{\mathrm{E}}^{\dagger} |0\rangle . \end{aligned} \tag{10.83}$$

Projecting with the states $\langle 0|\hat{d}_{\rm R}$ and $\langle 0|\hat{\tilde{d}}_{\rm E}$, respectively, a set of coupled differential equations for the functions $y(t)$ and $w_E(t)$ arises:

$$\frac{dy}{dt} = \mathrm{i}(E_{\rm R}-E_{\rm vac})y(t)+\mathrm{i}\int_{-\infty}^{-m_{\rm e}} dE\,\tilde{V}_E w_E(t)\,, \tag{10.84a}$$

$$\frac{dw_E}{dt} = \mathrm{i}(E-E_{\rm vac})\,w_E(t)+\mathrm{i}\,\tilde{V}_E^* y(t)\,. \tag{10.84b}$$

According to (10.81) the initial condition is

$$y(0)=1\,, \qquad w_E(0)=0\,. \tag{10.85}$$

The easiest way to solve the system of equations (10.84) is to apply a Fourier transformation to the functions $y(t)$ and $w_E(t)$. Since the vacuum energy enters both parts of (10.84) in the same way, it can be eliminated by

$$y(t) = \mathrm{e}^{-\mathrm{i}E_{\rm vac}t}\int_{-\infty}^{\infty} dp\,\mathrm{e}^{\mathrm{i}pt}\bar{y}(p)\,, \tag{10.86a}$$

$$w_E(t) = \mathrm{e}^{-\mathrm{i}E_{\rm vac}t}\int_{-\infty}^{\infty} dp\,\mathrm{e}^{\mathrm{i}pt}\,\bar{w}_E(p)\,. \tag{10.86b}$$

The Fourier transformation replaces all time derivatives by $(\mathrm{i}p)$, and converts the differential equations (10.84) into a set of algebraic equations

$$(p-E_{\rm R})\bar{y}(p) = \int_{-\infty}^{-m_{\rm e}} dE\,\tilde{V}_E\bar{w}_E(p)\,, \tag{10.87a}$$

$$(p-E)\,\bar{w}_E(p) = \tilde{V}_E^*\bar{y}(p)\,. \tag{10.87b}$$

Because this system of equations is homogeneous, any particular solution $\bar{y}'$, $\bar{w}_E'$ may be multiplied by an arbitrary function of p, i.e.

$$\bar{y}(p)=f(p)\bar{y}'(p)\,, \qquad \bar{w}_E(p)=f(p)\,\bar{w}_E'(p)\,, \tag{10.88}$$

are also solutions of (10.87). The function $f(p)$ must be determined from the initial conditions (10.85).

A particular solution of the system (10.87) is easily found, because a comparison indicates that it is just the complex conjugate of the system of eigenvalue equations encountered in autoionisation theory, (6.20). Hence

$$\bar{y}'(p)=a_{\rm R}(p)^*\,, \qquad \bar{w}_E'(p)=\tilde{h}_E(p)^* \tag{10.89}$$

must form a solution of our equations. The multiplicative factor must now be determined from the initial conditions. Combining (10.85, 86, 88, 89) gives the conditions

$$1=y(0)=\int_{-\infty}^{\infty} dp\,\bar{y}(p)=\int_{-\infty}^{\infty} dp\,f(p)\,a_{\rm R}(p)^*\,, \tag{10.90a}$$

$$0 = w_E(0) = \int_{-\infty}^{\infty} dp\, \bar{w}_E(p) = \int_{-\infty}^{\infty} dp f(p)\, \tilde{h}_E(p)^* \tag{10.90b}$$

for the unknown function $f(p)$.

This set of equations can be solved with the help of the orthogonality relation for the negative continuum eigenmodes Φ_p. Substituting the expansion (10.61) of Φ_p in terms of the projected basis functions

$$\Phi_\mathrm{p} = a_\mathrm{R}(p)\, \Phi_\mathrm{R} + \int_{-\infty}^{-m_\mathrm{e}} dE\, \tilde{h}_E(p)\, \tilde{\Phi}_\mathrm{E}\,, \tag{10.91}$$

and using the orthogonality of Φ_R and $\tilde{\Phi}_\mathrm{E}$ gives the completeness relation for the expansion coefficients [cf. (6.20c)]

$$\delta(p-p') = \langle \Phi_\mathrm{p} | \Phi_{\mathrm{p}'} \rangle = a_\mathrm{R}(p)^*\, a_\mathrm{R}(p') + \int_{-\infty}^{-m_\mathrm{e}} dE\, \tilde{h}_E(p)^*\, \tilde{h}_E(p')\,. \tag{10.92}$$

Hence, multiplying (10.90b) by $\tilde{h}_E(p')$ and integrating over E yields

$$\begin{aligned} 0 = \int_{-\infty}^{-m} dE\, \tilde{h}_E(p')\, w_E(0) &= \int_{-\infty}^{\infty} dp f(p)[\delta(p-p') - a_\mathrm{R}(p)^*\, a_\mathrm{R}(p')] \\ &= f(p') - a_\mathrm{R}(p') \int_{-\infty}^{\infty} dp f(p)\, a_\mathrm{R}(p)^*\,. \end{aligned} \tag{10.93}$$

The first boundary condition (10.90a) immediately yields

$$f(p') = a_\mathrm{R}(p')\,. \tag{10.94}$$

The correct solution of the algebraic equations (10.87) is therefore

$$\bar{y}(p) = |a_\mathrm{R}(p)|^2\,, \qquad \bar{w}_E(p) = a_\mathrm{R}(p)\, \tilde{h}_E(p)^*\,. \tag{10.95}$$

These functions must now be Fourier transformed according to (10.86) to obtain the functions $y(t)$ and $w_E(t)$, which describe the time evolution of the neutral vacuum in a supercritical field:

$$y(t) = \mathrm{e}^{-\mathrm{i}E_\mathrm{vac}t} \int_{-\infty}^{\infty} dp\, \mathrm{e}^{\mathrm{i}pt} |a_\mathrm{R}(p)|^2\,, \tag{10.96a}$$

$$w_E(t) = \mathrm{e}^{-\mathrm{i}E_\mathrm{vac}t} \int_{-\infty}^{\infty} dp\, \mathrm{e}^{\mathrm{i}pt} a_\mathrm{R}(p)\, \tilde{h}_E(p)^*\,. \tag{10.96b}$$

Again, these integrals are well known from the calculations in the Heisenberg picture, namely (10.63, 64), and even earlier (10.27, 31). Observe, however, that here we formally deal with the complex conjugate expressions because (10.87) differed from the original autoionisation equation by complex conjugation. Copying the previously obtained results, for $t > 0$

$$y(t) = \exp[-\mathrm{i}(E_{\mathrm{vac}} - E_{\mathrm{R}})t - \tfrac{1}{2}\tilde{\Gamma}t] , \tag{10.97a}$$

$$w_E(t) = \frac{\tilde{V}_E^*}{E - E_{\mathrm{R}} - \frac{\mathrm{i}}{2}\tilde{\Gamma}} \exp(-\mathrm{i}E_{\mathrm{vac}}t)[\mathrm{e}^{\mathrm{i}Et} - \exp(\mathrm{i}E_{\mathrm{R}}t - \tfrac{1}{2}\tilde{\Gamma}t)] . \tag{10.97b}$$

The amplitude $y(t)$, belonging to the K-shell component of the state $|\Omega(t)\rangle$, propagates like an excited state of energy $(-E_{\mathrm{R}}) = |E_{\mathrm{R}}|$ above the (charged) vacuum and decays in time. The decaying amplitude ends up in that of the positron states $w_E(t)$, yielding a resonance-type distribution of positrons with energy $(-E) = |E|$ after a sufficient time span has passed.

To summarize, we have confirmed within the framework of field theory that the neutral vacuum decays exponentially, giving rise to a distribution of antiparticles. The width of this distribution is determined either by the natural lifetime of the neutral vacuum state or by the time period for which the external field is supercritical, whichever is shorter.

10.5 The Vacuum in a Constant Electromagnetic Field

The simplest field configuration is when the entire space is filled with a constant, homogeneous electromagnetic field, characterized by the two constant vectors $\boldsymbol{E}, \boldsymbol{B}$. The relativistic expression of this situation is $\partial F^{\mu\nu}/\partial x^{\alpha} = 0$. Of course, this situation is an idealization, because the infinitely extended field contains an infinite amount of energy. It is, however, a very useful approximation if one deals with electromagnetic fields that vary little over regions of the order of the electron Compton wavelength $m_{\mathrm{e}}^{-1} \sim 386$ fm (Sect. 10.6)

$$\frac{|\nabla E|}{E} \ll m_{\mathrm{e}} . \tag{10.98}$$

A simple calculation reveals that these conditions are not satisfied for Coulomb fields $E \sim Z\alpha/r^2$, especially for super-high charges $Z\alpha \gg 1$. The condition may be fulfilled, however, in electromagnetic fields created by high-powered lasers or in astrophysical situations.

In any case, the example of a constant external field is extremely useful because it can be solved exactly, as done in the 1930s by the successive work of *Sauter* [Sa 31], *Heisenberg* and *Euler* [He 36] and *Weisskopf* [We 36]. Here we follow the derivation given by *Schwinger* [Schw 51] because it introduces an elegant and powerful method of dealing with the infinities encountered in QED and has the important virtue of conserving formal gauge invariance throughout the calculation. It starts from the expression of the vacuum energy in terms of the Feynman propagator. This, in turn, can be given by an integral representation with respect to a scalar variable that has many similarities with the proper time. The method is therefore also known as the "proper time technique".

It is possible to start directly from the expression (9.114) for the vacuum energy:

$$E_{\mathrm{vac}} = \int d^3x \, \mathrm{tr} \left[\gamma^0 \frac{\partial}{\partial t} S_{\mathrm{F}}(x, x' | A_\mu) \right]_{x' \to x} .$$

However, this expression does not form a useful basis for a covariant treatment, because the energy is not a relativistic invariant. We can obtain an invariant quantity by looking at the amplitude for the time evolution of the vacuum state. In the Schrödinger picture the vacuum state evolves as an eigenstate of the Hamiltonian with energy E_{vac} and therefore the vacuum–vacuum amplitude is given by

$$e^{\mathrm{i}W} = \langle 0(t) | 0(t+T) \rangle = \exp(-\mathrm{i}TE_{\mathrm{vac}}) , \tag{10.99}$$

if the external potential is time independent. Here W can be regarded as the vacuum action, and it is a relativistic invariant because the vacuum state is invariant under Lorentz transformations. In another Lorentz system, where the external potential is explicitly time dependent because the sources move with constant non-zero velocity, the vacuum state also has a non-zero momentum, and W contains a contribution from this. That W is indeed a relativistic invariant has been shown by *Tomonaga* and *Schwinger* [To 46, Schw 49].

A different representation of W that is manifestly Lorentz and gauge invariant can be obtained in the following way. The Lagrangian of the Dirac field has $-j^\mu A_\mu$ as coupling term to the external potential. An infinitesimal change δA_μ in the external potential therefore produces a change in the W action by

$$\begin{aligned} \delta W &= -\int d^4x \langle 0 | \hat{j}^\mu(x) | 0 \rangle \, \delta A_\mu(x) \\ &= \mathrm{i}e \int d^4x \, \delta A_\mu(x) \, \mathrm{tr} \, [\gamma^\mu S_{\mathrm{F}}(x; x' | A_\mu)]_{x' \to x} \end{aligned} \tag{10.100}$$

where we expressed the vacuum polarization density by (9.110). (In the absence of a non-zero real vacuum charge, the first-order change $\int d^4x \, A_\mu \delta j^\mu$ vanishes [Sect. 9.7, especially (9.131) and its derivation].) A direct comparison with (10.99) shows that the associated change in the vacuum energy follows (9.110) and (9.131):

$$\delta E_{\mathrm{vac}} = -\delta W/T = -\mathrm{i}e \int d^3x \, \delta A_\mu(x) \, \mathrm{tr} [\gamma^\mu S_{\mathrm{F}}(x; x' | A_\mu)]_{x' \to x} . \tag{10.101}$$

The basic idea is now to express δA_μ in terms of the propagator S_{F} itself. According to (8.53), $S_{\mathrm{F}}(x, x')$ is the inverse of the operator kernel

$$\left[\gamma^\mu \left(\mathrm{i} \frac{\partial}{\partial x^\mu} - eA_\mu \right) - m_{\mathrm{e}} + \mathrm{i}\varepsilon \right] \delta(x - x') \equiv S_{\mathrm{F}}^{-1}(x, x' | A_\mu) , \tag{10.102}$$

where the infinitesimal imaginary part determines the correct boundary conditions. The term "inverse" is meant here in the operator sense, i.e.,

$$\int d^4x'' \, S_{\mathrm{F}}^{-1}(x; x'' | A_\mu) \, S_{\mathrm{F}}(x''; x' | A_\mu) = \delta(x - x') . \tag{10.103}$$

Since S_{F}^{-1} depends on A_μ very simply, then

$$-e\gamma^\mu \delta A_\mu(x)\,\delta(x-x') = \delta S_F^{-1}(x;x'\,|A_\mu)\,. \tag{10.104}$$

This relation enables us to express the vacuum amplitude (10.100) solely in terms of the Feynman propagator through the infinitesimal equation

$$\delta W = -\mathrm{i}\int d^4x\, d^4x''\ \mathrm{tr}\,[\delta S_F^{-1}(x;x''\,|A_\mu)\, S_F(x'',x'\,|A_\mu)]_{x'\to x}\,. \tag{10.105}$$

This equation is the starting point for Schwinger's proper time method. The following formal manipulations are greatly facilitated if one regards $S_F(x;x')$ as the coordinate-space matrix element $\langle x|S_F|x'\rangle$ of an operator $\hat{S}_F$. (In the following the coordinate space operators $\hat{S}_F$, $\hat{\Pi}_\mu$, etc. are denoted for clarity by the operator circumflex symbol, although they are not Fock space operators.) The double space-time integral in (10.105) can then be understood as the coordinate space trace of the product of the operators $\delta\hat{S}_F^{-1}$ and $\hat{S}_F$. The full trace over space-time and spinor indices is denoted by "tr":

$$\begin{aligned}\delta W &= -\mathrm{i}\,\mathrm{tr}(\delta\hat{S}_F^{-1}\cdot\hat{S}_F) = -\mathrm{i}\,\mathrm{tr}(\delta\ln\hat{S}_F^{-1})\\ &= \mathrm{i}\,\delta[\mathrm{tr}(\ln\hat{S}_F)]\,.\end{aligned} \tag{10.106}$$

We furthermore introduce the gauge-covariant momentum operator $\hat{\Pi}_\mu$ with matrix elements

$$\langle x|\hat{\Pi}_\mu|x'\rangle = \left(\mathrm{i}\frac{\partial}{\partial x^\mu} - eA_\mu(x)\right)\delta(x-x')\,. \tag{10.107}$$

The propagator and its inverse then take the symbolic form:

$$\hat{S}_F^{-1} = \gamma^\mu\hat{\Pi}_\mu - m_e + \mathrm{i}\varepsilon \tag{10.108a}$$

$$\begin{aligned}\hat{S}_F = (\gamma^\mu\hat{\Pi}_\mu - m_e+\mathrm{i}\varepsilon)^{-1} &= (\gamma^\mu\hat{\Pi}_\mu + m_e)\,[(\gamma^\mu\hat{\Pi}_\mu)^2-(m_e-\mathrm{i}\varepsilon)^2]^{-1}\\ &= [(\gamma^\mu\hat{\Pi}_\mu)^2-(m_e-\mathrm{i}\varepsilon)^2]^{-1}(\gamma^\mu\hat{\Pi}_\mu+m_e)\,.\end{aligned} \tag{10.108b}$$

Since the operator $[(\gamma^\mu\hat{\Pi}_\mu)^2 - m_e^2]$ is self-adjoint (for not too singular potentials), it has only real eigenvalues, the inverse operator can be expressed by an integral representation:

$$\hat{D} \equiv [(\gamma^\mu\hat{\Pi}_\mu)^2-(m_e-\mathrm{i}\varepsilon)^2]^{-1} = -\mathrm{i}\int_0^\infty ds\,\exp[\mathrm{i}s((\gamma^\mu\hat{\Pi}_\mu)^2-(m_e-\mathrm{i}\varepsilon)^2)]\,. \tag{10.109}$$

The infinitesimal imaginary part ensures the convergence of the integral at the upper bound $s=\infty$. Because operators under the trace sign can be cyclically permuted, the vacuum amplitude (10.105) takes the form

$$\begin{aligned}\delta W &= -\mathrm{i}\,\mathrm{tr}\left\{\delta(\gamma^\mu\hat{\Pi}_\mu)\left[\frac{1}{2}(\gamma^\mu\hat{\Pi}_\mu+m_e)\hat{D} + \frac{1}{2}\hat{D}(\gamma^\mu\hat{\Pi}_\mu+m_e)\right]\right\}\\ &= -\frac{\mathrm{i}}{2}\mathrm{tr}\{[\delta(\gamma^\mu\hat{\Pi}_\mu)\,\gamma^\mu\hat{\Pi}_\mu + \gamma^\mu\hat{\Pi}_\mu\,\delta(\gamma^\mu\hat{\Pi}_\mu)]\,\hat{D}\}\end{aligned}$$

$$= -\frac{1}{2}\operatorname{tr}\left\{\delta[(\gamma^\mu\hat{\Pi}_\mu)^2 - m_e^2]\int_0^\infty ds\,\exp[\mathrm{i}s((\gamma^\mu\hat{\Pi}_\mu)^2-(m_e-\mathrm{i}\varepsilon)^2)]\right\}$$

$$= \frac{\mathrm{i}}{2}\delta\left\{\int_0^\infty \frac{ds}{s}\exp[-\mathrm{i}(m_e-\mathrm{i}\varepsilon)^2 s]\operatorname{tr}[\exp(\mathrm{i}s(\gamma^\mu\hat{\Pi}_\mu)^2)]\right\}. \qquad (10.110)$$

In the step from the first to the second line we used the property that the trace of any product of an odd number of γ matrices vanishes [Bj 64, Gr 81b].

Having expressed the rhs as a total differential, we can integrate and obtain an expression for the finite vacuum amplitude. The term in the exponent can be rewritten as

$$(\gamma^\mu\hat{\Pi}_\mu)^2 = \frac{1}{4}\{\gamma^\mu,\gamma^\nu\}\cdot\{\hat{\Pi}_\mu,\hat{\Pi}_\nu\} + \frac{1}{4}[\gamma^\mu,\gamma^\nu]\cdot[\hat{\Pi}_\mu,\hat{\Pi}_\nu]$$

$$= \frac{1}{2}g^{\mu\nu}\{\hat{\Pi}_\mu,\hat{\Pi}_\nu\} - \frac{\mathrm{i}}{2}\sigma^{\mu\nu}\left(-\mathrm{i}e\frac{\partial A_\nu}{\partial x^\mu} + \mathrm{i}e\frac{\partial A_\mu}{\partial x^\nu}\right)$$

$$= \hat{\Pi}_\mu\hat{\Pi}^\mu - \frac{1}{2}e\sigma^{\mu\nu}F_{\mu\nu} \equiv \hat{\Pi}^2 - \frac{1}{2}e\sigma\cdot F, \qquad (10.111)$$

giving

$$W = \frac{\mathrm{i}}{2}\int_0^\infty \frac{ds}{s}\exp[-\mathrm{i}(m-\mathrm{i}\varepsilon)^2 s]\operatorname{tr}\left\{\exp\left[\mathrm{i}s\left(\hat{\Pi}^2 - \frac{1}{2}e\sigma\cdot F\right)\right] - \exp(\mathrm{i}s\hat{p}^2)\right\}, \qquad (10.112)$$

where $\exp(\mathrm{i}s\hat{p}^2)$ comes from the lower limit $A_\mu \equiv 0$ of the variation after the external potential, with $\hat{p}_\mu = \hat{\Pi}_\mu|_{A=0}$.

Interestingly enough, (10.112) as it stands holds for any external potential, whether it is constant or not. Only for constant external field strengths $F_{\mu\nu}$, however, can the trace operations with respect to space-time indices and spinor indices be separated and carried out independently. This is because under this circumstance $\hat{\Pi}_\mu$ and $F_{\mu\nu}$ commute: $[\hat{\Pi}_\mu, F_{\lambda\nu}] = \mathrm{i}(\partial F_{\lambda\nu}/\partial x^\mu) = 0$, and $\hat{\Pi}_\mu$ is diagonal in spin space, $\sigma^{\mu\nu}F_{\mu\nu}$ diagonal in coordinate space. Therefore for constant $F_{\mu\nu}$

$$\operatorname{tr}\left\{\exp\left[\mathrm{i}s\left(\hat{\Pi}^2 - \frac{1}{2}e\sigma\cdot F\right)\right] - \mathrm{e}^{\mathrm{i}s\hat{p}^2}\right\}$$

$$= \operatorname{tr}\left[\exp\left(-\frac{\mathrm{i}}{2}e\sigma\cdot Fs\right)\right]\operatorname{tr}_x(\mathrm{e}^{\mathrm{i}s\hat{\Pi}^2}) - 4\operatorname{tr}_x(\mathrm{e}^{\mathrm{i}s\hat{p}^2}), \qquad (10.113)$$

where tr_x means the trace only in coordinate space, and the factor 4 comes from the trace over spinor variables.

The last term is easily evaluated with the help of the eigenstates $\exp(\mathrm{i}px)$ of the momentum operator:

$$\operatorname{tr}_x(\mathrm{e}^{\mathrm{i}s\hat{p}^2}) = \int d^4x\,\langle x|\mathrm{e}^{\mathrm{i}s\hat{p}^2}|x\rangle = \int d^4x\int\frac{d^4p}{(2\pi)^4}\mathrm{e}^{-\mathrm{i}p\cdot x}\mathrm{e}^{\mathrm{i}sp^2}\mathrm{e}^{\mathrm{i}p\cdot x}$$

$$= \int d^4x\int\frac{d^4p}{(2\pi)^4}\mathrm{e}^{\mathrm{i}sp^2} = \int d^4x\left(\int_{-\infty}^{\infty}\frac{dp_1}{2\pi}\mathrm{e}^{-\mathrm{i}sp_1^2}\right)^3\int_{-\infty}^{\infty}\frac{dp_0}{2\pi}\mathrm{e}^{\mathrm{i}sp_0^2}$$

$$= \int d^4x \left(\frac{\sqrt{\pi/\mathrm{i}s}}{2\pi} \right)^3 \frac{\sqrt{-\pi/\mathrm{i}s}}{2\pi} = \frac{1}{(4\pi)^2 \mathrm{i}s^2} \int d^4x \,. \tag{10.114}$$

The trace is obviously divergent, if taken over all of space-time, but for the contribution of a finite volume to the vacuum amplitude the expression is well defined.

The spinor trace in (10.113) can be readily evaluated, when one remembers that the trace does not change under coordinate transformations, hence is equal to the sum of eigenvalues. First consider

$$\begin{aligned}(\tfrac{1}{2}\sigma^{\mu\nu}F_{\mu\nu})^2 &= (\boldsymbol{\sigma}\cdot\boldsymbol{B} - \mathrm{i}\boldsymbol{\alpha}\cdot\boldsymbol{E})^2 = B^2 - E^2 - \mathrm{i}\{\boldsymbol{\sigma}\cdot\boldsymbol{B}, \boldsymbol{\alpha}\cdot\boldsymbol{E}\} \\ &= (B^2 - E^2) - 2\mathrm{i}\gamma^5(\boldsymbol{B}\cdot\boldsymbol{E}) \,. \end{aligned} \tag{10.115}$$

Since

$$\mathscr{F} = \tfrac{1}{2}(B^2 - E^2) = \tfrac{1}{4}F_{\mu\nu}F^{\mu\nu} \,, \qquad \mathscr{G} = \boldsymbol{B}\cdot\boldsymbol{E} = \tfrac{1}{8}\varepsilon^{\alpha\beta\mu\nu}F_{\alpha\beta}F_{\mu\nu} \tag{10.116}$$

are just the relativistic invariants of the electromagnetic field, then in a Lorentz-invariant way

$$(\tfrac{1}{2}\sigma^{\mu\nu}F_{\mu\nu})^2 = 2(\mathscr{F}\mathbb{1} - \mathrm{i}\gamma^5\mathscr{G}) = 2\begin{pmatrix} \mathscr{F} & 0 & -\mathrm{i}\mathscr{G} & 0 \\ 0 & \mathscr{F} & 0 & -\mathrm{i}\mathscr{G} \\ -\mathrm{i}\mathscr{G} & 0 & \mathscr{F} & 0 \\ 0 & -\mathrm{i}\mathscr{G} & 0 & \mathscr{F} \end{pmatrix} . \tag{10.117}$$

This 4×4 matrix has eigenvalues $2(\mathscr{F} \pm \mathrm{i}\mathscr{G}) = (\boldsymbol{B} \pm \mathrm{i}\boldsymbol{E})^2$, hence the eigenvalues of the original matrix $\sigma^{\mu\nu}F_{\mu\nu}/2$ must be

$$\pm\sqrt{2(\mathscr{F} \pm \mathrm{i}\mathscr{G})} = \pm\sqrt{(\boldsymbol{B} \pm \mathrm{i}\boldsymbol{E})^2} \tag{10.118}$$

with all four possible combinations of the signs. Now it is straightforward to calculate the trace

$$\begin{aligned}\mathrm{tr}[\exp(-\tfrac{\mathrm{i}}{2}e\sigma\cdot Fs)] &= \exp(-\mathrm{i}es\sqrt{2(\mathscr{F}+\mathrm{i}\mathscr{G})}) + \exp(\mathrm{i}es\sqrt{2(\mathscr{F}+\mathrm{i}\mathscr{G})}) \\ &\quad + \exp(-\mathrm{i}es\sqrt{2(\mathscr{F}-\mathrm{i}\mathscr{G})}) + \exp(\mathrm{i}es\sqrt{2(\mathscr{F}-\mathrm{i}\mathscr{G})}) \\ &= 2[\cos(es\sqrt{2(\mathscr{F}+\mathrm{i}\mathscr{G})}) + \cos(es\sqrt{2(\mathscr{F}-\mathrm{i}\mathscr{G})})] \\ &= 4\,\mathrm{Re}\{\cos(es\sqrt{2(\mathscr{F}+\mathrm{i}\mathscr{G})})\} \,. \end{aligned} \tag{10.119}$$

All that is left to do is to evaluate the space-time integral $\mathrm{tr}_x(\exp(\mathrm{i}s\hat{\Pi}^2))$. Incidentally, this is exactly the same expression we would have obtained for the vacuum amplitude of a spin-0 field in an electromagnetic background field. Thus this term contains the coupling between the electron charge and the external field, whereas the trace (10.119) describes the coupling between magnetic moment and field.

For a constant pure magnetic field $\boldsymbol{B}$ the spatial trace over the operator $\exp(\mathrm{i}s\hat{\Pi}^2)$ can be evaluated by elementary methods, because then the four-vector potential A_μ has only space-like components, e.g. $A_\mu = (0, \tfrac{1}{2}\boldsymbol{B}\times\boldsymbol{r})$, and the operator $\hat{\Pi}^2$ has a simple spectrum for a constant magnetic field. Actually, this spectrum is well known from non-relativistic quantum mechanics, where the Hamiltonian of a charged particle in an external magnetic field has the form

$$\hat{H} = \frac{1}{2m_e}\left(\hat{p} - \frac{e}{c}A\right)^2 = \frac{1}{2m_e}\hat{\Pi}^2 . \tag{10.120}$$

Its eigenvalues are the Landau energies [Ga 74]

$$E_n = \left(n + \frac{1}{2}\right)\frac{\hbar e B}{m_e c} + \frac{p_\parallel^2}{2m_e}, \qquad n = 0, 1, 2, \dots , \tag{10.121}$$

where $p_\parallel$ is the momentum component parallel to the magnetic field. Setting again $\hbar = c = 1$, the eigenvalues of the operator $\hat{\Pi}^2$ must therefore be

$$\xi_n = (2n+1)eB + p_\parallel^2 , \qquad n = 0, 1, 2, \dots . \tag{10.122}$$

Except for a normalization factor N which involves the density of states, the trace is given by the sum over the eigenvalues of the operator, as in (10.114),

$$\begin{aligned}
Z(s) &\equiv \mathrm{tr}_x(\mathrm{e}^{\mathrm{i}s\hat{\Pi}^2}) = \mathrm{tr}_{x^0}(\mathrm{e}^{\mathrm{i}s(\hat{p}^0)^2})\,\mathrm{tr}_{\boldsymbol{x}}(\mathrm{e}^{\mathrm{i}s\hat{\boldsymbol{\Pi}}^2}) \\
&= N\int dx^0 dx_\parallel \int \frac{dp^0}{2\pi}\frac{dp_\parallel}{2\pi}\,\mathrm{e}^{\mathrm{i}s(p^0)^2}\sum_{n=0}^{\infty}\mathrm{e}^{-\mathrm{i}s\xi_n} \\
&= N\int dx^0 dx_\parallel \frac{(-\pi/\mathrm{i}s)^{1/2}}{2\pi}\frac{(\pi/\mathrm{i}s)^{1/2}}{2\pi}\,\mathrm{e}^{-\mathrm{i}eBs}\sum_{n=0}^{\infty}\mathrm{e}^{-2\mathrm{i}eBsn} \\
&= N\int dx^0 dx_\parallel \frac{1}{4\pi s}\,\mathrm{e}^{-\mathrm{i}eBs}(1-\mathrm{e}^{-2\mathrm{i}eBs})^{-1} \\
&= \frac{N}{8\pi\mathrm{i}s}\int dx^0 dx_\parallel \frac{1}{\sin(eBs)} .
\end{aligned} \tag{10.123}$$

The normalization factor N can be determined by comparing the limit $B \to 0$ with the result (10.114) obtained in the field-free case. Agreement is found by choosing

$$N = \frac{eB}{2\pi}\int d^2x_\perp , \tag{10.124}$$

where $\int d^2x_\perp$ means integration over the spatial coordinates normal to the direction of the magnetic field. For the pure magnetic field

$$Z(s) \equiv \mathrm{tr}_x(\mathrm{e}^{\mathrm{i}s\hat{\Pi}^2}) = \frac{1}{(4\pi)^2\mathrm{i}s^2}\int d^4x \frac{eBs}{\sin(eBs)} . \tag{10.125}$$

Generally, when both electric and magnetic fields are present, the spatial trace can be evaluated, following *Schwinger* [Schw 51], by deriving a differential equation for $\mathrm{tr}_x[\exp(\mathrm{i}s\hat{\Pi}^2)]$ in the "proper time" variable s. This name derives from the observation that the operator $\exp(-\mathrm{i}s\hat{\Pi}^2)$ can be regarded as a "proper time" evolution operator with $\hat{\Pi}^2$ representing the Lorentz-invariant "super" Hamiltonian. Thus we are led to introduce the proper time-dependent operators

$$\hat{x}^\mu(s) = \mathrm{e}^{\mathrm{i}s\hat{\Pi}^2}\hat{x}^\mu \mathrm{e}^{-\mathrm{i}s\hat{\Pi}^2}, \quad \hat{\Pi}_\mu(s) = \mathrm{e}^{\mathrm{i}s\hat{\Pi}^2}\hat{\Pi}_\mu \mathrm{e}^{-\mathrm{i}s\hat{\Pi}^2} \tag{10.126}$$

and derive equations of motion for them. If eigenstates of the operator $\hat{x}^\mu(s)$ are denoted by

$$|x(s)\rangle = \mathrm{e}^{-\mathrm{i}s\hat{\Pi}^2}|x(0)\rangle, \quad |x(0)\rangle = |x\rangle, \tag{10.127}$$

the trace over space-time coordinates takes the very simple form

$$Z(s) = \mathrm{tr}_x(\mathrm{e}^{\mathrm{i}s\hat{\Pi}^2}) = \int d^4x \langle x(0)|\mathrm{e}^{\mathrm{i}s\hat{\Pi}^2}|x(0)\rangle = \int d^4x \langle x(s)|x(0)\rangle . \tag{10.128}$$

The usefulness of the proper time method is based on this relation. These preliminary remarks suffice, because this formal development is unnecessary to evaluate the trace. Differentiating $Z(s)$ with respect to s gives

$$\mathrm{i}\frac{d}{ds}Z(s) = -\mathrm{tr}_x(\hat{\Pi}^2\mathrm{e}^{\mathrm{i}s\hat{\Pi}^2}) . \tag{10.129}$$

To find a useful form of the rhs we start from the second rank tensor

$$\Theta_{\mu\nu} \equiv \mathrm{tr}_x(\hat{\Pi}_\mu\hat{\Pi}_\nu\mathrm{e}^{\mathrm{i}s\hat{\Pi}^2}) \tag{10.130}$$

regarded as a 4×4 matrix in the Lorentz indices μ, ν. From the definition (10.107) one easily verifies

$$[\hat{\Pi}_\mu, \hat{\Pi}_\nu] = -\mathrm{i}eF_{\mu\nu}, \tag{10.131}$$

and therefore

$$[\hat{\Pi}_\mu, \hat{\Pi}^2] = [\hat{\Pi}_\mu, \hat{\Pi}_\nu]\hat{\Pi}^\nu + \hat{\Pi}_\nu[\hat{\Pi}_\mu, \hat{\Pi}^\nu] = -\mathrm{i}e[\hat{\Pi}^\nu, F_{\mu\nu}] = -2\mathrm{i}eF_{\mu\nu}\hat{\Pi}^\nu . \tag{10.132}$$

In deriving (10.132) we have made use of the assumed constancy of $F_{\mu\nu}$, i.e. $[\Pi_\alpha, F_{\mu\nu}] = 0$. Expanding in a power series in $\hat{\Pi}^2$ yields

$$\hat{\Pi}_\nu\mathrm{e}^{\mathrm{i}s\hat{\Pi}^2} = \{\exp[\mathrm{i}s(\hat{\Pi}^2 - 2\mathrm{i}eF)]\}_{\nu\lambda}\hat{\Pi}^\lambda = \mathrm{e}^{\mathrm{i}s\hat{\Pi}^2}(\mathrm{e}^{2eFs})_\nu^{\ \lambda}\hat{\Pi}_\lambda \tag{10.133}$$

where the exponentiation of $F_{\nu\lambda}$ is a matrix operation in the Lorentz indices. Since products under the trace operation can be rotated, $\Theta_{\mu\nu}$ can be transformed as

$$\begin{aligned}\Theta_{\mu\nu} &= \mathrm{tr}_x[\hat{\Pi}_\mu\mathrm{e}^{\mathrm{i}s\hat{\Pi}^2}(\mathrm{e}^{2eFs})_\nu^{\ \lambda}\hat{\Pi}_\lambda] \\ &= \Theta_{\lambda\mu}(\mathrm{e}^{2eFs})_\nu^{\ \lambda} \\ &= \Theta_{\mu\lambda}(\mathrm{e}^{2eFs})_\nu^{\ \lambda} + \mathrm{i}eZ(s)F_{\mu\lambda}(\mathrm{e}^{2eFs})_\nu^{\ \lambda},\end{aligned} \tag{10.134}$$

where (10.131) was used. We now solve (10.134) for $\Theta_{\mu\nu}$ which appears on both sides:

$$\Theta_{\mu\lambda} = \mathrm{tr}_x[\hat{\Pi}_\mu \hat{\Pi}_\lambda \mathrm{e}^{\mathrm{i}s\hat{\Pi}^2}] = \mathrm{i}eZ(s)F_{\mu\sigma}(\mathrm{e}^{2eFs})^{\nu\sigma}[(1-\mathrm{e}^{-2eFs})^{-1}]_{\nu\lambda}, \tag{10.135}$$

where $1_{\nu\lambda} = g_{\nu\lambda}$ is the Minkowski metric $(1, -1, -1, -1)$. Through the antisymmetry of $F_{\nu\lambda}$ we have

$$(\mathrm{e}^{2eFs})^{\nu\sigma} = (\mathrm{e}^{-2eFs})^{\sigma\nu}\,. \tag{10.136}$$

which was used to derive (10.135). Combining the exponentials finally gives

$$\Theta_{\mu\lambda} = \mathrm{i}eZ(s)F_{\mu\sigma}[(\mathrm{e}^{2eFs}-1)^{-1}]^{\sigma}_{\lambda}\,. \tag{10.137}$$

Contracting the indices μ, λ, and combining this expression with (10.129, 130) yields the differential equation

$$\begin{aligned} \mathrm{i}\frac{d}{ds}Z(s) &= -\mathrm{i}eF_{\mu\sigma}[(\mathrm{e}^{2eFs}-1)^{-1}]^{\sigma\mu}Z(s) \\ &\equiv -\mathrm{i}e\left(\frac{F}{\mathrm{e}^{2eFs}-1}\right)^{\mu}_{\mu}Z(s)\,. \end{aligned} \tag{10.138}$$

This equation can be integrated:

$$\begin{aligned} \ln Z(s) &= -\int ds\left(\frac{eF\mathrm{e}^{-eFs}}{2\sinh(eFs)}\right)^{\mu}_{\mu} \\ &= \int ds\left(\frac{eF}{2}\coth(eFs)\right)^{\mu}_{\mu} = -\frac{1}{2}\left(\ln\frac{\sinh(eFs)}{eFs}\right)^{\mu}_{\mu} + \mathrm{const}\,. \end{aligned} \tag{10.139}$$

Here we used the fact that $F_{\nu\lambda}$ is antisymmetric in its indices and hence $(F^n)^{\mu}_{\mu} = 0$ for all odd exponents n. A comparison with the free-field expression (10.114) fixes the integration constant and gives

$$Z(s) = \mathrm{tr}_x(\mathrm{e}^{\mathrm{i}s\hat{\Pi}^2}) = \frac{1}{(4\pi)^2\mathrm{i}s^2}\int d^4x \exp\left[\frac{1}{2}\left(\ln\frac{eFs}{\sinh(eFs)}\right)^{\mu}_{\mu}\right]. \tag{10.140}$$

The complicated matrix expression is most easily evaluated with the eigenvalues of the 4×4 matrix $F_{\nu\lambda}$. A straightforward calculation yields as eigenvalues

$$\pm\frac{\mathrm{i}}{\sqrt{2}}(\sqrt{\mathscr{F}+\mathrm{i}\mathscr{G}} \pm \sqrt{\mathscr{F}-\mathrm{i}\mathscr{G}}) \equiv \pm\mathrm{i}f_{\pm}\,, \quad \text{whence} \tag{10.141}$$

$$\begin{aligned} \exp\left[\frac{1}{2}\left(\ln\frac{eFs}{\sinh(eFs)}\right)^{\mu}_{\mu}\right] &= \frac{ef_{+}s}{\sin(ef_{+}s)}\cdot\frac{ef_{-}s}{\sin(ef_{-}s)} \\ &= \frac{2(es)^2 f_{+}f_{-}}{\cos[es(f_{+}-f_{-})]-\cos[es(f_{+}+f_{-})]} \\ &= -\frac{(es)^2\,\mathscr{G}}{\mathrm{Im}\{\cos[es\sqrt{2(\mathscr{F}+\mathrm{i}\mathscr{G})}]\}}\,. \end{aligned} \tag{10.142}$$

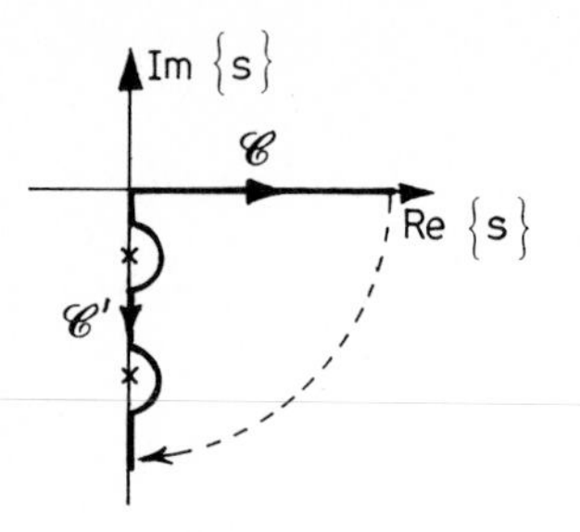

Fig. 10.11. Rotation of the integration contour in the complex s plane. The contribution of the quarter circle at infinity vanishes due to the exponential factor $\exp(-\mathrm{i}sm_e^2-\varepsilon s)$. Possible poles on the imaginary axis must be circumvented as indicated

In the special case of a pure magnetic field, considered previously, $\mathcal{F}=B^2/2$, $\mathcal{G}=0$, and therefore $f_+=\sqrt{2\mathcal{F}}=B$, $f_-=0$ in (10.141). Substituting these values into ((10.140, 142), the result (10.125) is recovered. This provides an excellent check for the formal manipulations which led to (10.140).

Collecting now the various terms (10.114, 119, 142) and inserting them into (10.112) gives for the vacuum amplitude

$$W=\frac{\mathrm{i}}{2}\int_0^\infty\frac{ds}{s}\exp[-\mathrm{i}s(m_e^2-\mathrm{i}\varepsilon)]\frac{4}{(4\pi)^2\mathrm{i}s^2}\int d^4x \times\left[-(es)^2\mathcal{G}\frac{\mathrm{Re}\{\cos[es\sqrt{2(\mathcal{F}+\mathrm{i}\,\mathcal{G})}]\}}{\mathrm{Im}\{\cos[es\sqrt{2(\mathcal{F}+\mathrm{i}\,\mathcal{G})}]\}}-1\right]. \tag{10.143}$$

Since the exponential mass factor always enforces convergence for $s\to\infty$, the contour can be deformed from the positive real axis to the negative imaginary axis (Fig. 10.11) making the substitution $s\to(-\mathrm{i}s)$:

$$W=-\frac{1}{8\pi^2}\int d^4x\int_0^\infty\frac{ds}{s^3}\mathrm{e}^{-sm_e^2}\left[(es)^2\mathcal{G}\frac{\mathrm{Re}\{\cosh[es\sqrt{2(\mathcal{F}+\mathrm{i}\,\mathcal{G})}]\}}{\mathrm{Im}\{\cosh[es\sqrt{2(\mathcal{F}+\mathrm{i}\,\mathcal{G})}]\}}-1\right]. \tag{10.144}$$

As expected from the general arguments given in Sect. 9.8, this integral is divergent due to its behaviour at $s\to 0$, the annoying term being quadratic in the external field strength. To isolate the divergence, the hyperbolic cosine function is expanded in a power series, yielding

$$\begin{aligned}W&=-\frac{1}{8\pi^2}\int d^4x\int_0^\infty\frac{ds}{s^3}\mathrm{e}^{-m_e^2s}\left[\frac{1+(es)^2\mathcal{F}+\frac{1}{6}(es)^4(\mathcal{F}^2-\mathcal{G}^2)+\dots}{1+\frac{1}{3}(es)^2\mathcal{F}+\frac{1}{90}(es)^4(3\mathcal{F}^2-\mathcal{G}^2)+\dots}-1\right]\\&=-\frac{e^2}{8\pi^2}\int d^4x\int_0^\infty\frac{ds}{s}\mathrm{e}^{-m_e^2s}[\tfrac{2}{3}\mathcal{F}-\tfrac{1}{45}(es)^2(4\mathcal{F}^2+7\mathcal{G}^2)+\dots].\end{aligned} \tag{10.145}$$

Going from the total action W to the associated Lagrangian density $\mathcal{L}$, $W=-\int d^4x\,\mathcal{L}(x)$, and combining this with the Lagrangian of the external field gives

$$\mathscr{L} = \mathscr{L}_{\text{ext}} + \mathscr{L}_{\text{vac}} = \left[1 + \frac{e^2}{12\pi^2}\int_0^\infty \frac{ds}{s}\,\mathrm{e}^{-m_e^2 s}\right]\mathscr{F}$$
$$- \frac{2e^4}{45(4\pi)^2}\int_0^\infty s\,ds\,\mathrm{e}^{-m_e^2 s}(4\mathscr{F}^2 + 7\mathscr{G}^2) + \dots \,. \tag{10.146}$$

We now argue as before that a contribution from the vacuum action which is exactly proportional to the external action cannot be observed and therefore must not be counted. Formally this may be done by defining a renormalized charge e_R by

$$\frac{1}{e_R^2} = \frac{1}{e^2} + \frac{1}{12\pi^2}\int_0^\infty \frac{ds}{s}\,\mathrm{e}^{-m_e^2 s} \tag{10.147}$$

and renormalized field strengths $\mathscr{F}_R$, $\mathscr{G}_R$ by

$$\mathscr{F}_R = \frac{e^2}{e_R^2}\mathscr{F} = \left(1 + \frac{e^2}{12\pi^2}\int_0^\infty \frac{ds}{s}\,\mathrm{e}^{-m_e^2 s}\right)\mathscr{F}\,, \qquad \mathscr{G}_R = \frac{e^2}{e_R^2}\mathscr{G}\,. \tag{10.148}$$

We set $e_R^2/4\pi = 1/137\ldots = \alpha$ and state that $\mathscr{F}_R$, $\mathscr{G}_R$ are the observable field quantities. The renormalized effective Lagrangian for the electromagnetic field in the polarizable electron–positron vacuum therefore is

$$\mathscr{L}_{\text{vac}} = \frac{1}{8\pi^2}\int_0^\infty \frac{ds}{s^3}\,\mathrm{e}^{-sm_e^2}\left[(es)^2\mathscr{G}\,\frac{\mathrm{Re}\{\cosh[es\sqrt{2(\mathscr{F}+\mathrm{i}\,\mathscr{G})}]\}}{\mathrm{Im}\{\cosh[es\sqrt{2(\mathscr{F}+\mathrm{i}\,\mathscr{G})}]\}} - 1 - \frac{2}{3}(es)^2\mathscr{F}\right]$$
$$= -\frac{2\alpha^2}{45}[(B^2-E^2)^2 + 7(\boldsymbol{E}\cdot\boldsymbol{B})] + \dots\,, \tag{10.149}$$

where we have dropped the subscript "R".

The most interesting special cases are obtained for a pure electric or magnetic field, when $\mathscr{G} = \boldsymbol{E}\cdot\boldsymbol{B} = 0$, and $\mathscr{F} = -1/2E^2$ or $\mathscr{F} = 1/2B^2$, and

$$\mathrm{Re}\{\cosh[es\sqrt{2(\mathscr{F}+\mathrm{i}\,\mathscr{G})}]\} = \begin{cases}\cos(esE)\\ \cosh(esB)\end{cases}. \tag{10.150}$$

The remaining part is an indefinite expression under these circumstances, and it is best to go back to (10.142), remembering that s has been replaced by $(-\mathrm{i}s)$. With $f_+ = \sqrt{2\mathscr{F}}$ and $f_- = 0$

$$\frac{(es)^2\mathscr{G}}{\mathrm{Im}\{\cosh[es\sqrt{2(\mathscr{F}+\mathrm{i}\,\mathscr{G})}]\}} = \frac{es\sqrt{2\mathscr{F}}}{\sinh(es\sqrt{2\mathscr{F}})}\cdot\lim_{f_-\to 0}\frac{esf_-}{\sinh(esf_-)}$$
$$= \begin{cases} esE/\sin(esE)\\ esB/\sinh(esB)\end{cases}. \tag{10.151}$$

The effective vacuum Lagrangian for a magnetic field becomes

$$\mathscr{L}(B) = \frac{B^2}{2} + \frac{1}{8\pi^2}\int_0^\infty \frac{ds}{s^3}\,\mathrm{e}^{-sm_\mathrm{e}^2}\left[(esB)\coth(esB) - 1 - \frac{1}{3}(esB)^2\right]. \quad (10.152)$$

The integral can be evaluated exactly, yielding [Di 76]

$$\mathscr{L}(B) = \frac{B^2}{2} + \frac{(eB)^2}{8\pi^2}\left[4\zeta'\left(-1, \frac{m_\mathrm{e}^2}{2eB}\right) + \left(\frac{m_\mathrm{e}^2}{2eB}\right)^2 - \frac{1}{3} + \left(\frac{m_\mathrm{e}^4}{2(eB)^2} - \frac{m_\mathrm{e}^2}{eB} + \frac{1}{3}\right)\ln\frac{2eB}{m_\mathrm{e}^2}\right]. \quad (10.153)$$

For a pure electric field, the integrand has poles at the locations $s = n\pi/(eE)$, as can be seen from (10.151). To resolve this, remember that the integration path has been deformed by clock-wise rotation by 90° in the complex plane. This means that the poles above must be circumvented, with the precise prescription (Fig. 10.11)

$$\frac{1}{\sin(eEs)} = P\frac{1}{\sin(eEs)} + \sum_{n=1}^\infty \mathrm{i}\pi\frac{(-1)^{n+1}}{eE}\delta\left(s - \frac{n\pi}{eE}\right); \quad (10.154)$$

hence

$$\mathscr{L}(E) = -\frac{1}{2}E^2 + \frac{1}{8\pi^2}P\int_0^\infty \frac{ds}{s^3}\,\mathrm{e}^{-sm_\mathrm{e}^2}\left[(esE)\cot(eEs) - 1 + \frac{1}{3}(eEs)^2\right] - \frac{\mathrm{i}}{8\pi}\sum_{n=1}^\infty\left(\frac{eE}{n\pi}\right)^2\exp\left(-n\frac{\pi m_\mathrm{e}^2}{eE}\right). \quad (10.155)$$

A different useful representation of the effective Lagrangian is [Mü 77b]

$$\mathscr{L}_\mathrm{eff} = -\frac{1}{2}E^2 - \frac{m_\mathrm{e}^4}{8\pi^2}\int_0^\infty ds\frac{f(s)}{\mathrm{e}^{\beta s} - 1} \quad (10.156a)$$

with the spectral function

$$f(s) = (s-1)\ln(1 - s + \mathrm{i}\varepsilon) + (s+1)\ln(s+1) - 2s, \quad (10.156b)$$

and $\beta = \pi m_\mathrm{e}^2/eE$. The integrand in (10.156a) bears some resemblance to a thermal distribution, but note that it has the wrong sign for fermions! The imaginary part of the Lagrangian signals that the vacuum state will not remain a vacuum as time goes by, but due to pair production will decay according to

$$|\langle 0(t)|0(t+T)\rangle|^2 = \exp(-2\,\mathrm{Im}\{W\}) = \exp\left(-\int_0^T dt\int d^3x\frac{\alpha E^2}{\pi^2}\sum_{n=1}^\infty\frac{1}{n^2}\mathrm{e}^{-n\pi m_\mathrm{e}^2/eE}\right). \quad (10.157)$$

In the limit $E \to 0$ the imaginary part of $\mathscr{L}(E)$ vanishes to all orders in E. This phenomenon indicates that pair production in a constant electric field cannot be found using perturbation theory in any finite order. Going back to (9.133), the lowest order polarization function has an imaginary part

$$\mathrm{Im}\{\Pi(p)\} = \frac{e^2}{12\pi}\,\mathrm{sgn}(p^0)\,\Theta(p^2-4m_\mathrm{e}^2)\left(1+\frac{2m_\mathrm{e}^2}{p^2}\right)\sqrt{1-\frac{4m_\mathrm{e}^2}{p^2}}\,. \qquad (10.158)$$

To create pairs in lowest order perturbation theory, an external potential A_μ must contain frequencies, or energy quanta, of a least $|p^0| = 2m_\mathrm{e}$, in complete agreement with the intuitive understanding that the creation of two particles with mass m_e each must take an energy of a least $2m_\mathrm{e}$ from the field. In higher order perturbation theory, the total energy necessary to produce a pair can be divided into several smaller quanta; for an Nth order process the energy threshold becomes $|p^0| = 2m_\mathrm{e}/N$. A constant field has only Fourier components with $p^0 = 0$, hence an infinite number of perturbative interactions is required. *Chiu* and *Nussinov* have, indeed, shown that the effective Lagrangian $\mathscr{L}(E)$ is obtained in the limit of very high-order perturbation theory for almost constant external fields [Ch 79].

10.6 Quantum Electrodynamics in Strong Macroscopic Fields

The situation encountered in the previous section is closely analogous to that discussed in Sect. 10.3, describing how time-dependent supercritical fields lead to (induced) pair creation which goes over into the spontaneous creation mechanism for strictly time-independent supercritical fields, Fig. 10.12. The spatially constant electric field is always supercritical, no matter how weak, but (10.158) states that the field strength must be of the order $E = \pi m_\mathrm{e}^2/e \sim 4\times10^{16}$ V/cm or larger to produce a significant effect.

Macroscopic electric fields of such strength are not known to exist anywhere in nature. If they could be created, they would dissipate very fast by attracting charged particles from their neighbourhood. This is different for magnetic fields, because there are no magnetic monopole charges (or they are extremely rare). Very strong macroscopic, permanent magnetic fields exist on the surface of neutron stars. From the observation of Larmor transitions of 58 keV in the x-ray spectrum of Hercules X-1, magnetic field strengths of the order of 5×10^8 T (5×10^{12} G) have been inferred [Tr 78], Fig. 10.13.

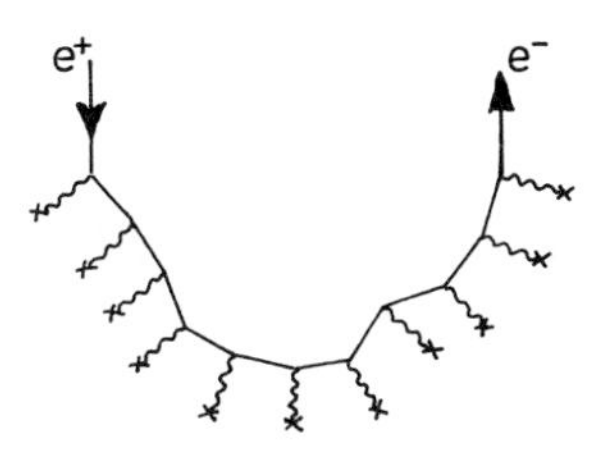

Fig. 10.12. In a very weakly time-dependent electric field pair creation through Feynman diagrams is of very high order. In the limit of vanishing time dependence the non-polynomial expression (10.155) is obtained

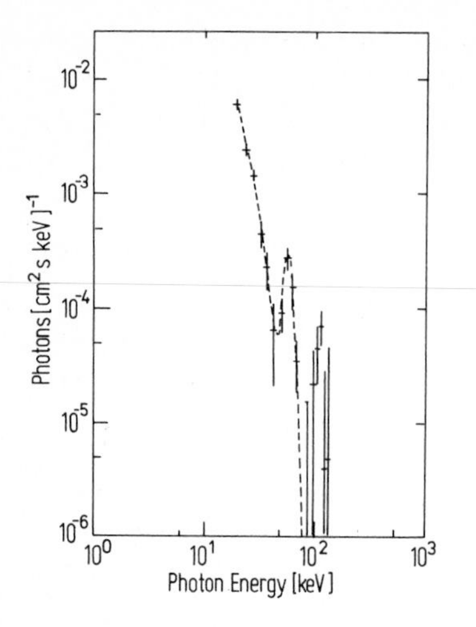

Fig. 10.13. Measurement showing the spectrum of the Hercules X-1 pulses. The peak at 58 keV is by the Larmor transition of electrons in a magnetic field of 5×10^8 T. Indications of the second transition at 115 keV are also seen [Tr 78]

Strong electric and magnetic fields in the laboratory can be produced by laser beams. Typically attainable field strengths are of the order of $E \approx 10^{11}\,\mathrm{Vm}^{-1}$ and $B \approx 10^3$T [La 79a]. As laser beams are essentially plane electromagnetic waves, the electric and magnetic field strength vectors are perpendicular to each other, $\boldsymbol{E} \cdot \boldsymbol{B} = 0$, and both are normal to the vector $\boldsymbol{k}$. Thus the second invariant $\mathscr{G} = \boldsymbol{E} \cdot \boldsymbol{B} = 0$, and the simplified forms (10.152 and 155) of the effective Lagrangian apply. However, because of the spatial and temporal variability of the electromagnetic fields in a laser wave,

$$E(x), B(x) \sim \sin kx\,; \qquad E(t), B(t) \sim \sin \omega t\,,$$

the question arises whether the approximation of a constant field is valid here. A detailed investigation shows [Ma 77c, Mo 79, Ma 79c] that the lowest order correction term to the effective Lagrangian is of the form

$$\Delta \mathscr{L}^{(2)} = \frac{\alpha}{15\pi m_\mathrm{e}^2} \Box \mathscr{F} + \mathrm{terms}(\alpha^2)\,. \tag{10.159a}$$

This term must be compared to the first correction term in the Heisenberg-Euler Lagrangian (10.155)

$$\mathscr{L}^{(4)} = -\frac{8\alpha^2}{45 m_\mathrm{e}^2} \mathscr{F}^2\,. \tag{10.159b}$$

Inserting $\mathscr{F} = (B^2 - E^2)/2$ from (10.116), the requirement that $|\Delta \mathscr{L}^{(2)}|$ be much smaller than $|\mathscr{L}^{(4)}|$ yields the conditions

$$\begin{aligned} |\dot{E}|,\ |\nabla E| &\ll \left(\frac{eE}{m_\mathrm{e}^2}\right) E m_\mathrm{e}\,, \\ |\dot{B}|,\ |\nabla B| &\ll \left(\frac{eB}{m_\mathrm{e}^2}\right) B m_\mathrm{e}\,, \end{aligned} \tag{10.160a}$$

which must be satisfied so that the virtual vacuum polarization effects are correctly described by the Heisenberg-Euler Lagrangian. For the imaginary part of the Lagrangian, i.e. for real pair production by the external field, the smoothness condition is slightly different [Sp 82]:

$$|\dot{E}|,\ |\nabla E| \ll \left(\frac{eE}{m_e^2}\right)^{3/2} E m_e\,. \tag{10.160b}$$

Inserting the present experimental limit $E \sim 10^{11}\,\mathrm{Vm}^{-1}$, condition (10.160a) yields that $|\dot{E}/E|$, $|\nabla E/E|$ should be much smaller than $(10^{-4}\mathrm{m})^{-1}$. For typical laser wavelengths (10^{-6}m), the condition is obviously not satisfied. However, because of the non-linear dependence of the rhs on the field strength, the condition may be satisfied in the future, if more powerful lasers are developed. Such a development will be needed anyway for testing QED, because the expansion parameter in the Heisenberg-Euler Lagrangian eE/m_e^2 becomes of order unity only when

$$E \approx E_0 = \frac{m_e^2}{e} \approx 10^{18}\,\mathrm{Vm}^{-1}\,, \tag{10.161}$$

i.e. at field strength 10^7 times higher than attainable today.

Besides pair creation [Br 70b, Po 73b, Ma 77c], other interesting QED phenomena associated with the vacuum polarization, i.e. the real part of the Heisenberg-Euler Lagrangian, can occur in strong macroscopic fields. Because it is possible to derive an explicit expression for the propagator of electrons and positrons in a constant magnetic field by essentially the same methods as discussed in Sect. 10.5 [Ts 74], QED processes in strong magnetic fields can be studied non-perturbatively. Thus the propagation of photons through the polarized vacuum has been investigated by *Brezin* and *Itzykson* [Br 71], *Bialynicka-Birula* and *Bialynicki-Birula* [Bi 70], *Adler* et al. [Ad 70], *Papanyan* and *Ritus* [Pa 72] and others. They found that the propagation of an electromagnetic wave in constant external field depends on the direction of propagation and on the polarization of the wave. The index of refraction of the vacuum is not one (except for propagation strictly parallel to the external field) and is direction dependent, so the vacuum has all the properties of a birefringent medium. Thus, in principle, rotation of the polarization plane can be observed (Faraday effect). Attempts to measure this effect experimentally in strong magnetic fields of 10^5G are underway [Ia 79]. Another interesting effect, also known from solid state physics, is photon splitting. This process is described by the Feynman diagrams

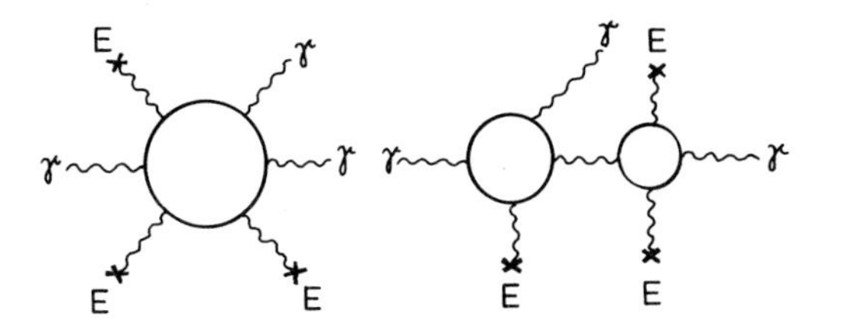

Fig. 10.14. Feynman diagrams describing photon splitting in a strong electromagnetic field

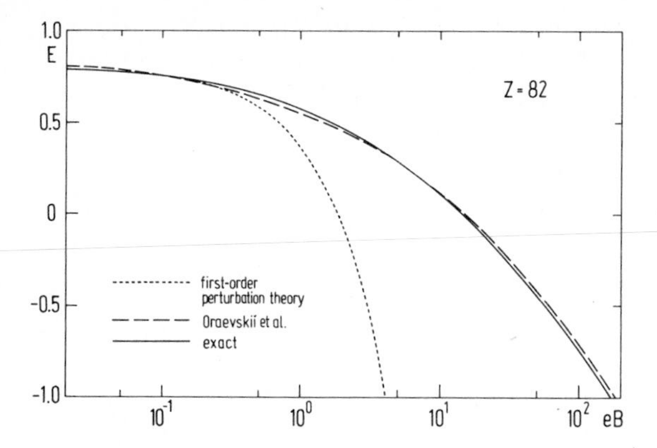

Fig. 10.15. Energy of a K-shell electron in Pb as function of the surrounding magnetic field strength. The energy is measured in the magnetic field in units of 4.4×10^9 T. The curve labelled exact is after *Schlüter* et al. [Schl 84b]

in Fig. 10.14. Unfortunately, it turns out that in lowest order the splitting probability depends on the sixth power of the external field, and is therefore very strongly suppressed for laboratory fields $E, B \ll m_e^2/e$, (10.161). In the magnetosphere of neutron stars, however, these phenomena may play an important role in the dynamics of the electron – positron plasma created by the high temperatures existing there [He 81b].

Another interesting frequently discussed question is whether a strong magnetic field can lead to spontaneous pair creation. The absence of an imaginary part in the Heisenberg-Euler effective Lagrangian (10.152) for a pure magnetic field shows that the vacuum is stable in a constant magnetic field. On the other hand, when inhomogeneous magnetic fields are considered, the vacuum may in principle become unstable, although only under conditions probably not satisfied anywhere in nature [Ac 79]. Superficial considerations indicate that the vacuum may become unstable at about 10^{13}T, when the interaction of the anomalous magnetic moment $\alpha(e\hbar/m_e c)/2\pi$ of the electron with the magnetic field is taken into account [O'Co 68, Ch 68]. This reasoning is not justified, however, because the anomalous magnetic moment of the electron itself is modified in the presence of a strong magnetic field [De 53, Ne 54, Ja 69, Ne 71, Ha 83], leading to an overall logarithmic dependence on the magnetic field strength.

Even for a homogeneous magnetic field $\boldsymbol{B}$, spontaneous pair production may occur in the vicinity of a nucleus. This is interesting, because heavy nuclei are synthesized in the supernova explosion that leads to the formation of a neutron star [Tr 75]. Heavy nuclei are, therefore probably present at the surface of a neutron star. The strength of the Coulomb field of the nucleus with charge Ze, that effectively acts on a K electron, is amplified by the presence of the magnetic field. Due to the Lorentz force the transverse motion of the electron is confined to a circle, with a typical radius $r_L \approx (eB)^{-1/2}$ in the limit of very large field strength. On the average, therefore, the electron is much closer to the nucleus than in a free atom, and feels a much stronger Coulomb field. In other words, the magnetic field acts as an agent to confine the electronic motion, and the Coulomb potential provides the strong binding necessary to produce a supercritical situation. Precise calculations by *Oraevskii* et al. [Or 77] and *Schlüter* et al. [Schl 84b] were made of the energy of a K-shell electron in Pb ($Z = 82$) as function of the magnetic field strength (eB). As shown in Fig. 10.15, the K shell becomes supercritically bound when the magnetic field strength reaches 10^{12} T. Due to the magnetic spin interaction, only the magnetic substate of the K shell with spin parallel to the magnetic field becomes supercritical. The physics of atoms in strong magnetic fields involves many other interesting phenomena which lie, however, outside the scope of this book. We refer the interested reader to [He 81b].

10.7 Klein's Paradox Revisited

We shall now discuss Klein's paradox, which was already studied in the context of Dirac's hole theory in Chap. 5, by means of the in/out formalism. As the potential in Klein's original version is a time-independent potential step, the in and out basis of wave functions is not distinguished by temporal boundary conditions (stationary states in the past or future), but instead by spatial boundary conditions corresponding to in-going or out-going particles. Remember that there are right- and left-moving plane wave solutions to the left (region I) and to the right (region II) of the potential barrier. Properly normalized, these solutions are:

$$z<0: \quad \psi_{\mathrm{I}}(z) = \sqrt{k_1}\begin{pmatrix}1\\0\\1/k_1\\0\end{pmatrix}\mathrm{e}^{\mathrm{i}p_1 z} \tag{10.162a}$$

$$z>0: \quad \psi_{\mathrm{II}}(z) = \sqrt{k_2}\begin{pmatrix}1\\0\\-1/k_2\\0\end{pmatrix}\mathrm{e}^{\mathrm{i}p_2 z} \tag{10.162b}$$

(10.162a,b): right moving

$$z<0: \quad \psi_{\mathrm{I}}^{\mathrm{refl}}(z) = \sqrt{k_1}\begin{pmatrix}1\\0\\-1/k_1\\0\end{pmatrix}\mathrm{e}^{\mathrm{i}p_1 z} \tag{10.162c}$$

$$z>0: \quad \psi_{\mathrm{II}}^{\mathrm{refl}}(z) = \sqrt{k_2}\begin{pmatrix}1\\0\\1/k_2\\0\end{pmatrix}\mathrm{e}^{\mathrm{i}p_2 z} \tag{10.162d}$$

(10.162c,d): left moving

where

$$p_1 = \sqrt{E^2 - m_{\mathrm{e}}^2}\,, \qquad k_1 = \frac{E+m_{\mathrm{e}}}{p_1}\,, \tag{10.163a}$$

$$p_2 = \sqrt{(E-V_0)^2 - m_{\mathrm{e}}^2}\,, \qquad k_2 = \frac{|E-V_0+m_{\mathrm{e}}|}{p_2}\,. \tag{10.163b}$$

Assume that $V_0 > 2m_e$ and $m_e < E < V_0 - m_e$, so that we are in the region of the upper continuum on the left and in the region of the lower continuum on the right of the barriers.

As in Sect. 5.2, the wave functions (10.162) can be combined to give solutions of the Dirac equation which are continuous everywhere (especially at $z=0$) and which are solely left- or right-moving solutions in one of the two regions I or II. Let us show, once more, explicitly how this works for a solely right-moving, i.e. ingoing, solution in region I. The most general ansatz is:

$$\psi_{\mathrm{I}}^{(+)}(z) = \psi_{\mathrm{I}}(z)\,\Theta(-z) + [b\,\psi_{\mathrm{II}}(z) + d\,\psi_{\mathrm{II}}^{\mathrm{refl}}(z)]\,\Theta(z)\,. \tag{10.164}$$

Requiring continuity at $z=0$, two equations for the coefficients b and d arise

$$\sqrt{k_1} = \sqrt{k_2}(b+d)\,, \tag{10.165a}$$

$$\frac{1}{\sqrt{k_1}} = \frac{1}{\sqrt{k_2}}(-b+d)\,, \quad \text{or} \tag{10.165b}$$

$$\sqrt{\gamma} = b+d\,, \quad \frac{1}{\sqrt{\gamma}} = -b+d\,, \tag{10.166}$$

where $\gamma = k_1/k_2$ was defined in (5.36). Adding and subtracting, then

$$b = \frac{\gamma-1}{2\sqrt{\gamma}}\,, \quad d = \frac{\gamma+1}{2\sqrt{\gamma}}\,. \tag{10.167}$$

The solution (10.164) is therefore

$$\psi_{\mathrm{I}}^{(+)}(z) = N\psi_{\mathrm{I}}(z)\,\Theta(-z) + \frac{N}{2\sqrt{\gamma}}[(\gamma-1)\,\psi_{\mathrm{II}}(z) + (\gamma+1)\,\psi_{\mathrm{II}}^{\mathrm{refl}}(z)]\,\Theta(z)\,, \tag{10.168}$$

where we have allowed for a normalization factor N. Let us check the orthogonality of the function $\psi_{\mathrm{I},E}(z)$ for different energies E, E' and calculate the factor N. Then

$$\begin{aligned}\langle\psi_{\mathrm{I},E'}^{(+)}|\psi_{\mathrm{I},E}^{(+)}\rangle &= \int_{-\infty}^{\infty} dz\,\psi_{\mathrm{I},E'}^{(+)}(z)^\dagger\psi_{\mathrm{I},E}^{(+)}(z)\\ &= NN'\int_{-\infty}^{0} dz\,\psi_{\mathrm{I},E'}(z)^\dagger\psi_{\mathrm{I},E}(z)\\ &\quad+\frac{NN'}{4\sqrt{\gamma\gamma'}}\int_0^\infty dz\,[(\gamma'-1)\,\psi_{\mathrm{II},E'}(z)+(\gamma'+1)\,\psi_{\mathrm{II},E'}^{\mathrm{refl}}(z)]^\dagger\\ &\quad\times[(\gamma-1)\,\psi_{\mathrm{II},E}(z)+(\gamma+1)\,\psi_{\mathrm{II},E}^{\mathrm{refl}}(z)]\\ &= NN'\frac{k_1'k_1+1}{\sqrt{k_1'k_1}}\int_{-\infty}^{0} dz\,\mathrm{e}^{\mathrm{i}(p_1-p_1')z}\end{aligned}$$

$$+\frac{NN'}{4\sqrt{\gamma'\gamma k_2' k_2}}\,[(\gamma'-1)(\gamma-1)(k_2' k_2+1)\int_0^\infty dz\,\mathrm{e}^{\mathrm{i}(p_2-p_2')z}$$

$$+(\gamma'+1)(\gamma+1)(k_2' k_2+1)\int_0^\infty dz\,\mathrm{e}^{-\mathrm{i}(p_2-p_2')z}$$

$$+(\gamma'-1)(\gamma+1)(k_2' k_2-1)\int_0^\infty dz\,\mathrm{e}^{-\mathrm{i}(p_2+p_2')z}$$

$$\left.+(\gamma'+1)(\gamma-1)(k_2' k_2-1)\int_0^\infty dz\,\mathrm{e}^{\mathrm{i}(p_2+p_2')z}\right]\,. \tag{10.169}$$

The integrals are really distributions and can be calculated by adding an infinitesimal imaginary part in the exponent

$$\int_0^\infty dz\,\mathrm{e}^{\mathrm{i}pz}\equiv\int_0^\infty dz\,\mathrm{e}^{\mathrm{i}(p+\mathrm{i}\varepsilon)z}=\frac{1}{\mathrm{i}(p+\mathrm{i}\varepsilon)}=\pi\delta(p)+\mathrm{i}P\left(\frac{1}{p}\right), \tag{10.170}$$

where 'P' means taking the Cauchy principal value. All contributions from the principal value (10.169) must cancel, and the delta-function part must be properly normalized. With (10.170) we obtain

$$\begin{aligned}\langle\psi_{1,E'}^{(+)}|\psi_{1,E}^{(+)}\rangle &= \pi\frac{NN'}{\sqrt{k_1' k_1}}\,[(k_2' k_2+1)\,\delta(p_1-p_1')\\ &\quad+\tfrac{1}{2}(\gamma\gamma'+1)(k_2' k_2+1)\,\delta(p_2-p_2')]\\ &\quad+\mathrm{i}\frac{NN'}{\sqrt{k_1' k_1}}P\left[\frac{k_1' k_1+1}{p_1'-p_1}-\frac{(\gamma+\gamma')(k_2' k_2+1)}{2(p_2-p_2')}\right.\\ &\quad\left.+\frac{(\gamma-\gamma')(k_2' k_2-1)}{2(p_2+p_2')}\right]\,.\end{aligned} \tag{10.171}$$

Substituting the explicit expressions for k_i, k_i', γ, γ', p_i, p_i' in terms of E, E', a rather lengthy but trivial calculation proves that the terms under the principal value sign cancel identically for $E \neq E'$. To evaluate the delta-function part, we use the chain rule for the delta function:

$$\delta(p_1-p_1')=\left|\frac{\partial E}{\partial p_1}\right|\delta(E-E')=\frac{p_1}{E}\,\delta(E-E')\,, \tag{10.172a}$$

$$\delta(p_2-p_2')=\frac{p_2}{|E-V_0|}\,\delta(E-E')\,. \tag{10.172b}$$

Then with (10.163)

$$\langle \psi_{\mathrm{I},E'}^{(+)} | \psi_{\mathrm{I},E}^{(+)} \rangle = \pi \frac{N^2}{k_1} \left[(k_1^2+1) \frac{p_1}{E} + \frac{1}{2}(\gamma^2+1)(k_2^2+1) \frac{p_2}{|E-V_0|} \right] \delta(E-E')$$

$$= \pi \frac{N^2}{k_1} [2k_1 + (\gamma^2+1)k_2] \, \delta(E-E')$$

$$= \pi N^2 \frac{(\gamma^2+1)}{\gamma} \delta(E-E') \,. \tag{10.173}$$

The correct normalization is obviously obtained by choosing

$$N = \frac{1}{\gamma+1} \sqrt{\frac{\gamma}{\pi}} \,. \tag{10.174}$$

By exactly the same manipulations we can derive the other three linearly independent solutions. Solutions purely right (left, reflected) moving in one region are denoted as "ingoing" ("outgoing") with index (+) [(−)]. Hence

$$\psi_{\mathrm{I}}^{(+)}(z) = \sqrt{\frac{\gamma}{\pi}} \frac{1}{\gamma+1} \psi_{\mathrm{I}}(z) \, \Theta(-z) + \frac{1}{2\sqrt{\pi}} \left[\frac{\gamma-1}{\gamma+1} \psi_{\mathrm{II}}(z) + \psi_{\mathrm{II}}^{\mathrm{refl}}(z) \right] \Theta(z) \,, \tag{10.175a}$$

$$\psi_{\mathrm{II}}^{(+)}(z) = \sqrt{\frac{\gamma}{\pi}} \frac{1}{\gamma+1} \psi_{\mathrm{II}}(z) \, \Theta(z) + \frac{1}{2\sqrt{\pi}} \left[-\frac{\gamma-1}{\gamma+1} \psi_{\mathrm{I}}(z) + \psi_{\mathrm{I}}^{\mathrm{refl}}(z) \right] \Theta(-z) \,, \tag{10.175b}$$

$$\psi_{\mathrm{I}}^{(-)}(z) = \sqrt{\frac{\gamma}{\pi}} \frac{1}{\gamma+1} \psi_{\mathrm{I}}^{\mathrm{refl}}(z) \, \Theta(-z) + \frac{1}{2\sqrt{\pi}} \left[\psi_{\mathrm{II}}(z) + \frac{\gamma-1}{\gamma+1} \psi_{\mathrm{II}}^{\mathrm{refl}}(z) \right] \Theta(z) \,, \tag{10.175c}$$

$$\psi_{\mathrm{II}}^{(-)}(z) = \sqrt{\frac{\gamma}{\pi}} \frac{1}{\gamma+1} \psi_{\mathrm{II}}^{\mathrm{refl}}(z) \, \Theta(z) + \frac{1}{2\sqrt{\pi}} \left[\psi_{\mathrm{I}}(z) - \frac{\gamma-1}{\gamma+1} \psi_{\mathrm{I}}^{\mathrm{refl}}(z) \right] \Theta(-z) \,. \tag{10.175d}$$

We can now decompose the field operator $\hat{\psi}(z)$ either with the solutions representing ingoing particles $\psi_{\mathrm{I/II}}^{(+)}$ or with the outgoing solutions $\psi_{\mathrm{I/II}}^{(-)}$. As is obvious from Fig. 5.2, the solutions in region II lie in the lower continuum, i.e. they describe antiparticle states. This explains why $\psi_{\mathrm{II}}^{(+)}$, which is a right-moving wave, represents an ingoing particle with respect to the potential barrier: a *hole* in this state moves toward the barrier. The two equivalent decompositions are:

$$\hat{\psi}(z) = \sum_E [\hat{b}_E^{(\mathrm{in})} \psi_{\mathrm{I},E}^{(+)}(z) + \hat{d}_E^{(\mathrm{in})\dagger} \psi_{\mathrm{II},E}^{(+)}(z)] \,, \tag{10.176a}$$

$$\hat{\psi}(z) = \sum_E [\hat{b}_E^{(\mathrm{out})} \psi_{\mathrm{I},E}^{(-)}(z) + \hat{d}_E^{(\mathrm{out})\dagger} \psi_{\mathrm{II},E}^{(-)}(z)] \,. \tag{10.176b}$$

The in vacuum is, as usual, defined as the state where there is no ingoing particle:

$$\hat{b}_E^{(\text{in})}|0,\text{in}\rangle = \hat{d}_E^{(\text{in})}|0,\text{in}\rangle = 0\,. \tag{10.177}$$

To find the relation between the in and out operators the single-particle S-matrix elements according to (10.5) are needed:

$$\begin{aligned}
S^{11}_{E'E} &= \langle \psi^{(-)}_{\text{I},E'} | \psi^{(+)}_{\text{I},E} \rangle = \frac{\gamma}{\pi(\gamma+1)^2} \int_{-\infty}^{0} dz\, \psi^{\text{refl}}_{\text{I},E'}(z)^\dagger \psi_{\text{I},E}(z) \\
&\quad + \frac{1}{4\pi}\int_0^\infty dz \left[\frac{\gamma-1}{\gamma+1}\psi_{\text{II},E'}(z) + \psi^{\text{refl}}_{\text{II},E'}(z)\right]^\dagger \left[\psi_{\text{II},E}(z) + \frac{\gamma-1}{\gamma+1}\psi^{\text{refl}}_{\text{II},E}(z)\right] \\
&= \frac{1}{4\pi}\cdot\frac{\gamma-1}{\gamma+1}\cdot\frac{k_2^2+1}{k_2}\, 2\pi\delta(p_2-p_2') \\
&= \frac{\gamma-1}{\gamma+1}\cdot\frac{k_2^2+1}{2k_2}\cdot\frac{p}{|E-V_0|}\,\delta(E-E') = \frac{\gamma-1}{\gamma+1}\,\delta(E-E')\,,
\end{aligned} \tag{10.178a}$$

again using (10.170). As before, the terms involving a principal value cancel. The remaining overlap matrix elements are

$$\begin{aligned}
S^{12}_{E'E} &= \langle \psi^{(-)}_{\text{I},E'} | \psi^{(+)}_{\text{II},E} \rangle = \frac{\sqrt{\gamma}}{2\pi(\gamma+1)} \left\{ \int_{-\infty}^{0} dz\, \psi^{\text{refl}}_{\text{I},E'}(z)^\dagger \left[\psi^{\text{refl}}_{\text{I},E}(z) - \frac{\gamma-1}{\gamma+1}\psi_{\text{I},E}(z)\right] \right. \\
&\quad \left. + \int_0^\infty dz \left[\psi_{\text{II},E'}(z) + \frac{\gamma-1}{\gamma+1}\psi^{\text{refl}}_{\text{II},E'}(z)\right]^\dagger \psi_{\text{II},E}(z) \right\} \\
&= \frac{\sqrt{\gamma}}{2\pi(\gamma+1)}\left[\frac{k_1^2+1}{k_1}\pi\delta(p_1-p_1') + \frac{k_2^2+1}{k_2}\pi\delta(p_2-p_2')\right] \\
&= \frac{\sqrt{\gamma}}{2(\gamma+1)}\left[\frac{(k_1^2+1)p_1}{k_1 E} + \frac{(k_2^2+1)p_2}{k_2|E-V_0|}\right]\delta(E-E') \\
&= \frac{2\sqrt{\gamma}}{\gamma+1}\,\delta(E-E')\,,
\end{aligned} \tag{10.178b}$$

$$S^{21}_{E'E} = \langle \psi^{(-)}_{\text{II},E'} | \psi^{(+)}_{\text{I},E} \rangle = S^{12}_{E'E}\,, \tag{10.178c}$$

$$S^{22}_{E',E} = \langle \psi^{(-)}_{\text{II},E'} | \psi^{(+)}_{\text{II},E} \rangle = -S^{11}_{E'E}\,. \tag{10.178d}$$

One easily verifies that the matrix $\hat{S}$ is unitary:

$$\begin{aligned}
\hat{S} &= \frac{1}{\gamma+1}\begin{pmatrix} \gamma-1 & 2\sqrt{\gamma} \\ 2\sqrt{\gamma} & -(\gamma-1)\end{pmatrix} = \hat{S}^\dagger\,, \\
\hat{S}^\dagger\hat{S} &= \frac{1}{(\gamma+1)^2}\begin{pmatrix} (\gamma-1)^2+4\gamma & 0 \\ 0 & (\gamma-1)^2+4\gamma\end{pmatrix} = \mathbb{1}\,.
\end{aligned} \tag{10.179}$$

The out operators can now be related to the in operators by means of (10.5):

$$\hat{b}_E^{(\text{out})} = \int dE' \left(S_{EE'}^{11} \hat{b}_{E'}^{(\text{in})} + S_{EE'}^{12} \hat{d}_{E'}^{(\text{in})\dagger} \right)$$

$$= \frac{\gamma-1}{\gamma+1} \hat{b}_E^{(\text{in})} + \frac{2\sqrt{\gamma}}{\gamma+1} \hat{d}_E^{(\text{in})\dagger} , \tag{10.180a}$$

$$\hat{d}_E^{(\text{out})} = \int dE' \left(S_{EE'}^{21\,*} \hat{b}_{E'}^{(\text{in})\dagger} + S_{EE'}^{22\,*} \hat{d}_{E'}^{(\text{in})} \right)$$

$$= \frac{2\sqrt{\gamma}}{\gamma+1} \hat{b}_E^{(\text{in})\dagger} + \frac{\gamma-1}{\gamma+1} \hat{d}_E^{(\text{in})} . \tag{10.180b}$$

From the in vacuum definition (10.177), the number of created electrons or positrons can now be calculated:

$$N_E = \langle 0, \text{in} | \hat{b}_E^{(\text{out})\dagger} \hat{b}_E^{(\text{out})} | 0, \text{in} \rangle$$

$$= \frac{4\gamma}{(\gamma+1)^2} \langle 0, \text{in} | \hat{d}_E^{(\text{in})} \hat{d}_E^{(\text{in})\dagger} | 0, \text{in} \rangle = \frac{4\gamma}{(\gamma+1)^2} \delta(E-E) , \tag{10.181a}$$

$$\bar{N}_E = \langle 0, \text{in} | \hat{d}_E^{(\text{out})\dagger} \hat{d}_E^{(\text{out})} | 0, \text{in} \rangle = \frac{4\gamma}{(\gamma+1)^2} \delta(E-E) = N_E . \tag{10.181b}$$

The delta functions at zero argument are understood as usual [Bj 64] as time integrals

$$\text{“}\delta(0)\text{”} \equiv \lim_{T\to\infty} \lim_{\varepsilon\to 0} \int_{-T/2}^{T/2} \frac{dt}{2\pi} e^{i\varepsilon t} = \lim_{T\to\infty} \frac{T}{2\pi} . \tag{10.182}$$

Dividing (10.181) by the total time T gives the spectral emission rate of electrons and positrons per unit time by the potential barrier:

$$\frac{dN_E}{dt} = \frac{d\bar{N}_E}{dt} = \frac{4\gamma}{2\pi(\gamma+1)^2} . \tag{10.183}$$

There is a steady flow of particles from the barrier region, electrons moving left, and positrons right.

We see that Klein's paradox really is no paradox at all! The formalism of second quantization clearly shows that pair creation occurs at the potential barrier with an equal number of outgoing particles and antiparticles contained in the in vacuum.

Let us compare (10.183) with the result of the old hole theory calculation (5.42). The ratio of positron flux in region II to incoming electron flux in region I is

$$w = \frac{|j_{\text{II}}|}{|j_{\text{I}}|} = \frac{4\gamma}{(\gamma-1)^2} . \tag{10.184}$$

Here w is the *relative* probability for production of an electron – positron pair, if the probability that nothing happens is equal to one. The correctly normalized pair-creation probability is then [Ni 70]

$$\frac{w}{1+w} = \frac{4\gamma}{(\gamma-1)^2}\left(1+\frac{4\gamma}{(\gamma-1)^2}\right)^{-1} = \frac{4\gamma}{(\gamma+1)^2}, \tag{10.185}$$

in perfect agreement with (10.183).

Bibliographical Notes

The in/out formulation for pair creation in strong external fields has been discussed in the framework of the Heisenberg picture by *Reinhardt* et al. [Re 80b] and by *Kirsch* et al. [Ki 80]. An analogous discussion in the context of the interaction picture can be found in [Bi 75].

The projection method for constructing an approximate supercritical bound state and an associated orthogonal continuum was developed by *Reinhardt* et al. [Re 80b, 81b].

The decay of a supercritical K vacancy by spontaneous positron emission was treated in second quantization (in the Schrödinger picture) by *Rafelski* et al. [Ra 74].

Our discussion of the vacuum energy in a constant electromagnetic field is based on the work of *Schwinger* [Schw 51]. Presentations of the "proper time" method are also found in [Schw 73, It 80].

A review of effects and possible tests of QED in strong, macroscopic fields is contained in [Ma 77c].

Our treatment of the Klein paradox in the framework of the in/out formalism is largely based on the work of *Hansen* and *Ravndal* [Ha 81a].

11. Superheavy Quasimolecules

Lacking the existence of stable nuclei with charge $Z > 173$ there is no immediate way to test the ideas about supercritical fields in QED. However, in a collision of two very heavy ions for a short period of time the electrons will experience the combined potential of both nuclei. If the nuclear motion is not too fast a superheavy quasimolecule may be formed. As a prerequisite for dynamical calculations, in this chapter the electron states of a hypothetical stationary molecule are discussed by solving the two center Dirac equation. Special attention is devoted to the critical internuclear distance R_{cr}, where the bound levels reach the lower continuum.

11.1 Heavy-Ion Collisions: General Remarks

The discussion of the binding energy of the *K*-shell electrons showed that instability of the neutral vacuum state occurs for nuclei with more than 170 protons. As this is far beyond the charge of stable nuclei and even the transactinides, a new kind of atoms, the so-called superheavy elements, would be required before laboratory tests of the theory of strong fields could be undertaken. For a considerable time there was much hope that certain (lighter) superheavy nuclei around $Z = 114$ could be stable and be produced in the laboratory [Ni 69a, Mo 68]. *Mosel* and *Greiner* even suggested that a second island of truly superheavy ("giant") nuclei around $Z = 164$ might be quasistable [Mo 69]. However, the recent unsuccessful attempts all over the world to synthesize superheavy elements (i.e. elements around $Z = 114$) through nuclear reactions have made such hopes slim.

Nevertheless, supercritical fields can be created temporarily in a collision of very heavy atoms, such as uranium and uranium. When the two nuclei have approached each other to a distance of, say 20 fm, in a head-on or nearly head-on collision, the electric field surrounding them very much resembles that which would surround a giant nucleus with $92 + 92 = 184$ protons. In fact, the multipole expansion of the Coulomb potential (in the Coulomb gauge) for two equal point-like nuclei with charge Z separated by a distance R is (Fig. 11.1) [Ra 76b, So 79a]

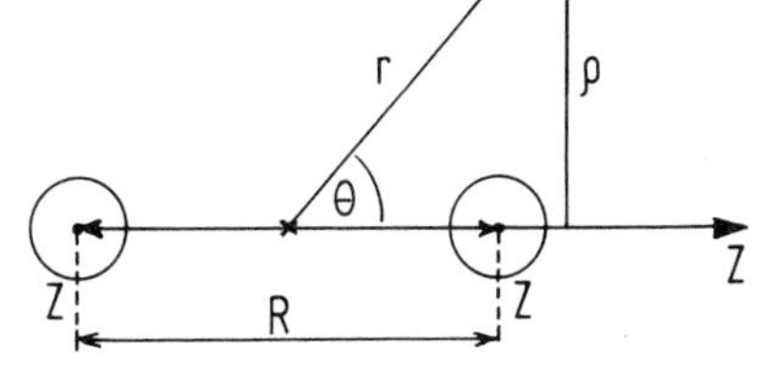

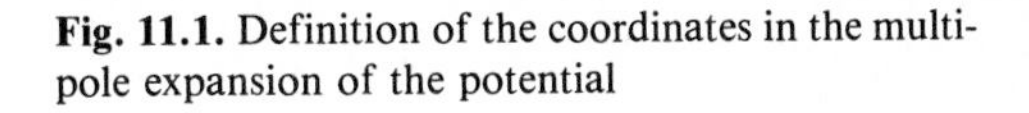
Fig. 11.1. Definition of the coordinates in the multipole expansion of the potential

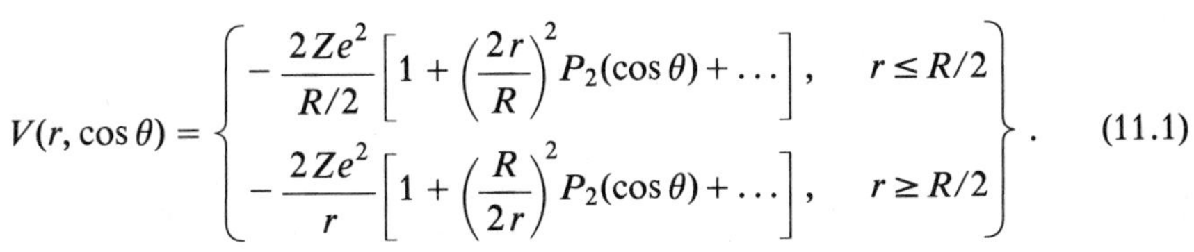

$$V(r, \cos\theta) = \begin{cases} -\dfrac{2Ze^2}{R/2}\left[1 + \left(\dfrac{2r}{R}\right)^2 P_2(\cos\theta) + \ldots\right], & r \leq R/2 \\ -\dfrac{2Ze^2}{r}\left[1 + \left(\dfrac{R}{2r}\right)^2 P_2(\cos\theta) + \ldots\right], & r \geq R/2 \end{cases}. \tag{11.1}$$

Clearly, at distances $r > R$ the potential differs little (to be precise, by less than one third) from the Coulomb potential of a nucleus with charge $(2Z)$. It is also seen that the potential difference comes from the quadrupole and higher multipole terms and falls off rapidly with increasing distance from the nuclei.

That the strong electric fields temporarily created in heavy-ion collisions could be utilized for laboratory tests of the theoretical predictions of QED of supercritical fields was proposed about 15 years ago independently by the two groups at Frankfurt [Gr 69b, Ra 71, Mü 72a – c] and Moscow [Ge 70, Ze 72]. They also argued why these transient fields should be able to produce supercritical binding, reviewed in the rest of this chapter. Chapters 12 and 13 then deal with the refined treatment of electronic phenomena in heavy-ion collisions and with the present status of experimental efforts in this field.

Three basic questions have to be addressed before concluding that the transient strong electric fields in a heavy-ion collision actually provide supercritical binding to the innermost electrons.

1) Is the collision slow enough so electrons can adjust to the increasing strength of the potential as the two nuclei approach? A first answer to this question compares the velocity of the scattering nuclei with the "orbiting" velocity of the electrons. The K-shell electrons in an uranium atom are bound by more than 100 keV. Due to the virial theorem, the binding energy in a Coulomb potential equals the average kinetic energy of the particle. Hence the ratio of the K-electron velocity v_{el} to the speed of light c in an uranium atom is estimated as $v_{\mathrm{el}}/c \sim (2|E|/m_e c^2)^{1/2} > 0.6$. For the nuclei to move at this relative speed would require a laboratory bombarding energy of approximately 200 MeV/nucleon. This is much more than one likes to have, because at projectile energies above 6 MeV/nucleon violent nuclear reactions set in that could provide a formidable background, e.g. in positron spectra. At energies up to 6 MeV/nucleon the Coulomb barrier slows down the approaching nuclei, thereby considerably reducing the relative velocity at close internuclear distances. We conclude that the nuclear motion is slower by a factor of 5 – 10 than the electronic motion in relevant collisions.

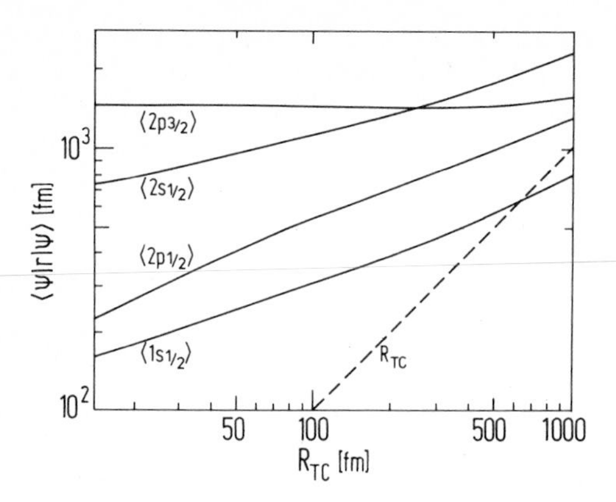

Fig. 11.2. Change of the average radii $\langle r\rangle$ of selected inner-shell wave functions in the Pb + Pb quasimolecule with varying internuclear distance. Observe that the $1s_{1/2}$ electron moves effectively around both nuclei up to $R_{TC} \gtrsim 500$ fm

2) Are the electrons moving in the region of very strong electric fields? Stated otherwise: under what circumstances will the nuclei approach so closely that the innermost electrons effectively orbit around both nuclei, seeing one giant quasinucleus? As we are primarily interested in the change of the vacuum state, we estimate the average radius of the electronic charge distribution when the binding energy reaches the threshold $E = 2m_e$. Again, we use of the virial theorem: $|E|$ must be equal to the average kinetic energy $\langle T\rangle$ of the electron, which, in turn, is related to the average squared momentum $\langle p^2\rangle$ by $\langle T\rangle = (m_e^2 + \langle p^2\rangle)^{1/2} - m_e$. From the uncertainty relation, for the ground state one has $\langle p^2\rangle\langle r^2\rangle \sim 1$.

Putting everything together

$$\langle r^2\rangle^{1/2} \sim (E^2 + 2|E|m_e)^{-1/2} = (2\sqrt{2}\,m_e)^{-1} \sim 135 \text{ fm} . \tag{11.2}$$

The validity of this estimate is shown in Fig. 11.2, where the average distance $\langle\psi|r|\psi\rangle$ of the strongly bound electrons from the centre of the Pb + Pb quasimolecule is shown as function of the nuclear separation. When the two Pb nuclei touch, the average extension of the K shell is about 150 fm.

Thus, when the nuclei are much closer together than this distance, the electrons will feel essentially the monopole part of the two-centre potential (11.1). When this is the case, the system comprising the two nuclei and the electrons attached to them is usually referred to as a *quasiatom*. A more general term is *quasimolecule*, which applies to a much longer phase of the collision, as long as the electrons have a substantial chance of being found at both nuclei. The quasimolecular states may be strongly deformed, whereas quasi-atomic states should have approximately good angular momentum, indicating that they exist in an almost spherical potential.

3) Even when the electron orbit becomes supercritical, is there enough time for the neutral vacuum state to decay? Because the dived electronic bound state is rather weakly coupled to the positron scattering states, the resonance width Γ is relatively small, typically a few keV, Chap. 2. This corresponds to a vacuum decay time of $\tau_{vac} \sim \Gamma^{-1} \sim 10^{-19}$ s, following Chaps. 6, 10. The normal collision time is much shorter, a reasonable estimate being $\tau_{coll} = R_{cr}/v_{ion}$, where R_{cr} is the internuclear distance at which the bound state becomes supercritical, and v_{ion} the

asymptotic ion velocity estimated before. From (11.2) follows that R_{cr} must be considerably smaller than 135 fm (for U+U the true value is about 30 fm, see below), hence the collision time τ_{coll} is less than 10^{-21} s. This result indicates that a positron would be spontaneously emitted in not more than one out of a hundred normal collisions of two uranium atoms. In addition, the positron distribution must be expected to be quite broad, typically $\tau_{coll}^{-1} \sim 1$ MeV, according to Sect. 10.2, making the identification of the positrons from the spontaneous vacuum decay difficult.

This may change radically when the two nuclei touch and stick together for a certain period of time under the action of the attractive nuclear force, forming a *nuclear molecule.* Such molecular configurations are well known from the scattering of light nuclei, e.g. C+C, O+O, and evidence is mounting for their presence even in collisions of the heaviest nuclei [Al 60, Sche 70, Pa 81]. If a composite nuclear system forms the nuclear sticking time T must be added to the collision time for Coulomb scattering, τ_{coll}. When T becomes comparable with the characteristic time for spontaneous positron emission τ_{vac}, the superheavy nuclear molecule has for our purposes essentially the same properties as a stable supercritical nucleus – vacancies decay by spontaneous emission of positrons of well-defined energy. How the shape of the positron line approaches the static limit with increasing T was discussed in Sects. 10.2, 3.

The qualitative picture is thus as follows. During the heavy-ion collision the deeply bound electrons adjust almost adiabatically to the momentary position of the nuclei on their scattering trajectory. When two uranium nuclei approach closer than ca. 50 fm, their electric field reaches the critical threshold. If a vacancy is brought into a temporarily supercritical bound state, it has a large chance of being emitted as a positron only if the nuclei are kept together by nuclear forces for about 10^{-19} s. These crude estimates are made quantitative in the following sections dealing with the collision dynamics. For the moment, we accept the adiabaticity assumption as a working hypothesis, but wish to emphasize one very important point: the K-shell vacancies that are absolutely necessary for the experimental observation of the vacuum decay can originate only when the collision process deviates from the strict adiabatic model. Solely because the nuclei move can we expect that sometimes an inner-shell electron is excited out of its quasi-molecular state. Therefore, there must be a delicate balance between adiabaticity and dynamics of the heavy-ion collision. That we find ourselves precisely in the favourable region must be considered a very fortunate circumstance.

11.2 The Two-Centre Dirac Equation

In the following two sections we leave aside all dynamical processes occurring during a heavy-ion – atom collision. The question asked is: if some "deus ex machina" kept two nuclei with charges Z_1 and Z_2, respectively, fixed at a distance R apart, what would the motion of the electrons in the electrostatic field of these two nuclei be? Since the potential would be static we could search for stationary states that are solutions of the *two-centre Dirac equation* [Mü 73a, 76b]

$$E\psi(\boldsymbol{r}) = H_{\mathrm{TCD}}(\boldsymbol{r})\,\psi \equiv \{c\boldsymbol{\alpha}\cdot\boldsymbol{p}+\beta m_{\mathrm{e}}c^2+V_1(|\boldsymbol{r}+\eta\boldsymbol{R}|)+V_2[\,|\boldsymbol{r}-(1-\eta)\boldsymbol{R}\,|]\}\,\psi(\boldsymbol{r})$$
$$= [c\boldsymbol{\alpha}\cdot\boldsymbol{p}+\beta m_{\mathrm{e}}c^2+V_1(r_1)+V_2(r_2)]\,\psi(\boldsymbol{r})\,, \tag{11.3}$$

where V_1 and V_2 are the potentials due to the two nuclei. Outside the nuclear radius ϱ_i $(i = 1,2)$ there are pure Coulomb potentials

$$V_i(|\boldsymbol{x}|) = -\frac{Z_i e^2}{|\boldsymbol{x}|}\,(|x| > \varrho_i)\,. \tag{11.4}$$

In (11.3) the origin of the coordinate system is conveniently chosen at the centre of mass of the two nuclei, which gives the value $M_1/(M_1+M_2)$ for the parameter η if $\boldsymbol{R}$ is taken to point from nucleus 1 to nucleus 2. We have neglected the mutual interaction between electrons in (11.3). The effect of this interaction – mainly screening the Coulomb potential – is discussed below.

Before discussing the properties of the solutions of the two-centre Dirac equation (11.3), it is useful to elaborate on the analogous non-relativistic problem, the two-centre Schrödinger equation

$$E\psi = \left(\frac{p^2}{2m} + V_1 + V_2\right)\psi\,. \tag{11.5}$$

This problem was solved by *Teller* [Te 30] and *Hylleraas* [Hy 31] after the first approximate results from *Heitler* and *London* [He 27]. The two-centre Schrödinger equation must have two constants of motion. Trivially, one of them is the projection L_z of angular momentum on the symmetry axis. A second operator commuting with the Hamiltonian was constructed by *Erickson* and *Hill* [Er 49] for point-like nuclei:

$$\Lambda = \tfrac{1}{2}(\boldsymbol{L}_1\cdot\boldsymbol{L}_2+\boldsymbol{L}_2\cdot\boldsymbol{L}_1) - m\mathrm{Re}^2\{Z_1\cos\theta_1 - Z_2\cos\theta_2\}\,,$$

where θ_i are the angles between $\boldsymbol{r}_i$ and the internuclear axis, whereas $\boldsymbol{L}_1$ and $\boldsymbol{L}_2$ are the angular momentum operators with respect to the two centres. In prolate spheroidal coordinates $\xi = (r_1+r_2)/R$, $\eta = (r_1-r_2)/R$, φ, the Schrödinger equation separates, producing two coupled eigenvalue equations which can be solved for the energy. For nuclear molecules the two-centre shell model has been developed at Frankfurt (1969 – 1973) to describe similar phenomena in nuclear fission and nuclear reactions at low energy. For a review of this active field see [Ci 81] and for the theoretical aspects, [Pa 85].

The two-centre Dirac equation is more difficult to handle since no second constant of motion exists besides the projection of the total angular momentum [Co 67]. A complete proof for the non-existence has recently been given by

Schlüter et al. [Wi 83] and the statement is supported by the numerical results (discussed below) that the relativistic two-centre states do not cross as a function of R. This would not happen if a second constant of motion existed, according to the famous non-crossing theorem of *von Neumann* and *Wigner* [Ne 29].

Let us next discuss the symmetries and common nomenclature of the relativistiv two-centre problem. Whereas the total angular momentum of the electron orbital is not conserved, its projection J_z along the axis that joins the nuclei commutes with the Dirac operator H_{TCD} since the azimuthal symmetry around this axis has been maintained. Therefore, letting the z coordinate pass through the nuclei as shown in Fig. 11.1,

$$J_z = L_z + \frac{\sigma_z}{2}\hbar = -\mathrm{i}\hbar\frac{\partial}{\partial\varphi} + \frac{\sigma_z}{2}\hbar\,, \tag{11.6}$$

an operator that commutes with H_{TCD}. Every electronic orbital is characterized therefore by a unique eigenvalue μ

$$J_z\psi_\mu = \mu\hbar\psi_\mu\,. \tag{11.7}$$

Further, μ can assume the values

$$\mu = \pm 1/2,\ \pm 3/2,\ \pm 5/2, \dots\,.$$

The solutions with $\pm|\mu|$ are degenerate in energy since there is no difference in rotation around the internuclear axis. The value of μ is commonly denoted by $\sigma, \pi, \delta, \dots$ for $J_z = \pm 1/2,\ \pm 3/2,\ \pm 5/2, \dots$.

When the two-centre potential V is, in addition, parity invariant, i.e. if both nuclei are identical

$$V(\varrho, z) = V(\varrho, -z)\,, \tag{11.8}$$

(ϱ is the cylindrical radial coordinate as shown in Fig. 11.1) then the eigensolutions have good parity and are referred to either as "gerade" (even parity) or "ungerade" (odd parity) states, abbreviated by "g" and "u", respectively.

To specify a complete set of solutions of (11.3), the two-centre states must be characterized by an additional label, usually corresponding to the quantum numbers of the asymptotic atomic states continuously reached in the united atom limit, e.g. a full description of the bound electron states in a heavy-ion collision would be $1s_{1/2}\sigma_g$, $2p_{3/3}\pi_g$, etc. This procedure is unique for the discrete spectrum, i.e. for the bound states, but not for the molecular continuum states, because there an infinite number of degenerate states exists for every energy.

We should, however, point out that the united atom label (e.g. $1s_{1/2}$) is not very useful from a physical point of view, except in a few cases. This is illustrated in Fig. 11.3 where a selected region of the two-centre diagram (binding energies versus internuclear distance) is shown for the system Pb + Cm. Due to the forbidden crossing of states with the same symmetry, the properties of the wave function often change abruptly along a continuous line. For example, at the avoided

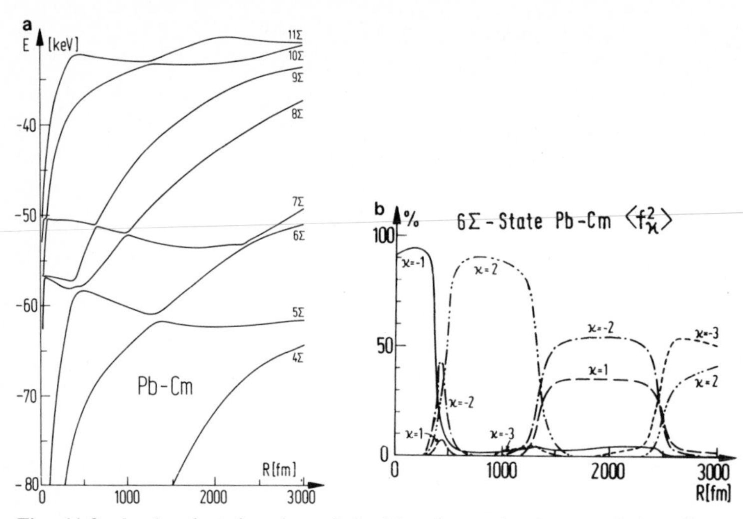

Fig. 11.3a, b. A selected region of the Pb+Cm molecular correlation diagram. At the avoided crossings seen in (**a**), the nature of the adiabatic wave functions changes abruptly, shown in (**b**) for the 6Σ state as a function of R

crossing at 1300 fm the continuous state labelled 6Σ changes from an approximate $d_{3/2}$ ($\varkappa=2$) configuration to an approximate $p_{1/2}-p_{3/2}$ ($\varkappa=+1,-2$) symmetry (Fig. 11.3). The opposite is true for the other state, labelled 5Σ, that participates in the avoided crossing. One therefore labels the states often only by consecutive numbers, $1\Sigma_{(g)}$, $2\Sigma_{(g)}$, $3\Sigma_{(g)}$, etc. The lowest states of each exact symmetry (of J_z and, possibly, parity) do not participate in avoided crossings, so that for these states the united atom nomenclature is commonly used and meaningful ($1s_{1/2}\sigma$, $2p_{3/2}\pi$, etc.).

The two-centre Dirac equation (11.3) was first solved by *Müller* et al. [Mü 73a, 76b, Fr 76b] by diagonalization in a basis of functions of the prolate spheroidal coordinates ξ, η and φ:

$$\psi_{nl\mu}(\xi,\eta,\varphi)=(\xi^2-1)^{\frac{1}{2}(|\mu|\pm\frac{1}{2})}\exp\left(\frac{\xi-1}{2a}\right)L_n^{|\mu|\pm\frac{1}{2}}\left(\frac{\xi-1}{a}\right)P_l(\eta)\,\mathrm{e}^{\mathrm{i}(\mu\pm\frac{1}{2})\varphi}\,. \tag{11.9}$$

Due to the presence of the negative energy continuum states, however, this approach, through correctly predicting the critical distance R_{cr}, was not successful for good convergence of the wave functions [Ra 76c]. Similar attempts by *Marinov* et al. encountered the same difficulties and even gave much too large values for R_{cr} [Ma 74, 75a, b].

Satisfactory results were obtained later by *Rafelski* and *Müller* [Ra 76b] with a numerical integration technique, described in the following. The dependence of the wave function on the angular variables is discretized by passing to the conjugate variable, i.e. angular momentum. This is achieved by expanding the wave function $\psi(r,\theta,\varphi)$ into an infinite series of multipole components

$$\psi_\mu(r,\theta,\varphi) = \sum_\varkappa \psi_{\varkappa\mu} = \sum_{\varkappa=\pm 1}^{\pm\infty} \begin{pmatrix} g_\varkappa(r) & \chi_\varkappa^\mu(\theta,\varphi) \\ \mathrm{i} f_\varkappa(r) & \chi_{-\varkappa}^\mu(\theta,\varphi) \end{pmatrix}. \tag{11.10}$$

Here $\chi_\varkappa^\mu$ are the spinor spherical harmonics, Sect. 3.3. We now also introduce a multipole expansion of the two-centre potential

$$V_1(r_1) + V_2(r_2) = \sum_{l=0}^{\infty} V_l(r,R) P_l(\cos\theta) \tag{11.11}$$

and substitute (11.10, 11) into the two-centre Dirac equation (11.3). Because the spinor spherical harmonics $\chi_\varkappa^\mu$ form a complete basis of two-component angular functions, we derive an equivalent set of coupled differential equations for the radial functions $g_\varkappa(r), f_\varkappa(r)$ by projecting with the $\chi_\varkappa^\mu$:

$$\begin{aligned} \frac{d}{dr} f_\varkappa(r) &= \frac{\varkappa-1}{r} f_\varkappa(r) - (E-m_e)\, g_\varkappa(r) + \sum_{\lambda l} g_\lambda(r)\, V_l(r,R) A_{\varkappa l\lambda} \\ \frac{d}{dr} g_\varkappa(r) &= (E+m_e) f_\varkappa(r) - \frac{\varkappa+1}{r} g_\varkappa(r) - \sum_{\lambda l} f_\lambda(r)\, V_l(r,R) A_{-\varkappa l-\lambda}\,. \end{aligned} \tag{11.12}$$

The coefficients $A_{\varkappa l\lambda} = \int d\Omega\, \chi_\varkappa^\mu(\Omega)^\dagger P_l(\cos\theta)\, \chi_\lambda^\mu(\Omega)$ can be evaluated by angular momentum algebra. After the sum in (11.11) has been truncated at a sufficiently large angular momentum $j_{\max}$, the $2(2j_{\max}+1)$ coupled differential equations are integrated. The energy eigenvalue E is determined by an iteration to make the wave function vanish at a large distance from the two Coulomb centres.

This highly accurate numerical integration method is also convenient since it easily allows the use of modified potentials (11.11) which, for instance, may contain electron screening effects. The influence of finite nuclear radii is readily calculated.

Calculations using this method have been systematically carried out on a broad scale during the last few years by *Betz* et al. [Be 76a, So 79a, Be 80b]. As a prototype we discuss the diagram for the uranium – uranium (U + U) system in some detail (Fig. 11.4). As is obvious from the diagram, the various states can be classified in two types, the first made up by those states with symmetry $s_{1/2}$ or $p_{1/2}$ ($\varkappa = \pm 1$) in the united atom limit. These states continuously gain binding energy when the internuclear separation decreases below several hundred fermis. This is typical for all superheavy quasimolecules ($Z_1 + Z_2 > 137$) and is caused by the "collapse to the centre" for $|\varkappa| < Z\alpha$, making the wave function very sensitive to the extension of the nuclear charge distribution.

The states with $|\varkappa| > Z\alpha$ approach a constant value of the binding energy at rather large internuclear separation (300 fm). This behaviour is known as the "runway" effect. Throughout this region the wave functions very closely resemble those of the united atom ($Z_1 + Z_2$). The difference between the solid and the dashed-dotted lines in Fig. 11.4 is caused by the finite extension of the individual nuclei. As may be expected, this effect considerably lowers the binding

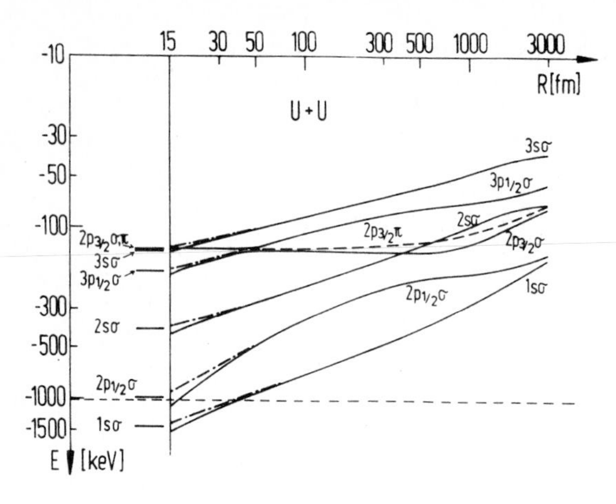

Fig. 11.4. The U+U correlation diagram ($Z_1 = Z_2 = 92$). (−·−) include the effect of the finite nuclear radius, (——) for point nuclei [So 79a, Be 80b]

energy for distances below several times the nuclear radii ($R \lesssim 50$ fm). It is felt only for the $s_{1/2}$ and $p_{1/2}$ states.

It is possible to show [Mü 76b] that the angular symmetry of the molecular states is profoundly changed at a typical nuclear separation of 500 fm. For larger distances the quadrupole part of the two-centre potential $V_2(r)$ becomes stronger than the spin-orbit force, changing the angular momentum coupling, rearranging the states of approximately good total angular momentum J into states of approximately good orbital angular momentum L and spin parallel or antiparallel to the molecular axis. For very large nuclear separations another recoupling occurs, resulting in (atomic) states with good angular momentum J_1, J_2 with respect to the individual nuclei.

We finally discuss an asymmetric system, Pb+Cm ($Z_1 = 82$, $Z_2 = 96$). A superficial look at Fig. 11.5 shows that the details are more complex due to the loss of parity symmetry. The behaviour of the strongest bound states, however, is practically unaffected by this; it is dominated by the continuous gain in binding energy upon approach of the nuclei. It is quite illuminating to follow the density distribution of some of the wave functions as a function of variable internuclear separation. The $1s_{1/2}\sigma$ wave function (Fig. 11.6), e.g., starts out as the K shell of the isolated Cm atom and becomes strongly polarized as the Pb nucleus approaches. For distances below ca. 500 fm, however, the wave function very closely resembles the K shell of the united atom, $Z = 178$. This would also happen for a lighter system ($Z_1 + Z_2 < 137$), but then the united atom wave function would not behave characteristically for a superheavy system, namely the radial extension of the density distribution shrinks further with decreasing distance between the nuclei (the "fall to the centre", compare the configurations at $R = 18$ fm and $R = 100$ fm). Similar behaviour is found for the $2p_{1/2}\sigma$ and $2s_{1/2}\sigma$ states, with a rather complicated region of intermediate distances, and a simple shape of the density distribution for distances below $R = 100$ fm.

A new method for solving the two-centre Dirac equation has recently been developed by *Wietschorke* et al. [Wi 83]. This method is similar in spirit to that employed by *Rafelski* and *Müller* [Ra 76b], i.e. the dependence of the wave

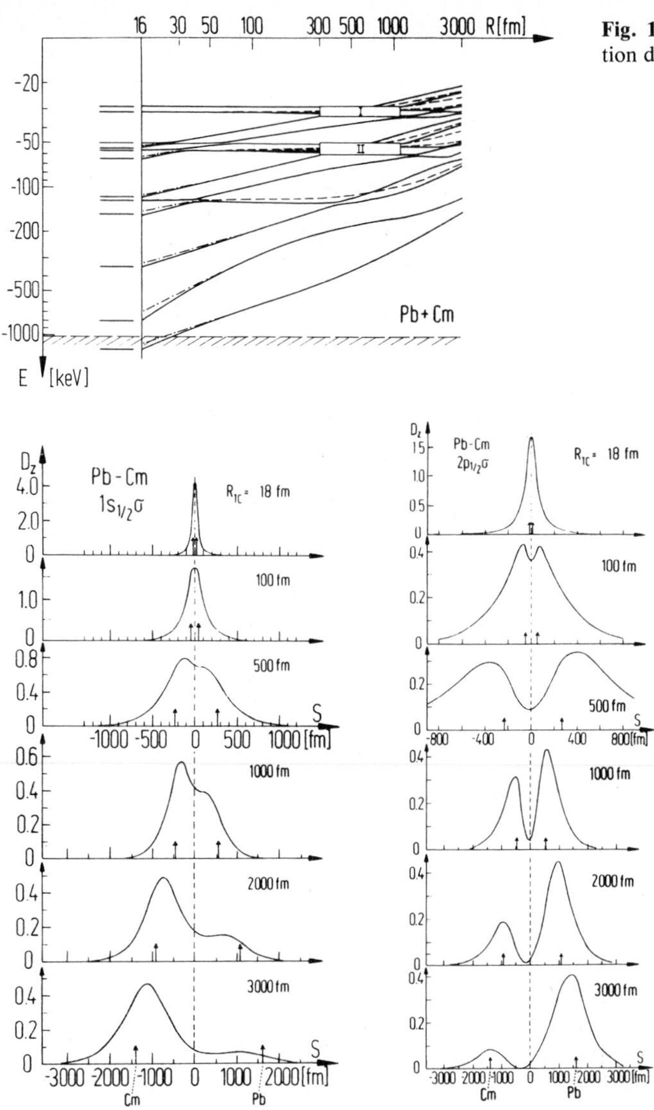

Fig. 11.5. The Pb + Cm correlation diagram ($Z_1 = 82$, $Z_2 = 96$)

Fig. 11.6. Evolution of the density distribution of the $1s_{1/2}$ and $2p_{1/2}$ wave functions in the Pb + Cm system as a function of nuclear separation R

function on the angular variable is discretized by expansion on a set of angular functions, and the remaining coupled radial equations are solved by numerical integration. However, *Wietschorke* et al. use the finite element technique with spline functions for the angular variable, which allows them so treat the Dirac equation also in other orthogonal coordinate systems (prolate spheroidal coordinates, Cassini coordinates, etc.). This method is therefore much more flexible, especially in the limit of large nuclear separation. Their method has also recently given numerical solutions for the two-centre Dirac continuum states for the first time [Wi 84].

11.3 The Critical Distance R_{cr}

The internuclear distance at which the binding energy of the lowest quasi-molecular state (the $1s\sigma$ state) reaches $2m_ec^2$ is of greatest immediate concern to us. Only when this *critical distance* R_{cr} exists and when the collision is sufficiently energetic to let the nuclei approach closer than R_{cr} can spontaneous pair creation occur: for $R < R_{cr}$ the quasi-molecular $1s\sigma$ state is supercritical. (For very large values of $Z_1 + Z_2$ the $2p_{1/2}\sigma$ state may also become supercritical.) This explains the importance of precisely determining R_{cr} as a function of the nuclear charges Z_1 and Z_2.

Figure 11.4 shows that the critical distance in the U + U system is of the order of 30 fm, agreeing with our estimate in Sect. 11.1 ($R_{cr} \ll 135$ fm). The precise value, however, depends on a number of assumptions. For example, for two point-like uranium nuclei $R_{cr} = 37 \pm 1$ fm [Mü 73a, Li 77] [1]. Inclusion of finite nuclear size reduces the critical distance to 35 fm, meaning that for such small internuclear separation the quadrupole and higher multipole contributions to the two-centre potential are unimportant (they affect R_{cr} by only 2 fm). This makes it possible to include also electron screening effects in the approximation by taking only the spherically symmetric part of the potential (the so-called monopole approximation) [Be 76b]. Assuming identical nuclei with radius R_N, the monopole part of the two-centre potential is ($Z_1 = Z_2 = Z/2$)

$$V_0(r) = \begin{cases} -Ze^2/r & \left(r > \frac{1}{2}R + R_N\right) \\ -\frac{2Ze^2}{R_N^3}\left[\frac{1}{16r}\left(\frac{1}{2}R - R_N\right)^3\left(\frac{1}{2}R + 3R_N\right) - \frac{1}{4}\left(\frac{1}{2}R + R_N\right)^2 \right. & \\ \quad \times\left(\frac{1}{2}R - 2R_N\right) + \frac{3}{8}r\left(\frac{1}{4}R^2 - R_N^2\right) & \\ \quad \left. -\frac{1}{8}Rr^2 + \frac{1}{16}r^3\right] & \left(\frac{1}{2}R + R_N > r > \frac{1}{2}R - R_N\right) \\ -\frac{2Ze^2}{R} & \left(r < \frac{1}{2}R - R_N\right). \end{cases} \tag{11.13}$$

[1] *Popov* [Po 73a] estimated $R_{cr} \approx 45$ fm for the double uranium system. Later, *Marinov* et al. [Ma 74, 75a, b] claimed that improved calculations gave a critical distance of about 51 fm, in contrast to the results of *Müller* et al. [Mü 73a, Ra 76c], who predicted $R_{cr} = 36.8$ fm. *Lisin* et al. [Li 77] finally obtained $R_{cr} \approx 38.5$ fm, which is in reasonable agreement considering numerical uncertainties.

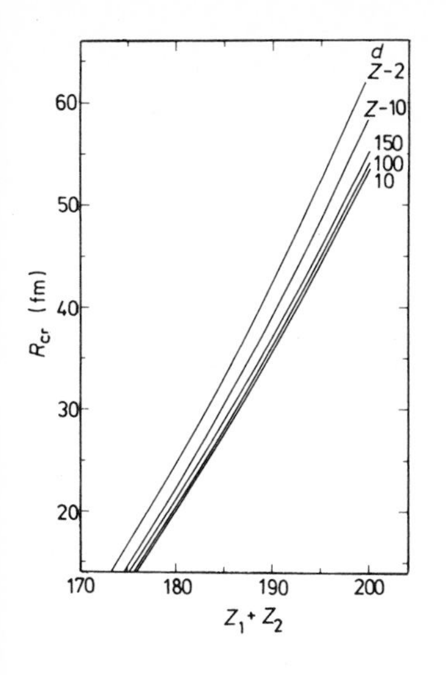

Fig. 11.7. Dependence of the critical distance R_{cr} as function of combined nuclear charge (Z_1+Z_2) in the monopole approximation. The different lines correspond to various degrees of ionization of the quasiatom [Wi 79]

Modifications of this potential due to screening have been considered within two models: the Thomas-Fermi and the Hartree-Fock methods.

The Thomas-Fermi method (density functional method) for molecular configurations has been investigated in great detail by *Gross* and *Dreizler*, who also attempted to include relativistic effects [Gr 79b, 83d]. Taking into account the influence of 30 electrons, the Thomas-Fermi method gives a critical distance of 30 fm [deR 81]. However, for such strong potentials the (non-relativistic) Thomas-Fermi method is known to underestimate the influence of the innermost electrons considerably. This is corroborated by Hartree-Fock-Slater calculations by *Wietschorke* et al. [Wi 79], who find a critical distance of 26 fm when all 184 electrons are present. However, in a violent heavy-ion collision a large number of electrons from the inner shells are also ionized, considerably reducing the screening effect. Taking this into account by allowing for various degrees of ionization, the results of *Wietschorke* and *Soff* predict a possible range of 26 – 31 fm for the critical distance in the U + U system. About 2 fm should be added to these values to account for the influence of the higher multipole terms of the two-centre potential. Allowing for further corrections due to vacuum polarization, self-energy, etc., $R_{cr}(\mathrm{U}+\mathrm{U}) = 30 \pm 2$ fm seems reasonable.

The variation of R_{cr} with combined nuclear charge $Z = Z_1 + Z_2$ is shown in Fig. 11.7. The curves are calculated within the monopole approximation, but with variable degrees of ionization, by Wietschorke and Soff. The values should be taken with an uncertainty of 2 – 3 fm due to the corrections discussed above. One may conclude that the system Pb + Cm $(Z_1 + Z_2 = 178)$ is just barely supercritical $(R_{cr} \sim 18$ fm), and one has to go to heavier systems to penetrate into the region where spontaneous positron emission is possible during a heavy collision.

Bibliographical Notes

The formation of superheavy quasimolecules in heavy-ion collisions and their applications in the physics of strong fields was discussed by *Müller* et al. [Mü 72b, c]. The solution of the two-centre Dirac equation and its properties are contained in [Mü 76b].

The validity of the use of an adiabatic basis was investigated in detail by [Th 79, 81].

12. The Dynamics of Heavy-Ion Collisions

To obtain information on the behaviour of electron states in strong fields using heavy-ion collisions as a tool, a full understanding of the scattering dynamics is essential. In this chapter a comprehensive discussion of the formalism required to describe electronic transitions in heavy-ion scattering is presented. It is first based on a classical picture of the nuclear motion, subsequently the language of quantum mechanics is used. A special section is devoted to the problem of collisions with nuclear contact. Collisions of this kind which lead to a prolonged interaction time are found to have the desired effect of enhancing selectively the process of spontaneous positron creation.

12.1 Rutherford Scattering

As discussed in Sect. 11.1, the time dependence of the collision is an important factor because, on the one hand, it enables vacancies to be created in the deeply bound electronic states (this is the "good news") and, on the other hand, it severely limits the time during which spontaneous pair creation can occur (the "bad news"). It is clear that increasing the production of vacancies during the collision generally results in a shorter time available for spontaneous positron creation. A thorough understanding of the consequences of the collision dynamics is therefore essential to calculate reliably the observability of spontaneous vacuum decay in supercritical heavy-ion collisions.

In heavy-ion collisions the two-centre Dirac Hamiltonian H_{TCD} of (7.3) becomes explicitly dependent on time, because the internuclear separation vector $\boldsymbol{R}$ changes with time. In general, the length of $\boldsymbol{R}$ as well as its direction is time dependent. In a collision below the Coulomb barrier of the two nuclei, the variation $\boldsymbol{R}(t)$ is governed by the Coulomb repulsion between the nuclei

$$\frac{d^2\boldsymbol{R}}{dt^2} = \frac{Z_1 Z_2 e^2}{M_{\mathrm{r}} R^2} \cdot \frac{\boldsymbol{R}}{R}, \tag{12.1}$$

where $M_{\mathrm{r}} = M_1 M_2/(M_1 + M_2)$ is the reduced nuclear mass and $R = |\boldsymbol{R}|$. Except for the change of sign this equation is identical with the equation of motion in the Kepler problem of planetary motion. Due to the central potential, angular momentum is conserved and scattering occurs in a plane, taken as the $y-z$ plane

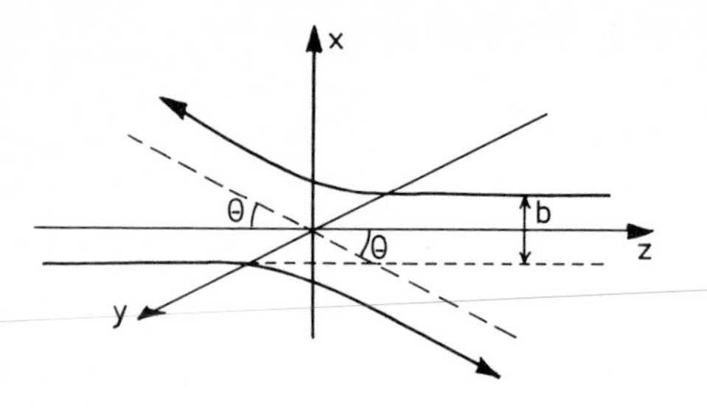

Fig. 12.1. Relative motion of the nuclei in Rutherford scattering and definition of the coordinate system

with the nuclei initially moving in the direction of the z axis. The angular momentum then points in the x direction:

$$\boldsymbol{L} = L\boldsymbol{e}_x = M_\mathrm{r} v_\infty b \boldsymbol{e}_x . \tag{12.2}$$

Here v_∞ is the asymptotic relative velocity of the nuclei and b is the impact parameter (Fig. 12.1). The radial part of (12.1) may be integrated once to an equation describing conservation of energy during the collision

$$\dot{R}^2 + \frac{2 Z_1 Z_2 e^2}{M_\mathrm{r} R} + \frac{v_\infty^2 b^2}{R^2} = v_\infty^2 . \tag{12.3}$$

The explicit solution of this differential equation cannot be given in terms of elementary analytic functions, but a very convenient parametric representation exists

$$R = a(\varepsilon \cosh \xi + 1) \tag{12.4a}$$

$$t = \frac{a}{v_\infty}(\varepsilon \sinh \xi + \xi) , \quad \text{where} \tag{12.4b}$$

$$a = \frac{Z_1 Z_2 e^2}{M_\mathrm{r} v_\infty^2} , \quad \varepsilon = \left(1 + \frac{b^2}{a^2}\right)^{1/2} . \tag{12.4c}$$

The parameter a is half the distance of closet approach in a head-on collision ($b = 0$). That (12.4) represent a solution of (12.3) may be easily proven by substitution, noting that $dt/d\xi = R/v_\infty$.

Note that the asymptotic scattering angle θ_∞ (Fig. 12.1) is related to the impact parameter b through

$$b = a \cot(\Theta_\infty/2) , \quad \varepsilon = \frac{1}{\sin^2(\Theta_\infty/2)} . \tag{12.5}$$

This is easily shown by noting that the variation of the angle between $\boldsymbol{R}$ and the z axis is connected with the angular momentum by

$$L = M_\mathrm{r} \dot{\Theta} R^2 . \tag{12.6}$$

The relative angle after the collision by integration [Gr 65] is

$$\begin{aligned}\Theta(\infty) &= \int_{-\infty}^{\infty} dt\,\dot{\Theta}(t) = \int_{-\infty}^{\infty} dt\,\frac{v_\infty b}{R(t)^2} = b\int_{-\infty}^{\infty}\frac{d\xi}{R}\\ &= 2\frac{b}{a}\int_0^{\infty}\frac{d\xi}{\varepsilon\cosh\xi+1} = 4\frac{b}{a}(\varepsilon^2-1)^{1/2}\arctan\left(\frac{(\varepsilon^2-1)^{1/2}}{\varepsilon+1}\right)\\ &= 4\arctan\left(\frac{b}{a+(a^2+b^2)^{1/2}}\right). \end{aligned} \tag{12.7}$$

Using the tangent addition theorem, then

$$\tan[\Theta(\infty)/2] = \frac{2\tan[\Theta(\infty)/4]}{1-\tan^2[\Theta(\infty)/4]} = \frac{b}{a}, \tag{12.8}$$

which is equivalent to (12.5) because the scattering angle is $\theta_\infty = \pi - \theta(\infty)$.

The relation between impact parameter and scattering angle is very important, because it allows direct experimental determination of the impact parameter. It relies on the assumption of a pure Coulomb force between the nuclei and therefore loses its validity when nuclear forces come into play. In the presence of nuclear interactions, (12.5) may be modified or, worse, a simple one-to-one correspondence between b and θ_∞ may no longer exist. This happens for orbiting, i.e. when the nuclei stick together and rotate around each other several times, or in the presence of inelastic scattering processes.

12.2 Expansion in the Quasi-Molecular Basis

To describe the heavy-ion collision in the framework of the in/out formalism of Sect. 10.1, we have to know the solutions of the time-dependent two-centre Dirac equation

$$\mathrm{i}\frac{\partial}{\partial t}\psi_k(\boldsymbol{r},t) = H_{\mathrm{TCD}}(\boldsymbol{r},\boldsymbol{R}(t))\,\psi_k(\boldsymbol{r},t)\,. \tag{12.9}$$

As initial condition the electron must occupy a stationary state in one of the two atoms before the collision.

As the quasi-molecular states $\Phi_k(\boldsymbol{r},\boldsymbol{R})$ of (11.3), which are the eigenstates of the instantaneous two-centre Hamiltonian $H_{\mathrm{TCD}}(\boldsymbol{r},\boldsymbol{R})$, tend to the stationary states of the separated atoms when $R = |\boldsymbol{R}| \to \infty$, we can require equivalently that

$$\psi_k^{(+)}(\boldsymbol{r},t) \xrightarrow{t\to-\infty} \Phi_k(\boldsymbol{r},\boldsymbol{R}(t))\exp(-\mathrm{i}\varepsilon_k t)\,, \tag{12.10}$$

where ε_k is the energy eigenvalue of the atomic state. Self-adjointness of the Hamiltonian H_{TCD} guarantees that the wave functions $\psi_k^{(+)}(\boldsymbol{r},t)$ are orthogonal at each instant of time and form a complete basis of Hilbert space.

If the collision proceeded infinitely slowly, the electrons would remain in their quasi-molecular state $\Phi_k(\boldsymbol{r}, \boldsymbol{R}(t))$ forever. The finite time dependence causes excitations into other states Φ_l, $l \neq k$. Since we have argued before (Sect. 11.1) that the velocity of the nuclei is much smaller than that of the electrons in states of interest, it is a useful starting point to expand the true solutions of the time-dependent Dirac equation (12.9) at each moment in terms of the instantaneous eigenstates Φ_l:

$$\psi_k^{(+)}(\boldsymbol{r}, t) = \sum_l c_{kl}(t)\, \Phi_l(\boldsymbol{r}, \boldsymbol{R}(t)) . \tag{12.11}$$

The initial condition (12.10) means that the coefficients $c_{kl}(t)$ must satisfy

$$c_{kl}(t) \xrightarrow{t \to -\infty} \exp(-\mathrm{i}\varepsilon_k t)\, \delta_{kl} . \tag{12.12}$$

This condition is somewhat inconvenient in practical calculations, because the amplitudes do not approach constant values for $t \to -\infty$. It is, therefore, useful to extract the phase factor,

$$c_{kl}(t) = a_{kl}(t) \exp[-\mathrm{i}\chi_l(t)] , \tag{12.13}$$

with the initial conditions

$$a_{kl}(t) \xrightarrow{t \to -\infty} \delta_{kl} , \tag{12.14a}$$

$$\dot{\chi}_l(t) \xrightarrow{t \to -\infty} \varepsilon_l . \tag{12.14b}$$

When the ansatz (12.11, 13) is inserted into the time-dependent Dirac equation (12.9), a set of coupled differential equations for the amplitudes $a_{kl}(t)$ is obtained after projection with the instantaneous stationary states Φ_k. We shall show that the still undetermined phases $\chi_l(t)$ may be utilized to eliminate all diagonal couplings in these equations. Start by substituting (12.11) into (12.9), keeping in mind that, by definition,

$$H_{\mathrm{TCD}}(\boldsymbol{r}, \boldsymbol{R})\, \Phi_l(\boldsymbol{r}, \boldsymbol{R}) = E_l(R)\, \Phi_l(\boldsymbol{r}, \boldsymbol{R}) . \tag{12.15}$$

Then

$$\sum_l (\mathrm{i}\dot{a}_{kl} + a_{kl}\dot{\chi}_l) \mathrm{e}^{-\mathrm{i}\chi_l} \Phi_l + \mathrm{i} \sum_l a_{kl} \mathrm{e}^{-\mathrm{i}\chi_l} \dot{\Phi}_l = \sum_l a_{kl} \mathrm{e}^{-\mathrm{i}\chi_l} E_l \Phi_l . \tag{12.16}$$

The expansion on the rhs is cancelled when

$$\dot{\chi}_l(t) = E_l(R(t)) , \tag{12.17}$$

which is clearly compatible with the initial condition (12.14b) since $\lim_{R \to \infty} E_l(R) = \varepsilon_l$. The set of equations (12.16) then reduces to

$$\sum_l \dot{a}_{kl} \mathrm{e}^{-\mathrm{i}\chi_l} \Phi_l = -\sum_l a_{kl} \mathrm{e}^{-\mathrm{i}\chi_l} \dot{\Phi}_l , \tag{12.18}$$

which obviously relates the change in the occupation amplitudes a_{kl} to the time variation of the basis of instantaneous eigenstates. Making use of the orthogonality of the wave functions Φ_l, by projection

$$\dot{a}_{kn}(t) = -\sum_l a_{kl}(t)\,\mathrm{e}^{\mathrm{i}(\chi_n-\chi_l)}\langle \Phi_n(\boldsymbol{R}(t))\,|\,\dot{\Phi}_l(\boldsymbol{R}(t))\rangle \,. \tag{12.19}$$

These are the coupled equations for the amplitudes $a_{kl}(t)$. Taken together for all n, they are completely equivalent to the original time-dependent Dirac equation (12.9), and no approximation is involved so far: using the quasi-molecular states Φ_l in the expansion (12.11) does not constitute an approximation in itself. The loss of exactness occurs when the infinite set of equations (12.19) is truncated to enable a practical numerical solution. It must then be tested carefully whether a sufficient number of coupled equations has been retained to obtain the required accuracy of the solution. This is entirely a question of the practical usefulness of (12.19), not of their principal validity, Sect. 12.6.

To return to the coupled equations (12.19), there is an interesting way to rewrite the matrix element on the rhs, which also shows the absence of diagonal coupling elements. Starting from the orthogonality relation for the quasi-molecular wave functions Φ_l, combined with their property of being eigenstates of H_{TCD}

$$E_n\,\delta_{nl} = \langle \Phi_n|H_{\mathrm{TCD}}|\Phi_l\rangle \,. \tag{12.20}$$

Taking the derivative with respect to time,

$$\dot{E}_n\,\delta_{nl} = \langle \dot{\Phi}_n|E_l\,\Phi_l\rangle + \langle E_n\,\Phi_n|\dot{\Phi}_l\rangle + \langle \Phi_n|\dot{H}_{\mathrm{TCD}}|\Phi_l\rangle \,, \tag{12.21}$$

where we made use of the self-adjointness of H_{TCD}. In the same way, the orthogonality relation itself yields

$$0 = \langle \dot{\Phi}_n|\Phi_l\rangle + \langle \Phi_n|\dot{\Phi}_l\rangle \,. \tag{12.22}$$

Combining this result with (12.21) for off-diagonal matrix elements, to eliminate $\langle \Phi_n|\Phi_l\rangle$,

$$l \neq n: \quad \langle \Phi_n|\dot{\Phi}_l\rangle = (E_l-E_n)^{-1}\left\langle \Phi_n\left|\frac{\partial H_{\mathrm{TCD}}}{\partial t}\right|\Phi_l\right\rangle . \tag{12.23}$$

The change of the quasi-molecular state Φ_l has been expressed through the change of the two-centre Dirac Hamiltonian itself. The sole part of H_{TCD} which depends on time is the potential $V_{\mathrm{TC}} = V_1 + V_2$, (10.3).

The potential depends on time only via the nuclear relative distance $\boldsymbol{R}(t)$. The final expression for the off-diagonal coupling elements is therefore

$$\langle \Phi_n|\dot{\Phi}_l\rangle = (E_l-E_n)^{-1}\dot{\boldsymbol{R}}\cdot\langle \Phi_n|\nabla_R V_{\mathrm{TC}}|\Phi_l\rangle \,, \tag{12.24}$$

which is well suited to numerical evaluation. For the diagonal elements, (12.22) yields

$$\langle \Phi_n|\dot{\Phi}_n\rangle = -\langle \dot{\Phi}_n|\Phi_n\rangle = -\langle \Phi_n|\dot{\Phi}_n\rangle^* \,, \tag{12.25}$$

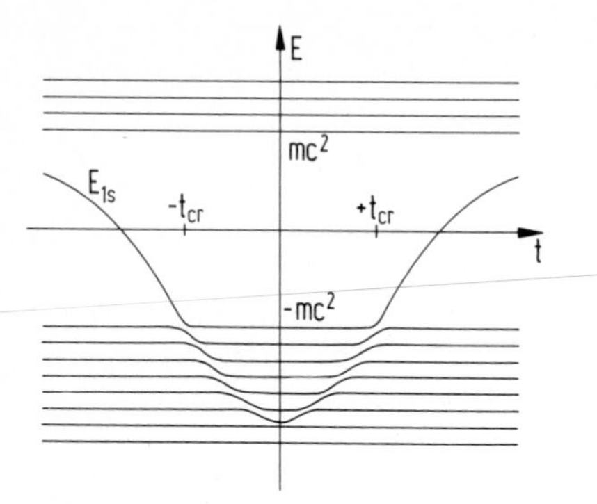

Fig. 12.2. Schema of how the supercritical resonance moves through the discretized continuum. Strong and very much localized couplings occur between neighbouring continuum states

implying that the matrix element is purely imaginary. For real, stationary functions Φ_n it must therefore vanish, i.e. diagonal couplings are absent in this expansion scheme. For $n=1$, (12.21) yields the *Hellmann-Feynman* theorem [He 37, Fe 39]

$$\dot{E}_n = \langle \Phi_n | \dot{H}_{\mathrm{TCD}} | \Phi_n \rangle \,, \tag{12.26}$$

Sect. 9.7, (9.135). The theoretical discussion up to now was limited to subcritical collisions, where the energy eigenvalues of the adiabatic electronic bound states are confined to the gap region $-m_e c^2 > R > m_e c^2$. In supercritical collisions $(Z_1+Z_2) > Z_{\mathrm{cr}}$ the deepest bound state joins the lower continuum at R_{cr} and becomes a resonance. In the static limit, a hole brought into this state decays spontaneously by positron emission, leaving a stable filled atomic K shell. As discussed in Sect. 11.1, the lifetime of the resonance is of the order 10^{-19} s and therefore considerably longer than the collision time ($\simeq 2\times 10^{-21}$ s for U – U collisions). Excitations induced by nuclear motion are eminently important. Therefore a formalism is required which describes dynamical excitations and at the same time accounts for the resonance character of the supercritical state.

The coupled differential equations (12.19) are not directly applicable to this situation. In the region $R < R_{\mathrm{cr}}$ the $1s\sigma$ state together with its amplitude disappears from the set of discrete states. Instead, the radial coupling matrix elements involving the lower continuum develop very strong, narrow maxima near the (time-dependent) position of the resonance. The situation is made even worse since the continuum must be discretized in any practical calculation and only a finite number of scattering states can be taken to represent the whole continuum. The narrow resonance would then only accidentally coincide with one of the chosen energy grid points, making a straightforward solution of the coupled channel equations impossible (Fig. 12.2).

These problems are solved at one stroke, if one uses the projection method presented in Sect. 10.3, whereby the supercritical resonance state is represented by a localized wave function Φ_R which is not an eigenstate of the Hamiltonian.

The modified negative energy continuum states $\tilde{\Phi}_E$ are orthogonal to Φ_R and contain no resonance influence. Therefore no strong variation is felt in the wave function $\tilde{\Phi}_E$ when the resonance state just passes through this particular energy region during the heavy-ion collision. In other words, coupling among the modified continuum states remains weak.

The price paid is that Φ_R and $\tilde{\Phi}_E$ are not eigenstates of the two-centre Dirac Hamiltonian, so that off-diagonal matrix elements like (10.59)

$$\tilde{V}_E = \langle \tilde{\Phi}_E | H_{\mathrm{TCD}} | \Phi_R \rangle \tag{12.27}$$

exist. This modifies the coupled amplitude equations (12.19). The expansion for the time-dependent wave function in the supercritical region $R < R_{\mathrm{cr}}$ reads

$$\psi_k^{(+)}(\boldsymbol{r}, t) = \sum_{E_l > -m_e} c_{kl}\, \Phi_l(\boldsymbol{r}, \boldsymbol{R}(t)) + c_{k,1s}\, \Phi_R(\boldsymbol{r}, \boldsymbol{R}(t)) + \int_{-\infty}^{-m_e} dE\, c_{kE}\, \tilde{\Phi}_E(\boldsymbol{r}, \boldsymbol{R}(t)) , \tag{12.28}$$

where the resonance function and the modified continuum functions now depend on the nuclear relative distance $\boldsymbol{R}(t)$. The initial conditions are not modified, because at $t = -\infty$ the potential is subcritical.

Inserting the ansatz (12.28) into the time-dependent Dirac equation, the rhs of (12.16) changes slightly because Φ_R and $\tilde{\Phi}_E$ are not eigenstates of H_{TCD}:

$$\cdots = \sum_{E_l > -m_e} a_{kl}\, \mathrm{e}^{-\mathrm{i}\chi_l} E_l\, \Phi_l + a_{k,1s}\, \mathrm{e}^{-\mathrm{i}\chi_{1s}} H_{\mathrm{TCD}}\, \Phi_R + \int_{-\infty}^{-m_e} dE\, a_{kE}\, \mathrm{e}^{-\mathrm{i}\chi_E} H_{\mathrm{TCD}}\, \tilde{\Phi}_E . \tag{12.29}$$

There is now no way to choose the phases χ_l to eliminate the rhs completely. All that can be achieved is to cancel the diagonal matrix elements of the two-centre Hamiltonian by setting

$$\dot{\chi}_l(t) = E_l(R(t)) , \quad E_l > -m_e , \tag{12.30a}$$

$$\dot{\chi}_{1s}(t) = E_R(R(t)) \equiv \langle \Phi_R | H_{\mathrm{TCD}} | \Phi_R \rangle , \tag{12.30b}$$

$$\dot{\chi}_E(t) = E , \quad E < -m_e , \tag{12.30c}$$

where E_R is the expectation value of the resonance energy as defined in (10.60). The new equation is

$$\sum_l \dot{a}_{kl}\, \mathrm{e}^{-\mathrm{i}\chi_l} \Phi_l = -\sum_l a_{kl}\, \mathrm{e}^{-\mathrm{i}\chi_l} \dot{\Phi}_l + a_{k,1s}\, \mathrm{e}^{-\mathrm{i}\chi_{1s}} (H_{\mathrm{TCD}} - E_R)\, \Phi_R + \int_{-\infty}^{-m_e} dE\, a_{kE}\, \mathrm{e}^{-\mathrm{i}\chi_E} (H_{\mathrm{TCD}} - E)\, \tilde{\Phi}_E . \tag{12.31}$$

Now projecting with a wave function Φ_n, the additional terms produce no diagonal matrix elements, but couplings like (12.27) occur. There are no additional couplings among the modified continuum states, because the Hamiltonian is diagonal within the subspace of the functions $\tilde{\Phi}_E$, (10.58). The new equations are (for $R < R_{\mathrm{cr}}$)

$$\dot{a}_{k,1s} = -\sum_l a_{kl}\, \mathrm{e}^{\mathrm{i}(\chi_{1s} - \chi_l)} \langle \Phi_R | \dot{\Phi}_l \rangle + \int_{-\infty}^{-m_e} dE\, a_{kE}\, \mathrm{e}^{\mathrm{i}(\chi_{1s} - Et)}\, \tilde{V}_E^* , \tag{12.32a}$$

$$\dot{a}_{kE} = -\sum_l a_{kl} e^{i(Et-\chi_l)} \langle \tilde{\Phi}_E | \dot{\Phi}_l \rangle + a_{k,1s} e^{i(Et-\chi_{1s})} \tilde{V}_E , \tag{12.32b}$$

while the equations for n with $E_n > -m_e$ remain formally unchanged. The additional terms in (12.32) describe the decay of a vacancy in the supercritical resonance state into the positron continuum. This proceeds in much the same way as discussed in Sect. 10.4, where we described the decay of a supercritical K vacancy, using the projected basis in the Schrödinger picture. Note, however, that (12.32) refer to the time evolution of single-particle amplitudes, whereas we described the evolution of Fock space states in Sect. 10.4. The coupling $\langle \tilde{\Phi}_E | \dot{\Phi}_R \rangle$ describes the dynamic process of induced positron creation due to the nuclear motion [Sm 74], whereas the matrix element $\tilde{V}_E = \langle \tilde{\Phi}_E | H | \Phi_R \rangle$ is responsible for the spontaneous positron emission from the vacant supercritical K shell [Re 80b, Re 81b]. In the limit of infinitely slow motion of the nuclei only the spontaneous process would survive.

Since Φ_R and $\tilde{\Phi}_E$ go over continuously into the unprojected eigenstates Φ_{1s} and Φ_E at the critical distance, the differential equations (12.32) smoothly join (12.19), describing the time evolution in the subcritical region. This is also clear from the Fano theory in Chap. 6, whereby the matrix element V_E (here $\tilde{V}_E$) is proportional to the strength of the supercritical part of the potential. At $R = R_{cr}$ hence $\tilde{V}_E = 0$.

Positron creation in heavy-ion collisions with supercritical binding can, therefore, be described by integrating the coupled channel equations (12.19, 32). When the final amplitudes $a_{kl}(t=\infty)$ are known, observable physical quantities in the final state may be calculated with the in/out formalism of Sect. 10.1. The in-basis of wave functions is provided by the time-dependent solutions $\psi_k^{(+)}(r,t)$ of (12.11, 28), while for the out-basis knowing the asymptotic properties is sufficient:

$$\psi_i^{(-)}(r,t) \xrightarrow{t\to+\infty} \Phi_i(r, R(t)) \exp[-i\chi_i(t)] , \tag{12.33}$$

in analogy to (12.10). The single-particle scattering matrix element is conveniently evaluated long after the collision, say at $t = t_f$; from (8.48)

$$\begin{aligned} S_{ik} &= \int d^3r\, \psi_i^{(-)}(r,t_f)^\dagger \psi_k^{(+)}(r,t_f) \\ &= \sum_l c_{kl}(t_f) e^{i\chi_l(t_f)} \langle \Phi_i(r, R(t_f)) | \Phi_l(r, R(t_f)) \rangle \\ &= c_{ki}(t_f) e^{i\chi_i(t_f)} = a_{ki}(t_f) \approx a_{ki}(\infty) , \end{aligned} \tag{12.34}$$

where the orthogonality of the functions Φ_k was used. According to (10.8) the number of positrons with energy $E < -m_e$ after the heavy-ion collision is given by

$$\begin{aligned} \frac{d\bar{N}}{dE} &= \sum_{k>F} |S_{Ek}|^2 = \sum_{k>F} |a_{kE}(\infty)|^2 \\ &= \sum_{-m_e<E_k<m_e} |a_{kE}(\infty)|^2 + \int_{m_e}^{\infty} dE' \, |a_{E'E}(\infty)|^2 . \end{aligned} \tag{12.35}$$

Electrons in excited states, inner-shell vacancies and possible coincidences between particles and holes are also given by (10.7–9) when the amplitudes $a_{ik}(\infty)$ are substituted for S_{ki}.

We conclude that knowing the final single-particle amplitudes is sufficient to describe all observable phenomena relating to electronic motion in heavy-ion collisions. Although this conclusion has been based on a description of the collision process where only the interaction between the nuclei and the electrons is taken into account, it may be carried over to a large extent into a description that also includes the electrostatic interaction among many electrons. The formalism remains essentially unchanged [Ki 80] if the interaction among the electrons is represented by an average screening potential that is the same for all electrons. All that has to be done is to amend the two-centre Hamiltonian (11.3) by the additional term

$$H_{\text{TCD}}^{\text{screen}} = \boldsymbol{\alpha} \cdot \boldsymbol{p} + \beta m_e + V_1 + V_2 + V_{\text{screen}}(\boldsymbol{r}, t) . \tag{12.36}$$

In the (time-dependent) Hartree approximation the self-consistent screening potential is derived from the density distribution of the electrons:

$$V_{\text{screen}}(\boldsymbol{r}, t) = V_{\text{TDH}}(\boldsymbol{r}, t) = e^2 \sum_{n<F} \int d^3r' \, \frac{\psi_n^{(+)}(\boldsymbol{r}', t)^\dagger \psi_n^{(+)}(\boldsymbol{r}', t)}{|\boldsymbol{r} - \boldsymbol{r}'|} . \tag{12.37}$$

Often one uses instead the so-called adiabatic time-dependent Hartree approximation, where the quasi-molecular state $\Phi_n(\boldsymbol{r}, \boldsymbol{R}(t))$ is used in place of the fully time-dependent wave function $\psi_n^{(+)}(\boldsymbol{r}, t)$:

$$V_{\text{ATDH}}(\boldsymbol{r}, t) = e^2 \sum_{n<F} \int d^3r' \, \frac{\Phi_n(\boldsymbol{r}', \boldsymbol{R}(t))^\dagger \Phi_n(\boldsymbol{r}', \boldsymbol{R}(t))}{|\boldsymbol{r} - \boldsymbol{r}'|} . \tag{12.38}$$

Here the influence of electronic excitation on the screening potential is neglected. Due to the Pauli principle a non-local exchange term must be added to the screening potential. Usually this exchange term, discussed further in the context of many-body theory of supercritical fields in Sect. 16.1, is replaced by an effective local interaction (the so-called Hartree-Fock-Slater approximation). The Hamiltonian then also has the form of (12.36). Before discussing the present experimental situation in the next section, we discuss in some detail how the time-dependent two-centre Dirac equation (12.9) can be derived from a quantum mechanical scattering theory of the heavy-ion collision when the semiclassical approximation is applied. This also enables us to account for the effect of nuclear forces.

12.3 Heavy-Ion Collisions: A Quantal Description

Although they are much heavier than electrons, the nuclei are still particles of atomic size properly described in the framework of quantum mechanics. A more detailed inquiry into the validity of transforming the two-centre Dirac equation

(11.3) into a dynamical equation (12.9) by replacing the energy eigenvalue E by the time-derivative $\mathrm{i}\,\partial/\partial t$ and the parameter $\boldsymbol{R}$ by the trajectory function $\boldsymbol{R}(t)$ is therefore in order. We start from a fully quantum mechanical formulation of the scattering problem for the many-body system of two nuclei (masses $M_\mathrm{A}, M_\mathrm{B}$; charges $Z_\mathrm{A}, Z_\mathrm{B}$) and Z' electrons interacting electromagnetically. Assuming that the nuclear motion can be treated non-relativistically, the total Hamiltonian is in the nuclear centre-of-mass system:

$$
\begin{aligned}
H &= \frac{P^2}{2\mu} + V^{\mathrm{AB}}(\boldsymbol{R}) + \sum_{i=1}^{Z'} [\boldsymbol{\alpha}^{(i)} \cdot \boldsymbol{p}_i c + \beta^{(i)} m_\mathrm{e} c^2 + V^{\mathrm{eA}}(\boldsymbol{r}_i, \boldsymbol{R}) + V^{\mathrm{eB}}(\boldsymbol{r}_i, \boldsymbol{R})] \\
&\qquad + \frac{1}{2} \sum_{i \neq j}^{Z'} V^{\mathrm{ee}}(|\boldsymbol{r}_i - \boldsymbol{r}_j|) \\
&\equiv \frac{P^2}{2\mu} + V^{\mathrm{AB}} + H^{\mathrm{el}}(\boldsymbol{r}_1, \ldots, \boldsymbol{r}_{Z'}; \boldsymbol{R}) .
\end{aligned}
\tag{12.39}
$$

The first two terms denote the relative kinetic energy of the nuclei [$\mu = M_\mathrm{A} M_\mathrm{B}/(M_\mathrm{A} + M_\mathrm{B})$ is the nuclear reduced mass] and the nuclear Coulomb barrier. The single sum contains the relativistic energy of the electrons and their interaction with the nuclei, whereas the double sum expresses the electron – electron interaction.

A number of approximations have been made: we have neglected retardation [Ko 78] and relativistic reduced mass effects, and also the magnetic field created by the nuclear current [Ra 76b, So 81b], and we have dropped a term describing the collective motion of *all* electrons with respect to the nuclear centre of mass. Last but not least, the nuclei are treated as indestructible, rigid particles having no internal degress of freedom. This is not really warranted since it is known that collective states may be excited in the colliding nuclei by the long-range Coulomb potential ("Coulomb excitation") and also by nuclear interaction and nucleon transfer or exchange. For our purposes, however, it is sufficient to include such effects here in a phenomenological way, e.g. by including a friction term in the potential V^{AB}. Internal excitations of the nuclei are discussed to some extent in Sect. 12.5, but a full investigation of possible consequences of nuclear reactions does still not exist.

As the total energy is conserved, the system (two nuclei and Z' electrons) is described by a stationary state of H:

$$
H(\boldsymbol{r}_i, \boldsymbol{R})\, \Psi(\boldsymbol{r}_i, \boldsymbol{R}) = E\, \Psi(\boldsymbol{r}_i, \boldsymbol{R}) . \tag{12.40}
$$

The wave function Ψ contains the electronic as well as the relative nuclear motion. It may be expanded with respect to a set of eigenstates of some electronic Hamiltonian, e.g. of the Hamiltonian $H^{\mathrm{el}}(\boldsymbol{r}_i, \boldsymbol{R})$, from (12.39),

$$
H^{\mathrm{el}}(\boldsymbol{r}_i, \boldsymbol{R})\, \Phi_n(\boldsymbol{r}_i, \boldsymbol{R}) = \varepsilon_n(R)\, \Phi_n(\boldsymbol{r}_i, \boldsymbol{R}) . \tag{12.41}
$$

The many-electron energy eigenvalues ε_n depend only upon the nuclear distance $R = |\boldsymbol{R}|$.

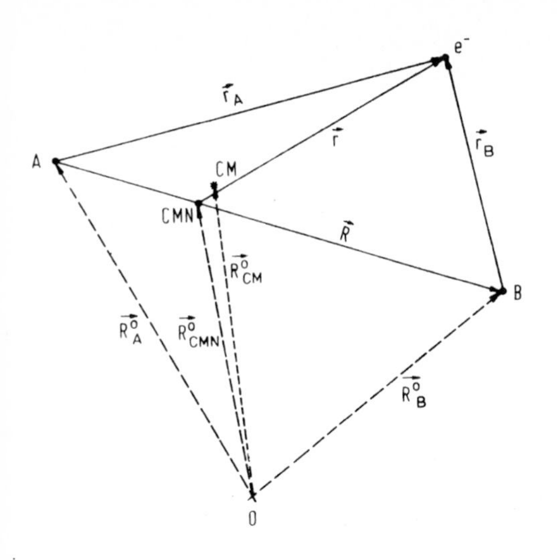

Fig. 12.3. Definition of the coordinates [He 81a]

When the electron–electron interaction is neglected, the electronic Hamiltonian is simply a sum of the single-electron two-centre Hamiltonians (11.3)

$$H^{\mathrm{el}}(\boldsymbol{r}_i, \boldsymbol{R}) = \sum_{i=1}^{Z'} H_{\mathrm{TCD}}(\boldsymbol{r}_i, \boldsymbol{R}) , \qquad (12.42)$$

and the many-electron wave functions Φ_n can be taken as Slater determinants of two-centre Dirac eigenfunctions. For asymmetric systems ($A \neq B$) the bound states of the molecular basis (12.41) factorize into wave functions localized around one of the nuclei, when the nuclear separation becomes very large:

$$\Phi_n(\boldsymbol{r}_i, \boldsymbol{R}) \xrightarrow{R\to\infty} \Phi_{n_A}(\boldsymbol{r}_{Ai})\, \Phi_{nB}(\boldsymbol{r}_{Bi}) , \qquad (12.43)$$

$\boldsymbol{r}_{Ai}$, $\boldsymbol{r}_{Bi}$ are the coordinates of electron i with respect to nucleus A, B (Fig. 12.3).

For symmetrical systems the $\Phi_n(\boldsymbol{r}, \boldsymbol{R})$ have good parity, but an appropriate combination of even and odd states has the clustering property (12.43). This property is due to the fact that for $r \to \infty$ an electron in ion A no longer feels ion B. This is not true for continuum electrons: continuum energy eigenstates are spread over all space and always feel the potential of both nuclei. *Heinz* et al. discussed how the formalism must be modified to apply to the continuum wave functions [He 81a]. However, since we treat the continuum only in a very restricted approximation, we forgo this discussion and assume that (12.43) is satisfied.

Expanding $\Psi(\boldsymbol{r}, \boldsymbol{R})$ with respect to the molecular basis $\Phi_n(\boldsymbol{r}, \boldsymbol{R})$, then

$$\Psi(\boldsymbol{r}, \boldsymbol{R}) = \sum_n \chi_n(\boldsymbol{R})\, \Phi_n(\boldsymbol{r}, \boldsymbol{R}) \qquad (12.44)$$

where $\chi_n(\boldsymbol{R})$ contain the nuclear relative motion. Projecting out the molecular states Φ_n yields a set of coupled differential equations for χ_n. Unfortunately, this leads to very long-range couplings among the functions $\chi_n(R)$ describing the various channels, because for $R \to \infty$ ($V^{AB} \to 0$), $\chi_n(R)$ become eigenfunctions of $p^2/2\mu$, i.e. plane waves with good momentum $\boldsymbol{P}$:

$$\chi_n(\boldsymbol{R}) \xrightarrow{R\to\infty} \exp\left(\frac{\mathrm{i}}{\hbar}\boldsymbol{P}\cdot\boldsymbol{R}\right). \tag{12.45}$$

Physically, however, the asymptotic relative motion should be that of two ions (nuclei plus electrons) moving apart, described by eigenfunctions of the momentum $\boldsymbol{P}_{\mathrm{AB}}$ canonically conjugate to the inter*atomic* distance $\boldsymbol{R}_{\mathrm{AB}}$:

$$\chi_n(\boldsymbol{R}) \xrightarrow{R\to\infty} \exp\left(\frac{\mathrm{i}}{\hbar}\boldsymbol{P}_{\mathrm{AB}}\cdot\boldsymbol{R}_{\mathrm{AB}}\right). \tag{12.46}$$

That is, the appropriate relative coordinate in the asymptotic region is the separation between the centres of mass of the two ions $\boldsymbol{R}_{\mathrm{AB}} = \boldsymbol{R}_{\mathrm{CMA}} - \boldsymbol{R}_{\mathrm{CMB}}$, and not between the two nuclei as such [Ba 58, Th 65, Mi 73b, Th 78, Gr 81a, He 81a].

Therefore the molecular basis $\Phi_n(\boldsymbol{r},\boldsymbol{R})$ is changed to $\Phi_n(\boldsymbol{r},\tilde{\boldsymbol{R}})$ so that in the expansion

$$\Psi(\boldsymbol{r},\tilde{\boldsymbol{R}}) = \sum_n \chi_n(\tilde{\boldsymbol{R}})\,\Phi_n(\boldsymbol{r},\tilde{\boldsymbol{R}}) \tag{12.47}$$

the new functions $\chi_n(\tilde{\boldsymbol{R}})$ behave according to (12.46) in the asymptotic region.

Instead of working with the new basis functions $\Phi_n(\boldsymbol{r},\tilde{\boldsymbol{R}})$, it is more convenient to express the transformation from $\boldsymbol{R}$ to $\tilde{\boldsymbol{R}}$ by a translation operator $T(\boldsymbol{X})$ which shifts $\boldsymbol{R}$ by $\boldsymbol{X} = \tilde{\boldsymbol{R}} - \boldsymbol{R}$. Thus

$$\Psi(\boldsymbol{r},\boldsymbol{R}) = \sum_n T(\boldsymbol{X})\,\chi_n(\boldsymbol{R})\,\Phi_n(\boldsymbol{r},\boldsymbol{R}), \tag{12.48}$$

where $T(\boldsymbol{X})$ is a translation operator defined by

$$T(\boldsymbol{X})f(\boldsymbol{R}) = f(\boldsymbol{R}+\boldsymbol{X}) = f(\tilde{\boldsymbol{R}}). \tag{12.49}$$

For constant $\boldsymbol{X}$

$$T(\boldsymbol{X}) = \exp(\boldsymbol{X}\cdot\nabla_R) = \exp\left(\frac{\mathrm{i}}{\hbar}\boldsymbol{X}\cdot\boldsymbol{P}\right); \tag{12.50}$$

when $\boldsymbol{X}$ itself depends on $\boldsymbol{R}$, the exponential is defined by the series expansion where all factors $\boldsymbol{X}$ are to the left of all operators $\boldsymbol{P}$.

The translation vector $\boldsymbol{X}$ is defined by two asymptotic conditions.

i) For $R \to 0$ we want to work in the molecular framework. Hence in the limit the internuclear distance $\boldsymbol{R}$ is the correct scattering coordinate.
ii) For $R \to \infty$, $\tilde{\boldsymbol{R}} = \boldsymbol{R} + \boldsymbol{X}$ should become $\boldsymbol{R}_{\mathrm{AB}}$, i.e. the interatomic distance. This scattering coordinate takes into account the asymptotic translation of the electron with the two nuclei.

After some algebra the following vector satisfies these requirements:

$$X = \frac{m_e}{\mu} \sum_{i=1}^{Z'} S_i \quad \text{with} \tag{12.51}$$

$$S_i = \tfrac{1}{2}[f(r_i, R) + \eta][r_i - \tfrac{1}{4}R(f(r_i, R) + \eta)] . \tag{12.52}$$

Here $\eta = (M_A - M_B)/(M_A + M_B)$ is the mass asymmetry and $f(r_i, R)$ is a so-called switching function which can be chosen freely up to the boundary conditions for $R \to 0$ and $R \to \infty$ [Schn 69]

$$f(r, R) \to \eta \quad (r \to \infty \text{ at finite } R) \quad \text{and} \tag{12.53a}$$

$$f(r, R) \to \begin{cases} +1 & (R \to \infty,\ r_B/R \to 0) \\ -1 & (R \to \infty,\ r_A/R \to 0) \end{cases} . \tag{12.53b}$$

Instead of working with the transformed wave functions (12.48), we can work with the transformed Hamiltonian (12.39).

$$H'(r_i, R) = T(X)^{-1} H(r_i, R)\, T(X) . \tag{12.54}$$

Since X and P do not commute in (12.12), $T(X)$ is not exactly unitary. Therefore the transformation of H should properly include factors of the form $(T^\dagger T)^{-1/2}$ correcting for the change in volume element from d^3R to $d^3\tilde{R}$. (For details, see [He 81a].)

Here H' cannot be given in closed form, but $H'(r, R)$ is expanded with respect to m_e/μ, as T is generated by the small vector $X = (m_e/\mu) \sum_i S_i$. Truncating after the first-order term gives a consistent expression for H', since the operator H is correct only in lowest order m_e/μ due to the non-relativistic approximation for the nuclear relative motion. After a lengthy but straightforward calculation [He 81a] one obtains in order m_e/μ

$$H' = \frac{P^2}{2\mu} + V^{AB}(R) + H^{el}(r_i, R) + \Delta + \frac{1}{2}(\dot{X} \cdot P + P \cdot \dot{X}) , \tag{12.55}$$

where Δ is the difference between transformed and old potential:

$$\Delta = V^{AB}(|R - X|) - V^{AB}(R) + \sum_{i=1}^{Z'} [V^{eA}(r_i, R - X) - V^{eA}(r_i, R) + V^{eB}(r_i, R - X) - V^{eB}(r_i, R)] , \tag{12.56}$$

and $\dot{X}$ is the time variation of X caused by the electronic part of the Hamiltonian:

$$\dot{X} = \frac{i}{\hbar}[H^{el}_{mol}(r, R), X] = \frac{m_e}{\mu} \sum_{i=1}^{Z'} \frac{1}{2}[f(r_i, R) + \eta]\, \alpha^{(i)} c . \tag{12.57}$$

Now we can write the Schrödinger equation for the transformed Hamiltonian H', and project on all possible states of the electronic molecular basis:

$$
\begin{aligned}
0 &= \sum_n \langle \Phi_m | H' - E | \Phi_n \rangle \chi_n \\
&= \sum_n \left[\frac{1}{2\mu} (-\mathrm{i}\hbar \nabla + \langle \boldsymbol{P} \rangle + \mu \langle \dot{\boldsymbol{X}} \rangle)^2_{mn} + (V^{\mathrm{AB}} + \varepsilon_n - E)\, \delta_{mn} \right. \\
&\quad \left. + \left(\langle \Delta \rangle - \frac{1}{2} \mu \langle \dot{\boldsymbol{X}} \rangle^2 \right)_{mn} \right] \chi_n(\boldsymbol{R}) \,. \qquad (12.58)
\end{aligned}
$$

Here $\langle \boldsymbol{P} \rangle$, $\langle \dot{\boldsymbol{X}} \rangle$, $\langle \Delta \rangle$ mean matrix elements taken with the electronic basis functions. It is possible to show [He 81a] that for large separations

$$
\boldsymbol{P} + \mu \dot{\boldsymbol{X}} \xrightarrow{R \to \infty} \boldsymbol{P}_{\mathrm{AB}} \left(\frac{\mu}{\mu_{\mathrm{AB}}} \right)^{1/2} \left[1 + O\left(\frac{m_{\mathrm{e}}}{\mu} \right) \right] , \qquad (12.59)
$$

where μ_{AB} is the reduced mass of the two *ions*. Since the electronic wave functions are constructed as eigenstates of $\boldsymbol{P}_{\mathrm{AB}}$, (12.45), all relevant couplings in (12.21) vanish in the limit $R \to \infty$, except for terms of the order of m_{e}/μ, which we neglect.

Since the terms quadratic in $\langle \boldsymbol{P} \rangle$ and $\langle \dot{\boldsymbol{X}} \rangle$ and the term $\langle \Delta \rangle$ are of higher order in m_{e}/μ, they can be neglected in (12.58), giving the coupled channel equations

$$
-\frac{\hbar^2}{2\mu} (\nabla_R^2 + k(R)^2)\, \chi_n(\boldsymbol{R}) = \frac{\hbar^2}{\mu} \sum_m \boldsymbol{D}_{nm} \cdot \nabla_R \chi_m(\boldsymbol{R}) - [\varepsilon_n(R) - \bar{\varepsilon}(R)]\, \chi_n(\boldsymbol{R}) \,, \qquad (12.60)
$$

where

$$
k(R)^2 = \frac{2\mu}{\hbar^2} [E - \bar{\varepsilon}(R) - V^{\mathrm{AB}}(R)] \qquad (12.61)
$$

and

$$
\boldsymbol{D}_{nm}(\boldsymbol{R}) = \langle \Phi_n | \nabla_R + \frac{\mathrm{i}\mu}{\hbar} \dot{\boldsymbol{X}} | \Phi_m \rangle \,. \qquad (12.62)
$$

Here $\bar{\varepsilon}(R)$ takes into account the average total electronic binding energy (i.e. the energy of all electrons added together) independent of the state of excitation. In many cases, this molecular contribution to the internuclear potential can be neglected, and one sets $\bar{\varepsilon}(R) = 0$. Below the nuclear Coulomb barrier, $k(R)$ then corresponds to a Rutherford hyperbola. We still have the free choice of the switching function $f(\boldsymbol{r}, \boldsymbol{R})$ introduced in (12.52, 53). Most authors choose $f(\boldsymbol{r}, \boldsymbol{R})$ ad hoc as an analytical expression containing several parameters that can be varied to make the matrix elements $\boldsymbol{D}_{nm}(R)$ as small as possible, Recently, however, *Taulbjerg* and *Vaaben* [Va 80] proposed a switching function based on an intuitive physical picture and giving results of equal quality.

Let $\boldsymbol{F}_{\mathrm{A}}$, $\boldsymbol{F}_{\mathrm{B}}$ be the forces acting on the electron from nuclei A and B; $\boldsymbol{F} = \boldsymbol{F}_{\mathrm{A}} + \boldsymbol{F}_{\mathrm{B}}$ is the resulting force. If $\boldsymbol{F}$ points directly to A, the electron belongs to A and $f = -1$; analogously $f = +1$ if $\boldsymbol{F}$ points to B. When $\boldsymbol{F}$ points to the nuclear centre of mass, $f = \eta$. Generalizing this argument, $1/2\,(\eta + f)$ equals the fraction of the internuclear distance spanned between the centre of mass and the basis of the projection of F on the internuclear axis (Fig. 12.4):

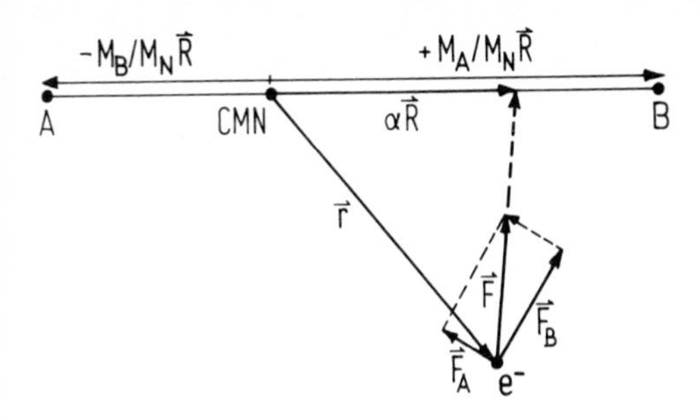

Fig. 12.4. Definition of the switching function via the centre of force

$$f(r,R) = -\eta + 2\,\frac{M_A+M_B}{M_A M_B}\cdot\frac{Z_B M_A r_A^3 - Z_A M_B r_B^3}{Z_B r_A^3 - Z_A r_B^3}\,. \tag{12.63}$$

One easily verifies that this definition obeys the correct boundary conditions (12.53), except for the minor difference that for $R \to 0$ the origin of the coordinate system is associated with the centre of charge and not with the centre of mass. For all practical purposes this is irrelevant for heavy systems.

12.4 The Semiclassical Approximation

Any numerical solution of the quantum mechanical equation (12.60) requires taking many electronic states and a huge number of values for the total angular momentum into account. For heavy systems such an approach would be completely intractable. On the other hand, for heavy systems the nuclei should move along classical Rutherford trajectories with very little influence from electronic excitations. Thus the nuclear trajectory is determined by the total energy and total angular momentum only. This description of the nuclear trajectory is obtained by applying the WKB approximation to (12.60). This method leads to a set of equations equivalent – except for minor modifications – to the time-dependent two-centre Dirac equation. Once this is achieved, we have proven that the methods developed in Sect. 10.1 for time-dependent external fields are applicable to heavy-ion collisions. The discussion can then proceed along the lines set out in Sect. 12.1.

Consequently assume that the nuclear motion is determined by the lhs of (12.60). The WKB solutions are then

$$\chi_n^{\mathrm{WKB}}(\boldsymbol{R}) = c_n k(R)^{-\frac{1}{2}} \exp\left[\mathrm{i}\int_{R_0}^{R} d\boldsymbol{R}\cdot\boldsymbol{k}(R)\right], \tag{12.64}$$

where c_n is a normalization factor. To allow for slight modifications of the nuclear motion caused by electronic transitions, apply the method of variation of constants, permitting c_n to be R dependent: $c_n(R)$. Neglecting $\nabla_R^2 c_n/\mu$ and $\nabla_R k(R)$ and dividing by the common factors gives a set of coupled equations for the amplitudes c_n:

$$\frac{\hbar k(R)}{\mu} \nabla_R c_n(R) = -\frac{\hbar k(R)}{\mu} \sum_m D_{nm} c_m(R) - \mathrm{i}\,\varepsilon_n(R) c_n(R) . \tag{12.65}$$

All we have to do now is to remember that $\hbar k(R)$ is the classical relative momentum of the nuclei on their scattering trajectories, which can be equated with $\mu\dot{\boldsymbol{R}}$. As $\boldsymbol{R}$ is a uniquely specified function of time t along the classical trajectories, the amplitudes c_n can be considered to be functions of t alone. This thus gives a set of differential equations for the time evolution of the amplitudes:

$$\dot{c}_n(t) = -\dot{\boldsymbol{R}} \sum_m \boldsymbol{D}_{nm}(\boldsymbol{R}(t))\, c_m(t) - \mathrm{i}\,\varepsilon_n c_n(t)/\hbar , \tag{12.66}$$

where the coupling matrix elements $\boldsymbol{D}_{nm}$ are given by (12.62).

The amplitudes $c_n(t)$, being related to the many-electron wave functions $\Phi_n(\boldsymbol{r}_i, \boldsymbol{R})$, (12.41), denote the probability amplitudes for finding the electronic system in the configuration Φ_n at time t. If one is willing to replace the true electron – electron interaction

$$\frac{1}{2} \sum_{i \neq j}^{Z'} V^{\mathrm{ee}}(|\boldsymbol{r}_i - \boldsymbol{r}_j|)$$

by an effective single-particle potential [cf. (12.36) ff.]

$$V_{\mathrm{eff}}^{\mathrm{ee}} = \sum_{i=1}^{Z'} V_{\mathrm{eff}}(\boldsymbol{r}_i, \boldsymbol{R}) \tag{12.67}$$

or to neglect it altogether, the wave functions Φ_n can be taken as Slater determinants built out of single-particle wave functions $\tilde{\Phi}_n(\boldsymbol{r}, \boldsymbol{R})$ satisfying the Dirac equation

$$\tilde{\varepsilon}_n \tilde{\Phi}_n = [\boldsymbol{\alpha} \cdot \boldsymbol{p} c + \beta m_{\mathrm{e}} c^2 + V^{\mathrm{eA}}(\boldsymbol{r}, \boldsymbol{R}) + V^{\mathrm{eB}}(\boldsymbol{r}, \boldsymbol{R}) + V_{\mathrm{eff}}(\boldsymbol{r}, \boldsymbol{R})]\, \tilde{\Phi}_n . \tag{12.68}$$

Since ∇_R and $\dot{X}$ are effectively single-particle operators, (12.57), the differential equations (12.66) for the configuration amplitudes are easily seen to split into equations for amplitudes associated with single-particle wave functions

$$\dot{\tilde{c}}_n = -\dot{\boldsymbol{R}} \cdot \sum_m \tilde{\boldsymbol{D}}_{nm} \tilde{c}_m - \frac{\mathrm{i}}{\hbar} \tilde{\varepsilon}_n \tilde{c}_n , \tag{12.69}$$

where

$$\varepsilon_n = \sum_{i=1}^{Z'} \tilde{\varepsilon}_{n_i} , \qquad c_n = \prod_{i=1}^{Z'} \tilde{c}_{n_i} \tag{12.70}$$

and

$$\tilde{\boldsymbol{D}}_{nm} = \langle \tilde{\Phi}_n | \nabla_R + \frac{\mathrm{i} m_{\mathrm{e}}}{2\hbar} [f(\boldsymbol{r}, \boldsymbol{R}) + \eta]\, \boldsymbol{\alpha} c | \tilde{\Phi}_m \rangle . \tag{12.71}$$

The final step in this reduction is to absorb the diagonal coupling into the amplitudes as in (12.13) by a phase term:

$$\tilde{c}_n(t) = a_n(t) \exp[-\mathrm{i}\chi_n(t)/\hbar] \quad \text{with} \tag{12.72}$$

$$\dot{\chi}_n(t) = \tilde{\varepsilon}_n(R(t)) \, . \tag{12.73}$$

Substituting (12.72) into (12.69) gives the coupled equations

$$\dot{a}_n(t) = -\sum_m \dot{\boldsymbol{R}} \cdot \tilde{\boldsymbol{D}}_{nm}(R(t))\, a_m(t) \exp\left[\frac{\mathrm{i}}{\hbar}(\chi_n - \chi_m)\right] \tag{12.74}$$

which are identical in form to (12.19), derived directly from the time-dependent two-centre equation. As already mentioned in connection with (12.24), any change in $\tilde{\Phi}_n$ is caused by variation of the internuclear distance vector $\boldsymbol{R}(t)$, whence the matrix element is

$$\langle \tilde{\Phi}_n | \dot{\tilde{\Phi}}_m \rangle = \dot{\boldsymbol{R}} \cdot \langle \tilde{\Phi}_n | \nabla_R | \tilde{\Phi}_m \rangle \, . \tag{12.75}$$

Although (12.19) are almost identical to the coupled channel equations (12.74), the coupling matrix elements are slightly different. For $R \to \infty$, i.e. for $t \to \pm\infty$, the matrix elements (12.75) in general do not approach zero, causing long-range couplings among the stationary quasi-molecular states $\tilde{\Phi}_n$. These couplings reflect that the expansion (12.11) is not appropriate for describing the asymptotic scattering region. To correct for this behaviour the basis functions $\tilde{\Phi}_n$ should rather be taken at the point $\tilde{\boldsymbol{R}} = \boldsymbol{R} + \boldsymbol{X} = \boldsymbol{R} + m_e/\mu\, \boldsymbol{S}$, with $\boldsymbol{S}$ defined in (12.52). Equivalently, we can take the corrected matrix elements $\tilde{\boldsymbol{D}}_{nm}$ of (12.71) in place of the simple form (12.75). The correction is commonly known as the "translation" correction since it is caused by the asymptotic relative motion of the two individual ions [Ba 58].

In the following we always assume that the electron – electron interaction is approximated by an effective potential. We therefore drop the tilde signs in the rest of this chapter and simply denote the one-electron wave functions, energies and matrix elements by Φ_n, ε_n and D_{nm}, respectively. The wave functions $\psi(\boldsymbol{r}, t)$ with the corrected amplitudes a_n defined by (12.74) satisfy the modified time-dependent two-centre Dirac equation

$$\begin{aligned} \mathrm{i}\hbar \frac{\partial}{\partial t} \psi &= [H_{\mathrm{TCD}} + V_{\mathrm{eff}} + \mu \dot{\boldsymbol{R}} \cdot \dot{\boldsymbol{X}}]\, \psi \\ &= \left[c\boldsymbol{\alpha} \cdot \left(\boldsymbol{p} + m_e \dot{\boldsymbol{R}} \frac{f+\eta}{2} \right) + \beta m_e c^2 + V^{\mathrm{eA}} + V^{\mathrm{eB}} + V_{\mathrm{eff}} \right] \psi \, , \end{aligned} \tag{12.76}$$

where the electronic momentum is corrected for the "translational" momentum $m\dot{\boldsymbol{R}}(f+\eta)/2$. This is a small correction whenever the electronic "velocity" p/m_e is small compared with the nuclear velocity $\dot{R}$, or when $(f+\eta) \approx 0$, i.e. for small internuclear separations.

12.5 Collisions with Nuclear Interaction

As already mentioned in Sect. 11.1, the characteristic duration of a heavy-ion collision in Coulomb scattering, i.e. the time the nuclei spend together closer than the critical distance, is about two orders of magnitude shorter than the lifetime of a supercritical K vacancy with respect to spontaneous positron emission. This led *Rafelski, Müller* and *Greiner* et al. [Ra 78c] to suggest that it might be useful to consider collisions at somewhat higher energy where the nuclei can come into contact and nuclear reactions occur.

Let us estimate the threshold for nuclear reactions, the so-called Coulomb barrier. Equating the Coulomb energy at the nuclear touching point with the initial kinetic energy in the centre-of-mass system, then

$$\frac{Z_1 Z_2 e^2}{R_1+R_2} = \frac{A_1 A_2}{A_1+A_2}\left(\frac{E_l}{A_1}\right)_{\mathrm{CB}}, \qquad (12.77)$$

where the index "1" refers to the projectile and "2" to the target, R_i stand for the nuclear radii and E_l is the beam energy in the laboratory system. For identical nuclei, e.g. uranium and uranium,

$$(E/A)_{\mathrm{CB}} = Z^2 e^2 / A R_{\mathrm{N}}\,. \qquad (12.78)$$

The uranium nucleus is, as all transactinide nuclei, strongly deformed, Fig. 12.5. Its nuclear radius R_{N} is therefore not well defined, but the threshold for nuclear reactions can be estimated by taking the length of the large major axis of the nuclear ellipsoid, yielding $R_{\mathrm{N}} \approx 9\,\mathrm{fm}$. With $Z=92$ and $A=238$ the Coulomb barrier corresponds to a beam energy of roughly 5.7 MeV per nucleon. Above this threshold nuclear reactions set in, first gently, then more violently as the energy is increased.

Crudely speaking, two scenarios could be envisaged in the energy region from the Coulomb barrier up to energies of about 10 MeV per nucleon. In the first, more conservative, scenario, one would try to make use of the prolonged interaction time between two nuclei in a violent, so-called deep inelastic collision, Fig. 12.6. The transfer of a very large number of nucleons observed in such collisions is probably associated with an extended period of nuclear contact that may reach 10^{-20} s for medium heavy nuclei. These estimates of the reaction time are based on macroscopic, statistical models for the nucleon exchange [Ag 77, 79, Ko 79a]. In these models the available reaction time is determined by the balance between the attractive nuclear interaction and the repulsive Coulomb force, in conjunction with a frictional force that slows the nuclear motion down. Unfortunately, the Coulomb force is almost overwhelmingly dominant in the double uranium system because of the large nuclear charge. Models that explain the gross features of mass transfer and kinetic energy dissipation in the U – U reaction at energies between 6.5 and 10 MeV per nucleon yield contact times of the order of 10^{-21} s. According to our prior estimate this is close to the collision time on a Rutherford trajectory in a collision below the Coulomb barrier, which may therefore be doubled by the nuclear reaction.

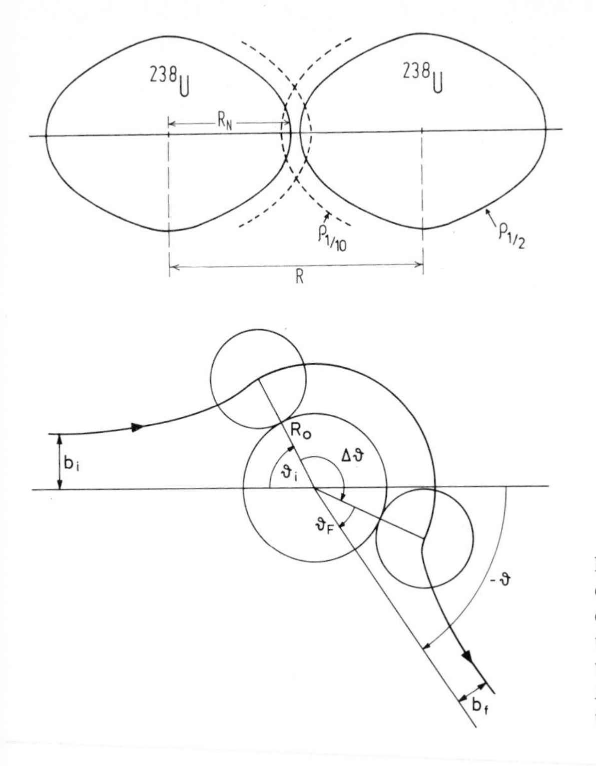

Fig. 12.5. Actual shape of two uranium nuclei in an oriented head-on collision [Re 81a]

Fig. 12.6. Trajectory in a deep inelastic nuclear collision. Nucleons are exchanged during the period of contact of the nuclei. Such trajectories were already discussed in 1976 by *Reinhardt* et al. [Re 76b] in the context of nuclear bremsstrahlung

The second, more spectacular, scenario assumes that much longer contact times are possible in collisions right at the Coulomb barrier, if the Coulomb force is locally cancelled by the sudden onset of nuclear attraction. This would lead to the existence of a (small) pocket in the nuclear potential $V(R)$ in which the relative nuclear motion could be captured for some time (Fig. 12.7). Such a phenomenon is well known from collisions of light and medium heavy nuclei ($^{12}C + {}^{12}C$, $^{16}O + {}^{16}O$, etc. up to $^{56}Ni + {}^{56}Ni$), where the temporary existence of quasi-stable dinuclear systems of very large deformation is firmly established [Al 60, Pa 81, Pa 86]. These "nuclear molecules" determine a large part of the reaction cross-section in those collisions (Fig. 12.8).

The possibility of forming a nuclear molecular configuration with two nuclei as heavy as uranium is not so firmly established and was, in fact, not seriously expected from the standpoint of standard heavy-ion theory. However, recent calculations [Rh 83, Se 84] in the context of accepted models of the interaction between complex nuclei have shown that there may be a narrow region of effective attraction just where the nuclear surfaces touch. The energy required to reach this molecular configuration is estimated to be about 5.9 MeV per nucleon in a collision of two uranium nuclei.

It is clear that for a very long nuclear contact time $T \gg 10^{-19}$ s, the spontaneous decay of a K vacancy by positron emission would proceed essentially as in a static situation. The spectrum of spontaneously created positrons would have

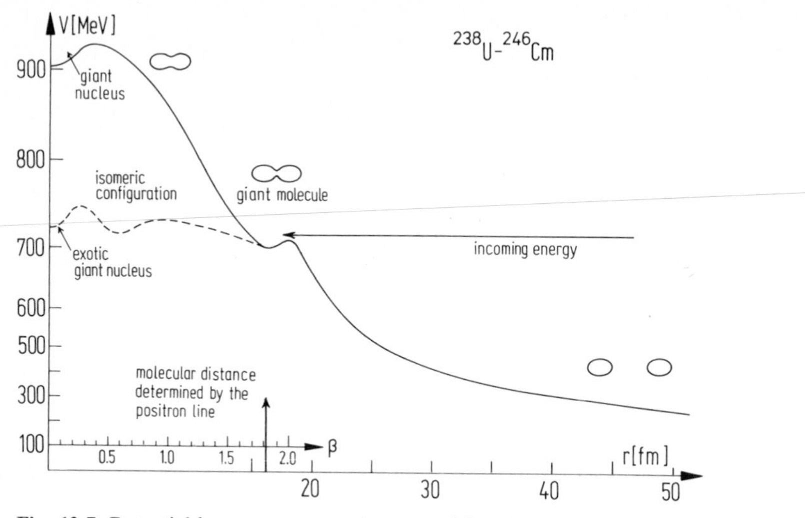

Fig. 12.7. Potential between two very heavy nuclei such as U+U or U+Cm. When the nuclei just touch, a small pocket may be formed in the potential, due to the sudden onset of the nuclear force [Se 84]

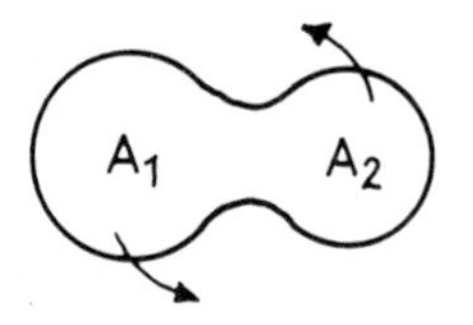

Fig. 12.8. A nuclear molecule in rotation

the shape of a Lorentzian, with a width given by the width of the bound state resonance embedded in the negative energy continuum, Sect. 10.3. Since the nuclear collision process is still present, leading to the initial formation of the K vacancy, a broad dynamically generated background would be superimposed on the far tails of this Lorentzian. If, however, the contact time T is comparable to, or shorter than, the spontaneous decay time of the neutral vacuum, contributions from spontaneous positron creation described by the coupling $V_E = \langle \tilde{\Phi}_E | H_{\text{TCD}} | \Phi_R \rangle$ and from dynamically induced positron creation [Sm 74, Re 81a] described by $\langle \tilde{\Phi}_E | \dot{\Phi}_R \rangle$ become thoroughly mixed, (12.32). In fact, their amplitudes must be added coherently because there is no way of experimentally distinguishing in a given collision whether a positron has been created by the action of the spontaneous or the dynamical coupling.

In the framework of the semiclassical approximation, where the nuclei move on a classical trajectory $R(t)$, a nuclear reaction modifies the function $R(t)$ due to the presence of nuclear forces. In general, it is possible to distinguish three regions, Fig. 12.9: (i) the part of the trajectory when the nuclei approach is described by a Rutherford hyperbola (12.4); (ii) the period during which nuclear forces delay the nuclear relative motion; (iii) the final part when the two nuclei fly apart is again given by Rutherford motion, but possibly with different

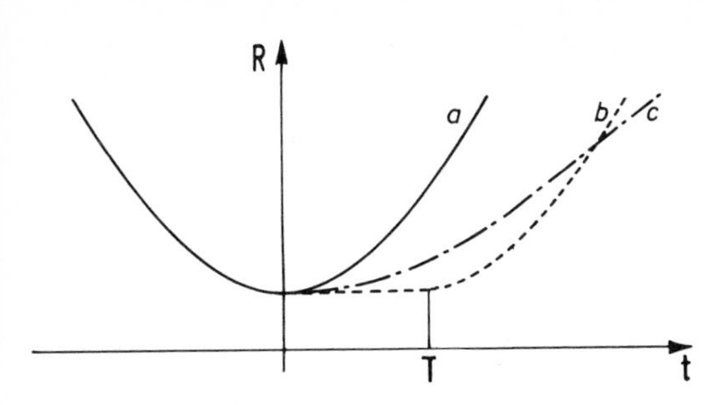

Fig. 12.9. Various trajectories $R(t)$ in a heavy-ion collision: (*a*) Rutherford trajectory, (*b*) Rutherford trajectory with finite contact time T, (*c*) trajectory with friction and energy loss

parameters owing to the transfer of energy, angular momentum and mass due to internal nuclear degrees of freedom. These regions are denoted separately by the superscripts (in), (nuc) and (out), respectively. Assume that the nuclei come into contact at $t = 0$ and break apart at $t = T$. The evolution of the solutions $\psi_k^{(+)}(\boldsymbol{r}, t)$ of the time-dependent two-centre Dirac equation can then be described by the product of the time-evolution operators in the three separate regions:

$$\psi_k^{(+)}(\boldsymbol{r}, t_\mathrm{f}) = U^{(\mathrm{out})}(t_\mathrm{f}, T)\, U^{(\mathrm{nuc})}(T, 0)\, U^{(\mathrm{in})}(0, t_\mathrm{i})\, \Phi_k(\boldsymbol{r}, \boldsymbol{R}(t_\mathrm{i}))\, \mathrm{e}^{-\mathrm{i}\chi_k(t_\mathrm{i})} . \qquad (12.79)$$

In each region, U is the solution of the time-dependent Dirac equation (8.45) with the initial conditions

$$U^{(\mathrm{out})}(T, T) = U^{(\mathrm{nuc})}(0, 0) = U^{(\mathrm{in})}(t_\mathrm{i}, t_\mathrm{i}) = 1 . \qquad (12.80)$$

Because of the completeness relation for the quasi-molecular states, these conditions can be written as

$$U^{(\mathrm{in})}(t_\mathrm{i}, t_\mathrm{i}) = \sum_k |\Phi_k(\boldsymbol{r}, \boldsymbol{R}(t_\mathrm{i}))\rangle \langle \Phi_k(\boldsymbol{r}, \boldsymbol{R}(t_\mathrm{i}))| \qquad (12.81\mathrm{a})$$

$$U^{(\mathrm{nuc})}(0, 0) = \sum_k |\Phi_k(\boldsymbol{r}, \boldsymbol{R}(0)\rangle \langle \Phi_k(\boldsymbol{r}, \boldsymbol{R}(0))| \qquad (12.81\mathrm{b})$$

$$U^{(\mathrm{out})}(T, T) = \sum_k |\Phi_k(\boldsymbol{r}, \boldsymbol{R}(T))\rangle \langle \Phi_k(\boldsymbol{r}, \boldsymbol{R})(T))| . \qquad (12.81\mathrm{c})$$

The time-evolution operator at unequal times is obtained by propagating the ket function according to the two-centre Dirac equation. If these time-dependent functions are represented by expansions in the instantaneous quasi-molecular states like (12.11), then

$$U^{(\mathrm{in})}(t, t_\mathrm{i}) = \sum_{kl} c_{kl}^{(\mathrm{in})}(t)\, |\Phi_l(\boldsymbol{r}, \boldsymbol{R}(t))\rangle \langle \Phi_k(\boldsymbol{r}, \boldsymbol{R}(t_\mathrm{i}))| , \qquad (12.82\mathrm{a})$$

$$U^{(\mathrm{nuc})}(t, 0) = \sum_{kl} c_{kl}^{(\mathrm{nuc})}(t)\, |\Phi_l(\boldsymbol{r}, \boldsymbol{R}(t))\rangle \langle \Phi_k(\boldsymbol{r}, \boldsymbol{R}(0))| , \qquad (12.82\mathrm{b})$$

$$U^{(\mathrm{out})}(t, T) = \sum_{kl} c_{kl}^{(\mathrm{out})}(t)\, |\Phi_l(\boldsymbol{r}, \boldsymbol{R}(t))\rangle \langle \Phi_k(\boldsymbol{r}, \boldsymbol{R}(T))| , \qquad (12.82\mathrm{c})$$

with the initial conditions

$$c_{kl}^{(\mathrm{in})}(t_\mathrm{i}) = c_{kl}^{(\mathrm{nuc})}(0) = c_{kl}^{(\mathrm{out})}(T) = \delta_{kl} . \qquad (12.83)$$

Substituting these expressions into (12.79) gives the time-dependent wave function at the final time t_f in terms of the amplitudes c_{kl}:

$$\psi_k^{(+)}(\boldsymbol{r}, t_f) = \sum_{lmn} c_{kl}^{(in)}(0)\, c_{lm}^{(nuc)}(T)\, c_{mn}^{(out)}(t_f)\, \Phi_n(\boldsymbol{r}, \boldsymbol{R}(t_f))\, e^{-i\chi_n(t_i)} . \tag{12.84}$$

Here the amplitudes are in reverse order to bring the indices in line for matrix multiplication.

It is now useful to replace the amplitudes c_{kl} by the amplitudes a_{kl} according to (12.13). Owing to the initial conditions (12.83) that differ slightly from (12.12), we choose

$$c_{kl}^{(in)}(0) = a_{kl}^{(in)}\, e^{-i[\chi_l(0) - \chi_k(t_i)]} , \tag{12.85a}$$

$$c_{lm}^{(nuc)}(T) = a_{lm}^{(nuc)}(T)\, e^{-i[\chi_m(T) - \chi_l(0)]} , \tag{12.85b}$$

$$c_{mn}^{(out)}(t_f) = a_{mn}^{(out)}(t_f)\, e^{-i\chi_n(t_f)} . \tag{12.85c}$$

The different choice of phase in (12.85c) is motivated by the desire to make the differential equations for the amplitudes $a_{mn}^{(out)}$ independent of the delay time T. We can now calculate the single-particle S-matrix elements as in (12.34), obtaining the final result

$$\begin{aligned} S_{ik}^{(T)} &= \int d^3r\, \psi_i^{(-)}(\boldsymbol{r}, t_f)^\dagger\, \psi_k^{(+)}(\boldsymbol{r}, t_f) \\ &= \sum_{lm} e^{i\chi_i(t_f)}\, c_{kl}^{(in)}(0)\, c_{lm}^{(nuc)}(T)\, c_{mi}^{(out)}(t_f)\, e^{-i\chi_k(t_i)} \\ &= \sum_{lm} a_{kl}^{(in)}(0)\, a_{lm}^{(nuc)}(T)\, a_{mi}^{(out)}(t_f)\, e^{-i\chi_m(T)} . \end{aligned} \tag{12.86}$$

It is now possible to substantiate our claim that spontaneous and dynamically induced positron production occur side by side in collisions with nuclear contact. To see this, assume that the nuclear molecular configuration remains completely frozen during the time of contact, i.e. $R(t) = 0$ for $0 \leqslant t \leqslant T$. Then $a_{lm}^{(nuc)}$ does not contain dynamical contributions because the matrix elements $\langle \Phi_l | \dot{\Phi}_m \rangle$ vanish during that time. The time-development operator $U^{(nuc)}(T, 0)$ for this case was derived in Sect. 10.3, (10.72), from which we can easily read off the amplitudes $a_{lm}^{(nuc)}(T)$:

$$a_{lm}^{(nuc)}(T) = \begin{cases} \delta_{lm} & (E_l, E_m > -m_e) \\ \exp(-\frac{1}{2}\tilde{\Gamma}T) & (l = m = R) \\ \delta(E_l - E_m) + \tilde{a}_{E_m, E_l}\, e^{iE_m T} & (E_l, E_m < -m_e) \\ \dfrac{1 - \exp[i(E_m - E_R + \frac{i}{2}\tilde{\Gamma})T]}{E_m - E_R + \frac{i}{2}\tilde{\Gamma}}\, \tilde{V}_{E_m} & (l = R, E_m < -m_e) \\ \dfrac{\exp[i(E_R - E_l)T] - \exp(-\frac{1}{2}\tilde{\Gamma}T)}{E_l - E_R + \frac{i}{2}\tilde{\Gamma}}\, \tilde{V}_{E_l}^* & (m = R, E_l < -m_e) . \end{cases} \tag{12.87}$$

Here E_R is the resonance energy in the nuclear molecular configuration. Taking $l = R$, clearly a vacancy in the supercritical K shell leaks out into the negative energy continuum states, causing a positron distribution

$$\frac{|\tilde{V}_E|^2}{(E-E_R)^2 + \frac{1}{4}\tilde{\Gamma}^2} \equiv |a_R(E)|^2 \tag{12.88}$$

for nuclear contact times $T \gg \Gamma^{-1}$. No matter how long T is, $a_{kl}^{(in)}$ and $a_{mn}^{(out)}$ are still present in (12.86), which describes positron production during the approach and departure phases of the heavy-ion collision. As already mentioned, the spontaneous mechanism contributes only negligibly to these amplitudes, even in a supercritical collision system, because the collision time on a Rutherford trajectory is too short.

Also note that the sum over intermediate states l and m in (12.86) involves the amplitudes and not the probabilities, i.e. the amplitudes for positron creation due to nuclear motion and due to spontaneous decay of the supercritical K vacancy are added coherently. Although this makes it impossible to determine whether a given positron is of dynamical or spontaneous origin by looking at a single collision, the spontaneous decay leaves a clear narrow line in the positron spectrum, if the nuclear contact time T is long enough.

Figure 12.10a illustrates this effect, where the positron spectra calculated with (12.86) from uranium–uranium collisions with various nuclear contact times

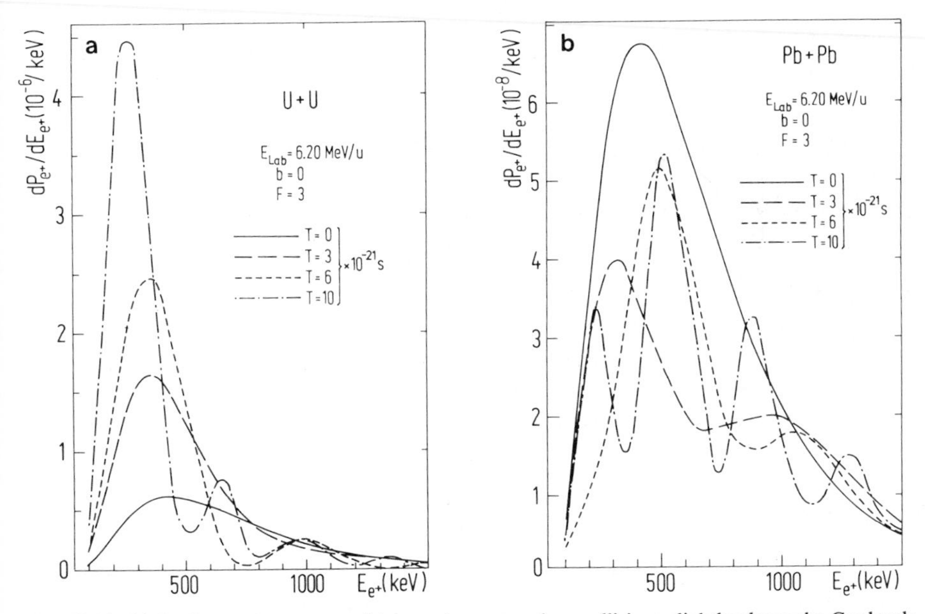

Fig. 12.10. **(a)** Positron spectra created in uranium–uranium collisions slightly above the Coulomb barrier with different nuclear delay times ranging up to $T = 10^{-20}$ s. For long delay times the spontaneous emission line emerges clearly. **(b)** Same for subcritical lead–lead collisions ($Z_1 + Z_2 = 164 < Z_{cr}$) [Re 81a]. Note the different scales in figures **(a)** and **(b)**

from $T=0$ to $T=10^{-20}$ s are seen. When the delay time T is much longer than the collision time on a Rutherford trajectory ($\tau_{\text{coll}} \sim 10^{-21}$ s), the spontaneously produced positrons show up as a conspicuous line in the spectrum. The positron of the line peak is determined by the resonance energy E_{R}, according to (12.88). The origin of the line structure from spontaneous decay is emphasized by Fig. 12.10b, which shows the same spectra for a collision of two lead atoms. As expected, no line structure grows out of the dynamically induced spectrum, because the Pb+Pb system with combined charge $Z_1+Z_2=164$ is not supercritical.

For long nuclear contact times we can obtain an approximate expression for the height and shape of the spontaneous line in the spectrum by considering only the dominant intermediate states in (12.35) for the positron energy distribution. Combining (12.35, 86) gives, generally ($E < -m_{\text{e}}$),

$$\frac{d\bar{N}(T)}{dE} = \sum_{k>F} \left| \sum_{lm} a_{kl}^{(\text{in})}(0)\, a_{lm}^{(\text{nuc})}(T)\, a_{mE}^{(\text{out})}(t_{\text{f}})\, \mathrm{e}^{-\mathrm{i}E_m T} \right|^2 . \tag{12.89}$$

Neglecting dynamic positron production or rescattering of positrons during the outgoing part of the collision means $m=E$ and $a_{EE}^{(\text{out})} \approx 1$. (Numerical calculations show that the dynamical coupling among positron states $\langle \Phi_{E'} | \dot{\Phi}_E \rangle$ is small [Re 81a, Mü 83c].) The spontaneous decay during the period of nuclear contact is then given by the amplitude $a_{lm}^{(\text{nuc})}$ for $l=R$, $m=E$ from (12.87). Since the sums over intermediate states have broken down,

$$\frac{d\bar{N}(T)}{dE} \approx \left(\sum_{k>F} |a_{k,1s}^{(\text{in})}(0)|^2 \right) |a_R(E)|^2 [1 - 2\mathrm{e}^{-\frac{1}{2}\tilde{\Gamma}T} \cos(E-E_{\text{R}})T + \mathrm{e}^{-\tilde{\Gamma}T}] \tag{12.90}$$

with $|a_R(E)|^2$ given by (12.88). The first factor is easily identified as the probability of forming a vacancy in the $1s\sigma$ state during the approach of the nuclei, i.e. the K-shell ionization probability at the moment the nuclei touch and spontaneous positron creation sets in. The next two factors are just the positron spectrum that would result had we started with a K vacancy right away, given by (10.66).

We have thus derived the (approximate) result that the positron spectrum in the line region factors for collisions with long nuclear contact time into the probability for making a K vacancy and the probability for its subsequent decay during the lifetime of the nuclear molecule:

$$\frac{d\bar{N}(T)}{dE} \approx P_k \cdot \bar{N}_E(T) . \tag{12.91}$$

We emphasize once more that this formula is only approximately valid and does not hold outside the region of the spontaneous positron line.

The discussion so far has been of academic nature because nuclear reactions are not classical deterministic processes with a definite time of duration T. In principle, nuclear reactions should be described in the framework of quantum mechanics, where the relevant quantity is not the reaction time but the scattering amplitude S_{fi} which is a function of the bombarding energy.

For deep inelastic collisions between heavy, complex nuclei such a treatment is impractical because of the vast number of open final channels. Here the quantal description is usually replaced by a classical or semiclassical statistical treatment. The duration time of the reaction is then not a sharply defined quantity, but has a certain statistical distribution denoted by a function $f(T)$. Some typical parameterizations of $f(T)$ are [To 83]

$$f(T) = \Gamma_0 e^{-\Gamma_0 T}, \tag{12.92}$$

which corresponds to the exponential decay of an isolated nuclear resonance state with lifetime $T_0 = \Gamma_0^{-1}$, or a Gaussian centred at $T = T_0$,

$$f(T) \sim \left(\frac{2}{\pi \Delta T}\right)^{1/2} \exp\left(-\frac{(T-T_0)^2}{2\Delta T^2}\right), \tag{12.93}$$

which might be the result of a statistical model of the nuclear reaction. With this distribution function the full expression for the positron spectrum in a heavy-ion collision is

$$\frac{d\bar{N}}{dE} = \int_0^\infty dT f(T) \frac{d\bar{N}(T)}{dE}, \tag{12.94}$$

where $d\bar{N}(T)/dE$ is given by (12.89).

One may wonder how this result can be derived from the quantum mechanical formulation of the heavy-ion collision presented in Sects. 12.3, 4. The influence of nuclear reactions on the dynamical processes occurring in the collisions has been studied by several authors [Ei 60, Ci 65, Fe 81, Be 82, McV 82], for positron production in particular by *Tomoda* and *Weidenmüller* [To 83, 84] and *Heinz* et al. [He 83a]. The equivalence of their results to (12.94) was shown by *Reinhardt* et al. [Re 83]. Let us sketch the derivation briefly, starting from (12.64) for the WKB wave function of nuclear relative motion. When a nuclear reaction occurs, the expressions for the incoming and outgoing wave cannot be matched directly at the turning point, since one has to take into account the additional phase shift in the outgoing wave due to the nuclear reaction. Equation (12.64) must therefore be replaced by

$$\begin{aligned} \chi_n^{\mathrm{WKB}}(R) = {} & c_n^{(\mathrm{in})}(R)\, k(R)^{-1/2} \exp\left[\mathrm{i}\int k(R)\, dR\right] \\ & + c_n^{(\mathrm{out})}(R)\, k(R)^{-1/2} S_{\mathrm{N}}(E - \varepsilon_{\mathrm{n}}(R_0)) \exp\left[-\mathrm{i}\int k(R)\, dr\right], \end{aligned} \tag{12.95}$$

where the nuclear S matrix S_{N} is taken at the energy available for the nuclear reaction, when the energy $\varepsilon_{\mathrm{n}}(R_0)$ is absorbed in the electronic configuration at the classical point R_0.

For elastic scattering, the S matrix is related to the scattering phase shift δ by $S_{\mathrm{N}}(E) = \exp[2\mathrm{i}\delta(E)]$. To arrive at (12.95) assume that the nuclear S matrix is unaffected by the electronic processes, expect for the available energy ($E = -\varepsilon_{\mathrm{n}}$).

More complicated processes, such as transfer of energy between nuclei and electrons (or positrons) during the reaction, are not covered by this expression.

For $c_n^{(\mathrm{in})}(R)$ and $c_n^{(\mathrm{out})}(R)$ (12.66) arises, which led to the single-particle equations (12.74) identical in form to (12.19), derived from the time-dependent Dirac equation. Now, however, from (12.95) the correct expression for the amplitude on the outgoing branch of the trajectory is not $c_n^{(\mathrm{out})}$ but $c_n^{(\mathrm{out})} S(E-\varepsilon_n(R_0))$. Solving (12.74) for the appropriate initial conditions as above, the result (12.86) is replaced by

$$S'_{ik}(E) = \sum_l a_{kl}^{(\mathrm{in})}(0)\, S_{\mathrm{N}}(E-\varepsilon_l(R_0))\, a_{li}^{(\mathrm{out})}(t_{\mathrm{f}}) \tag{12.96}$$

in quantum mechanical theory. To relate this to our result (12.94), remember that in a realistic experimental situation the beam energy E is never precisely determined, but can vary in a certain range. Then compute the averaged square of the S-matrix element $S'_{ik}(T)$ integrated over the reaction time distribution function $f(T)$

$$\begin{aligned}
&\sum_{ll'} a_{kl}^{(\mathrm{in})}(0)\, a_{kl'}^{(\mathrm{in})}(0)^*\, a_{li}^{(\mathrm{out})}(t_{\mathrm{f}})\, a_{l'i}^{(\mathrm{out})}(t_{\mathrm{f}})^* \,\langle S_{\mathrm{N}}(E-\varepsilon_l)\, S_{\mathrm{N}}^*(E-\varepsilon_{l'})\rangle_{\mathrm{av}} \\
&\quad = \sum_{lm}\sum_{l'm'} a_{kl}^{(\mathrm{in})}(0)\, a_{kl'}^{(\mathrm{in})}(0)^*\, a_{mi}^{(\mathrm{out})}(t_{\mathrm{f}})\, a_{m'i}^{(\mathrm{out})}(t_{\mathrm{f}}) \\
&\quad\quad \cdot \int_0^\infty dT\, a_{lm}^{(\mathrm{nuc})}(T)\, a_{l'm'}^{(\mathrm{nuc})}(T)^* \exp[\mathrm{i}(\varepsilon_{m'}-\varepsilon_m)T]\, f(T)\,.
\end{aligned} \tag{12.97}$$

Recall that the derivation of the semiclassical approximation in Sects. 12.3, 4 was done only for the case when the exact eigenstates of the two-centre Dirac Hamiltonian could be used everywhere. In that case the expressions in (12.87) simply reduce to

$$a_{lm}^{(\mathrm{nuc})}(T) = \delta_{lm}\,. \tag{12.98}$$

Taking this into account, (12.97) simplifies considerably, leading to

$$\langle S_{\mathrm{N}}(E-\varepsilon_l)\, S_{\mathrm{N}}^*(E-\varepsilon_{l'})\rangle_{\mathrm{av}} = \int_0^\infty dT f(T) \exp[\mathrm{i}(\varepsilon_{l'}-\varepsilon_l)T]\,, \tag{12.99}$$

which may be easily inverted to yield the quantum mechanical expression for the distribution function of nuclear reaction times:

$$\begin{aligned}
f(T) &= \int_{-\infty}^{\infty} d\omega\, \mathrm{e}^{\mathrm{i}\omega T} \langle S_{\mathrm{N}}(E-\varepsilon_l)\, S_{\mathrm{N}}^*(E-\varepsilon_l+\omega)\rangle_{\mathrm{av}} \\
&\approx \int_{-\infty}^{\infty} d\omega\, \mathrm{e}^{\mathrm{i}\omega T} \langle S_{\mathrm{N}}(E)\, S_{\mathrm{N}}^*(E+\omega)\rangle_{\mathrm{av}}\,,
\end{aligned} \tag{12.100}$$

where ε_l has been neglected, because in practice the scattering energy is not known to that accuracy, anyway. The average of the quantity in brackets is known as the autocorrelation function of the nuclear scattering matrix. Our result therefore shows that the classical time distribution must be identified with the Fourier transform of this quantity. If the nuclear scattering occurs in a region with many overlapping resonances, statistical resonance theory predicts that the autocorrelation function is of the form [Ma 79b]

$$\langle S_{\rm N}(E)\, S_{\rm N}^{*}(E+\omega)\rangle_{\rm av} \approx \frac{-\dfrac{\mathrm{i}}{2\pi}\,\Gamma_{\rm N}}{\omega-\mathrm{i}\Gamma_{\rm N}}\,, \tag{12.101}$$

yielding an experimental reaction time distribution like (12.92):

$$f(T) \approx \int\limits_{-\infty}^{\infty} d\omega\, \mathrm{e}^{\mathrm{i}\omega T} \frac{\left(\dfrac{-\mathrm{i}\Gamma_{\rm N}}{2\pi}\right)}{(\omega-\mathrm{i}\Gamma_{\rm N})} = 2\pi\mathrm{i}\left(\frac{-\mathrm{i}\Gamma_{\rm N}}{2\pi}\right)\mathrm{e}^{-\Gamma_{\rm N}T} = \Gamma_{\rm N}\mathrm{e}^{-\Gamma_n T}\,. \tag{12.102}$$

Another way to obtain the autocorrelation function of the nuclear S matrix is to study (12.99) for small values of $(\varepsilon_{l'}-\varepsilon_l)\,T$. Leaving aside the energy average and using the relation between S matrix and phase shift, $S_{\rm N}=\exp(2\mathrm{i}\delta)$, we expand in lowest order of a Taylor series. Because of unitarity $S_{\rm N}^{*}(E)\,S_{\rm N}(E)=1$, and

$$\begin{aligned} S_{\rm N}(E-\varepsilon_l)\,S_{\rm N}^{*}(E-\varepsilon_{l'}) &\approx S_{\rm N}(E)\,S_{\rm N}^{*}(E) - \varepsilon_l S_{\rm N}'(E)\,S_{\rm N}^{*}(E) - \varepsilon_{l'}\,S_{\rm N}(E)\,S_{\rm N}'(E)^{*} \\ &= 1-\varepsilon_l\frac{\partial}{\partial E}\ln S_{\rm N}(E) - \varepsilon_{l'}\frac{\partial}{\partial E}\ln S_{\rm N}^{*}(E) \\ &= 1+2\mathrm{i}\,(\varepsilon_{l'}-\varepsilon_l)\,\delta'(E)\,, \end{aligned} \tag{12.103}$$

where a prime indicates differentiation with respect to energy. Comparing (12.103) to the series expansion of the exponential on the rhs of (12.99), we identify

$$T \approx 2\frac{\partial\delta(E)}{\partial E}\,, \tag{12.104}$$

which is the standard expression for the time delay of a scattered wave in quantum mechanics [Bo 51, Ei 48, Wi 55, Sm 60]. Equation (12.100) can be understood as the generalization of this simple relation [Yo 74].

These remarks conclude the formal exposition of the theory of dynamical electronic processes in heavy-ion collision. Let us now discuss the state of numerical calculations and the experimental efforts to prove the existence of spontaneous positron creation in supercritical heavy-ion collisions.

12.6 Status of Numerical Calculations

The formalism developed in Sects, 12.1, 2 has been applied to calculate positron emission in various heavy-ion collision processes, without and with time delay due to a nuclear reaction. The amplitudes entering (12.35 or 89) were calculated

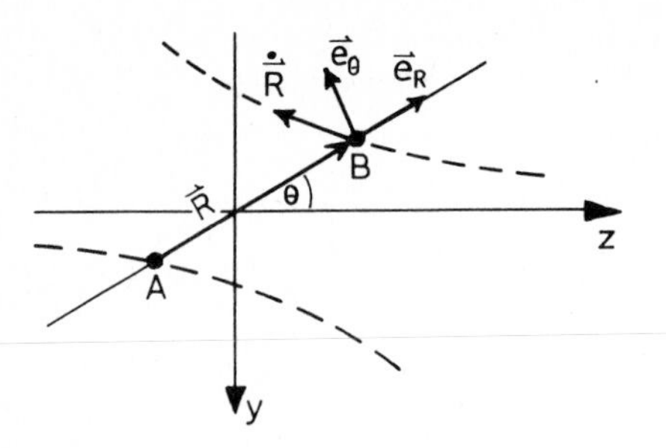

Fig. 12.11. Definition of vectors in the quasimolecule

by numerical integration of the coupled differential equations (12.19, 32). In the following we shall discuss the approximations inherent in these numerical calculations and present some results for both sub- and supercritical collisions.

As the energy eigenvalues of the two-centre Dirac equation (11.3) depend only on the relative nuclear distance $R = |\boldsymbol{R}|$, but not on the orientation of the vector $\boldsymbol{R}$, it is useful to separate the nuclear relative motion into a radial and an angular component, Fig. 12.11. The velocity vector can be decomposed as

$$\dot{\boldsymbol{R}} = v_{\rm R}\boldsymbol{e}_{\rm R} + v_{\Theta}\boldsymbol{e}_{\Theta} \quad \text{with} \tag{12.105}$$

$$v_{\rm R} = \dot{R}\,, \tag{12.106a}$$

$$v_{\Theta} = \dot{\Theta}R\,. \tag{12.106b}$$

The time-derivative operator entering the matrix element (12.75) is therefore

$$\begin{aligned}\dot{\boldsymbol{R}}\cdot\nabla_{\rm R} &= v_{\rm R}(\boldsymbol{e}_{\rm R}\cdot\nabla_{\rm R}) + v_{\Theta}(\boldsymbol{e}_{\Theta}\cdot\nabla_{\rm R})\\ &= \dot{R}\frac{\partial}{\partial R} + \dot{\Theta}R(\boldsymbol{e}_{\Theta}\cdot\nabla_{\rm R})\,.\end{aligned} \tag{12.107}$$

If the x axis is normal to the scattering plane,

$$R\boldsymbol{e}_{\Theta} = \boldsymbol{e}_x \times \boldsymbol{R}\,, \tag{12.108}$$

and the angular velocity vector of the rotation of the internuclear axis is

$$\boldsymbol{\Omega} = \dot{\Theta}\boldsymbol{e}_x\,. \tag{12.109}$$

Combining these relations, the angular part (12.107) becomes

$$\begin{aligned}\dot{\Theta}R(\boldsymbol{e}_{\Theta}\cdot\nabla_{\rm R}) &= \dot{\Theta}(\boldsymbol{e}_x\times\boldsymbol{R})\cdot\nabla_{\rm R} = \dot{\Theta}\boldsymbol{e}_x\cdot(\boldsymbol{R}\times\nabla_{\rm R})\\ &= \mathrm{i}\boldsymbol{\Omega}\cdot\boldsymbol{L}_{\rm R}\,,\end{aligned} \tag{12.110}$$

where $\boldsymbol{L}_{\rm R}$ is the operator of relative angular momentum of the nuclear motion. Of course $\boldsymbol{L}_{\rm R}$ is also the generator of an infinitesimal rotation of the internuclear axis. Figure 12.12 immediately shows that a small rotation of the internuclear axis is equivalent to a small rotation, but in the opposite direction, of the elec-

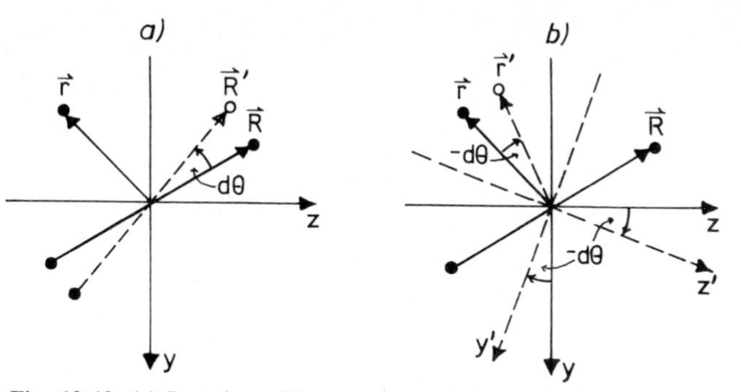

Fig. 12.12. **(a)** Rotation of internuclear axis by angle $d\theta$; **(b)** Rotation of electronic coordinate by $-d\theta$

tronic coordinate $\boldsymbol{r}$, except for an overall rotation of the coordinate system around the x axis. The overall rotation does not effect the scalar product in (12.110), hence, equivalently

$$\dot{\Theta} R (\boldsymbol{e}_\Theta \cdot \nabla_\mathrm{R})\, \Phi_l(\boldsymbol{r},\boldsymbol{R}) = -\mathrm{i}\,\boldsymbol{\Omega} \cdot \boldsymbol{j}\, \Phi_l(\boldsymbol{r},\boldsymbol{R}) = -\mathrm{i}\,\dot{\Theta} j_x\, \Phi_l(\boldsymbol{r},\boldsymbol{R})\,, \tag{12.111}$$

where $\boldsymbol{j}$ is the angular momentum operator of the electrons. Thus instead of (12.75),

$$\langle \Phi_n | \dot{\Phi}_m \rangle = \dot{R} \left\langle \Phi_n \left| \frac{\partial}{\partial R} \right| \Phi_m \right\rangle - \mathrm{i}\,\dot{\Theta} \langle \Phi_n | j_x | \Phi_m \rangle \quad . \tag{12.112}$$

The two contributions to the matrix element of the time-variation operator are known as *radial* and *rotational* coupling, respectively. The operator of radial coupling $\partial/\partial R$ connects only electronic states having the same projection of angular momentum on the axis, i.e. the same eigenvalue μ of j_z. On the other hand, j_x connects states which differ by ± 1 in the eigenvalue μ. In the united quasi-atomic limit $R \to 0$, the two-centre wave functions $\Phi_l(\boldsymbol{r},\boldsymbol{R})$ become wave functions of good angular momentum. The operator j_x then leads only to coupling among the magnetic substates of a given shell, but does not cause any excitations in this limit.

To obtain numerical results, several approximations have been applied to the semiclassical model.

1) While the (non-separable) two-centre Dirac equation has been solved for bound states [Mü 73a, 76b], no solutions have been available up to now for the relativistic molecular continuum. Detailed comparisons of binding energies and coupling matrix elements have shown [Be 76b, Be 80b], however, that up to internuclear distances of at least 500 fm, the inner-shell states are well described by restriction to the $l = 0$ part in the multipole expansion of the two-centre potential, (11.11), Fig. 12.13. Even for not too asymmetric heavy-ion systems, $|Z_1 - Z_2|/(Z_1 + Z_2) < 0.2$, the monopole terms are dominant. For the spherically symmetric problem both bound and continuum states have been derived in detail

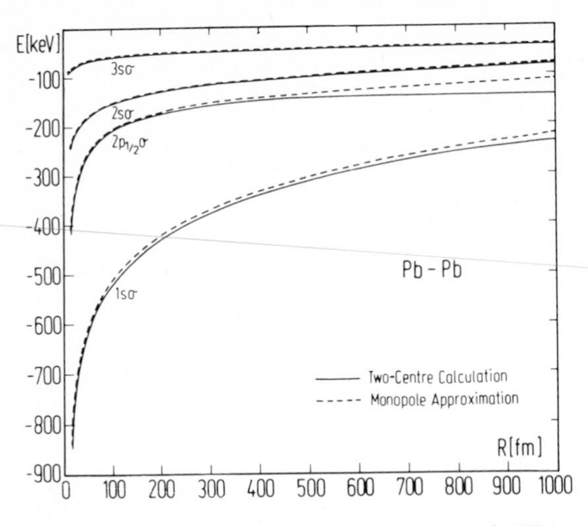

Fig. 12.13. Binding energies in Pb+Pb quasimolecules. The monopole approximation (– – –) is compared with exact two-centre Dirac results (———)

in Sects. 3.4, 5, and are easily generated. Therefore all calculations presented here have been done with monopole approximation.

2) The calculations are restricted to $\varkappa = -1$ and $\varkappa = +1$ states ($ns_{1/2}$ and $np_{1/2}$). Both sets of states are decoupled in (12.19) since they have different parity. They are expected to be the dominant channels on theoretical grounds, since in the superheavy systems under consideration, the wave functions with $|\varkappa| = 1$ are severely distorted by the strong potential, leading to large coupling matrix elements. Rotational coupling can be neglected in this approximation.

3) In Sect. 12.3 we discussed that the quasi-molecular model suffers from spurious asymptotic $\partial/\partial R$ couplings. Since the basis states are calculated assuming fixed nuclei, they do not satisfy the correct boundary conditions. With respect to this basis the non-vanishing nuclear velocity $\dot{R}$ induces transitions at arbitrarily large distance. As explained in Sect. 12.3, this problem may be avoided by introducing electron translation factors, which asymptotically switch over the basis to "travelling orbitals" correlated with either of the moving nuclei.

In these calculations we have simulated translation effects crudely: all coupling matrix elements are damped at separations $R \sim 1500 \dots 2000$ fm using a Gaussian factor. Compared with the non-relativistic case, relativistic quasi-molecular systems exhibit a strong maximum of the radial coupling matrix elements at small R where most of the excitation takes place (Fig. 12.14). Therefore translational effects should be less critical here.

4) Except where indicated otherwise, the electron – electron interaction has been included in the framework of the adiabatic time-dependent Hartree-Fock approximation (12.38), with Slater's expression for the exchange interaction [deR 84]. As most of the screening of an electron in the K shell is done by the electron in the other $1s$ state with opposite spin, and the probability of K-shell excitation is of the order of 10%, neglecting excitations in the screening potential should be allowed, at least for the $1s$ state.

The calculations from [Re 79] showed that it is not sufficient to solve the coupled channel equations (12.19) in perturbation theory. In fact, the probability

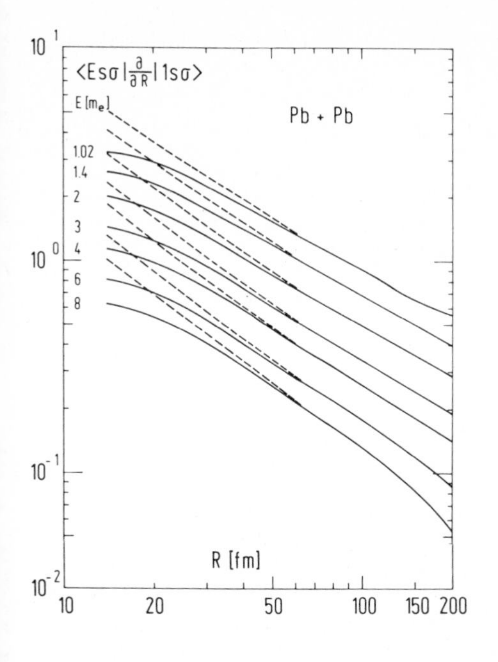

Fig. 12.14. Radial coupling matrix elements in the Pb+Pb quasimolecule between the $1s\sigma$ state and the $s\sigma$ continuum. (– – –) are for point nuclei, (———) take into account the finite nuclear radius

of ionization of the quasi-molecular $1s\sigma$ state comes out about a factor 5 too small in a perturbative calculation [So 79a], and similarly incorrect results are obtained for the positron spectrum. A complete numerical solution of the coupled equations for the amplitudes a_{kn} is therefore crucial. For each parity of the states with angular momentum one-half ($\varkappa = \pm 1$), it was found that 8 bound states suffice to obtain convergence of the calculation. The positive and negative energy continua were replaced by several suitably chosen representative states weighted with the corresponding energy interval

$$\int dE \rightarrow \sum_{i=1}^{N} \Delta \varepsilon_i ,$$

with $\Delta \varepsilon_i = \frac{1}{2}(\varepsilon_{i+1} + \varepsilon_{i-1}) - \varepsilon_i$ [Re 79]. The number and location of continuum states were varied until proper convergence was reached. For details see [So 80a, dRe 84]. The Fermi level of initially occupied quasi-molecular states was generally chosen above the three lowest bound states, i.e. above the $3s\sigma$ state in the $K = -1$ channel, and above the $4p_{1/2}\sigma$ state in the $K = +1$ channel, since the velocity of an electron in the next higher state, i.e. the $4s\sigma$ and $5p_{1/2}\sigma$ levels, respectively, is equal to, or lower than, the velocity of the projectile ion at beam energy of 5 MeV per nucleon. These states can, therefore, be considered as completely ionized during the passage of the projectile through the target. Other choices of the Fermi level and even a smeared distribution of Fermi levels have been studied [deR 81], with the result that a relative uncertainty of about 20% exists in the absolute probabilities. As this equals the possible normalization error in the present experimental data, we shall not discuss this effect further.

The next chapter discusses experimental results and compares then with theoretical calculations.

Bibliographical Notes

The theory of atomic collisions is the subject of several textbooks, including [Mo 33, Br 70a]. Research articles [Sm 75, He 81a, 83a] apply and extend the general theory to collisions involving superheavy quasimolecules. For the more generally interested reader [Re 76a, Re 77, Re 85] may be useful.

13. Experimental Test of Supercritical Fields in Heavy-Ion Collisions

This chapter describes the highlights of experimental research undertaken in recent years to get information on the physics of superheavy quasimolecules and, ultimately, to detect the decay of the neutral vacuum. Measurements of x-ray radiation, inner-shell hole production and electron emission have established the formation of superheavy quasimolecules and demonstrated the special properties of relativistic electron states in strong Coulomb potentials. To set the stage for a description of the latest development we introduce the experimental detectors used to measure positron spectra and subsequently discuss conversion processes producing background positrons and methods to deal with them. Results from experiments designed to study the gross features of positron emission are found to be in excellent agreement with theoretical predictions. In addition, however, several independent experiments have discovered remarkable narrow line structures in the emission spectrum. Further evidence indicates that the source of emission moves with the center of mass velocity and does not coincide with the individual nuclei. This supports the following interpretation: The lines are due to spontaneous positron emission in a supercritical potential. Their narrow width is due to the formation of longlived ($\gtrsim 10^{-19}$ s) "giant" nuclear compound systems of charge $Z \sim 182 \ldots 190$. Successes and remaining problems of this interpretation will be discussed in detail. At the close of this chapter attempts will be described to construct a reaction model, based on the idea of capture in a pocket of the internuclear potential, which gives a coherent explanation of giant nucleus formation and spontaneous positron formation.

13.1 Establishing Superheavy Quasimolecules

As mentioned in the introduction, the idea that quasi-molecular states are formed in slow atomic collisions dates back to the early 1930s [Hu 27, Mu 28, Mo 31, 33, Ma 33, Ba 53]. It was extended in the late 1960s to the idea of nuclear quasimolecules [Sche 69, Sche 70, Pr 70], which again had its feedback to atomic physics by triggering the suggestion that superheavy quasimolecules could exist in heavy-ion collisions [Gr 69a, b, Ge 70, Ra 71, Mü 72c]. First experimental indications for the existence of quasimolecules were found quite early [Co 34], but not pursued further because of lack of adequate experimental equipment. While more indirect evidence came forward from measurements of total ionization

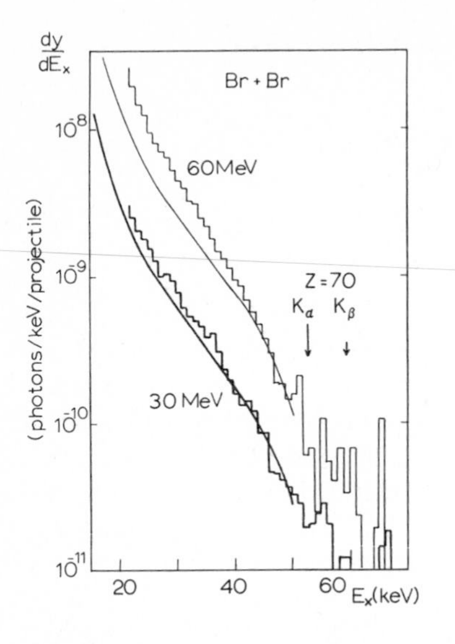

Fig. 13.1. Quasi-molecular K x-ray spectrum of the Br + Br system measured by *Meyerhof* and collaborators [Me 73b]. The spectrum extends up to the transition energy of the united atom Yb ($Z = 70$)

cross-sections in the 1960s [Lo 62, Fa 65], the work of *Saris* et al. [Sa 72] was a real step forward, who, for the first time, observed x-ray transitions between L quasi-molecular states directly. *Mokler* et al. [Mo 72] at the same time found similar results for the M shell. *Saris'* work, which involved L-shell transitions in fairly light systems, was extended to x-ray transitions into the quasi-molecular K-shell by *Meyerhof* et al. [Me 73b], who investigated the Br + Br system ($Z_1 + Z_2 = 70$). *Meyerhof* found a spectrum of non-characteristic x-rays that extended up to the K transition energy of the atom $Z = 70$, Fig. 13.1.

A breakthrough was made by *Davies* et al. [Da 74b, Gr 74], who showed the existence of a broad peak in the directional anisotropy of the quasi-molecular K radiation close to the transition energy of the united quasiatom, Fig. 13.2. This anisotropy had been predicted by theoretical work [Mü 74a, b, Gr 76] on the basis of the effect of rotation on the radiative emission process. A similar anisotropy was found by *Kraft* et al. [Kr 74] in M-shell transitions of the I + Au system. The energy-dependent anisotropy, once established, proved useful in two respects. For one, it enabled fairly systematic determination of the transition energy in the united quasiatom. This fact was exploited by *Wölfli* and collaborators [St 77, 78], who systematically studied the peak up to $Z_1 + Z_2 \sim 100$, Fig. 13.3. Unfortunately, the method lost its usefulness in the region of superheavy quasiatoms, as clearly shown by *Meyerhof* and collaborators [Me 79] in their measurement of the Pb + Pb system, where no anisotropy peak is seen (Fig. 13.4) in the quasi-molecular K x-ray spectrum.

The second, more important, application was *Meyerhof's* unambiguous proof of the quasi-molecular origin of the x-radiation [Me 75]. He showed that the angular asymmetry of emission was only forward-backward symmetric if viewed from the centre-of-mass system. In both the target (laboratory) and the projectile frames of reference, this symmetry was lost due to the Doppler shift, Fig. 13.5. Much further work on quasi-molecular x-rays has since been done and

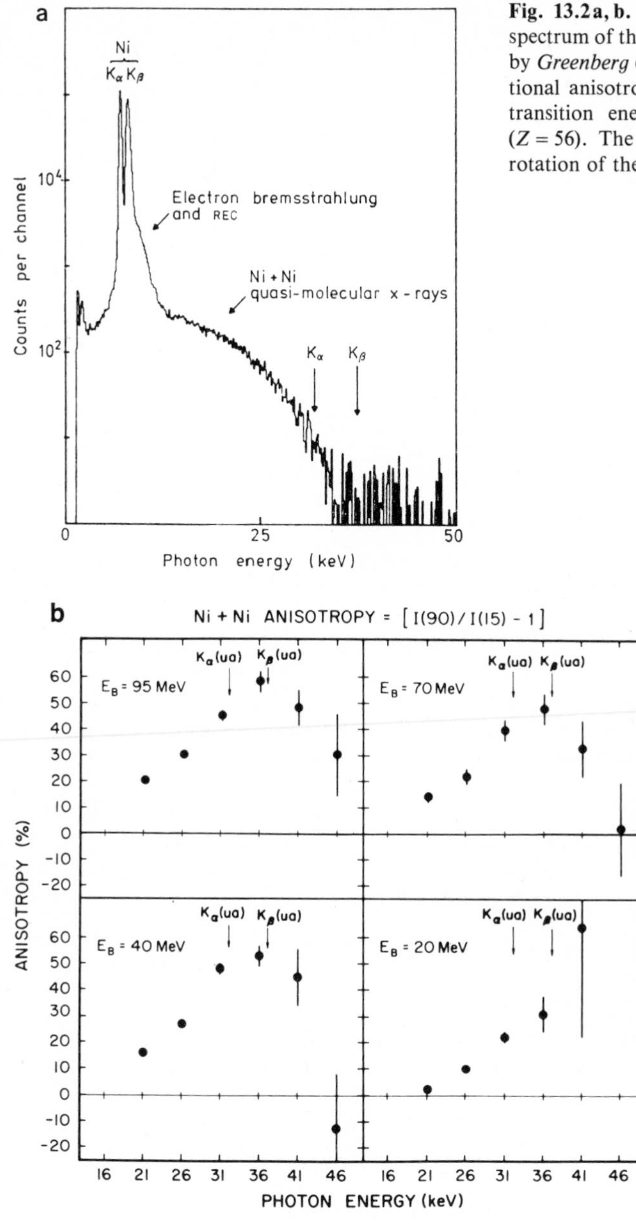

Fig. 13.2a, b. The quasi-molecular K x-ray spectrum of the Ni + Ni system (**a**), measured by *Greenberg* et al. [Gr 74], exhibits a directional anisotropy (**b**) which peaks at the K transition energy of he united atom Ba ($Z = 56$). The anisotropy is caused by the rotation of the internuclear axis

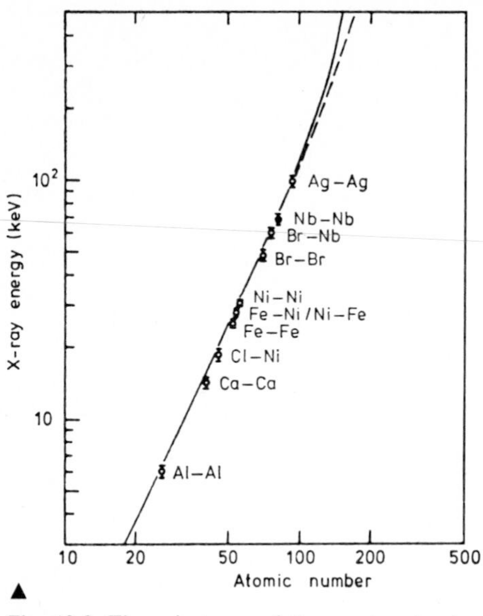

▲ **Fig. 13.3.** The anisotropy of the quasi-molecular spectrum was utilized by *Stoller* et al. [St 77] to measure the K x-ray energies of the united quasi-atoms

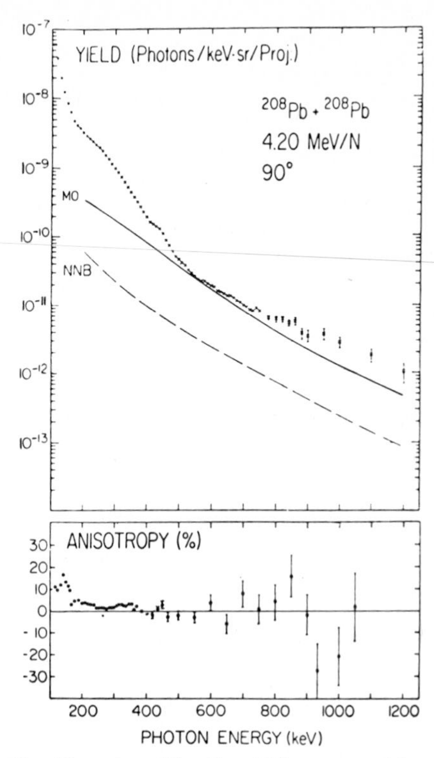

Fig. 13.4. Molecular orbital x-ray spectrum of the Pb–Pb system ($Z_1 + Z_2 = 164$), measured by *Meyerhof* et al. [Me 79]. The solid curve labelled MO reproduces the calculations of *Kirsch* et al. [Ki 78] for molecular x-rays, NNB stands for nuclear bremsstrahlung

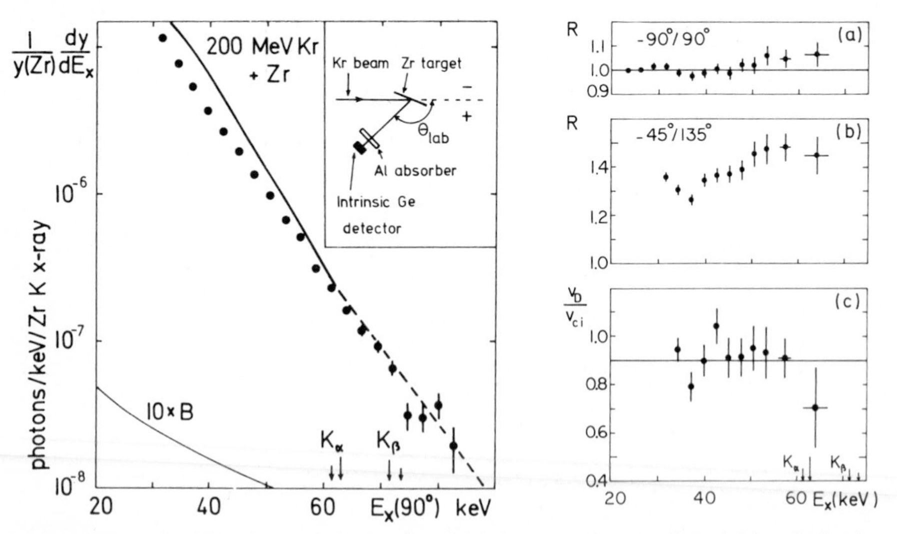

Fig. 13.5a–c. The anisotropy of the quasi-molecular x-ray spectrum allowed *Meyerhof* and co-workers to determine the velocity of the emitting system. As **(c)** shows, the emitted velocity V_D is very close to the velocity of the centre of mass V_{ci} [Me 75]

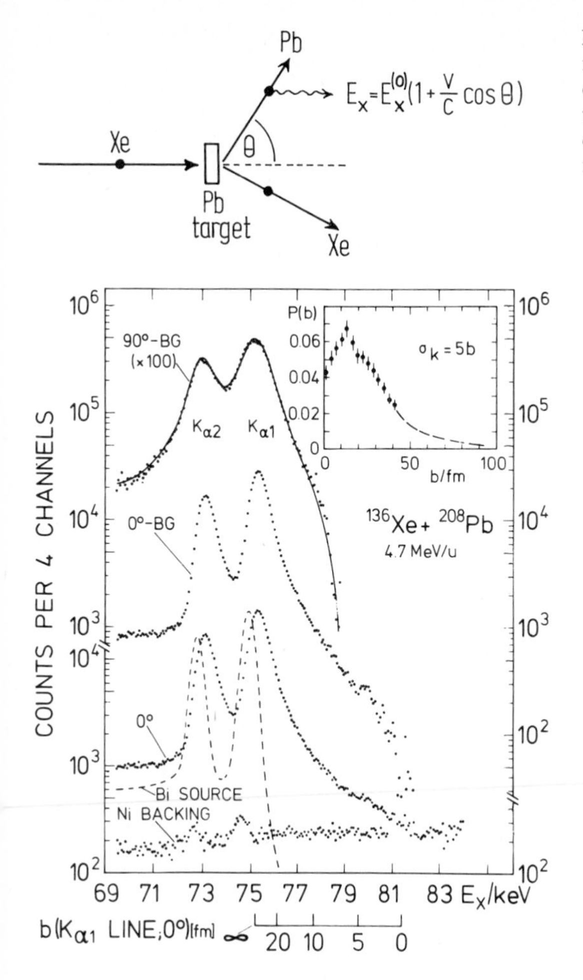

Fig. 13.6. The Doppler shift of a K x-ray observed along the beam axis depends on the direction of motion of the emitting atom. There is a unique correspondence between measured x-ray energy E_x and scattering angle θ

Fig. 13.7. Making use of the Doppler shift method explained in Fig. 13.6, *Greenberg* et al. [Gr 77] determined the impact parameter dependence $P(b)$ of K-shell ionization in Xe – Pb collisions. The insert shows $P(b)$ for K vacancies in the Pb atoms, which was found to reach several percent. The dip at $b < 15$ fm has not been confirmed by later measurements

is still going on, but has little direct bearing on our central theme, i.e. spontaneous positron production. We mention in passing the work of *Kaun's* group in Dubna, who were the first to investigate the quasi-molecular x-rays from a superheavy system (La + La) [Gi 74, Fr 76a]. *Anholt* [An 78a] quantitatively described the anisotropy in medium heavy systems, and *Kirsch* et al. [Ki 78] calculated the spectrum for Pb + Pb collisions. *O'Brien* et al. [O'Br 80] introduced an x-ray coincidence method to single out individual transitions in the quasimolecule, and *Tserruya* et al. [Ts 83] succeeded in finding interference structure in molecular orbital spectra.

Experimental investigation of the truly superheavy region of quasi-molecular systems ($Z_1 + Z_2 = 140 \dots 190$) started with the implementation in 1976 of the UNILAC accelerator at the GSI laboratory near Frankfurt in West Germany. This facility was constructed to accelerate all ions up to uranium to energies of about 10 MeV per nucleon (it has since been upgraded to provide energies up to

15 – 20 MeV per nucleon). The first experiment to yield evidence of large ionization probabilities at small impact parameters, which had been theoretically predicted [Be 76b], was conducted by *Greenberg* and co-workers [Gr 77]. They made rather ingenious use of the Doppler shift of characteristic K x-ray lines due to the motion of the projectile ion in the laboratory system. As discussed in Sect. 12.1, the scattering angle is a unique function of the impact parameter, and the Doppler shift of an x-ray emitted in beam direction from the scattered projectile depends on the cosine of the scattering angle (Fig. 13.6). *Greenberg* et al. found that the ionization probability of the $1s\sigma$ state is of the order of several percent, while that of the $2p_{1/2}\sigma$ state is 50% or even more, Fig. 13.7.

These results have been confirmed and extended by coincidence experiments by *Armbruster* et al. [Ma 78a, Be 78, Li 78, Be 79, Li 82], in which the scattering angle of the projectile is measured directly. Figures 13.8 – 10, compare some of the results with calculations, including screening. Clearly, the absolute agreement is impressive, as there are no adjustable parameters in either the experimental data or theory. In the heaviest system (Fig. 13.10) that has been studied in detail, Pb + Cm with $Z_1 + Z_2 = 82 + 96 = 178$, K-shell conversion of Coulomb-excited nuclear states of curium forms a major background in collisions with impact parameter $b \leq 30$ fm. For some time this seemed to disagree with theory [Li 80]. *Schwalm* and colleagues succeeded in separating the background of nuclear origin by utilizing the different scales of the lifetime of primary K vacancies ($\approx 10^{-17}$ s) and of excited nuclear states ($\sim 10^{-13} - 10^{-11}$ s), so ending the apparent discrepancy with the calculations [It 82], Fig. 13.11. The importance of these experi-

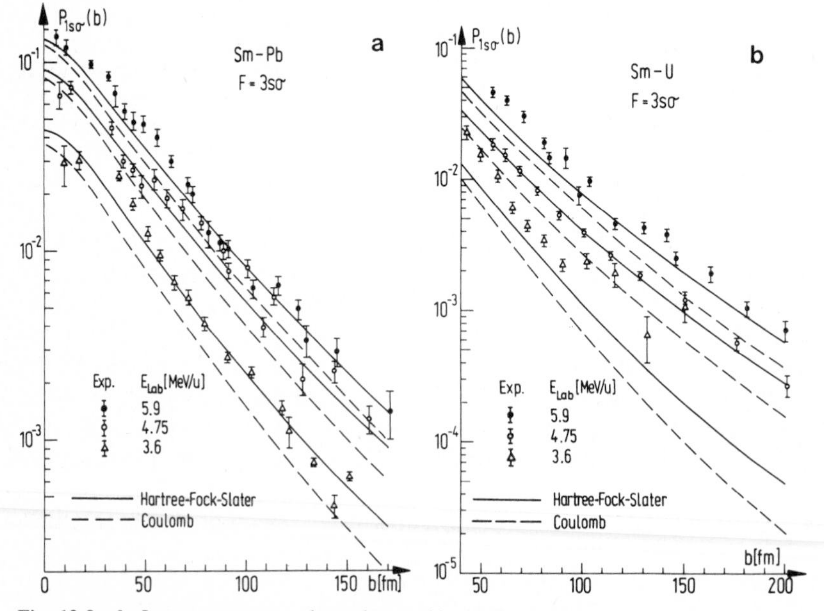

Fig. 13.8 a, b. Impact parameter dependence of ionization of the $1s\sigma$ orbital in Sm + Pb and Sm + U collisions compared with coupled channel calculations [deR 84]

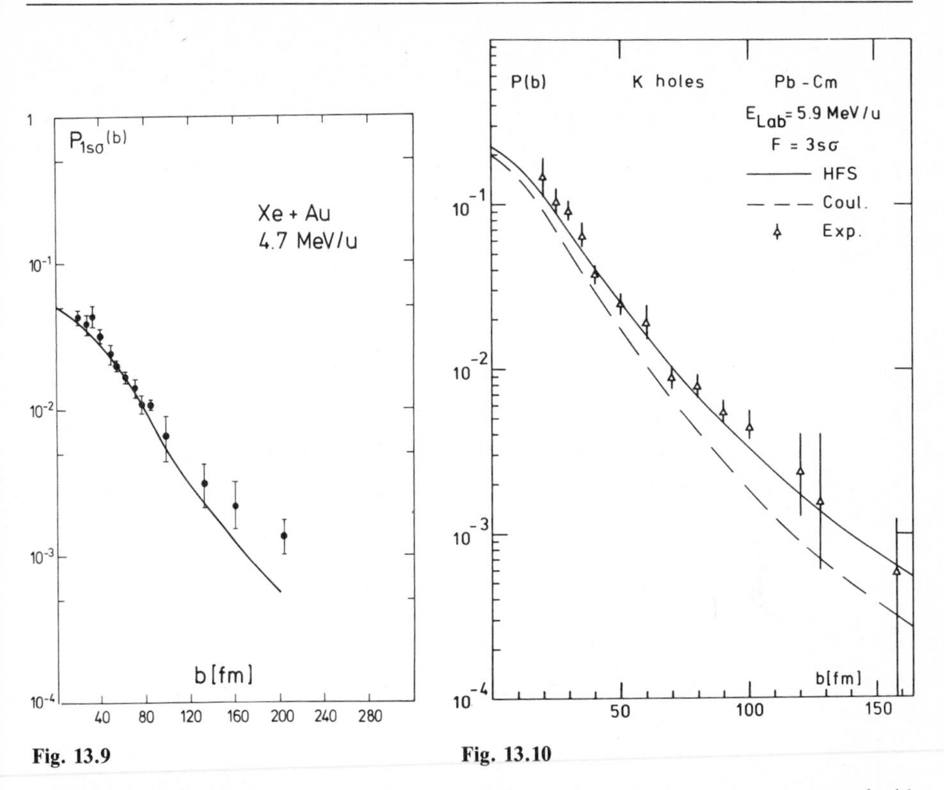

Fig. 13.9 **Fig. 13.10**

Fig. 13.9. Impact parameter dependence of $1s\sigma$ ionization in the Xe + Au system compared with theoretical results [So 81a]

Fig. 13.10. *K*-shell ionization of Cm after collisions with Pb atoms ($Z_1 Z_2 = 178$, [Li 80]). Calculations with screened molecular states give good agreement [deR 84]

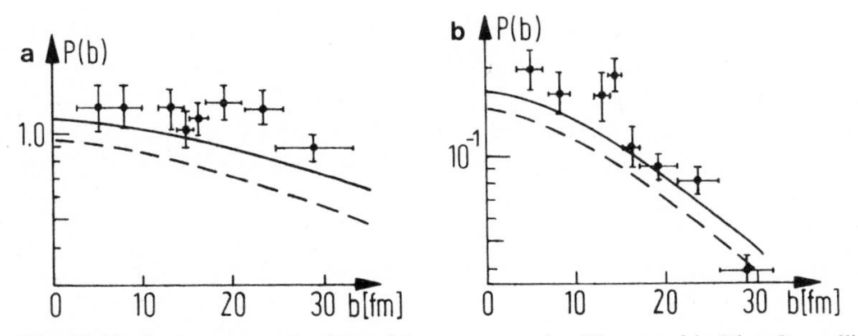

Fig. 13.11 a, b. $1s\sigma$ vacancies (**a**) and $2_{p1/2}\sigma$ vacancies (**b**) created in Pb + Cm collisions ($Z_u = 178$) at very small impact parameters [It 82] compared with theoretical calculations (see also Fig. 13.10)

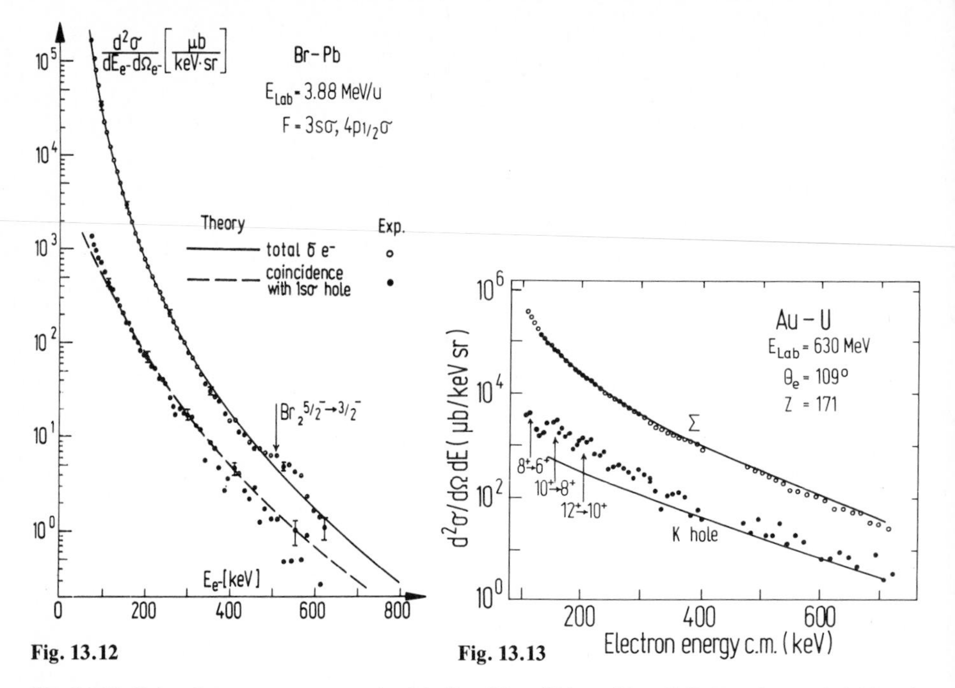

Fig. 13.12. Delta-electron spectrum emitted in Br + Pb collisions ($Z_u = 117$). For the total yield and for electrons originating from the $1s\sigma$ state (*lower line*) there is excellent agreement with theory [deR 84]

Fig. 13.13. Delta-electron spectrum from the Au + U system ($Z_2 + Z_2 = 171$) measured by *Koenig* et al. [He 84d]. The calculations are from *Mehler* et al. [Me 84a]

ments is twofold: firstly they test and confirm the dynamical theory in a non-trivial way, and secondly the rate of K-hole production is extremely important for spontaneous positron creation. The latter process depends crucially on the number of K holes present before the "diving process" starts. Even though the K-vacancy measurements as a function of impact parameter and ion energy yield information only about the holes present after the collision process is over, because of the astounding agreement between theory and experiment, it is plausible that K-hole production along the heavy-ion trajectory is also described correctly. This likelihood is increased by different types of experiments described in the following.

A second class of experiments that tests the strong localization of the quasi-atomic wave functions is spectroscopy of the energy distribution of ejected electrons, the so-called *delta electrons*. These have been measured for various collision systems up to uranium on uranium, $Z_1 + Z_2 = 184$ [Ko 77, Bo 78, Gü 82, Ko 82, He 84d]. Measurements including single spectra, coincidences with K vacancies, or scattered projectiles, or both, generally agree excellently with the theoretical predictions especially if screening is taken into account [So 78a, So 82a, deR 84, Me 84a]. Figures 13.12, 13 show several examples of this agreement.

It is quite instructive to view the ionization process, somewhat idealized, as an elastic collision between the Coulomb field of the projectile nucleus and the ejected electron. The classical picture requires energy and momentum conservation in this collision [Ge 26, Ko 77]. If ε_i, $\varepsilon_f(\boldsymbol{k}_i, \boldsymbol{k}_f)$ are the initial and final energies (momenta) of the electron, $\boldsymbol{P}_i, \boldsymbol{P}_f$ the initial and final momenta of the nucleus and M the nuclear mass, the conservation laws yield

$$\frac{\boldsymbol{P}_i^2}{2M} + \varepsilon_i = \frac{\boldsymbol{P}_f^2}{2M} + \varepsilon_f\,, \qquad \boldsymbol{P}_i + \boldsymbol{k}_i = \boldsymbol{P}_f + \boldsymbol{k}_f\,. \tag{13.1}$$

As $|\boldsymbol{P}_{f,i}| \gg |\boldsymbol{k}_{f,i}|$ due to the large mass of a heavy nucleus, the transfer of energy to the electron can be related to the momentum transfer

$$\varepsilon_f - \varepsilon_i = \frac{1}{2M}[\boldsymbol{P}_i^2 - (\boldsymbol{P}_i + \boldsymbol{k}_i - \boldsymbol{k}_f)^2] \approx \frac{\boldsymbol{P}_i}{M} \cdot (\boldsymbol{k}_f - \boldsymbol{k}_i) = \boldsymbol{V}_i \cdot (\boldsymbol{k}_f - \boldsymbol{k}_i)\,. \tag{13.2}$$

Here we have neglected terms of order k^2/M. The incident velocity of the projectile $\boldsymbol{V}_i$ is of the order of 1/10 of the speed of light c. For a typical ionized electron with several hundred keV kinetic energy, the relativistic relation

$$\varepsilon_f = (m_e^2 + k_f^2)^{1/2} \gtrsim |\boldsymbol{k}_f| c\,, \tag{13.3}$$

arises, so that $\boldsymbol{V}_i \cdot \boldsymbol{k}_f$ may be neglected. Hence we derive the condition that the momentum of the electron in its initial state must have been at least of magnitude

$$|\boldsymbol{k}_i| \gtrsim \frac{\varepsilon_f - \varepsilon_i}{|\boldsymbol{V}_i|}\,. \tag{13.4}$$

For typical values $(\varepsilon_f - \varepsilon_i) \sim 1$ MeV and $V_i/c \sim 0.1$, then $|\boldsymbol{k}_i| \gtrsim 10$ MeV/c, corresponding to distances of order $\hbar/|\boldsymbol{k}_i| \lesssim 20$ fm.

The conclusion we can draw from this greatly simplified argument is that the high-energy part of the spectrum of ionized electrons provides direct evidence of the high-momentum component contained in the initial bound state wave function. For collision systems with $Z_1 + Z_2 < \alpha^{-1} = 137$, the quasi-molecular wave functions become essentially independent of the nuclear separation R when R is much less than the K-shell radius $a_K = a_B/(Z_1 + Z_2)$, where $a_0 = \hbar^2/(e^2 m_e)$ is the Bohr radius. This is the *"run-way" effect* mentioned in Sect. 11.2. Therefore the high-energy fall off of the delta-electron spectra in light collision system is characterized by the parameter $V_i k_i \approx h\, V_i/a_K$. For superheavy collision systems with $Z_1 + Z_2 > 137$, the quasi-molecular wave functions change continuously as R decreases down to the touching point of the nuclei. This is possible only if the quasi-molecular wave function contains a sufficiently large contribution of momenta of size $|\boldsymbol{k}_i| \sim \hbar/R$. The high-energy fall off of the delta-electron spectrum is therefore characterized by the parameter $\hbar\, V_i/R_0$, where R_0 is the distance of closest approach of the nuclei on a given trajectory, e.g. given in (12.4).

The classical picture does not take into account that the initial state of the electron changes during the collision as a function of nuclear separation $R(t)$.

It turns out, however, that a similar result can be derived within the framework of quantum mechanics, restricted to first-order perturbation theory [Mü 78, Bo 80, Mü 83b]. The argument is based on the observation that the radial coupling matrix elements $\langle \varphi_k | \partial \varphi_l / \partial R \rangle$ between states of almost good angular momentum $j = 1/2$ behave like R^{-1} in superheavy quasiatoms. One can then demonstrate [Mü 83b] that the ionization probability in perturbation theory is dominated by the smallest complex zero $t_0 = \mathrm{i}\,\tau_0$ of the function $R(t)$:

$$|a_{\mathrm{fi}}|^2 \sim \left| \int_{-\infty}^{\infty} dt \frac{\dot{R}}{R} \exp[\mathrm{i}(\chi_{\mathrm{f}} - \chi_{\mathrm{i}})] \right|^2 \sim \exp\left[-2 \frac{\tau_0}{\hbar} (\varepsilon_{\mathrm{f}} - \varepsilon_{\mathrm{i}}) \right] . \tag{13.5}$$

As τ_0 is related to the distance of closest approach by $\tau_0 = \alpha R_0 / V_{\mathrm{i}}$, with the numerical constant α ranging between 1 and $\pi/2$, then

$$|a_{\mathrm{fi}}|^2 \sim \exp\left[-\frac{2\alpha R_0}{\hbar V_{\mathrm{i}}} (\varepsilon_{\mathrm{f}} - \varepsilon_{\mathrm{i}}) \right] , \tag{13.6}$$

relating the delta-electron spectrum to the distance of closest approach between the nuclei. In (13.6) the electronic wave function is sufficiently localized in the

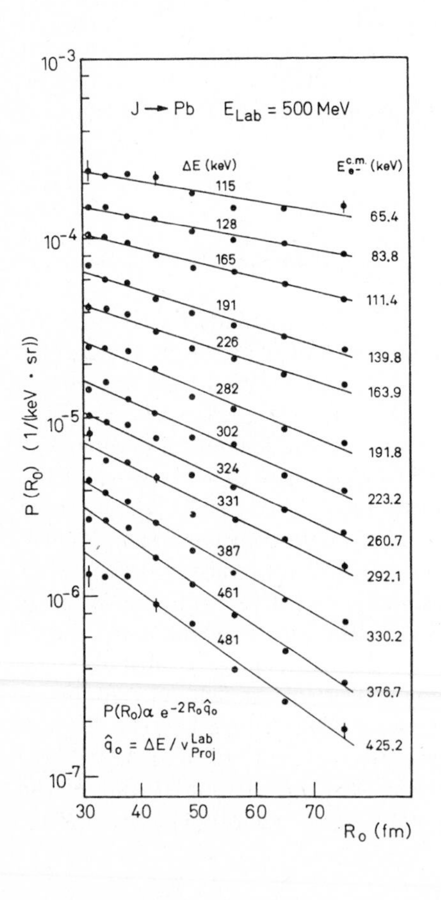

Fig. 13.14. Scaling plot for delta-electron creation in I + Pb collisions ($Z_{\mathrm{u}} = 135$), against distance of closest approach. ΔE is the transition energy obtained from a fit to (13.6)

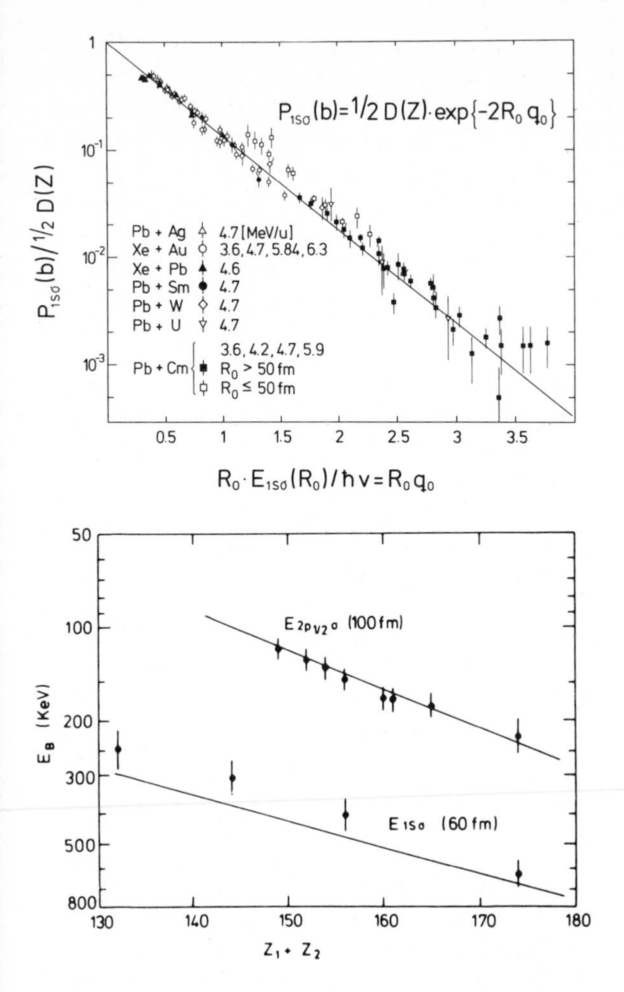

Fig. 13.15. Universal scaling plot of $1s\sigma$ ionization against the variable R_0q_0. A very nearly exponential distribution is found [Li 80]

Fig. 13.16. The scaling law (13.6) can be utilized to determine the binding energies in the quasimolecular system. The energies of the $1s\sigma$ states were determined at fixed nuclear distance R_0 for systems up to $Z_u = 174$ [Ar 78]

superheavy quasiatom to resolve very small nuclear separations of the order of 20 fm.

The scaling law (13.6), which is the generalization of *Bang* and *Hansteen's* formula for ionization by light-ion impact [Ba 59, 79b, 81b], is to some extent also valid for dynamically induced positron production [Ar 79, Ba 79a, Ka 80].

The simple formula (13.6) is somewhat obscured by the importance of multistep excitation mechanisms in the numerical calculations (Sect. 12.6). These higher order processes considerably increase the absolute magnitude of the ionization probability $|a_{\mathrm{fi}}|^2$, but cause little change in the kinematical behaviour of the exponential in (13.6). The main conclusion about the strong localization of the inner-shell states in superheavy quasiatoms therefore remains untouched.

Experimentally, the scaling law (13.6) is rather well established. Figure 13.14 exemplifies the probability for delta-electron emission in I + Pb collisions ($Z_1+Z_2=135$) plotted versus distance of closest approach for various electron energies [Ko 82]. The data can be fitted consistently with an exponential distribution like (13.6), where $\alpha = 1$ and $(\varepsilon_i - m_e c^2) \sim -55$ keV. A systematic evaluation of K-shell ionization data as a function of impact parameter was made by *Bosch* et al. [Bo 80, 82]. When these data are collected in one figure and plotted versus the scaling variable $2R_0 |E_{1s}^{B}(R_0)| / \hbar v_i \equiv 2qR_0$, a universal exponential distribution is found (Fig. 13.15). Being experimentally established, the scaling law can be applied to determine the transition energy $(\varepsilon_f - \varepsilon_i)$ [So 78b]. If $|a_{fi}|^2$ is measured for collision systems with various combined charges $Z_u = Z_1 + Z_2$ in collisions at fixed distance of closest approach, the increase in the binding energies of the quasi-molecular $1s\sigma$ and $2p_{1/2}$ states can be determined as a function of Z_u, Fig. 13.16 [Ar 78, Be 78]. As can be seen, binding energies of the quasi-molecular $1s\sigma$ state at $R = 60$ fm of up to 650 keV have been observed.

The wealth of experimental data and their excellent agreement with calculations based on the time-dependent two-centre Dirac equation and the quasi-molecular picture provide conclusive evidence that quasi-molecular and quasi-atomic states are formed in heavy-ion collisions at energies below and slightly above the Coulomb barrier. The large ionization probabilities observed and the abundance of high-energy delta electrons can only be understood by a dramatic increase of the binding energies of the inner-shell electrons during collisions, leading to strong localization of their wave functions. Therefore, the existence of superheavy quasiatoms is an experimentally established fact. This knowledge forms the basis for the experimental search for spontaneous positron creation leading to the charged vacuum, dealt with in the following sections.

13.2 Positron Spectrometers

Experimental configurations to detect positrons emitted in heavy-ion collisions have been designed to satisfy a number of needs [Ba 84]: (1) since the positron production cross-sections are fairly small (of the order of hundred μb/sr), the spectrometers have to accept positrons in a wide range of solid angles with a large detection efficiency; (2) they have to detect positrons among an overwhelming background of radiation composed of γ-rays, delta electrons and neutrons; (3) they must detect the scattered nuclei so as to rule out or, alternatively, select nuclear reaction processes. To meet this challenge, detection systems have been developed which combine a large sensitivity for primary positrons with good suppression characteristics of secondary positron production processes, such as external pair creation by electrons of γ-rays.

Before introducing these systems, a few remarks are pertinent concerning the transformation of the positron production probabilities (per scattered projectile ion) or cross-sections to the centre-of-mass system. This complication, which arises because the primary positrons are emitted from fast-moving systems, is aggravated by the fact that the positrons can originate either from the quasi-molecular system or from one of the scattered nuclei. The quasi-molecular or quasi-atomic positrons, including those resulting from spontaneous transition to the charged vacuum, are emitted from the centre-of-mass system moving for-

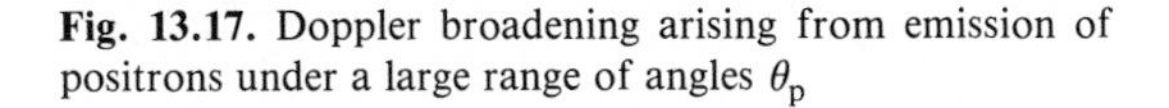

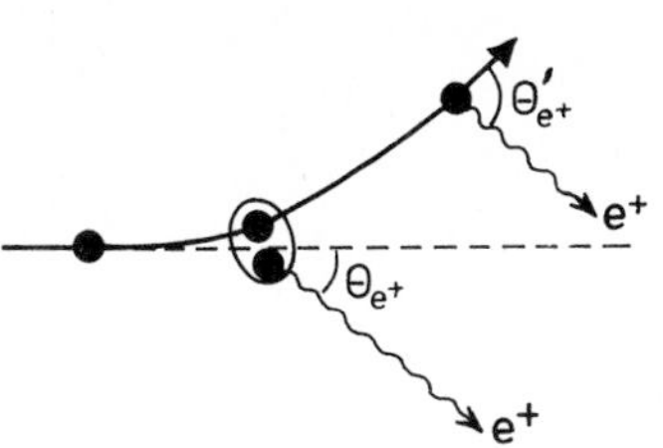

Fig. 13.17. Doppler broadening arising from emission of positrons under a large range of angles θ_p

ward with velocity v along the beam direction. Positrons produced by internal conversion of nuclear γ-rays, on the other hand, originate from scattered projectiles or target recoils, which have velocity components perpendicular to the beam direction and move with various velocities.

But even if positrons with discrete energies were emitted in the centre-of-mass system alone, large Doppler shift effects are expected in the laboratory system depending on the emission angle θ_p of the positron, Fig. 13.17. To first order in (v/c) the energy shift is given by

$$E_p^{\text{Lab}} = E_p^{\text{cm}}\left[1 + \frac{v}{c}\sqrt{1 + \left(\frac{m_e c^2}{E_p^{\text{cm}}}\right)}\cos\theta_p\right], \tag{13.7}$$

where E_p^{cm} is the kinetic energy of the positron line in the centre-of-mass system. If the positron emission angle is not observed, a Doppler broadening of the line results. Following the original experiment of *Meyerhof* et al. [Me 75], the Doppler shift may be turned from a curse into a blessing, if the magnitude of the observed broadening can be related to the velocity of the emitting source with (13.7). By this method one can, in principle, determine the origin of an observed positron line, Sect. 13.6.

All the positron spectrometers used so far have employed a magnetic transport system to collect the positrons emitted from the target and to guide them to the detector. In one of them, the "orange"-type β spectrometer, the magnetic field is also used for charge and momentum analysis of the detected particles, whereas energy determination is achieved solely by solid state detectors in the case of solenoidal transport systems. We shall start our discussion with the β spectrometer.

13.2.1 The "Orange"-Type β Spectrometer

This instrument was developed by *Moll* and *Kankeleit* [Mo 65] and perfected by *Kienle's* group [Ko 79b, Be 80a, Ki 83, Cl 83] as a detection system for positrons, Fig. 13.18. It is based on the fact that the Lorentz force acting on charged particles moving in a magnetic field is perpendicular both to the field lines and the direction of motion of the particles. A toroidal magnetic field is therefore able to focus positrons originating from a point on the axis (the target) onto a point on the axis on the other side of the field torus (Fig. 13.19). The β spectrometer used in the GSI experiments has 60 coils producing the magnetic field and accepts positrons emitted between 30° and 70° relative to the beam direction. As the

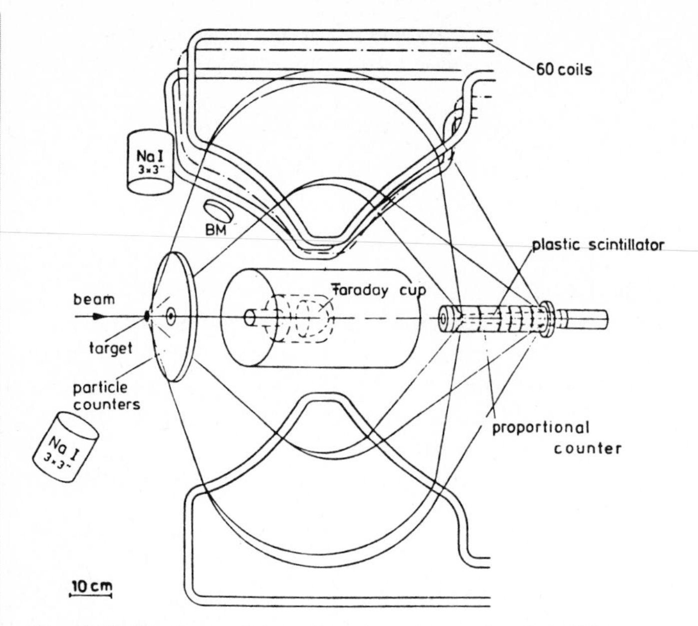

Fig. 13.18. The "orange"-type β spectrometer employed at GSI to measure positron spectra

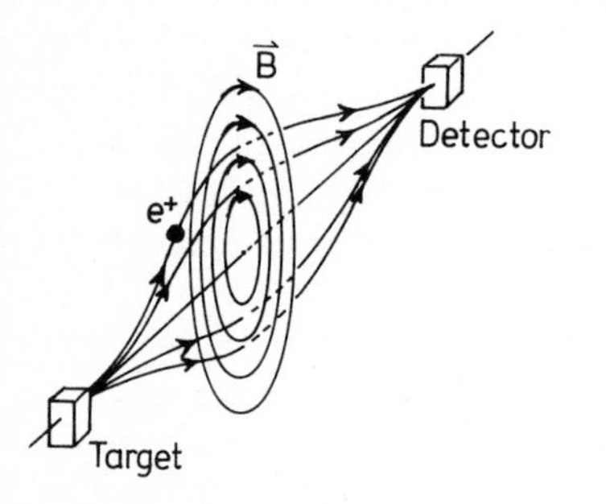

Fig. 13.19. Working principle of an "orange"-type β spectrometer. The toroidal magnetic field focusses positrons from the target onto the detector, both located on the axis of the torus

focal point depends on the momentum of the positron, the positron energy can be determined by a position-sensitive detector. At a given strength of the magnetic field only a narrow band of positron momenta is accepted ($\Delta p/p \sim 0.15$), so that the field strength must be varied in steps to measure a full positron spectrum.

The positron detector consisted of a plastic scintillator surrounded by a position-sensitive proportional counter to suppress low energy background radiation (mainly γ-rays) and to improve the energy resolution of the detection system. The scattered heavy ions were detected by an annular parallel plate avalanche counter placed in front of the β spectrometer. The counter was divided into concentric rings to allow a (rather crude) determination of the ion scattering angles.

Recently the detection system has been considerably improved. The "orange" spectrometer has been turned around, and the beam is shot through a hole in the centre of the spectrometer onto the target situated behind. Positrons emitted in backward direction are then focussed on the detector which is now placed in front of the spectrometer and also contains a hole to let the beam pass. The positron detector is formed by a large number of Si(Li) counters arranged in a pagoda-like configuration, enabling crude discrimination between different azimuthal angles (six segments covering 60° each). Also the counter for the scattered ions has been improved as there is sufficient space available behind the β spectrometer.

13.2.2 Solenoidal Transport Systems

The solenoidal magnetic transport systems [De 56, Bu 66]) employed at GSI for detecting positrons have undergone several stages of development. Here we concentrate on the detection system "EPOS" [Ba 80a] that led to the observation of a narrow line structure in the positron spectrum of the U + Cm system, Fig. 13.20. The magnetic field produced by the solenoid focusses positrons from the target onto the cylindrical Si(Li) counter which measures the positron energy. The spectrometer makes clever use of two basic features of a solenoidal transport system, namely the opposite helicity of the spiralling trajectories described by electrons and positrons, and the property that the particles always return to the axis if they are emitted from a point (the target) located on the axis.

Let us discuss these two points in some detail. As the Lorentz force on a charged particle in a magnetic field always points perpendicular to the field lines,

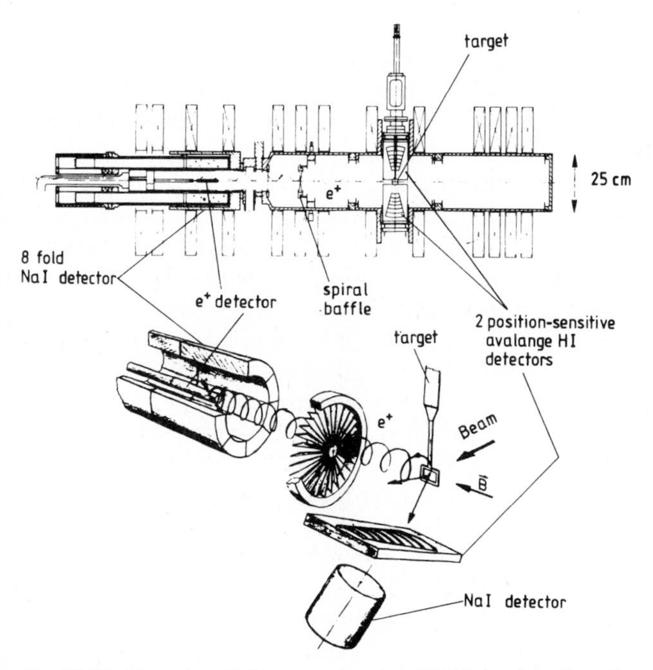

Fig. 13.20. The solenoidal transport system EPOS for detection of positrons at GSI

the path of a particle in a homogeneous magnetic field is a spiral. The motion along the field lines remains undisturbed, while the transverse motion is forced into a circular orbit. Since the direction of the Lorentz force depends on the sign of the charge of the particle, the spirals for electrons and positrons are just inverted. Also, because the transverse motion is a circle, the particle must return after each revolution to exactly that field line from which it started. These two properties are exploited to separate electrons very effectively from positrons by a helical baffle (shaped like a turbine wheel) positioned roughly in the middle of the solenoid. The orientation of the baffle lets positrons pass, but scatters the electrons that spiral in the opposite direction. When these particles are scattered at the baffle, their new point of origin is no longer on the axis of the solenoid, and they usually do not return to the axis, thus missing the detector. The transport efficiency of the solenoid is maximized by using a magnetic mirror on the side of the target opposite the positron detector, and a field depression at the position of the spiral baffle.

Despite these precautions many more electrons than positrons reach the Si(Li) detector. To discriminate against electrons, the detector is therefore surrounded by a ring of NaI counters which observe the characteristic positron annihilation radiation of two 511 keV γ-ray quanta. By requiring a coincidence between the particle and one of the annihilation quanta, total suppression of electrons and other unwanted particles is achieved. The total efficiency of the system for detection of a positron is between 10% and 15% for positrons below 1 MeV energy. The energy resolution is of the order of 10 keV.

The detection system for the scattered heavy ions is designed to determine the nuclear reaction kinematics as closely as possible. This is particularly important for collisions at or above the Coulomb barrier where unwanted reaction channels must be discriminated against, especially against fission of one or both nuclei. In fission exotic nuclear excited states are easily populated and can, by internal conversion, lead to serious contamination of the measured positron spectrum. For this reason, both the scattered projectile and the recoiling target nucleus are detected in two symmetrically arranged parallel-plate avalanche counters with delay-line lead out to determine the scattering angle. The detectors cover between $20° - 70°$ in an azimuthal sector to 60°. Some crude information about the mass and charge of the detected nuclei is obtained by time-of-flight and energy loss measurements. With the present system, however, it is not possible to distinguish between two heavy nuclei, say between uranium and lead, i.e. the amount of possible mass transfer between the nuclei cannot be determined.

Coincident detection of both particles also has the advantage that for the asymmetric systems, e.g. U+Pb ($Z_1 = 92$; $Z_2 = 82$) and to some degree even U+Cm ($Z_1 = 92$, $Z_2 = 96$), the impact parameter b can be determined uniquely from the scattering angle correlation for Coulomb scattering. This separation between forward and backward scattering events is seen clearly in Fig. 13.21, except in the region around 45° scattering angle.

The problem of electron suppression can be very efficiently solved when the electrons and positrons are transported in a curved magnetic field as produced by a solenoid of toroidal shape. In this slightly inhomogeneous field configuration

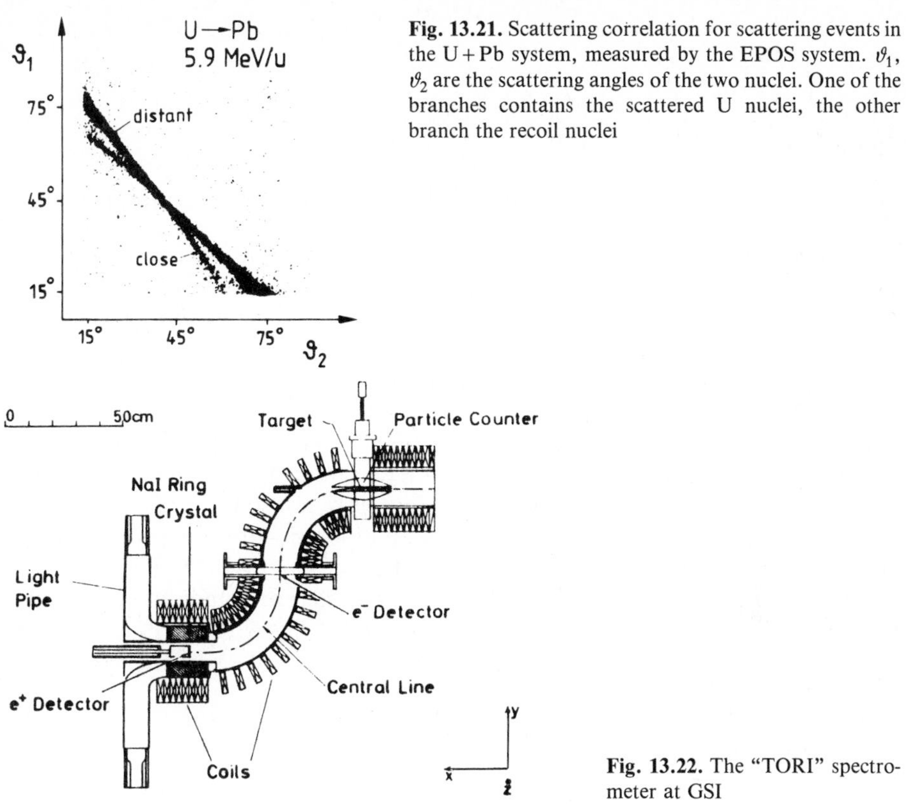

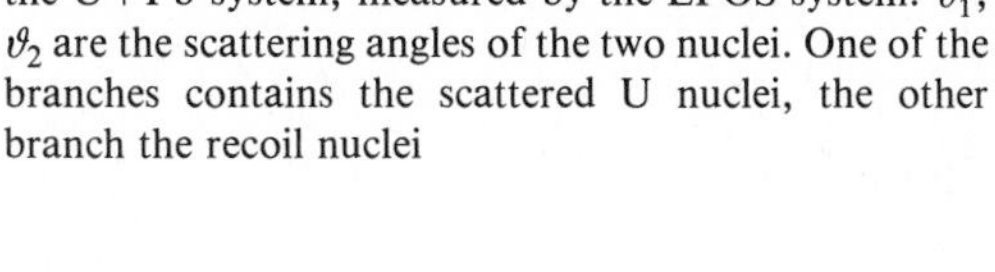

Fig. 13.21. Scattering correlation for scattering events in the U + Pb system, measured by the EPOS system. ϑ_1, ϑ_2 are the scattering angles of the two nuclei. One of the branches contains the scattered U nuclei, the other branch the recoil nuclei

Fig. 13.22. The "TORI" spectrometer at GSI

the spirals of electrons and positrons not only have opposite orientation, but their centres slowly drift apart perpendicular to the plane of the torus. The new "TORI" spectrometer of *Kankeleit's* group of [Ka 81a] employs this effect by joining two quarters of a toroidal solenoid to produce an S-shaped magnetic field (Fig. 13.22). In the first quarter electrons and positrons are separated, and the electrons are absorbed (and possibly detected). The sideward drift of the positrons is exactly reversed in the second quarter torus, with the result that the "TORI" spectrometer has imaging characteristics similar to a straight solenoidal transport system.

13.3 Background Effects Creating Positrons

A variety of background processes produce positrons in a heavy-ion collision. Most of these processes involve the conversion of a real or virtual γ-ray into an electron – positron pair. (Positrons can also be created by weak beta-decay processes, but these are so slow that they can be suppressed by coincidence requirements.) The processes differ in the origin of the γ quant and in the place where the conversion occurs. The most important of these background processes are (in that order):

i) internal conversion of γ-rays from highly excited nuclear states;
ii) external conversion of γ-rays in the target;
iii) external conversion of γ-rays in the transport system or in the detector;
iv) conversion of x-rays produced by nuclear or electronic bremsstrahlung.

Experimentally by far the most serious of these is the first, because excited nuclear states can be populated even in collisions far below the Coulomb barrier by the action of the multipole components of the long-range field on the individual nuclei. This effect is particularly strong for highly deformed nuclei which generally have a low-lying band of rotational states. All actinide nuclei (U, Th, Cm, etc.) belong to this class. When an excited nuclear state decays with a transition energy E_x larger than $2m_ec^2 = 1022$ keV, the excitation energy can be released by emission of an electron – positron pair instead of a γ-ray (Fig. 13.23). The probability for this conversion depends on the excess of the transition energy $E_x - 2m_ec^2$, and – more critically – on the multipolarity of the transition. Most serious are electric monopole transitions (E0) which can occur only via internal conversion or by ejection of a bound atomic electron. For higher multipolarities (E1, M1, E2, etc.) the γ-ray spectrum emitted by the nuclei reflects the positron spectrum from this source:

$$\frac{dN(e^+)}{dE_p} = N_\gamma(E_x) \cdot \frac{d\beta_{M\lambda}(E_x)}{dE_p}, \tag{13.8}$$

where $(M\lambda)$ denotes the multipolarity of the nuclear transition. Fortunately, the differential conversion coefficients $d\beta/dE_p$ can be calculated very reliably, because details of nuclear structure influence them negligibly for all existing nuclei [Schl 81]. Given the conversion coefficients it is possible to deduce the contribution to the positron spectrum from nuclear transitions, with the exception of monopole transitions, from the measured γ-ray spectrum:

$$\frac{dN(e^+)}{dE_p} = \int_{2m_ec^2}^{\infty} dE_x \cdot \frac{dN_\gamma(E_x)}{dE_x} \cdot \frac{d\beta_{M\lambda}(E_x)}{dE_p}. \tag{13.9}$$

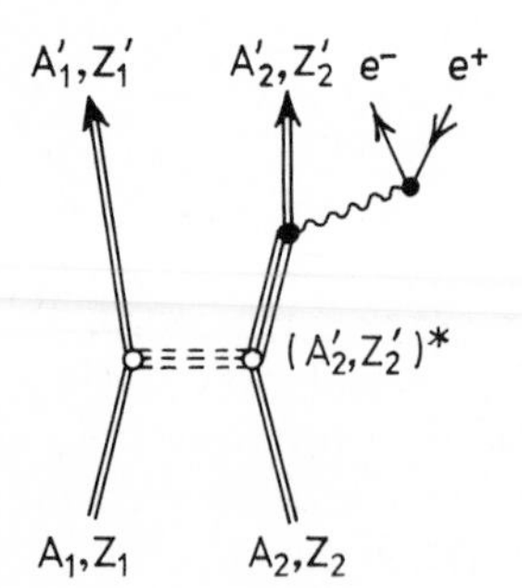

Fig. 13.23. Pair conversion of an excited nuclear state populated in the nuclear collision process. The time between excitation and de-excitation is typically $10^{-11} - 10^{-13}$ s. This is too short to allow for separation of this process by electronic timing methods

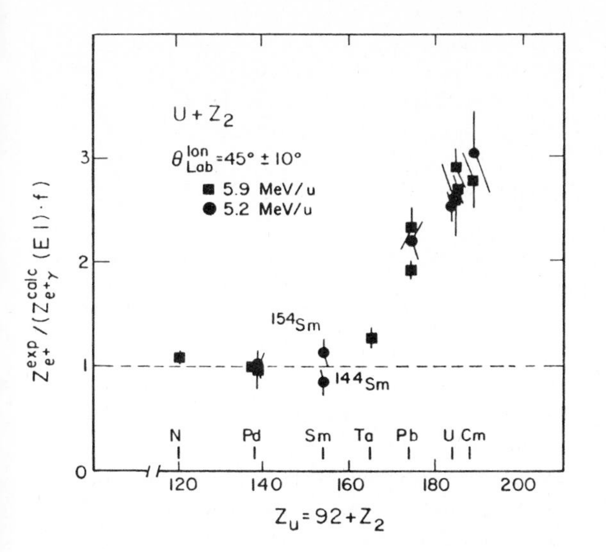

Fig. 13.24. Observed positron yield in uranium + Z_2 collisions compared with the yield expected from conversion of the measured γ-ray spectra. The steep rise beyond $Z_1+Z_2=160$ is due to atomic positron production ("shake off" of the vacuum polarization cloud)

The only unknown here is the multipolarity ($M\lambda$). In principle, ($M\lambda$) can be determined for an isolated γ-ray line by measuring the angular distribution of the γ-ray. However, this is almost impossible in our case, because the γ-ray spectrum above 1 MeV emitted from the nuclei after a heavy-ion collision is continuous and intrinsically Doppler shifted. Sole exceptions to this are collisions between two doubly magic nuclei, e.g. $^{208}Pb + ^{208}Pb$, which have just a very few excited states below 5 MeV energy.

A method to determine the average multipolarity required for the conversion of the γ-ray spectrum into a positron background spectrum was proposed by *Meyerhof* et al. [Me 77]. It involves measuring the positron yield as a function of combined nuclear charge $Z_u = Z_1 + Z_2$ for collisions of nuclei with similar structure. For systems with $Z_u < 160$ a negligible contribution from positrons of atomic origin is expected. The result of such a systematic survey is shown in Fig. 13.24. Uranium ($Z_1 = 92$) was used throughout as one collision partner. The data show that the ratio of the positron yield to the yield of nuclear γ-rays is constant up to $Z_u \approx 160$, in agreement with the underlying assumption that the average multipolarity is the same for nuclei of related level structure, in this case a certain mixture of electric dipole (E1) and electric quadrupole (E2). The effective multipolarity is now used to determine the background of positrons of nuclear origin for systems with combined charge $Z_u > 160$. The steep rise of the observed positron yield for growing excess ($Z_u - 160$) is firm evidence that atomic positrons are measured in these collision systems. The figure shows clearly that the signal-to-background ratio is most favourable in the heaviest systems U + U ($Z_u = 184$) and U + Cm ($Z_u = 188$). For these cases, the inherent uncertainty of this method of background subtraction (ca. 20% – 30%) results in a negligible uncertainty of the deduced spectrum of positrons produced by QED processes.

As mentioned above, positrons originating from pair conversion of nuclear electric monopole (E0) transitions escape this method of background subtraction, because these transitions do not occur by γ-ray but instead by electron emission. Fortunately, E0 transitions occur relatively rarely in the excitation spectra of most nuclei, especially in the spectra of the actinide nuclei which are all strongly deformed with a low-energy spectrum of rotational and vibrational states. Positrons from E0 transitions cannot, therefore, constitute a major contribution to the bulk of the nuclear positron background. Single, isolated E0 transitions, however, could produce structures in the measured positron spectra, if they are strongly populated in the nuclear collision. Whenever structures are found in the measured positron spectra, a careful search for associated conversion lines must be undertaken. This is explained in more detail below, when the possible origin of the line structures seen in positron spectra of U + U and U + Cm collisions is discussed.

We only mention briefly here a process that can lead to the emission of monochromatic positrons. When the atom that surrounds the decaying nucleus contains vacant electronic states, the electron associated with the positron in the pair creation may be captured into one of the vacant bound states [Sl 53]. The energy of the positron is then well defined and equal to $E_X - 2m_e c^2 + |E_B|$, where E_B is the binding energy of the vacant atomic orbital. The probability for this process to occur depends on the presence of a vacant bound state whose wave function has a large overlap with the nuclear volume. The vacancy also must be available for a time comparable to the lifetime of the nuclear excited state. Since inner-shell vacancies in heavy atoms are filled within 10^{-17} s by radiative transitions of electrons from outer shells, whereas the typical lifetime of an excited nuclear state is of the order $10^{-11} - 10^{-13}$ s, this process is normally suppressed by several orders of magnitude.

External pair conversion of γ-rays in the target, i.e. conversion in the Coulomb field of another nucleus, can be calculated by the Bethe-Heitler formula [Be 34, Ja 55], if the γ-ray spectrum and the target geometry are known. For typical thin targets less than 1 mg/cm^2 thick, this contribution yields less than 10% correction to the positron spectrum [Ba 84]. On the other hand, external pair conversion outside the target, i.e. in the solenoid or the detector, represents a technical problem that can be reduced by appropriate apparatus design. For the detection systems used in the present experiments at GSI this background effect was found to be negligible. Finally we mention that positrons from conversion of x-rays produced by nuclear or electronic bremsstrahlung constitute a negligible background [Re 76b].

13.4 Positron Experiments I: Gross Features

The objective of the first, exploratory experiments between 1977 – 1979 was to show that positron production could be used, *in practice*, as a signal to investigate this physics of very strong electric fields. The main questions that were asked, were:

i) Do atomic positron creation processes account for the major part of the total positron production cross-section in collisions of very heavy ions (with, say,

$Z_u = Z_1 + Z_2 > 160$), or is the background of positrons of nuclear origin overwhelming (as many physicists argued!)?

ii) Are there indications that positron creation in a strong electric field of two heavy nuclei is a non-perturbative phenomenon, as theory predicts (Sect. 12.6), that culminates in spontaneous pair creation in supercritical fields?

iii) Are the positrons created mainly in the moment when the nuclei are close together, and the electric field has maximal strength, or is the production process distributed all along the collisional trajectory?

Last but not least, the exploratory experiments should pave the way to the experimental detection of spontaneous pair creation, i.e. the decay of the neutral vacuum in strong electric fields.

The collision system ^{208}Pb + ^{208}Pb enables the correctness of theoretical predictions concerning positron production in strong fields to be tested and verified unambiguously. This is so because the nucleus ^{208}Pb is doubly magic and has only two excited states that are low enough to be populated in collisions below the nuclear Coulomb barrier, the 3^- state at 2.615 MeV, and the 2^+ state at 4.086 MeV excitation energy. The probabilities for Coulomb excitation of these states can be calculated and compared with the measured γ-ray yield from the decay of these states [Ob 76]. The pair-conversion coefficients for the states are well known, so the expected contribution to positron production from these nuclear states can be determined with great accuracy [Schl 81].

The experimental results, Fig. 13.25, showed clearly that the major fraction of the observed positron must originate from QED production processes. When plotted as a function of distance of closest approach R_{min}, the measured differential positron cross-section exhibited a much flatter fall off than nuclear Coulomb excitation. Moreover, when the theoretically deduced nuclear positron background is added to the predicted cross-section for quasi-molecular positron production, there is excellent quantitative agreement with the experimental data. That R_{min} should be a relevant parameter for the representation of the data is not surpris-

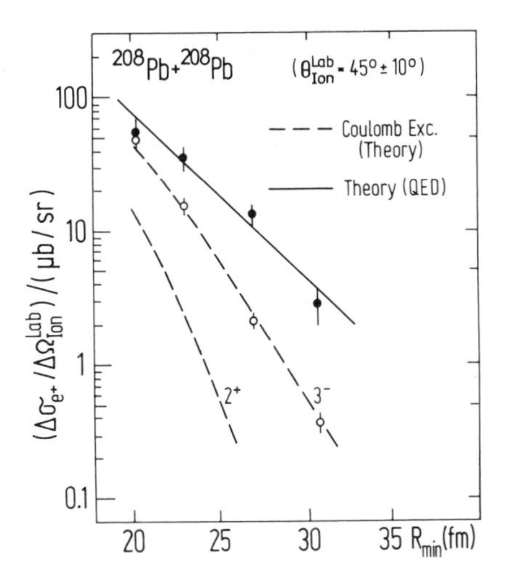

Fig. 13.25. Positron production in the ^{208}Pb + ^{208}Pb collision system as a function of the distance of closest approach between the nuclei reached in the collision, R_{min} [Ba 78]

ing, recalling the scaling law (13.6) for $1s\sigma$ vacancy creation. Intuitively, we expect that the scaling law may be applied to describe positron creation by dynamic excitation processes, if we replace the energy difference $(\varepsilon_f - \varepsilon_i)$ by the sum of the energy of the produced positron and the rest mass of the associated electron. That such a scaling law indeed holds empirically was shown by *Armbruster* and *Kienle* [Ar 79].

A detailed analysis by *Reinhardt* et al. [Re 81b] of the coupled channel calculations for positron production in the Pb + Pb system ($Z_u = 164$) showed that the main part of the positron cross-section is due to direct excitation of a pair of free electrons and positrons. Processes involving bound levels as intermediate states are less important. Therefore positron creation in the Pb + Pb system is not caused by strong binding of quasi-molecular electrons, but instead is a signal for the large degree of polarization of the electron – positron vacuum. In the strong electric field exerted by the combined nuclear charge, virtual electron – positron pairs are formed in much larger numbers than in normal atoms, and some fraction of these pairs are broken up, i.e. converted into real pairs, due to the time variation of the Coulomb field. This process has been called "shake-off of the vacuum-polarization cloud" [So 77a] (see Fig. 3b).

To some extent this term applies to all cases of positron production by time-dependent electric fields, e.g. in proton – proton collisions at high energy. However, in this long known example the virtual pairs broken up in the collision existed all the time in the Coulomb fields of the individual protons. The temporarily combined field of both protons acts only as a perturbation that excites the virtual particles onto the mass shell. In other words, electron – positron pair production in collisions of elementary particles is a perturbative process describable by the Feynman diagram in Fig. 13.26. Its cross-section is proportional to the square of the product of the charges of the two colliding particles. This is completely different for positron production in subcritical collisions of very heavy atoms. Because the effective coupling constant $(Z_u \alpha)$ is greater than one, the virtual vacuum polarization cloud around the combined nuclear charge differs very much from the simple superposition of the polarization clouds surrounding the separated nuclei. The virtual electron – positron pairs broken up due to the collision dynamics are produced only by the combined strong Coulomb field of both nuclei, in the first place.

This essentially non-perturbative character of dynamical positron production in collision systems with $Z_u \alpha > 1$, predicted theoretically by *Soff* et al. [So 77a], has found spectacular experimental proof. Figure 13.27 compares the probabilities for positron creation by atomic processes in Pb + Pb, U + Pb, and U + U

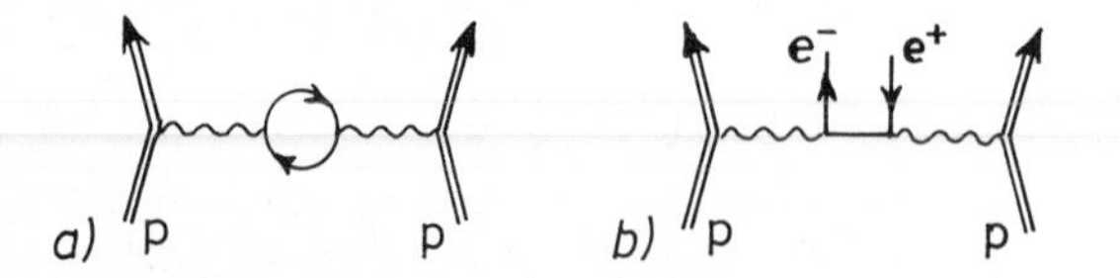

Fig. 13.26a, b. Feynman diagrams describing (**a**) virtual vacuum polarization and (**b**) real pair production in proton – proton collisions

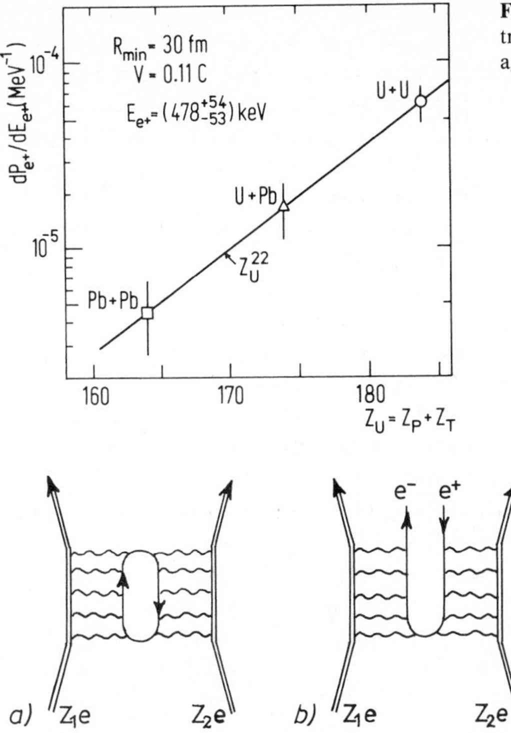

Fig. 13.27. Z dependence of induced positron production. The power law $\sim Z_u^{22}$ agrees with theoretical predictions [Gr 82b]

Fig. 13.28 a, b. Typical Feynman diagrams describing **(a)** virtual vacuum polarization and **(b)** real pair production ("shake off of the vacuum polarization cloud") in collisions of very heavy atoms. Experimentally the cross-section was found to scale approximately as $Z_u^{22} = (Z_1 + Z_2)^{22}$, i.e. the contributing diagrams contain 10 virtual photon lines on the average

systems ($Z_u = 164, 174, 184$, resp.) with the relevant kinematic variables R_{min} and velocity v fixed. From $Z_u = 164$ to $Z_u = 184$ the probability for pair creation increases by a factor of 13, corresponding to the highly non-linear growth law $dP/dE_p \sim Z_u^{22}$, shown in Fig. 13.27 by the solid straight line. This result is perhaps the most direct and convincing evidence that pair creation in strong Coulomb fields with coupling constant $Z\alpha > 1$ is an essentially non-perturbative phenomenon.

To represent such a process by Feynman diagrams, the typical diagram would contain more than 10 virtual photon lines which together describe the effect of the strong Coulomb field on the electron–positron vacuum (Fig. 13.28). The measured production cross-section would not be described by essentially a single Feynman diagram but by the coherent sum of a large number of diagrams similar to the one shown in the figure. It is therefore correct to say that pair production by dynamical processes in heavy-ion collisions is a process which tests the validity of QED in the non-perturbative regime of this theory.

Since the ultimate goal of experimental efforts is the search for spontaneous positron production in supercritically strong Coulomb fields, it is important to establish in more detail agreement between theory and data for other mechanisms of positron production, Figs. 13.29, 30. Figure 13.29 compares the measured positron spectra, integrated over ion scattering angle $25° < \theta_{Lab} < 65°$, with theory for four different systems all involving a uranium projectile. Part (a), representing the $^{238}U + ^{154}Sm$ system ($Z_u = 154$), shows that the nuclear background can be successfully described by the method outlined in the previous section. The agreement is non-trivial, because the average multipolarity of the γ-ray spectrum was adjusted to the data obtained in a different collision system, namely $^{238}U + ^{165}Ho$ ($Z_u = 159$). The nuclear background contribution in the following three systems is calculated in the same way. When the calculated positron spectra for dynamic atomic processes [Mü 83c] are added, excellent agreement is found also in the other systems (parts b – d). Note that the nuclear background constitutes a rapidly decreasing fraction of the total positron spectrum with increasing combined charge Z_u. This behaviour reflects the fast growth of the atomic pair production cross-section with Z_u, discussed above. In other words, our ability to predict the positron spectra without the help of semi-empirical methods for background subtraction becomes better with higher charge. This is a very important conclusion considering that we want to find modifications in the spectra of supercritical systems that signal spontaneous positron creation.

The same conclusion is borne out by Fig. 13.30, which shows the energy-integrated positron probability as a function of nuclear scattering angle. The

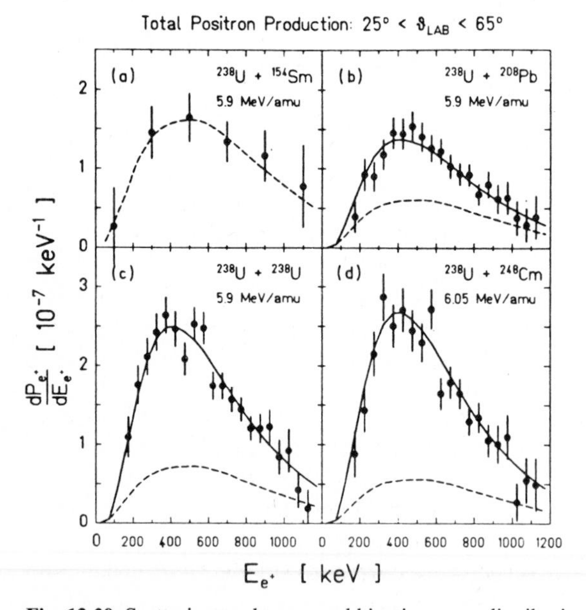

Fig. 13.29. Scattering-angle averaged kinetic energy distribution of positron production probabilities for four different scattering systems and projectile energies around the Coulomb barrier [Schw 84]. (– – –): nuclear background, (———): strong field QED assuming Rutherford trajectories, plus nuclear background

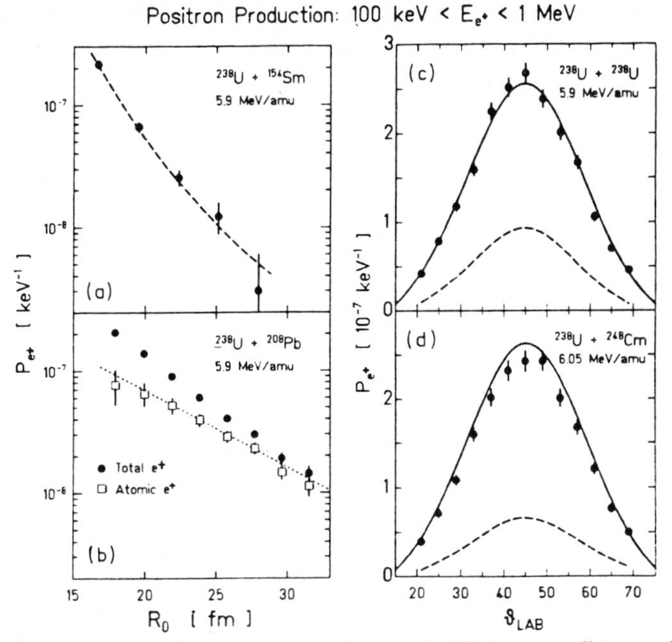

Fig. 13.30 a – d. Energy-averaged positron probabilities versus distance of closest approach R_0 (*left panel*) respectively lab. scattering angle (*right panel*) for the same systems. (– – –, ———) as in Fig. 13.29. (□□□): nuclear background corrected probabilities; (·········): strong field QED assuming Rutherford trajectories [Schw 84]

systems U + Sm and U + Pb are sufficiently asymmetric that forward and backward scattering events can be separated, allowing unambiguous determination of the impact parameter or, equivalently, the distance of closest approach in the collision, R_0. For the systems U + U and U + Cm this was not possible, so that every scattering angle θ_{Lab} corresponds to two impact parameters, assuming Rutherford scattering. The agreement between theory and experimental data is again excellent. In particular, the correct prediction of the slope of the atomic contribution for the U + Pb system, where the data can be plotted versus minimal distance R_0, proves that the positron production must occur predominantly when the two nuclei are very closet together. Otherwise the rapid and almost exponential fall off with R_0 would be hard to understand.

We have thus answered all the questions put forward at the beginning of this section. To summarize:

i) Atomic positron creation accounts for the major fraction of observed positrons in the very heaviest collision systems U + U and U + Cm ($Z_{\mathrm{u}} = 184, 188$). The nuclear background can be calculated reliably from the measured γ-ray spectrum.

ii) Atomic positron creation in heavy-ion collisions with $Z_{\mathrm{u}}\alpha > 1$ is an entirely non-perturbative process, evidenced by the increase according to Z_{u}^{22}. The agreement with theoretical predictions tests and confirms QED for the first time outside the perturbative sector.

iii) The positrons created by QED processes are produced mainly when the two nuclei are close together. These positrons constitute a probe of the strong Coulomb field in the united quasiatom.

iv) For the heaviest systems U + U and U + Cm, which are supercritical, the nuclear background contributes only 20% – 25% to the total positron cross-section. This forms an excellent basis to observe spontaneous positron emission in these collisions.

13.5 Positron Experiments II: Deep Inelastic Collisions

So far, we have discussed experiments where the energy of the projectile ion was chosen below the nuclear Coulomb barrier. When the energy of the projectile is increased, the two nuclei can come into contact in head-on collisions and start to interact, dissipating kinetic energy and relative angular momentum into internal nuclear degrees of freedom. In contrast to elastic or quasi-elastic reactions, these deep inelastic collisions are expected to be associated with finite reaction times, during which the two nuclei stick together at the distance of closest approach. Although reaction times of 10^{-20} s and more have been deduced for deep inelastic collisions of medium heavy nuclei [Schr 77, Go 80, Wo 82], sticking times of only about 10^{-21} s are expected for systems like U + U, due to the influence of the strong Coulomb repulsion between the two nuclei [Schm 78]. However, these reaction times have not been directly measured in a way that is independent of nuclear model assumptions. According to the discussion in Sect. 12.5, the nuclear sticking time should modify the observed positron spectrum, even for times as short as 10^{-21} s. It is, therefore, tempting to look at positron spectra from deep inelastic collisions in more detail.

Experiments of this kind have been performed by *Backe* and collaborators [Ba 83a, b], and more recently by the group working with the "TORI" spectrometer [Bo 83b], for the U + U system, using the subsequent fission of one or both uranium nuclei to identify deep inelastic collisions. The ^{238}U targets were bombarded with ^{238}U projectiles at energies between 5.9 and 10 MeV/u, and the resulting positron spectra were recorded in coincidence with fission by the solenoid detection system. In the presence of nuclear fission the separation of atomic and nuclear pair production processes is a major problem. The technique discussed in Sect. 13.3, involving conversion of measured γ-ray spectra into nuclear positron spectra, was augmented by measuring the spectra of electrons coincident with the positrons. It was found that the high-energy tails of these coincident electron spectra are almost entirely of nuclear origin. Thus the correctness of the γ-ray conversion procedure could be checked.

Figure 13.31 shows positron spectra corrected for nuclear background as obtained by the "TORI" group for U + U collisions at 8.4 MeV/u bombarding energy. The spectrum reproduced in Fig. 13.31a was recorded in coincidence with binary, i.e. elastic or quasi-elastic, events, while the spectrum in Fig. 13.31b was measured in coincidence with fission events. It is obvious that the drop of the

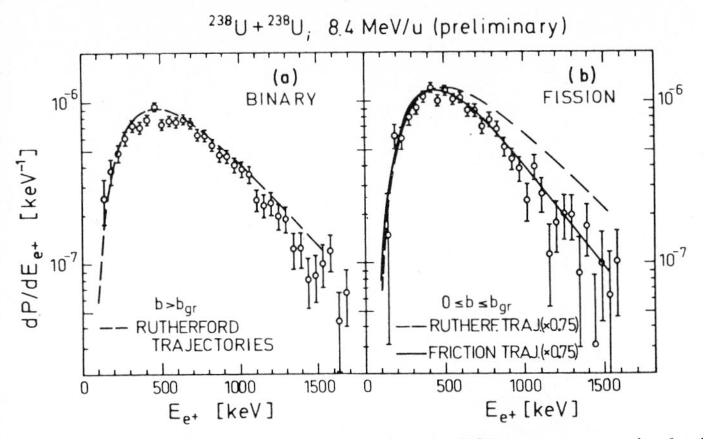

Fig. 13.31 a, b. Positron spectra measured at the TORI spectrometer in elastic (**a**) and deep inelastic (**b**) collisions of U + U at 8.4 MeV per nucleon [Bo 83b]. The steeper fall off found in coincidence with fission events can be related to a nuclear interaction time of about 10^{-21} s

high-energy tail of the positron spectrum observed in coincidence with fission is considerably faster than that observed for binary events, just as expected if fission events are associated with the time-delay caused by a massive nuclear reaction. The change in the slope corresponds to a sticking time of approximately 10^{-21} s. The difference is explained more quantitatively by comparing the data with theoretical predictions by *U. Müller* et al. [Mü 83c], assuming (i) Rutherford trajectories or (ii) trajectories following a macroscopic friction model for deep inelastic collisions [Schm 78]. Whereas the spectrum coincident with binary events is consistent with the result for Rutherford trajectories, the fission coincident spectrum is reproduced only by calculations involving trajectories with friction. These data, which form the first direct measurement of the reaction time in deep inelastic collisions (in fact, the first absolute measurement of such delay-times – "atomic clock"), are supported by a measurement of the *Q*-value dependence of the *K*-shell ionization probability by *Stoller* et al. [St 84b]. On the other hand, Fig. 13.31 also shows that delay times of the order of 10^{-21} s are too short to produce a clear signature for spontaneous positron creation.

13.6 Positron Experiments III: Narrow Structures in the Positron Spectrum

As the delay times observed in deep inelastic U + U collisions are obviously too short for that purpose, the search for spontaneous positron emission concentrated on measurements of positron spectra from supercritical collision systems at energies close to the Coulomb barrier (5.7 – 6.2 MeV/u). The reader should realize that, to some extent, this search implied the hope of a surprise. The calculations predicted that no sign of spontaneous positron production could be

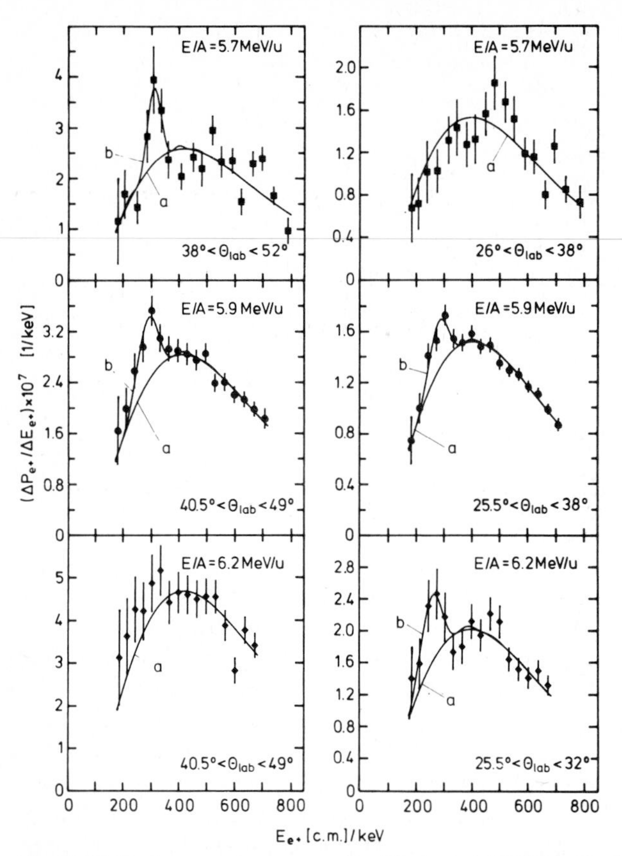

Fig. 13.32. Positron spectra measured by *Kienle* et al. [Cl 84] in U + U collisions close to the Coulomb barrier. The *smooth solid lines* (*a*) describe the nuclear background and dynamically induced positron production. The peaks that are fitted by Lorentz curves (*b*) are attributed to spontaneous positron creation

found in spectra originating from collisions on Rutherford trajectories, and standard nuclear reaction theory predicted that reaction times much longer than 10^{-21} s could not be expected for these collision systems. Nevertheless, the surprise came when narrow, peak-like structures were observed first in positron spectra from U + U ($Z_u = 184$) and U + Cm ($Z_u = 188$) collisions, and later also in spectra from the U + Th ($Z_u = 182$) and Th + Cm ($Z_u = 186$) systems.

First indications of these structures were obtained independently by the groups working with the orange spectrometer [Be 81] and with the EPOS detection system [Bo 81]. More positrons than expected were found with energies below 500 keV in the systems U + U and U + Cm. Subsequent experiments have shown beyond doubt that this surplus is contained in narrow peak-like structures in the positron spectrum at approximately 300 keV energy. It turned out that the presence or absence of the line structures in the spectrum depends critically on kinematic coincidence requirements for the heavy-ion scattering angles.

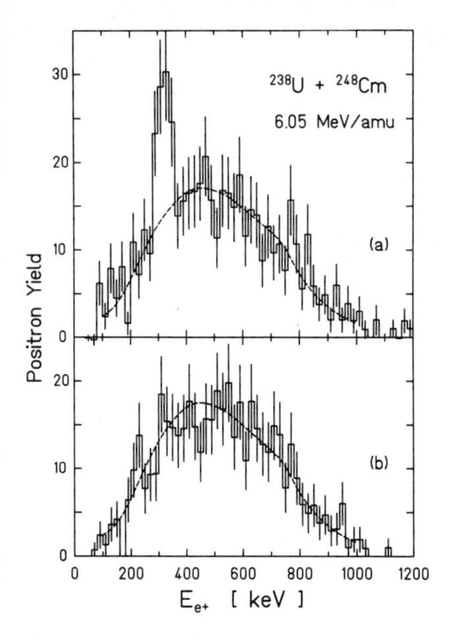

Fig. 13.33 a, b. Positron energy spectra observed for $^{238}U + ^{248}Cm$ collisions at a projectile energy of 6.05 MeV/u. Kinematic selections overlap preferentially with elastic scattering angles of **(a)** $100° < \theta_{cm} < 130°$ and **(b)** $50° < \theta_{cm} < 80°$. (– – –) represent the theoretical distribution for dynamic positron production based on Rutherford trajectories together with the nuclear background deduced from γ-ray spectra [Schw 83]

Figure 13.32 presents positron spectra resulting from $^{238}U + ^{238}U$ collisions measured by the orange β spectrometer [Cl 84]. The spectra were recorded at bombarding energies from 5.7 MeV/u up to 6.2 MeV/u. The line structure is most clearly seen in the spectrum corresponding to the lowest energy (5.7 MeV/u) and scattering angles centred around 45° in the laboratory system (90° in the centre-of-mass system), but the structure is also present in less pronounced form in spectra taken at higher energies. The smooth line (a) in Fig. 13.32 is the spectrum predicted by theoretical coupled-channel calculations including electron screening [Mü 83c], scaled by an appropriate normalization factor (equal to 0.94) to fit the spectra above 400 keV. As the nuclear background has not been subtracted from the spectra, the scaling factor takes background effects partly into account. The (b) lines are Lorentzian curves where position, width and intensity have been adjusted to fit the peak-line structures present in the data. The parameters for the Lorentzian in the upper left spectrum are, for example, a position at 308 ± 15 keV, a width of 56 ± 9 keV and an integrated intensity of $(8.9 \pm 2.0) \times 10^{-6}$ positrons per collision. The parameters for the other curves are similar [Cl 84].

Even more impressive is the line structure observed in $^{238}U + ^{238}Cm$ collisions at 6.05 MeV/u bombarding energy by the EPOS collaboration [Schw 83]. As shown in Fig. 13.33, this peak is present only under kinematic coincidence conditions corresponding preferentially to backward scattering (100° – 130° in the cm system) but not in spectra corresponding to forward scattering conditions. The position of the line structure has been determined to be 316 ± 10 keV, its width approximately 80 keV. Similar line structures have been observed in collisions of $^{238}U + ^{238}U$ and $^{232}Th + ^{248}Cm$ by the same experimental arrangement. The existence of these line structures is experimentally established. As they have

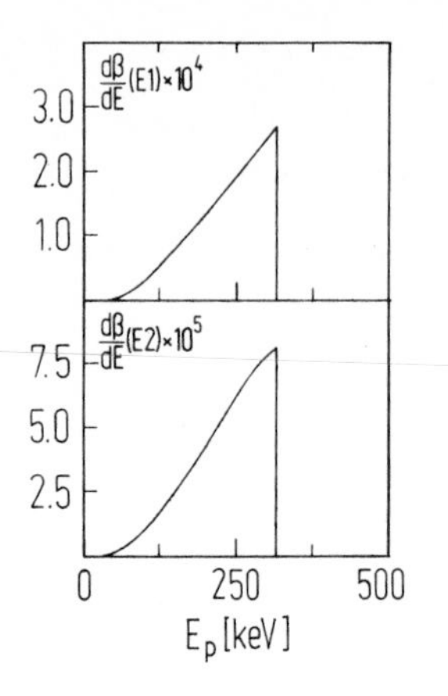

Fig. 13.34. Positron spectra expected from pair conversion of a transition in a uranium atom with energy 1.34 MeV. The multipolarities E1 and E2 are considered, the total conversion coefficients being $\beta = 6.4 \times 10^{-5}$ and 2×10^{-5}, respectively [Schl 83]

been observed only in spectra from collision systems with supercritical combined charge, more precisely with $Z_u \geq 182$, it is tempting to conjecture that they are caused by, or connected with, spontaneous positron production. To prove this conjecture, all other possible sources of the line structure must be ruled out conclusively. From the observed narrow widths of the peaks it follows immediately, by virtue of Heisenberg's uncertainty relation, that the emitting system must live for more than a period of $\hbar/\Gamma \gtrsim 4 \times 10^{-20}$ s, assuming $\Gamma \lesssim 100$ keV.

Let us first discuss the possibility that the positrons in the peak structures are emitted by the excited nuclei after the collision. Basically two different mechanisms have been considered in this connection. The first process is ordinary pair conversion of an excited nuclear state with a transition energy of 1.30 – 1.35 MeV, where the electron and positron are both emitted in continuum states. As the energy exceeding the combined rest mass of the pair can be arbitrarily distributed over the two particles in this case, the energy distribution of the emitted positrons is expected to be triangular with a peak at the maximal energy and a sudden drop beyond. The predicted saw-tooth-like shape of such a distribution, shown in Fig. 13.34, does not bear much resemblance to the shape of the observed line structure. The second process considered is monoenergetic pair conversion [Sl 53], where the electron is captured into an atomic bound state around the emitting nucleus [Schl 83]. Calculations show that only capture into the K shell is likely to be of interest. Since the binding energy of the K electron is gained in this case, the nuclear transition energy would have to be about 1.2 MeV to lead to a sharp positron line at about 300 keV.

For one of these processes to form the basis of a consistent explanation of the origin of our line structures, it must be possible to fit the observed line shape and the required intensity for the primary nuclear transition without contradicting other experimental observations. Let us start with the line shape. The argument here is based on the fact that the energy of a positron emitted from one of the scattered nuclei is subject to a Doppler shift which depends linearly on the velocity of that nucleus.

As discussed in Sect. 13.2, this effect causes line broadening that can be calculated for any of the processes under consideration. Figure 13.35 shows

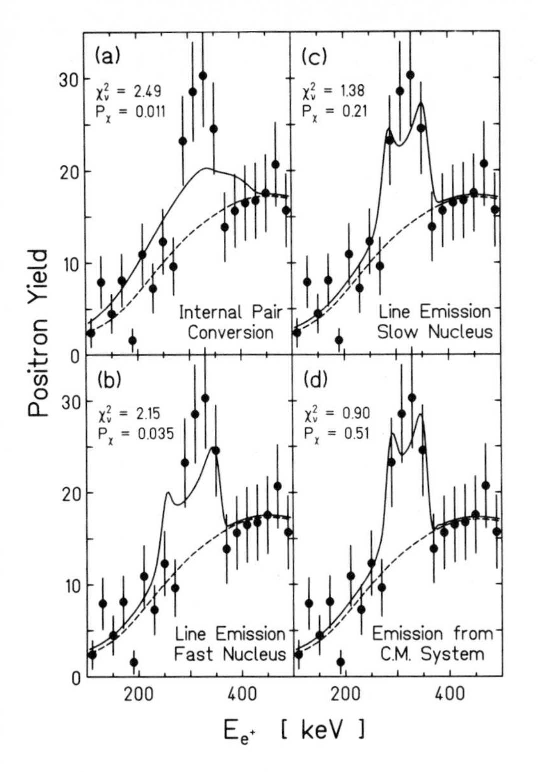

Fig. 13.35 a – d. Maximum likelihood fits to the entire energy spectrum in Fig. 13.33, assuming origins for the peak structure as indicated. The fit involves varying the peak intensity, its position and the continuous distributions of Fig. 13.34. The χ^2 is quoted for the 200 keV region around the peak [Schw 83]

the four possibilities for the U + Cm spectrum. (a) The line is due to ordinary pair conversion of a transition in the U-like nuclear fragment, or that the line is caused by monoenergetic pair conversion of a transition in (b) the Cm-like or (c) the U-like nucleus, and for comparison, (d) that a sharp positron line is emitted from a source moving with the velocity of the centre-of-mass. The chi-squared fits indicate that only processes (c) and (d) have a reasonable probability of describing the data, with a preference for the quasi-molecular origin of the line. A similar conclusion is reached when the velocity of the emitting source is deduced from the amount of Doppler broadening, assuming that the line structure has zero intrinsic width, i.e. that the width is totally due to the Doppler shift. At a 45° degree scattering angle in the laboratory system the velocity of the scattered nuclei is $\sqrt{2}\, v_{cm}$, where v_{cm} is the velocity of the centre-of-mass (Fig. 13.36). As Fig. 13.37 shows, the observed width of the line structure in the U + Cm system corresponds to $v = v_{cm}$, whereas the nuclear velocity $v = \sqrt{2}\, v_{cm}$ is far outside the experimental error bars. Similar results are obtained from an analysis of the U + U spectra measured by the β spectrometer group, but they are less conclusive at present because of the different acceptance properties of this detection system.

Leaving aside this evidence against a nuclear origin of the positron lines for a moment, let us ask whether their intensity is compatible with other observations. The argument exploits the fact that nuclear transitions that convert into electron positron pairs can also proceed by other mechanisms. When the multi-

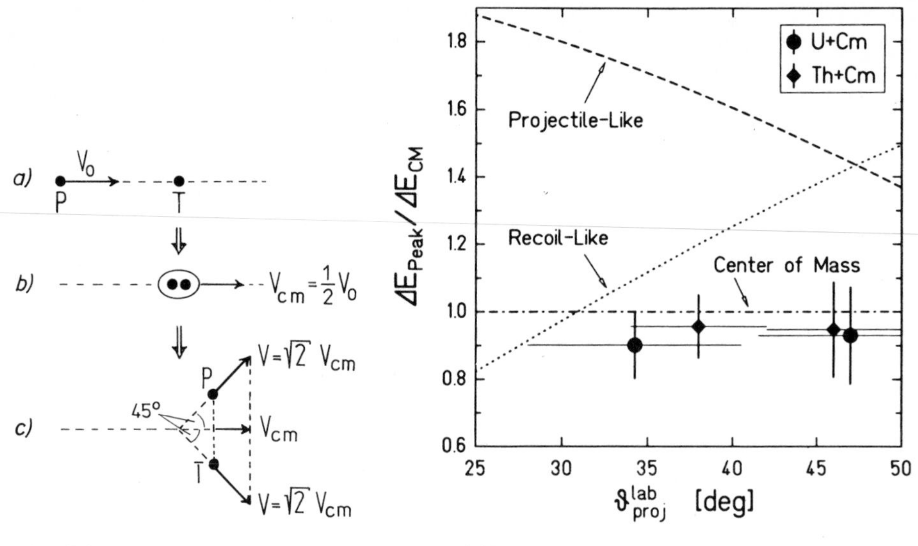

Fig. 13.36a–c **Fig. 13.37**

Fig. 13.36a–c. Kinematic relations in a collision of (nearly) equal nuclei leading to 45° scattering in the laboratory system: **(a)** before, **(b)** during, **(c)** after the collision

Fig. 13.37. Velocity of the source of emission of the line structure in the U+Cm and Th+Cm positron spectra, as determined from attributing the linewidth to Doppler broadening [Co 85]

polarity of the transition is not E0, the main de-excitation channel is by emission of a γ-ray. As the conversion coefficient, i.e. the branching ratio of the two decay channels, can be calculated precisely without knowing nuclear structure details, the intensity of the γ-ray line corresponding to the positron peak can be calculated assuming that the peak is caused by a nuclear transition:

$$P(\gamma) = \beta(E_\gamma)^{-1} \int dE_{\mathrm{p}} \frac{dP(\mathrm{e}^+)}{dE_{\mathrm{p}}} . \tag{13.10}$$

Taking into account the expected Doppler broadening of the γ line due to nuclear motion, we can compare the predicted intensity with the γ-ray spectrum that is routinely measured for nuclear scattering events. The results of this comparison are shown in the upper part of Figs. 13.38, 39 for the systems U+Cm and U+U, respectively, for assumed multipolarities of character E1 and E2. In both cases no indication for the presence of a line in the γ-ray spectrum is found, and its presence seems to be definitely excluded in the case of U+Cm. Other multipolarities would lead to even larger disagreement. A similar argument can be construed, if the nuclear transitions producing the positron lines had multipolarity E0. Then the transition could also occur by ejection of a bound electron, particularly from the K shell. As the ratio between K conversion and pair conversion can be calculated precisely without knowing nuclear structure,

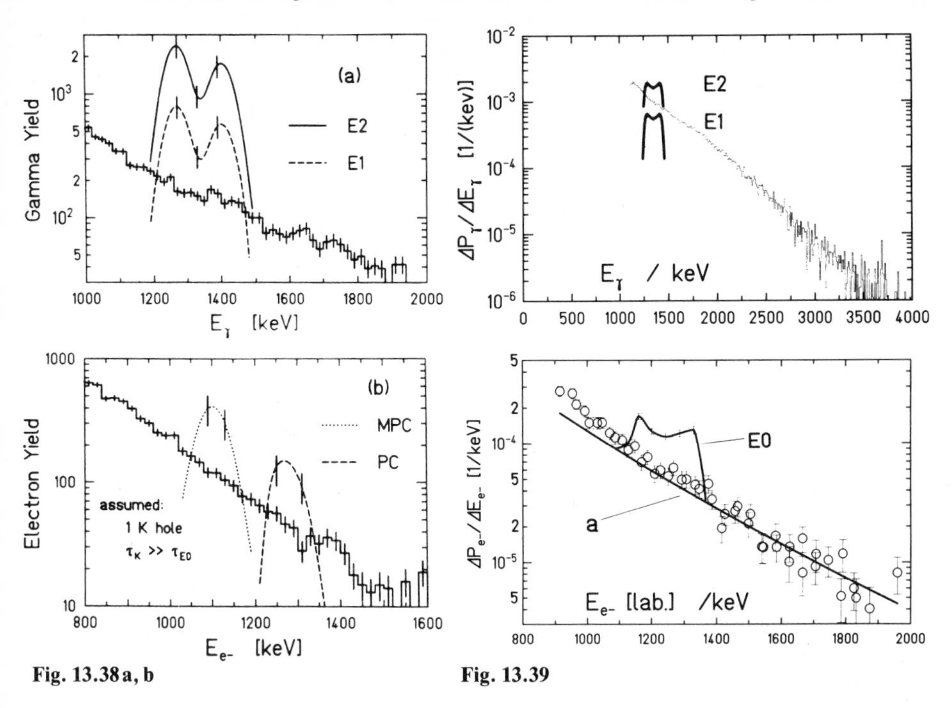

Fig. 13.38 a, b

Fig. 13.39

Fig. 13.38 a, b. Measured γ and e^- spectra compared **(a)** with γ-ray lines expected from the e^+-line intensity, assuming E1 or E2 transitions in the U or Cm nucleus, and **(b)** with internal conversion lines expected from normal (PC) or monoenergetic (MPC) internal pair conversion [Bo 84]

Fig. 13.39. The same figures for the γ-ray and δ-electron spectra measured in U + U collisions [Cl 84]

definite predictions concerning the intensity of the corresponding peak in the electron spectrum can be made [Schl 83]. When the predicted lines are compared with the measured δ-electron spectra, the results shown in the lower parts of Figs. 13.38, 39 are found. Again no indication of the presence of these lines is seen. We can conclude that also E0 multipolarity can be ruled out, if normal pair conversion is assumed to produce the line structures in the positron spectra.

The argument must be slightly modified if monoenergetic pair conversion is assumed. This process can occur only if vacancies are present in the *K* shell, and its intensity is proportional to the probability of finding a *K* vacancy during the nuclear transition. As mentioned in Sect. 13.3, the presence of a *K* vacancy in, say, uranium, at the moment of the transition is very unlikely due to the mismatch in transition times; whereas a *K* vacancy lives only about 10^{-17} s, a nuclear transition takes about $10^{-11}-10^{-13}$ s to occur. (This argument would not apply if all electrons in the *L* and *M* shells were simultaneously stripped away in the collision, so that no rapid transition to the *K* shell occurs. However, such a possibility must be considered as *extremely* improbable, as one-electron uranium ions are normally produced only in collisions at several hundred MeV/*u* energy.) Again, the branching ratio for monoenergetic pair conversion (MPC) relative to

K shell conversion does not depend on nuclear structure details. It does depend, however, on the number of vacancies available in the K shell. If there is no K vacancy available, monoenergetic pair conversion into the K shell is forbidden; on the other hand, if the K shell is completely empty, K conversion cannot take place. Denoting the primary conversion coefficient by β_{MPC}, and the average number of electrons contained in the K shell by n_K, the actual branching ratio is

$$\frac{P^{\mathrm{MPC}}(\mathrm{e}^+)}{P_K(\mathrm{e}^-)} = \frac{(2-n_K)}{n_K}\,\beta_{\mathrm{MPC}}(E_\mathrm{x})\,. \tag{13.11}$$

Thus, assumptions about the average number of available K vacancies enter crucially when trying to predict the intensity of the line in the electron spectrum that must be associated with the positron peak, if it is caused by monoenergetic pair conversion. As remarked above, one expects that the probability for a K vacancy to be present during the pair conversion process is very, very small.

Nevertheless, to be on the safe side, an average number of *one* K vacancy living indefinitely was assumed in the experimental analysis also included in the lower part of Fig. 13.30 (dotted line). The expected electron line is shifted by about 150 keV, because the required nuclear transition energy must be lower due to the binding energy gain of the electron from the pair captured into the K shell. Also this peak is clearly ruled out by the experimental data. In fact, an average number of more than 1.85 K vacancies (there can be at most 2!) would have to be assumed for this model [Co 84]. In view of the mismatch of the time scales this is a totally unreasonable assumption.

We conclude, therefore, that – on the basis of present experimental evidence – *a nuclear transition in the final fragments can be ruled out by shape and intensity as source of the observed line structures in the positron spectra.* Instead, line shape analysis favours a quasi-molecular origin, as shown in Fig. 13.37. Hence the possibility that the line structures are produced by spontaneous positron creation from the vacant supercritical K-shell in these systems must be seriously considered. This is now done for the line seen at 320 keV in the U + Cm spectrum [Schw 83]. Thereafter we discuss how the lines observed in the other collision systems fit into this model.

13.7 Giant Nuclear Systems and Spontaneous Positron Emission

We pointed out in the beginning of Sect. 13.6 that the source which emits the positron lines found in spectra of the U + Cm system and other supercritical collision systems must have a lifetime of at least several times 10^{-20} s by virtue of the uncertainty relation. Assuming that the line is caused by spontaneous pair creation, its position ($\sim$320 keV) must correspond to the location of the supercritical K-shell resonance in the negative energy continuum. The solutions of the two-centre Dirac equation indicate that the binding energy of the $1s\sigma$ state depends strongly on the distance of separation between nuclei, changing by about $\partial E_{1s}/\partial R \approx 20$ keV/fm for distances below the critical separation R_{cr}.

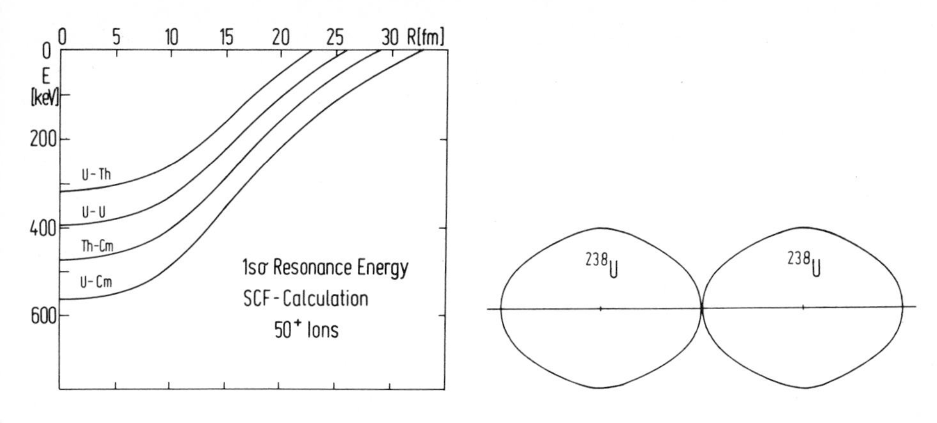

Fig. 13.40 **Fig. 13.41**

Fig. 13.40. Location of the supercritical $1s\sigma$ resonance as function of nuclear separation R for the systems U + Th ($Z_u = 182$), U + U ($Z_u = 184$), Th + Cm ($Z_u = 186$) and U + Cm ($Z_u = 188$) [Re 81a]

Fig. 13.41. Nuclear molecule formed by two oriented uranium nuclei. The distance between the nuclear centres is about 16.5 fm

The resonance location is shown in Fig. 13.40 for the systems U + Cm ($Z_u = 188$), U + U ($Z_u = 184$) and U + Th ($Z_u = 182$) as a function of R. For distances smaller than the nuclear touching point the nuclear charge distribution was taken in the form of two overlapping spheres, their radius adjusted to ensure volume conservation. In the limit $R \to 0$ one obtains a giant, spherical compound nucleus. The observed peak position corresponds to a distance $R = 16.5$ fm, just the separation at which the deformed nuclei ^{238}U and ^{248}Cm touch with their half-density radii, when they are oriented toward each other with their major axes, Fig. 13.41. In other words, the peak energy is aptly explained by the spontaneous emission line, if the two nuclei form a molecular configuration for some time, their surfaces just touching.

We shall name such a system a *giant* nuclear molecule to distinguish it from the "normal" superheavy nuclei conjectured to exist for proton numbers around $Z = 114$.

We can now make a model for the mechanism that produces the observed positron line in the U + Cm spectrum. In collisions with small impact parameter ($b \lesssim 5$ fm) the nuclei touch and are temporarily bound together by the attractive nuclear force. As long as the giant molecule lives, the existing K vacancies can be emitted as positrons. After some time T the system breaks up again, leaving very little excitation energy in the nuclei. The required length of contact can be determined by comparing the lines obtained in coupled-channel calculations, including T, with the measured linewidth. When the full observed width (~ 80 keV) is taken in this comparison, a contact time of 6.5×10^{-20} s is obtained. If part of the observed width is explained as Doppler broadening, the contact time must be accordingly longer. (Remember that the Doppler shift analysis in Fig. 13.37

was compatible with zero intrinsic width of the line!) The total intrinsic width $\Gamma = \Gamma_{\rm nucl} + \Gamma_{\rm e^+}$ is composed of the width (lifetime) of the giant nuclear system ($\Gamma_{\rm nucl}$) and the spontaneous decay width $\Gamma_{\rm e^+}$ of the positrons. For very long times T, the intrinsic linewidth Γ approaches the natural width $\Gamma_{\rm e^+}$ of the spontaneous positron line, here about 3 keV.

When the positron spectrum is measured together with scattered ions, these very special cases of giant nuclear molecule formation constitute only a minute fraction of all observed scattering events. At a given scattering angle most of the detected ions come from Rutherford scattering at larger impact parameters ($b \gtrsim 10$ fm), in which only the broad, dynamically generated spectrum of positrons is emitted. But because the height of the spontaneous emission line grows as T^2, even a very small reaction cross-section can show up in the positron spectrum if it is associated with sufficiently long contact time T. The combined spectrum is calculated according to

$$\begin{aligned} \frac{dP(\mathrm{e}^+)}{dE_\mathrm{p}} &= \frac{dP(\mathrm{e}^+, T=0)}{dE_\mathrm{p}} + \left(\frac{d\sigma}{d\Omega}\right)_{\rm react} \cdot \left(\frac{d\sigma}{d\Omega}\right)_{\rm Ruth}^{-1} \cdot \frac{dP(\mathrm{e}^+, T)}{dE_\mathrm{p}} \\ &\equiv \frac{dP(\mathrm{e}^+, T=0)}{dE_\mathrm{p}} + q \cdot \frac{dP(\mathrm{e}^+, T)}{dE_\mathrm{p}} \, . \end{aligned} \tag{13.12}$$

Taking $T = 1.0 \times 10^{-19}$ s, an excellent fit to the data from Fig. 13.33 is obtained for an admixture ratio $q = 0.9 \times 10^{-3}$ of reaction events at scattering angles around 45° (lab.). This fit, together with the experimental data, is reproduced in Fig. 13.42. Assuming that the angular distribution of the nuclear

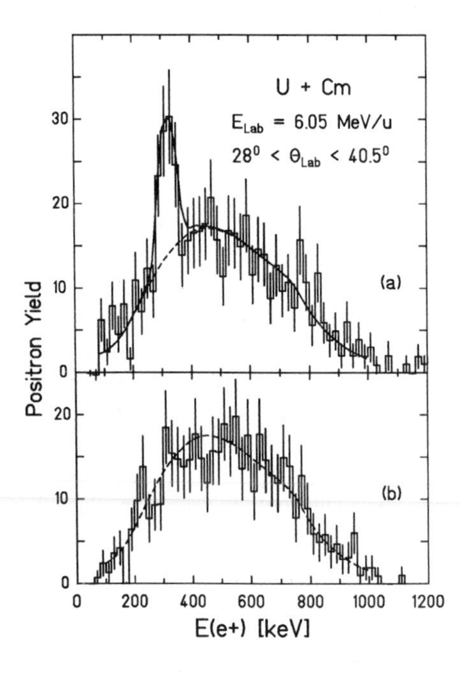

Fig. 13.42 a, b. Theoretical fit to the positron spectra observed in U + Cm collisions (Fig. 13.33). (– – –) describe positron spectra from Rutherford collisions, while (———) contains a contribution from collisions involving the formation of a giant nuclear molecule with average lifetime of 10^{-19} s and a cross-section of 15 mb

reaction products is isotropic, the value of q obtained in the fit implies a total reaction cross-section of about 15 mb for the nuclear molecule formation. In a sharp cut-off model this would correspond to a cut-off angular momentum of $l_c \approx 40\,\hbar$. (This means, in a semiclassical collision picture, that nuclei stick to form a giant nuclear system if the impact parameter $b < b_c \approx 0.7$ fm!) For longer values of the contact time the admixture coefficient q, and therefore σ_{react}, can be correspondingly smaller (approximately like $1/T$).

The success of this attempt to understand the origin of the line structure at 320 keV in the positron spectrum of the U + Cm system in terms of spontaneous positron emission motivates the search for a more microscopic model that explains the stability of the nuclear molecular configuration and allows its expected lifetime T to be calculated. The stability of the giant molecule requires the existence of an attractive pocket in the internuclear potential. One way to calculate this potential is by folding the density distributions of the two nuclei with a Yukawa-type potential containing phenomenologically determined parameters, such as the "M3Y" potential [Be 77, Mo 77, Sa 79]:

$$V(R, \alpha_\mu^{(1)}, \alpha_\mu^{(2)}) = \int d^3r_1\, d^3r_2\, \varrho_{\mathrm{U}}(\boldsymbol{r}_1)\, V_{\mathrm{eff}}(|\boldsymbol{r}_1 - \boldsymbol{r}_2|)\, \varrho_{\mathrm{Cm}}(\boldsymbol{r}_2)\,. \tag{13.13}$$

Because of the deformed shape of the uranium and curium nuclei, the potential depends on the relative orientations $\alpha_\mu^{(1)}$, $\alpha_\mu^{(2)}$ of the nuclei as well as on the separation R of their centres. The potential $V(R)$ has been calculated for various orientations by *Seiwert* et al. [Rh 83, Se 84], Fig. 13.43.

Indeed, $V(R)$ exhibits a shallow (10 – 15 MeV deep) pocket with a barrier at about 720 MeV/u for the configuration where the major axes of the nuclei are aligned. The barrier energy corresponds to a beam energy of 5.9 MeV/u, which practically coincides with the energy at which the positron line was found

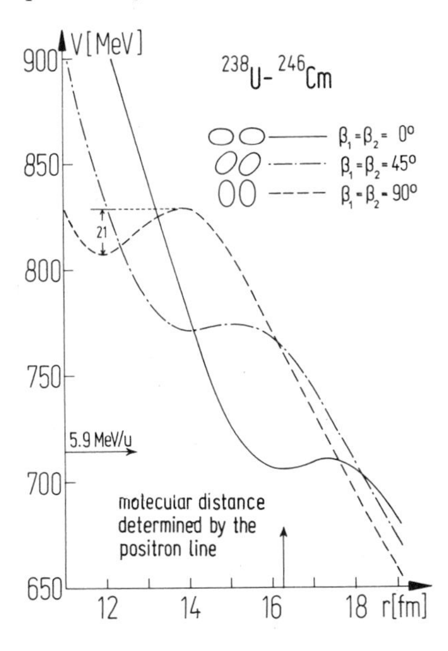

Fig. 13.43. Nucleus-nucleus potential for ^{238}U and ^{248}Cm with surface thickness corrections for various orientations of the nuclei. The lowest barrier is found for the "head-on-tails" configuration $\beta_1 = \beta_2 = 0$ at about 720 MeV, corresponding to 5.9 MeV/u bombarding energy. The molecular distance as determined from the energetical position of the positron line (vertical arrow, ↑) and the beam energy at which the positron line structure is experimentally observed (horizontal arrow, →) are indicated

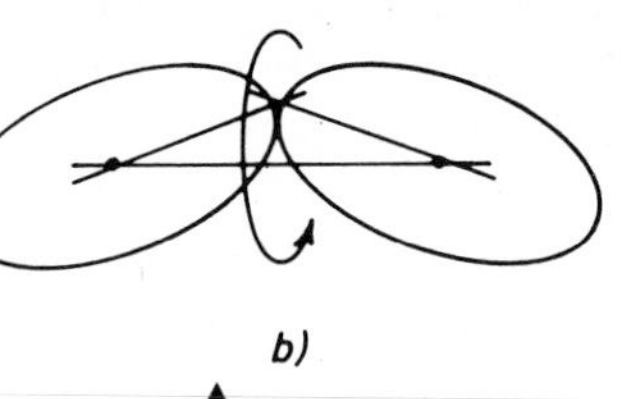

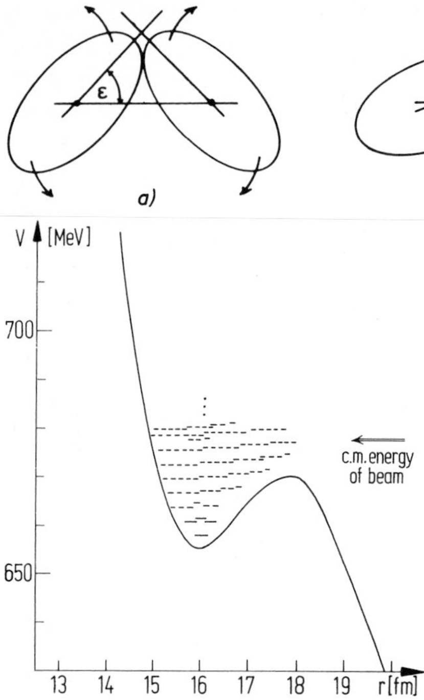

▲ **Fig. 13.44.** (**a**) Bending (or "butterfly") and (**b**) wriggling (or "belly dancer") modes of a giant nuclear molecule

Fig. 13.45. Resonances in the attractive pocket in the relative nuclear potential of a giant nuclear system. The beam energy of about 6 MeV/u corresponds to collisions at the barrier

experimentally. The minimum of the pocket lies between 16 and 16.5 fm, also in excellent agreement with our model assumptions for the giant nuclear molecule.

The giant dinuclear system is most probably not a static object. The nuclear molecule oscillates around the configuration that corresponds to the absolute minimum, i.e. around the configuration depicted in Fig. 13.41. Two typical modes, shown in Fig. 13.44, are the bending ("butterfly") and wriggling ("belly dancer") modes [Ni 65, Mo 80, He 84c, He 85]. There are also modes for vibrations in the relative separation R, and, of course, the vibrational modes of the individual nuclei. Combined, these modes form a multidimensional system of coupled oscillators, whose oscillator constants are determined by the relative nuclear potential $V(R, \alpha_\mu^{(1)}, \alpha_\mu^{(2)})$. The spectrum of eigenmodes of this system has been recently investigated in great detail by *Hess* et al. [He 84c, He 85], who predicted several hundred vibrational states in a 15 MeV deep pocket (Fig. 13.45). The spectrum is enriched further, because the nuclear molecule can rotate as a whole. Therefore, a complete rotational band is built on every single vibrational state, leading to a spectrum of the form

$$E_{nl} = E_n + \frac{\hbar^2 l(l+1)}{2\mu R_{\min}^2} \equiv E_n + \gamma l(l+1)\ . \tag{13.14}$$

For the U + Cm system at separation $R_{\min} \sim 16$ fm the rotational constant is only $\gamma \approx 0.7$ keV, so that the resulting density of states of the nuclear molecule is enormous.

Now the formalism developed at the end of Sect. 12.5 can be applied to describe spontaneous positron emission in U+Cm collisions leading to the formation of a nuclear molecular system. To evaluate (12.100) for the lifetime distribution of the molecule, we must know the energy dependence of the scattering matrix $S_l(E)$ in each partial wave l. For simplicity, we make the drastic assumption that the influence of the nuclear molecule on the scattering matrix can be fully described by elastic resonance scattering on the many vibrational and rotational states of the molecule. For each state of energy E_{nl} in (13.14) we assign a decay width Γ_{nl}, calculated with a barrier penetration formula [Hi 53, Br 59a]. We can then make a schematic, unitary ansatz for the nuclear scattering amplitude [He 84a, b]

$$S_l(E) = \prod_n \frac{E - E_{nl} - \frac{\mathrm{i}}{2}\Gamma_{nl}}{E - E_{nl} + \frac{\mathrm{i}}{2}\Gamma_{nl}} . \tag{13.15}$$

Here all interactions among the different molecular resonance states and all inelastic processes are neglected. Inserting (13.15) into (12.100), we can now calculate the distribution function $f(T)$ of the giant nuclear molecule lifetimes for any scattering angle, assuming a beam energy spread to 5 – 10 MeV. For numerical reasons, it has not yet been possible to perform a calculation with a realistic number of rotational bands (several hundred), but already the results obtained for 10 bands by *Heinz* et al. [He 83b, 84a, b] are very interesting. Figure 13.46 shows the nuclear scattering cross-section $d\sigma(T)/d\Omega$ associated with the lifetime T of two scattering angles (60° and 90° in the centre-of-mass system) as a function of T. After some threshold effects at small T the distribution falls off very slowly for delay times in the range $10^{-20}\,\mathrm{s} < T < 10^{-19}\,\mathrm{s}$. The slope corresponds to an average delay time of about 5×10^{-20} s, in almost perfect agreement with the value deduced in the macroscopic collision model (6.5×10^{-20} s). We stress that no parameters have been adjusted in the resonance formula (13.15) to obtain this agreement. A detailed analysis of the partial wave contributions to the long lifetime tail of the distribution shows that it originates from states with angular momentum around $l = 200\,\hbar$. This value should be reduced considerably by inclusion of more rotational bands.

The positron spectra calculated from the distributions are shown in Fig. 13.47. As expected from the presence of the long tails at high T, they exhibit striking peaks at supercritical $1s$ resonance. The width of the peak at half maximum is about 30 keV, again in excellent agreement with the value required to explain the experimentally observed structures. Unfortunately, the absolute intensity of the lines in Fig. 13.47 is about a factor 100 too small compared with the measured intensity of the peak structures in the U+Cm and U+U systems. This is probably an artefact of the oversimplified model for two reasons: (i) the intensity of the spectrum associated with long delay times is numerically found to increase roughly proportionally to the number of bands included. As mentioned above, a realistic number of bands is of the order of several hundred, i.e. a factor 20 – 50 larger than the 10 bands used to obtain the results in Figs. 13.46, 47. (ii) Inelastic processes may allow the nuclear molecule to dissipate energy away

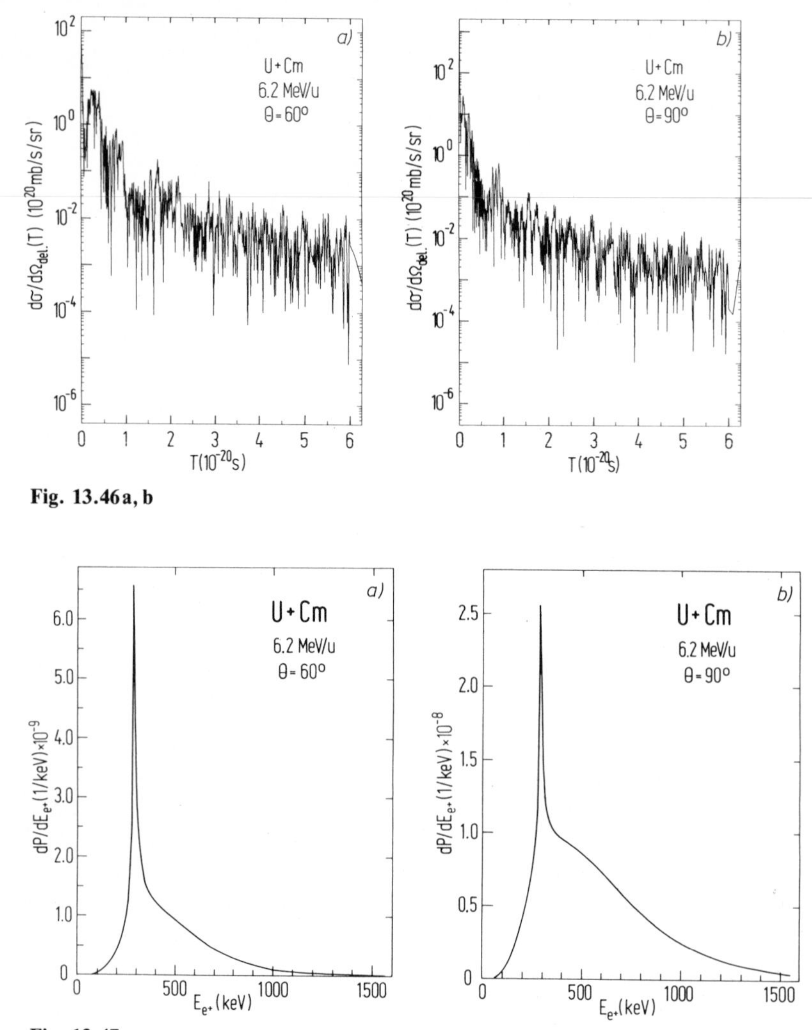

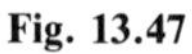

Fig. 13.47

Fig. 13.46a, b. Time distribution of the delayed nuclear cross-section for two scattering angles $\theta = 60°$ **(a)** and $90°$ **(b)** in the cm system, for 10 rotational molecular bands. The band heads were taken $1, 2, \ldots, 10$ MeV below the top of the potential barrier. After [He 84a]

Fig. 13.47. Positron spectra from delayed nuclear collisions, computed from the time-delay distributions shown in Fig. 13.46. The spectra are normalized to the nuclear Rutherford cross-section. The positrons from Rutherford scattering events are not added. After [He 84a]

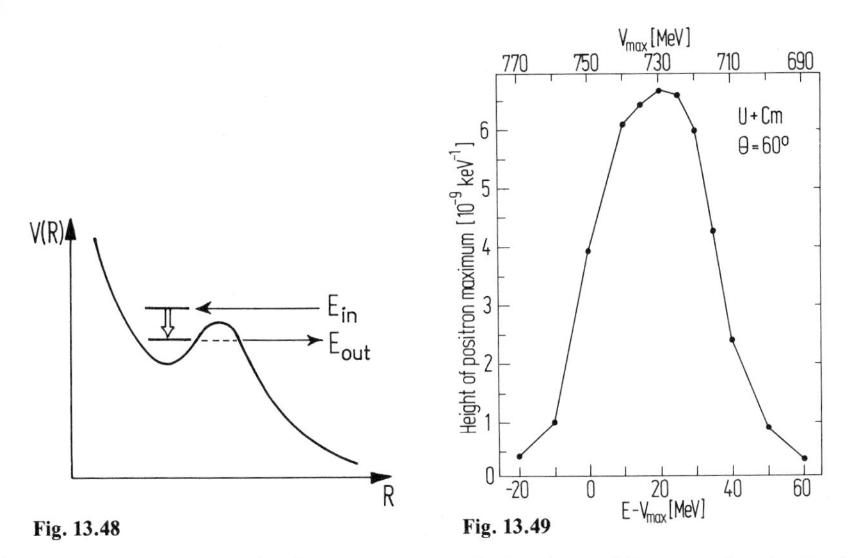

Fig. 13.48 **Fig. 13.49**

Fig. 13.48. Energy loss of inelastic processes may lead to the partial capture of projectiles that come in at energies above the barrier. This mechanism is expected to result in a drastic increase of the delayed scattering cross-section

Fig. 13.49. Height of the maximum in the positron spectrum in coincidence with scattering at 60° as a function of beam energy. V_{max} is the energy of the Coulomb barrier

from the degree of freedom of relative nuclear motion. Then the entrance width into a specific configuration can be much larger than the decay width, known in nuclear molecular physics as a "double-resonance mechanism" [Sche 70]. This effect, which is absent in our model of elastic scattering, is illustrated in Fig. 13.48. An estimate of the energy variability of the escape width indicates that an average energy loss of 2 MeV could lead to a gain of 3 orders of magnitude in the cross-section. This amount of inelasticity is already contained in the Coulomb excitation of the nuclei during the approach phase.

An interesting prediction of the microscopic model is the beam-energy dependence of the spontaneous positron line, shown in Fig. 13.49. At scattering energies below the top of the barrier the delayed cross-section is very small, because the projectiles are reflected off the barrier. A similar phenomenon occurs at energies high above the barrier where the widths of the molecular states are very wide, corresponding to short delay time. Only at beam energies slightly above the barrier is there a sizable contribution from long delay times. This causes the "resonance"-shaped excitation function of the positron line. Experiments indicated this behaviour [Schw 83, Cl 84, Gr 83c].

The present situation is then as follows. The model of giant nuclear molecule formation describes very nicely many details of the structure found in positron spectra of U + Cm collisions at energies close to the Coulomb barrier. The single remaining discrepancy concerns the absolute intensity and, as explained, there is good hope that an improved version of the model can correct this by including a realistic number of rotational bands and inelastic scattering processes. The astonishing feature of this microscopic model (which adds much to its credibility) is that it contains no parameter adjusted to fit the experimental data. The position of the barrier, the energy of the positron line and its width all come out correctly without parameter fiddling.

The problems of this interpretation, ascribing the positron lines to spontaneous pair creation, come when one looks at the other collision systems. The naive expectation would be that in all supercritical systems the same type of nuclear molecular configuration is formed, as the participating nuclei (Th, U, Cm) differ only minutely in their structure. On the basis of this expectation one would predict that the spontaneous positron emission line would shift as a function of $Z_u = Z_1 + Z_2$, because the binding energy of the supercritical K shell changes (Fig. 13.40). For each unit of charge less contained in the nuclei, the line should move by about 30 keV to lower energies: $dE_{1s}/dZ_u = -30$ keV. However, this is not what is found in the experiments: for all supercritical collision systems so far investigated the line structure is seen at approximately 280 – 360 keV [Co 85]. In principle, this behaviour could be explained if one assumes that the nuclear compound system changes in shape, i.e. that the average separation R in the molecule changes, so as to yield comparable K-shell binding energy always. From Fig. 13.40 it is clear that an almost spherical compound nucleus would have to be formed in the system U + Th ($Z_u = 182$). Although this explanation cannot be completely dismissed, it is – based on our present understanding of heavy nuclear composites – not very convincing.

Another possible explanation would be that the lines do not have the same origin. This possibility is made somewhat more likely since there are indications of further peaks in the positron spectra of the U + Cm and U + U systems at higher positron energies (Fig. 13.32). Some time ago already a mechanism was proposed by *Reinhardt* et al. that could produce several lines in the positron spectrum [Re 81a, Mü 83c]. The basic idea is simple: usually the giant nuclear molecule is produced in an excited state. When the K vacancy is spontaneously emitted as a positron, part of the excitation energy of the nuclear molecule can be transferred to the positron. After the emission, the nuclear molecule would be found in a state of lower excitation energy. This process, analogous to mono-energetic pair conversion in normal nuclei, is illustrated in Fig. 13.50. It becomes a fast process in supercritical systems because the conversion matrix elements increase very rapidly with nuclear charge Z. For $Z = 184$ the conversion probability is 10^5 times larger than for $Z = 92$ [Re 81a].

To find out how effective this mechanism to produce additional lines in the positron spectrum is, a simple classical analogue for the quantum process has been studied. In classical terms, an excited molecule means that the two nuclei oscillate back and forth around their equilibrium configuration. This oscillation may be described by a time-dependent separation $R(t)$

$$R(t) = R_0(1 - \alpha_0 \sin \omega t) . \tag{13.16}$$

By virtue of the correspondence principle, $\hbar\omega$ is identical with an excitation quantum in the potential pocket. Solving the coupled-channel equations for the

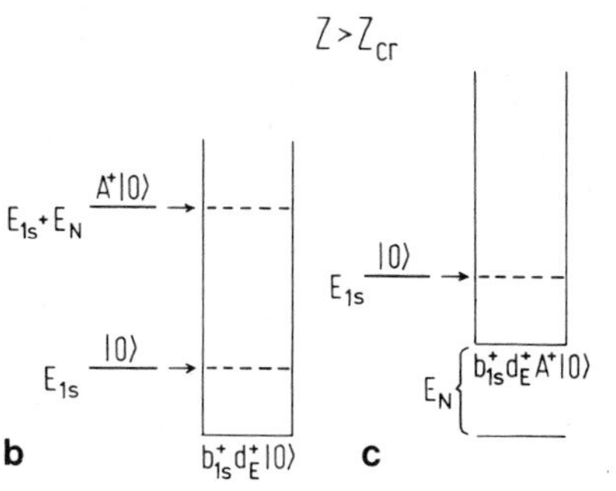

▲ **Fig. 13.50a–c.** Monoenergetic pair conversion filling a K vacancy in an atom associated with a giant nuclear system, can be induced by a nuclear transition with energy $E_N > m_e c^2 + E_{1s}$. In a subcritical system (**a**) the process would produce a single line; in supercritical systems (**b**) it can be understood as a Raman process associated with spontaneous pair production (two positron lines would appear). Also the inverse process is possible, where the nucleus becomes excited while a positron is emitted with reduced energy (**c**) [Re 81b]

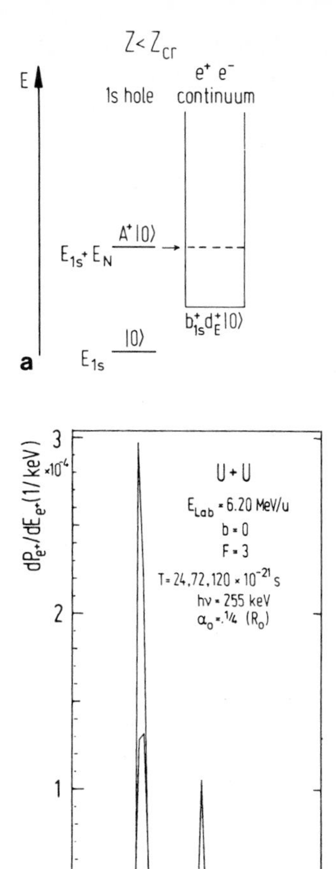

Fig. 13.51. Positron spectrum in delayed U+U collisions, assuming that the nuclear separation R oscillates during the period of contact T. The nuclear vibrational energy is taken as $\hbar\omega \approx 250$ keV. A Raman satellite of the spontaneous positron line is clearly visible [Mü 83c]

amplitudes $a_{kn}(t)$ in (12.32), one finds how the positron spectrum is modified under the influence of nuclear oscillations. Figure 13.51 shows the results of such calculations [Mü 83c] for an oscillation frequency corresponding to $\hbar\omega = 255$ keV and amplitude $\alpha_0 = 1/4$. A secondary peak (Raman line) is clearly present with about one third the intensity of the primary positron line, but shifted in energy by exactly one oscillator quantum. Obviously this mechanism could, indeed, be responsible for the appearance of further lines in the positron spectrum, which are not directly caused by, but are associated with, spontaneous positron emission.

This process was recently investigated by *Schramm* et al. [Schr 85] on the basis of a quantum mechanical model of nuclear resonance scattering. There it was found that the secondary line could be even more intense than expected from the classical model, because it is associated with an inherent inelasticity resulting in a prolonged period of nuclear contact (see Fig. 13.48 for a schematic explanation). Schramm et al. also considered the possible presence of more than one pocket in the internuclear potential, corresponding to several isomeric states of the giant nuclear molecule. In this case more than one line due to spontaneous positron emission results.

In principle, the nuclear Raman process discussed above can lead to narrow positron lines connected to the 1s-state of the giant atom even in subcritical systems. However, since the line energy remains related to the binding energy of the 1s-state, one would expect a shift in the line position with nuclear charge. The observed near-independence of this position has prompted speculations that, maybe, other mechanisms connected with the presence of strong fields are responsible for the narrow positron lines. E.g. *Schäfer, Müller* and *Greiner* [Scha 84] point out that a Bose condensate of Higgs particles might be established for much lower supercritical charge (see also Chap. 19 for Bose condensation). Most other explanations invoking the existence of new particles can be ruled out on the basis of present experimental evidence [Scha 85].

We conclude this section by repeating the fundamental results of the experimental efforts to prove the existence of spontaneous pair creation in supercritical electric fields.

i) Pair creation in systems with $Z_u\alpha > 1$ has been shown to be a highly non-perturbative phenomenon.
ii) Narrow lines have been observed in the positron spectra emitted by collision systems involving supercritical charge $Z_1 + Z_2 > Z_{cr}$.
iii) Nuclear pair conversion in normal nuclei has been conclusively ruled out as source of these line structures.
iv) A reaction model has been constructed that consistently explains the observed peak in the U + Cm systems as a *spontaneous positron emission* line from the K shell of a metastable giant nuclear molecular system.
v) Problems with the interpretation of lines in other systems persist, in particular, in understanding the Z-dependence of the line energy; their resolution must await further experimental and theoretical efforts. Most important in this context is a careful and systematic experimental study of subcritical collision systems ($Z_u \lesssim 180$).

Bibliographical Notes

Experimental results and their comparison with theory are discussed at length in the reviews [Re 77, 84]. Many details, especially concerning experimental aspects but also of theory, are found in the various articles contained in the *Proceedings of the Lahnstein School* [Gr 83a]. Pair conversion and internal electron conversion are summarized in the review article [Schl 81].

The present status of theory and experiment in the search for spontaneous positron emission is set forth in [Ba 84, Schw 84].

A comprehensive experimental review is forthcoming [Gr 85]. Limited aspects are reviewed in [Mo 84, Vi 84]. The experimental and theoretical aspects of nuclear reaction effects on electronic processes in quasiatoms are reviewed by *Meyerhof* and *Chemin* [Me 84b] and by *Anholt* [An 84].

14. Vacuum Polarization

Any comparison between theory and experiment hinges on precise understanding of the binding energies of electrons and the positron resonance energy. It is therefore of decisive importance to establish the role played by the effects of conventional radiative QED. Normally, such corrections, in particular vacuum polarization and self-energy (Lamb shift) effects, contribute at most 1% to the particle energies since they are expected to be one power of fine structure constant $\alpha = 1/137$ smaller. Nonetheless, these terms must be carefully scrutinized for influence of infinite renormalization of charge and mass. A further motivation to deepen this subject arises from the simple intention of properly understanding these effects in supercritical fields. In particular, it is interesting to see how for $Z \to Z_{\rm cr}$, the description of vacuum polarization and electron screening as being two different effects cannot be maintained anymore.

In the next three chapters we describe the radiative effects in supercritical fields – here we begin to consider the vacuum polarization in weak fields in a perturbative expansion. We continue to use the system of natural units in which $\hbar$ and c are equal to unity.

14.1 Vacuum-Current Density: Perturbative Expansion

The expectation value of the current operator of the Dirac field

$$\hat{j}_\mu(x) = \frac{e}{2}[\hat{\bar{\psi}}(x), \gamma_\mu \hat{\psi}(x)] \tag{14.1}$$

does not generally vanish when an external potential is present. First recall that (14.1) can be obtained as the symmetric limit $x' \to x$ of the time ordered product

$$\tfrac{1}{2}[\hat{\psi}(x), \hat{\bar{\psi}}(x)] = T(\hat{\psi}(x)\hat{\bar{\psi}}(x'))\,|_{x' \underset{s}{\to} x}\,, \tag{14.2}$$

where the symmetric limit ("s") implies that the limit $x' = x$ is approached symmetrically from future ($t' = t+0$) and past ($t' = t-0$) as indicated in (9.111).

Since the expectation value of the time-ordered product is the Feynman propagator, (9.106),

$$\mathrm{i}S_{\rm F}(x,x') = \langle OF|T(\hat{\psi}(x)\hat{\bar{\psi}}(x'))\,|OF\rangle\,, \tag{14.3}$$

$$\langle OF|\hat{j}_\mu(x)|OF\rangle = -\mathrm{i}e\,\mathrm{tr}\,[\gamma_\mu S_\mathrm{F}(x,x')]_{x' \underset{s}{\to} x}\,, \tag{14.4a}$$

where the symmetric limit here means explicitly

$$S(x,x')_{x \underset{s}{\to} x'} = \tfrac{1}{2} \lim_{\substack{\varepsilon'^2 \to 0+ \\ \varepsilon'_0 > 0}} [S(x,x+\varepsilon') + S(x,x-\varepsilon')]\,. \tag{14.4b}$$

Thus the induced displacement charge of the vacuum is, as discussed in Chap. 9, intimately connected with the propagator of the Dirac field. Depending on the actual application, the problem is to evaluate (14.4) either considering a non-perturbative representation of $S_\mathrm{F}(x,x')$ or expanding about a small perturbation $A_\mu^\mathrm{ex} \approx 0$ by the perturbative expansion method described in Chap. 7. Thus before turning to the more involved non-perturbative case for $Z \sim Z_\mathrm{cr}$, we first discuss the perturbative expansion and some of its classic pitfalls.

First, let us discuss the trivial case $A_\mu^\mathrm{ex} = 0$. At this point it is useful to know the explicit configuration space form of the free Feynman propagator. Beginning with (7.43),

$$S_0(x-x') = \int \frac{d^4p}{(2\pi)^4}\, \mathrm{e}^{-\mathrm{i}p(x-x')}\, \tilde{S}_\mathrm{SF}(p)\,, \tag{14.5a}$$

$$\tilde{S}_\mathrm{SF}(p) = \frac{\gamma \cdot p + m_0}{p^2 - m_0^2 + \mathrm{i}\varepsilon}\,. \tag{14.5b}$$

With (10.109)

$$S_0(x-x') = \int \frac{d^4p}{(2\pi)^4}\, \mathrm{e}^{-\mathrm{i}p(x-x')} (\gamma \cdot p + m_0)(-\mathrm{i}) \int_0^\infty ds\, \exp[\mathrm{i}s(p^2 - m_0^2 + \mathrm{i}\varepsilon)]\,,$$

which, interchanging the order of the integrations, leads first to a Gaussian integral:

$$\begin{aligned} S_0(x-x') = {} & -\mathrm{i} \int_0^\infty ds\, \exp[-\mathrm{i}s(m_0^2 - \mathrm{i}\varepsilon)] \int \frac{d^4p}{(2\pi)^4} (\gamma p + m_0) \\ & \times \exp[-\mathrm{i}p(x-x') + \mathrm{i}sp^2]\,. \end{aligned} \tag{14.6}$$

The last exponent is conveniently rewritten as

$$\mathrm{i}s\left(p^2 - 2p\frac{x-x'}{2s}\right) = \mathrm{i}s\left(p - \frac{x-x'}{2s}\right)^2 - \mathrm{i}\frac{(x-x')^2}{4s}\,. \tag{14.7}$$

This warrants the introduction of a new variable (four-vector)

$$l_\mu = p_\mu - \frac{x_\mu - x'_\mu}{2s}\,, \tag{14.8}$$

giving

$$\int \frac{d^4p}{(2\pi)^4} (\gamma p + m_0) \exp[-ip(x-x') + isp^2]$$

$$= \int \frac{d^4l}{(2\pi)^4} \left(\gamma l + \gamma \frac{(x-x')}{2s} + m_0 \right) e^{isl^2} \exp\left[-i \frac{(x-x')^2}{4s} \right]. \tag{14.9}$$

The 4-dimensional integral over l is dealt with in (10.114),

$$\int e^{isl^2} \frac{d^4l}{(2\pi)^4} = \frac{1}{(4\pi)^2} \frac{1}{is^2}, \tag{14.10}$$

while the linear term in l vanishes. Consequently

$$S_0(x-x') = -\frac{1}{(4\pi)^2} \int_0^\infty \exp[-is(m_0^2 - i\varepsilon) - i(x-x')^2/4s]$$

$$\times \left[\frac{\gamma(x-x')}{2s} + m_0 \right] \frac{ds}{s^2}. \tag{14.11}$$

This expression is highly singular as the so-called proper time parameter s approaches zero. To find the integral analytically, note that

$$S_0(x-x') = (\gamma \cdot p + m_0) D_0(x-x'), \tag{14.12a}$$

where $D_0(x-x')$ is simply the integral (14.11) without the square bracket factor. By definition of the Hankel path for Bessel functions, then

$$D_0(x) = -\frac{1}{4\pi} \delta(x^2) + \frac{m}{8\pi\sqrt{x^2}} \theta(x^2)(J_1(m\sqrt{x^2}) - i Y_1(m\sqrt{x^2}))$$

$$-\frac{im}{4\pi^2\sqrt{-x^2}} \theta(-x^2) K_1(m\sqrt{-x^2}). \tag{14.12b}$$

For $x^2 \to 0$ the highest (light cone: $x^2 = 0$) divergence is now identified. The Bessel functions contain a single pole at origin, i.e.

$$\left.\begin{aligned} K_1(z) &\to \frac{1}{z} \\ Y_1(z) &\to \frac{2}{\pi z} \end{aligned}\right\} z \to 0, \quad \text{and} \tag{14.13}$$

$$D_0(x-x') \xrightarrow[(x-x')^2 \to 0]{} \frac{i}{4\pi^2(x-x')^2 - i\varepsilon} + \ldots \tag{14.14}$$

which, when differentiated as required in (14.12a), leads to

$$S_0(x-x') \to \frac{1}{2\pi^2} \frac{(x-x')\gamma}{[(x-x')^2 - i\varepsilon]^2} . \tag{14.15}$$

Equipped with these analytic expressions we can now compute the induced vacuum charge density in the absence of external fields. When inserted in (14.4), (14.11) implies

$$\begin{aligned} \langle j_\mu \rangle_{(0)} &= \frac{4ie}{(4\pi)^2} \int_0^\infty \exp[-is(m_0^2 - i\varepsilon) - i(x-x')^2/4s] \frac{(x-x')_\mu}{2s} \frac{ds}{s^2} \bigg|_{x \underset{s}{\to} x'} \\ &= -(4ie)i \left[\frac{\partial}{\partial x^\mu} D_F(x-x') \right]_{x \underset{s}{\to} x'} \to 4e \frac{-i}{2\pi^2} \frac{(x-x')_\mu}{[(x-x')^2 - i\varepsilon]^2} \bigg|_{x \underset{s}{\to} x'} \end{aligned} \tag{14.16}$$

where, as required, the symmetric limit $x \to x'$ is reserved until the end of the calculation. It is important to appreciate that (14.16) is singular in the limit $x \to x'$ and only if this limit is taken symmetrically, i.e. by setting: $x = x' \pm \varepsilon'$, will (14.16) vanish, as expected. Clearly, we subtract infinity from infinity and the precise manner in which this is done matters. This problem is encountered several times in this chapter.

We would not have entered into this lengthy and seemingly fruitless discussion if the story was not entirely different once a finite, even arbitrarily small external potential is present. Consider the expansion (7.52):

$$S_{SF}(x,x') = S_0(x-x') + e\int d^4x_1 S_0(x-x_1)\gamma \cdot A_{ex}(x_1) S_{SF}(x_1,x') \tag{14.17}$$

which can be represented graphically (a thick line the full propagator S_{SF}) by

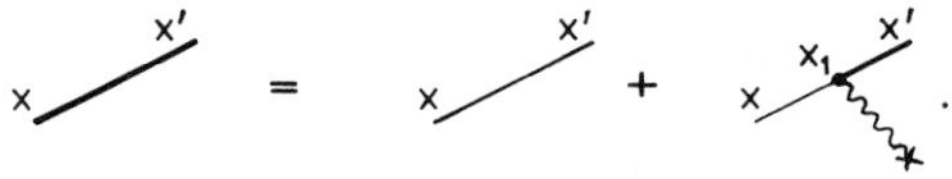

x — x′ = x — x′ + x — x_1 — x′ . (14.18)

Here the cross at the end of the wavy line stands for the external charge density. The limit $x \to x'$ may be obtained by closing the lines so that x and x' coincide. Then

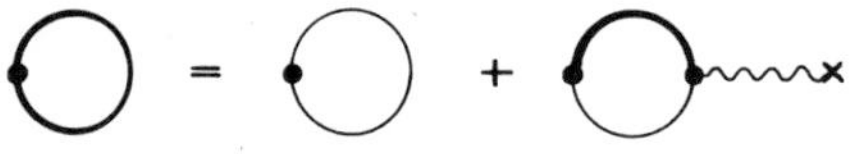

(14.19)

The first term in (14.19) is the free field term just considered. Inserting (14.17) into (14.4), the next term linear in $A_{ex,\nu}$ is

$$\langle OF|j^\mu(x)|OF\rangle_{(1)} = -ie^2 \int \mathrm{tr}[\gamma^\mu S_0(x-x_1)\gamma^\nu S_0(x_1-x')]\,|_{x \underset{s}{\to} x'} A_{ex,\nu}(x_1) d^4x_1 . \tag{14.20a}$$

Higher order terms are obtained accordingly. In nth order

$$\begin{aligned} \langle OF|j^\mu(x)|OF\rangle_{(n)} = &-ie^{n+1} \int \mathrm{tr}[\gamma^\mu S_0(x-x_1)\gamma^{\nu_1} \ldots \\ &\ldots S_0(x_{n-1}-x_n)\gamma^{\nu_n} S_0(x_n-x')]\,|_{x \underset{s}{\to} x'} A_{ex,\nu_1}(x_1) \ldots A_{ex,\nu_n}(x_n) d^4x_1 \ldots d^4x_n . \end{aligned} \tag{14.20b}$$

Note that for even n (even "power" of the external field) the induced current seems to depend on the convention chosen for the charge e in terms of its sign, since then the power of the coefficient e is odd. However, it is possible to show, quite generally (*Furry's* theorem [Fu 37]) that diagrams with a loop containing an odd number of "corners" vanish. Thus, only the odd powers $n=2k+1$ contribute, of which (14.20) is the first non-vanishing term.

It is often more convenient to consider the potential induced by the external charges, rather than the current. This is simply achieved by using the elementary solution of the Maxwell field equations, employing the Green's function $D_0^{\mu\nu}(x-x')$:

$$\langle A^\nu(x)\rangle = \int D_0^{\mu\nu}(x-x')\,\langle j_\mu(x')\rangle\, d^4x' \tag{14.21a}$$

which in the graphical presentation implies that a wiggly line is attached to the dot in (14.19):

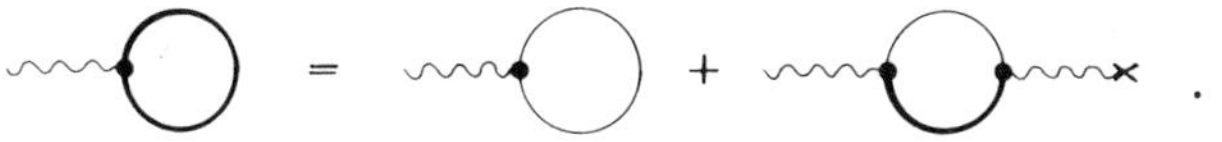

(14.21b)

Employing the momentum representation of the Feynman propagators in the form (14.5a) with variables p_1, p_2 and substituting

$$p=\frac{p_1+p_2}{2}\,, \qquad k=p_1-p_2 \tag{14.22}$$

we obtain for (14.20a)

$$\langle j^\mu(x)\rangle_{(1)} = -\mathrm{i}e^2\int\frac{d^4p}{(2\pi)^4}\,\frac{d^4k}{(2\pi)^4}\left\{\mathrm{tr}\left[\gamma^\mu \tilde{S}_{\mathrm{SF}}\left(p+\frac{k}{2}\right)\gamma^\nu \tilde{S}_{\mathrm{SF}}\left(p-\frac{k}{2}\right)\right]\right.$$
$$\left.\left.\cdot\,[A_{\mathrm{ex},\nu}(x_1)\,\mathrm{e}^{\mathrm{i}kx_1}d^4x_1]\,\exp[-\mathrm{i}p(x-x')]\,\exp\left[-\mathrm{i}k\frac{(x+x')}{2}\right]\right\}\right|_{x'\underset{\mathrm{s}}{\to}x} \tag{14.23}$$

where the Fourier transform of the external field appears

$$\tilde{A}_{\mathrm{ex},\nu}(k)=\int \mathrm{e}^{\mathrm{i}kx_1}A_{\mathrm{ex},\nu}(x_1)\,d^4x_1\,. \tag{14.24}$$

Note further that the limit $x'\to x$ can be taken with impunity in the factor $\exp[-\mathrm{i}k(x+x')/2]$, so

$$\langle j^\mu(x)\rangle_{(1)} = -e^2\int\frac{d^4k}{(2\pi)^4}\,\mathrm{e}^{-\mathrm{i}kx}\,\Pi_{(1)}^{\mu\alpha}(k)\,\tilde{A}_{\mathrm{ex},\alpha}(k) \tag{14.25a}$$

$$\langle A_\mu(x)\rangle_{(1)} = -e^2\int\frac{d^4k}{(2\pi)^4}\,\mathrm{e}^{-\mathrm{i}kx}\,\tilde{D}_{\mu\beta}(k)\,\Pi_{(1)}^{\beta\alpha}(k)\,\tilde{A}_{\mathrm{ex},\alpha}(k)\,, \tag{14.25b}$$

where $\tilde{D}_{\mu\beta}$ is the Fourier transform of $D_0^{\mu\beta}$ in (14.21a) and

$$\Pi^{\mu\nu}_{(1)}(k) = \lim_{\varepsilon' \to 0} \Pi^{\mu\nu}(k:\varepsilon')$$
$$= \lim_{\varepsilon' \to 0} \mathrm{i} \int \frac{d^4p}{(2\pi)^4} \mathrm{tr}\left[\gamma^\mu \tilde{S}_{\mathrm{SF}}\left(p+\frac{k}{2}\right)\gamma^\nu \tilde{S}_{\mathrm{SF}}\left(p-\frac{k}{2}\right)\right] \mathrm{e}^{\mathrm{i}p\varepsilon'} . \tag{14.26}$$

Inserting here the explicit momentum space form of the electron propagator (14.5b),

$$\Pi^{\mu\nu}(k,\varepsilon') = \int \frac{d^4p}{(2\pi)^4} \mathrm{e}^{\mathrm{i}p\varepsilon'} \frac{\mathrm{tr}\left\{\gamma^\mu\left[\gamma\left(p+\frac{k}{2}\right)+m_0\right]\gamma^\nu\left[\gamma\left(p-\frac{k}{2}\right)+m_0\right]\right\}}{\left[\left(p+\frac{k}{2}\right)^2 - m_0^2 + \mathrm{i}\varepsilon\right]\left[\left(p-\frac{k}{2}\right)^2 - m_0^2+\mathrm{i}\varepsilon\right]} . \tag{14.27}$$

First consider the trace in the numerator

$$\mathrm{tr}[\,] = \mathrm{tr}[\gamma^\mu\gamma^\alpha\gamma^\nu\gamma^\beta]\left(p_\alpha+\frac{k_\alpha}{2}\right)\left(p_\beta-\frac{k_\beta}{2}\right) + \mathrm{tr}[\gamma^\mu\gamma^\nu]\, m_0^2 , \tag{14.28}$$

where, as always, a sum is implied where two identical Greek indices are encountered, one as upper, the other as lower script. We have already ignored the terms with an odd number of γ matrices, since in general, the trace of a product of an odd number of γ matrices vanishes. Namely with $\gamma_5\gamma_5 = 1$

$$\mathrm{tr}[\gamma^{\alpha_1}\dots\gamma^{\alpha_{2n+1}}] = \mathrm{tr}[\gamma^{\alpha_1}\dots\gamma^{\alpha_{2n+1}}\gamma_5\gamma_5]$$
$$= (-)^{2n+1}\, \mathrm{tr}[\gamma_5\gamma^{\alpha_1}\dots\gamma^{\alpha_{2n+1}}\gamma_5] \tag{14.29}$$

with the last identity obtained upon $(2n+1)$-fold commutation of $\gamma_5 = \mathrm{i}\gamma^0\gamma^1\gamma^2\gamma^3$ with the γ matrices where

$$\gamma^\alpha\gamma_5 = -\gamma_5\gamma^\alpha . \tag{14.30}$$

Using the cyclic property of the trace

$$\mathrm{tr}(AB\dots C) = \mathrm{tr}(CAB\dots) ,$$

alternatively

$$\mathrm{tr}[\gamma^{\alpha_1}\dots\gamma^{\alpha_{2n+1}}] = \mathrm{tr}[\gamma_5\gamma^{\alpha_1}\dots\gamma^{\alpha_{2n+1}}\gamma_5] \tag{14.31}$$

and (14.29, 31) can both be true only if the trace of an odd number of γ matrices vanishes.

To compute the trace of the product of two γ matrices appearing in (14.28), we consider the trace of the anticommutation relation of the γ-matrices

$$\mathrm{tr}[\gamma^\mu\gamma^\nu] = -\mathrm{tr}[\gamma^\nu\gamma^\mu] + \mathrm{tr}[2g^{\mu\nu}] = -\mathrm{tr}[\gamma^\mu\gamma^\nu] + 4\cdot 2g^{\mu\nu}$$

and hence

$$\mathrm{tr}[\gamma^\mu \gamma^\nu] = 4 g^{\mu\nu} . \tag{14.32}$$

We proceed similarly with the trace of four γ matrices, where the basic anticommutator must be used 3 times in sequence:

$$\begin{aligned}\mathrm{tr}[\gamma^\mu\gamma^\alpha\gamma^\nu\gamma^\beta] &= \mathrm{tr}[2 g^{\mu\alpha}\gamma^\nu\gamma^\beta] - \mathrm{tr}[\gamma^\alpha\gamma^\mu\gamma^\nu\gamma^\beta] \\ &= 2 g^{\mu\alpha} 4 g^{\nu\beta} + \mathrm{tr}[\gamma^\alpha\gamma^\nu\gamma^\mu\gamma^\beta] - \mathrm{tr}[\gamma^\alpha 2 g^{\mu\nu}\gamma^\beta] \\ &= 8 g^{\mu\alpha} g^{\nu\beta} - 8 g^{\alpha\beta} g^{\mu\nu} + \mathrm{tr}[\gamma^\alpha\gamma^\nu 2 g^{\mu\beta}] - \mathrm{tr}[\gamma^\alpha\gamma^\nu\gamma^\beta\gamma^\mu] .\end{aligned}$$

Exploiting the cyclic property of the trace in the last term again gives

$$\mathrm{tr}[\gamma^\mu\gamma^\alpha\gamma^\nu\gamma^\beta] = 4 g^{\mu\alpha} g^{\nu\beta} - 4 g^{\mu\nu} g^{\alpha\beta} + 4 g^{\mu\beta} g^{\alpha\nu} . \tag{14.33}$$

This procedure may be continued further to yield the trace of an arbitrary even product of γ matrices.

Using (14.32, 33) it is now possible to evaluate (14.28)

$$\begin{aligned}&\frac{1}{4}\mathrm{tr}\left\{\gamma^\mu\left[\gamma\cdot\left(p+\frac{k}{2}\right)+m_0\right]\gamma^\nu\left[\gamma\cdot\left(p-\frac{k}{2}\right)+m_0\right]\right\} \\ &= \left(p^\mu+\frac{k^\mu}{2}\right)\left(p^\nu-\frac{k^\nu}{2}\right) - g^{\mu\nu}\left(p^2-\frac{k^2}{4}\right) + \left(p^\nu+\frac{k^\nu}{2}\right)\left(p^\mu-\frac{k^\mu}{2}\right) + g^{\mu\nu} m_0^2 \\ &= 2p^\mu p^\nu - \frac{1}{2}k^\mu k^\nu - g^{\mu\nu}\left(p^2 - m_0^2 - \frac{k^2}{4}\right)\end{aligned} \tag{14.34}$$

and the momentum integral (14.27) now becomes

$$\Pi^{\mu\nu}(k,\varepsilon') = 4\mathrm{i}\int \frac{d^4p}{(2\pi)^4}\,\mathrm{e}^{\mathrm{i}p\cdot\varepsilon'}\,\frac{2p^\mu p^\nu - \frac{1}{2}k^\mu k^\nu - g^{\mu\nu}\left(p^2-m_0^2-\frac{k^2}{4}\right)}{\left[\left(p+\frac{k}{2}\right)^2 - m_0^2 + \mathrm{i}\varepsilon\right]\left[\left(p-\frac{k}{2}\right)^2 - m_0^2 + \mathrm{i}\varepsilon\right]} . \tag{14.35}$$

14.2 Gauge Invariance and Vacuum Polarization

We proceed shortly to obtain the explicit form of the vacuum polarization tensor $\Pi^{\mu\nu}(k)$. At this point it is expedient to list certain general properties of the polarization tensor. Perhaps the most important arises from the requirement that the induced current density is conserved:

$$\frac{\partial}{\partial x_\mu}\langle j_\mu\rangle = 0 . \tag{14.36}$$

Since the perturbative expansion given in (14.20b)

$$\langle j_\mu \rangle = \sum_k \langle j_\mu \rangle_{(2k+1)} \tag{14.37}$$

is at the same time an (assumed) analytic expression in powers of e^2, the same condition (14.36) is valid for each coefficient in the power series expansion in e^{2k+1}. Hence each term satisfies separately

$$\frac{\partial}{\partial x_\mu} \langle j_\mu \rangle_{(2k+1)} = 0 \,. \tag{14.38}$$

Given (for $k = 0$) the form for the lowest non-vanishing displacement current (14.25), the explicit condition

$$0 = -\mathrm{i} e^2 \int \frac{d^4 k}{(2\pi)^4} A_{\mathrm{ex},\nu}(k)\, \Pi^{\mu\nu}_{(1)}(k) \frac{\partial}{\partial x^\mu} \mathrm{e}^{-\mathrm{i}kx} \tag{14.39}$$

arises and on differentiation gives

$$k_\mu \Pi^{\mu\nu}_{(1)}(k) = 0 \,. \tag{14.40}$$

However, in deriving (14.40) there are several tacit assumptions, in particular

a) the induced current $\langle j_\mu \rangle$ is regular and analytic at $e \to 0$ and hence the constructed power series expansion is identical to the function it is supposed to describe;

b) the weak external charge does not induce a *real* current. For supercritical fields this is not true. In principle, it is possible that a theory is supercritical for *any* strength of the coupling constant; imagine, in particular, that we were to consider vanishing elctron mass (i.e. the mass gap vanishes);

c) that the tensor $\Pi^{\mu\nu}_{(1)}$ is a well-behaved mathematical object: but note that $\Pi^{\mu\nu}_{(1)}$ arises from a product of two singular distributions, cf. (14.12), which can be written in the form

$$P^{\mu\nu}_{(1)}(x - x_1) = -e^2 \operatorname{tr}[\gamma^\mu S_0(x - x_1)\, \gamma^\nu S_0(x_1 - x')]\,|_{\substack{x \to x' \\ \mathrm{s}}} \,, \tag{14.41}$$

where $P^{\mu\nu}_{(1)}(x - x')$ is the Fourier transform of $\Pi^{\mu\nu}_{(1)}$

$$P^{\mu\nu}_{(1)}(x - x_1) = \int \frac{d^4 k}{(2\pi)^4} \mathrm{e}^{\mathrm{i}k(x - x_1)} \Pi^{\mu\nu}_{(1)}(k) \,. \tag{14.42}$$

Thus it is possible that $P^{\mu\nu}_{(1)}$ and the associated measure $d^4 x_1$ need to be reexamined in the light of the ill-defined properties along the light cone [Da 74a, 79];

d) that the considered theory be complete – but recall that electrons are not the only charged particles contributing to the vacuum polarization current. In principle *all charged particles* contribute and influence the so-called short-range behaviour of QED discussed here.

Thus condition (14.40) must be verified explicitly. However, it is useful to understand first what it implies in some detail.

Since the Lorentz tensor $\Pi_{\mu\nu}$ may be proportional only to the quantities $g^{\mu\nu}$, $k^\mu k^\nu$, we are free to write the general form as

$$\Pi^{\mu\nu}_{(1)}(k) = (g^{\mu\nu}k^2 - k^\mu k^\nu)\,\Pi_{(1)}(k^2) + \hat{\Pi}_{\rm sp}(k^2)\,g^{\mu\nu}\,, \tag{14.43}$$

where $\Pi_{(1)}$ is dimensionless. When $\Pi_{\rm sp}$ (sp for spurious) vanishes, (14.40) is satisfied for arbitrary functions $\Pi_{(1)}$; $\Pi_{(1)}$ must be determined by explicit calculations (see below). Inserting now (14.43) into (14.25), then

$$\begin{aligned}\langle j^\mu(x)\rangle_{(1)} = &-e^2 \int \frac{d^4k}{(2\pi)^4}\,\Pi_{(1)}(k^2)(g^{\mu\nu}k^2 - k^\mu k^\nu)\tilde{A}_{{\rm ex},\nu}(k)\,{\rm e}^{-{\rm i}kx}\\ &-e^2 \int \frac{d^4k}{(2\pi)^4}\,\Pi_{\rm sp}(k^2)\tilde{A}^\mu_{\rm ex}(k)\,{\rm e}^{-{\rm i}kx}\,.\end{aligned} \tag{14.44}$$

The first term does not contain explicitly the field $\tilde{A}_{{\rm ex},\nu}$, but instead the field strength $F_{\mu\nu}$:

$$F_{\mu\nu} = \frac{\partial}{\partial x^\mu}A_\nu - \frac{\partial}{\partial x^\nu}A_\mu\,. \tag{14.45a}$$

The Fourier transform of the field strength tensor is

$$\tilde{F}_{\mu\nu} = -{\rm i}(k_\mu\tilde{A}_\nu - k_\nu\tilde{A}_\mu) \tag{14.45b}$$

and hence the integrand of the first term in (14.44)

$${\rm i}k_\nu\tilde{F}^{\mu\nu} = -(g^{\mu\nu}k^2 - k^\mu k^\nu)\tilde{A}_{{\rm ex},\nu}(k)\,. \tag{14.46}$$

A further simplification is achieved recalling Maxwell's equation

$$\frac{\partial}{\partial x^\nu}F^{\mu\nu}_{\rm ex} = -j^\mu_{\rm ex}\,, \tag{14.47a}$$

and its Fourier transform

$$+{\rm i}k_\nu\tilde{F}^{\mu\nu}_{\rm ex} = \tilde{j}^\mu_{\rm ex}\,, \tag{14.47b}$$

which implies that the integrand of the first term in (14.44) actually is explicitly dependent on the applied external current

$$-(g^{\mu\nu}k^2 - k^\mu k^\nu)\tilde{A}_{{\rm ex},\nu}(k) = \tilde{j}^\mu_{\rm ex}(k)\,. \tag{14.48}$$

Thus, the first term in (14.44) is gauge invariant, as upon insertion of (14.46 or 48) it can be viewed as being dependent on the electromagnetic field strength tensor or the external current respectively, but independent of the gauge-dependent four potentials $A_{{\rm ex},\nu}$. Contrary to this, the second term in (14.44) induces a dependence of the induced current on the gauge selected by being a function of $A_{ex,\nu}$. Thus, since $\langle j_\mu \rangle_{(1)}$ is, in principle, observable, this latter term violates the gauge invariance of the theory. This would be a very severe contradiction to the experimentally confirmed gauge independence of QED. Thus, in a satisfactory formulation of the theory of charged particles, such a term (Π_{sp}) must not appear. We now explicitly show that it is indeed present, indicating that perturbative QED is not a complete theory. As one counter example or inconsistency suffices to prove a theory wrong, we should, in principle, spend the rest of this book searching for an improved theory. However, there is little active work on this today because: (1) there is a common belief that some artefact of exact mathematics is the source of this problem; (2) this problem may disappear when a properly generalized theory, including in its framework *all* charged Dirac particles, is achieved.

As this circumstance cannot be emphasized enough, we now proceed to calculate the spurious term in the vacuum polarization. To this end, consider first (14.35) in the $k \to 0$ limit, since, as shown, the only non-vanishing term is then the spurious polarization tensor Π_{sp}:

$$\begin{aligned}\Pi^{\mu\nu}(0;\varepsilon') &= 4\mathrm{i}\int \frac{d^4p}{(2\pi)^4}\,\mathrm{e}^{\mathrm{i}p\varepsilon'}\,\frac{[2p^\mu p^\nu - g^{\mu\nu}(p^2-m_0^2)]}{(p^2-m_0^2+\mathrm{i}\varepsilon)^2}\\ &= -4\mathrm{i}\int \frac{d^4p}{(2\pi)^4}\,\mathrm{e}^{\mathrm{i}p\varepsilon'}\,\frac{\partial}{\partial p_\mu}\,\frac{p^\nu}{p^2-m_0^2+\mathrm{i}\varepsilon}\\ &= -4\varepsilon'^{\mu}\int \frac{d^4p}{(2\pi)^4}\,\mathrm{e}^{\mathrm{i}p\varepsilon'}\,\frac{p^\nu}{p^2-m_0^2+\mathrm{i}\varepsilon}\,,\end{aligned} \tag{14.49}$$

with integration by parts. Further a derivative can replace p^ν

$$\begin{aligned}\Pi^{\mu\nu}(0;\varepsilon') &= \mathrm{i}4\varepsilon'^{\mu}\frac{\partial}{\partial\varepsilon'_\nu}\int \frac{d^4p}{(2\pi)^4}\,\mathrm{e}^{\mathrm{i}p\varepsilon'}\,\frac{1}{p^2-m_0^2+\mathrm{i}\varepsilon}\\ &= \mathrm{i}4\varepsilon'^{\mu}\frac{\partial}{\partial\varepsilon'_\nu}D_0(\varepsilon')\,,\end{aligned} \tag{14.50}$$

where we have used (14.5, 12a). The singularity of D_0 has been considered in (14.14), hence

$$\begin{aligned}\Pi^{\mu\nu}(0;\varepsilon') &= \mathrm{i}4\varepsilon'^{\mu}\frac{\partial}{\partial\varepsilon'_\nu}\left(\frac{\mathrm{i}}{4\pi^2}\,\frac{1}{\varepsilon'^2-\mathrm{i}\varepsilon}\right)\\ &= \frac{2}{\pi^2}\,\frac{\varepsilon'^{\mu}\varepsilon'^{\nu}}{(\varepsilon'^2-\mathrm{i}\varepsilon)^2}\,.\end{aligned} \tag{14.51}$$

Taking the symmetric limit with

$$\varepsilon'^{\mu} = \varepsilon'\,\delta^{\mu 0} \tag{14.52}$$

gives a highly singular term, quadratically divergent in the limit $\varepsilon' \to 0$,

$$\Pi^{\mu\nu}(0;\varepsilon') = g^{\mu 0} g^{\nu 0} \frac{2}{\pi^2} \frac{1}{\varepsilon'^2}, \tag{14.53}$$

which violates, as stated, the gauge invariance of the theory. Although explicitly present here, it must be omitted for either of the reasons spelled out previously. It is important to appreciate here that this term *does not arise* as a consequence of some illegitimate manipulations of Fourier transforms, as ε' has been kept finite throughout the derivation. It is possible to rederive it in configuration space without the use of Fourier transforms just using (14.15) for the Dirac propagator and (14.41) for the polarization tensor.

14.3 Charge Renormalization

Before computing the polarization tensor for arbitrary momenta k, we consider now how it influences the observable charge. As is well known from many areas of physics, polarizability of a medium changes the effective values of the constants involved. The remarkable difference now to the usual circumstance is that the bare charge cannot be taken out of the vacuum to observe its value. Hence, we must formulate the theory so that the observed physical electron charge arises on renormalization of the involved quantities, leaving a mathematical quantity e used in the Lagrange function as an unobserved bare charge.

Inserting (14.48) into (14.44) and dropping the gauge invariance breaking second term gives

$$\langle j^\mu(x)\rangle_{(1)} = e^2 \int \frac{d^4k}{(2\pi)^4} \Pi_{(1)}(k^2) \tilde{j}^\mu_{\rm ex}(k)\, e^{ikx}. \tag{14.54}$$

However, this expression does not explicitly contain the fact that $\langle j^\mu\rangle_{(1)}$ is a conserved quantity, cf (14.38). This can easily be remedied by noting the identity

$$g^{\mu\nu}k^2 - k^\mu k^\nu = (g^{\mu\alpha}k^2 - k^\mu k^\alpha)\left(g_\alpha^{\ \nu} - \frac{k_\alpha k^\nu}{k^2}\right). \tag{14.55}$$

Employing (14.55) in (14.44), observing (14.48) (and dropping $\Pi_{\rm sp}$) yields

$$\begin{aligned}\langle j^\mu(x)\rangle_{(1)} &= -e^2 \int \frac{d^4k}{(2\pi)^4} \Pi_{(1)}(k^2)\left(g^\mu_{\ \alpha} - \frac{k^\mu k_\alpha}{k^2}\right)(g^{\alpha\nu}k^2 - k^\alpha k^\nu)\tilde{A}_{{\rm ex},\nu}(k)\, e^{ikx} \\ &= e^2 \int \frac{d^4k}{(2\pi)^4} \Pi_{(1)}(k^2)\left(g^{\mu\nu} - \frac{k^\mu k^\nu}{k^2}\right)\tilde{j}_{\nu,{\rm ex}}(k)\, e^{ikx}. \end{aligned} \tag{14.56}$$

The Fourier transform of the induced current is even more appealing:

$$\langle \tilde{j}_\mu(k) \rangle_{(1)} = e^2 \Pi_{(1)}(k^2) \left(g_{\mu\nu} - \frac{k_\mu k_\nu}{k^2} \right) \tilde{j}^{\nu}_{\text{ex}}(k) \,. \tag{14.57}$$

Here $j_{\nu,\text{ex}}$ is the current which is physically felt from bringing in some external charge, thus it already contains the polarization current which is induced. It therefore can be decomposed into two terms: the bare current $j^{\text{bare}}_{\text{ex}}$ associated with some external source and the induced current

$$j_{\nu,\text{ex}}(k) = j^{\text{bare}}_{\nu,\text{ex}} + \langle j_\nu(k) \rangle \tag{14.58}$$

where ~ has been omitted as the use of the momentum space is self-evident here.

When inserted into (14.57)

$$\langle j_\mu(k) \rangle_{(1)} = e^2 \Pi_{(1)}(k^2) (j^{\text{bare}}_{\nu,\text{ex}}(k) + \langle j_\nu(k) \rangle) \left(g^{\nu}_{\mu} - \frac{k_\mu k^\nu}{k^2} \right) . \tag{14.59a}$$

Denoting the propagator with first order corrections in e by a double line, this can be denoted graphically as

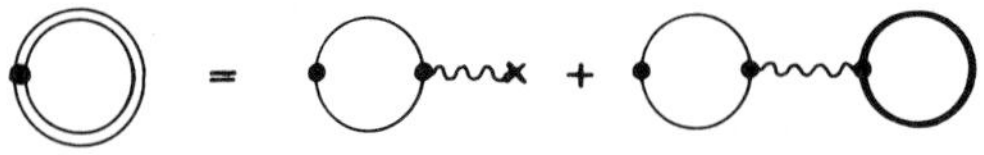

(14.59b)

which is self-explanatory in that the first term represents the effect of the (bare) external field, while the second is the contribution of the induced polarization charge. At this point it is worth returning to (14.19) which also contains higher order terms in the external potential contributing to the induced charge. In (14.59) these lead to

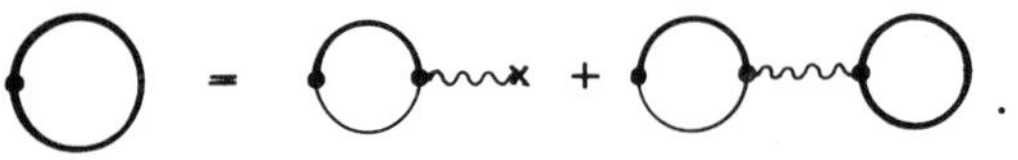

(14.59c)

Iterating (14.59) all diagrams that can be drawn in terms of *singly connected loops* are included [turning now to the induced potential, cf (14.21)] for symmetry:

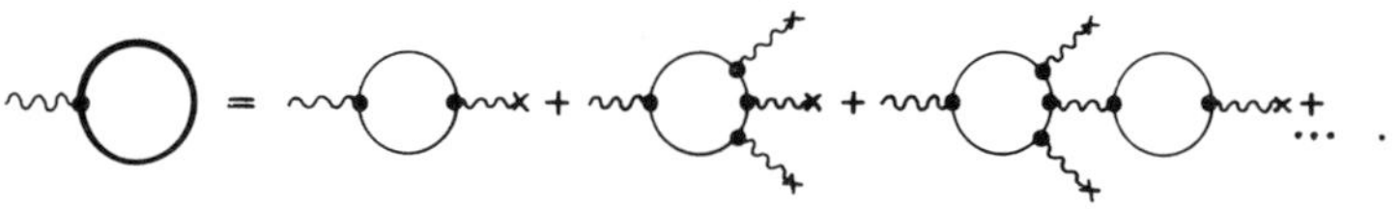

(14.59d)

It is obvious that singly connected here means "connected by one wavy line". However, the number of loops or the number of the external current crosses is arbitrary.

In the present context we constrain ourselves initially to diagrams with one external cross, as already partially done by considering the induced current in this approximation only. For this reason we approximate (14.58) by

$$j_{\nu,\text{ex}}(k) \approx j^{\text{bare}}_{\nu,\text{ex}}(k) + \langle j_\nu(k) \rangle_{(1)} \tag{14.60}$$

and solve (14.60) for $\langle j_\mu(k)\rangle_{(1)}$. Inserting this into (14.57) gives

$$j_{\mu,\mathrm{ex}}(k) - j^{(\mathrm{bare})}_{\mu,\mathrm{ex}}(k) = e^2 \Pi_{(1)}(k^2)\, j_{\nu,\mathrm{ex}}(k)\left(g^{\nu}_{\mu} - \frac{k_\mu k^\nu}{k^2}\right). \tag{14.61}$$

We now solve (14.61) for the physical current $j_{\nu,\mathrm{ex}}$:

$$j_{\nu,\mathrm{ex}}\left[g^{\nu}_{\mu} - e^2\Pi_{(1)}(k^2)\left(g^{\nu}_{\mu} - \frac{k^\nu k_\mu}{k^2}\right)\right] = j^{\mathrm{bare}}_{\mu,\mathrm{ex}}. \tag{14.62}$$

This can be multiplied with a tensor with unknown coefficients A, B such that

$$\left[g^{\nu}_{\mu} - e^2\Pi_{(1)}(k^2)\left(g^{\nu}_{\mu} - \frac{k^\nu k_\mu}{k^2}\right)\right]\left(A\, g^{\mu\alpha} + B\frac{k^\mu k^\alpha}{k^2}\right) = g^{\nu\alpha}. \tag{14.63}$$

Upon straightforward multiplication this condition is satisfied if

$$A = \frac{1}{1 - e^2\Pi_{(1)}(k^2)} \tag{14.64a}$$

$$B = -\frac{e^2\Pi_{(1)}(k^2)}{1 - e^2\Pi_{(1)}(k^2)} \tag{14.64b}$$

and hence

$$j_{\mu,\mathrm{ex}}(k) = \frac{1}{1 - e^2\Pi_{(1)}(k^2)}\, j^{\mathrm{bare}}_{\mu,\mathrm{ex}}(k) - \frac{e^2\Pi_1(k^2)}{1 - e^2\Pi_{(1)}(k^2)}\,\frac{k_\mu}{k^2}\, k^\alpha j^{\mathrm{bare}}_{\alpha,\mathrm{ex}}(k). \tag{14.65}$$

For all conserved bare external currents ($k^\alpha j^{\mathrm{bare}}_{\alpha,\mathrm{ex}} = 0$) there is a simple relation between the bare external current $j^{\mathrm{bare}}_{\mu,\mathrm{ex}}$ and the physical current $j_{\mu,\mathrm{ex}}$:

$$j_{\mu,\mathrm{ex}}(k) = \frac{1}{1 - e^2\Pi_{(1)}(k^2)}\, j^{\mathrm{bare}}_{\mu,\mathrm{ex}}(k). \tag{14.66}$$

Equation (14.66) implies that the bare current in momentum space is proportional to the observable physical current. In particular, for $k^2 \to 0$ the current is multiplicatively renormalized! We hence are led to define

$$j^{\mathrm{ren,b}}_{\mu}(k) = \frac{1}{1 - e^2\Pi_{(1)}(0)}\, \tilde{j}^{\mathrm{bare}}_{\mu}(k) \tag{14.67}$$

which allows (14.66) to be rewritten as

$$j_{\mu,\mathrm{ex}}(k) = \left[1 - \frac{e^2}{1 - e^2\Pi_{(1)}(0)}\left[\Pi_{(1)}(k^2) - \Pi_{(1)}(0)\right]\right]^{-1} j^{\mathrm{ren,b}}_{\mu,\mathrm{ex}}(k), \tag{14.68}$$

that is

$$j_{\mu,\mathrm{ex}}(k) = \frac{1}{1 - e_{\mathrm{r}}^2\Pi^{\mathrm{r}}_{(1)}(k^2)}\, j^{\mathrm{ren,b}}_{\mu}(k) \tag{14.69}$$

with the renormalized quantities defined by

$$e_{\mathrm{r}}^2 = \frac{e^2}{1 - e^2 \Pi_{(1)}(0)} \tag{14.70}$$

$$\Pi_{(1)}^{\mathrm{r}}(k^2) = \Pi_{(1)}(k^2) - \Pi_{(1)}(0) \,. \tag{14.71}$$

The essence of the above transformations is that in (14.69) they contain a polarization function $\Pi_{(1)}^{\mathrm{r}}(k^2)$ (14.71) which, by definition, vanishes in the limit $k^2 \to 0$. This implies that the potential we perceive at large distances is that of the renormalized current (14.67). The introduction of the renormalized charge (14.70) is logically consistent with the external current renormalization (14.67) and simplifies (14.69). Furthermore, as shown, this helps deal with the infinite character of $\Pi_{(1)}(0)$. However, it is important to appreciate here that the charge renormalization is not a consequence of the divergence of the polarization tensor in QED. It is the consequence of our definition of the charge as the strength of the Coulomb $1/r$ potential at infinite r, i.e. at $k \to 0$. Even if we were successful in constructing a theory with finite $\Pi_{(1)}(0)$, the renormalization would be necessary, except of course for the possible case where $\Pi_{(1)}(0) = 0$.

14.4 Explicit Form of the Polarization Function

Aside from the spurious divergence already discussed, the four-dimensional momentum integral (14.35) also contains in the limit $\varepsilon' \to 0$ the charge renormalization divergence. Thus we must evaluate the integral (14.35) so as to be able to separate easily the concerned terms from the finite and observable effect. We employ the same mathematical procedure as used to find the explicit form of the propagator (14.6). In particular, on a twofold application of (10.109), the denominator of (14.35) becomes

$$\begin{aligned} D &\equiv \frac{1}{\left(p + \dfrac{k}{2}\right)^2 - m_0^2 - \mathrm{i}\varepsilon} \, \frac{1}{\left(p - \dfrac{k}{2}\right)^2 - m_0^2 + \mathrm{i}\varepsilon} \\ &= (-\mathrm{i})^2 \int_0^\infty ds_1 \int_0^\infty ds_2 \exp\left[\mathrm{i}\left(p^2 + \frac{1}{4}k^2 - m_0^2 + \mathrm{i}\varepsilon\right)(s_1 + s_2) + \mathrm{i} p \cdot k (s_1 - s_2)\right] \end{aligned} \tag{14.72}$$

It is now advantageous to change to new variables λ and v so that

$$\left.\begin{aligned} s_1 &= \lambda\left(\frac{1+v}{2}\right) \\ s_2 &= \lambda\left(\frac{1-v}{2}\right) \end{aligned}\right\} \quad \lambda \in (0, \infty),\ v \in (-1, 1) \tag{14.73a}$$

$$ds_1 ds_2 = \tfrac{1}{2}\lambda\, d\lambda\, dv\,, \tag{14.73b}$$

which further simplifies our representation (14.72)

$$D = -\int_0^\infty \lambda\, d\lambda \frac{1}{2} \int_{-1}^{1} dv \exp\left[\mathrm{i}\left(p^2 - m_0^2 + \mathrm{i}\varepsilon + \frac{k^2}{4}\right)\lambda + \mathrm{i}pk\lambda v\right]. \tag{14.74}$$

Thus the following expression for the polarization function (14.35) results:

$$\Pi^{\mu\nu}(k;\varepsilon') = -2\mathrm{i}\int_0^\infty \lambda\, d\lambda \exp\left(\mathrm{i}\frac{k^2}{4}\lambda\right) \int_{-1}^{1} dv \int \frac{d^4p}{(2\pi)^4} \exp[\mathrm{i}(p^2 - m_0^2 + \mathrm{i}\varepsilon)\lambda]$$
$$\times \exp(\mathrm{i}p\varepsilon')\left[2p^\mu p^\nu - \frac{1}{2}k^\mu k^\nu - g^{\mu\nu}\left(p^2 - m_0^2 - \frac{k^2}{4}\right)\right] \mathrm{e}^{\mathrm{i}pk\lambda v}\,, \tag{14.75}$$

where the order of integration is exchanged: such manipulations are permitted when $\varepsilon' \neq 0$. It is possible to express the factor in square brackets entirely in terms of k^μ since

$$p^\mu p^\nu \mathrm{e}^{\mathrm{i}pk\lambda v} = -\frac{1}{\lambda^2 v^2} \frac{\partial}{\partial k_\mu} \frac{\partial}{\partial k_\nu} \mathrm{e}^{\mathrm{i}pk\lambda v}\,, \tag{14.76}$$

thus leading to

$$\Pi^{\mu\nu}(k;\varepsilon') = -2\mathrm{i}\int_0^\infty \lambda\, d\lambda \exp\left(\mathrm{i}\frac{k^2}{4}\lambda\right) \int_{-1}^{1} dv \left[\left(-\frac{2}{\lambda^2 v^2} \frac{\partial}{\partial k_\mu} \frac{\partial}{\partial k_\nu} - \frac{1}{2}k^\mu k^\nu\right)\right.$$
$$\left. - g^{\mu\nu}\left(-\frac{1}{\lambda^2 v^2} \frac{\partial}{\partial k^\alpha} \frac{\partial}{\partial k_\alpha} - m_0^2 - \frac{k^2}{4}\right)\right]$$
$$\times \int [d^4p/(2\pi)^4] \exp[\mathrm{i}(p^2 - m_0^2 + \mathrm{i}\varepsilon)\lambda + \mathrm{i}pk\lambda v + \mathrm{i}p\varepsilon']\,. \tag{14.77}$$

To evaluate the momentum integral, we complete the exponent, i.e.

$$\mathrm{i}p^2\lambda + \mathrm{i}pk\lambda v + \mathrm{i}p\varepsilon' = \mathrm{i}\left(p + \frac{v}{2}k + \frac{\varepsilon'}{2\lambda}\right)^2 \lambda - \mathrm{i}\frac{v^2}{4}k^2\lambda - \mathrm{i}\frac{\varepsilon'^2}{\lambda} - \mathrm{i}\frac{v}{2}k\varepsilon'\,. \tag{14.78}$$

Substituting $l_\mu = p_\mu + vk_\mu/2 + \varepsilon'_\mu/2\lambda$ we can then easily carry out the Gaussian integral over p_μ [cf. (10.114)] and obtain

$$\Pi^{\mu\nu}(k;\varepsilon') = -2\mathrm{i}\int_0^\infty \lambda\, d\lambda \frac{1}{(4\pi)^2 \mathrm{i}\lambda^2} \exp\left[-\mathrm{i}\frac{\varepsilon'^2}{4\lambda} - \mathrm{i}(m_0^2 - \mathrm{i}\varepsilon)\lambda + \mathrm{i}\frac{k^2}{4}\lambda\right] \int_{-1}^{+1} dv$$
$$\times \left[-\frac{2}{\lambda^2 v^2} \frac{\partial}{\partial k_\mu} \frac{\partial}{\partial k_\nu} - \frac{1}{2}k^\mu k^\nu - g^{\mu\nu}\left(-\frac{1}{\lambda^2 v^2} \frac{\partial}{\partial k^\alpha} \frac{\partial}{\partial k_\alpha}\right.\right.$$
$$\left.\left. - m_0^2 - \frac{k^2}{4}\right)\right] \exp\left(-\mathrm{i}\frac{v^2 k^2}{4}\lambda - \mathrm{i}\frac{v}{2}k\varepsilon'\right), \tag{14.79}$$

where the factor $\exp(-\mathrm{i}\varepsilon'^2/4\lambda)$ regularizes the singular point $\lambda \to 0$. The last factor in (14.79) which also contains ε' is unessential and may be omitted in future considerations. Carrying out the differentiations

$$
\begin{aligned}
\Pi^{\mu\nu}(k;\varepsilon') = &-\frac{1}{8\pi^2}\int_0^\infty \frac{d\lambda}{\lambda}\int_{-1}^{+1} dv \exp\left[\mathrm{i}\frac{k^2}{4}(1-v^2)\lambda - \mathrm{i}\frac{\varepsilon'^2}{4\lambda} - \mathrm{i}(m_0^2 - \mathrm{i}\varepsilon)\lambda\right] \\
&\times\left[\mathrm{i}\frac{g^{\mu\nu}}{\lambda} + \frac{v^2}{2}k^\mu k^\nu - \frac{1}{2}k^\mu k^\nu - \mathrm{i}g^{\mu\nu}\frac{2}{\lambda} - g^{\mu\nu}\frac{v^2}{4}k^2\right. \\
&\left.+ g^{\mu\nu}\left(\frac{k^2}{4} + m_0^2\right)\right]. \qquad (14.80)
\end{aligned}
$$

The $1/\lambda$ terms is square brackets $(-\mathrm{i}g^{\mu\nu}/\lambda)$ are now integrated by parts over λ:

$$
\begin{aligned}
&\int_0^\infty \frac{d\lambda}{\lambda}\left(-\mathrm{i}\frac{g^{\mu\nu}}{\lambda}\right)\exp\left[\mathrm{i}\frac{k^2}{4}(1-v^2)\lambda - \mathrm{i}\frac{\varepsilon'^2}{4\lambda} - \mathrm{i}(m_0^2-\mathrm{i}\varepsilon)\lambda\right] \\
&= +\mathrm{i}\frac{g^{\mu\nu}}{\lambda}\exp\left[\mathrm{i}\frac{k^2}{4}(1-v^2)\lambda - \mathrm{i}\frac{\varepsilon'^2}{4\lambda} - \mathrm{i}(m_0^2-\mathrm{i}\varepsilon)\lambda\right]\Bigg|_{\lambda=0}^{\infty} \\
&+\int_0^\infty \frac{d\lambda}{\lambda}(-\mathrm{i}g^{\mu\nu})\left(\mathrm{i}\frac{k^2}{4}(1-v^2) + \frac{\mathrm{i}\varepsilon'^2}{4\lambda^2} - \mathrm{i}m_0^2\right) \\
&\times\exp\left[\mathrm{i}\frac{k^2}{4}(1-v^2)\lambda - \mathrm{i}\frac{\varepsilon'^2}{4\lambda} - \mathrm{i}(m_0^2-\mathrm{i}\varepsilon)\lambda\right]. \qquad (14.81)
\end{aligned}
$$

The surface term in (14.81) vanishes at $\lambda = \infty$ because of the definition of the integration path through ε. For $\lambda \to 0$ the singularity of the surface term is reminiscent of the spurious contribution found previously, cf. (14.50). Subtracting this k-independent term, inserting the result (14.81) into (14.80) (notice that m_0^2 is cancelled) gives

$$
\begin{aligned}
\Pi^{\mu\nu}(k;\varepsilon') - \Pi^{\mu\nu}(0;\varepsilon') = &-\frac{1}{16\pi^2}(g^{\mu\nu}k^2 - k^\mu k^\nu)\int_0^\infty \frac{d\lambda}{\lambda}\int_{-1}^{+1} dv(1-v^2) \\
&\times\exp\left[\mathrm{i}\frac{k^2}{4}(1-v^2)\lambda - \mathrm{i}\frac{\varepsilon'^2}{4\lambda} - \mathrm{i}(m_0^2-\mathrm{i}\varepsilon)\lambda\right], \qquad (14.82)
\end{aligned}
$$

which, as expected, exhibits explicitly the gauge invariant form.

Schwinger "avoided" the spurious term in his paper [Schw 51]: in his proper time approach, he performs almost identical transforms, though he derives them from the proper time technique. However, he (1) sets $\varepsilon' = 0$, and (2) does not record the surface term at $\lambda \to 0$ (14.81) [see his equations (6.22, 23)]. Thus there are two cancelling errors.

Further significant separation of terms in (14.82) is achieved by a further partial integration with respect to v:

$$\begin{aligned}
&\Pi^{\mu\nu}(k;\varepsilon') - \Pi^{\mu\nu}(0;\varepsilon') \\
&= -\frac{1}{16\pi^2}(g^{\mu\nu}k^2 - k^\mu k^\nu)\int_0^\infty d\lambda \exp\left[-\mathrm{i}\frac{\varepsilon'^2}{4\lambda} - \mathrm{i}(m_0^2 - \mathrm{i}\varepsilon)\lambda\right] \\
&\quad \times \left\{\frac{1}{\lambda}\left(v - \frac{v^3}{3}\right)\exp\left[\mathrm{i}\frac{k^2}{4}(1-v^2)\lambda\right]\Bigg|_{v=-1}^{+1}\right. \\
&\quad \left. + \mathrm{i}\frac{k^2}{4}\int_{-1}^{+1} dv\left(v^2 - \frac{1}{3}v^4\right)\exp\left[+\mathrm{i}\frac{k^2}{4}(1-v^2)\lambda\right]\right\} .
\end{aligned} \tag{14.83}$$

For $\varepsilon'^2 \to 0$ the first term in curly brackets diverges logarithmically. Its value can be easily determined, as it is related to $\partial D_0(\varepsilon^2)/\partial m_0^2$, (14.11, 12),

$$\begin{aligned}
&\frac{1}{(4\pi)^2}\int\frac{d\lambda}{\lambda}\exp\left[-\mathrm{i}\frac{\varepsilon'^2}{4\lambda} - \mathrm{i}(m_0^2 - \mathrm{i}\varepsilon)\lambda\right] \\
&= \mathrm{i}\frac{\partial}{\partial m_0^2}\frac{1}{(4\pi)^2}\int\frac{d\lambda}{\lambda^2}\exp\left[-\mathrm{i}\frac{\varepsilon'^2}{4\lambda} - \mathrm{i}(m_0^2 - \mathrm{i}\varepsilon)\lambda\right] \\
&= -\mathrm{i}\frac{\partial}{\partial m_0^2}D_0(\varepsilon'^2, m_0^2) \xrightarrow[\varepsilon'^2 \to 0+]{} \frac{1}{16\pi^2}\ln(m_0^2\varepsilon'^2) ,
\end{aligned} \tag{14.84}$$

using (14.14) for the singular part of the D_0 function. Consequently,

$$\begin{aligned}
\Pi^{\mu\nu}(k,\varepsilon') - \Pi^{\mu\nu}(0,\varepsilon') = &-\frac{1}{(4\pi)^2}\ln(m_0^2\varepsilon'^2)\frac{4}{3}(g^{\mu\nu}k^2 - k^\mu k^\nu) \\
&+ (g^{\mu\nu}k^2 - k^\mu k^\nu)\,\Pi_{(1)}^{\mathrm{r}}(k^2) ,
\end{aligned} \tag{14.85}$$

where the renormalized part of the polarization function, i.e. the second term in (14.85) is finite, hence we take the limit $\varepsilon' \to 0$:

$$\begin{aligned}
\Pi_{(1)}^{\mathrm{r}}(k^2) = &\frac{\mathrm{i}}{16\pi^2}\frac{k^2}{2}\int_0^\infty d\lambda\int_0^{+1} dv\left(v^2 - \frac{1}{3}v^4\right) \\
&\times \exp\left\{-\mathrm{i}\lambda\left[-\frac{k^2}{4}(1-v^2) + m_0^2 - \mathrm{i}\varepsilon\right]\right\} .
\end{aligned} \tag{14.86}$$

The λ integral can be easily executed and for $\Pi_{\mathrm{ren}}^{(1)}$, introduced in (9.133), then

$$\Pi_{\mathrm{ren}}^{(1)}(k) \equiv e^2\Pi_{(1)}^{\mathrm{r}} = \frac{\alpha}{4\pi}\int_0^1 dv(v^2 - \tfrac{1}{3}v^4)\left(-\frac{k^2}{m_0^2}\right)\frac{1}{1 - \dfrac{k^2}{4m_0^2}(1-v^2)} . \tag{14.87}$$

Integrating by parts gives

$$\Pi_{\mathrm{ren}}^{(1)} = -\frac{\alpha}{2\pi}\int_0^1 dv\left(v-\tfrac{1}{3}v^3\right)\frac{\dfrac{k^2}{4m_0^2}\,2v}{1-\dfrac{k^2}{4m_0^2}(1-v^2)}$$

$$= \frac{\alpha}{2\pi}\int_0^1 dv(1-v^2)\ln\left[1-\frac{k^2}{4m_0^2}(1-v^2)\right]. \tag{14.88}$$

The integral of this expression is given below, (14.91, 92). Its limit for small k^2 can now easily be determined:

$$\Pi_{\mathrm{ren}}^{(1)} = \frac{\alpha}{2\pi}\int_0^1 dv(1-v^2)\left[-\frac{k^2}{4m_0^2}(1-v^2)-\frac{k^4}{32m_0^4}(1-v^2)^2+\dots\right]$$

$$= -\frac{\alpha}{\pi}\,\frac{1}{15}\,\frac{k^2}{m_0^2}-\frac{\alpha}{\pi}\,\frac{1}{140}\,\frac{k^4}{m_0^4}+\dots\,. \tag{14.89}$$

For $k^2 > 4m_0^2$ (14.87, 88) are indefinite as they contain a pole. Recall that in (14.86) m_0^2 can be thought of as containing a negative imaginary part, defining the way the pole is circumvented. Hence

$$\Pi_{\mathrm{ren}}^{(1)}(k^2) = \frac{\alpha}{4\pi}\int_0^1 dv\left(v^2-\tfrac{1}{3}v^4\right)\left(-\frac{k^2}{m_0^2}\right)\frac{1}{1-\dfrac{k^2}{4m_0^2}(1-v^2)-\mathrm{i}\varepsilon}$$

$$= \frac{\alpha}{4\pi}P\int_0^1 dv\left(v^2-\tfrac{1}{3}v^4\right)\left(-\frac{k^2}{m_0^2}\right)\frac{1}{1-\dfrac{k^2}{4m_0^2}(1-v^2)}$$

$$+\pi\mathrm{i}\frac{\alpha}{4\pi}\int_0^1 dv\left(v^2-\tfrac{1}{3}v^4\right)\left(-\frac{k^2}{m_0^2}\right)\delta\left(1-\frac{k^2}{4m_0^2}(1-v^2)\right), \tag{14.90}$$

where the second imaginary term can be easily reduced to

$$\mathrm{Im}\{\Pi_{\mathrm{ren}}^{(1)}(k^2)\} = \theta(k^2/4m_0^2-1)\,\frac{\alpha}{4}\left[1-\frac{4m_0^2}{k^2}-\frac{1}{3}\left(1-\frac{4m_0^2}{k^2}\right)^2\right]$$
$$\times\left(-\frac{k^2}{m_0^2}\right)\frac{1}{\dfrac{k^2}{2m_0^2}\sqrt{1-\dfrac{4m_0^2}{k^2}}},$$

since the integral contributes only where $v^2 = 1-(4m_0^2/k^2)$. Some simple algebra then leads to

$$\mathrm{Im}\{\Pi^{(1)}_{\mathrm{ren}}(k^2)\} = \frac{1}{3}\alpha\left(1+\frac{2m_0^2}{k^2}\right)\sqrt{1-\frac{4m_0^2}{k^2}}\,\theta(k^2-4m_0^2)\,. \tag{14.91}$$

The appearance of this imaginary term for $k^2 > 4m_0^2$ reflects the possibility of electron pair creation by an external current distribution, non-vanishing over $k^2 > 4m_0^2$. Then the induced current acquires an imaginary part, which can be related to (10.149) [Schw 51]. The real part can also be easily obtained using standard tables of integrals, as first obtained by *Uehling* [Ue 35]

$$\mathrm{Re}\{\Pi^{(1)}_{\mathrm{ren}}(k^2)\} = -\frac{\alpha}{3\pi}\left[\frac{5}{3}+\frac{4m_0^2}{k^2}-\left(1+\frac{2m_0^2}{k^2}\right)\sqrt{\left|1-\frac{4m_0^2}{k^2}\right|}\right.$$
$$\left.\times\ln\left(\frac{\sqrt{\left|1-\frac{4m_0^2}{k^2}\right|}+1}{\sqrt{\left|1-\frac{4m_0^2}{k^2}\right|}-1}\right)\right]. \tag{14.92}$$

Equation (14.91) further offers the opportunity to use the analytic property of $\Pi^{(1)}_{\mathrm{ren}}$ to find an alternate representation of the polarization function. Dispersion integral representation over the discontinuity of the polarization function (once subtracted) gives [Li 73]

$$\Pi^{(1)}_{\mathrm{ren}}(k^2) = -\int_{4m^2}^{\infty}\frac{1}{\pi}\mathrm{Im}\{\Pi^{(1)}(M^2)\}\left(\frac{1}{k^2-M^2+\mathrm{i}\varepsilon}+\frac{1}{M^2}\right)dM^2\,, \tag{14.93a}$$

that is

$$\Pi^{(1)}_{\mathrm{ren}}(k^2) = \int_{4m^2}^{\infty}\left(-\frac{\alpha}{3}\right)\left(1+\frac{2m_0^2}{M^2}\right)\sqrt{1-\frac{4m_0^2}{M^2}}\,\frac{k^2}{k^2-M^2+\mathrm{i}\varepsilon}\,\frac{dM^2}{M^2}\,, \tag{14.93b}$$

which can also be obtained by a suitable variable transformation directly from (14.90),

$$M^2 = \frac{4m_0^2}{1-v^2}\,.$$

It is useful to notice that the spectral factor

$$f(M^2) \equiv \left(1+\frac{2m_0^2}{M^2}\right)\sqrt{1-\frac{4m_0^2}{M^2}}\,\theta(M^2-4m_0^2) = \frac{3}{\alpha}\mathrm{Im}\{\Pi^{(1)}\} \tag{14.94a}$$

reaches quickly its asymptotic value of 1, Fig. 14.1.

For $k^2 \gg 4m_0^2$ the integral (14.93b) can be approximated by substituting for $f(M^2) \approx \theta(M^2/4m_0^2-1)$,

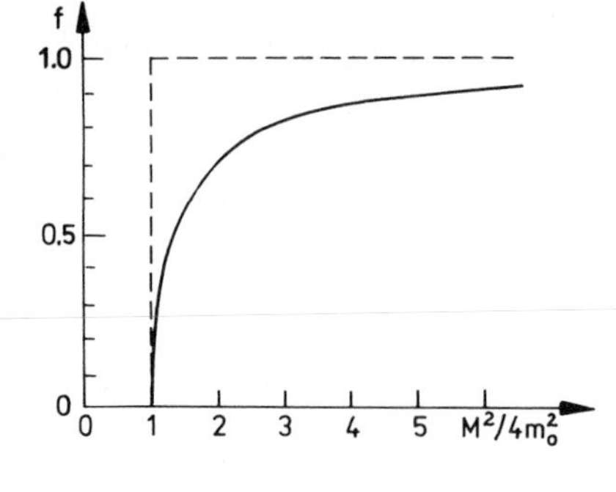

Fig. 14.1. Spectral function for vacuum polarization

$$\Pi_{\text{ren}}^{(1)}(k^2) \approx -\frac{\alpha}{3\pi}\int_{4m_0^2}^{\infty}\left(\frac{1}{k^2-M^2}+\frac{1}{M^2}\right)dM^2 = \frac{2\alpha}{3\pi}\ln\left(\frac{-M^2}{k^2-M^2}\right)\bigg|_{4m_0^2}^{\infty}$$

$$= \frac{\alpha}{3\pi}\ln\left(1-\frac{k^2}{4m_0^2}\right); \qquad k^2 > 4m_0^2\,, \tag{14.94b}$$

which surprisingly, has the correct behaviour even for $k^2 \to 0$, though the result is larger by a factor 5/4 than (14.89) in that limit.

It is interesting to note that while at $k^2 \lesssim 4m_0^2$ the deviation of f from a step function dominates the ultimate form of the polarization function, the high q^2 behaviour is always logarithmic, and the particular properties of the QED are contained in the coefficient $\alpha/3\pi$ of (14.94). The built-in requirement that $\Pi_{\text{ren}}^{(1)}(k^2 \to 0) \to 0$ must also be remembered here.

This discussion can be extended even further. As described in the previous section, the expression for the chain of loops in the polarization function after renormalization is simply, cf. (14.69, 87),

$$\Pi^{\text{chain}}(k^2) = \frac{1}{1-\Pi_{\text{ren}}^{(1)}} - 1 = \Pi_{\text{ren}}^{(1)} + (\Pi_{\text{ren}}^{(1)})^2 + \dots\,. \tag{14.95}$$

The real and imaginary parts in this expression can be separated, in particular

$$\text{Im}\{\Pi^{\text{chain}}(k^2)\} = \frac{\text{Im}\{\Pi_{\text{ren}}^{(1)}\}}{(1-\text{Re}\{\Pi_{\text{ren}}^{(1)}\})^2 + (\text{Im}\{\Pi_{\text{ren}}\})^2}\,, \tag{14.96}$$

and with (14.93) give

$$\Pi^{\text{chain}}(k^2) = -\int_{4m_0^2}^{\infty} \frac{\frac{1}{\pi}\,\text{Im}\{\Pi_{\text{ren}}^{(1)}(M^2)\}\,\frac{dM^2}{M^2}}{(1-\text{Re}\{\Pi_{\text{ren}}^{(1)}(M^2)\})^2 + (\text{Im}\{\Pi_{\text{ren}}^{(1)}(M^2)\})^2}\,\frac{k^2}{k^2-M^2+i\varepsilon}\,, \tag{14.97}$$

where we assumed that Π^{chain} has the same singularities as $\Pi_{\text{ren}}^{(1)}$. However, when $\Pi_{\text{ren}}^{(1)}$ exceeds 1, e.g. in large $(-k^2)$ limit, a new, previously ignored, pole appears in (14.95). This so-called Landau ghost is commonly believed to be an artefact of the first-order approximation to $\Pi_{\rm ren}$. Hence by omitting this term in the analytical representation of $\Pi^{\rm chain}$ (14.97), where the "bar" indicates that this omission has been made, we effectively account for any such higher order terms. Thus the dispersive representation (14.97) contains only contributions along the cut $k^2 > 4m_0^2$, generated by the two-particle continuum of electron – positron pairs produced at an invariant mass $s = M^2$.

At this point recall [Ad 70] that the quantity $\mathrm{Im}\{\Pi_{\rm ren}(q^2)\}$ is in principle a measurable object as it describes the density of electron and positron states at a given q^2. Thus the electron – positron production cross-sections by a photon are proportional to this quantity, in first-order in α as given by (14.92), and are known experimentally through all orders in α.

The approximate form (14.94) of $\Pi_{\rm ren}^{(1)}$ can give a useful approximation of (14.97): for $M_{\rm max}^2 \gg k^2 \gg 4m_0^2$

$$\bar{\Pi}^{\rm chain} \approx -\int\limits_{4m_0^2}^{M_{\rm max}^2} \frac{(\alpha/3\pi)\,dM^2/M^2}{\left[1-\dfrac{\alpha}{3\pi}\ln\left(\dfrac{M^2}{4m_0^2}-1\right)\right]^2+\left(\dfrac{\alpha}{3}\right)^2}\,\frac{k^2}{k^2-M^2+\mathrm{i}\varepsilon}\,, \tag{14.98}$$

where $M_{\rm max}^2$ is chosen not too large in order to avoid the Landau ghost: $M_{\rm max}^2 < 4m_0^2 e^{3\pi/\alpha}$. Equation (14.98), though of theoretical interest, is not applicable to the vacuum polarization effects to be discussed in the next section which are encountered mainly at $|k^2| \ll 4m_0^2$ in electronic systems and $|k^2| \gtrsim 4m_0^2$ in muonic atoms and, in both cases, it is necessary to use the correct form (14.91, 92) of the imaginary part of the polarization function.

14.5 Vacuum Polarization Effects in Atoms

We now compute the modification of the (Coulomb) potential due to the electronic vacuum polarization. As shown, the dispersive representation (14.94) of the polarization function is quite useful and leads straightforwardly to an integral representation for the vacuum polarization potential. We first return to (14.69) to separate the induced current from the bare, renormalized current:

$$\begin{aligned} j_{\mu,\rm ex}(k) &= j_{\mu,\rm ex}^{\rm ren,b}(k) + \left(\frac{1}{1-\Pi_{\rm ren}^{(1)}} - 1\right) j_{\mu,\rm ex}^{\rm ren,b} \\ &\equiv j_{\mu,\rm ex}^{\rm ren,b}(k) + j_\mu^{\rm in}(k)\,. \end{aligned} \tag{14.99}$$

The second term is the induced current, explicitly (in the chain approximation)

$$j_\mu^{\rm in} = \left(\frac{1}{1-\Pi_{\rm ren}^{(1)}} - 1\right) j_{\mu,\rm ex}^{\rm ren,b}(k) = [\Pi_{\rm ren}^{(1)} + (\Pi_{\rm ren}^{(1)})^2 + \ldots]\, j_{\mu,\rm ex}^{\rm ren,b}(k)\,. \tag{14.100}$$

Now, since the coefficient relating the renormalized external current $j_{\mu,\mathrm{ex}}^{\mathrm{ren,b}}$ to the induced current begins with a power k^2, it is more convenient to introduce the induced vacuum polarization potential

$$\begin{aligned}(\Box g_{\mu\nu} - \partial_\mu\partial_\nu)A_{\mathrm{in}}^{\nu}(x) &= j_\mu^{\mathrm{in}}(x)\\ -(k^2 g_{\mu\nu} - k_\mu k_\nu)A_{\mathrm{in}}^{\nu}(k) &= j_\mu^{\mathrm{in}}(k)\end{aligned} \tag{14.101}$$

[see (14.48) for the momentum form of Maxwell's equations].

As discussed in Sect. 14.2, the induced current is conserved up to a later omitted spurious term (see (14.36)). Hence

$$A_{\mathrm{in},\mu}(k) = \frac{1}{k^2} j_\mu^{\mathrm{in}}(k)\,; \qquad k^\mu A_\mu^{\mathrm{in}} = \frac{1}{k^2} k^\mu j_\mu^{\mathrm{in}} = 0\,. \tag{14.102}$$

In terms of Π^{chain}, (14.97) now reveals (we now set $m_0 = m_{\mathrm{e}}$)

$$\begin{aligned}A_{\mathrm{in},\mu} &= -\frac{1}{k^2}\Pi^{\mathrm{chain}}(k^2)\, j_\mu^{\mathrm{ren,b}}\\ &= \int_{4m_{\mathrm{e}}^2}^{\infty} \frac{\mathrm{Im}\{\Pi_{\mathrm{ren}}^{(1)}(M^2)\}}{(1-\mathrm{Re}\{\Pi_{\mathrm{ren}}^{(1)}\})^2 + (\mathrm{Im}\{\Pi_{\mathrm{ren}}^{(1)}(M^2)\})^2}\, \frac{dM^2/\pi M^2}{k^2 - M^2 + \mathrm{i}\varepsilon}\, j_\mu^{\mathrm{ren,b}}(k^2)\,.\end{aligned} \tag{14.103}$$

The advantage of this expression, aside from the parametric integration over M^2, is that the current appears multiplied by the propagator of a massive particle, cf. (14.12b),

$$D_0(x;M) = \int \frac{d^4k}{(2\pi)^4}\, \mathrm{e}^{-\mathrm{i}kx} \frac{1}{k^2 - M^2 + \mathrm{i}\varepsilon}\,, \tag{14.104}$$

which then allows us to write (14.103) in configuration space in a straightforward manner

$$\begin{aligned}A_{\mathrm{in},\mu}(x) = \int_{4m_{\mathrm{e}}^2}^{\infty} &\frac{\mathrm{Im}\{\Pi_{\mathrm{ren}}^{(1)}(M^2)\}}{(1-\mathrm{Re}\{\Pi_{\mathrm{ren}}^{(1)}\})^2 + (\mathrm{Im}\{\Pi_{\mathrm{ren}}^{(1)}(M^2)\})^2}\\ &\cdot \int D_{\mathrm{F}}(x-x';M)\, j_\mu^{\mathrm{ren,b}}(x')\, d^4x'\, \frac{dM^2}{\pi M^2}\,.\end{aligned} \tag{14.105}$$

If the bare renormalized current $j_\mu^{\mathrm{ren,b}}$ is independent of time, e.g. if it originates in a stationary charge distribution, then (14.105) simplifies further, since the dx_0' integration can be carried out above

$$\begin{aligned}&\int dx_0'\, D_0(x-x')\\ &= \int \frac{d^3k}{(2\pi)^3} \exp[-\mathrm{i}\boldsymbol{k}\cdot(\boldsymbol{x}-\boldsymbol{x})] \int \frac{dk_0}{2\pi} \frac{\exp(-\mathrm{i}k_0x_0)}{k_0^2 - (\boldsymbol{k}^2 + M^2) + \mathrm{i}\varepsilon} \int dx_0' \exp(\mathrm{i}k_0x_0')\\ &= -\int \frac{d^3k}{(2\pi)^3} \frac{\exp[-\mathrm{i}\boldsymbol{k}\cdot(\boldsymbol{x}-\boldsymbol{x}')]}{\boldsymbol{k}^2 + M^2}\,,\end{aligned} \tag{14.106}$$

whereby $+i\varepsilon$ is omitted as the integral is not singular any more. As is well known, the above integral leads to the Yukawa potential form

$$\int dx_0' \, D_0(x-x') = -\frac{1}{4\pi} \frac{\exp(-M|\boldsymbol{x}-\boldsymbol{x}'|)}{|\boldsymbol{x}-\boldsymbol{x}'|} . \tag{14.107}$$

Thus for (14.105)

$$A_{\text{in},\mu}(\boldsymbol{x}) = -\int_{4m_e^2}^{\infty} \frac{\text{Im}\{\Pi_{\text{ren}}^{(1)}(M^2)\}}{(1-\text{Re}\{\Pi_{\text{ren}}^{(1)}\})^2+(\text{Im}\{\Pi_{\text{ren}}^{(1)}(M^2)\})^2} \int \frac{\exp(-M|\boldsymbol{x}-\boldsymbol{x}'|)}{4\pi|\boldsymbol{x}-\boldsymbol{x}'|}$$

$$\times j_\mu^{\text{ren, b}}(\boldsymbol{x}') \, d^3x' \frac{dM^2}{\pi M^2} , \tag{14.108}$$

where $\text{Im}\{\Pi\}$ and $\text{Re}\{\Pi\}$ are given in (14.91, 92). Through first-order in $\alpha = e_r^2/4\pi$, i.e. keeping only the numerator in (14.108), this can be written as

$$A_\mu(\boldsymbol{x}) = -\int \frac{j_\mu^{\text{ren, b}}(\boldsymbol{x}')}{4\pi|\boldsymbol{x}-\boldsymbol{x}'|} \left(1 + \frac{\alpha}{3\pi} Z_0 |\boldsymbol{x}-\boldsymbol{x}'|\right) d^3x' , \tag{14.109}$$

which includes the non-induced part of the potential and

$$Z_0(|\boldsymbol{x}-\boldsymbol{x}'|) = \int_{4m_0^2}^{\infty} e^{-M|\boldsymbol{x}-\boldsymbol{x}'|} f(M^2) \frac{dM^2}{M^2} . \tag{14.110}$$

Where $f(M^2)$ is given by (14.94a) and shown in Fig. 14.1.

As the integral (14.110) normally needs to be evaluated numerically, including the chain of loop diagrams through the denominator of (14.108), which modifies the argument of (14.110) multiplicatively, presents no additional calculational problems. Explicitly,

$$Z_{\text{chain}}(|\boldsymbol{x}-\boldsymbol{x}'|) = \int_{4m_e^2}^{\infty} e^{-M|\boldsymbol{x}-\boldsymbol{x}'|} f_{\text{chain}}(M^2) \frac{dM^2}{M^2} \tag{14.111a}$$

$$f_{\text{chain}}(M^2) = \frac{f(M^2)}{[1-g(M^2)]^2 + \left[\frac{\alpha}{3} f(M^2)\right]^2} \tag{14.111b}$$

$$g(M^2) = -\frac{\alpha}{3\pi}\left[\frac{5}{3} + \frac{4m_e^2}{3\pi} - \left(1+\frac{2m_e^2}{M^2}\right)\sqrt{\left|1-\frac{4m_e^2}{M^2}\right|}\right.$$

$$\left. \times \ln\left(\frac{\sqrt{\left|1-\frac{4m_e^2}{M^2}\right|}+1}{\sqrt{\left|1-\frac{4m_e^2}{M^2}\right|}-1}\right)\right] . \tag{14.111c}$$

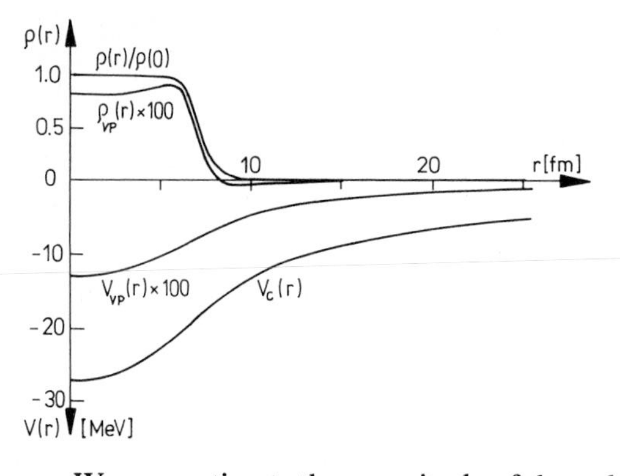

Fig. 14.2. Normalized nuclear and polarization ($\times 100$) charge density ϱ with the associated potentials for ^{238}U

We now estimate the magnitude of the polarization effect. It is important to appreciate that the range of the vacuum polarization potential is of the order $1/2m_e$, since $M \geqslant 2m_e$ in (14.110). Thus, two effects combine to limit the vacuum polarization effect, as (14.110) explicitly shows: (1) the coefficient $\alpha/3\pi$ limits the correction to about 1% of the Coulomb potential and (2) this potential dies off exponentially within the distance of $1/2m_e \approx 200$ fm for polarization due to virtual e^+e^- pairs. Consequently, only the atomic s-state electrons, which have an appreciable probability to be near the centre of the Coulomb potential, are affected by this polarization correction. Figure 14.2 shows the charge distributions and potentials of uranium: both Coulomb and vacuum polarization ($\times 100$) contributions are displayed. Note, as already mentioned, that the observable part of the vacuum polarization actually strengthens the total potential generated by a bare (renormalized) charge density. As shown below, the vacuum polarization potential is quite small compared to the Coulomb potential, so we are entitled to use 1st-order perturbation theory to calculate the change of energy in electronic atoms:

$$\Delta E_n = \langle \psi_n | e A_{\mathrm{in},0}(\boldsymbol{x}) | \psi_n \rangle = -\frac{\alpha}{3\pi} \int \frac{j_0^{\mathrm{ren,b}}(\boldsymbol{x}')}{4\pi|\boldsymbol{x}-\boldsymbol{x}'|} Z_0(|\boldsymbol{x}-\boldsymbol{x}'|)\, \varrho_n(\boldsymbol{x})\, d^3x\, d^3x' , \tag{14.112}$$

where $\varrho_n(\boldsymbol{x}) = |\psi_n(\boldsymbol{x})|^2$.

Taking a point charge at rest $j_0(\boldsymbol{x}') = Ze^2\delta^3(\boldsymbol{x}')$,

$$\begin{aligned} \Delta E_n &= \frac{\alpha}{3\pi} \frac{Ze^2}{4\pi} \int d^3x\, \varrho_n(\boldsymbol{x}) \frac{Z_0(|\boldsymbol{x}|)}{|\boldsymbol{x}|} \\ &= \frac{Z\alpha^2}{3\pi} \int_{4m_e^2}^{\infty} \frac{dM^2}{M^2} f(M^2) \left[\int d^3x \frac{\exp(-M|\boldsymbol{x}|)}{|\boldsymbol{x}|} \varrho_n(|\boldsymbol{x}|) \right] . \end{aligned} \tag{14.113}$$

Over the range of the polarization potential it is safe to assume initially that the s-state density does not change. Hence

$$\Delta E_{ns} = -\frac{Z\alpha^2}{3\pi} \left[\int_{4m^2}^{\infty} \frac{dM^2}{M^2} f(M^2) \frac{1}{M^2} \right] 4\pi |\psi_{ns}(0)|^2 , \tag{14.114}$$

where the square bracket is related to the polarization function at $k^2 = 0$, (14.93, 95),

$$
-\frac{\alpha}{3}\int_{4m_e^2}^{\infty}\frac{1}{\pi}f(M^2)\frac{1}{M^2}\frac{dM^2}{M^2}
$$

$$
=\int_{4m_e^2}^{\infty}\frac{1}{\pi}\,\mathrm{Im}\{\Pi_{\mathrm{ren}}^{(1)}(M^2)\}\frac{1}{k^2-M^2+\mathrm{i}\varepsilon}\frac{dM^2}{M^2}\bigg|_{k^2=0}=\frac{1}{k^2}\Pi_{\mathrm{ren}}^{(1)}\bigg|_{k^2=0}. \tag{14.115}
$$

Thus the lowest-order correction to the energy levels of electronic atoms is described by, cf. (14.89),

$$
\Delta E_{ns}=Z\alpha\pi|\psi_{ns}(0)|^2\left(-\frac{\alpha}{\pi}\frac{1}{15m_e^2}\right)
=\begin{cases}-\dfrac{4}{15\pi}\dfrac{Z^4\alpha^5}{n^3}m_e & \text{(electronic atoms)}\\[2ex] -\dfrac{4}{15\pi}\dfrac{Z^4\alpha^5}{n^3}\dfrac{m_\mu^3}{m_e^2} & \text{(muonic atoms)}\end{cases}. \tag{14.116}
$$

The expression given for muonic atoms must be taken only as a first approximation, as even for $Z=n=2$ it fails by a factor 2, since for sufficiently large Z, the electron (or muon for muonic atoms) has a substantial probability to be found near the finite size nucleus. Hence to evaluate the influence of the vacuum polarization on atomic energies, we must evaluate (14.112) more carefully, a task accomplished most readily by numerical methods [Pi 69]. It is important to appreciate that the effect always remains less than 1%, since even if the entire electronic wave function comes within the range of the vacuum polarization potential, for Z near Z_{cr}, the polarization effect remains small, being proportional to the fine structure coupling constant. In Fig. 14.3 the energy shift of K electrons in very heavy atoms is shown – as expected, it always remains a small function of the total binding energy, and is almost precisely 1% just at the critical point.

Since vacuum polarization effects in light electronic atoms usually remain very small, they cannot be tested independently of many other contributions. However, certain *muonic* orbits in atoms serve as an ideal and precise testing ground of the vacuum polarization effect.

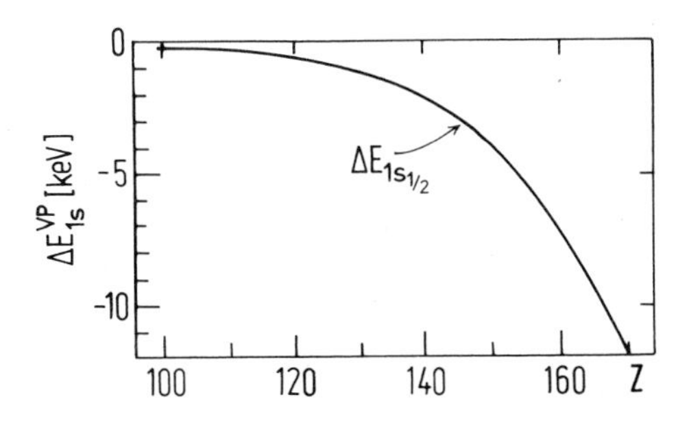

Fig. 14.3. First-order virtual vacuum polarization correction of the most tightly bound $1s_{1/2}$ electron as a function of nuclear charge

As regards electromagnetic interaction, the muon has identical properties to the electron, but a mass which is 105.6594 MeV, i.e. ca. 206.77 times larger than the electron's mass. Consequently, muonic Bohr orbits are 207 times smaller and bring the muon into the full range of the vacuum polarization potential. On the other hand, the heavy muon has almost a "classical" orbit as the uncertainty relation allows it to be much better localized. Certain muonic orbits, when selected properly, lie outside the finite-size nucleus but inside the electron's innermost orbit and within the range of the vacuum polarization potential. In these cases the vacuum polarization effect as described by (14.112) is the dominant correction to the Dirac equation eigenenergies with Coulombic ($1/r$) potential, where reduced mass of the muon is introduced normally to account for the dominant part of the (relativistic) recoil correction. A further remarkable point is that since the function Z_0 (14.112) smears the effect of finite nuclear size, the finite nuclear size influences such orbits through vacuum polarization potential rather than through the 100 times stronger Coulomb potential.

One of the favourite muonic transitions in this context is the $5g_{\frac{9}{2}} - 4f_{\frac{7}{2}}$ dipole transition in lead, which has a Dirac transition energy of 429344 eV. Table 14.1 [Br 78] lists all the different corrections to the Dirac eigenenergies in eV.

Table 14.1. Summary of contributions to energy levels [eV] in muonic lead ^{208}Pb

	Contributions	$4f_{\frac{7}{2}}$	$5g_{\frac{9}{2}}$
a)	*Static external potential*		
	Dirac Coulomb energy[a]	−1188314	−758970
	Finite nuclear size	4	0
b)	*Vacuum polarization*	−3594	−1509
c)	*Self-energy*	10	3
d)	*Nuclear motion*	−4	−1
e)	*Nuclear polarization*	−5	0
f)	*Atomic electrons*[b]	−89	−172
Total		−1191992	−760660

Transition energy ≅ 431.332 eV
Experimental energy 431.331 ± 5 eV [Du 78, Ha 77]

[a] Includes reduced mass correction $m_\mu \to Mm_\mu/(m_\mu + M)$.
[b] Constant term V_0 is not included in this screening correction [Vo 73].

As can be seen, the self-energy term (Chap. 16.3) is entirely negligible here and by far the dominant correction arises from vacuum polarization.

We thus conclude that vacuum polarization is successfully verified at the level of precision of 0.15% for $Z\alpha \lesssim 0.6$.

Bibliographical Notes

Vacuum polarization in a weak external field is discussed by *Källen* [Kä 58], while the photon propagation function is described by *Schwinger* [Schw 73]. Both give considerably more detailed description of perturbative effects in QED. A noteworthy detail is *Källen's* proof that the charge renormalization is infinite in QED, and the discussion of infrared difficulties in QED.

15. Vacuum Polarization: Arbitrarily Strong External Potentials

In the previous chapter we recorded the relation (14.4) between the vacuum polarization charge and Green's function. This relation is now exploited to derive the polarization effects for arbitrarily strong potential sources. Clearly, our interest here is directed particularly towards supercritical potentials. Our previously acquired knowledge of the single particle spectrum of the Dirac equation is extensively used and applied throughout this chapter.

15.1 Green's Function for Arbitrarily Strong External Potentials

A straightforward calculation leads from (14.3) for the Feynman propagator to an expression in terms of the single particle wave functions, assuming that the external potential is time independent. As this has been carefully discussed in Chap. 8, we summarize here the result by writing [cf. (8.63)]

$$\mathrm{i}S_{\mathrm{F}}(x,x') = \begin{cases} \sum\limits_{E_n>E_{\mathrm{F}}} \varphi_n(\boldsymbol{x})\,\bar{\varphi}_n(\boldsymbol{x}')\exp[-\mathrm{i}E_n(t-t')]\; t>t' \\ -\sum\limits_{E_n<E_{\mathrm{F}}} \varphi_n(\boldsymbol{x})\,\bar{\varphi}_n(\boldsymbol{x}')\exp[-\mathrm{i}E_n(t-t')]\; t<t', \end{cases} \tag{15.1}$$

which can also be written with the help of the step functions $\theta(t-t')$ and $\theta(t'-t)$, respectively.

In view of the well-known representations of the step functions in terms of contour integrals (Fig. 15.1)

$$\theta(s) = \int\limits_{-\infty}^{+\infty} \frac{\mathrm{e}^{\mathrm{i}zs}}{z-\mathrm{i}\varepsilon}\,\frac{dz}{2\pi\mathrm{i}}\,. \tag{15.2a}$$

Since equation (15.2a) leads to a real result its complex conjugate provides an alternative representation:

$$\theta(s) = \int\limits_{-\infty}^{+\infty} \frac{\mathrm{e}^{-\mathrm{i}zs}}{z+\mathrm{i}\varepsilon}\,\frac{dz}{2\pi}\,. \tag{15.2b}$$

We now rewrite (15.2) in terms of a contour integral. Respecting the time translation invariance implied by (15.2), we are led to consider the form

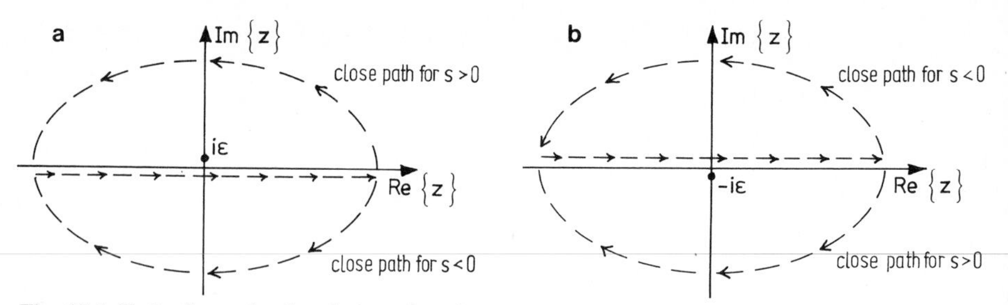

Fig. 15.1. Paths for evaluating the step function **(a)** for (15.2a), **(b)** for (15.2b)

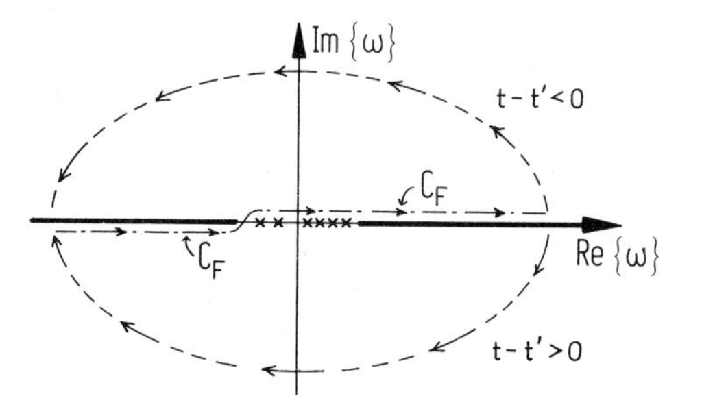

Fig. 15.2. (– · –) path C_F for the Feynman propagator (15.5). (– – –): possible continuations of the path for $t-t' \lessgtr 0$

$$\mathrm{i}S_F(x,x') = \int_{C_F} \frac{d\omega}{2\pi\mathrm{i}} \exp[-\mathrm{i}\omega(t-t')]\, G(\boldsymbol{x},\boldsymbol{x}':\omega)\,, \tag{15.3}$$

where the integration contour is related to the boundary conditions imposed on S_F. Equations (15.1, 2) imply that

$$G(x,x';\omega) = \sum_n \frac{\varphi_n(\boldsymbol{x})\,\bar{\varphi}_n(\boldsymbol{x}')}{E_n-\omega}\,, \tag{15.4}$$

where G is a 4×4 matrix. To prove (15.4), we insert it into (15.3)

$$\mathrm{i}S_F(x,x') = \sum_n \varphi_n(\boldsymbol{x})\,\bar{\varphi}_n(\boldsymbol{x}') \int_{C_F} \frac{d\omega}{2\pi\mathrm{i}} \frac{\exp[-\mathrm{i}\omega(t-t')]}{E_n-\omega}\,. \tag{15.5}$$

We now demonstrate that the path C_F must be as in Fig. 15.2 to reproduce (15.1). Here the singularities of $G(\boldsymbol{x},\boldsymbol{x}':\omega)$ for "weak" potentials are displayed, with crosses indicating simple poles associated with the discrete spectrum, and thick lines indicating the cuts associated with the continuum spectrum.

For $t-t'>0$ and $t-t'<0$ the paths can be closed as indicated by dashed lines in Fig. 15.2 and in absence of other singularities, they may be deformed to run along the singularities on the real ω axis, as shown in Fig. 15.3.

Thus for $t-t'<0$ we sum over the states $E_n<E_F$, and for $t-t'>0$, over the states $E_n>E_F$ but with negative orientation (i.e. clockwise path). Explicitly

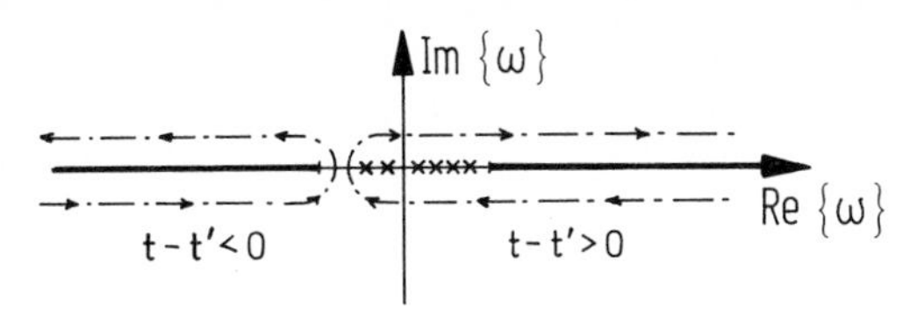

Fig. 15.3. Deformed path C_F for the Feynman propagator

$$\int_{C_F} \frac{d\omega}{2\pi i} \frac{e^{-i\omega(t-t')}}{E_n-\omega} = \begin{cases} -\theta(E_n-E_F)e^{-iE_n(t-t')}, & t-t'>0 \\ +\theta(E_F-E_n)e^{-iE_n(t-t')}, & t-t'<0 \end{cases}. \tag{15.6}$$

When inserted into (15.5), this leads straightforwardly to (15.1). Hence the distribution (15.4) leads together with the choice of the path C_F, Fig. 15.2, to an independent characterization of the Feynman propagator (15.1).

The particular advantage of G in (15.4) is that it is the time-independent Green's function of the Dirac equation. To see this apply the Dirac operator to (15.4) from the left:

$$\begin{aligned}[\boldsymbol{\alpha}\cdot\boldsymbol{p}+\beta m_e+V(\boldsymbol{x})]\,G(\boldsymbol{x},\boldsymbol{x}';\omega) &= \sum_n \frac{[\boldsymbol{\alpha}\cdot\boldsymbol{p}+\beta m_e+V(\boldsymbol{x})]\,\varphi_n(\boldsymbol{x})\,\bar{\varphi}_n(\boldsymbol{x}')}{(E_n-\omega)} \\ &= \sum_n \frac{(E_n-\omega)+\omega}{E_n-\omega}\,\varphi_n(\boldsymbol{x})\,\bar{\varphi}_n(\boldsymbol{x}') \\ &= \sum_n \varphi_n(\boldsymbol{x})\,\bar{\varphi}_n(\boldsymbol{x}')+\omega G(\boldsymbol{x},\boldsymbol{x}':\omega)\,.\end{aligned}$$

Hence, since the completeness relation for the Dirac equation reads

$$\gamma^0\delta^3(\boldsymbol{x}-\boldsymbol{x}') = \sum_n \varphi_n(\boldsymbol{x})\,\bar{\varphi}_n(\boldsymbol{x}')\,,$$

then

$$[\boldsymbol{\alpha}\cdot\boldsymbol{p}+\beta m_e+V(\boldsymbol{x})-\omega]\,G(\boldsymbol{x},\boldsymbol{x}';\omega) = \gamma^0\delta^3(\boldsymbol{x}-\boldsymbol{x}')\,. \tag{15.7a}$$

Similarly, the conjugate equation is given by applying the conjugate Dirac operator from the rhs

$$G(\boldsymbol{x},\boldsymbol{x}';\omega)\,[\boldsymbol{\alpha}\cdot\boldsymbol{p}'+\beta m_e+V(\boldsymbol{x}')-\omega] = \gamma^0\delta^3(\boldsymbol{x}-\boldsymbol{x}')\,. \tag{15.7b}$$

To study the particularly interesting case of supercritical potentials further, it is convenient to consider spherically symmetric potentials $V(r)$, $r=|\boldsymbol{x}|$. Then the Green's function G can be expanded in eigenfunctions of the Dirac angular momentum

$$K = \gamma^0(\boldsymbol{\sigma}\cdot\boldsymbol{L}+1) \tag{15.8}$$

with the eigenvalues $k = \pm(j+\frac{1}{2})$ in terms of the total angular momentum j. With the radial form of the Dirac equation given in Chap. 3, $G(\boldsymbol{x},\boldsymbol{x}';\omega)$ has a partial wave decomposition [Mo 74]

$$\begin{bmatrix} m_e + V(r) - \omega & -\frac{1}{r}\frac{d}{dr}r + \frac{\varkappa}{r} \\ \frac{1}{r}\frac{d}{dr}r + \frac{\varkappa}{r} & -m_e + V(r) - \omega \end{bmatrix} \cdot G_\varkappa(r, r'; \omega) = \frac{\delta(r-r')}{rr'}\gamma_0 . \tag{15.9}$$

We first notice a symmetry property of $G_\varkappa$ which is handy when considering the behaviour of the vacuum polarization: changing $V \to -V$, $\omega \to -\omega$, $\varkappa \to -\varkappa$ and the overall sign of G gives

$$\begin{bmatrix} -m_e + V(r) - \omega & \frac{1}{r}\frac{d}{dr}r + \frac{\varkappa}{r} \\ -\frac{1}{r}\frac{d}{dr}r + \frac{\varkappa}{r} & m_e + V(r) - \omega \end{bmatrix} \cdot [-G_{-\varkappa}(r, r'; -\omega, -V)] = \frac{\delta(r-r')}{rr'}\gamma_0 ,$$

which is almost the same as (15.9). With $\sigma_1 = \gamma_5 \alpha_1$ we can introduce

$$\bar{G}_\varkappa = \sigma_1 G_{-\varkappa}(r, r'; -\omega - V)\sigma_1 ,$$

and we easily recognize that $\bar{G}$ satisfies the *same* equation as G. Note that σ_1 anticommutes with γ_0. Consequently, with $\sigma_1\gamma_0\sigma_1 = -\gamma_0$,

$$\begin{aligned} G_\varkappa(r, r'; \omega, V) &= +\sigma_1 G_{-\varkappa}(r, r'; -\omega, -V)\sigma_1 , \\ G_\varkappa(r, r'; \omega, V)\gamma_0 &= -\sigma_1 G_{-\varkappa}(r, r'; -\omega, -V)\gamma_0\sigma_1 . \end{aligned} \tag{15.10}$$

The usefulness of this equation arises from

$$\operatorname{tr} G(\boldsymbol{x}, \boldsymbol{x}'; \omega)\Big|_{x \to x'} = \sum_{\substack{\varkappa = \pm(j+\frac{1}{2}) \\ j \in (0, \infty)}} \frac{2|\varkappa|}{4\pi} \operatorname{tr} G_\varkappa(r, r'; \omega)\Big|_{x \to x'} \tag{15.11}$$

in which the magnetic quantum numbers have been summed over, leading to the multiplicity factor $2|\varkappa| = 2j+1$. This relation can also be written more explicitly as

$$\operatorname{tr} G(\boldsymbol{x}, \boldsymbol{x}'; \omega)\Big|_{x \to x'} = \sum_{j=\frac{1}{2}, \frac{3}{2} \ldots} \frac{2j+1}{4\pi} \operatorname{tr}[G_{\varkappa=j+\frac{1}{2}}(r, r; \omega) + G_{\varkappa=-j-\frac{1}{2}}(r, r; \omega)] . \tag{15.12}$$

Equation (15.10) enables the second part of (15.12) to be rewritten as

$$\operatorname{tr} G(\boldsymbol{x}, \boldsymbol{x}'; \omega)|_{x \to x'} = \sum_{j=\frac{1}{2}, \frac{3}{2} \ldots} \frac{2j+1}{4\pi} \operatorname{tr}[G_{\varkappa=j+\frac{1}{2}}(r, r'; \omega, V) + G_{\varkappa=j+\frac{1}{2}}(r, r'; -\omega, -V)] . \tag{15.13}$$

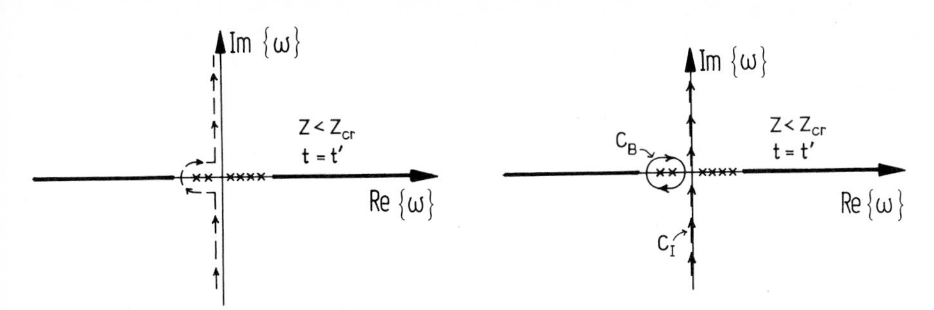

Fig. 15.4. Path C_F along the imaginary ω axis

Fig. 15.5. Path $C_F = C_B + C_I$

Thus (15.13) displays the particular symmetry

$$\operatorname{tr} G(\boldsymbol{x},\boldsymbol{x}';\omega,V) = +\operatorname{tr} G(\boldsymbol{x},\boldsymbol{x}';-\omega,-V) \tag{15.14a}$$

and similarly

$$\operatorname{tr}[G(\boldsymbol{x},\boldsymbol{x}';\omega,V)\gamma_0] = -\operatorname{tr}[G(\boldsymbol{x},\boldsymbol{x}';-\omega,-V)\gamma_0] \tag{15.14b}$$

proven here for spherically symmetric potentials. Relations (15.14) are needed to establish that the vacuum polarization is an odd function of $(Z\alpha)$. Furthermore, the symmetry displayed by (15.14) suggests that it is of certain advantage to deform the paths C_F so that they follow the imaginary ω axis, Fig. 15.4. This can be done only when $t = t'$, since then the integrals are not spoiled by exploding exponential factors.

Whenever some bound states descend below $E = 0$, it is of advantage to treat them separately, allowing us to perform an integral along the $\mathrm{Im}\{\omega\}$ axis (C_I), together with a circle (C_B) around the bound states' singularities with $-m_e < E < 0$, as shown in Fig. 15.5.

Inserting (15.4) into (15.3) separates the influence of strongly bound states from the other contributions quite simply: with $C_F = C_B + C_I$, for $t = t'$,

$$\mathrm{i} S_F(x,x')\bigg|_{t=t'} = \int_{C_B} \frac{d\omega}{(2\pi \mathrm{i})} \sum_{E_n<0} \frac{\varphi_n(\boldsymbol{x})\bar{\varphi}_n(\boldsymbol{x})}{E_n - \omega} + \int_{-\infty}^{+\infty} \frac{dy}{(2\pi)} G(\boldsymbol{x},\boldsymbol{x}';\mathrm{i}y)\,. \tag{15.15}$$

One can easily carry out, for $Z < Z_{cr}$, the first residuum integral which leaves a simple sum over the bound states in the interval $-m_e < E < 0$:

$$\mathrm{i} S_F(x,x')\bigg|_{\substack{t=t'\\ Z<Z_{cr}}} = \sum_{-m_e<E_n<0} \varphi_n(\boldsymbol{x})\bar{\varphi}_n(\boldsymbol{x}') + \int_{-\infty}^{+\infty} \frac{dy}{2\pi} G(\boldsymbol{x},\boldsymbol{x}':\mathrm{i}y)\,. \tag{15.16}$$

We can now consider $Z \to Z_{cr}$ and $Z > Z_{cr}$. Imagine that we are just below Z_{cr}. The second part of (15.16) is well behaved, as we introduce a small change in Z such that $Z > Z_{cr}$. But as we know very well, there is a sudden change in the

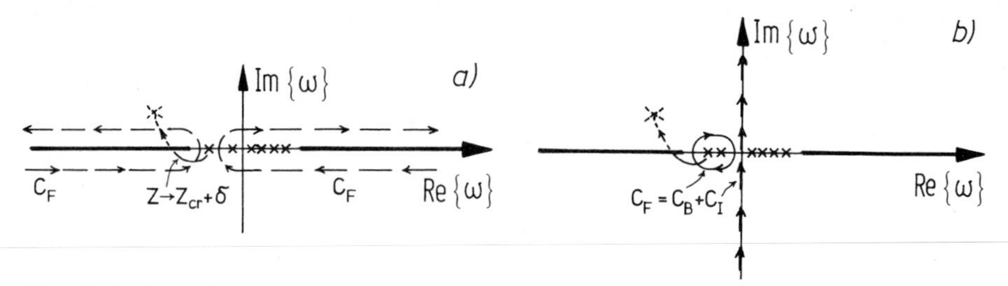

Fig. 15.6a, b. Motion of the supercritical bound state pole as Z increases beyond $Z_{\rm cr}$. **(a)** $C_{\rm F}$ path along $\mathrm{Re}\{\omega\}$, **(b)** $C_{\rm F}$ path along $\mathrm{Im}\{\omega\}$

spectrum: one of the bound states disappears and hence the first term in (15.16) changes quite drastically. Thus $S_{\rm F}(x,x')$ is non-continuous at $Z_{\rm cr}$. This behaviour is, of course, also apparent from the path $C_{\rm F}$ along the real ω axis (Fig. 15.3), where the bound state cross has to move through the path into the second sheet of the ω plane, Fig. 15.6.

The Feynman propagator has a discontinuity, as Fig. 15.6a indicates, for arbitrary $t-t'$. This discontinuity is solely associated with the disappearance of the bound state pole of G from the physical ω sheet: for $E<-m_{\rm e}$ only the continuous spectrum exists, but below $\mathrm{Re}\{\omega\}+\mathrm{i}\varepsilon$, there is a second sheet on which now the bound state pole resides. However, its presence is felt on the physical sheet as it influences e.g. the scattering phase shift of continuous states.

Thus, associated with the need to introduce a charged vacuum state for $Z=Z_{\rm cr}+\delta$, there is a discontinuity in the behaviour of the Feynman electron propagator due to the well-described "immobility" of the supercritical state, whose electronic charge is localized forever near the source of the supercritical field. Furthermore, S, as described, changes discontinuously at $Z=Z_{\rm cr}$. We reemphasize here that this behaviour is different from that observed by *Wichmann* and *Kroll* [Wi 56], where the $1/r$ potential was studied analytically, and the discontinuity of the propagator was associated with the disappearance of the s states from the physical spectrum (non-self adjointness of the Hamiltonian).

15.2 Vacuum Polarization Charge Density

The formal developments presented above enable the vacuum polarization charge as defined by (14.4) to be computed exactly for any strength of the external source, characterized by its charge Z.

To carry out this program explicitly, we first express the induced charge density in terms of G, (15.4). Inserting (15.3) into (14.4), for static external charge distributions (time-independent sources) gives

$$\langle \hat{j}_0\rangle = -e\int_{C_{\rm F}}\frac{d\omega}{2\pi\mathrm{i}}\,\mathrm{e}^{-\mathrm{i}\omega(t-t')}\,\mathrm{tr}[G(\boldsymbol{x},\boldsymbol{x}':\omega)\,\gamma_0]\Big|_{\boldsymbol{x}\underset{\mathrm{s}}{\to}\boldsymbol{x}'}\,. \tag{15.17}$$

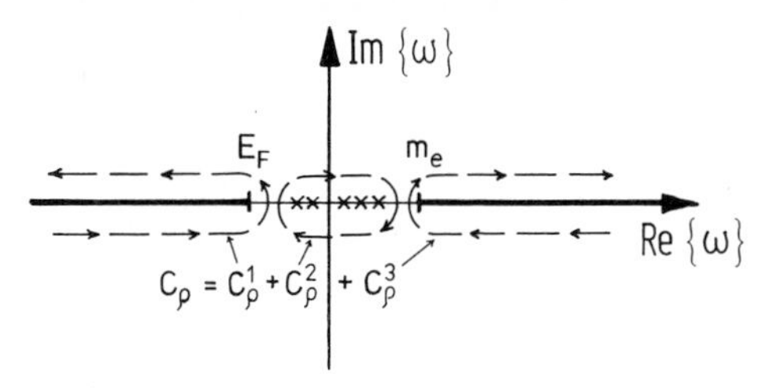

Fig. 15.7. Path C_ϱ for the vacuum polarization charge density

Using $\boldsymbol{x} = \boldsymbol{x}'$, $t - t' = \pm\delta$ for the symmetric limit

$$\langle \hat{j}_0 \rangle = -\frac{e}{2} \int_{C_F} \frac{d\omega}{2\pi i} e^{-i\omega\delta} \operatorname{tr}[G(\boldsymbol{x},\boldsymbol{x};\omega)\gamma_0]|_{\delta\to 0^+}$$

$$-\frac{e}{2} \int_{C_F} \frac{d\omega}{2\pi i} e^{i\omega\delta} \operatorname{tr}[G(\boldsymbol{x},\boldsymbol{x};\omega)\gamma_0]|_{\delta\to 0^+} . \tag{15.18}$$

Thus $t - t' < 0$ and $t - t' > 0$ contribute with the weight $\frac{1}{2}$ and *both* paths shown in Fig. 15.3 contribute. Omitting the convergence factor $\exp(\pm i\omega\delta)$

$$\langle j_0 \rangle = -\frac{e}{2} \int_{C_\varrho} \frac{d\omega}{2\pi i} \operatorname{tr}[G(\boldsymbol{x},\boldsymbol{x};\omega)\gamma_0] , \tag{15.19}$$

where C_ϱ is the path enclosing all the singularities along the real ω axis as separated by the choice of the Fermi surface E_F, Fig. 15.7. Furthermore, the integration path has been cut at $E = m_e$ to make certain symmetries visible.

The path C_ϱ can be decomposed into three contributions: C_ϱ^1 for the negative continuum, C_ϱ^2 for the bound states and C_ϱ^3 for the positive continuum. The C_ϱ^2 integration contributes only at the bound state poles:

$$\langle j_0 \rangle = -\frac{e}{2} \sum_{-m_e<E_n<m_e} \varphi_n^+(\boldsymbol{x})\varphi_n(\boldsymbol{x})$$

$$-\frac{e}{2} \int_{C_\varrho^1 + C_\varrho^3} \frac{d\omega}{2\pi i} \operatorname{tr}[G(\boldsymbol{x},\boldsymbol{x};\omega,V)\gamma_0] . \tag{15.20}$$

Another observation along these lines concerns the dependence of the induced current on the choice of the Fermi surface. Had we not chosen $E_F = -m_e$ but $E_F = +m_e$ (fully filled atomic shells) the identical calculation would arise, except that the bound state poles would be integrated around in the opposite direction. Thus (15.20) would read

$$\langle j_0 \rangle \Big|_{E_F = m_e} = +\frac{e}{2} \sum_{-m_e<E_n<m_e} \varphi_n^+(\boldsymbol{x})\varphi_n(\boldsymbol{x})$$

$$-\frac{e}{2} \int_{C_\varrho^1 + C_\varrho^3} \frac{d\omega}{2\pi i} \operatorname{tr}[G(\boldsymbol{x},\boldsymbol{x};\omega,V)\gamma_0] . \tag{15.21}$$

Thus

$$\langle j_0\rangle\Big|_{E_F=m_e} - \langle j_0\rangle\Big|_{E_F=-m_e} = e \sum_{-m_e<E_n<m_e} \varphi_n^+(x)\,\varphi_n(x)\,, \tag{15.22}$$

which is the usual charge density of electrons bound in a neutral atom.

Returning to the discussion of (15.20), note that the paths C_ϱ^1 and C_ϱ^3 can be transformed into each other. Hence, noting the orientation of the integration

$$\int_{C_\varrho^1} \frac{d\omega}{2\pi i}\,\mathrm{tr}\,[G(x,x;\omega,V)\,\gamma_0] = \int_{C_\varrho^3} \frac{d\omega}{2\pi i}\,\mathrm{tr}\,[G(x,x;-\omega,V)\,\gamma_0]\,. \tag{15.23}$$

Further, with (15.14b)

$$\int_{C_\varrho^1} \frac{d\omega}{2\pi i}\,\mathrm{tr}\,[G(x,x;\omega,V)\,\gamma_0] = -\int_{C_\varrho^3} \frac{d\omega}{2\pi i}\,\mathrm{tr}\,[G(x,x;\omega,-V)\,\gamma_0]\,. \tag{15.24}$$

Inserting (15.24) into (15.20) gives

$$\begin{aligned}\langle j_0\rangle = &-\frac{e}{2}\sum_{-m_e<E_n<m_e} \varphi_n^+(x)\,\varphi_n(x)\\ &-\frac{e}{2}\int_{C_\varrho^3} \frac{d\omega}{2\pi i}\,\mathrm{tr}\,[G(x,x;\omega,V)\,\gamma_0 - G(x,x;\omega,-V)\,\gamma_0]\end{aligned} \tag{15.25}$$

which is explicitly odd in V in its continuum part described by the integral over the path C_ϱ^3. It is interesting to observe that even for $V\to 0$ but $V\neq 0$ (15.25) contains the non-perturbative aspects introduced via the (non-analytic) appearance of bound states. It is important to appreciate that all contributions from bound states in (15.25) add up. Being positive definite, this contribution must remain positive, even though when $V\to -V$ the ordering of states in the sum is inverted. However, in the discrete sum this does not introduce an additional "orientation" sign, which made the continuum contribution odd in the external potential. This observation contradicts the general belief based on the behaviour of the continuum contribution to $\langle j_0\rangle$, which is an odd function in the external potential [Fu 37]. However, a perturbative series in V can be expanded only for the continuum contribution!

The charge conjugation invariance of (15.25) is only properly executed when E_F is transformed to $(-E_F)$ together with $V\to -V$. As (15.21) demonstrates convincingly, the bound state sum then picks up an additional sign required to make (15.25) odd under charge conjugation. This has already been stated more formally in (9.95).

For the mathematical convenience of carrying through a calculation, it is useful to deform the path C_3, Fig. 15.7, such that it turns around at $\omega = 0$. As shown in Fig. 15.8, this has the effect that then the residuum sum contains only the discrete states in the interval $-m_e < E_n < 0$.

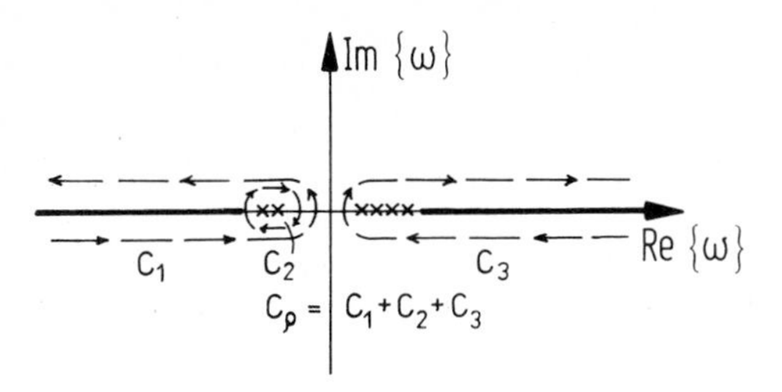

Fig. 15.8. Decomposition of the path C_3 for the vacuum polarization

Thus [cf. (15.20)]

$$\langle j_0 \rangle = -\mathrm{e} \sum_{-m<E_n<0} \varphi_n^+(\boldsymbol{x})\,\varphi_n(\boldsymbol{x}) - \frac{e}{2} \int_{C_1+C_3} \frac{d\omega}{2\pi\mathrm{i}} \operatorname{tr}\left[G(\boldsymbol{x},\boldsymbol{x}\colon \omega)\,\gamma_0\right]. \tag{15.26}$$

It is now possible to deform the paths C_1, C_3 to the imaginary ω axis:

$$\langle j_0 \rangle = -e \sum_{-m<E_n<0} \varphi_n^+(\boldsymbol{x})\,\varphi_n(\boldsymbol{x}) - e \int_{-\infty}^{-\infty} \frac{dy}{2\pi} \operatorname{tr}\left[G(\boldsymbol{x},\boldsymbol{x}\colon \omega = \mathrm{i}y)\,\gamma_0\right]. \tag{15.27}$$

Our next aim is to separate from (15.27) the terms contained in the perturbative approach used in Sect. 14.1. This is easily achieved by developing (formally) the integral equation for Green's function G. Denoting

$$\boldsymbol{\alpha}\cdot\boldsymbol{p} + \beta m - \omega = [G^{(0)}\gamma_0]^{-1}, \tag{15.28}$$

(15.7a) can now be written as

$$\{[G^{(0)}\gamma_0]^{-1} + V\}G = 1\,\gamma_0 \tag{15.29}$$

and hence corresponds to the integral equation

$$G = G^{(0)} - G^{(0)}\gamma_0 V G, \tag{15.30}$$

or once iterated

$$G = G^{(0)} - G^{(0)}\gamma_0 V G^{(0)} + G^{(0)}\gamma_0 V G^{(0)}\gamma_0 V G. \tag{15.31}$$

We define the Green function containing the first correction in V as

$$G^{(1)} = G^{(0)} - G^{(0)}\gamma_0 V G^{(0)}. \tag{15.32}$$

Thus (15.31) can also be written as

$$G - G^{(1)} = G^{(0)}\gamma_0 V G^{(0)}\gamma_0 V G \equiv G^{n>2}, \tag{15.33}$$

where the upper index $n>2$ indicates that only effects of higher power than $[V]^2$ have been retained – note that $[V]^2$ vanishes under the trace (*Furry*'s theorem for free propagators [Fu 37]). With (15.33, 27) can be written as [cf. (14.20)]

$$\langle j_0 \rangle = \langle j_0 \rangle_{(1)} - e \sum_{-m_e < E_n < 0} \varphi_n^+(x)\, \varphi_n(x)$$
$$- e \int_{-\infty}^{+\infty} \frac{dy}{2\pi} \operatorname{tr}[G(x,x,\mathrm{i}y)\,\gamma_0 - G^{(1)}(x,x,\mathrm{i}y)\,\gamma_0] \,. \tag{15.34}$$

The charge renormalization can now be carried out, as the first term in (15.34) contains the (infinite) term proportional to the external charge. Grouping this divergent term together with the bare charge on the lhs of (15.34) (as in Sect. 14.3), allows us to divide by a constant, divergent term, (14.67), changing e^2 on the rhs of (15.34) to e_r^2.

However, one difficulty worth mentioning here remains. The last term in (15.34) contains, in particular, the terms $[V]^3$ as shown by the iterative expansion of the vacuum charge density (the double line now refers to the exact propagator in the external field, but excluding vacuum corrections):

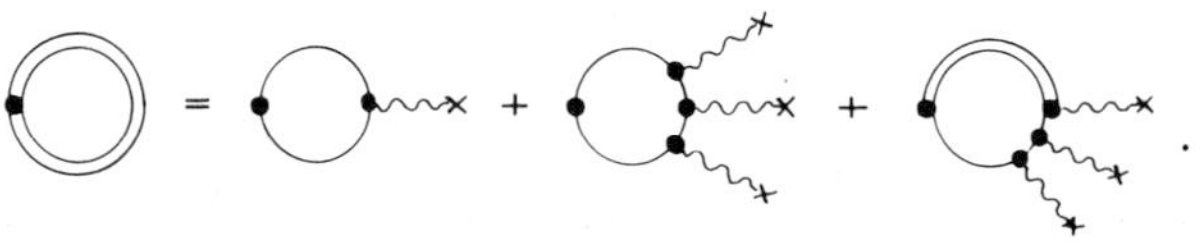

(15.35)

This term is not finite when evaluated as discussed here, (15.34). However, this is only a mathematical difficulty associated with our way of evaluating different terms, and, as it turns out, it can be dealt with simply by decomposing the vacuum polarization charge density (15.34) into partial waves, according to the angular quantum number $\varkappa$, see (15.11). This spurious term, arising from sloppy handling of not always absolutely convergent expressions, does not appear then [Gy 75].

15.3 Vacuum Polarization in External Fields of Arbitrary Strengths

To evaluate vacuum polarization effects in all orders in $(Z\alpha)$ for arbitrarily large Z, we extensively use the partial wave decomposition of the Green's function, (15.11). Restating this equation in slightly different form gives

$$\operatorname{tr}[G(x,x;\omega)\,\gamma^0]$$
$$= \sum_{j=\frac{1}{2},\frac{3}{2}\dots} \frac{2j+1}{4\pi} \operatorname{tr}[G_{\varkappa=j+\frac{1}{2}}(r,r;\omega)\,\gamma_0 + G_{\varkappa=-j-\frac{1}{2}}(r,r;\omega)\,\gamma_0] \,, \tag{15.36}$$

where $G_\varkappa(r,r';\omega)\,\gamma_0$ is the radial Green's function that satisfies (15.9). With (15.36), we can now define the vacuum polarization density due to a particular angular momentum j with $\varrho(r) = \sum_j \varrho_j(r)(2j+1)/4\pi$ as

$$\varrho_j(r) = -\frac{e}{2}\int_{C_\varrho}\frac{d\omega}{2\pi i}\,\mathrm{tr}\,[G_{j+\frac{1}{2}}(r,r;\omega)\,\gamma_0 + G_{-j-\frac{1}{2}}(r,r;\omega)\,\gamma_0]\,. \tag{15.37}$$

Next, decompose C_ϱ into C_1, C_2 and C_3 as in (15.26):

$$\varrho_j(r) = -e\sum_{-m_e<E_{jn}<0}\varphi^+_{jn}(r)\,\varphi_{jn}(r)$$
$$-e\int_{-\infty}^{+\infty}\frac{dy}{2\pi}\,\mathrm{tr}\,[G_{j+\frac{1}{2}}(r,r;\omega=\mathrm{i}y)\,\gamma_0 + G_{-j-\frac{1}{2}}(r,r;\omega=\mathrm{i}y)\,\gamma_0]\,. \tag{15.38}$$

To identify the effects of order $(Z\alpha)^{n>2}$ in (15.38) we must subtract the influence of $G^{(1)}$, (15.32), i.e. the $(Z\alpha)$ order vacuum polarization:

$$\varrho_j^{n>2}(r) = -e\sum_{-m_e<E_{jn}<0}\varphi^+_{jn}(r)\,\varphi_{jn}(r) - \mathrm{e}\int_{-\infty}^{+\infty}\frac{dy}{2\pi}\,\mathrm{tr}\,[G_{j+\frac{1}{2}}(r,r;\mathrm{i}y)\,\gamma_0$$
$$+G_{-j-\frac{1}{2}}(r,r;\mathrm{i}y)\,\gamma_0 - G^{(1)}_{j+\frac{1}{2}}(r,r;\mathrm{i}y)\,\gamma_0 - G^{(1)}_{-j-\frac{1}{2}}(r,r;\mathrm{i}y)\,\gamma_0]\,. \tag{15.39}$$

It is possible, though not essential, to simplify (15.39) further by observing that, (15.10),

$$\int_{-\infty}^{+\infty}\frac{dy}{2\pi}\,\mathrm{tr}\,[G^{(1)}_{-j-\frac{1}{2}}(r,r;\mathrm{i}y)\,\gamma_0] = \int_{-\infty}^{+\infty}\frac{dy}{2\pi}\,\mathrm{tr}\,[G^{(1)}_{j+\frac{1}{2}}(r,r;\mathrm{i}y)\,\gamma_0]\,, \tag{15.40}$$

where we have explicitly used the fact that the above expression is an odd function of the potential V. Thus

$$\varrho_j^{n>2}(r) = -e\sum_{-m_e<E_{jn}<0}\varphi^+_{jn}(r)\,\varphi_{jn}(r)$$
$$-\mathrm{e}\int_{-\infty}^{+\infty}\frac{dy}{2\pi}\,\mathrm{tr}\,[G_{j+\frac{1}{2}}(r,r;\mathrm{i}y,V)\,\gamma_0 + G_{j+\frac{1}{2}}(r,r;\mathrm{i}y,-V)\,\gamma^0$$
$$-2G^{(1)}_{j+\frac{1}{2}}(r,r;\mathrm{i}y)\,\gamma^0]\,. \tag{15.41}$$

The partial wave Green's functions are easily constructed, which justifies the approach taken here. Namely, it is possible [Wi 56, Gy 75] to derive explicitly the form of the Green's function, exploiting standard formal techniques developed for second-order differential equations.

Let $\psi_\mathrm{R}(r)$ and $\psi_\mathrm{I}(r)$ be the regular and irregular solutions of the radial Dirac equation $(H-\omega)\,\psi=0$, where $(H-\omega)$ is the matrix in (15.9). The regular solution is integrable near $r\to 0$, while the irregular solution is integrable at $r\to\infty$. The Green's function is then given by [Wi 56]

$$G_k(r,r';\omega)\,\gamma_0 = \frac{1}{W(\omega)}\,[\theta(r'-r)\,\psi_\mathrm{R}(r)\,\psi^+_\mathrm{I}(r') + \theta(r-r')\,\psi_\mathrm{I}(r)\,\psi^+_\mathrm{R}(r')]\,, \tag{15.42}$$

where the $W(\omega)$ is the (r-independent!) Wronskian given by

$$W(\omega) = r^2[\psi_{R2}(r)\,\psi_{I1}^\dagger(r) - \psi_{R1}(r)\,\psi_{I2}^\dagger(r)]\,. \tag{15.43}$$

(Here index 1 or 2 refers to the upper and lower components, respectively, of the four-spinor ψ.) It is easy to verify that (15.42) satisfies (15.9).

We need consider only the trace of $G_\varkappa$, (15.41), which takes the form

$$\mathrm{tr}\,[G_\varkappa(r,r;\omega)\,\gamma_0] = \psi_I^+(r)\,\psi_R(r)/W(\omega)\,. \tag{15.44}$$

An important example [Wi 56] of the above was obtained for a pure Coulomb potential $V(r) = -Z\alpha/r$. The solutions are $\psi_R(r) = M(r)$ and $\psi_I(r) = W(r)$, where M involves linear combinations of regular Whittacker functions $M_{\nu\pm\frac{1}{2},\gamma}(2pr)$ and W involves linear combinations of the irregular Whittacker functions $W_{\nu\pm\frac{1}{2},\gamma}(2pr)$.

The parameters on which M and W depend are $\gamma = \sqrt{\varkappa^2-(Z\alpha)^2}$, $p = \sqrt{m_e^2-\omega^2}$, and $\nu = Z\alpha\omega/c$. Whittacker functions are related to confluent hypergeometric functions. The most important parameter for strong fields $(Z\alpha\sim 1)$ is $\gamma = \sqrt{1-(Z\alpha)^2}$ for $\varkappa = \pm 1$, $j = 1/2$, states. At $Z\alpha = 1$, γ and consequently M have a branch point as a function of $Z\alpha$ for $j = 1/2$. Although it seems that W would also have a branch point at $Z\alpha = 1$, $W_{\alpha,\gamma}$ is an even function of γ and therefore non-singular at $Z\alpha = 1$. The non-analytic behaviour of the M function causes a singularity of $\mathrm{tr}\,G_{\varkappa=\pm 1}$ for this point nuclear charge case at $Z\alpha = 1$. Higher angular momentum states $(j\geq 3/2)$ are, on the other hand, well behaved near $Z\alpha = 1$. Therefore to extend $Z\alpha$ beyond 1, it is, for practical purposes, sufficient to include the effect of finite nuclear size in the above considerations of the $\varkappa = \pm 1$ partial waves only.

To learn about the role of finite nuclear size, a simple but semirealistic model may be considered. The simplest finite-size nuclear model is a surface distribution, for which $V(r)\sim -1/r$ for $r > R$, R being the nuclear radius. For $r < R$, the functions ψ_R, ψ_I, called $j(r)$ and $h(r)$, are simply related to spherical Bessel functions. For $r > R$, the Coulomb M and W solutions apply. Continuity at $r = R$ determines the particular linear combination of M and W that joins the interior $j(r)$ solution giving the regular solution. Continuity at $r = R$ also determines the linear combination of the j and h solutions that joins the exterior W function giving the irregular solution. With such solutions for a finite-size external charge, simply

$$\mathrm{tr}\,G_\varkappa(r,r;\omega) = \begin{cases} \mathrm{tr}\,G_\varkappa^0\left(r,r;\omega+\dfrac{Z\alpha}{R}\right) + \mathrm{tr}\,\Delta G_\varkappa^<\,, & r < R \\[2ex] \mathrm{tr}\,G_\varkappa^{\mathrm{coul}}(r,r;\omega) + \mathrm{tr}\,\Delta G_\varkappa^>\,, & r > R \end{cases}\,, \tag{15.45}$$

where $G_\varkappa^0$ is the free radial Green's function, $G_\varkappa^{\mathrm{coul}}$ is the point nucleus Coulomb Green's function, and $\Delta G_\varkappa^{<(>)}$ are finite-size correction functions. Explicit formulas are given by *Gyulassy* [Gy 75].

We first discuss the vacuum polarization charge density of order $n > 2$, evaluated in the limit of a point-like nucleus. As *Wichman* and *Kroll* have shown, [Wi 56], the non-linear vacuum polarization density assumes the form

$$e\varrho^{n>2} = \varphi^{n>2}\delta(r)/4\pi r^2 + \tilde{\varrho}(r)\,, \tag{15.46}$$

where $\varphi^{n>2}$ is the magnitude

$$\varphi^{n>2} = -e^2\{(Z\alpha)^3(0.021) + (Z\alpha)^5(0.007)F[(Z\alpha)^2]\} \tag{15.47}$$

of the point screening charge, where $F \sim 1$ except near $Z\alpha = 1$ where $d\varphi^{n>2}/dZ \to -\infty$ and in the limit $Z\alpha \to 1$ $\varphi^{n>2} \to \sim e^2/20$. Of course the compensating density $\tilde{\varrho}(r)$ spread out to $r \sim \lambda_e = \hbar c/m_e$ assures that no net charge is contained in (15.46).

An important observation is that 90% of the screening charge is due to the $j = 1/2$ contribution in (15.41) and for orders $n > 5$ this fraction is 99%. Thus the essence of the situation is described by studying the behaviour of the $j = 1/2$ term. In particular, for $Z\alpha \to 1$, the $1/r$ potential eigenenergies of $j = 1/2$ states of the Dirac equation become zero. This causes a sudden change of the properties of Green's function – actually for $Z\alpha \to 1$, as often stated, there is no proper theoretical framework, since the Hamiltonian is not self-adjoint. The proper resolution of this problem, the finite nuclear size, can be well studied in view of (15.45) and the above remarks, stressing the importance of few partial waves only. In particular, finite-size effects need to be included only in the $j = 1/2$ term, while use of the point charge form for $\varrho_j^{n>2}$ for $j > 3/2$ remains justified.

To evaluate $\varrho_{j=1/2}^{n>2}$ for large Z, the bound state wave functions with energies $E_{n1/2}$ between $-m_e < E_{n1/2} < 0$, see first term in (15.41), must be determined. As it happens, the $1s_{1/2}$ state is the only state in this interval for $146 < Z < 172$. Practically simultaneously with the $1s_{1/2}$ disappearing into the second ω sheet at $Z = 172$, the $2p_{1/2}$ state descends below $E = 0$. Thus practically at all times, there is only one member in the discrete sum of (15.41), except perhaps for $170 < Z < 172$, when there may be two bound states.

We wish to consider now the continuity of (15.41) as Z is increased from about 137 to supercritical values. The first discontinuity is expected at around $Z_0 \sim 146$ when the $1s_{1/2}$ state descends below $E = 0$ and must therefore be counted in the discrete sum of (15.41). However, since for $Z \lesssim Z_0$ the path

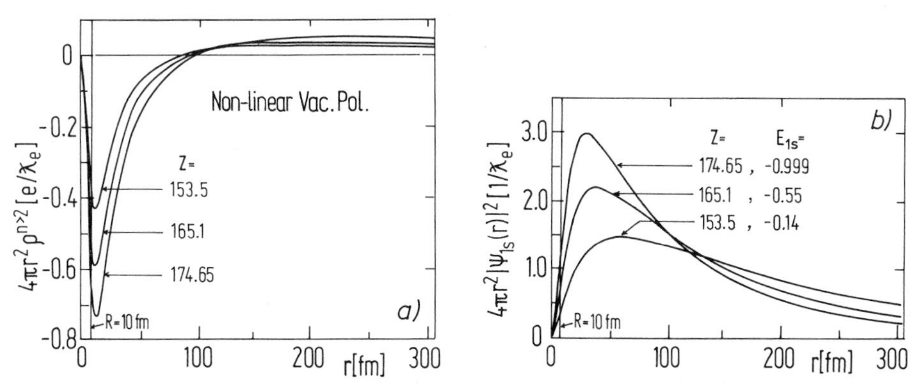

Fig. 15.9 a, b. Comparison for several values of Z of the charge densities of (**a**) non-linear vacuum polarization with (**b**) radial charge density of $1s$ states at $Z_0 < Z \lesssim Z_{cr}$ [Gy 75]

integral along the $\mathrm{Im}\{\omega\}$ axis feels an approaching pole in advance, there is no major change of G. However, a different situation appears when $Z = Z_{cr}$ is exceeded. There, the pole disappears onto another sheet without being near to the path of integration, hence the charge density (15.41) experiences a sudden change corresponding to the charge density of the disappearing s state. It is important to appreciate that for $Z < Z_{cr}$ the second term in (15.41) contains the compensating charge density to the first term, because these terms arise from a deformation of the path C_ϱ, Fig. 15.7, which does not have any sensitivity to the $1s$ energy descending below $E = 0$.

The approach to criticality of the non-linear vacuum polarization charge density is compared in Fig. 15.9 to the form of the $1s_{1/2}$ wave functions at different values of $Z > Z_0$. The results confirm the substantial cancellation of different terms in (15.41).

15.4 Real and Virtual Vacuum Polarization

When Z exceeds Z_{cr}, we expect, according to the above discussion, that the vacuum polarization charge density acquires a substantial contribution, that is the negative of the first term in (15.41) (of the charge density of the critical state), which has disappeared from the physical ω sheet. Of course, what happens now is that we cannot prepare a bare nucleus any more, as it dresses itself spontaneously with an electronic charge cloud. It is worth mentioning that had we in the first place taken a filled $1s$ electron shell, then another integration path C_q (Fig. 15.10) would be required to describe the vacuum polarization charge.

The additional pole, counted now as a "positron" state, leads to a negative *real* vacuum polarization charge density, usually called (for $Z < Z_{cr}$) the K-shell electron charge density. We can also choose to call this term "real vacuum polarization". This (for $Z < Z_{cr}$) quite normal situation becomes a necessary feature of the theory when the state concerned and its pole wander off the first ω sheet. Then there is no option of "choosing" the charged vacuum, but, as described at length, it must be accepted as the true ground state of the theory. Hence, what previously could also be called charge density of the bound electron must now be referred to as *real* vacuum polarization since its origin is not a fluctuating pair or a bound electron state, but the structured vacuum of supercritical QED. The remainder of the vacuum polarization charge density is hence termed "virtual".

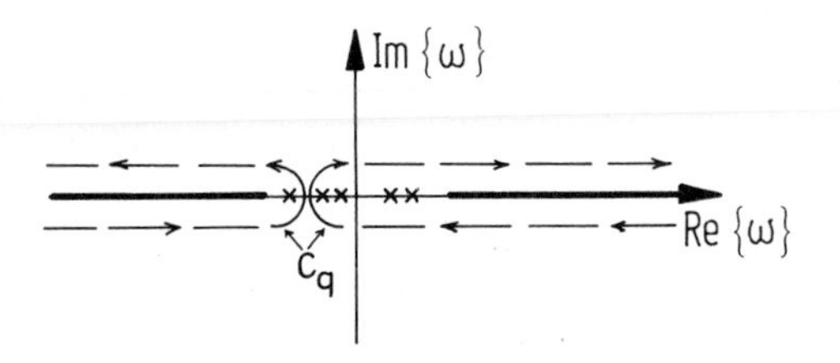

Fig. 15.10. Integration path C_q of the charged vacuum at $Z < Z_{cr}$

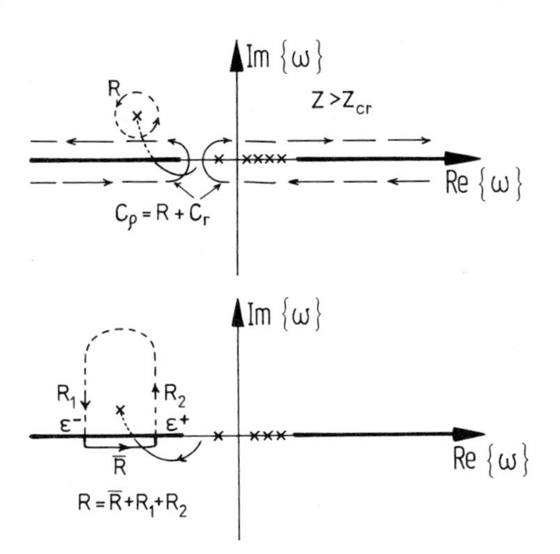

Fig. 15.11. Integration paths for real (R) and virtual (C_r) vacuum polarization

Fig. 15.12. Deformed path R for the real vacuum polarization

How can the division into "real" and "virtual" parts can be made uniquely? The simplest way to identify the *real* vacuum polarization charge density is to perform the integral around the pole on the second sheet

$$\langle j_0 \rangle^R \equiv -\frac{e}{2} \int_R \frac{d\omega}{2\pi \mathrm{i}} \operatorname{tr} G(\boldsymbol{x}, \boldsymbol{x}; \omega)\, \gamma_0 \,, \tag{15.48}$$

where the path R surrounds the singularity on the second sheet, Fig. 15.11. Together with the path C_r it forms the (deformed) path C_ϱ.

Let us now focus on the path R only. It can easily be deformed to an integral along the real axis for certain interval $\omega \in (\varepsilon^+, \varepsilon^-)$ and then the path is extended to $\omega = \mathrm{i}\infty$ on the second ω sheet, as shown in in Fig. 15.12. The $\varepsilon^\pm$ are chosen so that the paths $R_{1,2}$ to $\mathrm{i}\infty$ are relatively unaffected by the presence of the pole. But to avoid computing the contributions of R_1, R_2 at all, one may simply consider the difference of Green's functions:

$$\langle j_0 \rangle^R \approx -e \int_{\bar{R}} \frac{d\omega}{2\pi \mathrm{i}} \operatorname{tr} [G(\boldsymbol{x}, \boldsymbol{x}; \omega, V)\, \gamma_0 - G(\boldsymbol{x}, \boldsymbol{x}; \omega, V - \delta V)\, \gamma_0] \,, \tag{15.49}$$

where δV is chosen so conveniently that (i) there is no pole in the second term within the range of integration, and (ii) to a good approximation, the integrals over the paths R_1, R_2 cancel and only the contribution $\bar{R}$ along the $\mathrm{Re}\{\omega\}$ in the interval $(\varepsilon^-, \varepsilon^+)$ remains:

$$\langle j_0 \rangle^R \approx +e \int_{\varepsilon^-}^{\varepsilon^+} d\omega [\psi_\omega^+(\boldsymbol{x}, V)\, \psi_\omega(\boldsymbol{x}, V) - \psi_\omega^+(x, V - \delta V)\, \psi_\omega(\boldsymbol{x}, V - \delta V)] \,. \tag{15.50}$$

Here one can view the second term as effecting the subtraction of the effects of virtual vacuum polarization. The only currently available real vacuum polarization calculation uses (15.50). Its result was already discussed in Chap. 6. Fig. 6.14, on p. 164, shows the densities of the $1s_{1/2}$ charge for $Z = 184$ and $2p_{1/2}$ charge for $Z = 198$ with $2s_{1/2}$ contribution for $Z = 255$. The very strong localization of the electrons near the nuclear surface is noticeable, and also that the localization is more pronounced for the real vacuum polarization charge than for the $1s_{1/2}$ electron at $Z = 172$.

At this point, it is most useful to reconsider the perturbative approach of Chap. 6, where we expanded the supercritical states in undercritical basis. The supercritical continuum states are given by [cf. (6.10)]

$$\psi_\varepsilon = a(\varepsilon)\,\psi_{1s}^{\mathrm{cr}}(x) + \int\limits_{-\infty}^{-m_e c^2} h_E(\varepsilon)\,\psi_E^{\mathrm{cr}}(x)\,dE\,. \tag{15.51}$$

The Greens function contains the term associated with the former bound state:

$$G(x,x';\omega) = \int\limits_{-\infty}^{+\infty} d\varepsilon \frac{|a(\varepsilon)|^2}{\varepsilon-\omega+\mathrm{i}\eta}\,\psi_{1s}^{\mathrm{cr}}(x)\,\bar{\psi}_{1s}^{\mathrm{cr}}(x') + \dots\,, \tag{15.52}$$

where η is negative when $\omega < -m_e$ and positive when $\omega > m_e$, as required by the choice of contour C_F in the ω plane. From (15.52) it is apparent that $a(\varepsilon)$ carries the singularity associated with the resonance. This pole, however, occurs on the second sheet and the only contribution to the integral of (15.52) arises from the pole at $\varepsilon = \omega - \mathrm{i}\eta$ (provided that $\omega < -m_e$). Thus the result of the integration is

$$G(x,x';\omega) \sim \mathrm{i}\frac{\Gamma\theta(-m_e-\omega)}{(\omega-\varepsilon_r)^2+\frac{1}{4}\Gamma^2}\,\psi_0^{\mathrm{cr}}(x)\,\bar{\psi}_0^{\mathrm{cr}}(x')\,, \tag{15.53}$$

where the resonance has been treated approximately as discussed in Chap. 6, (6.34). A very different result would have been obtained had we chosen to enclose the supercritical resonance in the integration path, i.e. including the path R in the integral. Then the pole at $\varepsilon = \varepsilon_r + \mathrm{i}\Gamma/2$ contributes as

$$G_{C'}(x,x';\omega) = \frac{\psi_0^{\mathrm{cr}}(x)\,\bar{\psi}_0^{\mathrm{cr}}(x)}{\omega-\varepsilon_r-\mathrm{i}\Gamma/2}\,, \tag{15.54}$$

which is characteristic of a complex eigenvalue, a reflection of the lack of stability of the state of reference chosen, i.e. a neutral vacuum for $Z > Z_{\mathrm{cr}}$.

Equation (15.53) can be inserted into (15.48), revealing that the charge density of the critical state is an approximation to the space density of the real vacuum polarization. This is correct for moderate supercritical charges $Z - Z_{\mathrm{cr}} > 0$.

Bibliographical Notes

Further details may be found only in the original literature, especially in [Wi 56, Ra 74, Gy 75].

16. Many-Body Effects in QED of Strong Fields

Since this book emphasizes effects related to supercritical fields, there is not enough space for a comprehensive treatment of the different many-body effects in QED, including the proper treatment of interacting quantum fields. Fortunately, most of the general formalism developed for interacting quantized fields remains valid even when supercritical fields are encountered. Thus this chapter treats only the most important extensions of the theory specific to supercritical fields.

16.1 Self-Consistent Hartree-Fock Equations

This section addresses briefly how to formulate the self-consistent equations which will, aside of the screening effects of the real electrons, include effects of the vacuum polarization as well. As previously shown, both effects relate to the same physical phenomenon and it is the choice of the Fermi surface which separates these effects. The Hartree-Fock method is applied, which provides the generalization we are looking for. Rather than attempt the derivation, we propose and justify the following linearized equations for the electron field operator $\hat{\psi}$

$$\begin{aligned}(\gamma\cdot p-e\gamma\cdot A^{\mathrm{ex}}-m_0)\,\hat{\psi}(x)-e^2\!\int d^4y\,\gamma^\mu\langle\hat{j}^\nu(y)\rangle D^{\mathrm{F}}_{\mu\nu}(\boldsymbol{y}-\boldsymbol{x};t_y-t_x)\,\hat{\psi}(\boldsymbol{x},t_x)\\ +\mathrm{i}e^2\!\int d^4y\,\gamma^\mu S_{\mathrm{F}}(\boldsymbol{x},\boldsymbol{y};t_x-t_y)\,\gamma^\nu\hat{\psi}(\boldsymbol{y},t_y)\,D^{\mathrm{F}}_{\mu\nu}(\boldsymbol{y}-\boldsymbol{x};t_y-t_x)=0\,. \qquad (16.1)\end{aligned}$$

Here the first term is the Dirac equation in the external potential A^{ex}, the second term describes the induced potential generated by the induced charge. The propagator of the Maxwell equations (photon) is $D^{\mathrm{F}}_{\mu\nu}$, taken again with the Feynman choice of boundary conditions. The last term corresponds to the radiative (exchange) correction arising when the two-particle interactions are approximated by a product of an expectation value with an operator. Typically,

$$\hat{\psi}(y)\hat{\bar{\psi}}(y)\,\hat{\psi}(x)\sim\langle\hat{\psi}(y)\hat{\bar{\psi}}(y)\rangle\,\hat{\psi}(x)+\hat{\psi}(y)\,\langle\hat{\bar{\psi}}(y)\,\hat{\psi}(x)\rangle+\dots\,, \qquad (16.2)$$

which explains the appearance of the two different terms in (16.1). Equation (16.1) can be formally derived from the Dyson-Schwinger integral equations. To our knowledge, the first investigation of Hartree-Fock equations in QED following the Dyson-Schwinger equations for the electron propagator was by *Pratt* [Pr 63], though he did not obtain explicit expressions for the selfconsistent states.

An explicit equation has subsequently been found by *Reinhard* et al. [Re 70, 71]. A similar equation was later independently derived by *Gomberoff* and *Tolmachev* [Go 71]. *Reinhard* approximated the Dyson-Schwinger equations for the coupled fermion-photon propagators, neglecting higher-order contributions to the vertex function and the photon propagator. Thus (16.1) excludes these effects and contains only a certain subset of all Feynman diagrams, but its importance resides in that it reduces, upon neglecting vacuum polarization and fluctuation contributions, to the usual Hartree-Fock description of many-electron systems. We therefore now proceed to find the single-particle equation associated with the linearized quantum field, (16.1).

It is convenient to introduce the (self-consistent) amplitudes

$$\hat{\psi}(\boldsymbol{x},0) = \sum_{E_p>E_\mathrm{F}} \phi_p(\boldsymbol{x})\hat{b}_p + \sum_{E_q<E_\mathrm{F}} \phi_q(\boldsymbol{x})\hat{d}_q^+ \,. \tag{16.3}$$

We now require a diagonal form of the normal-ordered Hamiltonian

$$:H: = \sum_{E_p>E_\mathrm{F}} E_p\hat{b}_p^+\hat{b}_p - \sum_{E_q<E_\mathrm{F}} E_q\hat{d}_q^+\hat{d}_q \,, \tag{16.4}$$

which will force a self-consistent set of equations for ϕ_q and ϕ_p. The time dependence of the Heisenberg picture operator $\hat{\psi}(x,t)$ is consequently the usual one

$$\begin{aligned}\hat{\psi}(\boldsymbol{x},t) &= \mathrm{e}^{-\mathrm{i}\hat{H}t}\hat{\psi}(\boldsymbol{x},0)\,\mathrm{e}^{\mathrm{i}\hat{H}t}\\ &= \sum_{E_p>E_\mathrm{F}} \phi_p(\boldsymbol{x})\mathrm{e}^{-\mathrm{i}E_pt}\hat{b}_p + \sum_{E_q<E_\mathrm{F}} \phi_q(\boldsymbol{x})\mathrm{e}^{-\mathrm{i}E_qt}\hat{d}_q^+ \,.\end{aligned} \tag{16.5}$$

To determine a self-consistent equation for $\phi_p(\boldsymbol{x})$ consider the matrix elements of (16.1) between the single-particle states

$$\langle\Omega|\ldots\hat{b}_p^+|\Omega\rangle$$
$$\langle\Omega|\hat{d}_q\ldots|\Omega\rangle \,.$$

Consulting (16.5) (taking the set q, but equally valid for the set p), this yields

$$\begin{aligned}&(\gamma_0E_q-\boldsymbol{\gamma}\cdot\boldsymbol{p}-e\gamma A^{\mathrm{ex}}-m_0)\,\phi_q(\boldsymbol{x})-e^2\int d^3y\,\gamma^\mu\langle j^\nu(\boldsymbol{y})\rangle\int dt_yD_{\mu\nu}^\mathrm{F}(\boldsymbol{y}-\boldsymbol{x};t_y-t_x)\,\phi_q(\boldsymbol{x})\\ &\quad+\mathrm{i}e^2\int d^4y\,\gamma^\mu S_\mathrm{F}(\boldsymbol{x},\boldsymbol{y};t_x-t_y)\,\gamma^\nu\phi_q(\boldsymbol{y})\exp[-\mathrm{i}E_q(t_y-t_x)]\,D_{\mu\nu}^\mathrm{F}(\boldsymbol{y}-\boldsymbol{x};t_y-t_x)=0\,.\end{aligned} \tag{16.6}$$

The time dependence of the direct term involving the vacuum polarization charge is easily removed by noting that

$$\begin{aligned}\int dt_yD_{\mu\nu}^\mathrm{F}(\boldsymbol{y}-\boldsymbol{x};t_y-t_x) &= \int d(t_y-t_x)\,D_{\mu\nu}^\mathrm{F}(\boldsymbol{y}-\boldsymbol{x};t_y-t_x)\\ &= \int d(t_y-t_x)\exp[\mathrm{i}k_0(t_y-t_x)]\,D_{\mu\nu}^\mathrm{F}(\boldsymbol{y}-\boldsymbol{x};t_y-t_x)\big|_{k_0=0}\\ &= \tilde{D}_{\mu\nu}^\mathrm{F}(\boldsymbol{y}-\boldsymbol{x};k_0=0) = g_{\mu\nu}\left(-\frac{1}{4\pi r}\right),\end{aligned} \tag{16.7}$$

which in the Feynman gauge ($D_{\mu\nu} \sim g_{\mu\nu}$) simply becomes the Coulomb potential, as indicated, and where $r = |\boldsymbol{x} - \boldsymbol{y}|$.

The time dependence of the exchange term, the last term of (16.6), is more difficult to treat. Recall that the Fermion propagator can be concisely written as

$$\mathrm{i} S_{\mathrm{F}}(\boldsymbol{x}, \boldsymbol{y}; \tau) = \sum_{E_n} \theta[\tau \operatorname{sgn}(E_n - E_{\mathrm{F}})] \operatorname{sgn}(E_n - E_{\mathrm{F}}) \phi_n(\boldsymbol{x}) \bar{\phi}_n(\boldsymbol{y}) \mathrm{e}^{-\mathrm{i} E_n \tau}, \tag{16.8}$$

where $\operatorname{sgn}(x) = \theta(x) - \theta(-x)$. We further need

$$D^{\mathrm{F}}_{\mu\nu}(x) = g_{\mu\nu} \frac{\mathrm{i}}{4\pi^2} \frac{1}{x^2 - \mathrm{i}\varepsilon}, \tag{16.9}$$

again implying the Feynman gauge, which is the most suitable gauge whenever virtual effects dominate. (A transverse gauge may be of considerable advantage when regular screening effects are considered.) It is worth recording here that while QED is a gauge invariant theory, Hartree-Fock truncated theory may not be and, in particular, subsequent order by order expansion in the coupling constant e^2 is not a gauge-invariant approximation. Hence the choice of an appropriate gauge may be an important step towards obtaining the best approximation.

Taking (16.5, 8) into account, the integral appearing in the last term of (16.6) has the form

$$\begin{aligned} I &= \int_{-\infty}^{+\infty} \operatorname{sgn}(E_n - E_{\mathrm{F}}) \frac{\exp[-\mathrm{i}\tau(E_n - E_q)]}{\tau^2 - r^2 - \mathrm{i}\varepsilon} \theta(\operatorname{sgn}(E_n - E_{\mathrm{F}})\tau) \, d\tau \\ &= \operatorname{sgn}(E_n - E_{\mathrm{F}}) \int_0^{+\infty} \frac{\exp[-\mathrm{i}\tau(E_n - E_q) \operatorname{sgn}(E_n - E_{\mathrm{F}})]}{\tau^2 - r^2 - \mathrm{i}\varepsilon} \, d\tau \, . \end{aligned} \tag{16.10}$$

Figure 16.1 shows the deformation of the integration path in the complex τ plane, required to obtain the result

$$\begin{aligned} I = {} & \operatorname{sgn}(E_n - E_{\mathrm{F}}) \, \theta((E_n - E_{\mathrm{F}})(E_n - E_q)) \cdot \int_0^{-\mathrm{i}\infty} \frac{\exp[-\mathrm{i}\tau(E_n - E_q) \operatorname{sgn}(E_n - E_{\mathrm{F}})]}{\tau^2 - r^2} \, d\tau \\ & + \operatorname{sgn}(E_n - E_{\mathrm{F}}) \, \theta(-(E_n - E_{\mathrm{F}})(E_n - E_q)) \\ & \cdot \left\{ \int_0^{\mathrm{i}\infty} \frac{\exp[-\mathrm{i}\tau(E_n - E_q) \operatorname{sgn}(E_n - E_{\mathrm{F}})]}{\tau^2 - r^2} \, d\tau \right. \\ & \left. + \frac{\pi \mathrm{i}}{r} \exp[-\mathrm{i} r(E_n - E_q) \operatorname{sgn}(E_n - E_{\mathrm{F}})] \right\} . \end{aligned} \tag{16.11}$$

With the help of

$$\int_0^{\infty} dx \frac{\exp(-x|z|)}{x^2 + r^2} = \frac{1}{r} \{ \mathrm{Ci}(|zr|) \sin(|zr|) - [\mathrm{Si}(|zr|) - \pi/2] \cos(|zr|) \} , \tag{16.12}$$

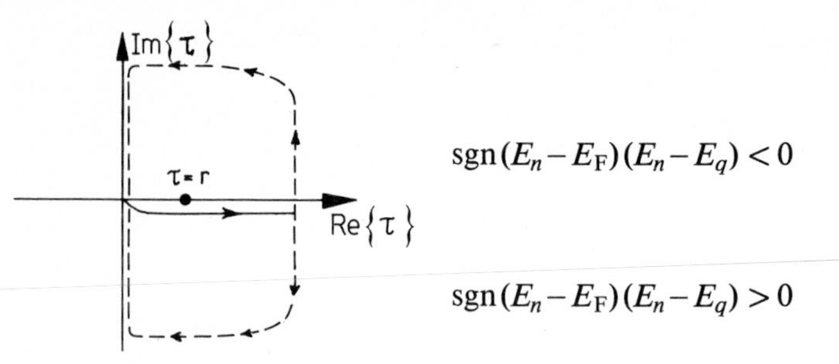

Fig. 16.1. Integration path of (16.10)

where Ci(z), Si(z) are the usual cosine and sine integrals as defined in [Ab 65], we find

$$
\begin{aligned}
I = & \frac{\mathrm{i}}{r}\{\mathrm{Ci}[r(E_n-E_q)]\sin[r(E_n-E_q)] - \mathrm{Si}[r(E_n-E_q)]\cos[r(E_n-E_q)]\} \\
& + \mathrm{sgn}(E_n-E_\mathrm{F})\frac{\mathrm{i}\pi}{2r}\cos[r(E_n-E_q)] \\
& + \theta((E_\mathrm{F}-E_n)(E_n-E_q))\frac{\pi}{r}\sin[r(E_n-E_q)] \, .
\end{aligned} \tag{16.13}
$$

It is convenient now to introduce the "tilde" sum defined by

$$
\tilde{\sum_n} = \sum_{E_n<\mathrm{F}} - \sum_{E_n>\mathrm{F}} = \sum_{E_n} \mathrm{sgn}(E_\mathrm{F}-E_n) \, , \tag{16.14}
$$

which results naturally from the integral C_ϱ enclosing the singularities of Green's function in the ω plane, Fig. 15.7. Then, for example,

$$
\langle j^\nu(y)\rangle = \frac{e}{2}\tilde{\sum_p}\bar{\phi}_p(y)\gamma^\nu\phi_p(y) \, , \tag{16.15}
$$

and we obtain for (16.6), using (16.13), but simplifying the notation using (16.14),

$$
\begin{aligned}
& (\gamma_0 E_q - \boldsymbol{\gamma}\cdot\boldsymbol{p} - e\boldsymbol{\gamma}\cdot\boldsymbol{A}^{\mathrm{ex}} - m)\phi_q(\boldsymbol{x}) - \frac{e^2}{8\pi}\gamma_\mu\int d^3y \frac{\tilde{\sum_n}\bar{\phi}_n(y)\gamma^\mu\phi_n(y)}{|\boldsymbol{x}-\boldsymbol{y}|}\phi_q(\boldsymbol{x}) \\
& + \frac{e^2}{8\pi}\gamma_\mu\int d^3y\tilde{\sum_n}\bar{\phi}_n(y)\gamma^\mu\phi_q(y)\frac{\cos[|x-y|(E_n-E_q)]}{|\boldsymbol{x}-\boldsymbol{y}|}\phi_n(\boldsymbol{x}) \\
& + \frac{e^2}{8\pi}\gamma_\mu\int d^3y\sum_n\bar{\phi}_n(y)\gamma^\mu\phi_q(y)V_{nq}(|\boldsymbol{x}-\boldsymbol{y}|)\phi_n(\boldsymbol{x}) = 0 \, ,
\end{aligned} \tag{16.16}
$$

where

$$
\begin{aligned}
V_{nq}(r) = & -\frac{2}{r\pi}\{\sin[r(E_n-E_q)]\,\mathrm{Ci}[r(E_n-E_q)] - \cos[r(E_n-E_q)]\,\mathrm{Si}[r(E_n-E_q)]\} \\
& + \frac{\mathrm{i}2}{r}\sin[r(E_n-E_q)]\cdot\theta[(E_\mathrm{F}-E_n)(E_n-E_q)] \, .
\end{aligned} \tag{16.17}
$$

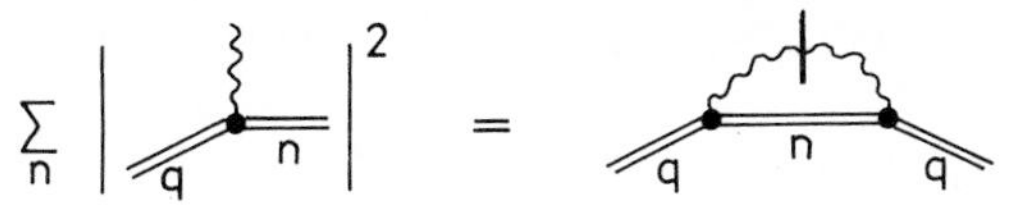

Fig. 16.2. The square of a radiative decay matrix element of a hole q is representable as a self-energy Feynman diagram with a mass-shell photon, denoted here by the cut on the photon line

The last term in (16.17) is imaginary and gives rise to a radiative width of unstable states, as obtained usually by perturbation theory. In our approach, this term emerges naturally and is required for consistency. This last remark can be illustrated by computing the square of the radiation term. Graphically, this is done in Fig. 16.2, where it is related to the self-energy diagram with the mass-shell photon. Thus when $E_q < E_n < E_F$, i.e. when there is a hole below the Fermi surface, the self-consistent equations allow any electron above it to fall into this hole, and hence they properly describe the total radiative width.

The first term in (16.17), the real part of V_{nq}, is a small (20%) correction to the *Breit retardation function*, the second last term in (16.16). It is independent of the choice of the Fermi surface and consequently can be entirely associated with the effect of the vacuum fluctuations. These can be more fully identified in (16.16) by realizing that a change of the Fermi energy implies

$$\tilde{\sum_{p|E_F}} = 2 \sum_{E_F > E_p > E_F'} + \tilde{\sum_{p|E_F'}} \tag{16.18}$$

which follows directly from the definition (16.14). Here the subscripts E_F and E_F' indicate the choice of the Fermi surface in the tilde sum. Taking $E_F = +m_0$ (neutral atom) and $E_F' = -m_0$ (fully ionized atom) separates terms in (16.16) so that the *real* particle effects are neatly separated from the virtual effects (the index bs denotes the occupied bound states):

$$\begin{aligned}
&(\gamma_0 E_q - \boldsymbol{\gamma}\cdot\boldsymbol{p} - e\boldsymbol{\gamma}\cdot\boldsymbol{A}^{\mathrm{ex}} - m)\,\phi_q(\boldsymbol{x}) - \frac{e^2}{4\pi}\gamma_\mu \int d^3y \frac{\sum\limits_{bs} \bar{\phi}_{bs}(y)\,\gamma^\mu \phi_{bs}(y)}{|\boldsymbol{x}-\boldsymbol{y}|}\,\phi_q(\boldsymbol{x})\\
&\quad + \frac{e^2}{4\pi}\gamma_\mu \int d^3y \sum_{bs} \bar{\phi}_{bs}(y)\,\gamma^\mu \phi_q(y) \frac{\cos[\,|\boldsymbol{x}-\boldsymbol{y}|(E_{bs}-E_q)]}{|\boldsymbol{x}-\boldsymbol{y}|}\,\phi_{bs}(\boldsymbol{x})\\
&\quad - \frac{e^2}{8\pi}\gamma_\mu \int d^3y \frac{\tilde{\sum}{}'\,\bar{\phi}_n(y)\,\gamma^\mu \phi_n(y)}{|\boldsymbol{x}-\boldsymbol{y}|}\,\phi_q(\boldsymbol{x})\\
&\quad + \frac{e^2}{8\pi}\gamma_\mu \int d^3y \frac{\tilde{\sum}{}'\,\bar{\phi}_n(y)\,\gamma^\mu \phi_q(y)}{|\boldsymbol{x}-\boldsymbol{y}|}\cos[\,|\boldsymbol{x}-\boldsymbol{y}|(E_n-E_q)]\,\phi_n(\boldsymbol{x})\\
&\quad + \frac{e^2}{8\pi}\gamma_\mu \int d^3y \sum_n \bar{\phi}_n(y)\,\gamma^\mu \phi_q(y)\, V_{nq}(|\boldsymbol{x}-\boldsymbol{y}|)\,\phi_n(\boldsymbol{x}) = 0\,,
\end{aligned} \tag{16.19}$$

where the last three terms are in the limit of weak fields reducing to the vacuum polarization and the so-called self-energy corrections, Sect. 16.2, while the first three terms are simply the Hartree-Fock equations of bound electrons in an atom.

We further recognize the vacuum polarization potential of a bare nucleus, in which the polarization charge density is now written as a sum $\tilde{\sum}'$ over all eigenstates of the self-consistent equation. This term is not discussed further here, as it has been described at length in Chaps. 14, 15. However, recall that the Hartree term for electron – electron interactions is just a part of the polarization of the vacuum, and in the limit of strong fields any conceivable difference between these two terms disappears, with real vacuum polarization having the same magnitude as the normal Hartree term. Similar remarks apply to the Fock term which contains the effect of the retardation. Further consideration of this term alone, without vacuum fluctuations, can be found in [Ma 71].

16.2 Self-Energy Effects in Atoms

The Fock term is intimately connected to the self-energy fluctuations, represented in (16.19) by the last two terms. As is well known, these terms lead to a noticeable splitting in atomic $2p_{1/2}-2s_{1/2}$ states called the Lamb shift. Of course, vacuum polarization also contributes to this splitting, as described previously. However, since the electrons in light atoms move far away from the range of the potential, vacuum polarization remains inferior to the self-energy effects up to quite large Z. These are associated with the electron itself and can therefore be larger than the short-range vacuum polarization.

Diagrammatically, this term is represented as the following diagram:

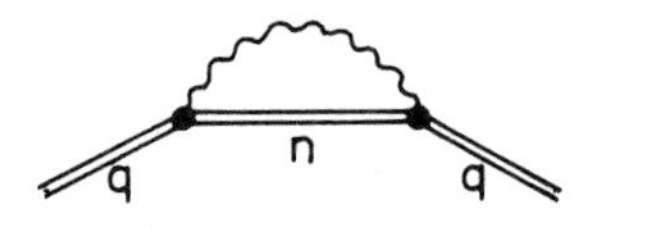

We recognize here that the electron (hole) q emits a photon and makes a virtual transition to an arbitrary state n from where it returns to q. As is evident, this diagram first appears in second-order perturbation theory, and some important contributions to this diagram arise when the transitions involve neighbouring $q-n$ states, that is nearly degenerate states as compared to the rest mass. For light atoms, this makes the evaluation of the self-energy correction difficult: then the energy separation of different states is $\sim m_e\alpha^2 \ll m_e$. However, for high Z the situation simplifies as the separation of the critical state from the rest of the spectrum is nearly m_e. Concentrating on q, being the lowest bound s state for $Z \lesssim Z_{cr}$, we can approximate the necessary sum over n by just taking the contribution of the very isolated single state: $n=q$. Then in (16.19), the two last terms simply reduce to a single term which, in lowest order, leads to an energy shift:

$$\Delta E_{SE} \underset{Z \lesssim Z_{cr}}{\approx} +\frac{e^2}{8\pi}\int d^3x\, d^3y \left[\bar{\Phi}_q(y)\,\gamma_\mu\,\Phi_q(y)\frac{1}{|\boldsymbol{x}-\boldsymbol{y}|}\bar{\Phi}_q(x)\,\gamma^\mu\,\Phi_q(x)\right]. \tag{16.20}$$

This term is dominated by the Coulomb interaction, i.e. $\mu=0$, and so its magnitude can be easily determined and is never more than the effect of the change of the nuclear charge by one unit.

However, perhaps more important is the observation that the same term appears with opposite sign in the vacuum polarization term, the third term in (16.19). These two terms, dominating the respective expressions for self-energy and vacuum polarization, cancel each other exactly. Therefore for $Z \to Z_{\mathrm{cr}}$, we expect vacuum polarization and the self-energy effect on the critical state nearly to cancel each other, and in any event to remain a small, controllable contribution to the energy of the $1s$ state, on the balance. For example, calculations by *Soff* et al. [So 82b], discussed in more detail in Sect. 16.3, indicate that at the critical point the individual shifts are ca. 11 keV, while the remainder of both radiative effects then is ca. 300 eV and for all practical purposes can be ignored in view of the 1 022 000 eV binding. Thus one concludes that it is tolerable to ignore the radiative corrections when the critical states are investigated. For further study of high Z self-energy terms, consult the work of *Desiderio* and *Johnson* [De 71], *Mohr* [Mo 74], *Soff* et al. [So 82b] and Sect. 16.3.

However, the self-energy term deserves, in principle, similar attention as the vacuum polarization effects, and to ensure that the physical mass of an electron coincides with the expected values, mass renormalization is required. We shall only sketch here the relevant points, as the essence of the development is similar to the previous extensive study of vacuum polarization.

We begin by calculating the free field electron self-energy. To this end $\Sigma(x,y)$ denotes the combination of electron and photon propagator:

$$-\mathrm{i}\,\Sigma(x,y) = $$

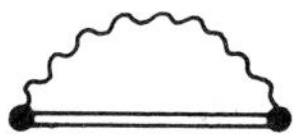

$$= (-\mathrm{i}e)^2 \gamma_\mu S_{\mathrm{F}}(x,y)\,\gamma_\nu D_{\mathrm{F}}^{\mu\nu}(y-x)\,,$$

which in the free field limit becomes:

$$-\mathrm{i}\,\Sigma^{(2)}(x-y) = (-\mathrm{i}e)^2 \frac{\mathrm{i}}{(2\pi)^4}\int d^4q\, \mathrm{e}^{-\mathrm{i}q(x-y)}\gamma_\mu \frac{1}{\gamma\cdot q - m_0 + \mathrm{i}\varepsilon}\gamma_\nu$$
$$\cdot\left(\frac{-\mathrm{i}}{(2\pi)^4}\right)\int d^4k\, \mathrm{e}^{-\mathrm{i}k(y-x)}\frac{\sum\limits_\lambda \varepsilon_\lambda^\mu \varepsilon_\lambda^\nu}{k^2+\mathrm{i}\varepsilon} \tag{16.21}$$

where the sum over all possible photon polarizations depends on the gauge. Taking the most convenient Feynman gauge

$$\sum_\lambda \varepsilon_\lambda^\mu \varepsilon_\lambda^\nu = g^{\mu\nu}\,,$$

for the Fourier transform of Σ then

$$-\mathrm{i}\,\tilde{\Sigma}^{(2)}(p) = -\mathrm{i}\int \mathrm{e}^{\mathrm{i}p(x-y)}\,\Sigma^{(2)}(x-y)\,d^4(x-y)$$
$$= -e^2\int \frac{d^4k}{(2\pi)^4}\gamma_\mu \frac{1}{\gamma(p-k)-m_0+\mathrm{i}\varepsilon}\gamma^\mu \frac{1}{k^2-\mu^2+\mathrm{i}\varepsilon}\,. \tag{16.22}$$

Evaluating (16.21) follows exactly the same pattern as for the calculation of vacuum polarization. An additional difficulty arises from the masslessness of the photon: there is an additional (infrared) divergence as $k \to 0$ in the integrand of (16.22). This we have already chosen to control in (16.22) by allowing the photon propagator to have a mass term μ. Of course all physical quantities are calculated in the limit $\mu \to 0$. But it is necessary to have this term in order to disentangle $k \to 0$ (infrared divergence) and $k \to \infty$ (ultraviolet) behaviour. The latter is dealt with by renormalization, while the former is an artefact of the formulation.

To appreciate this properly, summing the chain of self-energy diagrams gives a better understanding of the role of $\tilde{\Sigma}(p)$:

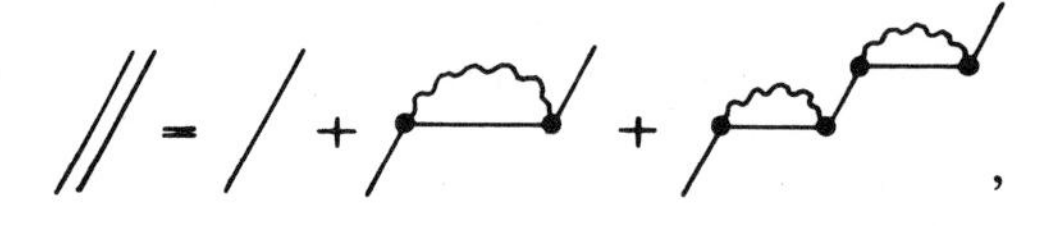

with the result that

$$\begin{aligned} S_F(p) &= S_F^0(p) + S_F^{(0)} \tilde{\Sigma}^{(2)}(p) S_F^{(0)} + \dots \\ &= \frac{\mathrm{i}}{\gamma \cdot p - m_0 - \Sigma^{(2)}(p) + \mathrm{i}\varepsilon} . \end{aligned} \tag{16.23}$$

Let us define the physical mass m_e of the electron as the pole $\gamma \cdot p = m_e$ of (16.23). This definition assures us that the physical properties of the electron correspond to those determined experimentally. Therefore, it is convenient to expand $\Sigma^{(2)}(p)$ in a power series in $(\gamma \cdot p - m_e)$

$$\tilde{\Sigma}^{(2)}(p) = A + [(B + C(p)](\gamma \cdot p - m_e) , \tag{16.24}$$

where $C(p)$ is of a higher order in $(\gamma \cdot p - m_e)$. Inserting this into (16.23) we determine

$$S_F(p) = \frac{\mathrm{i}}{\gamma \cdot p - (m_0 + A) + [B + C(p)](\gamma \cdot p - m_e)} ,$$

which requires that

$$m_e = m_0 + A .$$

Further denoting

$$Z_2 = 1/(1 + B) ,$$

then

$$S_F(p) = \frac{\mathrm{i} Z_2}{(\gamma \cdot p - m_e)} \frac{1}{1 + Z_2 C(p)} , \tag{16.25}$$

which is very similar in form to the renormalization of charge by vacuum polarization. Indeed Z_2 can be viewed as a further renormalization of the electric charge. However, this is not so as its effect is offset by the charge renormalization effect arising from other types of radiative corrections (vertex renormalization) [Kä 58].

Our analysis shows that quantities A and B are absorbed in the renormalization process, leaving us with function $C(p)$ such that $C(\gamma \cdot p = m_e) = 0$. Similar to the discussion on vacuum polarization, A, B can also be divergent quantities, the hope being that in a complete theory these would not appear.

Straightforward calculations [Gr 84b] lead to

$$\tilde{\Sigma}^{(2)}(p) \approx \frac{3\alpha}{4\pi} m_e \log \frac{\Lambda^2}{m_e^2} - \frac{\alpha}{4\pi}(\gamma \cdot p - m_e)\left(\log \frac{\Lambda^2}{m_e^2} + 4 \log \frac{m_e^2 - p^2}{m_e^2}\right), \tag{16.26}$$

where $\Lambda^2 \to \infty$ is the ultraviolet cut-off of the integrals valid for $p^2 \approx m_e^2$ but $p^2 - m_e^2 \gg m_e \mu$. The logarithmic singularity noticeable here originates in the physical process of an electron radiating a photon.

We now turn to the next order in the external field, which is simply given by the usual external field expansion

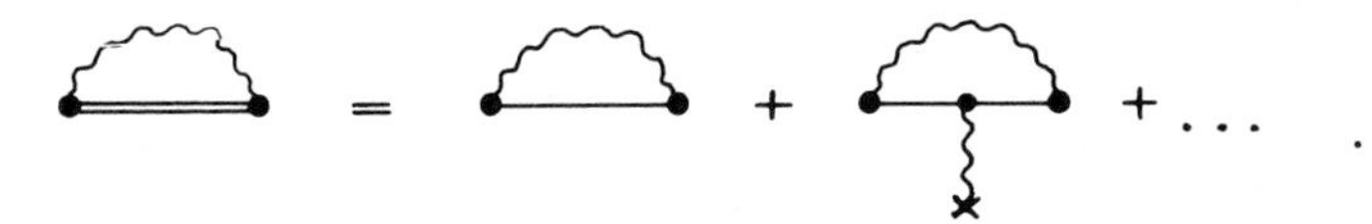

The most important effect is the term linear in Z, which played an important role in the development of QED. *Schwinger* [Schw 48] predicted that an electron would have an anomalous magnetic moment

$$\mu_e = \frac{e}{2m_e}\left(1 + \frac{\alpha}{2\pi}\right),$$

where $\alpha/2\pi$ originated in the diagram linear in Z.

All these terms influence the behaviour of the bound electron in an atom. However, as a brief calculation shows [Be 47], the effects for small Z are dominated by virtual transitions between the actual bound levels: the energy shift due to virtual excitation of a photon of momentum k and a state $|m\rangle$ of the atom is given in second-order perturbation theory by

$$\Delta E_n^< = e^2 \int_0^{k_{\min}} \frac{d^3k}{2k(2\pi)^3} \sum_{m,\varepsilon} \frac{\langle n|\alpha \cdot \varepsilon e^{ik \cdot r}|m\rangle \langle m|\alpha \cdot \varepsilon e^{-ik \cdot r}|n\rangle}{E_n - k - E_m}, \tag{16.27}$$

where we sum over two transverse polarizations of the photon and intermediate electron states. The upper cut off $k_{\min}$, proposed by *Bethe*, restricts this non-relativistic treatment and requires that one considers further the contributions $\Delta E_n^>$ of photon momenta $k > k_{\min}$. This calculational procedure is very convenient, although it does not manifestly respect Lorentz and gauge invariance. Then

$$\Delta E_{nlm}^< = \frac{4\alpha(Z\alpha)}{3m_e^2} \log(k_{\min}/\bar{E})\,|\psi_{nlm}(0)|^2$$

$$\Delta E^{>}_{nlm} = \frac{4\alpha}{3}\,\frac{Z\alpha}{m_e^2}\left(\log\frac{m_e}{2k_{\min}} + \frac{31}{120}\right)|\psi_{nlm}(0)|^2 .$$

Here $\bar{E}$ is an average of the excitation energy $|E_n - E_m|$, $\bar{E} \sim 8.9\,\alpha^2 m_e$ for hydrogen. This leads to a (self-)energy shift of the s states

$$\Delta E_{ns} = \frac{4\alpha(Z\alpha)^4}{3\pi n^3}\left(\log\frac{m_e}{2\bar{E}} + \frac{31}{120}\right)m_e . \tag{16.28}$$

As is well known, this self-energy effect is the dominant contribution to the Lamb splitting of otherwise degenerate $2p_{1/2} - 2s_{1/2}$ states in hydrogen. Its contribution in electronic hydrogen is ca. 40 times more relevant than that of vacuum polarization. However, as already explained, this changes completely at $Z = Z_{cr}$, where both effects nearly cancel.

16.3 Self-Energy in Superheavy Atoms

This section investigates the question whether the self-energy correction may prevent the extraordinary strong binding of K-shell electrons in superheavy systems close to criticality. The large coupling constant $Z\alpha \gtrsim 1$ of the external field obviously necessitates a different theoretical treatment of the self-energy compared with calculations for light atoms, where the traditional expansion in $Z\alpha$ may be still applied. In 1959, *Brown* et al. [Br 59b] showed how to avoid the $Z\alpha$ expansion entirely when evaluating self-energy shifts by treating the influence of the external field exactly. They expanded the electron propagator in terms of angular momentum states and performed the angular integration, leaving a complicated expression which required a numerical evaluation. However, they failed to predict correct numerical values for the binding energy shifts.

Desiderio and *Johnson* [De 71] generalized their calculation of the self-energy by including the effects of atomic screening and finite nuclear size on an arbitrary bound state. The first reliable results for the level shifts in heaviest elements of the periodic system were obtained. Three years later similar calculations were published by *Mohr* [Mo 74]. His precise analysis of self-energy corrections is based on the Coulomb potential for point-like nuclei, and is therefore restricted to nuclear charges below $Z = \alpha^{-1} \sim 137$. The calculations by *Desiderio* and *Johnson* were later refined and continued up to $Z = 160$, where a repulsive shift of about 1% of the K-shell binding energy was found [Ch 76].

The self-energy correction to be calculated is represented by the following Feynman diagram:

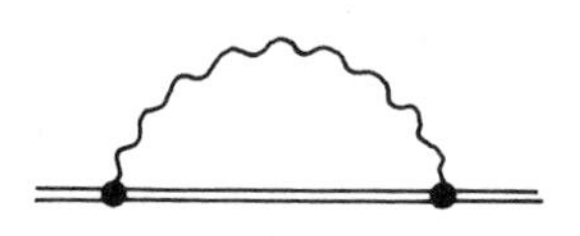

The double line indicates the exact electron propagator and wave function in the Coulomb field of a nucleus. According to this diagram the level shift follows from the *Gell-Mann – Low* theorem [Ge 51]

$$\Delta E = 4\pi \mathrm{i}\alpha \int dt\, d^3x\, d^3y\, \bar{\psi}_n(x)\gamma^\mu S_{\mathrm{F}}(x,y)\gamma^\nu \psi_n(y) D^{\mathrm{F}}_{\mu\nu}(x,y)\,, \tag{16.29}$$

where S_{F} denotes the Feynman propagator in the external field (Sects. 8.4, 9.6):

$$\begin{aligned} S_{\mathrm{F}}(x,y) &= -\mathrm{i}\langle 0|T\hat{\psi}(x)\hat{\bar{\psi}}(y)|0\rangle \\ &= -\mathrm{i}\left[\sum_{n>\mathrm{F}} \psi_n(x)\bar{\psi}_n(y)\theta(t_x-t_y) - \sum_{m<\mathrm{F}} \psi_m(x)\bar{\psi}_m(y)\theta(t_y-t_x)\right], \end{aligned} \tag{16.30}$$

which obeys the inhomogeneous Dirac equation

$$[\gamma\cdot(\hat{p}-eA^{\mathrm{ex}})-m_{\mathrm{e}}]_x S_{\mathrm{F}}(x,y) = \delta^4(x-y)\,.$$

In contrast to the definition of the free electron propagator this equation includes the external potential A^{ex}. The wave functions ψ_n are solutions of the Dirac equation

$$[\gamma(\hat{p}-eA^{\mathrm{ex}})-m_{\mathrm{e}}]\psi_n = 0\,.$$

The photon propagator in Feynman gauge is given by (cf. (16.21))

$$D^{\mathrm{F}}_{\mu\nu}(x,y) = -\mathrm{i}\langle 0|T\hat{A}_\mu(x)\hat{A}_\nu(y)|0\rangle = -g_{\mu\nu}\int \frac{d^4k}{(2\pi)^4}\,\frac{\exp[-\mathrm{i}k(x-y)]}{k^2+\mathrm{i}\varepsilon}\,. \tag{16.31}$$

The next step is to transform propagators and wave functions into momentum space. This admits a decomposition of the self-energy diagram, so enabling infinite mass terms to be identified and removed, leaving the finite observable part of the self-energy.

The theoretical justification of subtracting an infinite term from the self-energy shift ΔE is that the self-energy diagram for a free electron also exhibits this divergent term, as described in the previous section, and since only energy differences are measured in physics, we may take the self-energy of a free electron to be the zero point of energy for the self-energy of a bound electron.

From the point of view of mass renormalization, one may argue, as in Sect. 16.2, that the mass that appears in the Dirac equation is not the observable mass of the electron but the so-called bare mass. The observable mass is then composed of the bare mass and the mass shift due to a free electron's interaction with its own field. This mass shift is infinite. However, it is possible to combine this infinite mass shift with the bare mass that appears in the Feynman propagator for a free electron to form the observable mass, thus eliminating the divergence. This process of mass renormalization can be carried out successfully to all orders of perturbation theory. The ultimate justification of this technique is its ability to predict correctly the experimentally observed energy levels and energy level differences of electrons in atoms.

We introduce the following Fourier transformations

$$\psi(p) = \int d^4x\, \psi(x)\, e^{ipx} \tag{16.32a}$$

$$A_\mu^{ex}(p) = \int d^4x\, A_\mu^{ex}(x)\, e^{ipx} \tag{16.32b}$$

$$S_F(p_2, p_1) = \int d^4x_2\, d^4x_1\, S_F(x_2, x_1) \exp[-i(p_1 x_1 - p_2 x_2)] \,. \tag{16.32c}$$

The full Feynman propagator in momentum space obeys the integral equation (14.17). Iterating (14.18) and going over to momentum space:

$$\begin{aligned} S_F(p_2, p_1) = {} & (2\pi)^4 \delta^4(p_1 - p_2) S_F^{(0)}(p_2) + e S_F^{(0)}(p_2) A\!\!\!/^{ex}(p_2 - p_1) S_F^{(0)}(p_1) \\ & + e^2 S_F^{(0)}(p_2) \int d^4q\, d^4q'\, A\!\!\!/^{ex}(q)\, S_F(p_2 - q, p_1 + q')\, A\!\!\!/^{ex}(q')\, S_F^{(0)}(p_1) \,, \end{aligned} \tag{16.33}$$

where $S_F^{(0)}(p)$ denotes the free propagator

$$S_F^{(0)}(p) = \frac{1}{p\!\!\!/ - m_e} \,.$$

This result may be represented graphically, where a double line denotes S_F and a single line $S_F^{(0)}$:

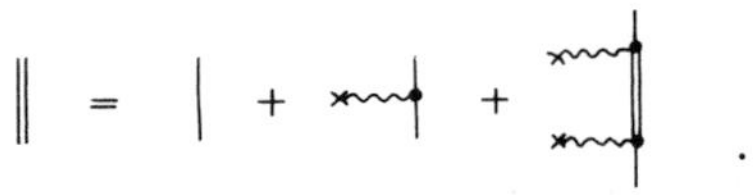

$$\| = | + \ldots$$

Iteration of this equation leads to

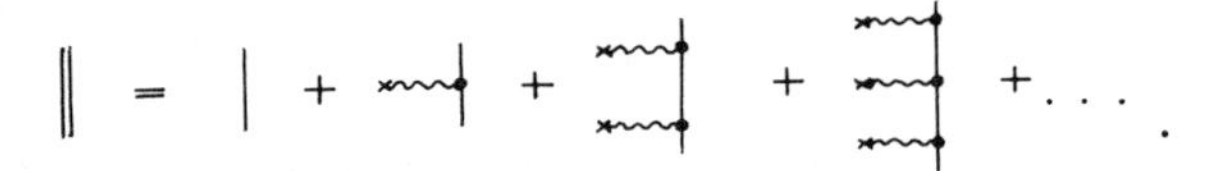

The decomposition of the Feynman propagator may be inserted into the self-energy graph, which yields

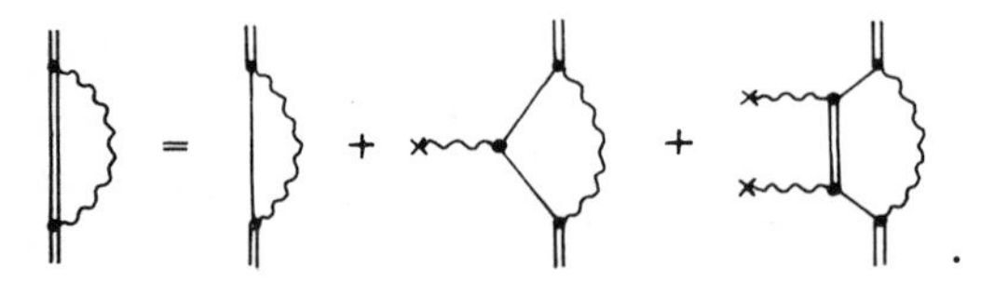

Calculating the various terms is rather lengthy and not very enlightening. Finally, for the renormalized self-energy of a K-shell electron:

$$\Delta E = \Delta E_0'(Z' = Z) - \Delta E_0'(Z' = 0) + i\pi R_0 + \Delta E^{(2)} + \Delta E_c \,. \tag{16.34}$$

This result is obtained from (16.29) by a very long and complicated derivation, which cannot be reproduced here. We refer the reader to [Schl 84a].

The term $\mathrm{i}\pi R_0$ is finite and is basically determined by the $1s$ wave function

$$\mathrm{i}\pi R_0 = -\alpha \int_0^\infty dr \int_0^r dx \left[\frac{2}{3} \frac{x}{r^2} Q(r) Q(x) - \frac{1}{r} P(r) P(x) \right] ,$$

with

$$Q(x) = 2 G_{1s}(x) F_{1s}(x) , \qquad P(x) = G_{1s}^2(x) + F_{1s}^2(x) .$$

Here $\Delta E^{(2)}$ is

$$\Delta E^{(2)} = -\frac{5\alpha}{4\pi} \langle V(x) \rangle_{1s} - \frac{2\alpha}{\pi^2} \int_0^\infty Q^-(p) \frac{\xi \ln \xi}{\xi - 1} p^2 dp$$

$$+ \frac{\alpha}{\pi^2} \int_0^\infty G_{1s}(p) F_{1s}(p) Z(\xi) p^3 dp + \frac{\alpha E_{1s}}{2\pi^2} \int_0^\infty Q^+(p) Z(\xi) p^2 dp$$

with $\xi = p^2 - E_{1s}^2 + 1$ and

$$Z(\xi) = \frac{\xi}{\xi - 1} \left(1 + \frac{\xi - 2}{\xi - 1} \ln \xi \right) .$$

Here $G_{1s}(p)$ and $F_{1s}(p)$ denote Bessel transforms of the radial component of the Dirac wave function

$$G_{1s}(p) = \int_0^\infty G_{1s}(x) j_0(px) x \, dx , \qquad F_{1s}(p) = \int_0^\infty F_{1s}(x) j_1(px) x \, dx ,$$

and

$$Q^\pm(p) = G_{1s}^2(p) \pm F_{1s}^2(p) ;$$

$\langle V(x) \rangle_{1s}$ is the expectation value of the potential energy and E_{1s} the energy eigenvalue of the K-shell electron. The counter term ΔE_c is determined by

$$\Delta E_c = -\frac{\alpha}{2\pi} \langle V(x) \rangle_{1s} \int_0^\infty \frac{d\omega}{(\omega^2 + 1)^{1/2}} .$$

The contribution of the main term $\Delta E_0'(Z')$ for a given nuclear charge Z' is

$$\Delta E_0' = -\frac{2\alpha}{\pi} \int_0^\infty \omega \, d\omega \int_0^\infty dr \int_0^r dx \sum_{\varkappa = \pm 1}^{\pm\infty} \mathrm{Re} \left\{ \frac{|\varkappa|}{\Delta_\varkappa (E_{1s} - \mathrm{i}\omega)} \right.$$

$$\times \left[\frac{(\varkappa - 1)^2}{\bar{l}(\bar{l} + 1)} B_{\bar{l}} Q^{\infty +}(r) Q^{0+}(x) \right.$$

$$+ \frac{l}{2l+1} B_{l-1} \left(Q^{\infty -}(r) + \frac{\varkappa + 1}{l} Q^{\infty +}(r) \right) \left(Q^{0-}(x) + \frac{\varkappa + 1}{l} Q^{0+}(x) \right)$$

$$+\frac{l+1}{2l+1}B_{l+1}\left(Q^{\infty-}(r)-\frac{\varkappa+1}{l+1}Q^{\infty+}(r)\right)$$
$$\times\left(Q^{0-}(x)-\frac{\varkappa+1}{l+1}Q^{0+}(x)\right)-B_l P^{\infty}(r)P^{0}(x)\Bigg]\Bigg\},$$

with l being the orbital angular momentum related to $\varkappa$ and $\bar{l}$ related to $-\varkappa$, respectively. Here we used the abbreviations

$$Q^{\infty,0\pm}(x)=G_{1s}(x,E_{1s})F_{\varkappa}^{\infty,0}(x,E_{1s}-\mathrm{i}\,\omega)\pm F_{1s}(x,E_{1s})G_{\varkappa}^{\infty,0}(x,E_{1s}-\mathrm{i}\,\omega)\,,$$
$$P^{\infty,0}(x)=G_{1s}(x,E_{1s})G_{\varkappa}^{\infty,0}(x,E_{1s}-\mathrm{i}\,\omega)+F_{1s}(x,E_{1s})F_{\varkappa}^{\infty,0}(x,E_{1s}-\mathrm{i}\,\omega)\,,$$
$$B_l=h_l^{(1)}(\mathrm{i}\,\omega r)\,j_l(\mathrm{i}\,\omega r)\,.$$

Further, $\Delta_{\varkappa}(E)$ denotes the Wronskian for a given complex energy E and angular momentum quantum number $\varkappa$:

$$\Delta_{\varkappa}(E)=F_{\varkappa}^{\infty}(x,E)\,G_{\varkappa}^{0}(x,E)-F_{\varkappa}^{0}(x,E)\,G_{\varkappa}^{\infty}(x,E)\,.$$

This $\Delta_{\varkappa}(E)$ is independent of x,and $F_{\varkappa}^{0,\infty}$ are solutions of the radial Dirac equation for complex energies which are regular either at the origin ($x=0$) or at infinity ($x=\infty$) (see also (15.42, 43):

$$dG_{\varkappa}/dx=-(\varkappa/x)\,G_{\varkappa}+[E+1-V(Z')]\,F_{\varkappa}\,,$$
$$dF_{\varkappa}/dx=-[E-1-V(Z')]\,G_{\varkappa}+(\varkappa/x)\,F_{\varkappa}\,.$$

Here j_l and $h_l^{(1)}$ are the spherical Bessel function and Hankel function of first kind for purely imaginary arguments. Notwithstanding their complex appearance, these expressions are in a form well suited for direct numerical evaluation.

The terms $\Delta E_0'(Z'=Z)$, $\Delta E_0'(Z'=0)$ and ΔE_{c}, if treated independently, would produce divergent results, but their combination in (16.34) yields a finite number. Integrating over the frequency ω of the intermediate photon and summing over angular momentum states $\varkappa$ when determining the electron propagator can be performed numerically only within finite limits. The extrapolation to infinity is accomplished analytically. If one neglects electron screening corrections, the external potential energy $V(x)$ is completely determined by the nuclear charge distribution, for which a homogeneously charged sphere with a radius $R=1.2A^{1/3}$ can be assumed.

The various existing calculations [Mo 74, Ch 76, So 82b] on the self-energy of K-shell electrons in high-Z system may be directly compared for mercury ($Z=80$). The self-energy contribution on the binding energy amounts to about 206 eV. The relative deviation between the different calculations was found to be less than 1%. The obtained energy shifts caused by the self-energy of the strongest bound electron are summarized in Fig. 16.3, where ΔE is plotted versus the nuclear charge number Z.

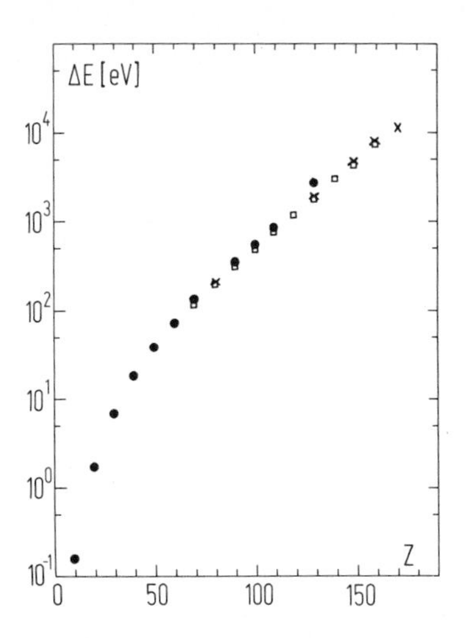

Fig. 16.3. The self-energy shift ΔE of K-shell electrons is plotted versus the nuclear charge number Z. (● ● ●) = numerical results of *Mohr* [Mo 74] for $1s$ electrons in the Coulomb field of point-like nuclei; (□□□) = the computed values of *Cheng* and *Johnson* [Ch 76]; (× × ×) = the results of *Soff* et al. [So 82b]

The apparent discrepancy between *Mohr's* calculation ($\Delta E = 2.586 \pm 0.156$ keV) and *Soff's* result ($\Delta E = 1.896$ keV) for $Z = 130$ is because *Mohr* neglected nuclear size effects.

The most important result of *Soff* et al. was the self-energy shift for $1s$ electrons in the superheavy atom with the critical nuclear charge number $Z = 170$. Here the nuclear radius was adjusted so that the K-electron energy eigenvalue differed only by 10^{-3} eV from the borderline of the negative energy continuum. A correction of $\Delta E = 10.989$ keV was obtained, which still represents only a 1% correction to the total K-electron binding energy. We thus conclude that the self-energy cannot prevent the K-shell binding energy from exceeding $2m_e c^2$ in superheavy systems with $Z > Z_{cr} \sim 170$.

One Last Remark: Due to the isolation of the $1s$ state in energy and in configuration space in almost critical systems one might be tempted to assume that $1s$ state dominates the partial wave expansion of the electron propagator. In this limit the QED self-energy would reduce to its classical counterpart. The classical energy shift is given by $\mathrm{i}\pi R_0$ in (16.34). However, this term accounts only for about 50% of the total energy correction. Consequently it is not sufficient to consider only this simple classical term.

16.4 Supercharged Vacuum

For very large external charge, the charged vacuum can be so highly charged that the effect of the self-consistent terms (16.19) becomes dominant. As shown in Sect. 15.4, the real vacuum polarization tends to neutralize the external applied charge. We now will investigate how this influences the observable quantities.

We begin by describing the neutralization of a large supercritical charge density such as that of a finite-size nucleus. When the nuclear charge is sufficiently large, the vacuum state acquires a very large, though finite charge.

When this charge is substantial, we can no longer neglect the screening effect of the electron – electron (real vacuum polarization) interactions within the vacuum and must approach this problem using the self-consistent equations.

We now focus on the effects of real vacuum polarization only, that is, we choose the Fermi surface as being that of a maximally ionized atom at $E_F = -m_e$, and furthermore neglect the effects of the virtual vacuum polarization and of electron self-energy.

Müller and *Rafelski* [Mü 75] proposed to treat this complex situation by the relativistic Thomas-Fermi approximation. The charge density of the vacuum is equal to the charge density carried by all the states that have joined the lower continuum. In the Thomas-Fermi model, the sum over all these states is represented by an integral over all states with momentum inside the Fermi sphere of radius p_F. The density of electrons is related to the Fermi momentum $p_F(x)$ by

$$\varrho_e = \frac{e}{3\pi^2} p_F^3 . \tag{16.35}$$

The effect of the spin degeneracy is included in (16.35). The relativistic relation between the Fermi energy E_F and Fermi momentum is

$$p_F^2 = [(E_F - eV)^2 - m_e^2]\,\theta(E_F - eV - m_e) . \tag{16.36}$$

The step function ensures that p_F^2 is positive.

From (16.35, 36), for the charge density of the ground state $|F\rangle$ characterized by a choice of E_F we now obtain

$$\langle F|\varrho_e|F\rangle = \frac{e}{3\pi^2}[(E_F - eV)^2 - m_e^2]^{3/2}\,\theta(E_F - eV - m_e) . \tag{16.37}$$

Introducing the total charge density ϱ_t, which is composed of the external "nuclear" part ϱ_N and the electronic part

$$\varrho_t = \varrho_N + \langle F|\varrho_e|F\rangle , \tag{16.38}$$

and using Coulomb's law

$$\Delta eV(r) = -e\varrho_t(r) , \tag{16.39}$$

we find a self-consistent non-linear differential equation for the average potential V that depends on the choice of the Fermi surface E_F, characterizing the ground state:

$$\Delta eV(r) = -e\varrho_N(r) - \frac{e^2}{3\pi^2}[(E_F - eV) - m_e]^{3/2}(E_F - eV - m_e) . \tag{16.40}$$

Consider only the real vacuum polarization and hence in (16.40) choose the Fermi energy at $E_F = -m_e$. This means that only the states accessible to *spontaneous* vacuum decay are filled. Inserting $E_F = -m_e$ into (16.40) yields

$$\Delta e V(r) = -e \varrho_N(r) - \frac{e^2}{3\pi^2}(2 m_e e V + e^2 V^2)^{3/2} \theta(-eV - 2m_e) . \tag{16.41}$$

We now proceed to discuss the solution of (16.41). Since the charge density of the vacuum must be confined to the vicinity of the external charge, we require a solution such that

$$e V(r) \xrightarrow[r\to\infty]{} -\frac{\gamma\alpha}{r} \tag{16.42a}$$

(α is the fine-structure constant). For every choice of Z, γ is determined by the boundary condition on the electrostatic potential at the origin

$$\left.\frac{\partial V}{\partial r}\right|_{r=0} = 0 . \tag{16.42b}$$

Equations (16.42a, b) are therefore eigenvalue equations for γ, the unscreened part of the nuclear charge, and $Z-\gamma$ gives the charge of the vacuum:

$$\int d^3x \langle F|\varrho_e|F\rangle = e(Z-\gamma) . \tag{16.43}$$

Neglecting at first the inhomogeneity of the solution, $V(0) = V_0$ is determined from

$$\varrho_t = \varrho_N + \langle F|\varrho_e|F\rangle = 0 \tag{16.44}$$

in the limit of large Z, i.e. when the distribution of nuclear charge is large compared with $1/m_e$; then

$$e V_0 = \{m_e - [m_e^2 + (3\pi^2 \varrho_N)^{2/3}]^{1/2}\} \to -(3\pi^2\varrho)^{1/3} . \tag{16.45}$$

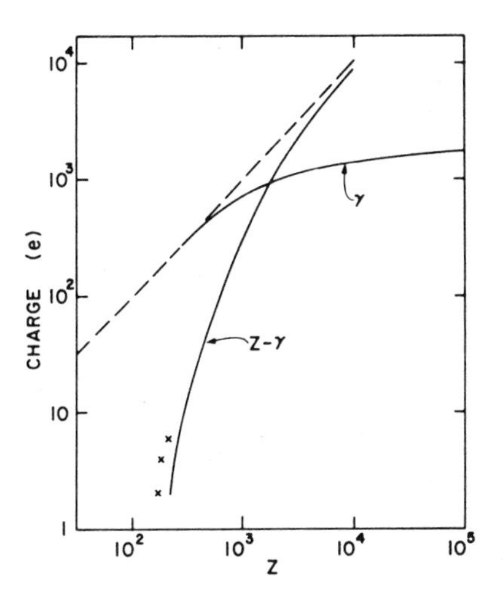

Fig. 16.4. The unscreened charge γ and the total charge of the vacuum $(Z-\gamma)$ as a function of Z. The crosses denote points from single-particle calculations, i.e. the stepwise increase of the vacuum charge as different electron shells descend below $E_F = -m_e$. (– – –) denotes the nuclear charge Z

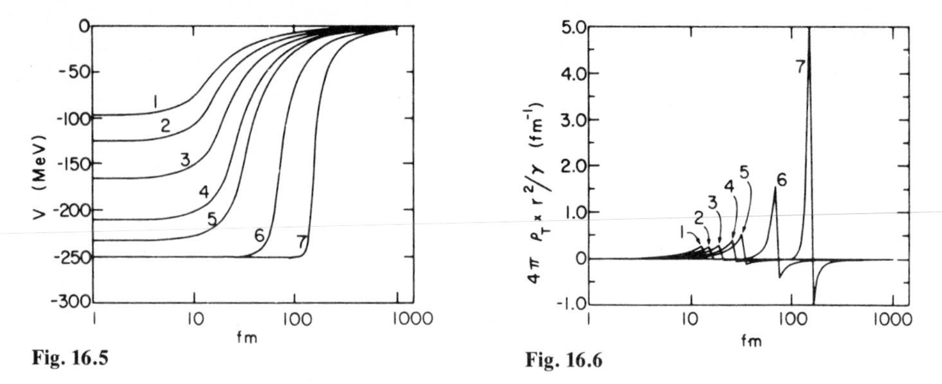

Fig. 16.5 **Fig. 16.6**

Fig. 16.5. Solutions for the self-consistent potential V of the relativistic Thomas-Fermi equation for selected values of the nuclear charge as a function of r. *Curve 1*, $Z = 600$; *curve 2*, 1000; *curve 3*, 2000; *curve 4*, 5000; *curve 5*, 10000; *curve 6*, 10^5, *curve 7*, 10^6

Fig. 16.6. The total charge densities, scaled with γ. Same selected values of nuclear charge as in Fig. 16.5

Numerical integration of (16.41) is straightforward. An equal number of protons and neutrons and normal nuclear density have been assumed for the nuclear charge distribution [Mü 75]. The results for γ are plotted in Fig. 16.4. The figure shows that γ increases monotonically with Z, and that γ/Z decreases as Z increases.

The single-particle results are denoted by crosses in Fig. 16.4 and agree reasonably well with an extrapolation of the Thomas-Fermi results into the realm of small values of $Z-\gamma \gtrsim 1$. Figure 16.5 considers the approach to infinite nuclear matter. The potential approaches the limit described by (16.45) as Z increases from 10^4 to 10^5. The radial total charge density, calculated from the rhs (16.41) is shown in Fig. 16.6. The results are scaled with γ so that each curve is normalized to unity. Evidently the charge density resembles more and more that of a surface dipole with clearly defined regions of positive and negative charge.

As expected, the charge generated by successive levels joining the lower continuum is sufficient to screen most of the bare nuclear charge as it increases without bound. This property is evident from the fact that for $Z > 10^5$ the self-consistent potential does not change within the nuclear matter distribution.

16.5 The Problem of a Supercritical Point Charge

The above results show that there is a limit to the coupling strength between electrons and charged matter. The boundary conditions chosen above – uniform density background charge – led to the finite self-consistent potential step V. But note that we could have equally well asked what happens when a different limit of these equations is considered: for a fixed, large external charge, the radius of the background charge distribution is taken to zero. Naturally, the self-consistent potential that emerges must be just subcritical, that is, no further state can become supercritical. This means, however, that the unscreened charge must be

$$Z-\gamma \xrightarrow[R\to 0]{} \alpha^{-1} \tag{16.46}$$

and the screening is entirely due to the s states. This latter statement can be qualitatively shown by considering the spectrum of the Dirac equation for point-like nuclei $V = -Z\alpha/r$, when $Z\alpha$ approaches one from below: at $Z\alpha = 1$ all s states disappear simultaneously. Considering a finite nuclear radius that tends to zero, all s states dive into the negative energy continuum when $Z\alpha > 1$. This supports the above argument. The intriguing interplay of the different limits has been studied in detail by *Gärtner* et al. [Gä 80, 81] and the above arguments have been explicitly demonstrated.

As discussed at length in Sect. 3.4, the Dirac equation with the Coulomb potential $V(r) = -Z\alpha/r$ of a point charge Ze leads to Sommerfeld's fine-structure formula (3.103), for the spectrum of electronic bound states:

$$E_{n\varkappa} = m_e\left(1+\frac{Z^2\alpha^2}{[n-|\varkappa|+(\varkappa^2-Z^2\alpha^2)^{1/2}]^2}\right)^{-1/2}. \tag{16.47}$$

Due to the term $(\varkappa^2-Z^2\alpha^2)^{1/2}$, this expression becomes imaginary for $Z\alpha > |\varkappa|$. For example, all states with total angular momentum $j = 1/2$ cease to exist beyond $Z = \alpha^{-1} = 137$, and the phase of the corresponding wave functions shows a logarithmic divergence at the origin, so that the wave functions become non-normalizable [see Sect. 6.4, especially (6.161)]. As can be proved [Ri 78a], this is due to the fact that the Dirac Hamiltonian with a pure Coulomb potential is self-adjoint only for $Z\alpha < 1$. A first attempt to solve this problem was made by *Case* [Ca 50]. He proposed that the wave functions have certain phase at the origin. This, however, led to the unphysical result that for $Z\alpha$ increasing beyond 1 the binding energy of the $1s_{1/2}$ state would decrease again, always remaining smaller than the electronic rest mass.

As discussed in Chap. 6, the physically meaningful resolution of the difficulty is obtained by taking into account the finite extension of the nucleus. This leads to the effect of the charged vacuum, where bound states have "dived" into the continuum and appear as resonances in the continuum wave functions, still bearing the electronic charge localized around the nucleus.

The question remains what happens when the nuclear radius R decreases and if a point nucleus limit exists. As shown in Fig. 6.15, the critical charge Z_{cr} approaches $\alpha^{-1} = 137$ for all states with $j = 1/2$ when the charge radius tends to zero. This means that for constant $Z > 137$ with shrinking nuclear radius an infinite number of states enter the negative energy continuum and reach an arbitrarily large binding energy. This corresponds to a result recently obtained for the two-centre Dirac Hamiltonian. It can be proved [Kl 80] that an infinite number of eigenvalues passes through any point in the gap $(-m_e, +m_e)$ when the distance between two point nuclei with $1/2 < Z\alpha < 1$ tends to zero.

The results described above are altered when one takes into account that the nuclear charge is screened by the vacuum charge located close to the nucleus. A similar effect occurs for a charged sphere in an infinite volume of electrolyte [La 80]. There the non-linear Poisson-Boltzmann equation has the property that the charge cloud of counterions around a point charge shrinks to zero radius, so that the point charge is electrically invisible from any finite distance away.

Investigating first the single-particle solutions of the Dirac equation in the field of a shrinking nucleus reveals that the vacuum charge in the vicinity of the nucleus consists of separate concentric shells that shrink at the same rate as the nucleus. Then we consider the screening effects in a quasi-classical approximation, showing that the screening prevents an infinite number of electrons from diving when the radius of a given charge distribution $Z > 137$ is decreased. In the limit of a vanishing nuclear radius the vacuum charge screens the nuclear charge to a value near $Z = 137$.

16.5.1 Overcritical Single-Particle States

The single-particle continuum solutions of the Dirac equation in the field of an extended nucleus with $Z\alpha > 1$ have been calculated in Sect. 6.4. The negative energy solutions are given by (6.141, 115):

$$\phi_1(x) = Nx^{-1/2}\left[e^{i\eta} M(x)_{-iy-1/2, i\gamma} + e^{-i\eta-\pi\gamma} \frac{i\gamma - iy}{\varkappa + iy m_0/E} M(x)_{-iy-1/2, -i\gamma} \right]$$

$$\phi_2(x) = \phi_1^*(x) . \tag{16.48}$$

Here abbreviations have been used (we write again m_0):

$$p = (E^2 - m_0^2)^{1/2} \tag{16.49}$$

$$x = 2ipr \tag{16.50}$$

$$y = Z\alpha E/p \tag{16.51}$$

$$\gamma = (Z^2\alpha^2 - \varkappa^2)^{1/2} \tag{16.52}$$

$$M(x)_{-iy-1/2, i\gamma} = e^{-x/2} x^{i\gamma+1/2} \cdot {}_1F_1(i\gamma + iy + 1, 2i\gamma + 1, x) . \tag{16.53}$$

Here $M(x)$ is a Whittaker function and η is the matching phase determined by the boundary condition on the charge surface. The behaviour of the wave function near the nucleus is determined by $x^{i\gamma}$, while the distant oscillations are described by the exponential function and the hypergeometric function ${}_1F_1$. Using the condition $2pr \ll 1$, which is fulfilled in the region where the electronic charge of the dived states is localized, we can set [Ab 65]

$$e^{-x/2} {}_1F_1 = 1 , \tag{16.54}$$

approximately. In this limit, the wave function becomes

$$\phi_1 = N\left(e^{i\eta}x^{i\gamma} + e^{-i\eta-\pi\gamma}x^{-i\gamma}\frac{i\gamma - iy}{\varkappa + iym/E}\right). \tag{16.55}$$

The transformation (3.113), which is

$$\begin{aligned} u_1 &= 2(-E-m_0)^{1/2}\mathrm{Re}\{\phi_1\} \\ u_2 &= 2(-E+m_0)^{1/2}\mathrm{Im}\{\phi_1\}, \end{aligned} \tag{16.56}$$

gives the components u_1 and u_2 of the Dirac spinor

$$\begin{aligned} u_1 &= 2Ne^{-\pi\gamma/2}(-E-m_0)^{1/2}\cdot[(1+\mathrm{Re}\{c\})\cos\alpha_r + \mathrm{Im}\{c\}\sin\alpha_r] \\ u_2 &= 2Ne^{-\pi\gamma/2}(-E+m_0)^{1/2}\cdot[(1-\mathrm{Re}\{c\})\sin\alpha_r + \mathrm{Im}\{c\}\cos\alpha_r] \end{aligned} \tag{16.57}$$

with

$$c = \frac{i\gamma - iy}{\varkappa + iym_0/E} \tag{16.58}$$

$$\alpha_r = \gamma\ln(2pr) + \eta\,. \tag{16.59}$$

Thus the wave functions can be written in the form

$$\begin{aligned} u_1 &= A_1\sin(\alpha_r + \varphi_1) \\ u_2 &= A_2\sin(\alpha_r + \varphi_2)\,, \end{aligned} \tag{16.60}$$

where the normalization constants A_i and the phases φ_i are determined by (16.57).

For $E \ll -m_0$, these constants read

$$A_1 = A_2 = 2Ne^{-\pi\gamma/2}\left\{-E\left[1+\left(\frac{\gamma+Z\alpha}{\varkappa}\right)^2\right]\right\}^{1/2} \tag{16.61}$$

$$\tan\varphi_1 = \frac{\varkappa}{\gamma+Z\alpha} \tag{16.62}$$

$$\tan\varphi_2 = (\tan\varphi_1)^{-1} = \frac{\gamma+Z\alpha}{\varkappa}\,. \tag{16.63}$$

The signs of the coefficients of the sine and cosine in (16.57) give the conditions that for $\varkappa > 0$

$$\varphi_1 \in [0, \pi/4] \quad \text{and} \quad \varphi_2 \in [\pi/4, \pi/2]\,,$$

for $\varkappa < 0$:

$$\varphi_1 \in [3\pi/4, \pi] \quad \text{and} \quad \varphi_2 \in [-\pi/2, -\pi/4]\,.$$

These formulae show that the relative phase shift $(\varphi_1 - \varphi_2)$ between u_1 and u_2 grows with Z, and that there is an additional relative phase shift of π for negative $\varkappa$.

We are interested mainly in the charge density of the supercritical electrons imbedded in the negative energy continuum. This charge density is given by (6.181) and can be written as

$$\varrho = \int_{\Delta E} [\Psi^+_{E,Z}\Psi_{E,Z} - \Psi^+_{E,Z+\delta Z}\Psi_{E,Z+\delta Z}]\,dE\,, \tag{16.64}$$

where ΔE is an energy interval containing the resonance, and δZ is chosen so that the potential $V = -(Z+\delta Z)\alpha/r$ does not generate a resonance in the interval ΔE. Thus, the second term subtracts the normal vacuum polarization charge in the energy interval ΔE. Because the real vacuum charge is localized close to the nucleus, the region of the asymptotic oscillations of $\Psi(r)$ need not be considered. In the vicinity of the nucleus, the resonant behaviour of the wave functions appears only in the normalization constant $N(E)$, which has a significant maximum at the resonance energy E_R. Hence at the resonance the normal background vacuum polarization is a small contribution, and the second term in (16.64) can be neglected.

For $E \ll -m_0$, the shape and localization of the wave functions are nearly independent of E so that

$$\varrho = \Psi^+_{E_R}\Psi_{E_R}N^{-2}(E_R)\int_{\Delta E} N^2(E)\,dE = N'^2\,\Psi^+_{E_R}\Psi_{E_R}\,. \tag{16.65}$$

This means that the charge distribution may be described by the inner part of the wave function at the resonance energy, and the new normalization constant N' (henceforth called N) is determined by the condition that each dived state is occupied by 2 electrons. To derive this formula, the condition $2pr \ll 1$ was used to cut off the distant oscillations which do not contribute to the charge density, so that the charge distribution (16.65) has to be cut off at a certain r determined by the number of oscillations of the wave function.

Thus the charge distribution of a resonant state can be derived from (16.57)

$$\begin{aligned} r^2\varrho &= u_1^2 + u_2^2 \\ &= 4N^2 e^{-\pi\gamma}\{-E(1+c^*c) - 2m_0\,\mathrm{Re}\{c\} - [2E\,\mathrm{Re}\{c\} + m_0(\mathrm{Re}^2\{c\} \\ &\quad -\mathrm{Im}^2\{c\}+1)]\cos 2\alpha_r - 2(E+m_0\,\mathrm{Re}\{c\})\,\mathrm{Im}\{c\}\sin 2\alpha_r\}\,. \end{aligned} \tag{16.66}$$

The matching phase η (contained in α_r) must be determined by the boundary conditions at the nuclear radius R. If, for simplicity, we assume the nuclear density to be distributed on a spherical shell, the potential inside the nucleus is constant: $V_N = -Z\alpha/R$. Then the wave function in the interior of the nucleus is given by the free spherical wave solutions of the Dirac equation [cf. (3.54)]

$$\begin{aligned} u_1^N &= Nrj_l(pr) \\ u_2^N &= N\,\mathrm{sgn}\,\varkappa\frac{pr}{E - V_N + m}j_{\bar{l}}(pr) \end{aligned} \tag{16.67}$$

with

$$p = [(E - V_N)^2 - m_0^2]^{1/2}$$

$$\bar{l} = l - \operatorname{sgn} \varkappa .$$

The matching condition on the nuclear surface requires

$$u_1^N(R)/u_2^N(R) = u_1(R)/u_2(R) . \tag{16.68}$$

We set

$$U_R = u_1^N(R)/u_2^N(R) = \operatorname{sgn} \varkappa \frac{E - V_N + m_0}{p} j_l(pR)/j_{\bar{l}}(pR) \tag{16.69}$$

and $\alpha_R = \gamma \ln(2pR) + \eta$ and calculate the ratio of the wave functions with the help of (16.57):

$$\tan \alpha_R = \frac{1 + \mathrm{Re}\{c\} - \mathrm{Im}\{c\} U_R(-E+m_0)^{1/2}(-E-m_0)^{-1/2}}{(1 - \mathrm{Re}\{c\}) U_R(-E+m_0)^{1/2}(-E-m_0)^{-1/2} - \mathrm{Im}\{c\}} . \tag{16.70}$$

Now, (16.66) becomes

$$r^2 \varrho = 4N^2 \mathrm{e}^{-\pi\gamma}[A - B \sin(2\gamma \ln r/R + 2\alpha_R + \varphi)] , \tag{16.71}$$

with

$$A = -E(1 + c^* c) - 2m_0 \mathrm{Re}\{c\} \tag{16.72}$$

$$B = \{[2E \,\mathrm{Re}\{c\} + m_0(\mathrm{Re}^2\{c\} - \mathrm{Im}^2\{c\} + 1)]^2 + 4(E + m_0 \mathrm{Re}\{c\})^2 \mathrm{Im}^2(c)\}^{1/2} \tag{16.73}$$

$$\tan \varphi = \frac{2E \,\mathrm{Re}\{c\} + m_0(\mathrm{Re}^2\{c\} - \mathrm{Im}^2\{c\} + 1)}{2(E + m_0 \mathrm{Re}\{c\}) \mathrm{Im}\{c\}} . \tag{16.74}$$

Formula (16.17) shows that the charge density outside the nucleus oscillates as a function of $\ln r/R$ with the frequency $2\gamma = 2(Z^2\alpha^2 - \varkappa^2)^{1/2}$. Since α_R does not depend on R (as shown below), R serves as a scaling parameter and the vacuum charge distribution shrinks together with the nuclear radius.

To understand the consequences of (16.71) better we discuss the case $E \ll -m_0$ in more detail. Then the density reads

$$r^2 \varrho = 4N^2 |E| \mathrm{e}^{-\pi\gamma}[1 + \varkappa^{-2}(\gamma + Z\alpha)^2 - 2|\varkappa|^{-1}(\gamma + Z\alpha) \sin(2\gamma \ln r/R + 2\alpha_R + \varphi)] . \tag{16.75}$$

If we can show that the phase of the argument of the sine function is independent of E and the sign of $\varkappa$, we can conclude that states with the same value of $(n-1)$ (where $l = j + \frac{1}{2} \operatorname{sgn} \varkappa$) and the same total angular momentum $j = |\varkappa| - 1/2$ have the same density distribution. For example, the $ns_{1/2}$ and $(n+1)p_{1/2}$ states could be treated as equivalent.

Indeed, with $V_N \ll E \ll -m_0$

$$U_R = \operatorname{sgn} \varkappa (j_{|\varkappa|}(Z\alpha)/j_{|\varkappa|-1}(Z\alpha))^{\operatorname{sgn} \varkappa} \tag{16.76}$$

$$\tan \alpha_R = \frac{1 - \operatorname{sgn} \varkappa (\gamma + Z\alpha) U_R / |\varkappa|}{U_R - \operatorname{sgn} \varkappa (\gamma + Z\alpha) / |\varkappa|} \tag{16.77}$$

$$\tan \varphi = \frac{m_0}{E} \left(\frac{|\varkappa|}{2(\gamma + Z\alpha)} - \frac{\gamma + 3Z\alpha}{2|\varkappa|} \right) \operatorname{sgn} \varkappa . \tag{16.78}$$

To investigate the dependence on the sign of $\varkappa$, we introduce the notation $\alpha_R^\pm$ for $\alpha_R(\varkappa = \pm 1)$ and $\varphi^\pm$ for $\varphi(\varkappa = \pm 1)$. Because of

$$\tan \alpha_r^- = -(\tan \alpha_R^+)^{-1}, \quad \text{then}$$

$$\alpha_R^- = \alpha_R^+ - \pi/2 .$$

In view of the signs of the sine and cosine in (16.66) we must choose

$$\varphi^+ \in [\pi, 3\pi/2] \quad \text{and} \quad \varphi^- \in [3\pi/2, 2\pi] ,$$

which together with $\tan \varphi^- = -\tan \varphi^+$ yield

$$\varphi^- = -\varphi^+ + 3\pi .$$

Thus for the total phase shift

$$2\alpha_R + \varphi = 2\alpha_R^+ - \varphi^+ + 2\pi = 2\alpha_R^+ + \varphi^+ - 2\varphi^+ + 2\pi . \tag{16.79}$$

In the limit $E \ll -m_0$, $\tan \varphi \approx 0$ and $\varphi^+ \approx \pi$. Then the phase is independent of the sign of $\varkappa$ and also of E, because of the periodicity of the sine function.

As mentioned before, the formula for the vacuum charge distribution is valid only in the region where $2pr \ll 1$ holds. In this region $(n-1)$ oscillations of the charge distribution are located, so that each state has the correct number of nodes, when the charge distribution is cut off at the point $2pr \approx 1$.

The results of this approximate analytic treatment can be illustrated by the numerical computation of the exact solutions of the Dirac equation [Gä 81]. The results are shown in Figs. 16.7, 8. The shape of the wave functions agrees well with the approximate analytic results derived above. Figure 16.8 shows that all maxima of the charge density of one state have almost the same height. However, since the plot in Fig. 16.8 is logarithmic, the outermost maximum is the broadest one, and nearly the total charge of the resonance $Q = 4\pi \int \varrho r^2 dr$ is localized in the latter. Therefore (concentrating on the region $1 < Z\alpha < 2$, where only $j = 1/2$-states can reach the lower continuum) the total vacuum charge distribution is very well described by a set of consecutive sine bumps (the sine depending on the logarithmic coordinate $\ln r/R$), of which each one is normalized so that it contains 4 charge units.

The analytic structure of the wave functions can be used to calculate the electronic charge located inside a shrinking nucleus. Evaluating the Bessel functions (16.76, 77) leads for $s_{1/2}$ electrons to

$$\tan \alpha_R = \frac{\gamma + Z\alpha - (Z\alpha)^{-1} + \cot Z\alpha}{(\gamma + Z\alpha) \cot Z\alpha - \gamma / Z\alpha} . \tag{16.80}$$

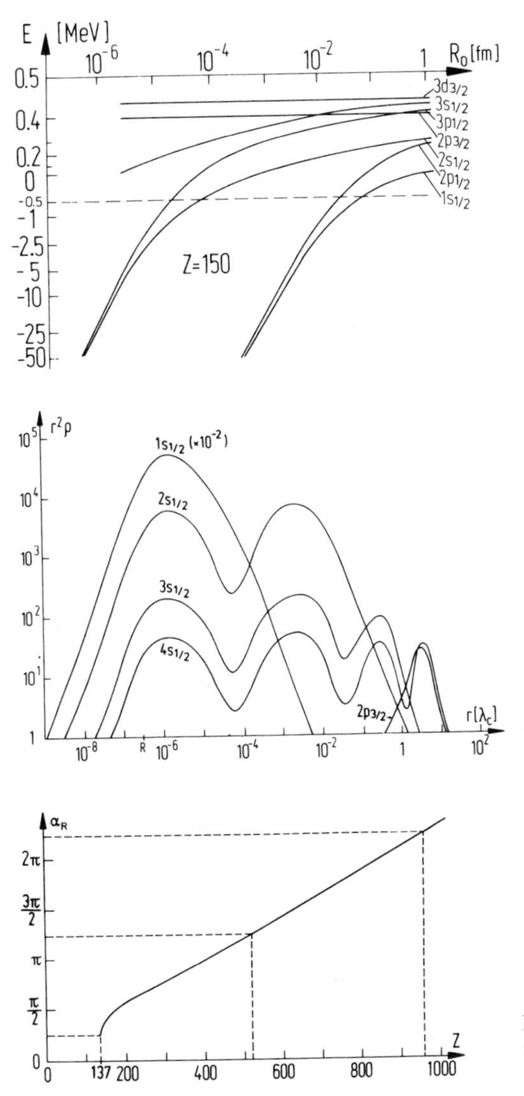

Fig. 16.7. Single-particle energies of electronic states in the field of a shrinking nucleus with charge $Z = 150$ and radius $R = R_0\,(2.5\,Z)^{1/3}$ (cf. Fig. 6.16)

Fig. 16.8. Single-particle densities (in arbitrary units) of some resonances and bound states for a nucleus with $Z = 150$ and $R_0 = 10^{-5}$ fm

Fig. 16.9. Phase of the wave function at the nuclear surface

The phase at the nuclear radius is given by $2\,\alpha_R$ because $\varphi = 0$ in this case. The phase α_R is independent of R, which means that the electronic charge located inside a shrinking nucleus is independent of the radius R for a given Z.

The $1s_{1/2}$ electron is located totally inside the nucleus if $\alpha_R \geqq 5\,\pi/4$ (see Fig. 16.9). This gives the condition

$$\cot Z_1\alpha = (Z_1\alpha)^{-1} + 1 \tag{16.81}$$

for the charge Z_1 at which the $1s_{1/2}$ electron is just inside the nucleus. This equation has the solution $Z_1 = 522$. For a nuclear charge that increases even more, the main maximum of the $2s_{1/2}$ state starts penetrating into the nucleus, until at $Z_2 = 960$ the $2s_{1/2}$ electrons are totally inside the nucleus too.

16.5.2 Screening Effects of the Vacuum Charge

Let us discuss the screening effect of the vacuum charge density, i.e. of those electrons that occupy the resonant states imbedded in the negative energy continuum. Obviously they will screen the nuclear charge. In fact, it could be strong enough to give an upper limit for the apparent charge of a point-like source. A possible method of investigation would be to perform Hartree-Fock calculations, e.g. based on the equations developed in Sect. 16.1. Because such calculations are very tedious and involve considerable numerical intricacies in the limit of a point-like source, we propose to study the vacuum charge distribution in the quasi-classical approximation. As shown, the simplest realization, the Thomas-Fermi approach, turns out to be inadequate for our problem. A better approach is the full WKB method which provides a correct description of the spatial structure of the wave functions and enables screening effects to be included satisfactorily.

The usual Thomas-Fermi model is obtained by assuming that the derivative of the potential can be entirely neglected. For particles with spin 1/2 and no other internal degrees of freedom, this gives (16.35) for the particle density, where the relativistic energy-momentum relation (16.36) has to be used. It is well known that this model gives a non-integrable particle density $\varrho \sim (Z\alpha/r)^3$ near the origin for a point source with a $1/r$ potential.

There are two points where this TF model contains inadequate simplifications. First, neglecting dV/dr is not allowed because $dV/dr \sim 1/r^2$ for a $1/r$ potential, so that dV/dr increases more strongly than V for small r. Second, the usual TF density includes states with all angular momenta. In the considered range $Z\alpha < 2$ only states with angular momentum $j = 1/2$ may become overcritical, so that our model must be restricted to include only such states. For that case, a TF density can be derived by summing approximated wave functions over all occupied states. Corresponding to the usual TF approximations, we used free spherical wave solutions of the Dirac equation. The summation over all states with $j = 1/2$ and $E < -m_0$ gives [Ra 76a, Gä 79]

$$\varrho_{1/2} = 2p_F/(\pi r)^2[1 - j_0^2(p_F r)] , \tag{16.82}$$

with ϱ_F taken from (16.36) and the Bessel function j_0. For small distances and $V = -Z\alpha/r$ this expression behaves like

$$\varrho_{1/2} = 2Z\alpha/(\pi^2 r^3)[1 - j_0^2(Z\alpha)] . \tag{16.83}$$

Again, we obtain a non-integrable density. Hence, this is obviously not a suitable expression for the vacuum charge around a point source. Another relativistic TF model was proposed in [Ru 52], using an effective momentum derived from the Dirac equation which includes the electron angular momentum. This model has the disadvantage that it is valid only for $Z < 137$ and $E_F > 0$. The extension of this model to the overcritical case is given by [Mi 77]

$$\varrho_j = |\varkappa|/(\pi r)^2 (V^2 + 2V - \varkappa^2/r^2)^{1/2} . \tag{16.84}$$

In our case this leads to

$$\varrho_{1/2} = (Z^2\alpha^2 - \varkappa^2)^{1/2}/(\pi^2 r^3) . \tag{16.85}$$

This model implies that the screening effects are described by the function $\gamma(r) = [Z^2(r)\,\alpha^2 - \varkappa^2]^{1/2}$, which has the important feature that it contains the electron angular momentum.

However, the shell structure of the vacuum charge is not reproduced, because replacing the function $\sin^2\theta$ (which describes the rapidly oscillating wave functions) by its average 1/2 in the derivation of this model is a good approximation only if the contributions from different j are summed up. Although this model shows some consequences of screening, the shape and location of the vacuum charge are not described correctly for a source that shrinks to a point.

To get approximations for the effective momentum and the charge density, *Gärtner* et al. studied the WKB solutions of the Dirac equation. The resulting expression for the charge density is inserted into the Poisson equation to calculate the screening effects self-consistently.

The radial Dirac equation can be reduced to a form analogous to the Schrödinger equation by using the following transformations [Ro 61] (here and in the following equations, the upper sign is valid for U_1, the lower sign for U_2)

$$U_{1/2} = u_{1/2}|E - V \pm m_0|^{-1/2} . \tag{16.86}$$

Then the Dirac equation reads

$$U''_{1/2} + p^2_{1/2}(r)\,U_{1/2} = 0 \quad \text{with} \tag{16.87}$$

$$p^2_{1/2}(r) = (E-V)^2 - m_0^2 - \frac{\varkappa(\varkappa \pm 1)}{r^2} \pm \frac{\varkappa}{r}\,\frac{dV/dr}{E - V \pm m_0}$$
$$- \frac{3}{4}\left(\frac{dV/dr}{E - V \pm m_0}\right)^2 - \frac{1}{2}\,\frac{d^2V/dr^2}{E - V \pm m_0} . \tag{16.88}$$

The structure of the solutions is determined by the shape of this effective momentum (Fig. 16.10). For $E < -m_0$ two different regions (I and III in Fig. 16.10) with $p^2 > 0$ exist, separated by a barrier with $p^2 < 0$. Thus the quasi-classical momentum reflects the resonance behaviour of the wave functions with $E < -m_0$. Since the oscillations in region III do not contribute to the charge density, the resonance can be handled like a bound state localized in region I.

To simplify the expression for p^2, we must insert an approximation for V. Outside the nucleus the potential can be written in the form

$$V(r) = -\frac{Z(r)\,\alpha}{r} . \tag{16.89}$$

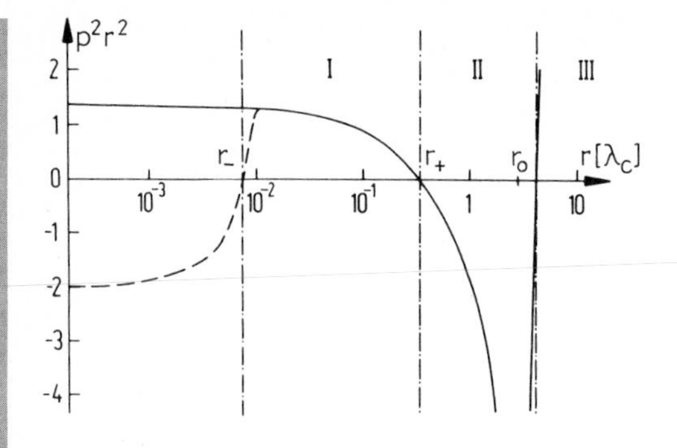

Fig. 16.10. Quasi-classical momentum of a nucleus with $Z=200$ at energy $E=-1.5\,m_e$ for the u_1 wave function with $\varkappa=+1$. r_0 denotes the point where $E-V+m=0$. (———) point nucleus; (– – –) extended nucleus with $R=10^{-2}\lambda_c$

In further considerations we neglect all derivatives of $Z(r)$, which is a better approximation than that used in the usual TF models, where the derivates of V are neglected. Thus

$$\begin{aligned} p_{1/2}^2(r) &= \left(E+\frac{Z\alpha}{r}\right)^2 - m_0^2 - \frac{\varkappa(\varkappa\pm 1)}{r^2} + \frac{(1\pm\varkappa)Z\alpha}{r^3}\left(E+\frac{Z\alpha}{r}\pm m_0\right)^{-1} \\ &\quad - \frac{3(Z\alpha)^2}{4r^4}\left(E+\frac{Z\alpha}{r}\pm m_0\right)^{-2} . \end{aligned} \tag{16.90}$$

For $p^2=Ar^{-2}$, the WKB solution with the Langer correction $q^2=p^2-1/4r^2$ [La 37] [which is equivalent to substituting $\varkappa(\varkappa\pm 1)$ by $(\varkappa\pm\frac{1}{2})^2$ is an exact solution of (16.87), as shown in [Fr 65]. Because the leading power in p^2 is r^{-2}, we use the Langer correction. The WKB solutions so modified should be a good approximation to the exact solutions outside the nucleus, not too close to the turning point, where $p^2\sim r^{-2}$ is not fulfilled. The behaviour in the vicinity of the turning point is discussed below.

The WKB solutions of (16.87) in the classically allowed region [where $q_i^2(r)>0$] between the turning points r_- and r_+ are given by

$$\begin{aligned} u_1(r) &= N\left[\frac{E-V+m_0}{q_1}\right]^{1/2} \sin\left(\int_{r_-}^{r} q_1\,dr' + \frac{\pi}{4}\right) \\ u_2(r) &= N\left[\frac{E-V-m_0}{q_2}\right]^{1/2} \operatorname{sgn}\varkappa \sin\left(\int_{r_-}^{r} q_2\,dr' + \frac{\pi}{4}\right) . \end{aligned} \tag{16.91}$$

The resonance energy E is determined by the Bohr-Sommerfeld quantization rule

$$\int_{r_-}^{r_+} q_1\,dr = \left(n+\frac{1}{2}\right)\pi . \tag{16.92}$$

In the following we discuss the turning points r_- and r_+. Inside the nucleus we use the potential of a homogeneously charged sphere with total charge Z_N

$$V_N = \frac{Z_N\alpha}{R}\left(\frac{1}{2}\left(\frac{r}{R}\right)^2 - \frac{3}{2}\right) , \tag{16.93}$$

and take into account that V_N is much deeper than the energy of the deepest bound state resonance, so that E and m_0 can be neglected. Thence

$$p_{1/2}^2(r) = V_N^2 - \frac{\varkappa(\varkappa\pm 1)}{r^2} - \frac{(\pm\varkappa-\frac{1}{2})Z_N\alpha}{R^3 V_N} - \frac{3}{4}\left(\frac{Z_N\alpha r}{R^3 V_N}\right)^2 . \tag{16.94}$$

Near the origin we can neglect the last term and set

$$V_N = V_0 = \frac{3}{2}\,\frac{Z_N\alpha}{R} . \tag{16.95}$$

Introducing the Langer correction yields

$$q_{1/2}^2(r) = \frac{1}{R^2}\left[\left(\frac{3}{2}Z_N\alpha\right)^2 \pm \frac{2}{3}\varkappa - \frac{1}{3}\right] - \left(\frac{\varkappa\pm\frac{1}{2}}{r}\right)^2 . \tag{16.96}$$

The classical turning point r_- is determined by the condition $q^2(r_-) = 0$, the solution to which is

$$r = R\,\frac{|\varkappa\pm\frac{1}{2}|}{[(\frac{3}{2}Z_N\alpha)^2 \pm \frac{2}{3}\varkappa - \frac{1}{3})^{1/2}} . \tag{16.97}$$

The outer turning point r_+ depends parametrically upon the energy E. We calculate the turning point r_+^F at the Fermi energy $E_F = -m_0$. This gives the spatial extension of the total vacuum charge density. At the Fermi surface (as in the previous chapter we set $\gamma^2 = Z^2\alpha^2 - \varkappa^2$)

$$q_{1F}^2(r) = -\frac{2Z(r)\,\alpha m_0}{r} + \frac{\gamma^2(r)}{r^2} . \tag{16.98}$$

The equation $q_{1F}^2(r_+^F) = 0$ is an implicit equation for r_+^F (the outer turning point at the Fermi energy), which cannot be solved explicitly. Only by neglecting the screening [i.e. setting $Z(r) = Z_N$] is an explicit solution possible:

$$r_+^F = \frac{Z_N^2\alpha^2 - \varkappa^2}{2Z_N\alpha m_0} . \tag{16.99}$$

The value of the exact turning point including the screening effects is always smaller than r_+^F. Thus (16.99) is an estimation for the spatial extension of the vacuum charge.

Outside the nucleus we introduce a further approximation: except close to the outer turning point, the potential V is much deeper than the energy $E \ll -m_0$, so that we can neglect all terms E and m_0 compared with $V = -Z\alpha/r$.

Then the quasi-classical momentum simplifies to

$$q_{1/2}^2(r) = V^2 - \frac{(\varkappa\pm\frac{1}{2})^2}{r^2} \mp \frac{\varkappa}{r}\,\frac{V'}{V} - \frac{3}{4}\left(\frac{V'}{V}\right)^2 + \frac{1}{2}\,\frac{V''}{V} = \frac{\gamma^2}{r^2} . \tag{16.100}$$

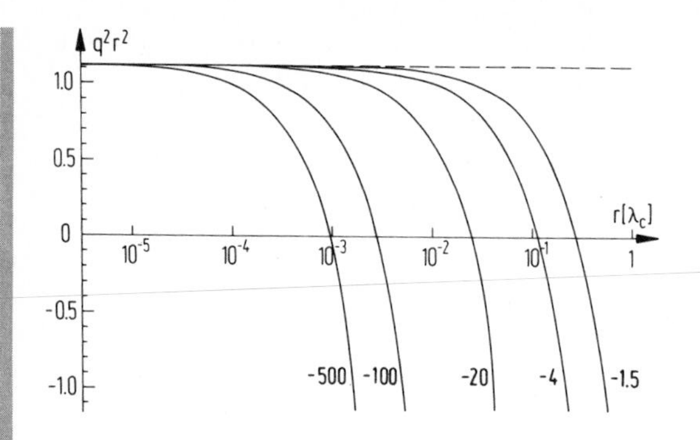

Fig. 16.11. Quasi-classical momentum for different energies (in units of m_0) for a point nucleus with $Z=200$. (– – –) represents (100)

This result can also be obtained by applying the approximation $V \ll E$ to the WKB solutions given in [Mu 78].

Since the condition $V \ll E$ is not fulfilled for $r \gtrsim r_+$, (16.100) is not valid for large r (Fig. 16.11). Thus we have to cut off q^2 for $r \gtrsim r_+$ in order to reproduce the behaviour of the exact q near the outer turning point.

In the classically allowed region the wave functions are

$$u_1(r) = N(Z\alpha/\gamma)^{1/2} \sin\left(\int_R^r \frac{\gamma}{r'} dr' + \eta_1\right)$$

$$u_2(r) = N \operatorname{sgn} \varkappa (Z\alpha/\gamma)^{1/2} \sin\left(\int_R^r \frac{\gamma}{r'} dr' + \eta_2\right) \quad \text{with} \tag{16.101}$$

$$\eta_i = \int_{r_-}^{R} q_i(r')\, dr' + \frac{\pi}{4} . \tag{16.102}$$

Neglecting the screening effects we can solve the phase integral

$$\int_R^r \frac{\gamma}{r'} dr' = \gamma \ln \frac{r}{R} , \tag{16.103}$$

i.e. the wave function then has the same structure as derived before, (16.60). In the following we neglect the phase shift [which is small when Z exceeds 137 slightly, according to (16.62, 63)] between the two components u_1 and u_2 determined by (16.96, 97, 102), so that the charge density simplifies to

$$r^2 \varrho_E = u_1^2 + u_2^2 = 2N_E^2 \frac{Z\alpha}{\gamma} \sin^2\left(\int_R^r \frac{\gamma}{r'} dr' + \eta\right) . \tag{16.104}$$

This formula depends on the energy only via the normalization constant N and the outer turning point r_+. It is independent of the sign of $\varkappa$, because inside the nucleus, u_1 and u_2 exchange their roles only when the sign of $\varkappa$ is changed. Thus we can label the states by $k = n - l$ and j, where k gives the number of oscillations. These results contain the shell structure of the total vacuum charge described in the previous section. From the formulae obtained it is obvious that screening does not destroy this shell structure; its effect is only to shift the outer shells of the vacuum charge further away from the nucleus.

To calculate the screening function $Z(r)$, we need an expression for the total charge density (considering states with $j = 1/2$, so that there are 4 electrons in each state):

$$r^2 \varrho = 4 \sum_{E<-m} r^2 \varrho_\varkappa = 8 \frac{Z\alpha}{\gamma} \sum_{E<-m} N_\varkappa^2 \sin^2 \left(\int_R^r \frac{\gamma}{r'} dr' + \eta \right) . \tag{16.105}$$

Applying the same arguments as before, the summation proceeds by introducing a new normalization $N(r)$ which is a step function determined by the condition that each oscillation of the charge density contains exactly 4 electrons:

$$r^2 \varrho = N^2(r) \frac{Z\alpha}{\gamma} \sin^2 \left(\int_R^r \frac{\gamma}{r'} dr' + \eta \right) \tag{16.106}$$

$$N(r) = N_i = \left[\pi \int_{r_i}^{r_i+1} \frac{Z\alpha}{\gamma} \sin^2 \left(\int_R^{r'} \frac{\gamma}{r''} dr'' + \eta \right) dr' \right]^{-1/2} \quad \text{for} \quad r \in [r_i, r_{i+1}] , \tag{16.107}$$

with r_i, r_{i+1} being two neighbouring zeros of the sine. Inserting the density (16.106) into the Poisson equation

$$d^2 Z / dr^2 = 4 \pi r \varrho \qquad (\text{with } r \geqq R) \tag{16.108}$$

gives a self-consistent equation for the screening function $Z(r)$

$$\frac{d^2 Z}{dr^2} = 4 \pi \alpha \frac{N^2(r) Z(r)}{r \gamma(r)} \sin^2 \left(\int_R^r \frac{\gamma(r')}{r'} dr' + \eta \right) , \tag{16.109}$$

where $N(r)$ is the step function defined by (16.106).

To simplify our calculations, we set $r_- = R$ (and hence $\eta = \pi/4$). This does not influence the screening effects because most of the electronic charge is located outside the nucleus (Fig. 16.9). The outer turning point r_+^{F} is given by the estimation of (16.99). This is allowed only if $\gamma(r) \neq 0$. When so many electrons have reached the continuum that $\gamma(r)$ becomes zero at a certain $r_0 < r_+^{\mathrm{F}}$, r_0 is the outer turning point. Determining the turning points yields a condition whether a state is dived or not: its wave function must be located entirely in the classically allowed region, as is expressed by the Bohr-Sommerfeld quantization condition (16.92).

The differential equation (16.109) has to be solved with the following boundary conditions. At the nuclear surface the Poisson equation requires

$$Z(r) = Z_{\mathrm{N}} + R \left. \frac{dZ}{dr} \right|_R , \tag{16.110}$$

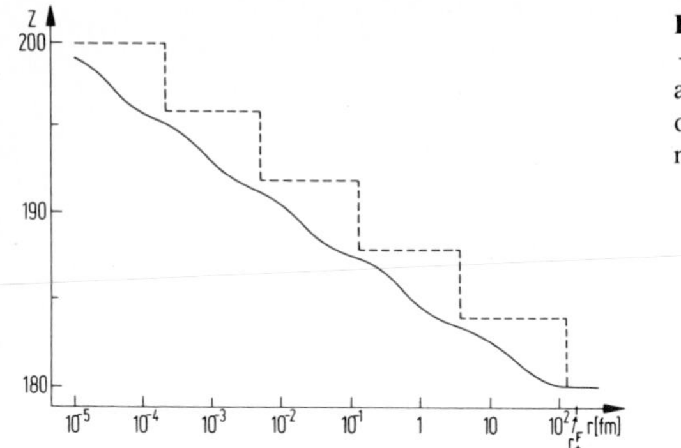

Fig. 16.12. Screening function $Z(r) = -Vr/\alpha$ for a nucleus with charge $Z_N = 200$ and radius $R = 10^{-5}$ fm. (———) solution of the Poisson equation; (– – –) approximate $Z_{\rm eff}$

where Z_N denotes the nuclear charge. We assume that our atom is so far ionized that it does not contain electrons in the gap $[-m_0, m_0]$. Then the boundary condition at the surface of the vacuum charge is

$$\left.\frac{dZ}{dr}\right|_{r_+} = 0 \quad \text{or equivalently:} \quad Z(r_+) = Z_N - Q_{\rm vac} .$$

Note that (16.109) combined with (16.107) is scale invariant under a transformation $r \to \lambda r$, $R \to \lambda R$. The scale invariance carries over to the boundary conditions if the outer turning point is determined by $\gamma(r_+) = 0$, but it is violated when condition (16.99) applies, which contains the electron mass as scale. We conclude that the solution is scale invariant when the vacuum screening is sufficiently strong to reduce the apparent charge to $1/\alpha$ so that $\gamma(r_+) = 0$. The scale is then set by the nuclear radius R alone, and the vacuum charge distribution is forced to shrink in line with the nuclear radius. For a point source the charge distribution must be point-like as well.

A typical solution of (16.108) is shown in Fig. 16.12. The wiggles in the screening function $Z(r)$ reflect the oscillating structure of the charge density. To investigate the limit $R \to 0$ we make a further simplification: we replace the screening function $Z(r)$ by a step function $Z_{\rm eff}(r)$ which is constant within every shell of the vacuum charge. We choose $Z_{\rm eff}$ such that the electrons in the nth shell are thought to be screened by the $4(n-1)$ electrons contained in the shells inside: $Z_{\rm eff}(r) = Z_N - 4(n-1)$ for $r_n \leqq r \leqq r_{n+1}$ (Fig. 16.12). This model enables all calculations to be performed analytically.

The positions s_n of the maxima of the screened charge density are given by

$$\ln s_n = \ln s_1 + \frac{\pi}{2}\left(\frac{1}{\gamma_1} + \frac{1}{\gamma_n}\right) + \pi \sum_{i=2}^{n-1} \frac{1}{\gamma_i} , \tag{16.111}$$

with

$$\gamma_n = [(Z_{\rm eff}(s_n)\,\alpha)^2 - \varkappa^2]^{1/2} . \tag{16.112}$$

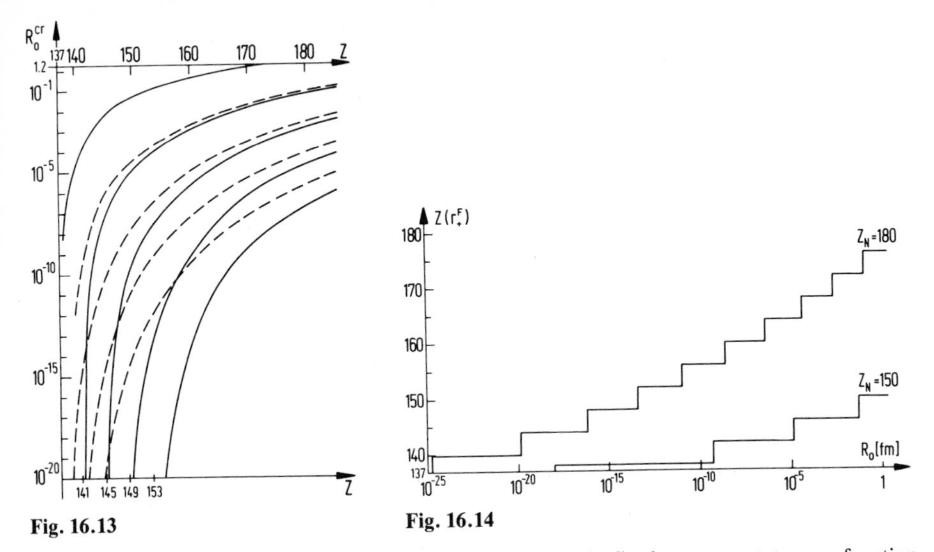

Fig. 16.13 **Fig. 16.14**

Fig. 16.13. Value of the critical radius parameter R_0^{cr} [fm] for the five lowest $s_{1/2}$ states as a function of the nuclear charge Z. The $1s_{1/2}$ state is not screened in our approximation. (———) including screening; (– – –) without screening

Fig. 16.14. Effective charge $Z(r_+^F) = Z_N - Q_{vac}$ as function of the radius parameter R_0 for two nuclei with charge $Z_N = 150$ and 200

Obviously the distance between two maxima becomes infinite when Z_{eff} tends to 137. This has to be interpreted as that the next level does not become supercritical for any value of R.

We can estimate within this approximation under which condition an electronic level can reach the lower continuum. The Bohr-Sommerfeld quantization rule determines the critical radius parameter R_0^{cr} [the nuclear radius is given by $R = R_0\,(2.5\,Z_N)^{1/3}$] at which the diving occurs, by the condition that the outer maximum passes the turning point r_+^F. This leads to

$$R_0^{cr}(n) = R_0^{cr}(1)\,s_1/s_n\,, \tag{16.113}$$

where the integer n denotes the label of the state.

Figure 16.13 shows that in this model the $ns_{1/2}$ and the $(n+1)p_{1/2}$ states never become supercritical for $Z_N < 137 + 4(n-1)$, even if the nucleus shrinks to a point. This means that in our simple model with the screening function approximated by a step function, the effective charge $Z(r_+^F) = Z_N - Q_{vac}$ at the vacuum charge surface cannot decrease below the limit 137-a, with $a \leqq 4$, when the nucleus shrinks to a point (Fig. 16.14). This is so because we have assumed that $s_{1/2}$ and $p_{1/2}$ electrons behave equally, so that Q_{vac} must be a multiple of 4. Thus this model enables only the limiting charge of a point nucleus with an uncertainty $\Delta Z = 4$ to be determined.

For a nucleus with $Z \neq 1/\alpha + 4n$, the question remains which shell of the electronic charge density is the last one that dives, and to what extent it contributes to the vacuum charge. Since $1/\alpha$ is not an integer, it is also interesting whether the vacuum charge may assume non-integer values. To investigate these questions, we have to discuss in more detail what happens at the outer turning point, trying to get a more precise statement of the limiting charge for a point nucleus.

Therefore we have to look at the simplifications we made. Taking the exact quasi-classical momentum defined by (16.88), it is clearly different for states with a different sign of $\varkappa$. This splits levels negligibly for most of the vacuum states, but it can become crucial for the decision whether a state screened by nearly $Z_N - 137$ electrons becomes overcritical or not [Mi 76]. It could happen that a $ns_{1/2}$ state dives, while the $(n+1)p_{1/2}$ state remains undercritical. Analogously, the magnetic moment of the nucleus causes hyperfine structure splitting, so that electrons with different quantum number μ separate. These considerations show that we must deal with a single electron and not with entire shells of four electrons, when determining the precise value of the vacuum charge.

We now tackle the question which is the last state that reaches the lower continuum in a shrinking nucleus? We know that the quasi-classical momentum for $E < -m_0$ has the shape shown in Fig. 16.10, and a resonance is defined by the Bohr-Sommerfeld condition applied to the classically allowed region I. Normalizable semiclassical solutions at other energies, which do not fulfil this condition, cannot be constructed, because they would be connected to an exponentially increasing part in the forbidden region II (the solutions must be connected by Airy functions in the region near the turning points, where WKB is not valid). The outer turning point is defined by the condition $q^2(r_+) = 0$, where r_+ is a function of E (Fig. 16.11). For all energies $E < -m_0$, the exact turning point lies closer to the origin than the point where $\gamma^2(r) = 0$, so that this condition gives an upper limit for the vacuum charge: $Q_{vac} < Z_N - 1/\alpha$. On the other hand, the vacuum charge will always be at least $Z_N - 1/\alpha - 1$, if the nuclear radius is chosen sufficiently small. Thus we can conclude that in this model, where $\gamma(r)$ is calculated with the Poisson equation, the limiting charge for a point nucleus is larger than $1/\alpha$, and its upper limit is $1/\alpha + 1$. The uncertainty comes from the difficulty to calculate exactly the charge located in the exponentially decreasing tail in region II.

The calculations in the last section yielded an upper limit of $1/\alpha$ for the charge of the point nucleus. The difference between these two results arises since the approximations used there subtract the self-interaction of the electrons, while the considerations leading to the limit $1/\alpha + 1$ include the self-interaction. The problem is that we did not consider the exchange interaction which decreases the self-interaction. Thus, in our semiclassical model, we cannot give a precise value of the limiting charge, which must be calculated including the exchange interaction.

The question remains as to what happens with the electrons that cannot become overcritical. Therefore we look at the quasi-classical momentum at energies $E > 0$ for an electron screened by more than $Z_N - 137$ electrons. Figure 16.15 shows that for small r the vacuum charge is located in the region with $q^2 > 0$. This part is scaled with the nuclear radius R. For large r, there is the region of the normal electronic states. Between these two regions, q^2 is nearly equal to zero.

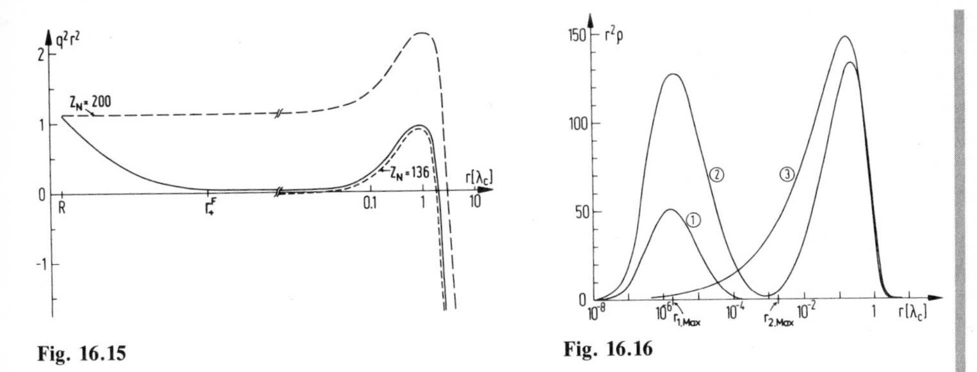

Fig. 16.15

Fig. 16.16

Fig. 16.15. The function q^2r^2 at the energy $E = 0.5\,m_0$. (– – –): computed without approximations in the single-particle case for $Z_N = 136$ and 200. (———): with screening for $Z_N = 200$. The wiggling structure according to Fig. 16.12 for $R < r < r_+^F$ is not shown

Fig. 16.16. Particle densities (in arbitrary units) for a nucleus with $R_0 = 10^{-5}$ fm: ① $Z = 150$, $1s_{1/2}$ state without screening, multiplied by the factor 10^{-5}, $E = -3.652\,m_0$; ② $Z = 150$, $2s_{1/2}$ state screened by 15 $1s_{1/2}$ electrons, $E = 0.217\,m_0$; ③ $Z = 135$, $1s_{1/2}$ state without screening, $E = 0.173\,m_0$. $r_{1.\max}$ and $r_{2.\max}$ are the radii where the maxima of the unscreened density of the $2s_{1/2}$ state would be located

This reveals the character of the vacuum function of a state that cannot reach the lower continuum due to the screening effects.

The innermost part of its charge density behaves like the dived states and shrinks together with the nucleus, while the outer maxima which contain nearly the total charge behave like a normal wave function at $Z < 1/\alpha$. Such a wave function computed with a Dirac-Hartree programme is shown in Fig. 16.16. To simulate this effect in a numerically practicable way, we computed an electronic wave function screened by 15 $1s$ electrons instead of a charge distribution described by (16.106).

One sees that only the outer maximum of the $2s_{1/2}$ state is affected by the screening, so that it resembles the charge density of a state with $Z = 135$, whereas the inner maximum shrinks with the nucleus like the unscreened wave function.

We conclude from this discussion that there is an upper limit for the charge of a point nucleus, because the screening charge of the vacuum degenerates to a point, too. The dived electrons reach an arbitrarily large binding energy, while the outer states remain undercritical, showing the behaviour described above.

16.5.3 Influence of Heavier Leptons

When discussing the problem of a point nucleus, we have to take into account that other leptons exist beside the electron. For $Z_N > 137$ the muon, for example, will also become overcritical in the field of a sufficiently small nucleus. Therefore we have to investigate how the Dirac equation behaves for particles with different masses.

The Dirac equation written in natural units $\hbar = c = m_0 = 1$ has the same solutions for all particles. Replacing m_e by another mass m_x in the Dirac equation is equivalent to a scale transformation. Only the external potential is independent of the mass, so that its value changes when the units changed: the nuclear radius R must be replaced by $R\, m_x/m_e$. This leads to a shift of the diving point R_0^{cr} by a factor m_e/m_x. In the limit $R \to 0$ for a fixed value of R the wave functions of all particles have the same position and the same binding energy. This is seen directly from the Dirac equation which becomes independent of the rest mass in the limit $|E| \to \infty$. To illustrate: with $m_\mu = 105.66$ MeV and $m_\tau = 1{,}800$ MeV, and neglecting recoil effects, the following critical radii for a nucleus with charge $Z_N = 150$ for the $1s_{1/2}$ single-particle state arise:

$$R_0^{cr}(\mu) = 2 \times 10^{-4}\,\text{fm}$$

$$R_0^{cr}(\tau) = 1.2 \times 10^{-5}\,\text{fm}\,.$$

Figure 16.7 shows that $R_0^{cr}(\mu)$ lies between the diving radii of the $2p_{1/2}$ and the $2s_{1/2}$ electron states. Therefore only four electrons have already become overcritical when the $1s_{1/2}$ muons dive, so that the diving of the muons is not prevented by electron screening. Adopting the results obtained for electrons in the last chapter, we can conclude that in general a nucleus with charge $Z_N > 1/\alpha$ in the limit $R_0 \to 0$ is surrounded by shells of overcritical leptons, where each shell contains the charge of 4 electrons, 4 muons, etc., until the shrinking nucleus is screened to the limiting charge of a point nucleus.

The question of the effects of virtual vacuum polarization remains. At first sight it seems reasonable to take this into account by replacing the coupling constant α of the classical theory by the "running" coupling constant $\alpha(r)$ found as solution of the renormalization group equations. For small distances r this is given by

$$\alpha(r) \sim \frac{\alpha}{1 + \dfrac{2\alpha}{3\alpha}\ln(m_0 r)}\,. \tag{16.114}$$

As r becomes very small, $\alpha(r)$ grows arbitrarily large. This would mean that for every source of strength Ze, even with $Z < \alpha^{-1}$ a radius R would exist such that the effective strength in its immediate vicinity, $Z\alpha(R)$, would exceed one. Thus every charge would become supercritical as its radius shrinks toward zero.

The validity of the foregoing argument is questionable for two reasons. The physical reason is that the correction in $\alpha(r)$ becomes as large as α itself only at distances smaller than 10^{-278} fm. This is a scale where effects of other interactions, in particular of quantum gravity, have long become important. It is certainly unreasonable to assume that QED can be extrapolated below distances corresponding to the Planck length 10^{-20} fm. The second reason is of a formal nature: if every charge shielded itself by real vacuum polarization, this would also be true of the elementary particles that are supposed to constitute the hypothetical source that shrinks to zero extension. For these, however, the bare charge cannot be observed and we must thus conclude that, in this case, the real vacuum charge would have to be renormalized. This would then also be true for the shielding charge of the nucleus. In addition, the "real" and "virtual" vacuum polarization can no longer be considered as linearly additive phenomena under these conditions. It is evident that this problem cannot be properly discussed without careful treatment of the renormalization process. Because of the unpleasant ultraviolet properties of QED it is possible that the virtual polarization effects cannot be rigorously included unless QED is imbedded in a unified (ultraviolet-free?) gauge theory of all interactions.

16.6 Klein's Paradox Revisited

We now have the necessary background to understand how to resolve Klein's paradox. The key to this resides in the observation made in Sect. 16.4 that a very large external charge density leads to a self-consistent potential step whose depth is determined by the charge density (16.45). Imagine for the purpose of this discussion that a large charged object is investigated. Inside the state illustrated in Fig. 16.17 pertains.

If tunnelling through the potential well the surface can be ignored, then one can view the inside of the charge distribution as a separate world, and the fact that there is a substantial inside potential may be simply gauged away. The inside can thus be viewed as a filled Fermi surface, beginning at the lowest energy m_0 (formerly $m_0 + V$). If there is communication with the outside world, then *even* if such a neutralizing charge is initialy absent, it would be spontaneouslly created up to the indicated energy $-m_0$ in Fig. 16.17, with reference to outside world.

Of course, for any macroscopic object the rise of the potential is too slow to produce any likelihood of overlap between inside and outside worlds. The criterion for this is that the wavelength of a given inside state can leak out through the barrier indicated by the two dots in Fig. 16.17:

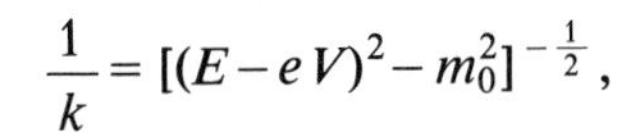

$$\frac{1}{k} = [(E - eV)^2 - m_0^2]^{-\frac{1}{2}}, \tag{16.115}$$

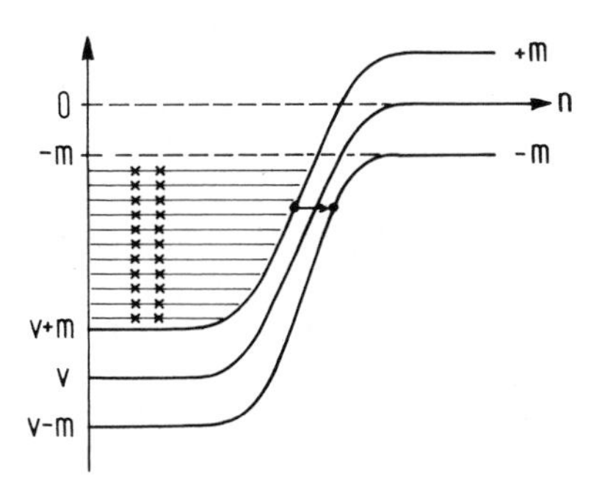

Fig. 16.17. Crosses indicate states for $-m_0 > E > V + m_0$ within the potential well filled by spontaneous vacuum decay

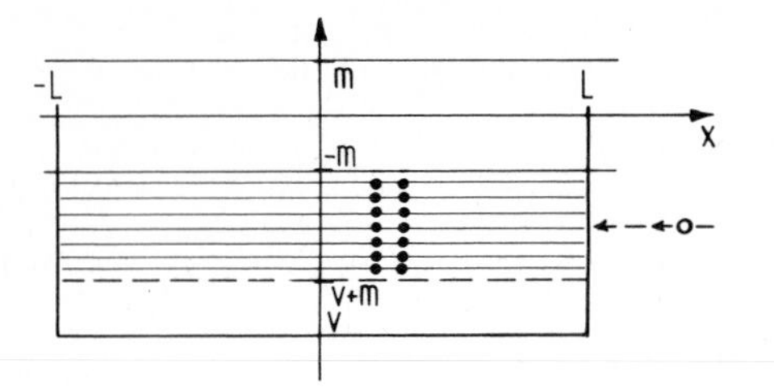

Fig. 16.18. Scattering of a positron of a negative potential step in the Klein framework

which as indicated can hardly be more than the electron Compton wavelength. But most real barriers are wider, dissociating the two worlds effectively.

Not so for a steep barrier considered in the context of Klein's paradox, Chap. 5. Such a suddenly rising barrier allows full communication between inside and outside, thus it can be established only self-consistently with the vacuum charge contributing to it. It is nonetheless worth recording that we actually cannot have a step as a solution to (16.40), except for highly singular surface charge distributions ϱ_N. However, given the popularity of Klein's paradox and what has been said and written about it, it is worthwhile even for such an unphysical potential scattering to understand it properly in the context of supercritical fields.

Before proceeding with our discussion of the Klein's paradox, we wish to emphasize that for didactic convenience, it is more appropriate to use a potential step with an opposite sign to that in Chap. 5, Fig. 5.1. Figure 16.18 presents a situation where all states below $E = -m_0$ are occupied and the ground state of reference is the charged vacuum.

We assume that the potential step is only in the x direction, i.e. that there are two parallel potential walls. Then the solutions of the Dirac equation are as determined in Sect. 3.1. In particular, recall that for certain energies there is substantial probability that a positron is transmitted through a barrier associated with the overcritical bound state. However, in the study of the one-step potential (Chap. 5), induced pair production is recorded. The difference in this physical behaviour arises from the mathematical detail that there is a second boundary of the potential – that for the properly redefined (charged) vacuum state, the positron either is resonantly transmitted through the barrier or is reflected. *Only* if the state of reference is the neutral vacuum will the incoming positron have an anomalous pattern of scattering characteristics of the Klein paradox, as it can stimulate the transition to the charged vacuum.

Hence the anomalous behaviour of single-particle scattering states of the Klein potential step is traced back to the improperly defined vacuum reference state.

The same remark applies to the phenomenon of pair production in a constant electric field as described by (14.49): in this instance of a linearly rising potential, the effective action has been computed with reference to the neutral ground state and spontaneous pair production is encountered. The vacuum state sparks on its way to the proper ground state and pair production continues until the true ground state is achieved.

Bibliographical Notes

Reinhard et al. [Re 71] give an extensive derivation of the self-consistent equations from Dyson-Schwinger equations, but do not evaluate the exchange integral explicitly. *Mohr* [Mo 74] applies the Wiechman-Kroll method [Wi 56] to compute the self-energy corrections in heavy atoms and this work is further extended to critical potentials by *Soff* et al. [So 82b]

Supercharged vacua have been discussed at length in [Re 77, Ra 78b, Gä 81]. Self-consistent supercritical states in the absence of an external potential have been discussed briefly in [Ra 76d].

17. Bosons Bound in Strong Potentials

This book would remain incomplete without a discussion of the behaviour of Bose particles in supercritical external fields. Indeed, the physics explored in the following sections differs fundamentally from the conclusions reached for fermions in arbitrarily strong external potentials. It is important to observe here that our intuition, sharpened by experience with classical physics, leads to a reasonably correct picture for bosons, contrary to the case for Fermi particles. Namely, the behaviour of condensed Bose particles is similar to that of classical, rather than quantum mechanical systems. In particular, we describe how in supercritical external potentials the stability of the vacuum state is assured by the appearance of a Bose condensate whose role is to screen the effective interaction down to a subcritical strength. When a very substantial screening of an extremely strong applied field occurs, a large number of condensed Bose particles form a macroscopic coherent state, similar in nature to the electromagnetic laser field. However, the validity of this analogy is limited, as in our case the condensed particles are bound and remain near the source of the supercritical interaction.

The development presented here is at first glance, for example for pions, of academic interest only. Namely, as is the case with muons, the relatively large mass of the pion compared to that of an electron makes it unlikely that pion supercritical fields are formed in a laboratory experiment. However, in quantum chromodynamics, the theory of strongly interacting particles, the carriers of the force – gluons – which like photons are spin-one bosons, are subject to such self-interaction that one is able to derive an identical set of field equations to that for spinless pions interacting with a Coulombic source. These developments deserve urgent attention.

As this chapter can be read omitting much of the prior developments and can be viewed at a fairly introductory level, we return to a notation with $\hbar$ and c given explicitly.

17.1 The Klein-Gordon Field

This Section treats the relativistic quantum theory of a charged spin-zero field. While the original motivation for considering this subject was the study of pions in the (Coulomb) field of a nucleus, we have learned since that the development is also a necessary step towards understanding the quantum theory of strongly interacting fields and particularly quantum chromodynamics (QCD). We proceed here in analogy to Chap. 2.

The Schrödinger wave equation is easily generalized to describe the relativistic motion. Recall the relativistic energy-momentum relation

$$(m_0c^2)^2 = E^2 - |\boldsymbol{p}|^2c^2 \tag{17.1}$$

and the quantization prescription

$$E \to \mathrm{i}\hbar\frac{\partial}{\partial t} \tag{17.2a}$$

$$\boldsymbol{p} \to -\mathrm{i}\hbar\nabla\,. \tag{17.2b}$$

Inserting (17.2a, b) into (17.1) gives the so-called Klein-Gordon [Kl 26, Go 26] wave equation. The Klein-Gordon equation was originally also proposed by Schrödinger [Schr 26], and independently by a number of other authors [Fo 26, Ku 26, deD 26]:

$$\left(\mathrm{i}\hbar\frac{\partial}{\partial t}\right)^2\phi(\boldsymbol{x},t) = (-\mathrm{i}\hbar c\nabla)^2\phi(\boldsymbol{x},t) + (m_0c^2)^2\phi(\boldsymbol{x},t)\,. \tag{17.3}$$

This form is significantly simpler in appearance than the Dirac wave equation studied in Chap. 2. Indeed, it was also the first relativistic wave equation to be considered and was initially discarded since the conserved probability current [see (17.29)] was seemingly not a positive definite quantity. Hence the wave function $\phi(\boldsymbol{x},t)$ could not always be normalized, the norm being the integral of the time-like component of the current:

$$N = \int d^3x\,\varrho_0[\phi] = \int d^3x\left[\phi^*\left(\mathrm{i}\hbar\frac{\partial}{\partial t}\phi\right) + \left(-\mathrm{i}\hbar\frac{\partial}{\partial t}\phi^*\right)\phi\right]. \tag{17.4}$$

For an eigensolution

$$\phi = \exp(-\mathrm{i}\,\omega_n t/\hbar)\,\varphi_n(\boldsymbol{x}) \tag{17.5}$$

the norm becomes

$$N_n = 2\,\omega_n\int d^3x\,|\varphi_n(\boldsymbol{x})|^2 \tag{17.6}$$

and explicitly depends on the signature of the frequency ω_n. To determine what values ω_n can assume, i.e. the spectrum of the free Klein-Gordon equation, insert (17.5) in (17.3), which yields

$$\omega_n^2\varphi_n(\boldsymbol{x}) = [(-\mathrm{i}\hbar c\nabla)^2 + (m_0c^2)^2]\,\varphi_n(\boldsymbol{x})\,. \tag{17.7}$$

The wave function $\varphi_n(\boldsymbol{x})$ is a plane wave,

$$\varphi_p(\boldsymbol{x}) = N_p^{1/2}\,\mathrm{e}^{\mathrm{i}\boldsymbol{p}\cdot\boldsymbol{x}/\hbar} \tag{17.8}$$

which, when inserted into (17.7), fulfils that equation, provided that the spectral condition for $\omega(\boldsymbol{p})$ is satisfied

$$\omega(p) = \pm\sqrt{|p|^2 c^2 + (m_0 c^2)^2}\,. \tag{17.9}$$

(Note that the lower index n has been replaced by the explicit momentum symbol p.) Of course this is expected in view of (17.1). However, we have now found that $\omega(p)$ can be negative and hence (17.6) has been shown to lead to negative "norm".

To determine the correct proportionality constant N_p in (17.8) for continuum normalization, consider the factor $2\,\omega_n$ in (17.6):

$$\varphi_p(x) = \frac{1}{\sqrt{2\,|\omega(p)|(2\,\pi)^3\hbar^3}}\,\mathrm{e}^{\mathrm{i}p\cdot x/\hbar}\,, \tag{17.10}$$

where the absolute value of the frequency enters for calculational convenience when dealing with modes with negative frequencies. For continuous states, the orthonormality relation can be found in (17.6). The wave function inner product is defined by

$$\pm\,\delta^3(p-p') = [\omega(p)+\omega(p')]\int d^3x\,\varphi_{p'}^*(x)\,\varphi_p(x)\,, \tag{17.11}$$

where the positive sign applies to positive frequencies, while the negative sign applies to negative frequencies in the spectrum.

Here a remark about the continuum-normalization is in order: for $p \to p'$ the lhs diverges as the volume V in which the plane wave is contained. By enclosing the entire system in a finite volume, the spectrum would be discrete, and we would have to include an additional factor $1/\sqrt{V}$ in the normalization of the wave function, which replaces the Fourier factor $1/\sqrt{(2\,\pi)^3}$ for continuum states.

Without a proper quantum field theoretical formulation, the sign ambiguity in (17.11) is not acceptable theoretically, as it seemingly contradicts the probability interpretation of quantum mechanics. This observation initially led to the discarding of the Klein-Gordon equation and subsequent development of the Dirac wave equation. The Klein-Gordon equation was later revived when the relation between spin and the form of the wave equation was recognized. While spin-$\frac{1}{2}$ particles are well described directly by the Dirac wave equation, the seemingly simpler case of spin-zero particles requires quantum field theory, in order to be able to interpret the wave equation *properly*. Indeed, *Pauli* and *Weisskopf* [Pa 34] showed that the signs in (17.11) correspond to two different possible charge states. The current is then not the probability current, but it is the charge current. Note that it is characteristic of the relativistic generalizations of particle equations of motion that they often involve both signs of the charge, and therefore, simultaneously both particle and antiparticle degrees of freedom.

Rather than proceed with the intuitive development, we consider the consequences of choosing the proper action for the Klein-Gordon field.

As usual, we assume that the action I is given by the covariant integral of the Lagrangian density $\mathscr{L}$ [cf. (8.2)]

$$I = \int d^4x\,\mathscr{L}\,[\phi, \phi^*, \partial_\mu\phi, \partial_\mu\phi^*]\,, \tag{17.12}$$

where the covariant derivatives have been introduced. Here, as before, $x^0 = x_0 = ct$ and $\partial_\mu = \partial/\partial x^\mu$, and hence

$$\partial_\mu \phi = \left(\frac{\partial}{c\partial t} \phi, \nabla \phi \right). \tag{17.13}$$

Recall that $x_\mu = (ct, -\boldsymbol{x})$, $x^\mu = (ct, \boldsymbol{x})$.

The Lagrangian density is a function of the fields and its first derivatives only. While higher order derivatives of the fields in $\mathscr{L}$ are in principle possible, they lead to the occurrence of several further massive fields simultaneously, a complication which is not essential for the developments we are pursuing [Da 72].

The requirement of least action under an arbitrary variation of the fields (Hamilton's principle) leads to the field equations, Chap. 8,

$$\frac{\partial}{\partial x_\mu} \frac{\partial \mathscr{L}}{\partial \left(\dfrac{\partial \phi^*}{\partial x_\mu} \right)} = \frac{\partial \mathscr{L}}{\partial \phi^*} \tag{17.14}$$

and similarly for the linearly independent field ϕ. Indeed, to recover (17.3) in covariant notation

$$\partial_\mu \partial^\mu \phi = -\left(\frac{m_0 c}{\hbar} \right)^2 \phi, \tag{17.15}$$

we must take

$$\mathscr{L} = (\hbar c \partial_\mu \phi^*)(\hbar c \partial^\mu \phi) - (m_0 c^2 \phi^*)(m_0 c^2 \phi). \tag{17.16}$$

Let us illustrate this in more detail. Consider the differentiation set out in (17.14), applied to the Lagrangian (17.16) with respect to ϕ^* and $\partial^\mu \phi^*$:

$$\frac{\partial \mathscr{L}}{\partial \phi^*} = -(m_0 c^2)^2 \phi, \tag{17.17a}$$

$$\partial_\mu \frac{\partial \mathscr{L}}{\partial(\partial_\mu \phi^*)} = c\hbar \partial_\mu c\hbar \partial^\mu \phi, \tag{17.17b}$$

which gives (17.15) when inserted in (17.14).

Next, consider the Hamiltonian which follows from the choice of the Lagrangian (17.16).

In non-covariant notation for the Lagrangian

$$\mathscr{L} = \left(-\mathrm{i}\hbar \frac{\partial}{\partial t} \phi^* \right)\left(\mathrm{i}\hbar \frac{\partial}{\partial t} \phi \right) - (\mathrm{i}\hbar c \nabla \phi^*) \cdot (-\mathrm{i}\hbar c \nabla \phi) - (m_0 c^2)^2 \phi^* \phi. \tag{17.18}$$

The fields π, π^* canonically conjugate to the complex and hence linearly independent fields ϕ, ϕ^* are

$$\pi = \frac{\partial \mathscr{L}}{\partial \dot{\phi}} = \hbar^2 \dot{\phi}^* \tag{17.19a}$$

$$\pi^* = \frac{\partial \mathscr{L}}{\partial \dot{\phi}^*} = \hbar^2 \dot{\phi} , \tag{17.19b}$$

and the Hamiltonian density becomes

$$\begin{aligned} \mathscr{H}[\pi, \pi^*, \phi, \phi^*] &= \pi \dot{\phi} + \pi^* \dot{\phi}^* - \mathscr{L} \\ &= \frac{1}{\hbar^2} \pi^* \pi + (\hbar c \nabla \phi^*) \cdot (\hbar c \nabla \phi) + (m_0 c^2)^2 \phi^* \phi , \end{aligned} \tag{17.20}$$

while the Hamiltonian is

$$H = \int d^3x \, \mathscr{H}(\boldsymbol{x}, t) . \tag{17.21}$$

It is useful to verify that the Hamiltonian equations of motion indeed correspond to the Lagrangian equations:

$$\dot{\phi} = \frac{\delta H}{\delta \pi} = \frac{1}{\hbar^2} \pi^* , \tag{17.22a}$$

$$\dot{\phi}^* = \frac{\delta H}{\delta \pi^*} = \frac{1}{\hbar^2} \pi , \tag{17.22b}$$

$$\dot{\pi} = -\frac{\delta H}{\delta \phi} = (\hbar c \nabla)^2 \phi^* - m_0 c^2 \phi^* , \tag{17.22c}$$

$$\dot{\pi}^* = -\frac{\delta H}{\delta \phi^*} = (\hbar c \nabla)^2 \phi - m_0 c^2 \phi . \tag{17.22d}$$

The respective pairs of equations a – d, b – c, are fully equivalent to the Lagrangian (17.15) or (17.3). Observe, however, the appearance of two first-order equations in place of one second-order equation, which is exploited thoroughly below.

The Klein-Gordon waves carry energy and momentum, which can be obtained from the general form of the energy momentum tensor discussed in Chap. 8:

$$T_{\mu\nu} = \mathscr{L} g_{\mu\nu} - \left(\frac{\partial \mathscr{L}}{\partial(\partial^\mu \phi)} \partial_\nu \phi + \frac{\partial \mathscr{L}}{\partial(\partial^\mu \phi^*)} \partial_\nu \phi^* \right) . \tag{17.23}$$

The conserved current of the waves is also obtained from the Lagrangian (cf. Chap. 9)

$$j_\mu = -\frac{\mathrm{i}e}{\hbar c}\left(\frac{\partial \mathcal{L}}{\partial(\partial^\mu \phi)}\phi - \frac{\partial \mathcal{L}}{\partial(\partial^\mu \phi^*)}\phi^*\right), \tag{17.24}$$

where e is an arbitrary, but constant unit of charge. We now proceed to verify that (17.24) defines a conserved four current. This requires

$$\begin{aligned} 0 = \partial^\mu j_\mu = &-\frac{\mathrm{i}e}{\hbar c}\left(\frac{\partial \mathcal{L}}{\partial(\partial^\mu \phi)}\partial^\mu \phi - \frac{\partial \mathcal{L}}{\partial(\partial^\mu \phi^*)}\partial^\mu \phi^*\right) \\ &-\frac{\mathrm{i}e}{\hbar c}\left[\phi\partial^\mu\left(\frac{\partial \mathcal{L}}{\partial(\partial^\mu \phi)}\right) - \phi^*\partial^\mu\left(\frac{\partial \mathcal{L}}{\partial(\partial^\mu \phi^*)}\right)\right]. \end{aligned} \tag{17.25}$$

Equation (17.14) and its conjugate, when multiplied with ϕ^* and, respectively ϕ, give

$$\phi^*\partial^\mu\left(\frac{\partial \mathcal{L}}{\partial(\partial^\mu \phi^*)}\right) = \frac{\partial \mathcal{L}}{\partial \phi^*}\phi^* \tag{17.26a}$$

$$\phi\partial^\mu\left(\frac{\partial \mathcal{L}}{\partial(\partial^\mu \phi)}\right) = \frac{\partial \mathcal{L}}{\partial \phi}\phi . \tag{17.26b}$$

Inserting this into (17.25) gives

$$\begin{aligned} \partial^\mu j_\mu = &-\frac{\mathrm{i}e}{\hbar c}\left(\frac{\partial \mathcal{L}}{\partial(\partial^\mu \phi)}\partial^\mu \phi + \frac{\partial \mathcal{L}}{\partial \phi}\phi\right) \\ &+\frac{\mathrm{i}e}{\hbar c}\left(\frac{\partial \mathcal{L}}{\partial(\partial^\mu \phi^*)}\partial^\mu \phi^* + \frac{\partial \mathcal{L}}{\partial \phi^*}\phi^*\right). \end{aligned} \tag{17.27}$$

Using the explicit form (17.16) of the Lagrangian we find that each of the square brackets is itself the Lagrangian, since $\mathcal{L}$ is a linear combination of ϕ and $\partial^\mu \phi$ as well as of ϕ^* and $\partial^\mu \phi^*$

$$\partial^\mu j_\mu = -\frac{\mathrm{i}e}{\hbar c}[\mathcal{L} - \mathcal{L}] = 0 . \tag{17.28}$$

The explicit form of the current (17.24) is, using (17.16),

$$j_\mu = -\mathrm{i}e\hbar c(\phi\partial_\mu\phi^* - \phi^*\partial_\mu\phi) , \tag{17.29}$$

which in non-covariant notation with $j_\mu = (\varrho, -j/c)$ becomes

$$\varrho = -\mathrm{i}e\hbar\left(\phi\frac{\partial}{\partial t}\phi^* - \phi^*\frac{\partial}{\partial t}\phi\right) \tag{17.30a}$$

$$j = e\hbar c^2(\phi\nabla\phi^* - \phi^*\nabla\phi) . \tag{17.30b}$$

We can now turn to the canonical quantization of the Klein-Gordon field.

In this conventional method to quantize a Bose field, assume that the fields and their conjugate momentum fields (17.19) satisfy the following equal time commutation relations:

$$[\hat{\phi}(\boldsymbol{x},t),\ \hat{\pi}(\boldsymbol{x}',t)] = \mathrm{i}\hbar\delta^3(\boldsymbol{x}-\boldsymbol{x}') \tag{17.31a}$$

$$[\phi^*(\boldsymbol{x},t),\ \hat{\pi}^*(\boldsymbol{x}',t)] = \mathrm{i}\hbar\delta^3(\boldsymbol{x}-\boldsymbol{x}') \tag{17.31b}$$

while all other commutators vanish at equal time. With this choice, the generalized Heisenberg equation for the time rate of change of any function of the fields and their conjugates $f[\hat{\phi},\hat{\phi}^*,\hat{\pi},\hat{\pi}^*]$

$$\mathrm{i}\hbar\dot{\hat{f}} = [\hat{f},\hat{H}]_- \equiv \hat{f}\hat{H}-\hat{H}\hat{f} \tag{17.32}$$

is consistent with the field equations in that the equations of motion for the quantum fields $\hat{\phi}$, $\hat{\phi}^*$, $\hat{\pi}$ and $\hat{\pi}^*$ following from (17.32) are identical to (17.22a – d). This is an important observation as it allows the Klein-Gordon equation to become, upon quantization, an operator-valued equation of the same analytic form.

Another important consequence is the conservation of charge carried by the Klein-Gordon field. With (17.30a, 19) the charge operator is

$$\hat{q} = \int d^3x\,\hat{\varrho} = -\frac{\mathrm{i}e}{\hbar}\int d^3x(\hat{\phi}\hat{\pi}-\hat{\pi}^*\hat{\phi}^*)\,, \tag{17.33a}$$

where the ordering of the fields in (17.33) ensures that $\hat{q}$ is a Hermitian operator

$$\hat{q}^+ = \hat{q}\,. \tag{17.33b}$$

The time development of the charge operator is obtained by evaluating (17.32)

$$\mathrm{i}\hbar\dot{\hat{q}} = [\hat{q},\hat{H}]\,. \tag{17.34a}$$

Our aim is now to show that like the case of the spin-$\frac{1}{2}$ particles [Chap. 9 and (9.93)]

$$[\hat{q},\hat{H}] = \mathrm{i}\hbar\oint dS\boldsymbol{n}\cdot\hat{\boldsymbol{j}}\,, \tag{17.34b}$$

where $\boldsymbol{n}$ is the normal surface S unit vector. The rhs of (17.34b) describes the outflow of charge from the volume considered. For stationary states, i.e. eigenstates of $\hat{H}$, there is no current, thus $\hat{q}$ and $\hat{H}$ can be diagonalised simultaneously.

To obtain (17.34b) consider the equal time commutator (dropping the argument t for convenience):

$$\begin{aligned}[q,\hat{H}] &= \frac{\mathrm{i}e}{\hbar}\int d^3x'([\hat{\phi}(\boldsymbol{x}')\hat{\pi}(\boldsymbol{x}'),\hat{H}]-[\hat{\pi}^*(\boldsymbol{x}')\hat{\phi}^*(\boldsymbol{x}'),\hat{H}])\\ &= \frac{\mathrm{i}e}{\hbar}\int d^3x'\int d^3x\left\{\frac{1}{\hbar^2}\hat{\pi}^*(\boldsymbol{x})[\hat{\phi}(\boldsymbol{x}'),\hat{\pi}(\boldsymbol{x})]\hat{\pi}(\boldsymbol{x}')\right.\end{aligned}$$

$$\left. + \hbar c \nabla_x \hat{\phi}^*(x)\, \hat{\phi}(x')\, [\hat{\pi}(x'), \mathrm{i}\hbar c \nabla_x \hat{\phi}(x)] + (m_0 c^2)^2 \hat{\phi}^*(x)\, \hat{\phi}(x')\, [\hat{\pi}(x'), \hat{\phi}(x)] + \text{h.c.} \right\} , \tag{17.35}$$

which is obviously an anti-Hermitian operator.

Aside from the usual commutators (17.31), one also needs here

$$\begin{aligned} [\hat{\pi}(x'), \hbar c \nabla_x \hat{\phi}(x)] &= \hbar c \nabla_x [\hat{\pi}(x'), \hat{\phi}(x)] = -\mathrm{i}\hbar^2 c \nabla_x \delta^3(x - x') \\ &= \mathrm{i}\hbar^2 c \nabla_{x'} \delta^3(x - x') . \end{aligned} \tag{17.36}$$

Thus

$$\begin{aligned} [\hat{q}, \hat{H}] &= \frac{\mathrm{i}e}{\hbar} \int d^3x \left\{ \frac{\mathrm{i}}{\hbar} \hat{\pi}^*(x)\, \hat{\pi}(x) - \mathrm{i}\hbar (m_0 c^2)^2 \hat{\phi}(x)\, \hat{\phi}^*(x) + \text{h.c.} \right\} \\ &+ \frac{\mathrm{i}e}{\hbar} \int d^3x \int d^3x'\, \hbar c \{ \nabla_x \hat{\phi}^*(x)\, \hat{\phi}(x')\, \mathrm{i}\hbar^2 c \nabla_{x'} \delta^3(x - x') + \text{h.c.} \} . \end{aligned} \tag{17.37}$$

The first term above is cancelled by its Hermitian conjugate (h.c.).

The second requires a more careful evaluation, recording the surface terms:

$$\begin{aligned} [\hat{q}, \hat{H}] &= -e\hbar^2 c^2 \int d^3x\, d^3x' \,[\nabla_x \hat{\phi}^*(x)\, \hat{\phi}(x') - \hat{\phi}^*(x')\, \nabla_x \hat{\phi}(x)]\, \nabla_{x'} \delta^3(x - x') \\ &= -e\hbar^2 c^2 \int d^3x [-\int d^3x' \,(\nabla_x \hat{\phi}^*(x)\, \nabla_{x'} \hat{\phi}(x') - \nabla_{x'} \hat{\phi}^*(x')\, \nabla_x \hat{\phi}(x))\, \delta^3(x - x') \\ &\quad - \int dS\, \boldsymbol{n} \cdot (\nabla_x \hat{\phi}^*(x)\, \hat{\phi}(x') - \hat{\phi}^*(x')\, \nabla_x \hat{\phi}(x))\, \delta^3(x - x')] ; \end{aligned} \tag{17.38}$$

where the last equality has been obtained by partial integration. Hence

$$[\hat{q}, \hat{H}] = -\mathrm{i}e\hbar c \int dS\, \boldsymbol{n} \cdot (\hat{\phi}(-\mathrm{i}\hbar c \nabla \hat{\phi}^*) + \hat{\phi}^*(\mathrm{i}\hbar c \nabla \hat{\phi})) , \tag{17.39}$$

which is the required relation (17.34b) in view of (17.30b). Here $\hat{H}$ and $\hat{q}$ are Hermitian operators which, because of (17.34), are simultaneously diagonalizable, with only real eigenvalues. The actual explicit determination of the spectrum using the Fock space is undertaken below. Before developing the quantum theory further, we first consider another form of the Klein-Gordon equation which is particularly suitable for the quantum treatment of bosons in the presence of an external potential.

17.2 Alternate Form of the Klein-Gordon Equation

Further developments can occasionally be greatly simplified if a formulation of the theory is employed which is dependent only on first derivatives in time. Such an approach is based on the hybrid two-component field $\hat{\chi}$:

$$\hat{\chi} \equiv \begin{pmatrix} \hat{\chi}_1 \\ \hat{\chi}_2 \end{pmatrix} \equiv \begin{bmatrix} k\hat{\phi} \\ \dfrac{\mathrm{i}}{k} \hat{\pi}^* \end{bmatrix} , \tag{17.40a}$$

where k is a real constant to be chosen appropriately. Note that the adjoint field is

$$\hat{\chi}^+ \equiv (\hat{\chi}_1^+, \hat{\chi}_2^+) = \left(k\hat{\phi}^*, -\frac{\mathrm{i}}{k}\hat{\pi}\right). \tag{17.40b}$$

We also introduce

$$\hat{\bar{\chi}} \equiv (\hat{\bar{\chi}}_1, \hat{\bar{\chi}}_2) = \left(-\frac{\mathrm{i}}{k}\hat{\pi}, k\hat{\phi}^*\right). \tag{17.40c}$$

Using (17.31), the equal time commutator of the hybrid is

$$[\hat{\chi}_i, \hat{\bar{\chi}}_j] = \begin{pmatrix} -\mathrm{i}[\phi, \hat{\pi}] & k^2[\phi, \phi^*] \\ \dfrac{1}{k^2}[\hat{\pi}^*, \hat{\pi}] & \mathrm{i}[\hat{\pi}^*, \phi^*] \end{pmatrix}_{ij}$$

$$= \hbar\delta^3(\boldsymbol{x}-\boldsymbol{x}')\,\delta_{ij}. \tag{17.41a}$$

Hence the new conjugate momentum is

$$(\hat{\pi}_\chi)_j = \mathrm{i}\hat{\bar{\chi}}_j \tag{17.41b}$$

and the description can be based on the new dynamical variables $(\hat{\chi}, \hat{\pi}_\chi)$.

We now describe the form of the Klein-Gordon equation which directly leads to (17.40, 41). We use this opportunity to introduce the electromagnetic field through the four-potential $A^\mu = (A^0, \boldsymbol{A})$. By virtue of the minimal coupling, the Klein-Gordon field equation now reads

$$\left(\mathrm{i}\hbar\frac{\partial}{\partial t} - eA_0\right)^2\phi = (-\mathrm{i}\hbar c\nabla - e\boldsymbol{A})^2\phi + (m_0c^2)^2\phi. \tag{17.42}$$

One often substitutes

$$V = eA^0 = eA_0. \tag{17.43}$$

Occasionally, but improperly, V is called the "scalar" electromagnetic potential, while $e\boldsymbol{A}$ is the electromagnetic vector potential.

Recall that many of the formal complications associated with the Klein-Gordon field are consequences of the explicit presence of second-order derivatives in time. We now present a modified *Feshbach-Villars* [Fe 58] form of (17.42) which avoids this difficulty. The price is loss of explicit relativistic covariance. We introduce two (complex) fields in a fashion motivated by (17.40):

$$\varphi^{(1)} = (m_0c^2)^{1/2}\phi \tag{17.44a}$$

$$\varphi^{(2)} = (m_0c^2)^{-1/2}\left(\mathrm{i}\hbar\frac{\partial}{\partial t} - eA_0\right)\phi, \tag{17.44b}$$

where the powers of (m_0c^2) have been chosen that the dimension of $\varphi^{(1)}$, $\varphi^{(2)}$ is $[L^{-3/2}]$. A trivial consequence of (17.44) is a relation between the components $\varphi^{(1)}$ and $\varphi^{(2)}$

$$\left(\mathrm{i}\hbar\frac{\partial}{\partial t}-eA_0\right)\varphi^{(1)}=m_0c^2\varphi^{(2)}. \tag{17.45a}$$

The Klein-Gordon equation (17.42) now assumes the form

$$\left(\mathrm{i}\hbar\frac{\partial}{\partial t}-eA_0\right)\varphi^{(2)}=\frac{(-\mathrm{i}\hbar c\nabla-e\boldsymbol{A})^2}{m_0c^2}\varphi^{(1)}+m_0c^2\varphi^{(1)}. \tag{17.45b}$$

Equations (17.45a, b) can be written more concisely when a two-component hybrid, cf. (17.40)

$$\chi=\begin{pmatrix}\varphi^{(1)}\\ \varphi^{(2)}\end{pmatrix} \tag{17.46a}$$

is introduced. Then

$$\left(\mathrm{i}\hbar\frac{\partial}{\partial t}-eA_0\right)\tau_1\chi=U\chi \tag{17.46b}$$

$$U\equiv(1+\tau_3)\frac{(-\mathrm{i}\hbar c\nabla-e\boldsymbol{A})^2}{2m_0c^2}+m_0c^2. \tag{17.46c}$$

Here the Pauli matrices in the standard representation have been employed:

$$\tau_1\equiv\sigma_1=\begin{pmatrix}0&1\\1&0\end{pmatrix};\quad \tau_2\equiv\sigma_2=\begin{pmatrix}0&-\mathrm{i}\\ \mathrm{i}&0\end{pmatrix};\quad \tau_3\equiv\sigma_3=\begin{pmatrix}1&0\\0&-1\end{pmatrix}. \tag{17.47}$$

Indeed, (17.46) is easily verified by using the above explicit representation of the Pauli matrices.

At this point (17.46) should be briefly reconsidered. Observe that this procedure corresponds to a "second" way of taking the root of $E^2=(cp)^2+m_0^2c^4$, discussed in the context of the derivation of the Dirac equation in Chap. 2. To see this, consider the square of both sides of the operator equality

$$\left(\mathrm{i}\hbar\frac{\partial}{\partial t}-eA_0\right)=\tau_1(1+\tau_3)\frac{(-\mathrm{i}\hbar c\nabla-e\boldsymbol{A})^2}{2m_0c^2}+m_0c^2\tau_1, \tag{17.48}$$

which becomes

$$\begin{aligned}\left(\mathrm{i}\hbar\frac{\partial}{\partial t}-eA_0\right)^2&=(m_0c^2)^2\tau_1^2+(-\mathrm{i}\hbar c\nabla-e\boldsymbol{A})^2\tfrac{1}{2}(\tau_1^2(1+\tau_3)+\tau_1(1+\tau_3)\tau_1)\\&=[(m_0c^2)^2+(-\mathrm{i}\hbar c\nabla-e\boldsymbol{A})^2]\tau_1^2,\end{aligned} \tag{17.49}$$

where we have already used

$$[\tau_1(1+\tau_3)]^2 = \tau_1(1+\tau_3)(1-\tau_3)\tau_1 = \tau_1(1-\tau_3^2)\tau_1 = 0\,. \tag{17.50}$$

Hence (17.46) is a root of the Klein-Gordon equation, if τ_1, τ_3 satisfy

$$\tau_1^2 = \tau_3^2 = 1 \tag{17.51a}$$

$$\tau_1\tau_3 = -\tau_3\tau_1\,. \tag{17.51b}$$

These equations define the Pauli matrices. Thus up to the choice of the representation of τ_1, τ_3 the form equation (17.46) is a unique and alternative way to the Dirac procedure of taking the root of $E^2 = (c\boldsymbol{p})^2 + m^2c^4$.

The Lagrangian from which (17.46) and its adjoint follow is easily worked out to be

$$\mathscr{L} = \chi^+\left(\mathrm{i}\hbar\frac{\partial}{\partial t} - eA_0\right)\tau_1\chi - \chi^+\left[(1+\tau_3)\frac{(-\mathrm{i}\hbar c\nabla - e\boldsymbol{A})^2}{2m_0c^2} + m_0c^2\right]\chi\,. \tag{17.52}$$

The conserved current $j^\mu = (\varrho, \boldsymbol{j}/c)$ can be derived by exploiting the implicit gauge invariance of (17.52). Consider the variation of the action under a change of A^μ:

$$\varrho(x) = -\frac{\delta}{\delta A_0(x)}\left[\int d^4x\,\mathscr{L}\right] = e\chi^+\tau_1\chi \tag{17.53a}$$

$$\begin{aligned}\boldsymbol{j}(x) &= \frac{c\delta}{\delta \boldsymbol{A}(x)}\left[\int d^4x\,\mathscr{L}\right]\\ &= ec\left[\chi^+(1+\tau_3)\frac{(-\mathrm{i}\hbar c\nabla - e\boldsymbol{A})}{2m_0c^2}\chi + \frac{(\mathrm{i}\hbar c\nabla - e\boldsymbol{A})}{2m_0c^2}\chi^+(1+\tau_3)\chi\right].\end{aligned} \tag{17.53b}$$

We have here exploited (9.131) and the general discussion presented in Chap. 9 relating to the changes under small variation of the electromagnetic potentials. It is a useful exercise to reinsert (17.44) in (17.53) to obtain the conserved current in terms of the original field φ:

$$\begin{aligned}\varrho(x) &= e\{\varphi^{*(1)}\varphi^{(2)} + \varphi^{*(2)}\varphi^{(1)}\}\\ &= e\left\{\phi^*\left(\mathrm{i}\hbar\frac{\partial}{\partial t} - eA_0\right)\phi + \left[\left(-\mathrm{i}\hbar\frac{\partial}{\partial t} - eA_0\right)\phi^*\right]\phi\right\}\end{aligned} \tag{17.54a}$$

$$\begin{aligned}\boldsymbol{j}(x) &= ec\left\{\varphi^{*(1)}\left[\frac{-\mathrm{i}\hbar c\nabla - e\boldsymbol{A}}{m_0c^2}\varphi^{(1)}\right] + \left[\frac{\mathrm{i}\hbar c\nabla - e\boldsymbol{A}}{m_0c^2}\varphi^{*(1)}\right]\varphi^{(1)}\right\}\\ &= ec\{\phi^*(-\mathrm{i}\hbar c\nabla - e\boldsymbol{A})\phi + [(\mathrm{i}\hbar c\nabla - e\boldsymbol{A})\phi^*]\phi\}\,.\end{aligned} \tag{17.54b}$$

This expression can be rewritten more concisely by introducing the differential operator

$$\frac{\overleftrightarrow{\partial}}{\partial x_\mu} = \overleftrightarrow{\partial}^\mu = \frac{\overrightarrow{\partial}}{\partial x_\mu} - \frac{\overleftarrow{\partial}}{\partial x_\mu} . \tag{17.55a}$$

Here the arrow indicates the direction in which the differential operates

$$A \overleftrightarrow{\partial}^\mu B = A(\partial^\mu B) - (\partial^\mu A) B . \tag{17.55b}$$

Thus

$$j^\mu = e[\phi^*(\mathrm{i}\hbar c \overleftrightarrow{\partial}^\mu)\phi - 2eA^\mu \phi^* \phi] \tag{17.56}$$

which for $A^\mu = 0$ is just (17.29), while (17.54) corresponds to (17.30) for ($A_0 = 0$, $\boldsymbol{A} = 0$). The only addition is the identification of the charge factor e, the coefficient of (17.56), with the charge of the Klein-Gordon field particle through (17.53).

Having seen the different forms of the conserved current, observe that (17.53a) introduces a measure

$$q = \int \varrho(x)\, d^3x = e \int \chi^+(x)\, \tau_1\, \chi(x)\, d^3x , \tag{17.57}$$

with is not positive definite, because of the appearance of the matrix τ_1. However, it is always real. Hence we can normalize the charge up to a sign

$$(q/e) = \pm 1 \tag{17.58}$$

with the exception of the (accidental) case $q = 0$. Recognize that the field χ contains both positively and negatively charged particle modes, once it is quantized.

For time-independent fields A_μ the posed problem can be considerably simplified by first understanding the stationary states of the wave equation (17.46). For a particular solution

$$\chi(\boldsymbol{x}, t) = \exp\left(-\frac{\mathrm{i}}{\hbar} E_n t\right) \chi_n(\boldsymbol{x}) \tag{17.59}$$

and the eigenfunctions χ_n satisfy the eigenvalue equation [cf. (17.46c)]

$$E_n \tau_1 \chi_1(\boldsymbol{x}) = \mathscr{H} \chi_n(\boldsymbol{x}) \tag{17.60a}$$

$$\mathscr{H} = eA_0 \tau_1 + (1 + \tau_3) \frac{(-\mathrm{i}\hbar c \nabla - e\boldsymbol{A})^2}{2m_0 c^2} + m_0 c^2 \tag{17.60b}$$

with the normalization condition following from (17.57):

$$\Theta_n = \int d^3x\, \chi_n^+ \tau_1 \chi_n , \tag{17.61}$$

where Θ_n is $+1$ for particles (negatively charged, observe that e is the electron charge), and -1 for antiparticles. In component form (17.60) becomes [cf. (17.45)]

$$(E_n - eA_0)\,\varphi^{(1)} = m_0c^2\varphi^{(2)} \tag{17.62a}$$

$$(E_n - eA_0)\,\varphi^{(2)} = \frac{(-\mathrm{i}\hbar c\nabla - e\boldsymbol{A})^2}{m_0c^2}\varphi^{(1)} + m_0c^2\varphi^{(1)}\,. \tag{17.62b}$$

At this point consider the simplest solutions, the plane waves of the Klein-Gordon field. In the free field limit, $A_\mu = 0$, we insert into (17.62) the ansatz

$$\varphi^{(j)} = \varphi^{(j)}_{(p)}\,\mathrm{e}^{\mathrm{i}p\cdot x/\hbar}\,, \tag{17.63}$$

and find that $\varphi^{(j)}_{(p)}$ satisfy

$$E(p)\,\varphi^{(1)}_{(p)} = m_0c^2\varphi^{(2)}_{(p)} \tag{17.64}$$

$$E(p)\,\varphi^{(2)}_{(p)} = \frac{(pc)^2}{m_0c^2}\varphi^{(1)}_{(p)} + m_0c^2\varphi^{(1)}_{(p)}\,. \tag{17.65}$$

Inserting (17.64) into (17.65) gives the (expected) relation for $E(p)$,

$$E^{\pm}(p) = \pm\sqrt{p^2c^2 + m_0^2c^4} = \pm\,\omega(p)\,, \qquad \text{where} \tag{17.66a}$$

$$\omega(p) = \sqrt{p^2c^2 + m_0^2c^4} = |E^{\pm}(p)|\,. \tag{17.66b}$$

Henceforth $E^{\pm}$ is the "frequency" of the mode which (contrary to the energy of the states) need not be positive definite. Of course there are no negative "frequencies" in a normal physics environment, but it is safe to refer to negative frequencies without being misunderstood; negative energies, e.g. (binding) energies of the Coulomb atoms, are already in use.

As can be seen, the spectrum of the Klein-Gordon equation is identical to that of the Dirac equation, but there is an important difference. Consider in more detail the plane wave solutions corresponding to the two branches of frequencies in (17.66a):

$$\chi_p^{\pm}(x) = N_p\begin{pmatrix} 1 \\ E^{\pm}(p)/m_0c^2 \end{pmatrix}\mathrm{e}^{\mathrm{i}p\cdot x/\hbar}\,. \tag{17.67}$$

We fix N_p by requiring continuum state orthonormality, i.e. in view of (17.61), we require

$$\int d^3x\,\chi_p^{(\pm)}(x)^\dagger\tau_1\chi_{p'}^{(\pm)}(x) = \pm\,\delta^3(p-p')\,, \tag{17.68}$$

which leads to

$$N_p = \sqrt{\frac{m_0c^2}{2\,\omega(p)(2\pi\hbar)^3}}\,. \tag{17.69}$$

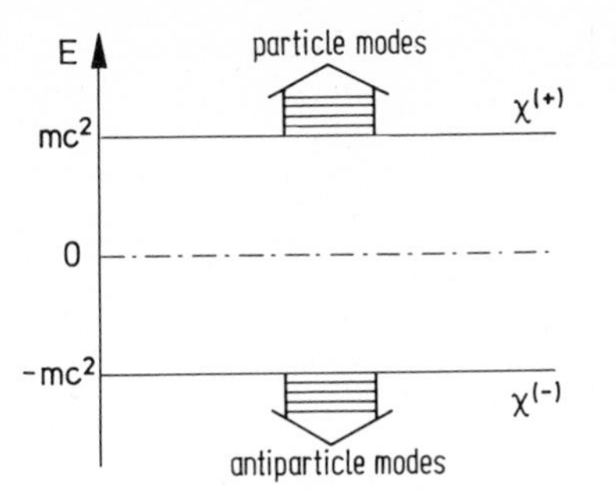

Fig. 17.1. Spectrum of free Klein-Gordon equation and signature of the norm of the plane wave solutions

Note here the existence of two sets of solutions, designated in (17.67) by the indices "$\pm$" for positive and negative values of the mode frequency $E(\boldsymbol{p})$, respectively. Also note that the solutions (17.67) of the Klein-Gordon equation have a very similar form to that of a free Dirac equation, and even the spectrum is identical, as shown in Fig. 17.1. However, the profound difference is the occurrence of the sign of the eigenvalue in (17.68) so that the positive frequency modes (particles) have a positive norm, while negative frequency modes (antiparticles) have a negative norm, Fig. 17.1.

We now discuss the completeness of the plane wave solutions and consider the quantity suggested by (17.68) ($i,j=1,2$):

$$
\begin{aligned}
&\int d^3p\,(\chi_p^{(\pm)}(x))_i(\chi_p^{(\pm)\dagger}(x')\,\tau_1)_j \\
&\quad=\int\frac{d^3p}{(2\pi\hbar)^3}\exp\left[\mathrm{i}\boldsymbol{p}\cdot(\boldsymbol{x}-\boldsymbol{x}')/\hbar\right]\left(\frac{m_0c^2}{2\,\omega(p)}\right) \\
&\quad\cdot\begin{pmatrix} E^{\pm}_{(p)}/m_0c^2 & 1 \\ E^2(p)/(m_0c^2)^2 & E^{\pm}(p)/m_0c^2 \end{pmatrix}_{ij}.
\end{aligned}
\tag{17.70}
$$

We still have to sum over the indices $(+,-)$. As (17.70) clearly shows, the desired result arises only if the difference between the $(+)$ and the $(-)$ contributions is considered

$$
\begin{aligned}
&\int d^3p(\chi_p^{(+)}(x)_i(\chi_p^{(+)\dagger}(x')\,\tau_1)_j-\int d^3p(\chi_p^{(-)}(x))_i(\chi_p^{(-)\dagger}(x)\,\tau_1)_j \\
&\quad=\int\frac{d^3p}{(2\pi\hbar)^3}\exp\left[\mathrm{i}\boldsymbol{p}\cdot(\boldsymbol{x}-\boldsymbol{x}')/\hbar\right]\begin{pmatrix}1&0\\0&1\end{pmatrix}=\delta_{ij}\,\delta^3(\boldsymbol{x}-\boldsymbol{x}')\,.
\end{aligned}
\tag{17.71}
$$

This result confirms that the form of the completeness relation is determined largely by the charge density of the field. Still, the form (17.71) is slightly surprising. It is used in discussions below.

Bibliographical Notes

Bjorken and *Drell* [Bj 64] discuss the Klein-Gordon field extensively. *Feshbach* and *Villars* [Fe 58] introduced the two component representation of the Klein-Gordon equation.

18. Subcritical External Potentials

To prepare for the discussion of overcritical phenomena in the case of spin-0 particles, in this chapter the quantized theory of bosons in external potentials will be treated. Starting from the Hamiltonian field equations and the commutation relations for the field operators, the standard procedure of canonical quantization is outlined. This leads to the construction of the many-particle Fock space and to diagonalized expressions for the charge operator and the Hamiltonian. Finally the energy spectrum of the Klein-Gordon equation with a point-like Coulomb potential is obtained. It is found to exhibit singular behaviour already at $Z\alpha = 1/2$.

18.1 Quantization of the Klein-Gordon Field with External Fields

The procedure of canonical quantization discussed in Chap. 17 can be adapted to suit the new form of the Lagrangian (17.52). First note that the canonically conjugate momentum is (17.41b)

$$\pi_\chi = \frac{\partial \mathscr{L}}{\partial \dot{\chi}} = \mathrm{i}\hbar\chi^+\tau_1 \equiv \mathrm{i}\bar{\chi} \tag{18.1a}$$

or equivalently

$$\chi^+ = -\mathrm{i}\frac{1}{\hbar}\pi_\chi\tau_1 \tag{18.1b}$$

$$\bar{\chi} \equiv -\mathrm{i}\pi_\chi = \hbar\chi^+\tau_1 \tag{18.1c}$$

and the Hamiltonian density $\mathscr{H}$ becomes

$$\mathscr{H} = \pi_\chi\chi - \mathscr{L} = eA_0\chi^+\tau_1\chi + \chi^+\left[(1+\tau_3)\frac{(-\mathrm{i}\hbar c\nabla - eA)^2}{2m_0c^2} + m_0c^2\right]\chi, \tag{18.2a}$$

while $$H = \int d^3x\,\mathscr{H}\,. \tag{18.2b}$$

In (18.2) we should have substituted (18.1b) for χ^+ to express the Hamiltonian density in terms of χ and π_χ only. The advantage of (18.2) is its apparent Hermiticity, even without recourse to the symmetrization normally needed between χ and π. Indeed, due to the introduction of the hybrid χ it is not necessary to consider χ^+ as an independent degree of freedom, cf. (17.40, 41). This is confirmed here by computing the explicit form of the Hamiltonian equations of motion. We find $\dot{\chi}$ and $\dot{\pi}_\chi$ in turn:

$$\dot{\chi} = \frac{\delta H}{\delta \pi_\chi} = \frac{1}{\mathrm{i}\hbar} \frac{\delta}{\delta(\chi^+ \tau_1)} \int d^3x\, \mathscr{H}$$

$$= \frac{\tau_1}{\mathrm{i}\hbar}\left[eA_0 \tau_1 \chi + (1+\tau_3)\frac{(-\mathrm{i}\hbar c\nabla - eA)^2}{2m_0c^2}\chi + m_0c^2\chi \right] \tag{18.3a}$$

$$-\dot{\pi}_\chi = -\mathrm{i}\hbar\dot{\chi}^+\tau_1 = \frac{\delta H}{\delta\chi} = \frac{\delta}{\delta\chi}\int d^3x\,\mathscr{H}$$

$$= eA_0\chi^+\tau_1 + \frac{(\mathrm{i}\hbar c\nabla - eA)^2}{2m_0c^2}\chi^+(1+\tau_3) + m_0c^2\chi^+ \,, \tag{18.3b}$$

which when written componentwise are equivalent to the four equations (17.22), except that now we have allowed for the presence of an electromagnetic field.

To quantize the field, assume now equal time commutation relations between $\hat{\chi}$ and its canonical conjugate $\hat{\pi}_\chi(\boldsymbol{x},0)$, (17.41b, 41),

$$[\hat{\chi}(\boldsymbol{x},0), \hat{\pi}_\chi(\boldsymbol{x}',0)] = \mathrm{i}\hbar\delta^3(\boldsymbol{x}-\boldsymbol{x}') \,, \tag{18.4a}$$

while all other commutators vanish at equal time. Inserting (17.41b) into the above gives

$$[\hat{\chi}_i(\boldsymbol{x},0), (\hat{\chi}^+(\boldsymbol{x}',0)\tau_1)_j] = \delta^3(\boldsymbol{x}-\boldsymbol{x}')\,\delta_{ij} \tag{18.4b}$$

where the indices $i,j \in (1,2)$ of the field χ are indicated. Hence the commutation relation (18.4) fully replaces both of those of (17.31). The explicit forms of (18.4) in terms of the original field φ, (17.44), are worth recording.

a) $i=1, j=2$; $\quad [\hat{\varphi}(\boldsymbol{x},0), \hat{\varphi}^*(\boldsymbol{x},0)] = 0$ (18.5a)

b) $i=j=1$; $\quad [\hat{\varphi}(\boldsymbol{x},0), (-\mathrm{i}\hbar\hat{\dot{\varphi}}^*(\boldsymbol{x},0) - eA_0\hat{\varphi}^*(\boldsymbol{x},0))] = \delta^3(\boldsymbol{x}-\boldsymbol{x}')$

$$[\hat{\varphi}(\boldsymbol{x},0), \hat{\dot{\varphi}}^*(\boldsymbol{x},0)] = \frac{\mathrm{i}}{\hbar}\delta^3(\boldsymbol{x}-\boldsymbol{x}') \tag{18.5b}$$

c) $i=j=2$; $\quad [(\mathrm{i}\hbar\hat{\dot{\varphi}}(\boldsymbol{x},0) - eA_0\hat{\varphi}(\boldsymbol{x},0)), \hat{\varphi}(\boldsymbol{x},0)] = \delta^3(\boldsymbol{x}-\boldsymbol{x}')\,.$

(18.5a) implies that (c) can be simplified to

$$[\hat{\varphi}^*(\boldsymbol{x},0), \hat{\dot{\varphi}}(\boldsymbol{x},0)] = \frac{\mathrm{i}}{\hbar}\delta^3(\boldsymbol{x}-\boldsymbol{x}') \,, \tag{18.5c}$$

which is the complex conjugate of (b). A *non-trivial* relation arises in the last case

d) $i=2, j=1$ $\quad [(\mathrm{i}\hbar\hat{\dot{\varphi}}(\boldsymbol{x},0) - eA_0\hat{\varphi}(\boldsymbol{x},0)),$
$(-\mathrm{i}\hbar\hat{\dot{\varphi}}^*(\boldsymbol{x},0) - eA_0\hat{\varphi}^*(\boldsymbol{x},0))] = 0\,.$

Exploiting (a, b, c),

$$[\hat{\varphi}(\boldsymbol{x},0),\hat{\varphi}^*(\boldsymbol{x},0)] = (2/\hbar^2)eA_0\delta^3(\boldsymbol{x}-\boldsymbol{x}') . \tag{18.5d}$$

As this commutator disappears in the limit $A_0 \to 0$, it can easily escape notice. This was first emphasized by *Synder* and *Weinberg* [Sn 40]. Although here the potential A_0 is an external, not quantized, hence "*c*-number" field, this relation nonetheless indicates the kind of trouble which arises in non-Abelian gauge theories with canonical quantization. In particular, when in such a case A_0 is a *q*-numbered quantum field, the rhs of (18.5) is operator-valued. The advantage of this first-order formalism is the lack of such complications, cf. (18.4). However, the most important aspect of working with the first-order Feshbach-Villars form of the Klein-Gordon equation is the possibility of using all the techniques and methods of quantum field theory developed previously, since in terms of the hybrid χ the quantum equations of motion are first order in the time derivative.

As discussed previously, the charge operator $\hat{q}$ is normally a conserved quantity. According to (17.53, 18.1)

$$\hat{q} = e\int d^3x\,\hat{\chi}^+\tau_1\hat{\chi} = -\frac{\mathrm{i}e}{\hbar}\int d^3x\,\hat{\pi}_\chi\hat{\chi} = \frac{e}{\hbar}\int d^3x\,\hat{\bar{\chi}}\hat{\chi} \tag{18.6}$$

and, just as for $A_0 = 0$ (17.34),

$$\dot{\hat{q}} = -(\mathrm{i}/\hbar)[\hat{q},\hat{H}] = \oint ds\,\boldsymbol{n}\cdot\hat{\boldsymbol{j}}\,, \tag{18.7}$$

where the rhs vanishes in the absence of a particle current. Then, since $\dot{\hat{q}} = 0$, eigenstates of $\hat{H}$ (18.2) can be characterized by eigenvalues of $\hat{q}$. Note that $\hat{H}$ from (18.2) is written here with a circumflex above the fields to indicate their operator character. Both $\hat{q}$ and $\hat{H}$ are Hermitian operators. Using (18.4) then

$$[\hat{\chi}(\boldsymbol{x},0),\hat{\chi}^+(\boldsymbol{x}',0)] = \delta^3(\boldsymbol{x}-\boldsymbol{x}')\,\tau_1\,. \tag{18.8}$$

Equation (18.7) may be verified pursuing the same approach as in (17.34 – 39). We refrain here from carrying this through again. Let us now determine the ground state and the excitation spectrum of the quantum field theory as defined by (18.6 – 8).

18.2 (Quasi) Particle Representation of the Operators

When considering representations of (18.4) first note that the satisfactory quantization (i.e. representation in terms of known quantities) of the quantum wave operator χ_i must somehow produce the necessary sign found in the completeness relation (17.71) when inserted into (18.4). This establishes an important constraint for the following explicit quantum representation of the wave operator in single-particle modes. Consider

$$\hat{\chi}(\boldsymbol{x},0) = \sum_n \chi_n^{(+)}(\boldsymbol{x})\,\hat{a}_n + \sum_k \chi_k^{(-)}(\boldsymbol{x})\,\hat{c}_k^+ , \tag{18.9}$$

introducing the so-called single-particle operators $\hat{a}_n$, $\hat{c}_k^+$. Their commutation properties are determined by attempting to satisfy (18.8). Equation (18.9) is written in terms of modes

$$\chi_n^{(\pm)} = \begin{pmatrix} \varphi_n^1 \\ \varphi_n^2 \end{pmatrix}$$

which are solutions of (17.62) with an external field. We distinguish explicitly between the positive $\{n, +\}$ and negative $\{k, -\}$ frequency solutions. In the limit of a vanishing external potential, (18.9) reduces explicitly (with continuum wave normalization) to

$$\hat{\chi}(\boldsymbol{x},0;A_\mu=0) = \int \frac{d^3p}{(2\pi\hbar)^{3/2}} \sqrt{\frac{m_0c^2}{2\,\omega(p)}}\, \mathrm{e}^{\mathrm{i}\boldsymbol{p}\cdot\boldsymbol{x}/\hbar} \cdot \left[\begin{pmatrix} 1 \\ \dfrac{E^+(p)}{m_0c^2} \end{pmatrix} \hat{a}_p + \begin{pmatrix} 1 \\ \dfrac{E^-(p)}{m_0c^2} \end{pmatrix} \hat{c}_p^+ \right] \quad \text{and} \tag{18.10a}$$

$$\hat{\chi}^\dagger(\boldsymbol{x},0{:}A_\mu=0)\,\tau_1 = \int \frac{d^3p}{(2\pi\hbar)^{3/2}} \sqrt{\frac{m_0c^2}{2\,\omega(p)}}\, \mathrm{e}^{-\mathrm{i}\boldsymbol{p}\cdot\boldsymbol{x}/\hbar} \cdot [(E^+(p)/m_0c^2;1)\,\hat{a}_p^+ + (E^-(p)/m_0c^2;1)\,\hat{c}_p] \,. \tag{18.10b}$$

The commutator (18.4b) now becomes $(i,j=1,2)$

$$[\hat{\chi}(\boldsymbol{x},0)_i, (\hat{\chi}^+(\boldsymbol{x}',0)\,\tau_1)_j] = \int \frac{d^3p}{(2\pi\hbar)^{3/2}} \int \frac{d^3p'}{(2\pi\hbar)^{3/2}} \frac{m_0c^2}{\sqrt{2\,\omega(p)\,2\,\omega(p')}}$$

$$\cdot \exp\left(-\mathrm{i}\frac{\boldsymbol{p}'\cdot\boldsymbol{x}'}{\hbar} + \mathrm{i}\frac{\boldsymbol{p}\cdot\boldsymbol{x}}{\hbar}\right)$$

$$\begin{pmatrix} \dfrac{E^+(p')}{m_0c^2}[\hat{a}_p,\hat{a}_{p'}^+] + \dfrac{E^-(p')}{m_0c^2}[\hat{c}_p^+,\hat{c}_{p'}] & [\hat{a}_p,\hat{a}_{p'}^+] + [\hat{c}_p^+,\hat{c}_{p'}] \\ \dfrac{E^+(p)}{m_0c^2}\dfrac{E^+(p')}{m_0c^2}[\hat{a}_p,\hat{a}_{p'}^+] + \dfrac{E^-(p)}{m_0c^2}\dfrac{E^-(p')}{m_0c^2}[\hat{c}_p^+,\hat{c}_{p'}] & \dfrac{E^+(p)}{m_0c^2}[\hat{a}_p,\hat{a}_{p'}^+] + \dfrac{E^-(p)}{m_0c^2}[\hat{c}_p^+,\hat{c}_{p'}] \end{pmatrix}_{ij} .$$

Here we have neglected to write the cross terms involving the commutators:

$$[\hat{a}_p, \hat{c}_{p'}] = [\hat{a}_p^+, \hat{c}_{p'}^+] = 0\,, \tag{18.11a}$$

which should vanish in a consistent quantization. Note that only if

$$[\hat{a}_p, \hat{a}_{p'}^+] = \delta^3(\boldsymbol{p}-\boldsymbol{p}') \tag{18.11b}$$

$$[\hat{c}_p, \hat{c}_{p'}^+] = \delta^3(\boldsymbol{p}-\boldsymbol{p}') \tag{18.11c}$$

will we get the desired answer, that is (18.4b). Assuming (18.11),

$$\begin{aligned}[\hat{\chi}(\boldsymbol{x},0), \hat{\chi}^+(\boldsymbol{x},0)\,\tau_1] &= \int \frac{d^3p}{(2\pi\hbar)^3}\,\frac{m_0c^2}{2\,\omega(p)} \exp[\mathrm{i}\boldsymbol{p}\cdot(\boldsymbol{x}-\boldsymbol{x}')/\hbar]\cdot\left(\frac{2\,\omega(p)}{m_0c^2}\,\delta_{ij}\right)\\ &= \delta^3(\boldsymbol{x}-\boldsymbol{x}')\,\delta_{ij}\end{aligned} \tag{18.12}$$

as desired.

It is now routine to prove explicitly that the other canonical commutations $[\hat{\chi}, \hat{\chi}]$ and $[\hat{\chi}^+, \hat{\chi}^+]$ vanish.

By first explicitly discussing the free fields, we ascertained that under the second quantization the spectrum of the Klein-Gordon equation naturally dissociates into two sectors, characterized by the sign of the norm, each of which can be associated entirely analogously to the Dirac field with respectively particle and antiparticle modes. It is worth noting, however, that the properties of solutions of the Klein-Gordon equation actually determine the nature of the mode by making that distinction in terms of the easily computable norm (17.57) (the only exception being the possible accidental case of a vanishing norm, discussed below).

Thus, unlike the case of the Dirac field, we can always, uniquely divide the spectrum into particle and antiparticle sectors, in the sense that once the potential is prescribed, the character of the solutions found is fixed. This is to be contrasted with the situation for Dirac fields, where this division is not unique.

Now we apply the developed formalism to the description of the fields in the presence of non-negligible external fields, and return to discuss (18.9) for arbitrary external fields. The indices (+) and (−) refer, respectively, to positive and negative norms, as in (17.61). For the canonical commutator (18.8), using the expansion (18.9) and the commutation relations (18.11),

$$\begin{aligned}&[\hat{\chi}(\boldsymbol{x},0), \hat{\chi}^\dagger(\boldsymbol{x},0)\,\tau_1]\\ &\quad= \sum_{\{n\}} \chi_n^{(+)}(\boldsymbol{x})(\chi_n^{(+)\dagger}(\boldsymbol{x}')\,\tau_1) - \sum_{\{k\}} \chi_k^{(-)}(\boldsymbol{x})(\chi_k^{(-)\dagger}(\boldsymbol{x}')\,\tau_1)\\ &\quad= \mathbb{1}\,\delta^3(\boldsymbol{x}-\boldsymbol{x}')\,,\end{aligned} \tag{18.13}$$

where $\mathbb{1}$ is the two by two (i,j) unit matrix. Denoting from now on the positive and negative norm solutions by the sets $\{n\}$ and $\{k\}$, the upper indices $(\pm)$ are henceforth dropped. We can now obtain the explicit form of the completeness relations (18.12) for the Klein-Gordon field $\hat{\phi}$, making use of (17.44, 62). For $i=j=1$; $i=j=2$; $i=1, j=2$; $i=2, j=1$, respectively,

$$\sum_{\{n\}} [E_n - V(\boldsymbol{x}')]\,\varphi_n^*(\boldsymbol{x}')\,\varphi_n(\boldsymbol{x}) - \sum_{\{k\}} [E_k - V(\boldsymbol{x}')]\,\varphi_k^*(\boldsymbol{x}')\,\varphi_k(\boldsymbol{x}) = \delta^3(\boldsymbol{x}-\boldsymbol{x}') \tag{18.14a}$$

$$\sum_{\{n\}} [E_n - V(\boldsymbol{x})]\,\varphi_n^*(\boldsymbol{x}')\,\varphi_n(\boldsymbol{x}) - \sum_{\{k\}} [E_k - V(\boldsymbol{x})]\,\varphi_k^*(\boldsymbol{x}')\,\varphi_k(\boldsymbol{x}) = \delta^3(\boldsymbol{x}-\boldsymbol{x}') \tag{18.14b}$$

$$m_0c^2\left[\sum_{\{n\}} \varphi_n^*(\boldsymbol{x}')\,\varphi_n(\boldsymbol{x}) - \sum_{\{k\}} \varphi_k^*(\boldsymbol{x}')\,\varphi_k(\boldsymbol{x})\right] = 0 \tag{18.14c}$$

$$\frac{1}{m_0c^2}\left\{\sum_{\{n\}} [E_n - V(\boldsymbol{x})][E_n - V(\boldsymbol{x}')]\,\varphi_n^*(\boldsymbol{x}')\,\varphi_n(\boldsymbol{x}) \right.$$
$$\left. - \sum_{\{k\}} (E_k - V(\boldsymbol{x}))(E_k - V(\boldsymbol{x}'))\,\varphi_k^*(\boldsymbol{x}')\,\varphi_k(\boldsymbol{x})\right\} = 0\,. \tag{18.14d}$$

Equation (18.14c) eliminates the $V(\boldsymbol{x}) \cdot V(\boldsymbol{x}')$ proportional term in (18.14d):

$$\sum_{\{n\}} E_n^2\varphi_n^*(\boldsymbol{x}')\,\varphi_n(\boldsymbol{x}) - \sum_{\{k\}} E_k^2\varphi_k^*(\boldsymbol{x}')\,\varphi_k(\boldsymbol{x})$$
$$= [V(\boldsymbol{x}) + V(\boldsymbol{x}')]\left(\sum_{\{n\}} E_n\varphi_n^*(\boldsymbol{x}')\,\varphi(\boldsymbol{x}) - \sum_{\{k\}} E_k\varphi_k^*(\boldsymbol{x}')\,\varphi(\boldsymbol{x})\right). \tag{18.14d$'$}$$

Now (18.14c) also allows us to write (18.14a) in the form

$$\sum_{\{n\}} E_n\,\varphi_n^*(\boldsymbol{x}')\,\varphi_n(\boldsymbol{x}) - \sum_{\{k\}} E_k\,\varphi_k^*(\boldsymbol{x}')\,\varphi_k(\boldsymbol{x}) = \delta^3(\boldsymbol{x}-\boldsymbol{x}')\,, \tag{18.15a}$$

so that (18.14d$'$) becomes

$$\sum_{\{n\}} E_n^2\varphi_n^*(\boldsymbol{x}')\,\varphi_n(\boldsymbol{x}) - \sum_{\{k\}} E_k^2\varphi_k^*(\boldsymbol{x}')\,\varphi_k(\boldsymbol{x}) = 2\,V(\boldsymbol{x})\,\delta^3(\boldsymbol{x}-\boldsymbol{x}')\,. \tag{18.15b}$$

Restating (18.14c)

$$\sum_{\{n\}} \varphi_n^*(\boldsymbol{x}')\,\varphi_n(\boldsymbol{x}) - \sum_{\{k\}} \varphi_k^*(\boldsymbol{x}')\,\varphi_k(\boldsymbol{x}) = 0\,, \tag{18.15c}$$

(18.15) gives the three completeness relations of the second-order differential equation required for consistent quantization. Indeed, these relations required by the quantization here have been proven, since in the equivalent Feshbach-Villars approach there is only (18.13), and its validity is self-evident, because only first-order time derivatives enter the description, allowing application of the theory of first-order differential equations.

Since the Feshbach-Villars representation (17.46) is first order in time, a propagator $G(\boldsymbol{x}, t; \boldsymbol{x}', t')$ exists, enabling the value of χ to be calculated at time t from the values of χ at t', subject to certain asymptotic boundary conditions

$$\chi(\boldsymbol{x}, t) = \int \mathrm{i}\,G(\boldsymbol{x}, t; \boldsymbol{x}', t')\,\chi(\boldsymbol{x}', t')\,d^3x'\,. \tag{18.16}$$

Such a propagator is in general [cf. (15.1)] given by the bilinear form

$$G(\boldsymbol{x}, t; \boldsymbol{x}', t') = \sum_{\{n\}} \exp[\mathrm{i}E_n(t'-t)/\hbar]\, \chi_n(\boldsymbol{x})\, \chi_n^{\dagger}(\boldsymbol{x}')\, \tau_1$$
$$- \sum_{\{k\}} \exp[\mathrm{i}E_n(t'-t)/\hbar]\, \chi_k(\boldsymbol{x})\, \chi_k^{\dagger}(\boldsymbol{x}')\, \tau_1 \,. \tag{18.17}$$

Equation (18.16) is now proven easily using (18.17). Consider a general choice of χ at time t',

$$\chi(\boldsymbol{x}', t') = \sum_j a_j^0 \exp(-\mathrm{i}E_j t'/\hbar)\, \chi_j(\boldsymbol{x}') \,, \tag{18.18}$$

where a_j^0 are (arbitrary) initial conditions on the entire set of modes j of the spin-0 particles. Inserting (18.17, 18) into (18.16),

$$\chi(\boldsymbol{x}, t) = \sum_{\{n\}} \exp[\mathrm{i}E_n(t'-t)/\hbar]\, \chi_n(\boldsymbol{x}) \sum_j a_j \exp(-\mathrm{i}E_j t'/\hbar) \int d^3x'\, \chi_n^{+}(\boldsymbol{x}')\, \tau_1\, \chi_j(\boldsymbol{x}')$$
$$- \sum_{\{k\}} \exp[\mathrm{i}E_k(t'-t)/\hbar]\, \chi_k(\boldsymbol{x}) \sum_j a_j \exp(-\mathrm{i}E_j t'/\hbar) \int d^3x'\, \chi_k^{+}(\boldsymbol{x}')\, \tau_1\, \chi_j(\boldsymbol{x}')$$
$$= \sum_{\{j\}} a_j \exp(-\mathrm{i}E_j t/\hbar)\, \chi_j(\boldsymbol{x}) \,,$$

as required. Here we have used (17.61)

$$\int d^3x'\, \chi_k^{+}(\boldsymbol{x}')\, \tau_1\, \chi_j(\boldsymbol{x}') = \delta_{kj}\, \theta_j \,,$$

where the values of the sign function $\theta_j = \pm 1$ compensate the sign in the expansion (18.17). This result was the reason for the above exercise: For spin-$\frac{1}{2}$ particles, this sign arose due to the anticommutation of the Fermi operators (Pauli principle). For spin-0 particles, it is obtained from the charge (normalization) of the single-particle modes.

In summary, a satisfactory representation of the quantum field operators has been achieved here, separating the associated set of single-particle operators into particle operators, $\hat{a}_n$ belonging to positive norm eigenmodes and antiparticle operators $\hat{c}_k$ to negative norm modes. The quantized fields satisfy all the canonical commutation relations, which, in turn, are consistent with the completeness relations of the spin-zero particles. Still untreated is the possible case of a potential supporting a solution of zero norm. Chapter 19 shows that this case must be discussed within the many-body theory, using a self-consistent potential.

18.3 The Fock Space and Diagonalization of the Hamiltonian

Equation (18.8) can be rewritten to clarify the nature of the single-particle operators $\hat{a}_n$, $\hat{c}_k^{+}$. Exploiting the orthonormalization of the solutions of the Klein-Gordon equations, then

$$\hat{a}_n = \int \chi_n^{(+)\dagger}(\boldsymbol{x})\, \tau_1\, \hat{\chi}(\boldsymbol{x}, 0)\, d^3x \tag{18.19a}$$

$$\hat{c}_k^{+} = -\int \chi_k^{(-)\dagger}(\boldsymbol{x})\, \tau_1\, \hat{\chi}(\boldsymbol{x}, 0)\, d^3x \tag{18.19b}$$

and the adjoint equations for $\hat{a}_n^+$ and $\hat{c}_k$. The Hamiltonian $\hat{H}$ (18.2) is a bilinear form in these operators. Our aim is to find a space in which the action of these operators can be better understood and the Hamiltonian diagonalized.

Hence we introduce a state $|0\rangle$ (as before) referred to as the *vacuum*, which is the eigenstate of the Hamiltonian with the *lowest* energy of all its eigenstates. An important property of $|0\rangle$ is

$$\begin{aligned} &\hat{a}_n\,|0\rangle = \hat{c}_k|0\rangle = 0 \\ &\langle 0|\hat{a}_n^+ = \langle 0|\hat{c}_k^+ = 0\,, \end{aligned} \tag{18.20}$$

which is identical to (9.35) for the Dirac field. Thus, $\hat{a}_n$ (respectively $\hat{c}_k$) are the particle (antiparticle) annihilation operators. Consider the commutation relation (18.11) for a set of discrete modes,

$$[\hat{a}_{n'}, \hat{a}_n^+] = \delta_{n'n}\,. \tag{18.21}$$

Hence

$$[\hat{a}_{n'}, a_n^+]\,|0\rangle = \delta_{n'n}|0\rangle\,. \tag{18.22}$$

In view of (18.20) it now follows that

$$\hat{a}_{n'}(\hat{a}_n^+\,|0\rangle) = \delta_{nn'}|0\rangle\,, \tag{18.23a}$$

and similarly

$$c_{k'}(c_k^+\,|0\rangle) = \delta_{k'k}|0\rangle\,. \tag{18.23b}$$

Thus the states

$$\hat{a}_n^+\,|0\rangle \equiv |n\rangle\,; \qquad \hat{c}_k^+\,|0\rangle \equiv |k\rangle \tag{18.24}$$

are qualitatively different from the vacuum states (18.20): the annihilation operators, when they match with their quantum numbers, convert these back to the vacuum state! Such states are called "single-particle states".

Consider now the slightly more general case:

$$\frac{1}{\sqrt{N!}}(\hat{a}_n^+)^N|0\rangle \equiv |Nn\rangle\,, \tag{18.25}$$

where the normalization factor $1/\sqrt{N!}$ has been introduced for later convenience. Applying the operator $\hat{a}_{n'}$ to (18.25), for $n' \neq n$ we get back to (18.20). Namely, if $n' \neq n$, the operator $\hat{a}_{n'}$ commutes with the operator coefficient in (18.25) and we can allow it to act directly on the vacuum:

$$\hat{a}_{n'}|Nn\rangle = \frac{1}{\sqrt{N!}}\hat{a}_{n'}(a_n^+)^N|0\rangle = \frac{1}{\sqrt{N!}}(\hat{a}_n^+)^N\hat{a}_{n'}|0\rangle = 0\,. \tag{18.26}$$

For $n' = n$ the commutation relation moves the annihilator $\hat{a}_n$ to the right of the creation operator $\hat{a}_n^+$:

$$\begin{aligned}
\hat{a}_n \frac{1}{\sqrt{N!}} (\hat{a}_n^+)^N &= \frac{1}{\sqrt{N!}} [\hat{a}_n, \hat{a}_n^+] (\hat{a}_n^+)^{N-1} + \frac{1}{\sqrt{N!}} \hat{a}_n^+ \hat{a}_n (\hat{a}_n^+)^{N-1} \\
&= \frac{1}{\sqrt{N!}} (\hat{a}_n^+)^{N-1} + \frac{1}{\sqrt{N!}} \hat{a}_n^+ [\hat{a}_n, \hat{a}_n^+] (\hat{a}_n^+)^{N-2} \\
&\quad + \frac{1}{\sqrt{N!}} (\hat{a}_n^+)^2 \hat{a}_n (\hat{a}_n^+)^{N-2} \\
&= \frac{2}{\sqrt{N!}} (\hat{a}_n^+)^{N-1} + \frac{1}{\sqrt{N!}} (\hat{a}_n^+)^2 [\hat{a}_n, \hat{a}_n^+] (\hat{a}_n^+)^{N-3} \\
&\quad + \frac{1}{\sqrt{N!}} (\hat{a}_n^+)^3 \hat{a}_n (\hat{a}_n^+)^{N-3} \\
&= \dots = \frac{N}{\sqrt{N!}} (\hat{a}_n^+)^{N-1} + \frac{1}{\sqrt{N!}} (\hat{a}_n^+)^N \hat{a}_n \, .
\end{aligned} \tag{18.27}$$

Hence, as $\hat{a}_n$ annihilates the vacuum state,

$$\hat{a}_n |Nn\rangle = \sqrt{N} \, |(N-1)\, n\rangle \, . \tag{18.28a}$$

Thus $\hat{a}_n$, in general, reduces the number of particles in the mode "n". Contrary to this, $\hat{a}_n^+$ increases the number of particles in the mode, since (18.25) implies

$$\hat{a}_n^+ \, |Nn\rangle = \sqrt{N+1} \, |(N+1)\, n\rangle \, . \tag{18.28b}$$

Combining (18.28a, b) gives the important relation

$$\hat{a}_n^+ \hat{a}_n |Nn\rangle = N |Nn\rangle \, . \tag{18.29}$$

Thus the operator

$$\hat{N}_n = \hat{a}_n^+ \hat{a}_n \tag{18.30}$$

counts the number of particles in the mode "n" of the state (18.25).

The set of states (18.25) is orthogonal and normalized. For $n' \neq n$, using the commutation relation as obtained in (18.23),

$$\langle Nn' | Nn\rangle = 0 \, , \qquad n' \neq n \, . \tag{18.31}$$

Similarly, for $n = n'$ but $N \neq N'$ the overlap vanishes. Hence, in general,

$$\langle N' n' | Nn\rangle = \delta_{N'N} \delta_{n'n} \, . \tag{18.32}$$

To derive this result we employ (18.27) and assume here that $N' \geq N$. (For $N' < N$ only the adjoint form need be considered. Since $n' \neq n$ is covered by (18.31), we also take $n' = n$.)

$$\begin{aligned}\langle N' n | N n\rangle &= \frac{1}{\sqrt{N!N'!}} \langle 0 | (\hat{a}_n)^{N'} (\hat{a}_n^+)^N | 0\rangle \\ &= \frac{1}{\sqrt{N!N'!}} \langle 0 | (\hat{a}_n)^{N'-1} [N(\hat{a}_n^+)^{N-1} + (\hat{a}_n^+)^N \hat{a}_n] | 0\rangle \\ &= \frac{1}{\sqrt{N!N'!}} \langle 0 | (N(N-1)\dots 1) \cdot \hat{a}_n^{N'-N} | 0\rangle \\ &= \begin{cases} 1 & N' = N \\ 0 & N' > N \end{cases}. \end{aligned} \tag{18.33}$$

In view of the above remark concerning $N' < N$ we have now verified the validity of (18.32). Further, note that our N-mode states are properly normalized, i.e. the normalization factor $(N!)^{-1/2}$ has been well chosen in (18.25).

So far we have put the particles into only one particular mode. But of course all we have said can be generalized to when different modes of particles and/or antiparticles are occupied. In general, we may construct a multiparticle state

$$\left(\frac{1}{\sqrt{N!}} \hat{a}_n^+\right)^N \dots \left(\frac{1}{\sqrt{N'!}} \hat{a}_{n'}^+\right)^{N'} \dots \left(\frac{1}{\sqrt{K!}} \hat{c}_k^+\right)^K \dots \left(\frac{1}{\sqrt{K'!}} \hat{c}_{k'}^+\right)^{K'} |0\rangle$$
$$= |\dots Nn; \dots N'n'; \dots Kk; \dots K'k'; \dots\rangle . \tag{18.34}$$

The set of all such states spans the so-called Fock space – each state is characterized by the set of numbers $\{N_n,\ n = 1\dots\infty;\ K_k,\ k = 1\dots\infty\}$, which are the occupation numbers of all particle and antiparticle modes. Without repeating the detailed proof (18.33), we point out here that this set of states is orthonormal. Further, this set is complete in the sense that any physical state containing spin-zero particles only can be described in the occupation number representation basis (18.34).

We are now ready to consider the Hamiltonian $\hat{H}$ and the charge operator $\hat{q}$ in terms of the Fock-space operators. However, through dividing the spectrum into particle and antiparticle operators, $\hat{H}$ and $\hat{q}$ are not symmetric any more when $\hat{\chi}$ and $\hat{\chi}^+$ are exchanged. Symmetrizing,

$$\begin{aligned}\hat{H}^{(s)} = \int d^3x \Bigg\{ & V\frac{1}{2}[\hat{\chi}^+ \tau_1 \hat{\chi} + \hat{\chi}(\hat{\chi}^+ \tau_1)] + m_0 c^2 \frac{1}{2}(\hat{\chi}^+ \hat{\chi} + \hat{\chi}\hat{\chi}^+) \\ & + \frac{1}{2m_0c^2}(\mathrm{i}\hbar c\nabla - eA)\hat{\chi}^+ \left(\frac{1+\tau_3}{2}\right)(-\mathrm{i}\hbar c\nabla - eA)\hat{\chi} \\ & + \frac{1}{2m_0c^2}(-\mathrm{i}\hbar c\nabla - eA)\hat{\chi}(\mathrm{i}\hbar c\nabla - eA)\hat{\chi}^+ \left(\frac{1+\tau_3}{2}\right)\Bigg\} ,\end{aligned} \tag{18.35}$$

and the charge operator becomes

$$(\hat{q}/e) = \int d^3x \tfrac{1}{2}[\hat{\chi}^+ \tau_1 \hat{\chi} + \hat{\chi}(\hat{\chi}^+ \tau_1)] \,. \tag{18.36}$$

It is important to note here that the inversion of the sequence of quantum operators does not imply a change in the summation over the indices of the hybrid $\hat{\chi}$, but is relevant only for the q-number elements of the quantum field operators.

We now consider explicitly first the simpler case of the charge operator $\hat{q}$. Inserting (18.9) and its adjoint into (18.36) yields

$$\begin{aligned}(\hat{q}/e) = &\sum_{nn'} \int d^3x\, \chi_n^+(\boldsymbol{x})\, \tau_1\, \chi_{n'}(\boldsymbol{x}) \frac{1}{2}(\hat{a}_n^+ \hat{a}_{n'} + \hat{a}_{n'} \hat{a}_n^+) \\ &+ \sum_{kk'} \int d^3x\, \chi_k^+(\boldsymbol{x})\, \tau_1\, \chi_{k'}(\boldsymbol{x}) \frac{1}{2}(\hat{c}_k \hat{c}_{k'}^+ + \hat{c}_{k'}^+ \hat{c}_k) \\ &+ \text{mixed } (\hat{c}_k, \hat{a}_n) \text{ terms} \,.\end{aligned}$$

In view of the orthogonality of the modes, integration over x eliminates the mixed terms and greatly simplifies the diagonal terms:

$$(\hat{q}/e) = \sum_{nn'} \delta_{nn'} \frac{1}{2}(\hat{a}_n^+ \hat{a}_{n'} + \hat{a}_{n'} \hat{a}_n^+) - \sum_{kk'} \delta_{kk'} \frac{1}{2}(\hat{c}_k \hat{c}_{k'}^+ + \hat{c}_{k'}^+ \hat{c}_k) \,.$$

Using the commutation relations (18.11) in the form (18.21) gives

$$(\hat{q}/e) = \sum_n \hat{a}_n^+ \hat{a}_n - \sum_k \hat{c}_k^+ \hat{c}_k + \frac{1}{2}\left(\sum_n 1 - \sum_k 1\right) \,. \tag{18.37}$$

Thus the charge operator contains two terms, the number operator (18.30) of particles less antiparticles and a *vacuum* term that, similar to Dirac particles, vanishes only when there are as many particle as antiparticle undercritical modes (the meaning of this term is determined more precisely in Chap. 19). For now we take it to vanish. Further, note that the $(-)$ sign in (18.37) originates in the sign of the norm, contrary to the Dirac field, where it stems from the commutation relations (see Chap. 9).

In the Fock-space basis, $(\hat{q}/e)$ is always diagonal

$$(\hat{q}/e)\,|\text{Fock space}\rangle = \text{Number of (particles} - \text{antiparticles)}\,|\text{Fock space}\rangle \tag{18.38}$$

with discrete eigenvalues, as expected. For the record, note that the operators

$$\hat{N} = \sum_n \hat{a}_n^+ \hat{a}_n = \sum_n \hat{N}_n \tag{18.39a}$$

$$\hat{K} = \sum_k \hat{c}_k^+ \hat{c}_k = \sum_k \hat{K}_k \tag{18.39b}$$

are the total number operators of particles and antiparticles, respectively.

The same approach must be employed to determine the form of $\hat{H}^{(s)}$ (18.35) in terms of the mode operators. But the (Feshbach-Villars) field equations must be used for the time-independent potential (17.60) so as to be able to replace the differential operators in (18.35) by the eigenvalues of individual modes. This done, we find the usual result, supplemented by an extra minus sign for the negative frequency modes, which again originate in the negative sign of the norm integral of these modes

$$\hat{H}^{(s)} = \sum_n E_n \hat{a}_n^+ \hat{a}_n + \sum_k (-E_k)\hat{c}_k^+ \hat{c}_k + \frac{1}{2}\left[\sum_n E_n + \sum_k (-E_k)\right] . \tag{18.40}$$

Here "weak potential" is defined such that for weak external potentials, all antiparticle modes have negative eigenenergy $-E_k > 0$. For weak potentials, (18.40) is explicitly positive definite, with a *positive* zero-point energy, last term in (18.40). In the presence of moderately strong external potentials, either some E_n become negative or alternatively, some E_k become positive. *Notably, even when only one of the modes "changes sign", $\hat{H}^{(s)}$ ceases to be explicitly bounded below*, Chap. 19.

The great advantage of the Fock states (18.34) is that the Hamiltonian (18.40) is diagonal in the Fock basis:

$$\hat{H}^{(s)}|\text{Fock state}\rangle = \left\{\left[\sum_n E_n N_n + \sum_k (-E_k) K_k + E_0\right]\right\}|\text{Fock state}\rangle , \tag{18.41}$$

where N_n and K_k are the occupation numbers of n or k modes and E_0 is the (state-independent) positive zero-point energy

$$E_0 = \frac{1}{2}\left[\sum_n E_n + \sum_k (-E_k)\right] . \tag{18.42}$$

Recall here that for the spin-$\frac{1}{2}$ particles the zero-point energy is negative, hence it appears that bosons and fermions have opposite influences on the vacuum state.

18.4 The Coulomb Problem for Spin-0 Particles

We have now developed a subcritical quantized theory of spin-0 particles in external weak potentials – its only immediate application is the description of the bound states of π-mesons in light nuclei, i.e. before the wave functions overlap too strongly with the nuclei, introducing essential modifications due to strong pion-nucleon interactions.

The charged π meson has a mass of 139.567 MeV, that is 273.12 times larger than the electron's mass. Accordingly, its non-relativistic Bohr orbit is much closer to the Coulomb centre. The charged pion lifetime of 2.60×10^{-8} s is substantially longer than the typical dipole transition times even in light π atoms, hence one can easily observe the transition radiation when pions cascade down to the $1s$ ground state. One actually can test the validity of the Klein-Gordon equation to great precision.

It turns out that to find explicit solutions, it is more convenient to employ the second-order form (17.42) of the Klein-Gordon equation. We are interested in $A = 0$, V being the spherically symmetric Coulomb potential. Thus

$$\varphi_{nlm} = \frac{1}{r} U_{nlm} Y_{lm}(\theta, \phi) \mathrm{e}^{-\mathrm{i}E_{nlm}t/\hbar} \tag{18.43}$$

which, when inserted into (17.42), leads to

$$[E_{nlm} - V(r)]^2 \left(\frac{U_{nlm}}{r}\right) = (-\hbar^2 c^2 \Delta_l + m_\pi^2 c^4) \left(\frac{U_{nlm}}{r}\right), \tag{18.44a}$$

where

$$\Delta_l = \frac{d^2}{dr^2} + \frac{2}{r}\frac{d}{dr} - \frac{l(l+1)}{r^2}, \tag{18.44b}$$

and for a point-like nucleus

$$V(r) = -\frac{Ze^2}{r} = -\frac{Z\alpha}{r}\hbar c = -\frac{Z\alpha}{s} 2pc. \tag{18.44c}$$

When searching for bound states it is convenient to introduce the momentum

$$p_{nlm} = \sqrt{m_\pi^2 c^2 - E_{nlm}^2/c^2} \tag{18.45a}$$

in terms of which the length is measured

$$s = 2 r p_{nlm}/\hbar ; \tag{18.45b}$$

the inconvenient indices n, l, m are dropped.

By explicit differentiation

$$r\Delta \frac{U(s)}{r} = 4p^2/\hbar^2 \left(\frac{d^2}{ds^2} - \frac{l(l+1)}{s^2}\right) U(s). \tag{18.46}$$

After some rearrangement of terms, for (18.44), using (18.46),

$$-2EVU(s) = -4p^2c^2 \left(\frac{d^2}{ds^2} - \frac{l(l+1)}{s^2}\right) U(s) + (c^2p^2 - V^2) U(s). \tag{18.47}$$

Now, using the explicit form (18.44c) of the potential, multiplication with $(-4p^2c^2)^{-1}$ gives

$$0 = \frac{d^2U(s)}{ds^2} + \left[-\frac{1}{4} + \left(\frac{Z\alpha E}{pc}\right)\frac{1}{s} + \frac{(Z\alpha)^2 - (l+1)l}{s^2}\right] U(s), \tag{18.48}$$

which has the form of Whittaker's differential equation [Ab 65, Eq. (13.1.31)] and hence the solution regular at origin is

$$U(s) = NW_{\varkappa,\mu}(s) = Ne^{-\frac{1}{2}s}s^{\frac{1}{2}+\mu}{}_1F_1(\tfrac{1}{2}+\mu-\varkappa, 1+2\mu; s) \,. \tag{18.49a}$$

Here ${}_1F_1(a, b; z) \equiv M(a, b, z)$ is the confluent hypergeometric function [Ab 65, Eq. (13.1.2)].

$$\varkappa = \frac{Z\alpha E}{pc} \quad \text{and} \tag{18.49b}$$

$$\tfrac{1}{4} - \mu^2 = (Z\alpha)^2 - (l+1)l\,; \quad \text{that is}$$

$$\mu = \sqrt{(l+\tfrac{1}{2})^2 - (Z\alpha)^2} \xrightarrow[Z\alpha\to 0]{} l + \tfrac{1}{2} \,. \tag{18.49c}$$

A solution irregular at origin exists, which must be considered whenever the asymptotic Coulomb $1/r$ potential is joined to a non-singular potential at $r \to 0$. It may simply be obtained by taking the negative root of (18.49c).

However, it must be certain that the solution (18.49a) is bounded at infinity, $s \to \infty$. This can be the case only if the series defining the ${}_1F_1$ function

$${}_1F_1(a, b; z) = 1 + \frac{az}{b} + \frac{(a)_2}{(b)_2}\frac{z^2}{2!} + \ldots \frac{(a)_n}{(b)_n}\frac{z^n}{n!} \tag{18.50a}$$

$$(a)_n = a \cdot (a+1) \cdot \ldots \cdot (a+n-1) \,, \quad (a)_0 = 1 \tag{18.50b}$$

terminates. In turn, in view of (18.50b), this occurs only when a is a negative integer

$$a \equiv \tfrac{1}{2} + \mu - \varkappa = -n' \,. \tag{18.51}$$

Then the series (18.50a) leads to generalized Laguerre polynomials

$${}_1F_1(-n', 1+2\mu; z) = \frac{n'!}{(1+2\mu)_{n'}} L_{n'}^{(2\mu)}(z) \,. \tag{18.52}$$

Equation (18.51), supplemented by (18.49b, c) allows a straightforward solution for E:

$$\varkappa = \frac{Z\alpha E}{\sqrt{m_\pi^2 c^4 - E^2}} = \frac{1}{2} + \mu + n' \,, \quad \text{and hence} \tag{18.53}$$

$$E = m_\pi c^2 \left(1 + \frac{(Z\alpha)^2}{(n'+\mu+\frac{1}{2})^2}\right)^{-1/2} . \tag{18.54}$$

The non-relativistic limit of this relation

$$E \to m_\pi c^2 - m_\pi c^2 \left[\frac{1}{2}(Z\alpha)^2 \frac{1}{(n'+l+1)^2}\right) + \ldots \Big] \tag{18.55}$$

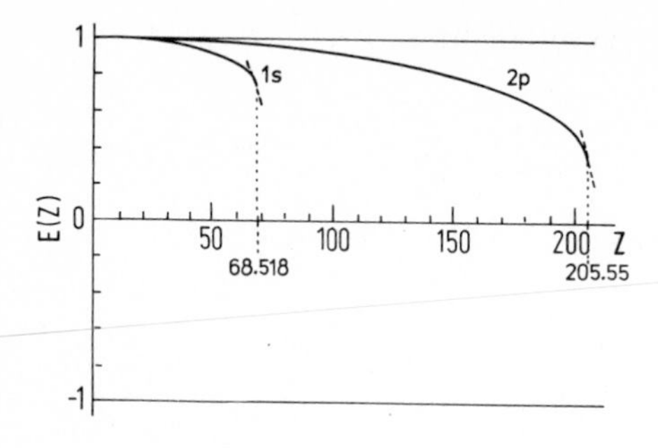

Fig. 18.1. $1s$ and $2p$ energies of the Klein-Gordon equation in units of $m_\pi c^2$, as a function of the nuclear charge Z. $V \sim Z/r$

allows us to identify

$$n = n' + l + 1\,, \tag{18.56}$$

where $n = 1, 2, 3 \ldots$ (where, of course $l = 0, 1, 2 \ldots$) is the principal quantum number.

Thus, for the solution φ_{nlm}

$$\varphi_{nlm} = N e^{-pr/\hbar} (2pr/\hbar)^{\mu - \frac{1}{2}} {}_1F_1(l + 1 - n, 1 + 2\mu; 2pr/\hbar) \tag{18.57}$$

and the normalization condition is

$$1 = 2 \int_0^\infty (E - V)\, \varphi_{nlm}^2 r^2 dr\,; \tag{18.58}$$

that is

$$N^{-2} = \frac{2\hbar^3}{(2p)^3} \int_0^\infty dz\, e^{-z} (E + 2pc Z\alpha/z)\, z^{2\mu+1} ({}_1F_1)^2\,. \tag{18.59}$$

We now consider what happens as Z is increased. As (18.54, 49c) indicate, the difficulty encountered here is that for $l = 0$, μ becomes 0 at $Z\alpha = 1/2$ and hence the bound state spectrum for $Z\alpha > 1/2$ misses certain states: for $Z\alpha > 1/2$ the $1/r$ potential becomes too singular, the Hamiltonian, though Hermitian, ceases to be self-adjoint. The situation here is, however, slightly different to that for the Dirac equation, since at the singular point the tangent to the energy is not singular, as shown qualitatively by dashed curves in Fig. 18.1.

Of course, even long before $Z = 68$, the nuclear finite size plays an important role. The charge density of the $1s$ state is practically completely confined to the inside of the nucleus $Z = 68$. Though this peculiar behaviour encountered for the $1/r$ potentials, viz. singularity at $r \to 0$, can be easily remedied, the critical and overcritical potentials for spin-0 fields must be thoroughly studied.

Bibliographical Notes

The Fock space concept is carefully developed by *Schweber* [Schw 61]. Details of bound Coulomb wave functions for spin-0 particles, however, cannot be found easily in other books. *Lurié* [Lu 68] notes, in particular, the presence of normal-dependent terms which we avoided in this Feshbach-Villars representation of the Klein-Gordon equation.

19. Overcritical Potential for Bose Fields

For a discussion of overcritical behaviour of bosons a full understanding of the single particle energy spectrum is required. The solutions of the Klein-Gordon equation can exhibit two distinct types of behaviour as the strength of an external potential exceeds a critical value: In the case of long-range potentials a bound state enters the lower continuum and becomes an antiresonance. For short range potentials the particle solution joins with a newly formed branch of bound anti-particles to form a state with zero norm and then vanishes from the spectrum. A new vacuum state is formed which is stabilized by the (Coulomb) self-interaction of spontaneously created boson pairs, instead of the Pauli principle which applied in the spin-1/2 case. A self consistent treatment of the newly formed Bose condensate will be developed and applied to the hypothetical case of very highly charged nuclei.

19.1 The Critical Potentials

We have excluded from our discussion the possibility that a mode of indefinite norm appears, i.e. one for which the norm integral (17.61) vanishes:

$$\int d^3x\, \chi_0^+ \tau_1 \chi_0 = 2\int d^3x (E_0 - V(\boldsymbol{x}))\, |\varphi(\boldsymbol{x})|^2 = 0\,. \tag{19.1}$$

Another case not considered is when, in a critical potential, the spectrum contains a particle mode imbedded in the antiparticle continuum (as for the Dirac equation). We first demonstrate that both these cases can occur, entailing a review of the consequences for the quantum field theory of Bose particles in strong fields.

We begin by separating the angular dependence. With $r = |\boldsymbol{x}|$ and Ω standing for the angular coordinates we set (18.43)

$$\varphi_{nlm} = \frac{u_{nlm}(r)}{r} Y_{lm}(\Omega) \tag{19.2}$$

and obtain for the radial wave function u the differential equation

$$[E_{nlm} - V(r)]^2 u(r) = -\hbar^2 c^2 u''(r) + \hbar^2 c^2 \frac{l(l+1)}{r^2} u(r) + (m_0 c^2)^2 u(r)\,. \tag{19.3}$$

We now solve this equation first for a short-range potential, taking here the example of the square well potential

$$V(r) = -V_0\theta(R-r)\,. \tag{19.4}$$

Then

$$(E_{nlm}+V_0)^2-(m_0c^2)^2>0$$

for the (bound) s states ($l=0$)

$$u(r)=\begin{cases}A\sin(\alpha r/R) & ; \quad r<R\\ A\sin\alpha\exp[-\beta(r/R-1)]\,; & \quad r>R\end{cases}, \tag{19.5a}$$

where

$$\alpha=\frac{R}{\hbar c}\sqrt{(E_n+V_0)^2-(m_0c^2)^2}\,, \tag{19.5b}$$

$$\beta=\frac{R}{\hbar c}\sqrt{(m_0c^2)^2-E_n^2}\,. \tag{19.5c}$$

Note that the wave function (19.5a) is continuous at the boundary. However, its derivative

$$u'(r)=\begin{cases}A(\alpha/R)\cos(\alpha r/R) & ; \quad r<R\\ A\sin\alpha(-\beta/R)\exp[-\beta(r/R-1)]\,; & \quad r>R\end{cases} \tag{19.6}$$

is continuous as required for it to be an eigensolution of (19.3) only at the point $r=R$, if

$$\alpha\cos\alpha=-\beta\sin\alpha\,, \tag{19.7}$$

which is the eigenvalue condition for the energies E_n. The explicit solutions of (19.7), giving E_n in terms of V_0, can be easily determined numerically. We are here interested only in bound states, i.e. we consider the domain

$$-m_0c^2<E_n<m_0c^2\,. \tag{19.8}$$

Considering (19.7) we ascertain that there are no bound state solutions in this domain for very small values of V_0. Some aspects of the spectrum of tightly bound states are shown qualitatively in Fig. 19.1 for the lowest s states as functions of V_0. Note that at some value of V_0, a *bound antiparticle* state appears, joins the bound particle state and they both *disappear* at $V_0=V_{cr}=2.97$. This is also seen by evaluating the normalization of the wave function (19.5a) explicitly:

$$\pm 1 = A^2 2\left\{(E_n+V_0)\int_0^R dr\sin^2(\alpha r/R)\right.$$
$$\left.+\sin^2\alpha\cdot E_n\int_R^\infty dr\exp[-2\beta(r/R-1)]\right\}$$

$$= A^2 R \left[(E_n + V_0)\left(1 - \frac{\sin 2\alpha}{2\alpha}\right) + E_n \frac{\sin^2\alpha}{\beta} \right]$$

$$= -A^2 \frac{\sin^2\alpha}{R} \frac{\partial}{\partial E}(\alpha \cot\alpha + \beta) , \qquad (19.9)$$

where the eigenvalue condition (19.7) is in parentheses. Thus plotting $Y = \alpha \cot\alpha + \beta$ in Fig. 19.2 as a function E for several values of V_0 at $R = \hbar c/m_0 c^2$ gives the approximate eigenvalues E_{1s} where the curves cross zero, and the same time the sign of the norm can be determined by simply inspecting the sign of the slope at these intersections.

Equation (19.9) implies further that the norm of the Klein-Gordon equation vanishes at $V_{\rm cr}$ where both possible solutions $E^{(+)}$ and $E^{(-)}$ meet. There, as shown in Fig. 19.2, two zeros coincide and a horizontal tangent arises as a function of E.

While the appearance of indefinite solution at $V_{\rm cr}$ is by itself somewhat surprising, at this level of discussion there is a lack of physical understanding of this binding of an antiparticle in a repulsive potential. This can be better understood by rewriting the Klein-Gordon equation so that it acquires a form similar to the Schrödinger equation. Defining

$$E = (E_n^2 - m_0^2 c^4)/2 m_0 c^2 \qquad (19.10a)$$

$$V' = V(E_n/m_0 c^2 - V/2 m_0 c^2) \qquad (19.10b)$$

gives for the Klein-Gordon equation the form

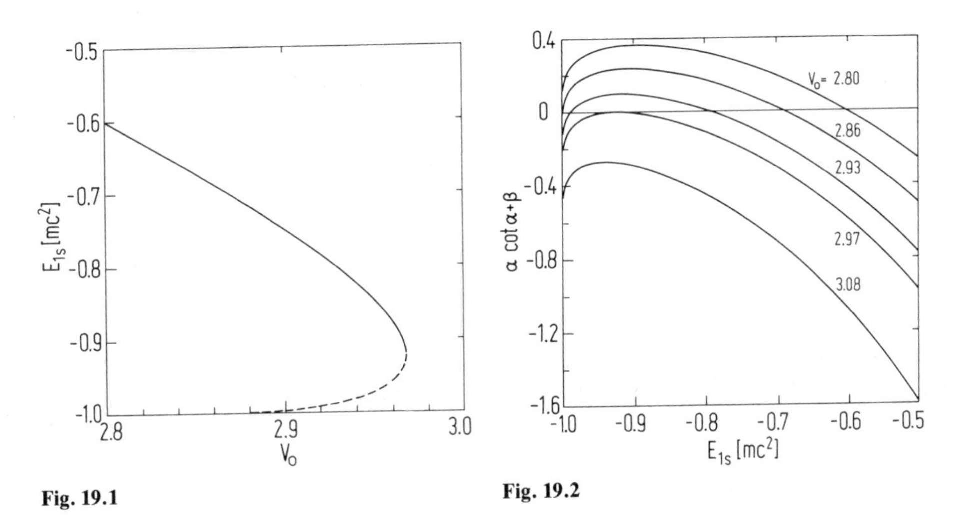

Fig. 19.1 **Fig. 19.2**

Fig. 19.1. The eigenfrequency E_{1s} of the Klein-Gordon equation. (———): positive norm, lowest $1s$ state solution; (– – –): negative norm, $1s$ state solution

Fig. 19.2. Eigenvalue condition for E_n with V_0 as parameter

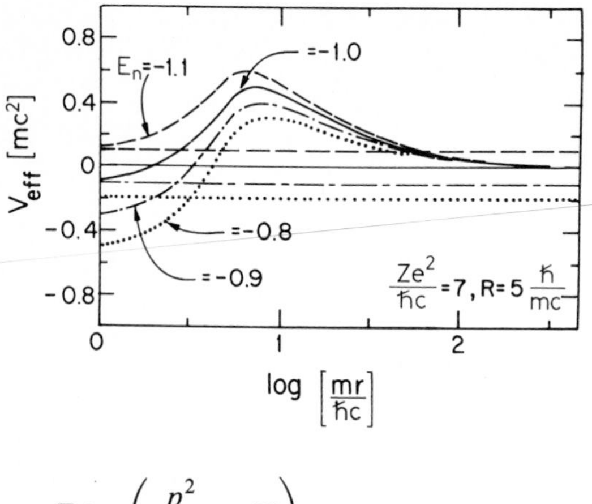

Fig. 19.3. Effective potential of homogeneous charge distribution of radius $R = 5(\hbar/mc)$ with $(Ze^2/\hbar c) = 7$ and $E_n = (-0.8; -0.9; -1; -1.1)\, m_0 c^2$ as a function of $(mr/\hbar c)$

$$E\phi = \left(\frac{p^2}{2m_0} + V'\right)\phi\,. \tag{19.11}$$

Except for the E_n dependence of the effective potential V', (19.11) has the appearance of a Schrödinger equation. We further note an always (independent of V) *attractive* term $-V^2/2m_0c^2$ in the effective potential V. Consider an eigenvalue near the lower, antiparticle, continuum. Then with $E_n = -m_0c^2 + \delta$

$$E = -\delta + 0(\delta^2/m_0c^2) \tag{19.12a}$$

$$V' = \left(-V - \frac{V^2}{2m_0c^2}\right) + \frac{\delta}{m_0c^2}V\,. \tag{19.12b}$$

For bound states of the effective Schrödinger equation (19.10b), $E<0$, i.e. $\delta>0$. Further, note that the effective potential changes sign near to $V = -2m_0c^2$, where the two first terms in (19.12b) cancel each other. For $V < -2m_0c^2$ the first term in (19.12) is *repulsive* (as it should be, assuming an antiparticle) but the relativistic term dominates, leading to a net attraction further helped by the last term in (19.12b). Thus a necessary condition on the potential to facilitate the appearance of a bound antiparticle state is that $V < -2m_0c^2$ in some important regions of space. Then the relativistic attraction $\sim V^2$ dominates the normal repulsion of the charges. Figure 19.3 illustrates this point quantitatively for a long-range $1/r$ potential. Near to the lower continuum the sign of the effective potential can change according to (19.12), so both positive and negative norm states may exist simultaneously, as discussed above. However, this can occur only for short-range potentials with a sudden drop of V to zero as a function of r. The precise conditions under which this behaviour is found have been discussed by *Bawin* and *Lavine* [Ba 79c].

In principal, different behaviour of the spectrum is possible, more similar to the case of the Dirac equation. For long-ranged potentials, like $V \sim 1/r$ (for large r), there is a large region in space where the effective potential (19.12b) shows a more usual behaviour as the relativistic term $|V^2/2mc^2| \ll |V|$is irrelevant [Ba 75].

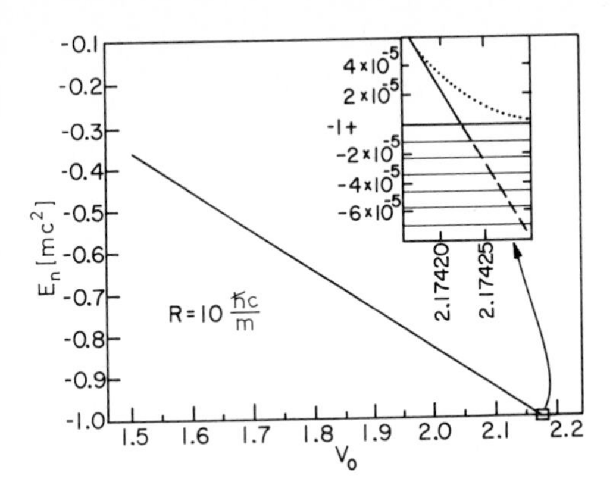

Fig. 19.4. Energy of the lowest bound state in the Coulomb potential of the nucleus with $V = -V_0[1-(r/R)^2/3]$, $R = 10\,\hbar/m_0c$. Insert is an enlargement of the critical area, showing (········) the behaviour including many-body effects. (– – –) is the (anti-)resonance in the continuum

We now consider in some detail the (academic) case of a very, very large nucleus with Z protons. For our purposes, the homogeneous charge distribution over a radius R is sufficient approximation of the proper charge density. Thus,

$$V = \begin{cases} -\dfrac{3}{2}\dfrac{Ze^2}{R}\left[1-\dfrac{1}{3}(r/R)^2\right]; & r<R \\ -\dfrac{Ze^2}{r} \qquad ; & r>R \end{cases} . \tag{19.13}$$

Figure 19.3 shows the effective potential as function of r in units of m_0c^2, with E being a parameter. For $E \sim -1$ there is a moderately attractive part of the potential for small r which is, however, exceeded by the repulsive part at large r.

Although analytical solutions of the Klein-Gordon equation with the potential (19.13) may be generated, it is perhaps simpler in these times of pocket computers to integrate the Klein-Gordon equation numerically to determine the behaviour of the spectrum near $E = -m_0c^2$. In Fig. 19.4 the lowest eigenvalue is given as a function of the potential depth $V_0 = -V(0) = (3Ze^2/2R)$, determined by varying Z for fixed $R = 10\,\hbar/m_0c$. The discrete state *joins* the negative energy (better frequency) continuum at $V_0 = 2.174\,m_0c^2$ (corresponding to $Z_{cr} = 1986$). Note that the value of Z_{cr} increases to 3007 for nuclei at normal nuclear density and equal number of protons and neutrons. The insert in Fig. 19.4 is an enlargement of the neighbourhood of the critical point. The dashed line (an *anti*-resonance in the continuum as shown further below) was obtained from actual calculations. The dotted line is the approximate behaviour of the solution after it is modified to include many-body effects, Sect. 19.3.

At this point it is necessary to consider in more detail the resonance shown in Fig. 19.4. We choose to do this in the Feshbach-Villars representation, as then the perturbative resonance formalism developed for fermions (cf. Chap. 6) can be easily adapted to our current problem. Suppose that V is long ranged and supercritical. Then

$$V = V_{cr} + V' , \tag{19.14}$$

where $V_{\rm cr}$ is just critical. Thus a complete basis is obtained by taking, cf. (17.60),

$$\mathscr{H}_{\rm c} = V_{\rm cr}\tau_1 + (1+\tau_3)\frac{(-{\rm i}\hbar c\nabla - eA)^2}{2m_0c^2} + m_0c^2\,, \tag{19.15a}$$

while

$$\mathscr{H} = \mathscr{H}_{\rm c} + \tau_1 V'\,. \tag{19.15b}$$

We study two types of eigenstates belonging to $\mathscr{H}$: (1) bound states near the top of the energy continuum and (2) negative frequency continuum states, both under the approximation that these can be represented as a superposition of (1) the bound state $\chi_-^{\rm c}$ at the top of the negative continuum

$$\mathscr{H}_{\rm c}\chi_-^{\rm c} = -m_0c^2\tau_1\chi_-^{\rm c}\,, \tag{19.16a}$$

and (2) of the negative energy continuum states $\chi_k^{\rm c}$

$$\mathscr{H}_{\rm c}\chi_k^{\rm c} = E_k\tau_1\chi_k^{\rm c}\,; \qquad E_k < -m_0c^2\,, \tag{19.16b}$$

both generated by $\mathscr{H}_{\rm c}$. We seek the solution of

$$\mathscr{H}\chi_k = E_k\tau_1\chi_k \tag{19.17a}$$

via the approximate expansion and following the methods developed in Chap. 6

$$\chi_k = a(E_k)\chi_-^{\rm c} + \int\limits_{-\infty}^{-m_0c^2} b_{E_{k'}}(E_k)\chi_{k'}^{\rm c}\,dE_{k'}\,. \tag{19.17b}$$

Neglecting the continuum – continuum matrix elements, but remembering the *signs* of the different norms, inserting (19.17b) in (19.17a) and projecting with $\chi_-^{\rm c}$ and $\chi_k^{\rm c}$, see (6.12, 13) give

$$[E_k - (-m_0c^2 + \Delta E_k)]\,a(E_n) = \int\limits_{-\infty}^{-m_0c^2} b_{E_{k'}}(E_k)\,V_{E_{k'}}dE_{k'} \tag{19.18a}$$

$$(E_k - E_k')\,b_{E_{k'}}(E_k) = -a(E_k)\,V_{E_{k'}}\,, \tag{19.18b}$$

where the sign on the rhs of (19.18b) arises through the negative norm of the lower continuum states. This sign is *not* present in (6.13). The different quantities are defined as before:

$$V_{E_{k'}} = \int \chi_k^{{\rm c}\dagger}(x)\,\tau_1\,V'(x)\,\chi_-^{\rm c}(x) \tag{19.18c}$$

$$\Delta E_k = \int \chi_-^{{\rm c}+}(x)\,\tau_1\,V'(x)\,\chi_-^{\rm c}(x)\,. \tag{19.18d}$$

First note that (19.17) possesses resonant continuum solutions. The solution of (19.18b) [see (6.33b) and discussion there] is

$$b_{E'_k} = \delta(E_k - E_{k'}) - \frac{a(E_k)\, V_{E_{k'}}}{E_k - E_{k'} + \mathrm{i}\delta} . \tag{19.19}$$

The negative sign in the second term is required by the negative norm of the negative frequency basis states (19.16b) and is consistent with the negative norm of (19.17b). This sign in (19.19) implies that for resonant $a(E_k)$ the phase shift falls by π (rather than increases). We describe this as *antiresonance* to contrast it with the resonant behaviour found for the Dirac equation in Chap. 6. Inserting (19.19) into (19.18a) gives a resonant $a(E_k)$:

$$a(E_k) = \frac{V^*_{E_k}}{E_k - (-m_0c^2 + \Delta E_k) + \mathrm{i}\delta + \int \frac{|V_{E_{k'}}|^2}{E_k - E_{k'} + \mathrm{i}\delta}\, dE_{k'}} . \tag{19.20}$$

The integral in the denominator may be approximated by its imaginary part, since

$$\begin{aligned} \int \frac{|V_{E_{k'}}|^2}{E_k - E_{k'} + \mathrm{i}\delta}\, dE_{k'} &= -\pi \mathrm{i}\, |V_{E_k}|^2 + P\int \frac{|V_{E_{k'}}|^2}{E_k - E_{k'}}\, dE_{k'} \\ &= -\pi \mathrm{i}\, |V_{E_k}|^2 + \delta E_k , \end{aligned} \tag{19.21}$$

where δE_k may be expected to be small but in any instance may be absorbed in the ΔE_k term in (19.20). By using (19.21) in (19.20), the explicit form of the behaviour of $a(E_k)$ determined

$$a(e_k) \approx \frac{V^*_{E_k}}{E_k - (-m_0c^2 + \Delta E_k + \delta E_k) - \pi \mathrm{i}\, |V_{E_k}|^2} . \tag{19.22}$$

In (19.22) the imaginary part is opposite in sign to the Dirac case, since it contains the extra sign found in (19.18b).

It is very important to appreciate that the modes χ_k, (19.17b), cannot, form a complete set of eigenstates, contrary to the Dirac case. More precisely, they cannot span the undercritical subspace formed by $\{\chi^{\mathrm{c}}_-, \chi^{\mathrm{c}}_{E_k}; E_k \in (-mc^2, -\infty)\}$. We prove this assertion by showing a contradiction in the assumption of (restricted) completeness as defined above. Inspecting the completeness relation (18.13), we require

$$-\sum_{E_k} \chi_k(\boldsymbol{x})[\chi^\dagger_k(\boldsymbol{x}')\tau_1] \stackrel{!}{=} \chi^{\mathrm{c}}_-(\boldsymbol{x})(\chi^{\mathrm{c}\dagger}_-(\boldsymbol{x}')\tau_1) - \sum_{E_k} \chi^{\mathrm{c}}_k(\boldsymbol{x})[\chi^{\mathrm{c}\dagger}_k(\boldsymbol{x}')\tau_1] . \tag{19.23}$$

Multiplying from the left with $\chi^{\mathrm{c}\dagger}_-(\boldsymbol{x})\tau_1$ and from right with $\chi^{\mathrm{c}}_-(\boldsymbol{x}')$ and integrating with respect to $\boldsymbol{x}, \boldsymbol{x}'$ yields

$$\begin{aligned} &-\sum_{E_k} \left[\int d^3x\, \chi^{\mathrm{c}\dagger}_-(\boldsymbol{x})\tau_1 \chi_k(\boldsymbol{x})\right]\left[\int d^3x'\, \chi^\dagger_k(\boldsymbol{x}')\tau_1 \chi^{\mathrm{c}}_-(\boldsymbol{x}')\right] \\ &= \left[\int \chi^{\mathrm{c}\dagger}_-(\boldsymbol{x})\tau_1 \chi^{\mathrm{c}}_-(\boldsymbol{x})\, d^3x\right]\left[\int \chi^{\mathrm{c}\dagger}_-(\boldsymbol{x}')\tau_1 \chi^{\mathrm{c}}_-(\boldsymbol{x}')\, d^3x'\right] \\ &\quad - \sum_{E_k} \left[\int d^3x\, \chi^{\mathrm{c}\dagger}_-(\boldsymbol{x})\tau_1 \chi^{\mathrm{c}}_k(\boldsymbol{x})\right]\left[\int d^3x'\, \chi^{\mathrm{c}\dagger}_k(\boldsymbol{x}')\tau_1 \chi^{\mathrm{c}}_-(\boldsymbol{x}')\right] . \end{aligned}$$

The last term vanishes due to the orthogonality of the basis set chosen. Hence

$$-\sum_{E_k} \left| \int \chi_k^{c\dagger}(\boldsymbol{x})\, \tau_1\, \chi_k(\boldsymbol{x})\, d^3x \right|^2 = +1\,, \tag{19.24}$$

which is impossible: note that similar arguments in the Dirac case do not hold, as both the right and left hand sides are positive definite there.

This means that we still have to find the missing state(s) which disappeared as the critical point was crossed. We therefore now look for *bound state* solutions of (19.18), non-existent in the Dirac case. Then $E_{k'}$ can never be E_k and another solution of (19.18b) is

$$b_{E_{k'}}(E_k) = -\frac{a(E_k)\, V_{E_{k'}}}{E_k - E_{k'}}\,. \tag{19.25}$$

Inserting this into (19.18a) gives the eigenvalue equation

$$E_k = (-m_0c^2 + \Delta E_k) - \int_{-\infty}^{-m_0c^2} dE_{k'} \frac{|V_{E_{k'}}|^2}{E_k - E_{k'}} \tag{19.26}$$

which is an integral equation for E_k. However, if $E_k > -mc^2$, (19.26) is inconsistent – the integral on the rhs is positive, since here $E_k - E_{k'} > 0$. Hence we must assume that E_k is complex, with

$$E_k = E_\mathrm{R} + \mathrm{i}E_\mathrm{I}\,. \tag{19.27a}$$

Taking the real and imaginary parts of (19.26),

$$E_k = (-m_0c^2 + \Delta E_k) - \int_{-\infty}^{-m_0c^2} \frac{|V_{E_{k'}}|(E_\mathrm{R} - E_{k'})}{(E_\mathrm{R} - E_{k'})^2 + E_\mathrm{I}^2}\, dE_{k'} \tag{19.27b}$$

$$1 = + \int_{-\infty}^{-m_0c^2} \frac{|V_{E_{k'}}|^2}{(E_\mathrm{R} - E_{k'})^2 + E_\mathrm{I}^2}\, dE_{k'}\,; \qquad E_\mathrm{I} \neq 0\,. \tag{19.27c}$$

Equations (19.27b, c) must be solved consistently; there is no apparent contradiction if $E_\mathrm{R} < -mc^2$, $E_\mathrm{I} \neq 0$. In view of the contradiction found with (19.24), there must be at least *one* pair of complex conjugate solutions $E_\mathrm{R} \pm \mathrm{i}E_\mathrm{I}$.

We now show that the corresponding states $\chi_\pm$ will have *vanishing* norm. Inserting (19.25) into (19.17b) gives for the wave functions

$$\chi_\pm = a(E_\mathrm{R} \pm \mathrm{i}E_\mathrm{I}) \left[\chi_-^\mathrm{c} - \int_{-\infty}^{-m_0c^2} \frac{V_{E_{k'}}}{E_\mathrm{R} \pm \mathrm{i}E_\mathrm{I} - E_{k'}}\, \chi_{E_{k'}}^\mathrm{c}\, dE_{k'} \right], \tag{19.28}$$

and hence the normalization integral becomes

$$\begin{aligned} &\int d^3x\, \chi_\pm^\dagger\, \tau_1\, \chi_\pm \\ &= |a(E_\mathrm{R} \pm \mathrm{i}E_\mathrm{I})|^2 \left[1 - \int_{-\infty}^{-m_0c^2} \frac{|V_{E_{k'}}|^2}{(E_\mathrm{R} - E_{k'})^2 + E_\mathrm{I}^2}\, dE_{k'} \right] = 0\,, \end{aligned} \tag{19.29}$$

which vanishes in view of (19.27c). Thus the paradox (19.24) is resolved. We have found two complex-conjugate bound state solutions with $E_R < -mc^2$ of vanishing norm. Appearance of such solutions beyond the critical point signifies that the Hamiltonian in question ceases to be self-adjoint. Consequently, efforts to construct a quantum theory using complex solutions found here would be ill advised. Rather than to search for self-adjoint extensions of the overcritical Hamiltonian without a natural physical motivation, we now discuss a simple physical resolution of the phenomena encountered here.

19.2 The True Ground State and Bose Condensation

For both long- and short-range overcritical potentials, we have just shown that one cannot find a consistent quantization: the real spectrum of the Hermitian Hamiltonian is not complete, meaning that the operator is not self-adjoint. This is turn means that we cannot exponentiate the Hamiltonian to obtain the time evolution operator for a physical state. Consequently, the ground state $|0\rangle$ of the Fock space, Sect. 18.3, must be modified to remove this deficiency of the theory. As it turns out, we succeed only when due consideration is given to small effects of the Hamiltonian, neglected in the discussion of bosons in external fields. Here a crucial difference from the case of the spin-1/2 particles becomes important. For fermions, a *consistent* and satisfactory theory can be developed even without reference to higher order in $\alpha = e^2/\hbar c$ effects such as self-energy or vacuum polarization. These terms, while important for the detailed description of the physical quantities, do not influence for fermions the qualitative nature of the approach to criticality or the supercritical behaviour. This is different for Bose fields. Without a change in the qualitative critical behaviour, we cannot obtain properly quantized Bose fields for strong potentials at all!

It is easy to pinpoint the particular microscopic problem for Bose fields as the critical potential is reached. Consider here the case of the short-range, square-well potential when $V = V_{cr}$, Fig. 19.1. Here two modes become degenerate, one particle mode ("+") and one antiparticle mode ("−"). Consider the Fock state

$$|N, +, -\rangle \equiv \frac{1}{N!}(\hat{a}^\dagger_{(+)})^N(\hat{c}^\dagger_{(-)})^N|0\rangle \tag{19.30}$$

made out of N particle−antiparticle pairs in the critical modes. This state is an eigenstate of the Hamiltonian (18.40) with the eigenenergy $E^{(N)}$ (E_0 is the zero-point energy):

$$E^{(N)} - E_0 = N[E_{(+)} + (-E_{(-)})] = N(\mu - \mu) = 0\,, \tag{19.31}$$

where $-m_0c^2 < \mu < m_0c^2$ is the point where both critical modes meet. In (19.31) there is an infinite degeneracy between all N-pair states, even for $N \to \infty$. It costs no energy to add a pair of particles at the critical point. Thus, in the absence of further interactions, the ground state becomes unstable in the sence that *a neutral condensate* containing an equal number of particles and antiparticles could develop, but with N being arbitrarily large. But as N becomes very large, we must consider effects which are small for $N \sim 1$ (Fermi case), but which will appear for $N \to \infty$. In particular, all Coulombic particle–particle interactions below have an expansion parameter which is not e^2 but $(Ne)^2$ (!) and hence cannot be neglected, even in a first approximation. It is helpful to appreciate here that by introducing the coupling of the Bose field to an external (electromagnetic) field, we have had to assign to it its proper (electric) charge, which, as we now recognize, may not be ignored anymore with respect to particle–particle interactions. The field generated by a large number of produced Bose pairs must become part of the "external" field.

To remedy this situation in the particular case of a *critical* Bose field, the mutual Coulomb interaction of the meson fields must now be included in the quantum theory. The effect of this step can be easily anticipated since the expectation value of the Coulomb self-interaction for *any* charge distribution is positive definite,

$$E_c = e^2 \int \varrho(\boldsymbol{x}) \frac{1}{4\pi|\boldsymbol{x}-\boldsymbol{x}'|} \varrho(\boldsymbol{x}') \, d^3x \, d^3x' > 0 \,, \tag{19.32}$$

where the latter inequality can be demonstrated in momentum space. Define

$$\tilde{\varrho}(\boldsymbol{p}) = \int e^{i\boldsymbol{p}\cdot\boldsymbol{x}} \varrho(\boldsymbol{x}) \, d^3x \,, \tag{19.33a}$$

and note that

$$\frac{1}{p^2} = \int e^{i\boldsymbol{p}\cdot(\boldsymbol{x}-\boldsymbol{x}')} \frac{1}{4\pi|\boldsymbol{x}-\boldsymbol{x}'|} \frac{d^3p}{(2\pi)^3} \,; \tag{19.33b}$$

hence as stated

$$E_c = e^2 \int |\tilde{\varrho}(\boldsymbol{p})|^2 \frac{1}{p^2} \frac{d^3p}{(2\pi)^3} > 0 \,. \tag{19.34}$$

Since E_c is quartic in the amplitudes (H is quadratic), for large amplitudes it can ultimately *dominate* all energies and substantially modify the critical behaviour. Here we can, in particular, anticipate effective screening of the applied external potential through particle production, ultimately leading to a subcritical behaviour of so-called quasiparticles. Creation and annihilation operators of these new particle modes generate a new and different physical ground state. We shall refer to the latter as the true ground state $|\mathrm{Vac}\rangle$ to contrast it with the perturbative vacuum state $|0\rangle$.

While the above discussion implied the existence of critical particle and antiparticle modes in a short-range potential, the same remarks apply to the slightly different case of long-ranged external potentials (including in particular the external Coulombic potential). Here, as shown, a particle mode approaches the antiparticle continuum when $V \to V_{\mathrm{cr}}$. We can proceed to construct an N particle state similar to (18.25), which now must contain only an arbitrary number of particles alone, as the entire energy necessary to create a pair is gained by placing a particle in a state with $E^+ \to -m_0 c^2$. Here the situation is formally similar to the Dirac case, with the antiparticle partner emitted to infinity and having no immediate importance except to ensure charge conservation. However, in the Bose case there is no Pauli principle to stop pair production! Thus

$$|N, +\rangle = \frac{1}{\sqrt{N!}} (\hat{a}^\dagger_{(+)})^N |0\rangle \tag{19.35}$$

and again this is an eigenstate of the Hamiltonian (18.40) with the eigenenergy $E^{(N)}_{(+)}$ (E_0 is again the zero point energy)

$$E^{(N)}_{(+)} - E_0 = N E_{(+)} = -N m_0 c^2 . \tag{19.36}$$

Here $E^N_{(+)}$ is the energy required to put N antiparticles at infinity. Thus it again costs no energy to add a pair, the particle remaining localized near the source of the external potential (and screening it), while the antiparticles go to infinity. A very similar degeneracy occurs for the values of N as in the case of short-range potentials, except that now the true ground state contains only (a macroscopic number of) particles. Thus the latter is a *charged* condensate, as compared with the *neutral* condensate of short-range potentials, described previously.

To formulate these ideas more precisely, we simply supplement the Hamiltonian (18.2) [omitting for the present explicit reference to the needed symmetrization (18.35)] by the (Coulomb) particle–particle interaction[1]:

$$\begin{aligned} \hat{H} = \int d^3x\, \hat{\chi}^\dagger \left[\tau_1 V(x) + m_0 c^2 + (1+\tau_3) \frac{(-i\hbar c \nabla)^2}{2 m_0 c^2} \right] \hat{\chi} \\ + \frac{1}{2} e^2 \int [\hat{\chi}^\dagger(\boldsymbol{x}')\, \tau_1 \hat{\chi}(\boldsymbol{x}')] \frac{1}{4\pi |\boldsymbol{x}-\boldsymbol{x}'|} [\hat{\chi}^\dagger(\boldsymbol{x})\, \tau_1 \hat{\chi}(\boldsymbol{x})]\, d^3x\, d^3x' , \end{aligned} \tag{19.37}$$

where $V(x)$ is by assumption an external (supercritical) potential.

Let us now assume that V is strongly supercritical and recall that in view of the two different types of degeneracies (19.31, 35) encountered, we need to consider separately the two cases of long- and short-range potentials.

We begin with the mathematically simpler case of a long-range potential. Assume that the true ground state has charge q and that it can be approximated for large q by the state vector [note that it is *not* the same state as (19.35)]

[1] The Hamiltonian (19.37) must be understood as a simplified model devised to describe the self-interaction of the condensate. *Dyson* has argued that the associated nonrelativistic Hamiltonian may not possess an absolute ground state [Dy 67]. Note, however, that Dyson's arguments are not conclusive when the current–current interaction is added to (19.37). In any case, the possible instability of the Hamiltonian at very high particle–antiparticle densities is an artefact because all known charged bosons have repulsive short range interactions which overwhelm the Coulomb attraction between bosons of opposite charge. A simple way of incorporating these effects into (19.37) is to add a self-interaction $H_4 = \lambda \int d^3x\, \varphi(x)^4$ with positive λ [Mi 72].

$$|q\rangle \cong \frac{1}{\sqrt{q!}}(\hat{\alpha}^{\dagger})^{q}|\mathrm{Vac}\rangle\,, \tag{19.38a}$$

where [cf. (18.9)]

$$\hat{\alpha}^{\dagger} = \int \hat{\chi}^{\dagger}(\boldsymbol{x})\,\tau_1\,\chi_0(\boldsymbol{x})\,d^3x \tag{19.38b}$$

is the creation operator associated with the particle wave function $\chi_0(\boldsymbol{x})$, which is to be *determined self-consistently*. The function $\chi_0(\boldsymbol{x})$ is normalized according to

$$\int \chi_0^{\dagger}\tau_1\,\chi_0\,d^3x = 1 \tag{19.39}$$

and the true vacuum state is characterized by

$$\hat{\alpha}|\mathrm{Vac}\rangle = 0\,. \tag{19.40}$$

We next calculate the expectation value of $\hat{H}$, (19.37) with respect to (19.38) as a trial function. The result contains a semiclassical part extractable by simply replacing $\hat{\chi} \rightarrow \hat{\alpha}\chi_0$ in (19.37) and calculating the resulting expectation value. (There are also quantum fluctuations, which are not further discussed here). The result is, with $\mathscr{H}$ being the differential operator of the first (single-particle) part of (19.37),

$$W(q) \equiv \langle q|\hat{H}|q\rangle \cong q\int \chi_0^{\dagger}\mathscr{H}\,\chi_0\,d^3x + \frac{1}{2}q^2e^2\int \frac{\varrho_0(\boldsymbol{x})\,\varrho_0(\boldsymbol{x}')}{4\pi\,|\boldsymbol{x}-\boldsymbol{x}'|}\,d^3x\,d^3x' \tag{19.41a}$$

$$\varrho_0(\boldsymbol{x}) = \chi_0^{\dagger}(\boldsymbol{x})\,\tau_1\,\chi_0(\boldsymbol{x})\,, \tag{19.41b}$$

where the second term in (19.41a) is the Coulomb energy (19.32). We shall determine $\chi_0(\boldsymbol{x})$ and q from the variational principle

$$\delta\left[W(q) - \mu q\int \varrho_0(\boldsymbol{x})\,d^3x\right] = 0\,, \tag{19.42}$$

where μq is the Lagrange multiplier for (19.39). Varying with respect to $\chi_0^{\dagger}$ at fixed q gives the "condensate" equation

$$\begin{aligned}\mu\tau_1\,\chi(\boldsymbol{x}) &= \mathscr{H}\,\chi(\boldsymbol{x}) + \tau_1\,V_{\mathrm{cond}}(\boldsymbol{x})\,\chi(\boldsymbol{x})\\ &\equiv \mathscr{H}_{\mathrm{eff}}\,\chi(\boldsymbol{x})\end{aligned} \tag{19.43a}$$

where we have introduced a condensate (macroscopic) wave function

$$\chi = \sqrt{q}\,\chi_0\,, \tag{19.43b}$$

and the potential generated by the condensed charge

$$V_{\mathrm{cond}}(\boldsymbol{x}) = \frac{e^2}{4\pi}\int \frac{\varrho(\boldsymbol{x}')}{|\boldsymbol{x}-\boldsymbol{x}'|}\,d^3x'\,. \tag{19.43c}$$

Here

$$\varrho(\boldsymbol{x}) = \chi^{\dagger}\tau_1\,\chi\,. \tag{19.43d}$$

Here the physical significance of the parameter μ can be established by computing $\langle q+1|\hat{H}|q+1\rangle$ in the same approximation that led to (19.41a):

$$\langle q+1|\hat{H}|q+1\rangle = W(q)+\mu\,. \tag{19.44}$$

Thus μ is the energy of the last particle added to the condensate. Therefore we must seek solutions of (19.43) for a particular choice of (a real) μ within the bound state interval

$$-m_0c^2 < \mu < m_0c^2$$

such that all physical constraints are respected. In the case just discussed, according to (19.44) we should choose $\mu = -m_0c^2$, as the condensate appears only after the associated antiparticle can be created. However, in the presence of other interactions, such as weak interactions, it can suffice to create a different, much lighter particle to accompany the condensing field. Then, e.g. $\mu = -m_e c^2 \sim 0$, as it costs almost nothing to balance the charge conservation with the mass m_e of an electron.

To obtain a physical solution (19.43) fix μ as dictated by the physics of the condensate and find a solution to (19.43) (as described below). In general then

$$\int \chi^\dagger \tau_1 \chi\, d^3x = q \tag{19.45}$$

will not be an integer. Now increase μ very slightly until q becomes, as required, the next larger integer. For large $(Z-Z_0)$ this step is of no practical importance and need not be carried out. At this point note that q may be considered a continuous function of Z with μ fixed. It is obvious physically that in addition to this correct (stable) solution, solutions with larger values of q exist – overscreened solutions. Therefore the minimum value of (an integer) $|q|$ obtainable should be sought, such that the spectrum of $\mathscr{H}_{\text{eff}}$, (19.43), remains complete. In Fig. 19.5a the charged condensate is illustrated. Further note that at the stationary point, using (19.43),

$$W(q)|_{\min} \equiv W_0 = \mu q - \frac{1}{2}\,\frac{e^2q^2}{4\pi}\int \frac{\varrho_0\varrho_0'}{|\boldsymbol{x}-\boldsymbol{x}'|}\,d^3x\,d^3x'\,. \tag{19.46}$$

In the limit $q\to\infty$ (which requires $Z\to\infty$), this may be viewed as a phase transition to a charged superfluid state with macroscopic occupation of the mode $\chi_0(\boldsymbol{x})$, since another definition of $\chi_0(\chi)$, equivalent to the one given, is, cf. (18.28),

$$\langle q|\hat{\chi}(\boldsymbol{x})|q+1\rangle = \sqrt{q}\,\chi_0(\boldsymbol{x}) = \chi(\boldsymbol{x})\,. \tag{19.47}$$

Even when the energy of the critical state has not reached $-m_0c^2$, there is a domain of strong vacuum fluctuations permitting the onset of the condensate to proceed (precritical fluctuations).

An independent derivation of (19.46) can be based on definition (19.47) and the Heisenberg equation for $\hat{\chi}(\chi)$, but we shall not pursue this approach further here.

Still another approach valid for large q replaces the charge eigenstate (19.38) by a coherent state

$$|\alpha_0\rangle = N\exp[\alpha_0\hat{\alpha}^\dagger]\,|\text{Vac}\rangle\,, \tag{19.48}$$

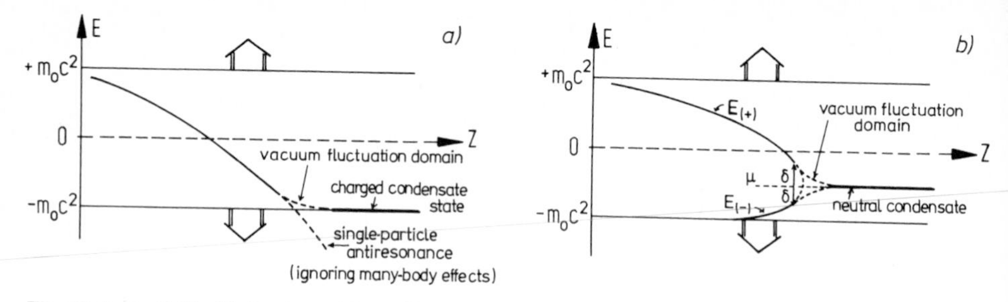

Fig. 19.5 a, b. Critical behaviour of Bose fields. **(a)** Long-range potential, $\mu = -m_0c^2$; **(b)** short-range potential

where $\alpha_0 = q^{1/2}$. Equation (19.18) differs from the state (19.38a) in that it is a superposition of such states of different q. Equation (19.48) is an eigenstate of the operator $\hat{\alpha}$, while (19.38) is an eigenstate of the operator $\hat{\alpha}^\dagger\hat{\alpha}$.

So far we have described the case of long-range potentials, i.e. when a bound particle mode descends to join the antiparticle continuum. Somewhat different is the case of short-range potentials, i.e. when particle and antiparticle modes meet in the gap $-m_0c^2 < E < m_0c^2$, i.e. the case discussed at the beginning of this section. Assume now that for $V(x) = V_{\text{cr}}(x)$ in (19.37), some particle and antiparticle modes χ_+ and χ_- become degenerate at $-m_0c^2 < \mu < m_0c^2$. For $V_0 \lesssim V_{\text{cr}}$ (where "<" is used as usual to indicate "of a strength less than critical"),

$$E_{(\pm)}\tau_1\chi_{(\pm)} = \mathscr{H}_0\chi_{(\pm)} \tag{19.49a}$$

$$\mathscr{H}_0 = \tau_1 V_0 + m_0c^2 + (1+\tau_3)\frac{(-\mathrm{i}\hbar c\nabla)^2}{2m_0c^2} \tag{19.49b}$$

and it is useful to parametrize the energies $E_{(\pm)}$ in terms of their average μ and a (small) deviation δ, Fig. 19.5,

$$E_{(\pm)} = \mu \pm \delta\,. \tag{19.49c}$$

The quantized field is [cf. (18.9)]

$$\hat{\chi} = \hat{a}\chi_{(+)} + \hat{c}^\dagger\chi_{(-)} + \hat{\chi}'\,, \tag{19.50}$$

where, for simplicity, we have not written out the indices $(\pm)$ for the operators $\hat{a}$, $\hat{c}$. Here $\hat{\chi}'$ describes the remainder of the modes, which are assumed to be relatively little affected by the approach to criticality.

We now proceed in a self-consistent fashion similar to our treatment of the charged condensates, except that two modes are present now. It is convenient to rewrite (19.50) as

$$\hat{\chi} = \hat{\chi}' + \tfrac{1}{2}(\chi_{(+)} + \chi_{(-)})(\hat{a} + \hat{c}^\dagger) + \tfrac{1}{2}(\chi_{(+)} - \chi_{(-)})(\hat{a} - \hat{c}^\dagger) \tag{19.51a}$$

$$\equiv \hat{\chi}' + \chi_{\text{e}}\hat{X} + \chi_{\text{o}}\hat{P}^\dagger \tag{19.51b}$$

with the implicit definitions

$$\hat{X} = \frac{1}{\sqrt{2}}(\hat{a} + \hat{c}^\dagger) \tag{19.51c}$$

$$\hat{P} = \frac{1}{\sqrt{2}}\mathrm{i}(\hat{a}^\dagger - \hat{c}) \tag{19.51d}$$

$$\chi_\mathrm{e} = \frac{1}{\sqrt{2}}(\chi_+ + \chi_-) \tag{19.51e}$$

$$\chi_\mathrm{o} = \frac{\mathrm{i}}{\sqrt{2}}(\chi_+ - \chi_-)\,, \tag{19.51f}$$

where

$$[\hat{X}, \hat{P}] = \mathrm{i}\,, \tag{19.51g}$$

while the critical part of the charge operator reads

$$\hat{q} = \hat{a}^\dagger \hat{a} - \hat{c}^\dagger \hat{c} = -\mathrm{i}(\hat{X}^\dagger \hat{P}^\dagger - \hat{P}\hat{X})\,. \tag{19.51h}$$

The Hamiltonian (19.37) can now be expressed in terms of $\hat{X}$, $\hat{P}$, and $\hat{q}$ only, so that it becomes a simple analogue of a two-dimensional classical anharmonic system, with a constant of motion $\hat{Q}$. Details of this tedious exercise [Kl 75a] are omitted, concentrating instead on the limit $\delta \to 0$, i.e. $\chi_\mathrm{o} \to 0$, since both $\chi_{(+)}$ and $\chi_{(-)}$ coincide at $E_{(\pm)} \to \mu$. Going at the same time to the classical limit, valid for macroscopic occupation of the modes, i.e. replacing $\hat{X}^2 \to q^2$, $\hat{P} \to 0$, in (19.37)

$$W(q^2) = q^2 \int \chi_\mathrm{e}^\dagger \left(\tau_1 V + m_0 c^2 + \frac{1+\tau_3}{2} \frac{(-\mathrm{i}\hbar\nabla)^2}{2 m_0 c^2} \right) \chi_\mathrm{e} d^3x$$
$$+ \frac{1}{2} q^4 e^2 \int \chi_\mathrm{e}^\dagger(\boldsymbol{x})\, \tau_1\, \chi_\mathrm{e}(\boldsymbol{x}) \frac{1}{4\pi|\boldsymbol{x}-\boldsymbol{x}'|} \chi_\mathrm{e}^\dagger(\boldsymbol{x}')\, \tau_1\, \chi_\mathrm{e}(\boldsymbol{x}')\, d^3x\, d^3x'\,, \tag{19.52}$$

where, because of the supercriticality, the coefficient of q^2 *is negative*, since V is a strongly attractive potential and without the last term in (19.52) $W_0(q^2) \sim -|E_{1s}| q^2$. Thus the minimum of (19.52) is found at large $q^2 \neq 0$, where $W(q^2)$ assumes a negative value.

Varying $W(q^2)$ subject to the normalization condition (remember χ_- has negative norm)

$$\int \chi_\mathrm{e}^\dagger \tau_1 \chi_\mathrm{e} d^3x = \tfrac{1}{2} \int \chi_+^\dagger \tau_1 \chi_+ + \tfrac{1}{2} \int \chi_-^\dagger \tau_1 \chi_- = 0\,, \tag{19.53}$$

the same condensate equations arise as for the Coulomb case, (19.43), except that now the norm of the solution must vanish. The self-consistent potential $V + V_\mathrm{cond}$ must be such to remain critically, i.e. the energies $E_{(\pm)} \approx \mu$, no matter what the (short-range) external field V does, Fig. 19.5b. The only complication, compared to the charged condensate case, is the lack of completeness, even in the condensed ground state, namely, while χ_e satisfies

$$(\mathscr{H}_{\text{eff}}-\mu\tau_1)\chi_e = 0 \quad \text{with} \tag{19.54a}$$

$$\mathscr{H}_{\text{eff}} = \mathscr{H}_0 + V_{\text{cond}} \quad \text{and} \tag{19.54b}$$

$$V_{\text{cond}}(\boldsymbol{x}) = \frac{q^2e^2}{4\pi}\int \chi_e^\dagger(\boldsymbol{x}')\tau_1\chi_e(\boldsymbol{x}')\frac{1}{|\boldsymbol{x}-\boldsymbol{x}'|}d^3x' \tag{19.54c}$$

we have apparently lost the mode χ_o from the spectrum of $\mathscr{H}$. To complement the spectrum of $\mathscr{H}$ suitably, consider the set χ_i such that

$$\mathscr{H}_{\text{eff}}\chi_i = E_i\tau_1\chi_i\,, \tag{19.55a}$$

with some i = e being just critical as in (19.54) and therefore of zero norm. Now multiply (19.55a) with $\chi_o^\dagger$, which satisfies

$$(\mathscr{H}_{\text{eff}}-\mu\tau_1)\chi_o \stackrel{!}{=} \lambda\tau_1\chi_e\,. \tag{19.55b}$$

Then

$$\int d^3x\,\chi_o^\dagger\mathscr{H}_{\text{eff}}\chi_i = E_i\int d^3x\,\chi_o\tau_1\chi_i\,. \tag{19.56a}$$

But using (19.55b) gives

$$\int d^3x\,\chi_o^\dagger\mathscr{H}_{\text{eff}}\chi_i = \mu\int d^3x\,\chi_o^\dagger\tau_1\chi_i + \lambda\int d^3x\,\chi_e^\dagger\tau_1\chi_i\,. \tag{19.56b}$$

For $E_i \neq \mu$, combining (19.56a, b), the required orthogonality relation between the state χ_o defined by (19.55) and the remainder of the spectrum is

$$\int d^3x\,\chi_o^\dagger\tau_1\chi_i = \frac{\lambda}{E_i-\mu}\int d^3x\,\chi_e^\dagger\tau_1\chi_i = 0\,, \quad i \neq e\,. \tag{19.57}$$

When i is the critical state, i = e, recalling that

$$\int d^3x\,\chi_e^\dagger\tau_1\chi_o = \tfrac{1}{2}\left(\int d^3x\,\chi_+^\dagger\tau_1\chi_+ - \int d^3x\,\chi_-^\dagger\tau_1\chi_-\right) = 1\,, \tag{19.58a}$$

$$\int d^3x\,\chi_e^\dagger\tau_1\chi_e = \tfrac{1}{2}\left(\int d^3x\,\chi_+^\dagger\tau_1\chi_+ + \int d^3x\,\chi_-^\dagger\tau_1\chi_-\right) = 0\,, \tag{19.58b}$$

then $E_e = \mu$, when inserting (19.57). Thus we have shown that (19.55b, 58) define a suitable function χ_o *orthogonal to all solutions of the Klein-Gordon equation* (19.54), which therefore provides the missing function to make up a complete set. (Actually, also one can show by construction, taking the limit $\delta \to 0$, that (19.56) arises explicitly [Kl 75a].) As we have not constructed this explicitly here, the proof of this statement still requires that we show the existence of a non-trivial solution. This is, however, certainly the case, since we have an *incomplete* set of states $\{\chi_i\}$ when the norm of χ_e, (19.58b), vanishes. Alternatively, when the norm does not vanish, (19.56) can be satisfied only for $\lambda = 0$, making χ_o a member of the then complete set of states. Thus (19.55b) is actually more general, applying to all cases in which a zero-norm wave function χ_e is found.

We conclude this section by indicating how one can proceed to show that the (charged) ground state is stable with respect to fluctuations of zero charge. The approximate, coherent ground state (19.48) is in fact an approximate eigenstate of the operator

$$\hat{H}' = \hat{H} - \mu\hat{L}\,, \quad \text{where} \tag{19.59a}$$

$$\hat{L} = \int \hat{\chi}^\dagger \tau_1 \hat{\chi}\, d^3x \tag{19.59b}$$

is proportional to the charge operator.

The coherent state is furthermore characterized by the condition

$$\langle \alpha_0 | \hat{\chi}(\boldsymbol{x}) | \alpha_0 \rangle = \chi(\boldsymbol{x})\,. \tag{19.60a}$$

Then

$$\hat{\chi}(\boldsymbol{x}) = \chi(\boldsymbol{x}) + \hat{\phi}(\boldsymbol{x})\,, \tag{19.60b}$$

where $\chi(\boldsymbol{x})$ is the macroscopic condensate wave function and $\hat{\phi}(\boldsymbol{x})$ describes the small quasiparticle oscillations around the new ground state. A standard procedure of variations with respect to $\hat{\phi}$ may then be followed to show that $|\alpha_0\rangle$ is locally stable under these fluctuations – the proof is straightforward but tedious as it requires that the linearized spectrum of $\hat{\phi}$ derived from

$$\left.\frac{\delta H[\hat{\chi} = \chi + \hat{\phi}]}{\delta\hat{\phi}(\boldsymbol{x})}\right|_{\hat{\phi}=0} U_n(\boldsymbol{x}) = \varepsilon_n U_n(\boldsymbol{x}) \tag{19.61}$$

contains only positive eigenvalues ε_n.

19.3 Solutions of the Condensate Equations

In this section we solve the charged condensate equations (19.43). It is clear from the outset that these complicated integro-differential equations must be approached numerically. To this end it is convenient to convert them back into the (original) second-order form of the Klein-Gordon equation (recall that this is often done with the Dirac equation as well). Introducing the effective potential,

$$\begin{aligned} V_{\text{eff}} &= V + V_{\text{cond}} \\ &\equiv V + q V_0\,, \end{aligned} \tag{19.62}$$

(19.43) yields

$$(\mu - V_{\text{eff}})^2 \varphi_0 = (c^2 p^2 + m_0^2 c^4)\, \varphi_0\,. \tag{19.63a}$$

Here the wave function φ_0 is normalized in the usual way

$$\int \varrho_0(\boldsymbol{x})\, d^3x = 1\,, \quad \text{where} \tag{19.63b}$$

$$\varrho_0 = 2\varphi_0(\mu - V_{\text{eff}})\,\varphi_0\,. \tag{19.63c}$$

The expression (19.62) for the effective potential, contains the potential generated by a unit condensed charge

$$V_0 = e^2 \int \frac{\varrho_0(x')}{|x-x'|} d^3x' . \tag{19.63d}$$

For the remainder of this discussion, the Coulomb potential V will be that of the uniformly charged sphere of radius R, (19.13).

Motivated by the (hypothetical) example of isospin zero nuclear matter $[f = (Z/A) = 1/2]$ at near normal density, the nuclear radius is given by

$$R = r_0 A^{1/3}, \quad \text{where} \tag{19.64a}$$

$$r_0 = 1.2\ \text{fm}\ (\varrho_N^0 | \varrho_N)^{1/3} \tag{19.64b}$$

and ϱ_N^0 is the normal nuclear density. Now $m_0 = m_\pi$ is the pion mass. We are interested in solutions of (19.63) in the limit $Z > Z_{cr} = 3006.5$, the critical value for $\varrho_N = \varrho_N^0$, $f = 1/2$.

When solving these equations, a numerical iterative procedure is quite helpful: given $V_0^{(n)}$ (in the nth iteration) we can solve (19.63a) for the wave function $\phi_0^{(n)}$ in the nth iteration:

$$(\mu - V - q^{(n)} V_0^{(n)})^2 \varphi_0^{(n)} = (c^2 p^2 + m_0^2 c^4) \varphi_0^{(n)} \tag{19.65}$$

with $\mu = -m_\pi$ fixed and $q^{(n)} > 0$. As explained in Sect. 19.2, we are interested in the smallest $q^{(n)}$ that yields a solution. Given the eigenvalue $q^{(n)}$, then

$$V_{\text{eff}}^{(n)} = V + q^{(n)} V_0^{(n)} \quad \text{and} \tag{19.66}$$

$$\varrho_0^{(n)} = 2 \varphi_0^{(n)} (\mu - V_{\text{eff}}^{(n)}) \varphi_0^{(n)} . \tag{19.67}$$

The iteration loop is closed by computing

$$V_0^{(n+1)}(x) = \frac{e^2}{4\pi} \int \frac{\varrho_0^{(n)}(x')}{|x-x'|} d^3x' . \tag{19.68}$$

The above procedure is then repeated until

$$\| V_0^{(n)} - V_0^{(n+1)} \| < \delta ,$$

with δ to be chosen sufficiently small.

In practice, the initial choice

$$V_0^{(0)} = \frac{1}{Z} V ,$$

i.e. just the part of the Coulomb potential when its charge is changed to one unit, turns out to be sufficient to make the iteration procedure converge rapidly.

We now discuss the properties of the numerical results. The free parameter in the solutions in the "nuclear charge" Z, and all quantities will be given in units of m_π. Consider arbitrarily chosen values of $Z = 3025$, 3500 and 3750 to encompass the essential features of the numerical calculations. While the smallest of these values is still close to the critical value 3006.5, the largest is significantly supercritical. Consider first the "large" supercritical charge $Z = 3750$. The most remarkable feature here is that with growing supercriticality the total potential approaches more and more closely the value $-(2+\varepsilon)m_\pi$, where $\varepsilon \sim (\pi/\langle r \rangle)$, Fig. 19.6. This result agrees with our expectation that the potential must be deeper than $-2m_\pi c^2$ to generate a bound state at $\mu = -m_\pi c^2$. Note that the potential shows a local minimum for high Z, an effect associated with the relatively high localization of the condensate charge distribution ϱ_0 as compared to the assumed nuclear charge, Fig. 19.7. Also the relative similarity of the total potentials at the origin for high Z, as shown in Fig. 19.8, can be understood in terms of the increased influence of the condensate potential on the form of the total potential at the origin.

A further expected qualitative aspect of the condensate – the flattening of the total potential with increasing Z, since the neutralization by the condensate charge becomes increasingly smooth – is displayed in Fig. 19.8, where the main point is the approach of the self-consistent $V_{\text{total}}\,(r = 0)$ to the expected value at $-2m_\pi c^2$. The associated increase in the width of the potential is reflected by the increasing delocalization of the charge density, Fig. 19.7.

Figure 19.9 completes this discussion of the properties of the condensate by illustrating the charge of the condensate q (and therefore that of the "true vacuum" state) as a function of the external source strength Z. As can be expected, q grows monotonically with $Z' = Z - Z_{\text{cr}}$, at first linearly, but soon Z'/q begins to decrease monotonically with increasing Z. It is worth recording for later use that at $Z = 3700$, q changes half as fast as Z', and that for two external charges, one new pair must be created to keep the potential just critical.

As alluded to previously, charged condensates need not appear only at $\mu = -m_\pi c^2$. Instead, if other interactions can be responsible for charge balancing, μ can be at any place in the interval $(-m_\pi c^2, 0)$ provided that the energy to create (1) the particle for the condensate, i.e. $m_\pi c^2$ and (2) the balancing charge is available. Thus the condition

$$E_{\text{binding}} = m_\pi c^2 - \mu = |E^{(+)}| + |E^{(-)}| \tag{19.69}$$

arises, where $E^{(\pm)}$ refer to (1) and (2), respectively.

If, instead of being emitted to infinity, the antiparticle (here called a positively charged pion) is converted immediately into a positron and a neutrino by the weak interaction, then $E^{(+)} = m_\pi c^2$ and $E^{(-)} = m_e c^2$. Hence, from (19.69) $\mu = -m_e c^2$ (which is practically zero). Everything said above for $\mu = -m_\pi c^2$ about the condensate solutions applies here, except that (a) the total potential now has a depth near to $-m_\pi c^2$ and (b) the critical point is at $Z \approx 1150$. Figure 19.10 shows the analogue of Fig. 19.6, which by comparing different potentials

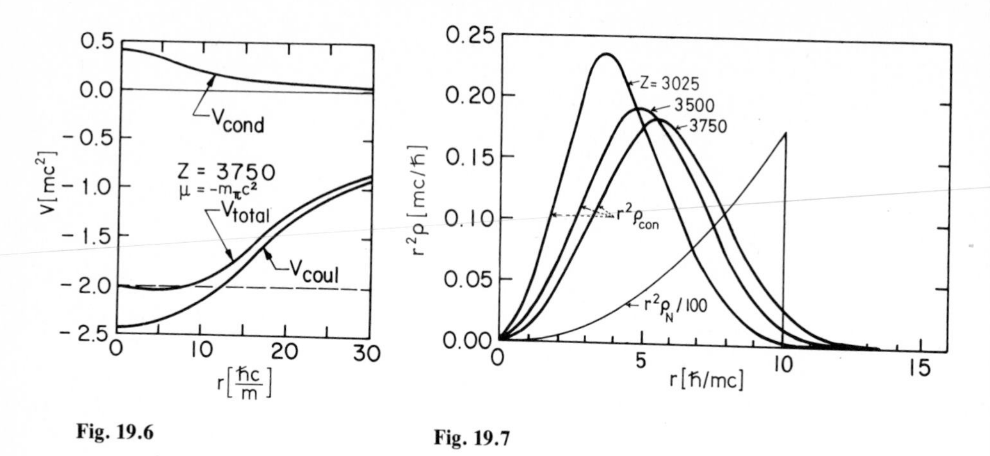

Fig. 19.6

Fig. 19.7

Fig. 19.6. Comparison of the total potential V_{total}, the condensate potential V_{cond}, and the Coulomb potential V_{coul}, for $Z' = 3750$

Fig. 19.7. Comparison of the condensate charges ϱ_{cond} for selected values of nuclear charge with the assumed homogeneous nuclear charge density $r^2 \varrho_N$

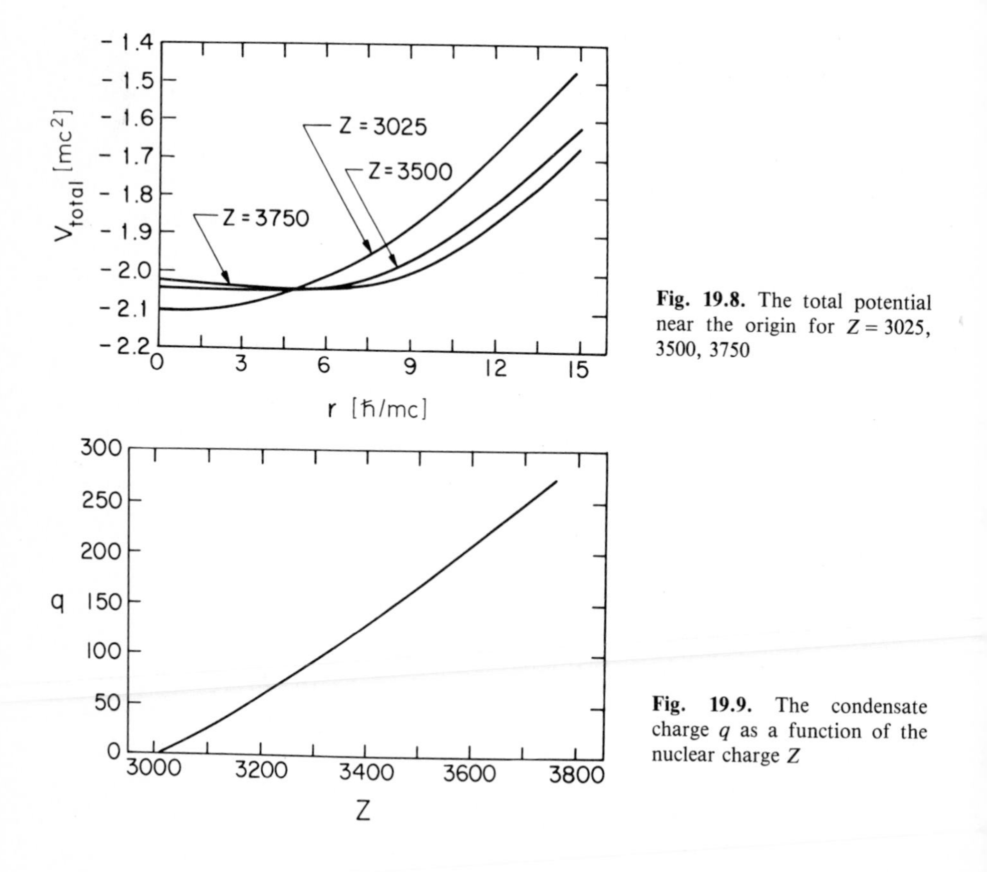

Fig. 19.8. The total potential near the origin for $Z = 3025$, 3500, 3750

Fig. 19.9. The condensate charge q as a function of the nuclear charge Z

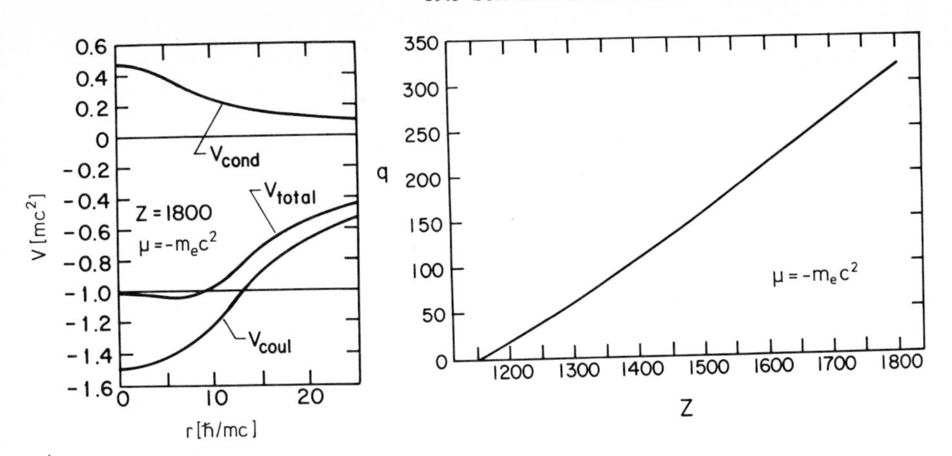

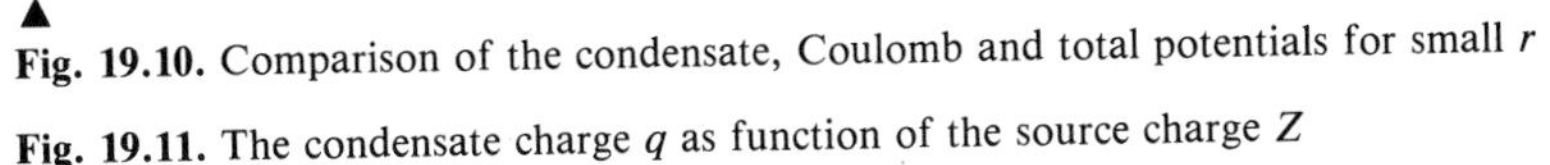

▲

Fig. 19.10. Comparison of the condensate, Coulomb and total potentials for small r

Fig. 19.11. The condensate charge q as function of the source charge Z

illustrates point (a) above, while in Fig. 19.11 the condensate charge of the true vacuum with $\mu = -m_e c^2$ illustrates (b).

In conclusion, the charge density of the condensate is such as to screen down the effect of a supercritical potential to a just critical one with depth corresponding to the chosen value of μ, i.e. $V_0 \sim -m_0 c^2 + \mu$. Thus, contrary to the case of fermions, where the self-consistent potential changed with the density of external charge, (16.45), the self-consistent potential for meson condensates generates *one* critical state only, since the Pauli exclusion principle is inoperative, and all condensed particles sit in one state. This also means that the limit of point-like nuclei of arbitrary charge for meson fields should be quite different from the Dirac case, (16.46). This behaviour has so far not been considered. Let us now turn to the more relevant case of strong interactions.

Bibliographical Notes

Bose condensation in strong fields was first discussed by *Migdal* [Mi 72]. The subject was further developed by *Klein* and *Rafelski* [Kl 75a, 77] and is reviewed by *Rafelski, Fulcher* and *Klein* [Ra 78b], and in [Ra 78d]. The different types of approach to criticality for scalar particles have been discussed by *Bawin* and *Lavine* [Ba 75, 79c]. So-called "pion condensation" in nuclear matter, discussed e.g. by *Migdal* [Mi 73c], *Sawyer* and *Scalapino* [Sa 73], *Brown* and *Weise* [Br 76], is a many-body effect and only marginally connected with the Bose condensate in the vacuum treated in this chapter.

20. Strong Yang-Mills Fields

The intention of this and the following chapters is to outline problems in other domains of physics related to the problems of QED of strong fields studied in this book. The two domains selected are the modern theory of strong interactions, i.e. quantum chromodynamics, that is treated here, and the theory of strong gravitational fields, i.e. the general theory of relativity, Chap. 21.

In principle, each subject deserves a book on its own. The need to condense some of the most interesting aspects into a single chapter obviously necessitates a change in our presentation both in style and detail. We refer the reader who develops a deeper interest in these subjects to the original literature. Still, we hope that these last two chapters fulfil their twofold purpose: to serve as introduction to these fields, and to show that the theory of QED of strong fields is pertinent to other areas of active research, and that the methods we developed in its context have many applications in these other areas.

20.1 Quantum Chromodynamics

Numerous physicists believe today that quantum chromodynamics (QCD) is the fundamental theory of strong interactions. What is QCD? From high-energy scattering experiments with leptons (electrons, muons or neutrinos) on hadrons, i.e. strongly interacting particles, we know that there must be almost point-like constituents inside the hadrons (Fig. 20.1). These constituents are commonly called partons, and are identified with the quarks first postulated when classifying hadrons according to the representations of SU(3). From the analysis of lepton-hadron scattering it is known that quarks have spin-1/2, i.e. they are Dirac particles, and that they carry non-integer electric charge.

Today we know that there are at least five different "flavours" of quarks, called u(p), d(own), s(trange), c(harm) and b(ottom), respectively, and there

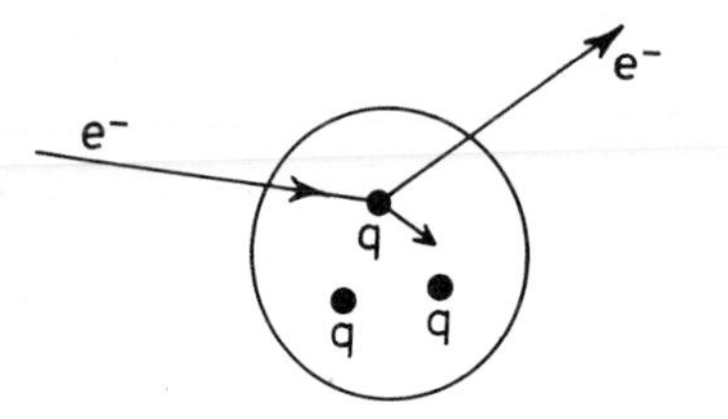

Fig. 20.1. Quarks have been discovered as point-like scattering centres in the interior of nucleons

are strong theoretical reasons to suspect the existence of one more flavour, called t(op). The quarks carry electric charge, which takes the value +2/3 for u, c, (t) and the value −1/3 for d, s, b in units of proton charge. Quarks of various flavour differ widely in their mass. Because the quarks are confined to the inside of hadrons – at least under normal circumstances – their mass cannot be measured directly. Much of the mass of the hadron itself does not originate in the intrinsic mass of the quark constituents but resides in the kinetic energy of the confined quarks and in the field that binds the quarks together, the so-called glue field. By current algebra one can extract something like bare quark masses (called masses of "current quarks"), for which typical values are [La 79b]

$$m_{\mathrm{u}} \simeq 5\ \mathrm{MeV}\,, \qquad m_{\mathrm{d}} \simeq 10\ \mathrm{MeV}\,, \qquad m_{\mathrm{s}} \simeq 150\ \mathrm{MeV}\,,$$
$$m_{\mathrm{c}} \simeq 1500\ \mathrm{MeV}\,, \qquad m_{\mathrm{b}} \simeq 5000\ \mathrm{MeV}\,, \qquad m_{\mathrm{t}} > 20\ \mathrm{GeV}\,.$$

The quarks must carry one further internal quantum number, called *colour*. This can be deduced from the existence of particles like the $\Delta^{++} = (\mathrm{uuu})$, the $\Delta^{-} = (\mathrm{ddd})$ and $\Omega^{-} = (\mathrm{sss})$ which contain three identical quarks with parallel spin in an s wave. The Pauli principle requires that the quarks differ from each other by an additional quantum number. An independent argument relies on the ratio of the production of hadrons from $\mathrm{e^+e^-}$ collisions as compared with the production of muon pairs. If the production of hadrons proceeds through creation of a quark – antiquark pair which subsequently fragments into hadrons, the ratio of the cross-sections can be related to the sum over the square of the charges of all sorts of quarks that can be created. Counting the different colours separately gives a factor N_{c} for the number of colours:

$$R = \sigma(\mathrm{e^+e^-} \to \mathrm{hadrons})/\sigma(\mathrm{e^+e^-} \to \mu^+\mu^-) = N_{\mathrm{c}} \sum_i \mathrm{e}_i^2/\mathrm{e}^2 = (11/9)\,N_{\mathrm{c}}\,, \quad (20.1)$$

for $i = \mathrm{u, d, s, c, b}$. The experiment yields $N_{\mathrm{c}} = 3$, Fig. 20.2.

We believe today that the quantum number of colour resembles very much in its properties electric charge, so much so that one also speaks of "colour charge". Like electric charge, colour charge is exactly conserved, and it acts as the source of a force field which is long range unless screened (Fig. 20.3). The difference is that there are three different colour degrees of freedom, say red, green and blue, whereas electric charge is a one-dimensional, and therefore additive, quantity. The theory of colour forces, i.e. QCD, is derived from the principle of gauge invariance against arbitrary rotations in colour space. Since the complex rotations of a three-dimensional vector are described by unitary 3 × 3 matrices $U(u_{ik})$ of unit determinant, the symmetry group of the gauge transformations is SU(3).

The precise form of the QCD Lagrangian is obtained by considering invariance under local colour rotations, i.e. by allowing the gauge matrix U to change from one point in space to another. To see how this principle of local gauge invariance works, let us go back to QED. In QED the gauge transformations correspond to changes in the phase of the wave function: $\psi \to \exp(\mathrm{i}\,\alpha)\,\psi$. If phase α varies from point to point, i.e. if $\alpha = \alpha(x)$, the derivative of a wave function changes by a non-trivial term ($\partial_\mu = \partial/\partial x^\mu$):

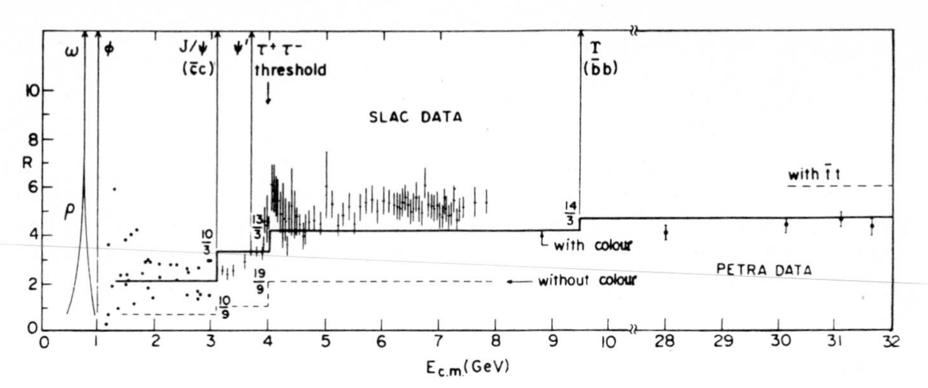

Fig. 20.2. The ratio R of hadronic versus electromagnetic cross-section in e^+e^- annihilations as function of energy

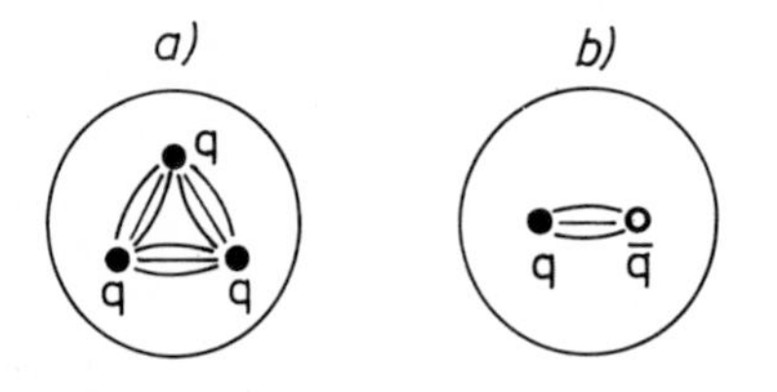

Fig. 20.3 a, b. Quarks are permanently confined in baryons (**a**) and mesons (**b**) due to strong "colour" fields

$$\partial_\mu \psi \to \partial_\mu (\mathrm{e}^{\mathrm{i}\alpha(x)} \psi) = \mathrm{e}^{\mathrm{i}\alpha}(\partial_\mu \psi) + \mathrm{e}^{\mathrm{i}\alpha}(\mathrm{i}\,\partial_\mu \alpha)\,\psi\,. \tag{20.2}$$

The unwanted second term is cancelled by the gauge change of the electromagnetic potential A_μ, if the potential is added to the derivative operator in minimal coupling $(\partial_\mu + \mathrm{i}eA_\mu)$. With

$$A_\mu(x) \to A_\mu(x) - e^{-1}(\partial_\mu \alpha)\,, \tag{20.3}$$

the minimally coupled derivative remains invariant:

$$(\partial_\mu + \mathrm{i}eA_\mu)\,\mathrm{e}^{\mathrm{i}\alpha(x)}\psi(x) = \mathrm{e}^{\mathrm{i}\alpha(x)}(\partial_\mu + \mathrm{i}eA_\mu)\,\psi(x)\,. \tag{20.4}$$

The electromagnetic field strength tensor $F_{\mu\nu} = \partial_\mu A_\nu - \partial_\nu A_\mu$ also being invariant against the gauge transformations (20.3), the Lagrangian of QED is

$$L_{\mathrm{QED}} = \mathrm{i}\,\bar{\psi}\gamma^\mu(\partial_\mu + \mathrm{i}eA_\mu)\,\psi - m\,\bar{\psi}\psi - \tfrac{1}{4}F^{\mu\nu}F_{\mu\nu}\,. \tag{20.5}$$

The form of this Lagrangian, and therefore the form of the electromagnetic interaction, is uniquely determined by the requirement of local gauge invariance.

Because there are three colours, the wave function of a quark has three components in colour space: $\Psi = (\psi_{\mathrm{r}}, \psi_{\mathrm{g}}, \psi_{\mathrm{b}})$, where the indices r, g, b stand for "red", "green", and "blue". As already said, a colour gauge transformation is described by a unitary 3×3 matrix U with $\det(U) = 1$, which rotates the colour components of the wave function:

$$\Psi \to U\Psi = \begin{pmatrix} U_{\rm rr} & U_{\rm rg} & U_{\rm rb} \\ U_{\rm gr} & U_{\rm gg} & U_{\rm gb} \\ U_{\rm br} & U_{\rm bg} & U_{\rm bb} \end{pmatrix} \begin{pmatrix} \psi_{\rm r} \\ \psi_{\rm g} \\ \psi_{\rm b} \end{pmatrix} . \tag{20.6}$$

Any unitary matrix can be written as the imaginary exponential of a Hermitian matrix: $U = \exp(\mathrm{i}L)$. We have $U^+ U = 1$ and $L^+ = L$, and $\det(U) = 1$ means $\mathrm{tr}(L) = 0$. All traceless Hermitian 3×3 matrices can be expressed as linear combination of the eight λ matrices of *Gell-Mann* [Ge 62]

$$L = \frac{1}{2} \sum_{a=1}^{8} \theta_a \lambda_a . \tag{20.7}$$

In the standard representation the *Gell-Mann* matrices are given by:

$$\lambda_1 = \begin{pmatrix} 0 & 1 & 0 \\ 1 & 0 & 0 \\ 0 & 0 & 0 \end{pmatrix} \quad \lambda_2 = \begin{pmatrix} 0 & -\mathrm{i} & 0 \\ 1 & 0 & 0 \\ 0 & 0 & 0 \end{pmatrix} \quad \lambda_3 = \begin{pmatrix} 1 & 0 & 0 \\ 0 & -1 & 0 \\ 0 & 0 & 0 \end{pmatrix}$$

$$\lambda_4 = \begin{pmatrix} 0 & 0 & 1 \\ 0 & 0 & 0 \\ 1 & 0 & 0 \end{pmatrix} \quad \lambda_5 = \begin{pmatrix} 0 & 0 & -\mathrm{i} \\ 0 & 0 & 0 \\ \mathrm{i} & 0 & 0 \end{pmatrix}$$

$$\lambda_6 = \begin{pmatrix} 0 & 0 & 0 \\ 0 & 0 & 1 \\ 0 & 1 & 0 \end{pmatrix} \quad \lambda_7 = \begin{pmatrix} 0 & 0 & 0 \\ 0 & 0 & -\mathrm{i} \\ 0 & \mathrm{i} & 0 \end{pmatrix} \quad \lambda_8 = \frac{1}{\sqrt{3}} \begin{pmatrix} 1 & 0 & 0 \\ 0 & 1 & 0 \\ 0 & 0 & -2 \end{pmatrix} ,$$

but in most cases we do not need the explicit form of the λ_a, only the commutation relations

$$[\lambda_a, \lambda_b]_- = 2\mathrm{i} f_{abc} \lambda_c , \qquad [\lambda_a, \lambda_b]_+ = (4/3)\,\delta_{ab} + 2\, d_{abc} \lambda_c , \tag{20.8}$$

where the f_{abc} and d_{abc} are the antisymmetric and symmetric structure constants of the Lie group SU(3), respectively, and summation over c is implied.

We now make the colour rotation U space dependent by allowing the eight real parameters θ_a to change from point to point $U(x) = \exp[\frac{\mathrm{i}}{2}\theta_a(x)\,\lambda_a]$ with summation over "a" implicitly understood. The partial derivative then acquires an unwanted additional term

$$\partial_\mu [U(x)\,\Psi] = U \partial_\mu \Psi + (\partial_\mu U)\,\Psi = U[\partial_\mu \Psi + U^\dagger (\partial_\mu U)\,\Psi] . \tag{20.9}$$

Instead of the term $(\mathrm{i}\partial_\mu \alpha)$ in (20.2) there is now a matrix term $U^\dagger(\partial_\mu U)$. To cancel this contribution we have to introduce a colour potential $\hat{A}_\mu$ which is a 3×3 matrix, indicated by the circumflex. As the potential must be Hermitian, we can represent it as a linear combination of the Gell-Mann matrices with eight real potential functions $A_\mu^a(x)$

$$\hat{A}_\mu(x) = \frac{1}{2} \sum_{a=1}^{8} A_\mu^a(x)\, \lambda_a . \tag{20.10}$$

The field $\hat{A}_\mu(x)$ is called a *Yang-Mills* field [Ya 54, Ut 56, Ge 61]. If the potential changes under a local colour rotation according to

$$\hat{A}_\mu \to U^\dagger \hat{A}_\mu U - \mathrm{i} g^{-1} U^\dagger (\partial_\mu U) , \tag{20.11}$$

the minimally coupled derivative $(\partial_\mu - \mathrm{i} g \hat{A}_\mu)$ remains invariant in form under the gauge transformation:

$$(\partial_\mu - \mathrm{i} g \hat{A}_\mu)\, U \Psi = U(\partial_\mu - \mathrm{i} g \hat{A}_\mu)\, \Psi . \tag{20.12}$$

That one needs eight colour potentials instead of one is not really a surprise. The three-component quark wave function forms a triplet of the colour group SU(3), while the wave function of an antiquark forms an antitriplet. From the product of a triplet and an antitriplet one can form a SU(3) singlet or an octet: $3 \times \bar{3} = 8 + 1$. Since the colour potential must have the same quantum numbers as a quark – antiquark pair and cannot be a colourless singlet, it must be described by an octet of fields. The eight-component field strength tensor

$$F_{\mu\nu}^a = \partial_\mu A_\nu^a - \partial_\nu A_\mu^a + g f_{abc} A_\mu^b A_\nu^c \tag{20.13}$$

remains form invariant under a local colour gauge transformation, i.e. $\hat{F}_{\mu\nu} \to U^+ \hat{F}_{\mu\nu} U$, where the matrix $\hat{F}$ is defined in the same way by its eight colour components as the matrix $\hat{A}$. The complete Lagrangian of QCD is then [Fr 72, We 73, Gr 73]

$$L_{\mathrm{QCD}} = \mathrm{i}\, \tilde{\Psi} \gamma^\mu (\partial_\mu - \mathrm{i} g \hat{A}_\mu)\, \Psi - m\, \tilde{\Psi} \Psi - \tfrac{1}{4} F_{\mu\nu}^a F_a^{\mu\nu} . \tag{20.14}$$

The similarity to the Lagrangian of QED (20.5) is conspicuous. The difference lies in the non-linear term entering the definition of $F_{\mu\nu}^a$, (20.13), which is quadratic in the colour potentials A_μ^a. Since the antisymmetric structure coefficients f_{abc} vanish when two indices are equal, the non-linear terms disappear if just a single colour component of A_μ^a is different from zero. It follows directly from this observation that all solutions of classical electrodynamics, where there is only one potential to start with, are also solutions of classical chromodynamics. There are, of course, further solutions to which the non-linear terms contribute.

In the quantum theory one cannot make certain components of the colour potential vanish at will, and therefore the non-linearities always show up. As a consequence, the properties of QCD are radically different from those of QED, in particular, free colour charges do not exist. This property, called *confinement of colour*, means that particles carrying colour charge always come in combinations which form total colour singlets, i.e. long-range colour forces are completely screened. The proof of these assertions is provided by complicated numerical calculations (see [Sa 82] for a review).

The analogue of the Maxwell equations may be derived for the colour field from the Lagrangian (20.14)

$$\partial_\nu F_a^{\mu\nu} = g j_a^\mu - g f_{abc} A_\nu^b F_c^{\mu\nu} , \tag{20.15}$$

where the colour current of the quarks is obtained from L_{QCD}

$$j_\mu^a = \tfrac{1}{2} \bar{\Psi} \gamma^\mu \lambda_a \Psi . \tag{20.16}$$

The non-linear term appearing on the rhs of (20.15) indicates that the colour field acts partially as its own source. In other words, the quanta of the colour field, which are called *gluons*, carry colour charge themselves. This is the origin of the differences between QCD and QED. From (20.15) one can see that j_μ^a does not obey a continuity equation, which means that the colour charge of the quarks alone is not conserved. This is not surprising since quarks can emit or absorb gluons which carry away colour. Only by adding the colour charge residing in the gluon field, represented by the second term on the rhs (20.15), is a conserved colour current obtained. Finally, note that one can derive an energy-momentum tensor

$$T_\mu^\nu = -L_{\mathrm{QCD}} \delta_\mu^\nu - F_{\mu\lambda}^a F_a^{\nu\lambda} \tag{20.17}$$

for the gluon field. Its divergence yields the analogue of the Lorentz force, now the force acting on a particle moving in a colour field:

$$\partial_\nu T^{\mu\nu} = g F_a^{\mu\nu} j_\nu^a . \tag{20.18}$$

The energy density has the familiar form $T_{00} = \frac{1}{2}(E^a)^2 + \frac{1}{2}(B^a)^2$, where E^a and B^a are the electric and magnetic components of the colour field strength tensor.

Perturbative quantization in QCD proceeds similarly as in QED (Chap. 14), because QCD can be shown to be renormalizable [tH 71]. The quadratic terms in the Lagrangian L_{QCD} define free quark and gluon fields described by propagators which have the same form as those for electrons and photons in QED. One difference arises because of the coupling of the gluon field to itself. To ensure gauge invariance of the quantum theory one has to introduce fictitious particles called (*Faddeev-Popov*) ghosts [Fa 67], which carry colour but behave like fermions although they propagate like spin-zero particles. These particles cancel the contributions from the unphysical degrees of freedom of the colour gauge field. The terms of third and fourth order in L_{QCD} give rise to interaction vertices among the free propagators of quarks, gluons and ghosts. There is a quark-gluon vertex, three-gluon and four-gluon vertices, and a gluon-ghost vertex. Propagators and vertices can be combined to Feynman diagrams in all possible ways, with the exception that ghosts, being fictitious particles, can occur only in intermediate states. The propagators and vertices of QCD in the Coulomb gauge are shown graphically and analytically in Fig. 20.4.

The free propagator of the gluon, which is proportional to $1/q^2$, suggests that the colour force falls off like $1/r$. The true gluon propagator, however, is modi-

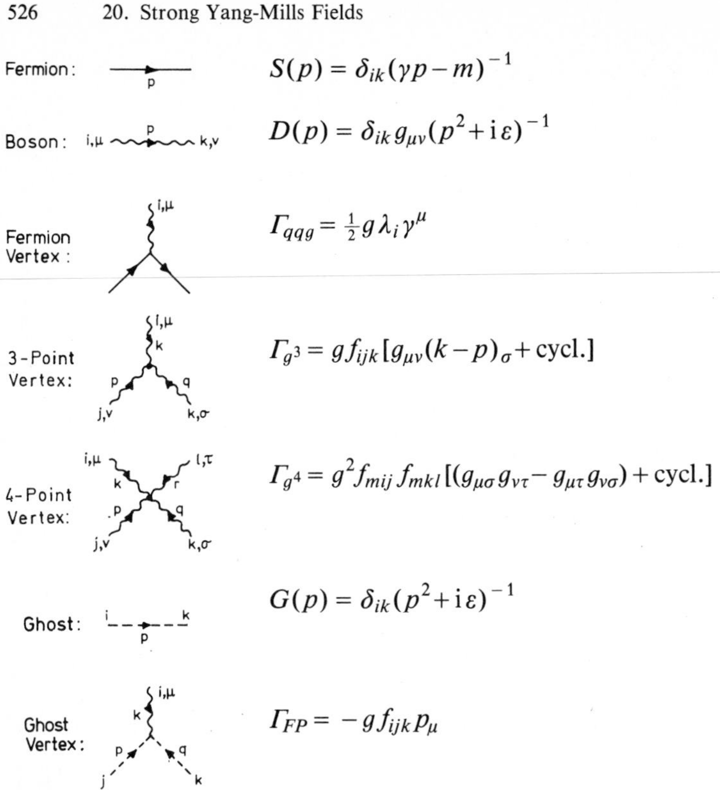

Fig. 20.4. Propagator and vertices of QCD. Ghost lines are not allowed to appear as external lines in Feynman diagrams

fied by vacuum polarization and shows completely different behaviour. To see how this arises, one has to evaluate the loop diagrams corresponding to polarization of the vacuum by the virtual creation of a pair of coloured particles from the vacuum:

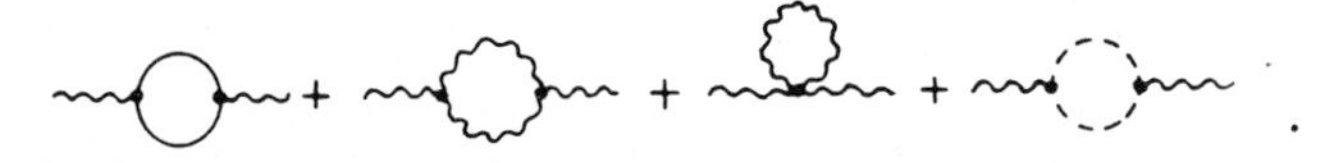

As in QED, the effect of these diagrams can be expressed in terms of a polarization function

$$\Pi(q^2) = -(33 - 2N_\mathrm{F})/48\,\pi^2 g^2 q^2 \ln(-q^2/\mu^2)\,, \tag{20.19}$$

where μ is a reference point introduced by the renormalization procedure and N_F counts the number of quark flavours with mass below $|q^2|^{1/2}$. The sign of Π is opposite to that of the polarization function in QED due to the presence of the diagram involving the gluon loop. The higher-order diagrams, in which the gluon interacts consecutively once, twice, three times with the vacuum polarization, and so on, can be summed into a geometric series for the full propagator (see Sect. 14.4):

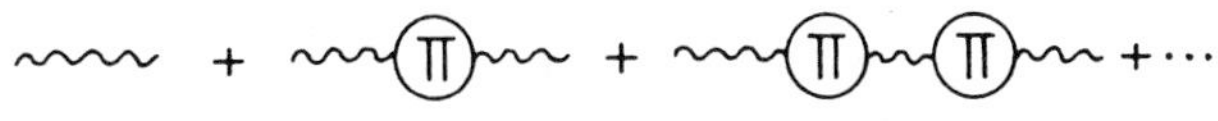

$$\begin{aligned} D(q^2) &= D_0(q^2) + \mathrm{i}^2 D_0(q^2)\,\Pi(q^2)\,D_0(q^2) \\ &\quad + \mathrm{i}^4 D_0(q^2)\,\Pi(q^2)\,D_0(q^2)\,\Pi(q^2)\,D_0(q^2) + \ldots \\ &= D_0(q^2)[1 - \Pi(q^2)\,D_0(q^2)]^{-1} \\ &= 1/q^2[1 + (33 - 2N_{\mathrm{F}})\,\alpha_{\mathrm{s}}/12\pi \ln(-q^2/\mu^2)]^{-1} \end{aligned} \tag{20.20}$$

where $\alpha_{\mathrm{s}} = g^2/4\pi$.

After renormalization the factor in brackets acts as a momentum-dependent modification of the strong coupling constant. Combined with α_{s}, it gives the "running" coupling constant

$$\alpha_{\mathrm{s}}(q^2) = 4\pi[(11 - 2N_{\mathrm{F}}/3)\ln(-q^2/\Lambda^2)]^{-1}, \tag{20.21}$$

where Λ is a dimensional parameter introduced in the renormalization process. Note that α_{s} has vanished altogether from (20.21), an effect known as "dimensional transmutation", commonly found in massless quantum field theories [Co 73].

To determine Λ from experimental data is somewhat tricky [St 81]. Besides other subtleties this has to do with the different behaviour of (20.21) for space-like and time-like momenta. For space-like momenta $q^2 < 0$, the logarithm in (20.21) is real, and the measured coupling constant can be identified with $\alpha_{\mathrm{s}}(q^2)$. For time-like momenta $q^2 > 0$, on the other hand, the logarithm becomes complex, $\ln(q^2/\Lambda^2) - \mathrm{i}\pi$, and the measured value must be identified with $\mathrm{Re}\{\alpha_{\mathrm{s}}\}$. The occurrence of an imaginary part in α_{s} corresponds to the possibility of a time-like gluon splitting into two gluons or into two light quarks. Analyses of experiments involving space-like momenta (in particular charmonium level fits) indicate a value for Λ in the range of 300 – 500 MeV/c [Vi 80, Mo 83]. The dependence of $\mathrm{Re}\{\alpha_{\mathrm{s}}\}$ for positive and negative q^2 is shown in Fig. 20.5.

At $q^2 = -\Lambda^2$ the running coupling constant has a pole. That this pole occurs at a finite value of q^2 is an artefact of the approximation (20.19) involving only the simplest loop diagram. More sophisticated approximate solutions of the Schwinger-Dyson equations for the gluon propagator [Ba 81a] indicate that the pole should really be at $q^2 = 0$, and that $\alpha_{\mathrm{s}}(q^2)$ should behave as $1/q^2$ in the limit $q^2 \to 0$. Converted into coordinate space this would mean that $\alpha_{\mathrm{s}}(r)$ grows like r^2 for large distances, corresponding to a linearly rising potential (Fig. 20.6). This implies permanent confinement of quarks and gluons, because an infinite amount of energy would be needed to separate two colour charges.

On the other hand, the logarithm in (20.21) gradually decreases the coupling strength between colour charges at large momenta or small distances. This is the characteristic feature of *asymptotic freedom* which makes QCD the prime can-

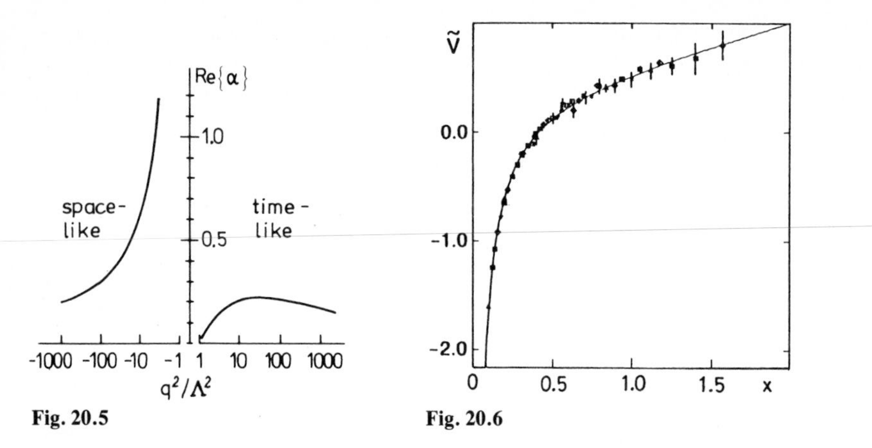

Fig. 20.5

Fig. 20.6

Fig. 20.5. Running coupling constant $\alpha_s(q^2)$ as a function of q^2/Λ^2. For time-like momenta only the real part is shown

Fig. 20.6. The quark – antiquark potential as obtained from Monte Carlo calculations on a lattice. Dynamical quark pairs are not included [St 84a]

didate for a theory of strong interaction, because it describes the approximate scaling of cross-sections at very high energy that led to the discovery of quarks as constituents of the nucleon in the first place.

20.2 Gluon Condensates in Strong Colour Fields

Originally, the field of Bose condensation presented in Chap. 19 was developed as an esoteric theoretical exercise to further understanding of supercritical fields. Meanwhile it has been realized that these ideas are highly relevant in understanding strongly interacting particles [Ma 76, 77a] and in particular, the structure of the QCD vacuum [Mi 78, Mü 82]. This arises through the strong self-interaction of the gluon field, believed to be the essential element responsible for the phenomenon of quark confinement.

We shall study two aspects of this very complex and difficult problem. In this section we consider what happens when a localized, almost point-like colour charge becomes so strong that the gluon vacuum is rendered unstable. In this case a gluon condensate forms which screens the strong colour source. (There is also the possibility that the quark field becomes supercritical. This would be analogous to supercritical fields in atoms [Ma 77c]. Supercritical quark fields are suspected to play a role in breaking chiral symmetry and in pion physics, in general [Go 81a, Fi 80].)

In the next section we shall show that a constant colour-magnetic field is also unstable against gluon condensation, resulting in domain-structured colour-magnetic configuration that has lower energy than the normal vacuum state.

Let us assume for the moment that we have a Coulomb-type solution for the potential $A_0 \sim Qg/4\pi r$. Since the f_{abc} are of order one, we can estimate crudely that the non-linear term in (20.13) for the gluon field strength becomes comparable to the gradient term when

$$\frac{Qg}{4\pi r^2} = \frac{\partial A_0}{\partial r} \sim g(A_0)^2 = g\frac{(Qg)^2}{16\pi^2 r^2} \quad \text{or} \tag{20.22}$$

$$Q\alpha_s = Qg^2/4\pi \sim 1 , \tag{20.23}$$

where we have introduced the strong (interaction) coupling constant α_s.

How the *self-coupling* of the gluon field affects the equation of motion, given a quark source, is treated in the remainder of this section. Our treatment is considerably simplified, if we pass from the QCD colour gauge group SU(3) to the group SU(2) which has only three generators and therefore only three gauge potentials A_μ^a $(a = 1,2,3)$. Note that SU(2) is a subgroup of SU(3). Therefore, even this simplification is directly relevant to the understanding of QCD.

We study the response of the gauge quantum field to the presence of a coloured charge of strength Q located near $r = 0$. We assume (without loss of generality) that it points in the colour "direction" $a = 3$:

$$j_a^\mu = Q\delta_{\mu 0}\delta_{a3}\,\delta(\boldsymbol{r}) = \varrho(\boldsymbol{r})\,\delta_{\mu 0}\delta_{a3} . \tag{20.24}$$

Since the structure constants $f_{abc} = \varepsilon_{abc}$ for SU(2) do not allow coupling between any same coloured direction in colour space, but only between different colours, the following approximation scheme is apt: the charge (20.24) is a source of a static Coulomb potential in its rest frame and we assume that the vector potential in the $a = 3$ direction vanishes,

$$A_\mu^3 = (\phi(\boldsymbol{r}), 0) . \tag{20.25}$$

In treating the colour components $a = 1,2$ it is useful to work in the "temporal" gauge $A_0^a = 0$ $(a = 1,2)$ which, being a gauge condition, is not an approximation. We then deal with three coupled fields, $\phi, \boldsymbol{A}^1, \boldsymbol{A}^2$:

$$A_\mu^a = \begin{pmatrix} 0 & \boldsymbol{A}_1 \\ 0 & \boldsymbol{A}_2 \\ \phi & 0 \end{pmatrix} . \tag{20.26}$$

The components of the field strength tensor $F_{\mu\nu}^a$, (20.13), are

$$\begin{aligned} \boldsymbol{E}^1 &= -\dot{\boldsymbol{A}}^1 + g\phi\boldsymbol{A}^2 \qquad & \boldsymbol{B}^1 &= \operatorname{curl}\boldsymbol{A}^1 \\ \boldsymbol{E}^2 &= -\dot{\boldsymbol{A}}^2 - g\phi\boldsymbol{A}^1 \qquad & \boldsymbol{B}^2 &= \operatorname{curl}\boldsymbol{A}^2 \\ \boldsymbol{E}^3 &= -\nabla\phi & \boldsymbol{B}^3 &= -g\boldsymbol{A}^1 \times \boldsymbol{A}^2 . \end{aligned} \tag{20.27}$$

We further note at this point that since an external charge has been prescribed, the local colour (isospin) gauge invariance of the theory was lost, a feature allowing us, in the first place, to single out ϕ among the fields A_μ^a. Thus, in the following, we do not discuss how non-Abelian neutrality is achieved, but concentrate on the part of the theory selected here. However, we select a charge density confined to a shell of fixed radius in order to retain essential features of local gauge invariance for non-Abelian fields [Hu 78].

It is advantageous to combine the neutral vector fields $\boldsymbol{A}^1$ and $\boldsymbol{A}^2$ into one complex field

$$W = \tfrac{1}{2}(A^1 + \mathrm{i}A^2)\,. \tag{20.28}$$

In terms of this field, the Lagrangian can be written after some elementary manipulations as

$$\begin{aligned} L = {}& \frac{1}{2}(\nabla\phi)^2 + \left(\frac{\partial}{\partial t} + \mathrm{i}g\phi\right) W^* \cdot \left(\frac{\partial}{\partial t} - \mathrm{i}g\phi\right) W \\ & - (\operatorname{curl} W^*) \cdot (\operatorname{curl} W) + \frac{g^2}{2}(W^* \times W)^2 + g\varrho\Phi\,. \end{aligned} \tag{20.29}$$

The non-linear classical field equations are obtained by variation with respect to ϕ and W^*, yielding

$$\nabla^2\phi = g\left[\varrho - W^* \cdot \left(\mathrm{i}\frac{\partial}{\partial t} - g\phi\right) W + \left(\mathrm{i}\frac{\partial}{\partial t} + g\phi\right) W^* \cdot W\right] \tag{20.30a}$$

$$\left[\left(\mathrm{i}\frac{\partial}{\partial t} - g\phi\right)^2 + \nabla^2 - \operatorname{grad}\operatorname{div} - g^2(W^* \times W)\right] W = 0\,. \tag{20.30b}$$

These equations have an extremely interesting structure. The first one, for the Coulomb potential ϕ, is essentially the Poisson equation, but with a crucial modification: the source term contains not only the external colour charge density but also the colour density generated by the complex vector field W, cf. (17.30),

$$\varrho_W = W^* \cdot \left[\left(\mathrm{i}\frac{\partial}{\partial t} - g\phi\right) W\right] - \left[\left(\mathrm{i}\frac{\partial}{\partial t} + g\phi\right) W^*\right] \cdot W\,. \tag{20.31}$$

The W field generally *screens* the external supercritical charge ϱ. That indeed it usually screens and does not antiscreen can be easily seen by neglecting the time dependence of W. Then (20.30a) can be written as

$$[-\nabla^2 + 2g^2(W^* \cdot W)]\,\phi = -g\varrho\,, \tag{20.32}$$

where the complex field acts as an effective mass term for the Coulomb-like field ϕ and reduces the range of ϕ. Equation (20.30b) for the W field does not contain a source term, but instead the third-order term $g^2(W^* \times W) \times W$. *Except for this term, it is identical to the usual wave equation for a charged vector boson in an external Coulomb field* ϕ. Hence the problem in gauge theory is the same as that treated in Chap. 17–19, ϕ standing now for the total potential. The similarity is even closer, because the W^3 term in (20.30b) is shown below to be negligible.

As described, the Coulomb problem on its own develops an instability when the external charge is too large. Our intention now, given (20.30), is to construct a variational ground state that will eliminate the Coulomb instability. As shown, the instability of the classical modes means that these modes of the *Yang-Mills* field W are probably present in the structured true vacuum state. Indeed, the true ground state is characterized by the presence of a *finite* number of W particles, called "condensate" state in this chapter.

Naturally, we are not able to give the "exact" new stable ground state, since this would imply the complete solution of a non-trivial interacting quantum field theory. Instead, we construct an approximate variational ground state similarly to our foregoing work with condensed Bose fields.

The path to the solution of the condensate problem has already been described at length. Here only the essential points are repeated. Quantization of the W field requires a complete set of solutions of the linearized equation (20.30b)

$$\left[\left(\mathrm{i}\frac{\partial}{\partial t} - g\phi\right)^2 + \nabla^2 - \operatorname{grad}\operatorname{div}\right] W_n^{(\pm)} = 0 \,. \tag{20.33}$$

Here the third-order term has been omitted, but it will be included later as a perturbation. The quantum field *operator* $\hat{W}$ is then decomposed into contributions from modes with positive and negative norms:

$$\hat{W} = \sum_n \hat{\alpha}_n W_n^{(+)}(x) + \sum_n \hat{\beta}_n^+ W_n^{(-)}(x) \,, \tag{20.34}$$

where

$$\int d^3x\, W_n^{(\varepsilon)*} \cdot \left(\mathrm{i}\frac{\overleftrightarrow{\partial}}{\partial t} - 2g\phi\right) W_m^{(\varepsilon')} = \varepsilon\,\delta_{nm}\,\delta_{\varepsilon\varepsilon'} \,. \tag{20.35}$$

The true vacuum state $|\mathrm{Vac}\rangle$ is defined to be annihilated by all particle destruction operators $\hat{\alpha}_n$, $\hat{\beta}_n$. When the potential ϕ is time independent and a function of the radial variable $r = |\boldsymbol{x}|$ only, then stationary modes can be introduced

$$W_n(\boldsymbol{x}, t) = W_n(\boldsymbol{x}) \exp(-\mathrm{i}\omega t) \,. \tag{20.36}$$

First consider the perturbative state $|0\rangle$ in which the charge density (20.32) generated by the charged field W is zero. Then the potential outside the source is $g(r) = -Q\alpha/r$. The strongest bound states of (20.33) belong to the positive parity modes for which $\operatorname{div} W_n$ vanishes, so avoiding an extra node in the wave function. Hence,

$$W_n(\boldsymbol{x}) = h_l(r)\, T_{lll_z}(\vartheta, \varphi) \tag{20.37a}$$

with the vector spherical harmonics [Ei 70]

$$T_{jlj_z} = \sum_{m,\mu} (l\,1\,j\,|\,m\mu j_z)\, Y_{lm}\, \xi_\mu \boldsymbol{e}_\mu\,, \tag{20.37b}$$

where the vector $\xi_\mu \boldsymbol{e}_\mu$ characterizes the spin-1 property of the gluon. This recovers, inserting (20.37b) into (20.33), the old Coulomb problem again, with the slight complication caused by the spin: the total angular momentum is now at least 1, but $l \geqslant 1$:

$$\left[\frac{1}{r}\frac{\partial^2}{\partial r^2} r + (\omega - g\,\Phi)^2 - \frac{l(l+1)}{r^2}\right] h_l(r) = 0\,, \qquad l \geqslant 1\,. \tag{20.38}$$

We already know that the naive perturbative state $|0\rangle$ becomes degenerate at critical strength ($Q\alpha = l + \frac{1}{2}$ for $1/r$ potentials) for charged states. The new condensate ground state can be developed using quasi-particle modes. With ($l_z = -1, 0, 1$)

$$\hat{\alpha}^\dagger_{l_z} = 2\int d^3x\, \hat{W}^*(x) \cdot (\omega - g\,\Phi)\, h(r)\, T_{1,1,l_z}(\vartheta, \varphi) \qquad \text{then} \tag{20.39}$$

$$|q\rangle \equiv (q!)^{-3/2} (\hat{\alpha}^\dagger_{-1}\hat{\alpha}^\dagger_0\hat{\alpha}^\dagger_1)^q |\mathrm{Vac}\rangle\,, \tag{20.40}$$

where $\hat{\alpha}|\mathrm{Vac}\rangle = 0$ and the integer q denotes the number of quanta in each of the magnetic substates. In (20.40) we have allowed only for equal occupation of the different magnetic submodes, which assures a spherically symmetric state. But we cannot, a priori and definitely, exclude dynamical breaking of rotational invariance, which would require uneven distribution of condensate strength among the modes in (20.40).

The condensate trial state (20.40) must be compared with (19.38), and all further developments that now follow duplicate the developments already presented following that equation. Thus we here follow the second-order differential form of the formalism for a change, rather than the Feshbach-Villars method used earlier. The total energy contained in the gauge field condensate is now

$$W(q) = \langle q|\hat{H}|q\rangle = \frac{\mathrm{tr}}{2}\int d^3x\, \langle q|\hat{\boldsymbol{E}}^2 + \hat{\boldsymbol{B}}^2|q\rangle\,, \tag{20.41}$$

with $\boldsymbol{E}$ and $\boldsymbol{B}$ as given implicitly by (20.13), while the trace sums the implicit colour components of the fields $\boldsymbol{E}$ and $\boldsymbol{B}$. Omitting the zero-point energy and off-diagonal self-interaction terms gives for $W(q)$

$$W(q) = \int d^3x \left[\frac{1}{2}(\nabla\,\Phi)^2 + q\sum_m |(\omega - g\,\Phi)\,\boldsymbol{W}_m|^2 + |\nabla \times \boldsymbol{W}_m|^2 \right.$$
$$\left. + g\varrho\,\Phi + \frac{1}{2}g^2 q^2 \sum_{m\bar{m}} |\boldsymbol{W}_{\bar{m}} \times \boldsymbol{W}_m|^2\right]. \tag{20.42}$$

Since the gluons are massless, $\omega = 0$ is the only possible choice, while the source ϱ, rather than the mass, provides a scale to the theory. In (20.42) then, in view of (20.37),

$$\sum_m |W_m|^2 = \frac{1}{4\pi} 3h(r)^2 , \tag{20.43a}$$

$$\sum_{m\bar{m}} |W_m^* \times W_{\bar{m}}|^2 = \frac{1}{32\pi^2} h^4(r) , \tag{20.43b}$$

which we insert into (20.42). Varying now (20.42) with respect to h and ϕ gives the condensate equations

$$\left[\frac{1}{r}\frac{d^2}{dr^2}r - \frac{2}{r^2} + (g\Phi)^2 + \frac{g^2q^2}{16\pi^2}h^2(r)\right]h(r) = 0 \tag{20.44a}$$

$$\nabla^2\Phi = g\left(\varrho + \frac{3q}{\pi}g\Phi h^2\right), \tag{20.44b}$$

with the normalization condition of the Klein-Gordon equation in potential $g\Phi$ for vanishing frequency ω:

$$1 = -2\int r^2 dr (g\Phi) h^2 . \tag{20.44c}$$

It is important to observe that the assumption in (20.25), $A^3 = 0$, is indeed satisfied in the sense that $\langle q|A^3|q\rangle = 0$.

The surprise one encounters when solving (20.44) is that the h^3 term in (20.44a) has negligible influence compared to the essential non-linear term, $\sim \Phi h^2$ in (20.44b), and which has the same structure as the non-linear term contributing to the total potential, (19.62, 63), in the previously described case of spin-0 mesons in an external ("Coulombic") potential. Thus the current problem is identical in structure to the case of a massless Bose condensate problem in an external field, with total angular momentum $l = 1$.

Let us illustrate the behaviour of the associated solutions by taking the coupling constant $\alpha_s = g^2/4\pi = 1/2$ and the strength of Q of the source to be 8, corresponding to eight "quarks" with parallel isospin colour, distributed on a shell with a radius R. The choice of a shell (rather than, e.g., a homogeneous sphere) is necessary to ensure gauge invariance of the source for non-Abelian theories [Hu 78]. Since the effective strength of the source is $8 \cdot 1/2 > 3/2$, an overcritical potential arises. For $q = 2$, i.e. altogether six condensed gluons, the source strength reduces to unity, i.e. below the supercritical value 3/2. Indeed this is the solution to the self-screening problem found numerically [Mü 82], and surprisingly it means that the condensate *overscreens* the source to below criticality (see Fig. 20.7). Overscreening of the source is quite peculiar to the massless character of the gluon field, which allows the quantum wave function to creep into the classically forbidden region. Further, the term in (20.43b) contributes only 3% to the condensate energy (i.e. 0.3% to the total energy), justifying its perturbative treatment.

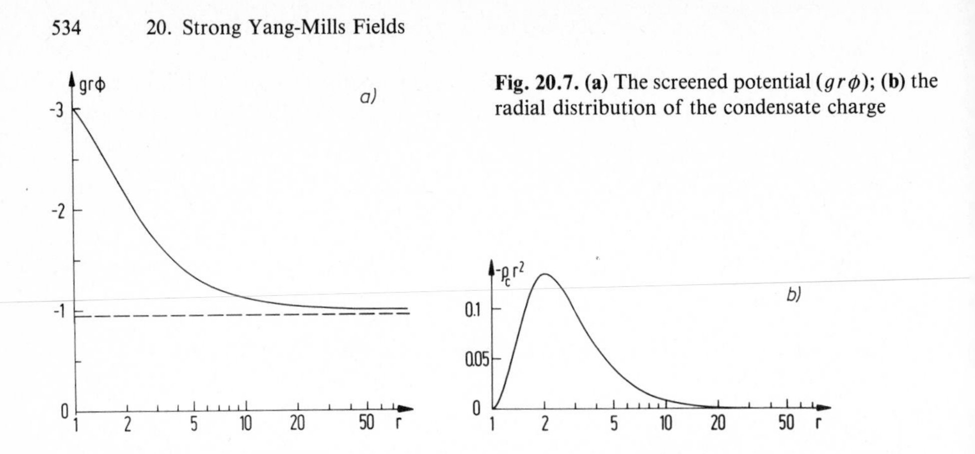

Fig. 20.7. **(a)** The screened potential $(gr\phi)$; **(b)** the radial distribution of the condensate charge

It is found that the appearance of the gluon condensate reduces the energy of the (unstable) Coulomb solution by about 10%. At this point one should nonetheless emphasize a slight distinction in the case of the non-Abelian gauge fields compared to Abelian charged particles. In the latter case, the supercritical Coulomb potential always exists, at least theoretically, as we may *disregard* the charge interaction of the different particles (pions) among themselves. In the non-Abelian case it is the *same* field that generates the Coulomb field and the condensate, cf. (20.30), and gluon–gluon interaction cannot possibly be ignored. Thus, while the Coulomb solution $\phi \sim 1/r$, $A_{1,2} = 0$ always exists, for strong sources it has higher energy than the condensate solution with $A_{1,2} \neq 0$ (i.e. $q \neq 0$), so the Coulomb case is part of the "false" vacuum of QCD.

20.3 The "Magnetic" Vacuum of QCD

The properties of baryons and mesons have been described very successfully by phenomenological bag models [Ch 74b, deG 75, Th 80, 84]. These models assume that the quarks (and gluons) are confined to a region of space of hadronic size, in which they can move almost freely, but from which they cannot escape [Bo 67]. The mechanism for this confinement is thought to reside in the nature of the *true vacuum state* of QCD which does not support the presence of colour charges. Inside the hadronic bag, the true vacuum is assumed to be destroyed and replaced by a perturbative state (similar in nature to the normal vacuum of QED) which allows quarks to be present (Fig. 20.8). This state, however, has a higher energy density than the true vacuum state, and is characterized by a negative pressure $P = -B$. The value of the bag constant $B^{1/4}$ may be somewhere between 150 MeV [deG 75] and 230 MeV [Fl 84]. The bag thus tends to contract as much as possible, and is kept stable by the balance between the pressure of the confined quarks and the pressure of the vacuum.

To have all these postulated properties the true ground state of QCD must have a complicated structure:

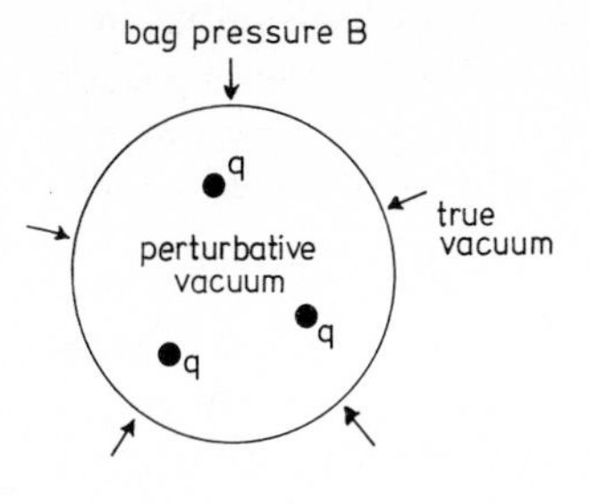

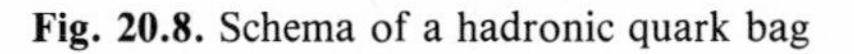
Fig. 20.8. Schema of a hadronic quark bag

a) its energy density must be lower than that of the perturbative vacuum;
b) it must be a perfect colour dielectric, i.e. $\varepsilon = 0$, $\mu = \infty$, to guarantee confinement of colour charges,
c) it must be Lorentz and gauge invariant.

Since we expect a phase transition into the perturbative vacuum, there must be an order parameter describing this phase transition. The order parameter must vanish in the perturbative phase and must be non-zero for the true ground state. What may this order parameter be? Because of the symmetry requirements it must be a gauge-invariant Lorentz scalar. For the gauge field alone, it can be composed only of scalar combinations of the field strength tensor, such as

$$\langle F^a_{\mu\nu}(x) F^{\mu\nu}_a(x) \rangle \, , \qquad \langle d_{abc} F^a_{\mu\nu}(x) F^{\nu\lambda}_b(x) F^{a\mu}_\lambda(x) \rangle \, , \tag{20.45}$$

or, if we also permit non-local expressions involving a structure function $S(x-y)$

$$\langle \int d^4y \, F^a_{\mu\nu}(x) S(x-y) F^{\mu\nu}_a(y) \rangle \, , \qquad \text{etc.} \tag{20.46}$$

For full QCD bilinear expressions involving quark fields may also occur as an order parameter, e.g. $\langle \bar{\psi}\psi \rangle$.

Although it is, in principle, possible to investigate the energy of a trial bound state as a function of these order parameters, this would be a very hard problem to solve. However, often a well-chosen trial state may yield a good estimate of the ground state energy and static properties of a physical system, even if it does not conserve all the required symmetries. Ample evidence for this is provided by the success of the Nilsson deformed shell model in explaining nuclear deformation parameters, magnetic moments, etc. Although the restoration of the artificially broken symmetry gives some, often small, corrections to the ground state energy, the real importance shows up only in transition matrix elements between different states.

In this spirit, one is led to consider the expectation value $\langle F^a_{\mu\nu}(x) \rangle$ itself as an order parameter. To be sure, the state described by a non-zero expectation value of $F^a_{\mu\nu}$ cannot be the true ground state, because it violates the invariance properties of the vacuum state, but its energy expectation value may nonetheless be a good approximation to the ground state energy [Pa 78].

For simplicity, let us again start by discussing SU(2) gauge theory instead of full SU(3). A constant colour-electric field is not a reasonable choice for the approximate ground state, because it accelerates coloured particles and leads to immediate pair creation. Therefore, let us study the energy of the gluon vacuum in the presence of a constant chromomagnetic field of strength H, which we may take, without loss of generality, to point in the colour-3 direction and into z direction in coordinate space:

$$\boldsymbol{H}_a = H\boldsymbol{e}_z\delta_{a3}\,, \qquad \boldsymbol{A}_a = Hx\boldsymbol{e}_y\delta_{a3}\,. \tag{20.47}$$

We shall now investigate the properties of small fluctuations around this field configuration. The non-linear terms in the Yang-Mills equations couple the three different colour components among each other through the SU(2) structure constants ε_{abc}, where (abc) must be a permutation of (123). Consequently, the fluctuations in the colour-3 direction are not aware of the average chromomagnetic field (in lowest order), while the fluctuations in the $a=1,2$ directions are influenced by the background field.

It is convenient to express the two modes by a single complex field $W_\mu = (A_\mu^1 + \mathrm{i}A_\mu^2)/\sqrt{2}$. This corresponds to a transition from Cartesian to spherical coordinates in colour space. Keeping only terms of second order in the field W, the part of the total SU(2) Lagrangian governing small oscillations in the colour-1,2 directions is [Ni 78]

$$L_2(W) = -\tfrac{1}{2}|(\partial_\mu - \mathrm{i}gA_\mu^3)\,W_\nu - (\partial_\nu - \mathrm{i}gA_\nu^3)\,W_\mu|^2 - \mathrm{i}g(\partial_\mu A_\nu^3 - \partial_\nu A_\mu^3)\,W_\mu^*\,W_\nu\,, \tag{20.48}$$

from which the equations for the W modes may be obtained by variation with respect to W_μ^*. The gauge freedom can be utilized to simplify these equations by imposing the co-called background field gauge

$$(\partial_\nu - \mathrm{i}gA_\nu^3)\,W^\nu = 0\,. \tag{20.49}$$

Furthermore, the expression involving A_μ^3 in the last term in (20.5) can be identified with the field strength tensor $F_{\mu\nu}^3$ of the background field, because the non-linear term vanishes for the special field configuration (20.47). Thus the linearized equation of motion for small amplitude oscillations of the W field is

$$[(\partial_\lambda - \mathrm{i}gA_\lambda^3)^2\delta_{\mu\nu} + 2\mathrm{i}gF_{\mu\nu}^3]\,W^\nu(x) = 0\,. \tag{20.50}$$

Exactly the same equation would be obtained in QED of a charged vector (spin 1) field in the presence of an electromagnetic background field. (The mass of the particle is zero, which makes the comparison academic since there is no known electrically charged massless particle in nature.) The first contribution describes motion due to the interaction of the charge of the particle with the field, while the second term contains the effect of the field on the magnetic moment of the spin-1 particle. The factor 2 is nothing else than the g factor for a point-like particle. This can be seen as follows: restricting to space-like values of the vector indices $\mu, \nu = 1,2,3$, the matrix $\mathrm{i}F_{\mu\nu}^3$ can be identified with the spin matrix for a spin-1 particle (the three-dimensional representation of the rotation group):

$$\mathrm{i}F^3_{\mu\nu} = H \begin{pmatrix} 0 & \mathrm{i} & 0 \\ -\mathrm{i} & 0 & 0 \\ 0 & 0 & 0 \end{pmatrix} = -HS_z\,. \tag{20.51}$$

For the space-like components of W_ν the second term in (20.50) can therefore be written in the suggestive form

$$(2gH \cdot S)W\,, \tag{20.52}$$

characteristic of a magnetic dipole interaction.

For stationary states

$$W(x) = \xi_s \exp(\mathrm{i}k_2 y + \mathrm{i}k_3 z - \mathrm{i}\varepsilon t)\, W_{s,k_2k_3}(x) \tag{20.53}$$

(20.50) reduces to the equation for the Landau orbits of a charged particle in a magnetic field

$$[-\varepsilon^2 - \partial^2/\partial x^2 - (k_2 - gHx)^2 + k_3^2 - 2gHs]\, W_{s,k_2k_3}(x) = 0\,. \tag{20.54}$$

Here ξ_s ($s = -1, 0, 1$) are the eigenvectors of the spin matrix S_z. Because gluons are massless, the eigenvector with $s = 0$ is ruled out, i.e. only the two transverse polarizations $s = \pm 1$ exist. The energy eigenvalues determined by (20.54) are well known:

$$\varepsilon_{ns}(k_3) = [gH(2n+1) + k_3^2 - 2gHs]^{1/2}\,, \tag{20.55}$$

where $n = 0, 1, 2, \ldots$, $s = \pm 1$ and $-\infty < k_3 < \infty$. Due to the absence of a mass term in (20.55), the radicand becomes negative for the mode $n = 0$, $s = +1$ when $k_3^2 < gH$. The interaction between the spin of the gluon and the chromomagnetic field is so strong that it overwhelms the zero-point energy of Landau motion. Mathematically speaking, the occurrence of imaginary energies means that the operator in (20.50) is not self-adjoint. In physical terms this means that a *constant chromomagnetic field is always supercritical.*

The reason for the unwanted behaviour is the homogeneity of the chromomagnetic field. If the field would, e.g., oscillate in the transverse direction, then the interaction with the spin of gluons in the lowest mode would at least partially cancel, because $2g\langle H\rangle s$ would involve the average of H over the entire gluon mode. It is clear that the effective spin interaction could be made arbitrarily small for sufficiently fast varying orientation of the background field. Now imagine the background field to become more and more homogeneous, starting out from a rapidly oscillating field configuration. The spin interaction would gradually increase, causing the energy eigenvalue of the lowest Landau mode to decrease until it finally reaches zero. At that moment, gluons would be spontaneously produced from the vacuum, forming a condensate in the supercritical lowest mode which, being spin-oriented, would provide an extra contribution to the average chromomagnetic field, Fig. 20.9. Since the lowest mode is localized in a region of size $(gH)^{1/2}$, the additional field from the condensate has structure of

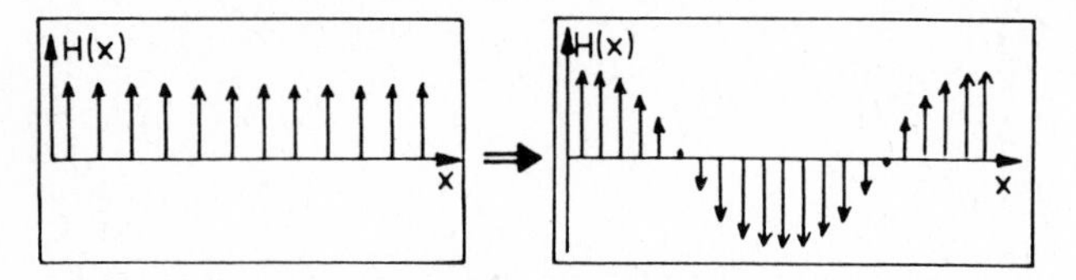

Fig. 20.9. A homogeneous chromomagnetic field is unstable against local fluctuations in the orientation of the field. A gluon condensate develops, causing a domain structure of the field configuration

that scale. We conclude that also the average chromomagnetic field must exhibit structure on the scale of the size of the lowest Landau orbit. Any attempt to make it more homogeneous would cause the structure to reappear via spontaneous gluon creation. Observe incidentally, that the mathematics is improved at the same time, because the linear approximation does not work for a condensate, so that the full non-linear eigenvalue problem must be studied.

This non-linear problem has not yet been solved, although some attempts have been made [Ni 78, 79a, b]. To keep our model as simple as possible – remember that this state is not expected to have good symmetries, anyway – we simply discard the troublesome modes ($n=0$, $s=1$, $k_3^2<gH$). We shall denote this "kitchen recipe" by a prime on the summation sign. The vacuum energy is then just the sum of the zero-point energies of all modes, where we subtract the energy of the perturbative vacuum (i.e. for $H=0$) as the reference point:

$$\begin{aligned} E_{\mathrm{v}}(H)-E_{\mathrm{v}}(0) &= \tfrac{1}{2}\sum_i{}'\,\varepsilon_i(H)-\tfrac{1}{2}\sum_i \varepsilon_i(0) \\ &= \tfrac{1}{2}V\sum_{\mathrm{C}}\sum_s\left[gH/\pi\sum_n{}'\int dk_3/2\pi\,\varepsilon_{ns}(k_3,H)-\int d^3k/(2\pi)^3\,|k|\right]. \end{aligned} \tag{20.56}$$

The additional sum $\sum_{\mathrm{C}}$ runs over the degrees of freedom of the gluon field in colour space. For SU(3) this contributes a factor 3. To get rid of the square roots, the trick is to introduce the Fock-Schwinger proper-time representation, already applied in Sect. 10.5, via the integral formula

$$\varepsilon^{\alpha}=\Gamma(-\alpha)^{-1}\int_0^{\infty} ds/s\,s^{-\alpha}\mathrm{e}^{-s\varepsilon^2} \tag{20.57}$$

for $\alpha=1/2$. The sums over n and s yield geometric series, while the momentum integrations reduce to Gaussian integrals (some care must be taken for the lowest mode). Substituting $x=gHs$, the final integration becomes

$$\begin{aligned} E_{\mathrm{v}}(H)-E_{\mathrm{v}}(0) = &-3\,V(gH)^2/8\pi^2\int_0^{\infty} dx/x^2[1/\sin(x) \\ &-1/x-\mathrm{e}^{-x}+\mathrm{e}^{x}\Gamma(\tfrac{1}{2},x)/\Gamma(\tfrac{1}{2})]\,. \end{aligned} \tag{20.58}$$

It turns out that the integral is divergent due to the $1/x$ behaviour of the integrand in the limit $x\to 0$. The coefficient of this term is found to be $-1/6+2=11/6$. The divergent part is easily separated by dimensional regular-ization, replacing $\alpha = 1/2$ by $\alpha = 1/2 - \varepsilon$ in (20.57). The integral then splits into a finite part (which is uninteresting in this connection) and an infinite part proportional to $(gH)^2$

$$\begin{aligned} \int dx/x^{1-\varepsilon}(gH)^{2-\varepsilon} &= (gH)^{2-\varepsilon}/\varepsilon = (gH)^2[1-\varepsilon\ln(gH)]/\varepsilon \\ &= (gH)^2/\varepsilon - (gH)^2\ln(gH) + 0(\varepsilon) . \end{aligned} \tag{20.59}$$

Since QCD is as renormalizable theory, all terms like $(gH)^2$ renormalize the coupling constant in the free Lagrangian $H^2/2$, and only the finite logarithmic contribution remains (see Chap. 14 for detailed discussion of renormalization in the context of QED)

$$E_{\rm v}'(H) = [E_{\rm v}(H) - E_{\rm v}(0)]_{\rm ren} = 3\,V/8\,\pi^2\,11/12(gH)^2\ln(H^2/H_0^2) , \tag{20.60}$$

where H_0 is a constant depending on the renormalization point. Since H_0 is introduced by the renormalization procedure, its value cannot be calculated but must be determined by comparison with measured quantities.

Despite its simplicity, which corresponds with the simplicity of our choice of the trial vacuum state, this result (first obtained by *Savvidy* et al. [Ba 77, Sa 77]) has interesting properties.[1] Because the logarithm is negative for small values of the chromomagnetic field H, the energy of our trial state has a non-trivial minimum given by

$$\partial E_{\rm v}'(H)/\partial H = 0 \rightarrow H_{\rm min} = H_0 \exp(-\tfrac{1}{2}) , \quad \text{where}$$

$$\Delta\varepsilon = E_{\rm v}'(H_{\rm min})/V = -11/32\,\pi^2(gH_{\rm min})^2 = -11/32\,\pi^2 e(gH_0)^2 . \tag{20.61}$$

The full curve $\varepsilon_{\rm vac}'(H)$ is shown in Fig. 20.10 where the existence of the non-trivial minimum is obvious. We conclude that the perturbative QCD vacuum is unstable against spontaneous formation of a chromomagnetic field H. The "true" vacuum state is characterized in this model by a non-vanishing expectation value of H, and is lower in energy by an amount $\Delta\varepsilon$ given in (20.61). Courageously identifying this energy difference with the bag constant B of the MIT bag model gives

$$\varepsilon_{\rm v}'(H) = 11/32\,\pi^2(gH)^2[\ln(gH)^2/B - \ln(32\,\pi^2/11) - 1] . \tag{20.62}$$

However, bear in mind that in deriving this result we have simply ignored the unstable gluon modes. The true ground state must have even lower energy. Details of this simple model for the QCD ground state, e.g. the question of confinement, have been widely discussed, and several attempts to improve on the

[1] The sign of $E_{\rm v}'(H)$ is opposite to that of the Euler-Heisenberg effective action in QED, Sect. 10.5, because of the asymptotic freedom of QCD and, indeed, the factor 11/12 in (20.60) is intimately related to the factor 11 in the running coupling constant (20.21) [Ba 77]. The "wrong" sign was found by *Vanyashin* and *Terent'ev* [Va 65] as early as 1965.

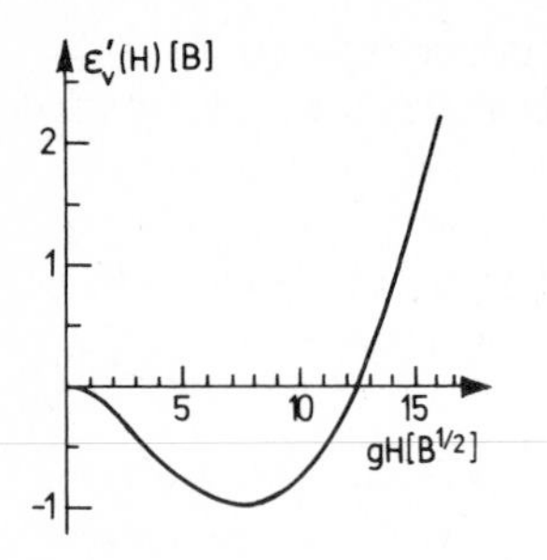

Fig. 20.10. Energy density of the chromomagnetic trial ground state as a function of the magnetic field strength. The minimum at a non-zero value of H corresponds to the non-trivial ground state of the gauge theory. The difference in the energy density to the state with $H = 0$ is identified with the bag constant B

trial state exist. The interested reader is referred to [Ni 78a, 79a, b, Am 80, Fl 80, Fu 80, Ad 81].

In conclusion, note that the "magnetic vacuum" of QCD predicts a phase transition to a deconfined phase at high temperature, in which quarks and gluons can exist as free particles [Mü 81, Ka 81b]. This so-called *quark-gluon plasma* would be a new phase of nuclear hadronic matter with entirely novel properties [Ja 82, Mü 85].

20.4 Spontaneous Quark Pair Production and Fission of Quark Bags

In the introduction of Sect. 20.3 we mentioned that the bag models very successfully describe the low-energy properties of hadronic particles, i.e. baryons and mesons. The simplest of these is the MIT bag model [Ch 74b], in which the quarks are confined by assuming that their (effective) mass, which is small inside the bag volume, becomes very large outside the bag. Neglecting short-range interactions among the quarks, the quark wave functions then must obey the Dirac equation

$$\mathrm{i}\gamma^\mu \partial_\mu \psi - M\psi + (M-m)\,\Theta_V \psi = 0\,, \tag{20.63}$$

where $\Theta_V = 1$ inside the bag and $\Theta_V = 0$ outside. In the limit $M \to \infty$ it is possible to show that the wave function vanishes outside the bag volume and satisfies a linear boundary condition at the bag surface:

$$\mathrm{i}\gamma^\mu n_\mu \psi|_\mathrm{S} = -\mathrm{i}(\gamma \cdot \boldsymbol{n})\,\psi|_\mathrm{S} = \psi|_\mathrm{S}\,. \tag{20.64}$$

Because $\gamma^{\mu\dagger} = \gamma^0\gamma^\mu\gamma^0$, (4.21), the adjoint equation reads $-\mathrm{i}\bar{\psi}\gamma^\mu n_\mu|_\mathrm{S} = \bar{\psi}|_\mathrm{S}$. It is now easily seen that the linear boundary condition ensures that the normal flow of quark current through the bag surface vanishes: $n_\mu j^\mu = n_\mu(\bar{\psi}\gamma^\mu\psi) = 0$. The boundary condition (20.64) thus guarantees confinement of quarks.

As explained in Sect. 20.3, bag models balance the internal pressure of the confined quarks by the external pressure from the complicated "true vacuum" of QCD which is destroyed by the presence of quarks in the bag's interior.

Because of Lorentz invariance the external vacuum pressure (which really is a negative internal pressure) must be characterized by a scalar constant B, the bag constant, with dimensions of an energy density. The original fit calculated by the MIT group to the spectrum of hadronic states [deG 75] gave the value $B^{1/4} = 145$ MeV, but other values up to $B^{1/4} = 235$ MeV have been reported [Ha 81b]. The requirement of pressure balance at the surface leads to the quadratic boundary condition

$$-\frac{1}{2} n^\mu \partial_\mu \left(\sum_i \bar{\psi}_i \psi_i \right) \bigg|_S = B \,, \tag{20.65}$$

where the sum runs over all quarks contained in the bag. For a spherical bag, condition (20.6) is equivalent to the requirement that the total energy contained in the bag volume is a minimum with respect to the bag radius R: $\partial M/\partial R = 0$. Due to the scalar nature of the bag constant B, the difference between the vacua in the interior and the exterior of the bag contributes an energy proportional to the bag volume V:

$$M(R) = \sum_i E_i(R) + \frac{4\pi}{3} BR^3 \,, \tag{20.66}$$

where $E_i(R)$ are the energy eigenvalues of the occupied quark states. For a spherical bag, $E_i(R) \sim R^{-1}$.

There is no obvious reason why a bag should not contain real gluons, although such states – called glueballs – have not yet been experimentally identified. The appropriate boundary conditions for the glue field are obtained from the requirement that the colour-electric field may not penetrate out the bag into the "true vacuum". This is analogous to the boundary conditions in classical electrodynamics for a medium with dielectric constant $\varepsilon = 0$ and magnetic permeability $\mu = \infty$:

$$\boldsymbol{n} \cdot \boldsymbol{E}^a = 0 \,, \quad \boldsymbol{n} \times \boldsymbol{H}^a = 0 \,, \quad \text{i.e.} \quad n_\mu F_a^{\mu\nu} = 0 \,. \tag{20.67}$$

From Gauss' theorem, an integration of the electric boundary condition over the full bag surface yields the result that the total colour charge contained in the bag volume must be zero, ensuring that all hadrons are colour singlets.

The following attempts to explain the microscopic process of fission of excited bags. It is generally believed that a quark with high kinetic energy, gained, e.g., in hadronic collisions, attempts to leave the colourless object within which it is confined, but is held back by a string-like bond of colour-electric field connecting it with the remaining quarks. Also heavy quarkonium resonances are supposed to form similar strings [Jo 76] with the heavy quarks attached to their ends. The energy stored in the colour-electric field forming the string increases with increasing separation of the colour charges until it becomes "supercritical". This means that the field is strong enough to allow for a light quark – antiquark pair to be spontaneously produced, since the energy eigenvalues of the quark and antiquark states approach zero. These additional quarks screen the original colour charges, and the string breaks into two [Ko 75], Fig. 20.11.

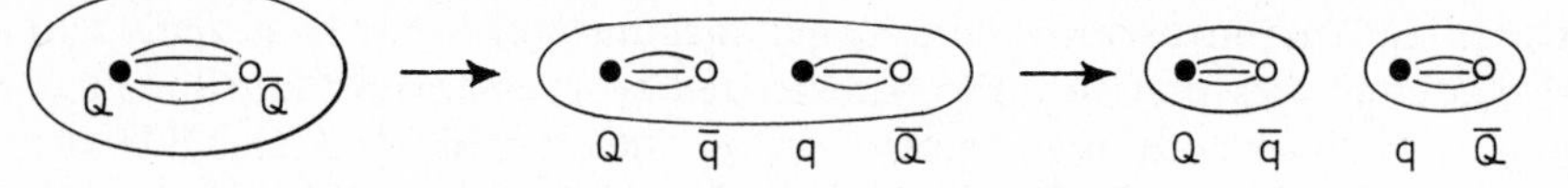

Fig. 20.11. Bag fission by glue string breaking due to spontaneous quark – antiquark pair production

As a first step towards understanding the fission of bags one can investigate the structure of single-particle states for light quarks in a deformed cavity containing a strong colour-electric field. To start with a simple adiabatic picture, assume the string to be an excited state of a heavy quarkonium system and view it as a static prolate ellipsoidal MIT bag with two point-like colour charges Q' and $\bar{Q}'$ held fixed at the two foci [Ha 78, 81b]. Its shape may be parametrized by the two half-axes a and b with $a \geqslant b$, Fig. 20.12. The potential produced by this configuration [with the condition $\boldsymbol{n} \cdot \boldsymbol{E} = 0$ on the surface as stated in (20.67)] is in the lowest order of the coupling constant equivalent to an Abelian field with charge $Q = 2Q'/\sqrt{3}$ and can be calculated analytically in this approximation. In prolate elliptical coordinates

$$
\begin{aligned}
x &= d \sinh \eta \sin \vartheta \cos \varphi \\
y &= d \sinh \eta \sin \vartheta \sin \varphi \\
z &= d \cosh \eta \cos \vartheta
\end{aligned}
\tag{20.68}
$$

the solution for the shape of the ellipsoid fixed by a and b reads [Wa 81]

$$
\begin{aligned}
\Phi(\eta, \vartheta, \varphi) = \frac{Q}{4\pi d} \bigg[& \frac{1}{\cosh \eta - \cos \vartheta} - \frac{1}{\cosh \eta + \cos \vartheta} \\
& - 2 \sum_l (4l+3) \frac{Q'_{2l+1}(\cosh \eta_0)}{P'_{2l+1}(\cosh \eta_0)} P_{2l+1}(\cosh \eta) P_{2l+1}(\cos \vartheta) \bigg]
\end{aligned}
\tag{20.69}
$$

with the focal length $d = (a^2 - b^2)^{1/2}$ and the eccentricity $1/\cosh \eta_0 = d/a$. The Legendre functions P_n and Q_n appear in the part of the potential produced by the confining boundary. Note that $\bar{Q} = -Q$. Here and in the following $Q(q)$ denotes the heavy (light) quarks as well as their charge. The field lines corresponding to the potential (20.69) are shown in Fig. 20.13.

We now ask at which separation $2d$ between the heavy quarks is the colour-electric field strong enough to create a pair of light quarks q spontaneously? The answer involves calculating the energies of quark and antiquark states in the potential Φ and seeing where the gap between them vanishes. The light quark [mass m, charge $(-q)$] wave function $\psi(r)$ is a solution of the stationary Dirac equation

$$
H(-q)\,\psi(\boldsymbol{r}) = \{-\mathrm{i}\boldsymbol{\alpha} \cdot \nabla + \beta m - q\,\Phi(\boldsymbol{r})\}\,\psi(\boldsymbol{r}) = E\,\psi(\boldsymbol{r}) \tag{20.70}
$$

within the bag volume V_{B}, satisfying the linear boundary condition

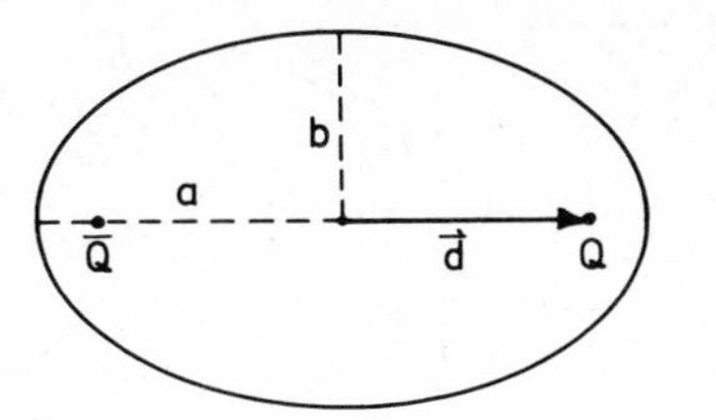

Fig. 20.12. The shape of the ellipsoidal bag is defined by the two half-axes a (long) and b (short). The vector $2\boldsymbol{d}$ points in the direction of Q and its length equals the distance between the charges

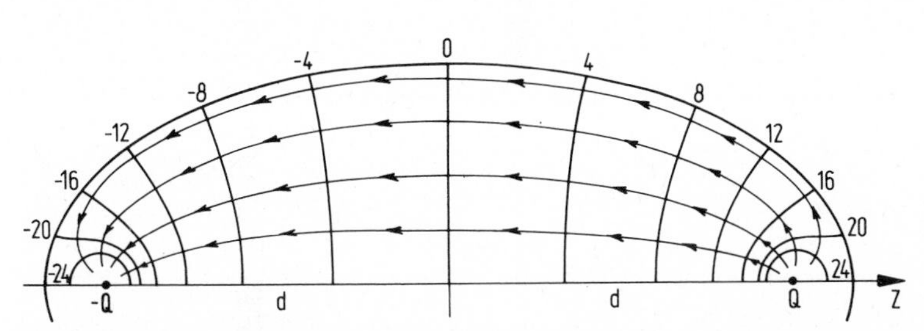

Fig. 20.13. The equipotential lines of $\Phi(\eta, \vartheta)$ in units $Q/4\pi d$ and the lines of force given by $-\nabla\Phi$ (*lines with arrows*). The potential $\Phi(\eta, \vartheta)$, produced by two charges Q and $-Q$ and modified by the confining boundary, is plotted here for deformation $a/b = x = 1.9$. The arrows indicate the direction of the colour-electric field

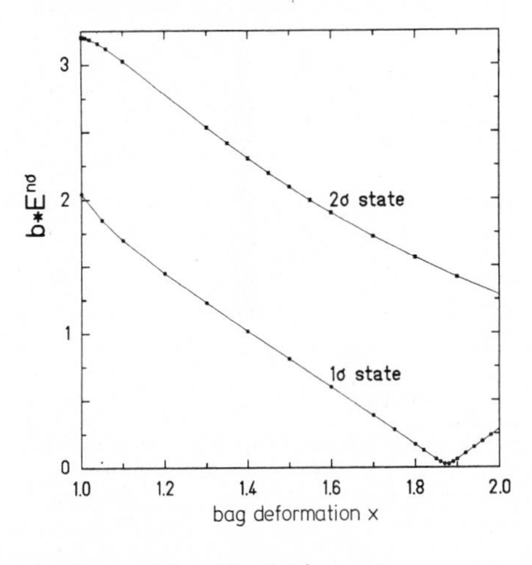

Fig. 20.14. The scale-independent energy eigenvalues (bE) for the value 1/2 (indicated by σ) of the z component of the total angular momentum, which is a good quantum number because of the axial symmetry of the bag. n is the principal quantum number

$$[\mathrm{i}\,\gamma\cdot \boldsymbol{n}(\boldsymbol{r})+1]\,\psi(\boldsymbol{r})\,|_{\mathrm{S_B}}=0 \tag{20.71}$$

on the surface $\mathrm{S_B}$, where $\boldsymbol{n}$ is the outward normal vector.

For massless quarks ($m = 0$), as long as the pressure balance condition of the MIT bag model is not considered, there is no scale parameter other than the bag size. We use the short half-axis of the prolate ellipsoid b as unit length, define $x = a/b$ and write $d = b\sqrt{(x^2-1)}$. In the limit of a very elongated tube-like ellipsoid, b is the tube radius.

The Dirac equation (20.70) has been solved numerically together with the boundary condition (20.71) by *Vasak* et al. [Va 83]. They found that the scale-independent eigenvalues $bE > 0$ decrease with increasing deformation, Fig. 20.14.

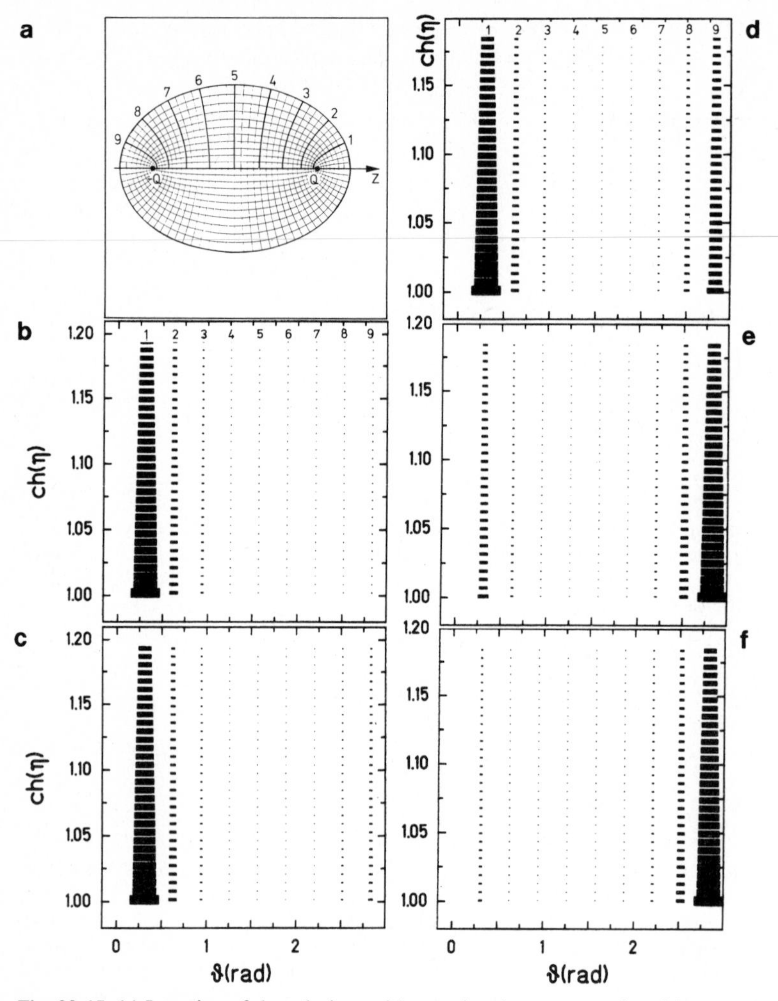

Fig. 20.15. (**a**) Location of the solutions with negative charge $-q$ and positive energies switches from the positive source charge Q to the negative one $-Q$ beyond the critical deformation, as shown in this sequence of cluster plots (**b**) $x = 1.85$, (**c**) $x = 1.87$, (**d**) $x = 1.88$, (**e**) $x = 1.89$, (**f**) $x = 1.90$. Large black rectangles correspond to high density

The energy of the lowest state [which is in the spherical case where the charges cancel each other has the value of the $1s_{1/2}$ state in a spherical bag, $bE(x=1) = 2.04$] reaches its minimum near zero at the "critical deformation" $x_{\rm cr}$. The position of this minimum depends on the coupling constant $\alpha_{\rm s} = qQ/4\pi$. A level crossing at zero energy is avoided [Ne 29] because the solution with negative energy, which rises with increasing deformation, has the same quantum numbers as the solution with positive energy falling with increasing x. Note that the influence of the boundary condition (20.67) on the colour-electric field is crucial for supercriticality. Without the boundary condition, only the first two terms in the bracket of (20.69) would be present. For these terms alone, and for the chosen value of the coupling constant $\alpha_s = 0.385$, the potential would not become supercritical at any deformation.

Figure 20.15 shows the density distribution of the lowest energy state 1σ. At the deformation x_{cr} the density $\psi^\dagger \psi$ of the "negatively" charged light-quark solution, peaked around the "positive" charge Q for $x < x_{cr}$, switches to the opposite charge $-Q$ for $x > x_{cr}$, Fig. 20.15. To understand this behaviour we first explore the symmetries of (20.2) and then reinterpret the results in the language of field theory.

The direction $\boldsymbol{d}$ (Fig. 20.12), defined to point from $-Q$ to Q, breaks not only the spherical but also the parity invariance. The parity transformation in effect reverses the sign of the potential, or equivalently, if the potential is kept unchanged, it reverses the charge of the light quark. This can be readily proved. If $\psi(\boldsymbol{r})$ is a solution of (20.70) with energy E and charge $-q$, then, (2.133),

$$\psi_P(\boldsymbol{r}) = \hat{P}\psi(\boldsymbol{r}) = \beta\psi(-\boldsymbol{r}) \tag{20.72}$$

is also a solution of (20.70) with $-q$ replaced by q:

$$\begin{aligned} E\psi_P(\boldsymbol{r}) &= E\hat{P}\psi(\boldsymbol{r}) = \hat{P}H(-q)\psi(\boldsymbol{r}) \\ &= \hat{P}[-\mathrm{i}\boldsymbol{\alpha}\cdot\nabla + \beta m - q\Phi(\boldsymbol{r})]\hat{P}^{-1}\hat{P}\psi(\boldsymbol{r}) \\ &= [-\mathrm{i}\boldsymbol{\alpha}\cdot\nabla + \beta m + q\Phi(\boldsymbol{r})]\psi_P(\boldsymbol{r}) = H(q)\psi_P(\boldsymbol{r}), \end{aligned} \tag{20.73}$$

$$[\mathrm{i}\boldsymbol{\gamma}\cdot\boldsymbol{n}(\boldsymbol{r}) + 1]\psi_P(\boldsymbol{r})|_{S_B} = P[\mathrm{i}\boldsymbol{\gamma}\cdot\boldsymbol{n}(\boldsymbol{r}) + 1]\psi(\boldsymbol{r})|_{S_B} = 0,$$

because $\Phi(-\boldsymbol{r}) = -\Phi(\boldsymbol{r})$ and $\boldsymbol{n}(-\boldsymbol{r}) = -\boldsymbol{n}(\boldsymbol{r})$. Hence, the wave functions ψ and ψ_P belong to particles with the same energy but different charges, and their densities are peaked at source charges with opposite signs. Similarly, the charge-conjugate wave function, (4.26),

$$\psi_C(\boldsymbol{r}) = \hat{C}\psi(\boldsymbol{r}) = \gamma_2\psi^*(\boldsymbol{r}) \tag{20.74}$$

can be shown to be an eigenfunction of $H(q)$ but with the negative eigenvalue $-E$ and with its density peaking around the "wrong" charge. Furthermore, the "time-reversed" wave function, (4.59),

$$\psi_T(\boldsymbol{r}) = \hat{T}\psi(\boldsymbol{r}) = \gamma_1\gamma_3\psi^*(\boldsymbol{r}) \tag{20.75}$$

is a solution of the original equation (20.70) with energy E. Both ψ_C and ψ_T also satisfy the boundary condition (20.71). Thus, if written in time-dependent form, (20.70) clearly possesses three symmetry operations: *CP*, *T* and, of course, *CPT*.

Next we discuss the solutions with negative energy in the framework of field theory. When quantizing, we have to choose a vacuum state with all negative energy modes occupied, and renormalize the vacuum energy to zero. The notation used is displayed in Table 20.1. The first subscript c of the wave function φ_{cp}^k indicates the charge of the light quark under consideration, p is the sign of the charge of the heavy quark at which this wave function is localized.

Table 20.1. Notation for the four different wave functions. The subscript c in φ^k_{cp} denotes the sign of the light-quark charge, and p denotes the sign of the external charge to which the corresponding wave function is attached. The superscript k summarizes the energy and all other quantum numbers

Wave function	Charge	Energy	Notation	Vacuum occupation
$\psi(\boldsymbol{r})$	$-q$	$+E$	$\varphi^k_{-+}(\boldsymbol{r})$	no
$PC\psi(\boldsymbol{r})$	$-q$	$-E$	$\varphi^k_{--}(\boldsymbol{r})$	yes
$C\psi(\boldsymbol{r})$	$+q$	$-E$	$\varphi^k_{++}(\boldsymbol{r})$	
$P\psi(\boldsymbol{r})$	$+q$	$+E$	$\varphi^k_{+-}(\boldsymbol{r})$	

The superscript k summarizes energy and all other quantum numbers. The modes φ^k_{-+} and φ^k_{+-} belong to positive, φ^k_{--} and φ^k_{++} to negative energy eigenvalues.

It must be realized that the two sorts of quarks differing in their charge are not independent particles. Consider, e.g., the field operator $\hat{\psi}_-(\boldsymbol{r})$ of the negatively charged light-quark field. It can be expanded in terms of the complete system $\{\varphi^k_{-+}, \varphi^k_{--}\}$ of wave functions with positive and negative energies:

$$\hat{\psi}_-(\boldsymbol{r}) = \sum_k [\hat{b}^k_{-+}\varphi^k_{-+}(\boldsymbol{r}) + \hat{b}^{k\dagger}_{--}\varphi^k_{--}(\boldsymbol{r})] \, . \tag{20.76}$$

The vacuum state $|0\rangle$ is defined by

$$\hat{b}^k_{-+}|0\rangle = \hat{b}^k_{--}|0\rangle = 0 \, . \tag{20.77}$$

All negative-energy modes are occupied. With the field-theoretical charge-conjugation operator $\hat{\mathscr{C}}$ we obtain from this (remember that charge conjugation contains complex conjugation)

$$\begin{aligned}\hat{\mathscr{C}}\,\hat{\psi}_-(\boldsymbol{r})\,\hat{\mathscr{C}}^{-1} &= \sum_k [\hat{b}^{k\dagger}_{-+}\hat{C}\varphi^k_{-+}(\boldsymbol{r}) + \hat{b}^k_{--}\hat{C}\varphi^k_{--}(\boldsymbol{r})] \\ &= \sum_k [b^{k\dagger}_{-+}\varphi^k_{++}(\boldsymbol{r}) + \hat{b}^k_{--}\varphi^k_{+-}(\boldsymbol{r})] \, .\end{aligned} \tag{20.78}$$

On the other hand, in terms of the complete set $\{\varphi^k_{+-}, \varphi^k_{++}\}$ of wave functions of the positively charged light quarks, the expansion

$$\hat{\mathscr{C}}\,\hat{\psi}_-(\boldsymbol{r})\,\hat{\mathscr{C}}^{-1} = \hat{\psi}_+(\boldsymbol{r}) = \sum_k [\hat{b}^k_{+-}\varphi^k_{+-}(\boldsymbol{r}) + \hat{b}^{k\dagger}_{++}\varphi^k_{++}(\boldsymbol{r})] \tag{20.79}$$

holds. Comparing (20.78) with (20.79) the Fock-space operators for quarks with negative energy and charge c, $\hat{b}^k_{cp}$ are identified with the Fock-space operators for quarks with positive energy and charge $-c$, $\hat{b}^{k\dagger}_{-c,p}$. An unoccupied state with charge c and negative energy (hole) is identified with an occupied state with charge $-c$ and positive energy (antiparticle). For a complete description of both positively and negatively charged particles, we thus need only one kind of charged field. As usual, $c = -q$ and states with charge $+q$ are attributed to antiquarks. This choice follows the patterns of QED (Chap. 4) where negatively charged electrons are considered (defined) as particles and the positrons as

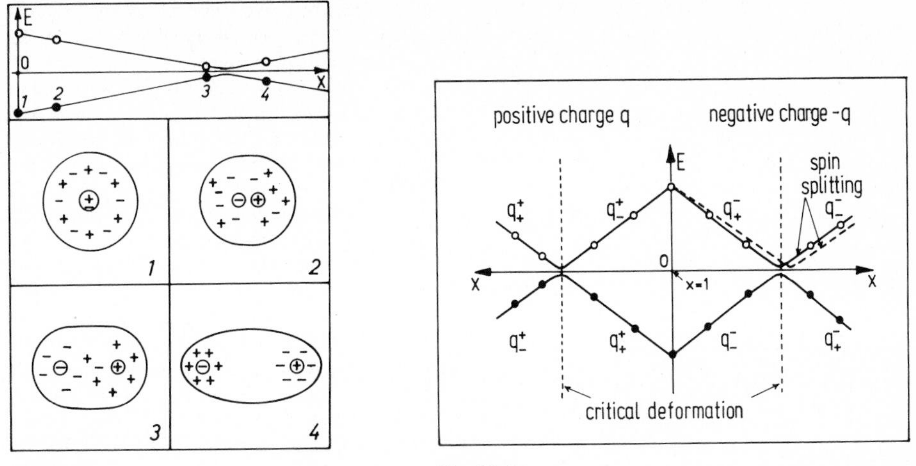

Fig. 20.16

Fig. 20.17

Fig. 20.16. In the spherical case the potential is zero, and the solutions with different charges degenerate (*1*). In the vacuum state (*top*) all negative energy modes are occupied (● ● ●). When pulling the source charges apart, the wave functions start localizing (*2* and *3*). Beyond the critical deformation the colour field is strong enough to pull the wave functions to the opposite side (*4*)

Fig. 20.17. In the vacuum state below the critical deformation an occupied light-quark mode with negative energy and a certain charge (● ● ●) is attached to the source charge with the same sign. Beyond the critical deformation, (– – –), the definition of the vacuum state has to be modified. The spin (and colour spin) splitting due to $q\bar{q}$ interaction is indicated. The *right* and *left parts* of the figure correspond to the two equivalent possibilities to describe the quantum field

antiparticles. Vacuum occupation remains symmetric if the current operator is antisymmetrized, (9.81).

This picture enables the "migration" of the wave function to be interpreted as spontaneous pair creation. Figure 20.16 shows how the density of occupied states with negative energy changes with deformation. In the spherical case the modes φ^k_{c+} and φ^k_{c-} are degenerate; by pulling the charges apart the spherical symmetry is broken and the degeneracy removed.

Between the subcritical and supercritical values of x the density of the 1σ state, which in the spherical case goes over into the $1s_{1/2}$ state, changes its location. In Fig. 20.17 interprets decay of the vacuum by spontaneous pair production: the previously occupied $\varphi^{1\sigma}_{--}$ state (or $\varphi^{1\sigma}_{++}$ in an equivalent description with $c=q$) becomes empty beyond the critical deformation. A quark $\varphi^{1\sigma}_{--}$ has been removed from the location around the charge $-Q$ (or vice versa, a quark $\varphi^{1\sigma}_{++}$ has been removed from the opposite location) and leaves a hole behind. The previously unoccupied mode $\varphi^{1\sigma}_{-+}(\varphi^{1\sigma}_{+-})$ is filled. The vacuum charge density within the bag changes. This can be viewed as the decay of the subcritical vacuum state into a new vacuum state by creation of a quark–antiquark pair.

Below the critical distance light quarks with positive energy are not present in the heavy deformed quarkonium. After the deformation has become super-

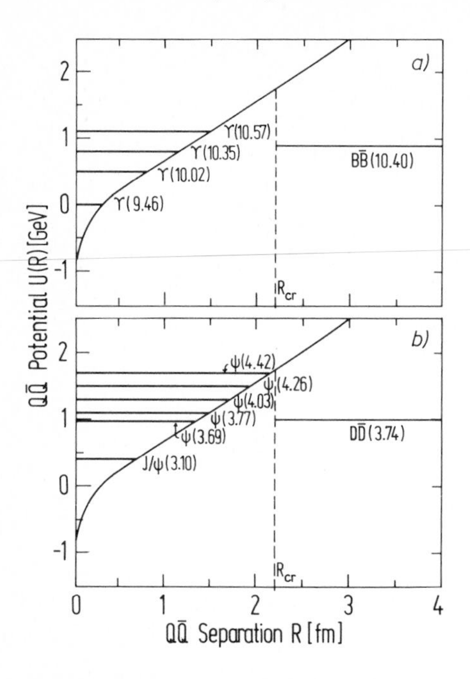

Fig. 20.18a, b. If the potential energy $U(R)$ is used in a Schrödinger equation to describe the radial motion of two heavy quarks Q and $\bar{Q}$, **(a)** the bottomonium and **(b)** the charmonium masses can be calculated. To obtain the actual mass of the meson, the masses of the constituent quarks have to be added to the eigenvalue of the Schrödinger equation. The results agree well with experimental data, indicated by horizontal lines. The potential is modified at R_{cr} since beyond this point the $Q\bar{Q}$ system prefers to exist as two atom-like mesons $Q\bar{q}+\bar{Q}q$. The masses of the $b\bar{u}$, $\bar{b}u$ (B mesons) and $c\bar{u}$, $\bar{c}u$ (D mesons) pairs with current masses of the constituent quarks subtracted fix the asymptotic value of $U(R)$ for $R \geq R_{\mathrm{cr}}$. Indeed, sharp resonances with binding energies above the barrier $U(R_{\mathrm{cr}}) \approx 1.8$ GeV have not been discovered

critical, the created pair screens the point charges, and the pressure of the colour-electric field is replaced by the pressure due to the light quarks. Because the light-quark wave functions are concentrated around the point charges, the bag develops a neck to "save" volume energy, and finally fission into two spherical atom-like bags occurs, each containing one heavy quark in its centre and one light quark around it. If the pair creation did not occur, the potential would keep on growing linearly. Pair creation modifies it at the fission point and it remains constant beyond. The dynamics of this fission process has been formulated by *Vasak* [Va 85].

As an example, consider the $b\bar{b}$ and $c\bar{c}$ mesons. The masses of the ground and several excited states have been calculated in the adiabatic approximation of the MIT bag model [Ha 81b], and agree with the experimental values [Ro 82]. Thus the value $R_{\mathrm{cr}} = 2.2$ fm is consistent with experimental data. Since the heavy quarks are non-relativistic, their mass enters the total quarkonium energy additively and has to be added to the energy eigenvalue of the Schrödinger equation to obtain the meson mass. The resonance $\gamma(10.57)$ [$\psi(3.77)$], which is found slightly above the threshold for production of pairs of atom-like mesons $B+\bar{B}(D+\bar{D})$, each consisting of a heavy quark b (c) and a light antiquark, is still well below the critical distance, Fig. 20.18. The $Q\bar{Q}$ potential energy $U(R)$ is modified at the critical distance R_{cr} and drops to the threshold value $2(m_B - m_b)$ or $2(m_D - m_c)$, respectively.

The main intent of this section has been to demonstrate how quark confinement is related to spontaneous pair creation in supercritical fields. We have seen how the change of the vacuum state leads to fission of deformed bags. Furthermore, the physics of strong fields, originally developed to describe novel phenomena occurring in super-heavy atomic systems, has equally important applications to the physics of quarks and gluons.

To recall the common characteristics of all these phenomena, let us summarize the basic results: when the field of force acting on some species of particles (electrons, pions, gluons, etc.) exceeds a certain critical strength, the vacuum state is suddenly forced to change its symmetry properties. For instance, in the original case of a supercritical atomic nucleus, the vacuum state becomes electrically charged, and positrons are emitted at the same time. In general, the vacuum state is rearranged so as to diminish the effect of the applied "external" force, i.e. the vacuum acts as a screening agent.

The phenomenon of gluon condensation and non-perturbative vacua discussed here may prove to be the clue for understanding the elusiveness of quarks. While the experimental evidence for the presence of quark-like constituents in the proton and other hadrons is overwhelming, the fact that these constituents seem to be permanently confined to the interior of elementary particles is not yet fully understood. Several recent investigations, however, have yielded preliminary evidence that the true vacuum of QCD is characterized by a global condensate of gluons interacting among themselves [Ni 79a, b, Ha 82, Ra 83]. It is likely that the methods developed to describe the charged electronic vacuum, pion condensates and gluon condensates, will also elucidate the vacuum state for strong interactions.

Bibliographical Notes

For an introduction to Yang-Mills theories and QCD, we particularly refer to [It 80, Ai 82].

The possible role of supercritical fields in understanding the vacuum state of QCD and quark confinement is discussed in [Ra 83].

Charged vacuum states have recently attracted considerable interest in connection with catalysis of baryon decay by magnetic monopoles [Gr 83e, Ya 83, Ni 83, Pa 83], with topological solitons [Ja 76a, 81, Go 81b] and with chiral bag models [Go 83]. A discussion of these applications, however, lies outside the scope of this book.

An introduction into the mathematics of non-Abelian symmetry groups and their applications in particle physics is found in [Gr 84a].

21. Strong Fields in General Relativity

This chapter treats the influence of strong gravitational fields on the vacuum. After a brief introduction to the Dirac equation in generally covariant form, it is shown that the gap between particle and antiparticle states is narrowed in the gravitational field of a collapsed star, such as a neutron star. When the star collapses to a black hole, the gap shrinks to zero at the Schwarzschild radius $r_s = 2GM/c^2$, and the vacuum becomes unstable.

To understand this instability better, we then discuss the vacuum in a uniform acceleration field (Rindler space), where particles are spontaneously created with a thermal spectrum. The temperature is related to the magnitude of acceleration g, namely $T \sim g\hbar/2\pi c k_B$. When this result is applied to the gravitational acceleration field of a black hole, it is found that the black hole radiates particles away, as first discovered by Hawking in 1974.

21.1 Dirac Particles in a Gravitational Field

In Einstein's general theory of relativity, the gravitational field is described by a metric tensor $g_{\mu\nu}(x)$ that defines the invariant distance between two infinitesimally separated space-time events

$$ds^2 = g_{\mu\nu}(x)\, dx^\mu dx^\nu . \tag{21.1}$$

According to the principle of equivalence the metric tensor describes not only the effects of gravitational forces but also those of (apparent) inertial forces through the geodesic equation of motion for a point-like mass

$$\frac{d^2x^\alpha}{ds^2} + \left\{ \begin{matrix} \alpha \\ \mu\nu \end{matrix} \right\} \frac{dx^\mu}{ds} \frac{dx^\nu}{ds} = 0 , \tag{21.2}$$

where affine connections are given by the Christoffel symbol

$$\left\{ \begin{matrix} \alpha \\ \mu\nu \end{matrix} \right\} = \frac{1}{2} g^{\alpha\beta} \left(\frac{\partial g_{\mu\beta}}{\partial x^\nu} + \frac{\partial g_{\beta\nu}}{\partial x^\mu} - \frac{\partial g_{\mu\nu}}{\partial x^\beta} \right) . \tag{21.3}$$

The Christoffel symbols also enter into the definition of the covariant derivative of a vector field ξ^μ

$$D_\nu \xi^\mu \equiv \frac{\partial \xi^\mu}{\partial x^\nu} + \left\{ \begin{matrix} \mu \\ \alpha\nu \end{matrix} \right\} \xi^\alpha , \tag{21.4}$$

which transforms like a two-component tensor (the partial derivative $\partial \xi^\mu / \partial x^\nu$ does not!).

Because the Christoffel symbol $\left\{ \begin{matrix} \alpha \\ \mu\nu \end{matrix} \right\}$ describes both gravitational and inertial forces, it is not invariant under a general coordinate transformation, i.e. it is not a tensor. Only the tidal forces of gravity can be given a coordinate system-independent meaning; according to Einstein they are related to the curvature of space-time through the Riemann tensor

$$R^{\alpha}_{\ \beta\mu\nu} = \frac{\partial}{\partial x^\nu} \left\{ \begin{matrix} \alpha \\ \mu\beta \end{matrix} \right\} - \frac{\partial}{\partial x^\mu} \left\{ \begin{matrix} \alpha \\ \beta\nu \end{matrix} \right\} + \left\{ \begin{matrix} \alpha \\ \lambda\nu \end{matrix} \right\} \left\{ \begin{matrix} \lambda \\ \mu\beta \end{matrix} \right\} - \left\{ \begin{matrix} \alpha \\ \lambda\mu \end{matrix} \right\} \left\{ \begin{matrix} \lambda \\ \nu\beta \end{matrix} \right\} , \tag{21.5}$$

which is defined as the commutator of the second covariant derivative:

$$R^{\alpha}_{\ \beta\mu\nu} \xi^\beta \equiv D_\nu D_\mu \xi^\alpha - D_\mu D_\nu \xi^\alpha . \tag{21.6}$$

In a general, non-Minkowskian, metric $g^{\mu\nu}$, the Dirac matrices have to satisfy the tensor equation

$$\gamma^\mu \gamma^\nu + \gamma^\nu \gamma^\mu = 2 g^{\mu\nu} \tag{21.7}$$

which is an obvious generalization of (2.73). When the metric depends on x, so do the γ matrices, and therefore the γ^μ do not commute with the derivative operator $\partial/\partial x^\mu$, in general. Consequently, the Dirac equation for the fermion field $\psi(x)$ in a general metric $g_{\mu\nu}$ must take a somewhat more complicated form:

$$\mathrm{i}\gamma^\mu \left(\frac{\partial}{\partial x^\mu} + \Gamma_\mu \right) \psi - m\psi \equiv \mathrm{i}\gamma^\mu D_\mu \psi - m\psi = 0 , \tag{21.8}$$

where the 4×4 spin matrices Γ_μ contain derivatives of the Dirac matrices. The precise form of matrices Γ_μ is obtained, after *Weyl*, and *Fock* and *Iwanenko* [We 29, Fo 29a, b], by the requirement that the Dirac operator be Hermitian. As a result, it is found that the Γ_μ can be represented as

$$\Gamma_\mu = \frac{1}{4} \gamma_\nu \left(\frac{\partial \gamma^\nu}{\partial x^\mu} + \left\{ \begin{matrix} \nu \\ \lambda\mu \end{matrix} \right\} \gamma^\lambda \right) = \frac{1}{4} \gamma_\nu D_\mu \gamma^\nu . \tag{21.9}$$

To be specific, let us restrict further considerations to spherically symmetric gravitational fields as they exist around dense, heavy stars, e.g. white dwarfs, neutron stars or black holes [So 77b]. Any static, spherically symmetric gravitational field can be described by the metric

$$g_{\mu\nu} = \begin{bmatrix} e^{\nu(r)} & & & \\ & -e^{\lambda(r)} & & \\ & & r^2 & \\ & & & r^2 \sin^2\Theta \end{bmatrix} . \tag{21.10}$$

Outside the matter distribution of the star the functions $\nu(r)$ and $\lambda(r)$ are given by Schwarzschild's solution of the Einstein equations:

$$e^{\nu(r)} = e^{-\lambda(r)} = 1 - \frac{2MG}{r} , \tag{21.11}$$

where G is the gravitational constant and M the stellar mass. Because the metric tensor (21.10) is diagonal, a simple solution of the defining equation (21.7) of the Dirac matrices can be found in the form

$$\begin{aligned} \gamma_0 &= e^{\nu/2}\tilde{\gamma}_0 , & \gamma^0 &= e^{-\nu/2}\tilde{\gamma}_0 , \\ \gamma_1 &= e^{\lambda/2}\tilde{\gamma}_1 , & \gamma^1 &= -e^{-\lambda/2}\tilde{\gamma}_1 , \\ \gamma_2 &= r\tilde{\gamma}_2 , & \gamma_2 &= -\frac{1}{r}\tilde{\gamma}_2 , \\ \gamma_3 &= r\sin\Theta\,\tilde{\gamma}_3 , & \gamma_3 &= -\frac{1}{r\sin\Theta}\tilde{\gamma}_3 , \end{aligned} \tag{21.12}$$

where the $\tilde{\gamma}_\mu$ are the usual Dirac matrices in Minkowski space.

Charge density and energy density of the Dirac field can be derived from the action principle

$$\delta\int d^4x \sqrt{-g}\,(\mathrm{i}\,\bar{\psi}\gamma^\mu D_\mu\psi - m\,\bar{\psi}\psi) = 0 \tag{21.13}$$

by Noether's theorem, giving

$$j^\mu(x) = \bar{\psi}(x)\,\gamma^\mu\psi(x) \quad \text{and} \tag{21.14}$$

$$\begin{aligned} T_{\mu\nu}(x) = \frac{\mathrm{i}}{4}[&\bar{\psi}(x)(\gamma_\mu D_\nu\psi(x) + \gamma_\nu D_\mu\psi(x)) \\ &- (D_\mu\bar{\psi}(x)\,\gamma_\nu + D_\nu\bar{\psi}(x)\,\gamma_\mu)\,\psi(x)] . \end{aligned} \tag{21.15}$$

For the spherically symmetric case

$$j^0(x) = \psi^\dagger(x)\,\psi(x)\,e^{-\nu/2} \quad \text{and} \tag{21.16}$$

$$T_0^0(x) = \frac{\mathrm{i}}{2}\left(\psi^\dagger\frac{\partial\psi}{\partial t} - \frac{\partial\psi^\dagger}{\partial t}\psi\right)e^{-\nu/2} . \tag{21.17}$$

In the spherically symmetric field (21.10) the four-component Dirac equation can be reduced to a two-component form as done for the normal hydrogen atom in Minkowski space (Chap. 2):

$$\psi = \mathrm{e}^{-\lambda/4}\frac{1}{r}\begin{pmatrix} \Phi_1(r,t) & \chi^\mu_\varkappa \\ \mathrm{i}\,\Phi_2(r,t) & \chi^\mu_{-\varkappa}\end{pmatrix}, \quad \Phi = \begin{pmatrix}\Phi_1 \\ \Phi_2\end{pmatrix} \tag{21.18}$$

with the spinor spherical harmonics

$$\chi^\mu_\varkappa(\Theta,\varphi) = \sum_{m=\pm\frac{1}{2}} (l\tfrac{1}{2}j\,|\mu-m\,m\,\mu)\, Y_{l,\mu-m}(\Theta,\varphi)\,\chi_m \tag{21.19}$$

which satisfy the eigenvalue equation

$$\hat{K}\chi^\mu_\varkappa \equiv \tilde{\gamma}_0(\boldsymbol{\sigma}\cdot\hat{\boldsymbol{L}}+1)\,\chi^\mu_\varkappa = -\varkappa\chi^\mu_\varkappa\,. \tag{21.20}$$

Inserting this into (21.8), after some elementary algebra

$$\begin{aligned}\mathrm{i}\frac{\partial\Phi}{\partial t} &= \left[-\mathrm{i}\,\alpha_r \exp\left(\frac{\nu-\lambda}{4}\right)\frac{\partial}{\partial r}\exp\left(\frac{\nu-\lambda}{4}\right) + \mathrm{i}\beta\alpha_r \mathrm{e}^{\nu/2}\frac{\varkappa}{r} + \beta\mathrm{e}^{\nu/2}m + V\right]\Phi \\ &\equiv (H_0+V)\,\Phi \end{aligned} \tag{21.21}$$

with the notations

$$\beta = \tilde{\gamma}_0 = \begin{pmatrix}1 & 0\\ 0 & -1\end{pmatrix}, \quad \alpha_r = \sigma_2 = \begin{pmatrix}0 & -\mathrm{i}\\ \mathrm{i} & 0\end{pmatrix}, \quad V = eA_0\,. \tag{21.22}$$

By partial integration it is immediately clear that the Hamiltonian H is explicitly Hermitian with respect to integration over r

$$\int dr\, \Phi_1^\dagger(H\Phi_2) = \int dr (H\Phi_1)^\dagger\,\Phi_2\,. \tag{21.23}$$

The radial density is

$$\varrho(r) = r^2\int d\Omega\, j^0(x)\exp\left(\frac{\nu+\lambda}{2}\right) = \Phi^\dagger(r)\,\Phi(r)\,, \tag{21.24}$$

so that the normalization condition becomes

$$\int_0^\infty dr\,\Phi^\dagger(r)\,\Phi(r) = 1\,. \tag{21.25}$$

For a *static* gravitational field, the Dirac equation (21.21) can be reduced further by solving for stationary states

$$\Phi(r,t) = \exp\left(\frac{\lambda-\nu}{4}\right)\begin{pmatrix}f(r)\\ g(r)\end{pmatrix}\mathrm{e}^{-\mathrm{i}Et}\,. \tag{21.26}$$

The radial equations are

$$\frac{d}{dr}\begin{pmatrix}f\\ g\end{pmatrix} = \mathrm{e}^{\lambda/2}\begin{pmatrix}\varkappa/r & m-\mathrm{e}^{-\nu/2}(E-V)\\ m+\mathrm{e}^{-\nu/2}(E-V) & -\varkappa/r\end{pmatrix}\begin{pmatrix}f\\ g\end{pmatrix}. \tag{21.27}$$

These equations have been solved [So 77b] for the metric field of a dense star composed of an incompressible fluid of density ϱ_0 [Ad 65]. If r_0 is the stellar radius, the mass of the star is given by

$$M = \frac{4\pi}{3}\varrho_0 r_0^3 . \tag{21.28}$$

Outside the star, i.e. for $r > r_0$, the metric has the Schwarzschild form (21.11), and the stellar interior it is

$$e^{\nu(r)} = \frac{3}{2}\sqrt{1 - \frac{r_0^2}{\hat{R}^2}} - \frac{1}{2}\sqrt{1 - \frac{r^2}{\hat{R}^2}} , \tag{21.29a}$$

$$e^{-\lambda(r)} = 1 - \frac{r^2}{\hat{R}^2} , \quad \text{where} \tag{21.29b}$$

$$\hat{R}^2 = \frac{r_0^3}{2MG} . \tag{21.30}$$

The radius r_0 of the star should be greater than its Schwarzschild radius

$$r_s = 2MG . \tag{21.31}$$

Indeed, the model breaks down already for $r_0 = 9/8\, r_s$, because the pressure becomes infinite. To study the limit $r_0 \to r_s$, the metric was analytically continued to $r_s < r_0 < 9/8\, r_s$ by choosing

$$\begin{aligned} &e^{\nu(r)} = a e^{br^2} \qquad r \le r_0 \\ &b = r_0^{-2}\left(\frac{r_0}{MG} - 2\right)^{-1} , \qquad a = \left(1 - \frac{2MG}{r_0}\right) e^{-br_0^2} . \end{aligned} \tag{21.32}$$

For $r_0 \le 9/8\, r_s$ this metric does not account for a reasonable physical situation but serves to define an energy eigenvalue of the Dirac equation in this region.

The radial Dirac equation (21.27) has been numerically integrated for $V = 0$ and the energy eigenvalues have been obtained, Fig. 21.1. Note that the energy eigenvalues of all bound states (with finite number of modes in their radial functions) tend to zero as $r_0 \to r_s$ and a quasicontinuum arises.

To show that this is not an artefact of the special metric (21.29, 32), a different, more schematic model has also been studied. The metric of a thin gravitating mass shell of mass M and radius r_0 is given by

$$\begin{aligned} &e^{\nu} = e^{-\lambda} = 1 - \frac{2MG}{r} \equiv 1 - \frac{r_s}{r} , \qquad (r > r_0) \\ &e^{\nu} = 1 - \frac{2MG}{r_0} \equiv 1 - \frac{r_s}{r_0} , \qquad e^{\lambda} = 1 \qquad (r < r_0) . \end{aligned} \tag{21.33}$$

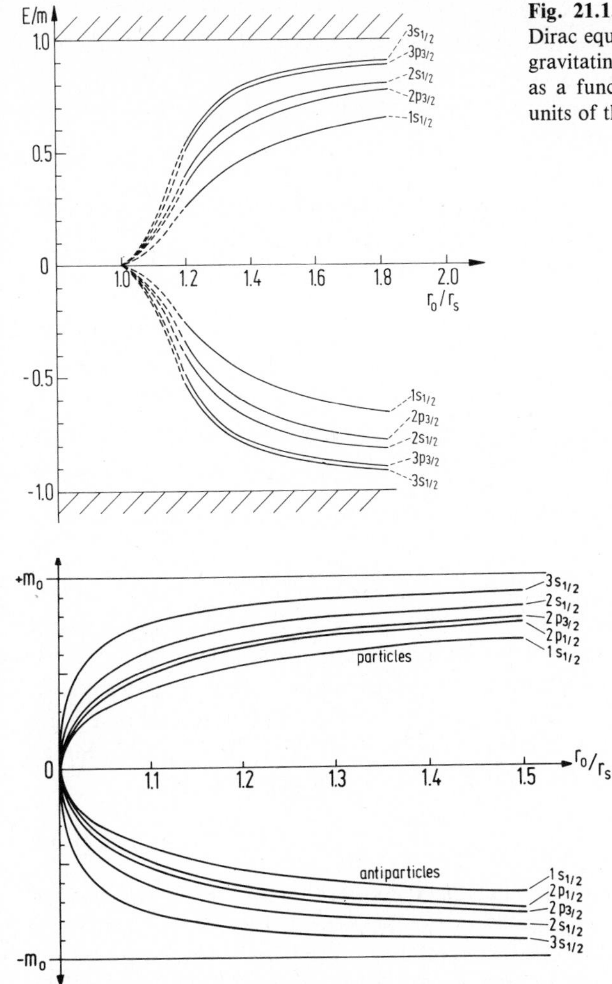

Fig. 21.1. The energy eigenvalues of the Dirac equation in the field of an extended gravitating source. The energies are shown as a function of the source radius r_0 (in units of the Schwarzschild radius r_s)

Fig. 21.2. The lowest energy eigenvalues of Dirac's equation in the field of a gravitating mass shell are shown as functions of the shell radius r_0

When the Dirac equation is solved with this metric, the same phenomenon is found: when r_0 approaches the Schwarzschild radius r_s, the energy eigenvalues E of all bound states tend to zero, Fig. 21.2; the gap between particle and antiparticle states vanishes in the limit $r_0 \to r_s$.

The mathematical reason for this behaviour is obvious: $e^{\nu(r)}$ approaches zero inside the mass shell when $r_0 \to r_s$, and therefore the term $e^{-\nu/2}E$ in the Dirac equation (21.27) can remain finite only if E goes to zero in this limit. Since $e^{\nu/2}$ is the red shift factor in a spherically symmetric gravitational field, this result can be understood in the following way. Asymptotic energies E are obtained by searching for stationary states in the time variable x^0. Since x^0 coincides with the

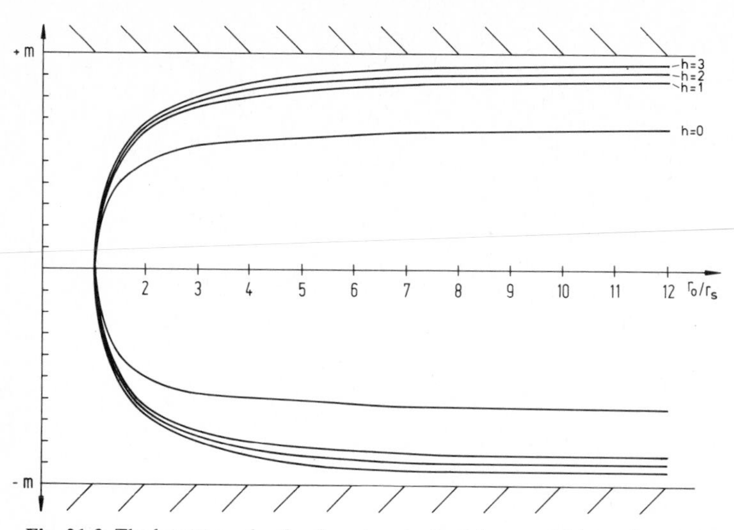

Fig. 21.3. The lowest $s_{1/2}$ levels of an atom in the Schwarzschild metric are plotted against the atomic position r_0/r_s, where r_s is the Schwarzschild radius

physical (proper) time only in the asymptotic region ($r \to \infty$), these energies represent quantities as measured by an asymptotically distant observer.

This interpretation is confirmed by a calculation first performed by *Papapetrou* [Pa 56]. If an atom (i.e. electron in the Coulomb potential of a point nucleus) is located at $r_0 > r_s$ in a Schwarzschild field, then the following energy spectrum is obtained when all derivatives of the metric are neglected:

$$E_i(r_0) = e^{\nu(r_0)/2} E_i \, . \tag{21.34}$$

Here E_i are the usual energy eigenvalues in Minkowski space. The behaviour of the $s_{1/2}$ energy levels as a function of r_0 can be seen in Fig. 21.3. Clearly the behaviour of $E_i(r_0)$ as a function of r_0 results from the red shift caused by the gravitational field between $r = r_0$ and $r = \infty$. Similarly, the behaviour of the energy levels in Fig. 21.2 essentially results from a mean gravitational red shift of the wave function $\langle e^{\nu/2}(r) \rangle$.

However, one should not conclude that this red shift is only an effect of coordinate transformations. The Newtonian gravitational potential Φ_N is related to the metric tensor through the approximate equation

$$\Phi_N(r) \approx m(1 - e^{\nu(r)/2}) \, . \tag{21.35}$$

This potential can be made to vanish only in a local, geodesic coordinate system in which the overall $g_{\mu\nu}$ field appears dynamical. Considering an electron in a central Newtonian potential, the binding of the electron results from the attractive Newtonian force, and one would speak of red shift only if a photon leaves the field of force. This point of view results from the existence of a global and absolute time coordinate representing physically measurable time intervals in

Newton's theory in contrast to Einstein's theory of gravity. In fact, a more detailed calculation [Pa 80] reveals that the energy levels of a hydrogen atom are also influenced by the local curvature. These effects cannot be transformed away by any choice of coordinate system (i.e. they are gauge invariant).

One might be tempted to conclude that a supercritical situation analogous to Klein's paradox appears for $r_0 \leq r_s$ in Figs. 21.1, 2. The case, however, is not so simple since for $r_0 \leq r_s$ the gravitational field is no longer globally static. This is obvious from the pure Schwarzschild field (21.11): in the "interior" of the black hole (for $r < r_s$) r is a time-like and t a space-like coordinate, i.e. the gravitational field is genuinely dynamic in this region. Pair creation, connected with the formation of an event horizon (the so-called *Hawking* effect [Ha 74], and the dynamics associated with it, are discussed in Sect. 21.3.

21.2 Limiting Charge of Black Holes

The decay of the vacuum, related to Klein's paradox, also appears in the study of charged black holes. When a charged stellar object of mass M collapses beyond the Schwarzschild event horizon, it leaves behind a spherically symmetric gravitational field given by the Reissner-Nordstrøm metric [Ad 65]

$$e^{\nu(r)} = e^{-\lambda(r)} = 1 - \frac{2MG}{r} + \frac{Q^2 G}{r^2}, \tag{21.36}$$

where Q is the charge of the black hole. The electrostatic potential $V(r)$ is simply given by

$$V(r) = \frac{eQ}{r}. \tag{21.37}$$

When the potential at the Schwarzschild radius $V(r_s) = eQ/r_s$ exceeds the electron rest mass, spontaneous pair production occurs. This process can be best understood and described using an *effective potential.* The effective potential was first derived from the classical Hamilton-Jacobi formalism by *Christodoulou* and *Ruffini* [Ch 71]. For spin-1/2 particles it must be derived by transforming the Dirac equation (21.27) into a second-order differential equation and applying the WKB approximation [So 82c]. First, squaring (21.27) yields the second-order equation

$$(E - V)^2 \Phi = e^{\nu}\left[-\Delta_r + m^2 + \frac{\varkappa^2}{r^2} + e^{-\lambda/2}\tilde{\gamma}_0\left(\frac{\varkappa}{r^2} - \frac{\nu'\varkappa}{2r}\right) - i\tilde{\gamma}_1 e^{-\lambda/2}\frac{\nu'}{2} m\right]\Phi \tag{21.38}$$

with

$$\Delta_r = \mathrm{e}^{-\frac{\lambda+\nu}{2}} \frac{\partial}{\partial r} \mathrm{e}^{\frac{\nu-\lambda}{2}} \frac{\partial}{\partial r}, \qquad \Phi(r) = \begin{pmatrix} f(r) \\ g(r) \end{pmatrix}, \tag{21.39}$$

the prime indicating a derivative with respect to the radial coordinate r.

If the derivatives of $g_{\mu\nu}$ are neglected in (21.38), the WKB approximation is obtained. The behaviour of the Dirac particle then is essentially determined by the effective potential V_{eff}

$$V_{\mathrm{eff}}(r) = V(r) \pm \mathrm{e}^{\nu(r)/2} \left(m^2 + \frac{\varkappa^2}{r^2} \right)^{1/2}. \tag{21.40}$$

The motion of classical test particles in a manifold with metric (21.10) is completely determined by V_{eff}, e.g. for classical test particles moving in the equatorial plane of a Schwarzschild black hole, the geodesic equation (21.2) yields

$$\left(\frac{dr}{ds} \right)^2 + V_{\mathrm{eff}}(r)^2 = E^2. \tag{21.41}$$

Figure 21.4 qualitatively shows the shape of V_{eff} for $\varkappa^2 \geq 2m^2 r_\mathrm{s}^2$ in the Schwarzschild field. Note that for $\varkappa^2 \geq 2m^2 r_\mathrm{s}^2$ classical bound states are possible between the turning points b and c. On the other hand, general relativity effects may destroy the angular momentum barrier for test particles moving around a black hole. Differentiating V_{eff} with respect to r causes bound states for classical test particles to appear only if $\varkappa^2 \geq 12 m^2 (GM)^2$ or $\varkappa^2 \geq 2m^2 r_\mathrm{s}^2$.

The form of the effective potential in the Schwarzschild field in Fig. 21.4 indicates that the spectrum of the Dirac equation in the pure Schwarzschild geometry is continuous with resonance solutions [So 77b]. They correspond to particles quasi bound in the potential pocket between b and c, but have a certain probability for tunnelling through the barrier to $r = a$ and then falling into the black hole. To exhibit these resonance solutions, it is useful to restrict this manifold to the exterior Schwarzschild geometry $r > r_\mathrm{s}$ by introducing a new radial coordinate r^* through

$$\frac{dr^*}{dr} = \left(1 - \frac{r_\mathrm{s}}{r} \right)^{-1} \equiv \mathrm{e}^{\lambda(r)}, \qquad r^* = r + r_\mathrm{s} \ln \left(\frac{r - r_\mathrm{s}}{r_\mathrm{s}} \right). \tag{21.42}$$

Introducing the r^* coordinate means that the Schwarzschild horizon r_s is projected to minus infinity. Wave functions in the interior of the black hole are then not taken into account. Dirac's equation (21.27) in r^* coordinates reads

$$\begin{aligned} \frac{dg}{dr^*} &= -\frac{\varkappa}{r} \mathrm{e}^{\nu/2} g + (E + \mathrm{e}^{\nu/2} m) f, \\ \frac{df}{dr^*} &= \frac{\varkappa}{r} \mathrm{e}^{\nu/2} f - (E - \mathrm{e}^{\nu/2} m) g, \end{aligned} \tag{21.43}$$

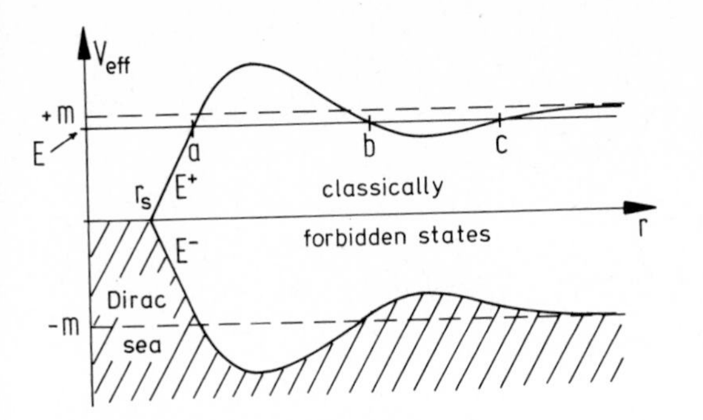

Fig. 21.4. The effective potential $V_{\mathrm{eff}}(r)$ in the Schwarzschild field is shown. The three roots of $E^2 - V_{\mathrm{eff}}^2$ are denoted by a, b and c. Between b and c classical bound states are possible

where $\mathrm{e}^{\nu(r)}$ is still given by (21.11).

In the sense of a WKB approximation then

$$\frac{d^2 g}{dr^{*2}} = [V_{\mathrm{eff}}(r)^2 - E^2]\, g(r^*) \equiv W(r^*)\, g(r^*)\,, \tag{21.44}$$

if spin-dependent terms are small and terms of order $O(1/r^2)$ are neglected. Let a, b and c be the zeros of the function W as shown in Fig. 21.4. The WKB approximation then yields the following solution near the horizon [De 74]

$$g(r^*) \sim \sqrt{-W} \exp\left(-\mathrm{i} \int\limits_{\mathrm{a}}^{r^*} dr^* \sqrt{-W}\right) \left[\mathrm{i}\, \frac{\sinh \mathrm{e}^{-\varepsilon}}{2} + 2\cosh \mathrm{e}^{\varepsilon}\right] + \mathrm{c.c.}\,, \tag{21.45}$$

where $\varepsilon \equiv \int_{\mathrm{a}}^{\mathrm{b}} dr^* \sqrt{W}$, and W is defined in (21.44).

The physical boundary condition that particles can only fall into the black hole but not emerge from it requires that there are only "ingoing" waves at the horizon, i.e. the c.c. term in (21.45) should vanish. Hence

$$-\mathrm{i} \sinh \mathrm{e}^{-\varepsilon} + 4 \cosh \mathrm{e}^{\varepsilon} = 0\,, \tag{21.46}$$

which has a solution only in the complex energy plane, i.e.

$$\bar{E}_n = E_n - \frac{i}{2}\Gamma_n\,, \tag{21.47}$$

clearly indicating the resonance character of these solutions. In the context of quantum mechanics, a particle bound in the region between b and c in Fig. 21.4 is able to tunnel through the potential barrier between a and b and be subsequently absorbed by the black hole. The absorption probability is given by the width Γ_n of the resonance corresponding to a lifetime of the quasi-bound state of $\tau_n = \Gamma_n^{-1}$.

The energy of a resonance may be obtained by the Bohr-Sommerfeld condition ($n = 0, 1, 2, \ldots$)

$$\int\limits_{\mathrm{b}}^{\mathrm{c}} dr^* \sqrt{-W(E_n)} = (n + \tfrac{1}{2})\,\pi\,, \tag{21.48}$$

its width Γ_n being given by

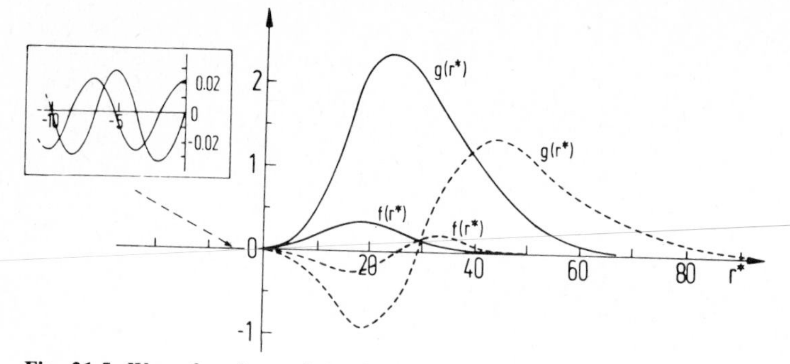

Fig. 21.5. Wave functions of the first [(——), $E_1 = 0.976\,m$] and the second [(– – –), $E_2 = 0.983\,m$] resonances of the Dirac equation in a microscopic Schwarzschild field ($r_s = 2$, $\varkappa = 4$) [So 82c]

$$\Gamma_n = 2\left|\frac{d}{dE}\cos\int_b^c dr^*\sqrt{-W}[1+\mathrm{e}^{2\varepsilon}]\right|^{-1}_{E=E_n} . \tag{21.49}$$

It is also possible to obtain exact solutions of (21.43) by numerically integrating from large positive to negative values of r^*. The resonances may be found by inspecting the phase shift $\delta(E)$ near the horizon. For $r^* \to \infty$ (21.43) yields

$$\begin{aligned} g(r^*) &\sim A_0 \sin(Er^* - \delta(E)) , \\ f(r^*) &\sim A_0 \cos(Er^* - \delta(E)) . \end{aligned} \tag{21.50}$$

Here $\delta(E)$ rapidly changes magnitude π at the resonance energy $E = E_n$. Figure 21.5 shows the wave function of the first (full curve, $E_1 = 0.976\,m$) and the second (broken curve, $E_2 = 0.983\,m$) resonance of the Dirac equation in a microscopic Schwarzschild field. The widths of these two lowest resonances are of the order $\Gamma \sim 10^{-6}$ corresponding to a lifetime of $\tau \sim 10^{-15}$ s.

We now return to our discussion of the spontaneous discharge by pair production of a charged black hole. According to (21.40), the effective potential describing the behaviour of test particles in the field of a charged (non-rotating) black hole of charge Q is approximately given by

$$V_{\mathrm{eff}}^{(\pm)}(r) = \frac{eQ}{r} \pm \left(1 - \frac{r_s}{r}\right)^{1/2}\left(m^2 + \frac{\varkappa^2}{r^2}\right)^{1/2} , \tag{21.51}$$

assuming that $\dfrac{Q^2G}{r_s^2} \ll 1$, i.e. $Q \ll MG^{1/2}$.

The effective potential for various charged black holes is shown in Fig. 21.6. From these it is clear that spontaneous pair creation and, therefore, a change of the vacuum state occurs for a highly charged black hole in complete analogy to the situation encountered in QED, Chap. 6.

Let r_+ be the outer event horizon of a charged black hole ($r_+ = MG + \sqrt{M^2G - Q^2G} \simeq 2MG$). Then pair creation occurs if $Qe/r_+ \gtrsim mc^2$, as is clear

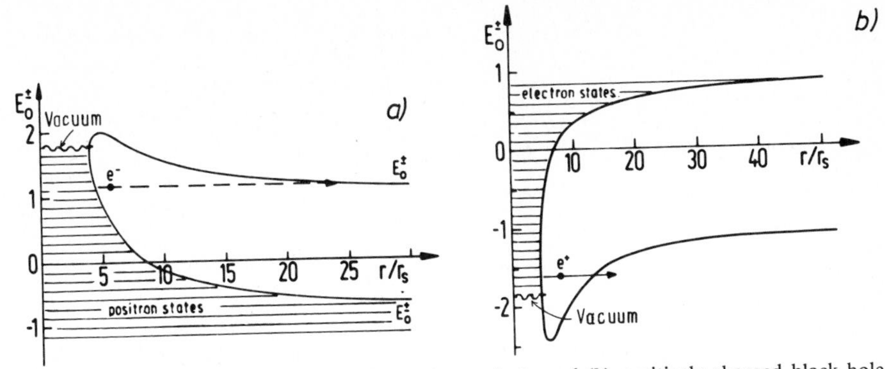

Fig. 21.6a, b. The effective potential for a **(a)** negatively and **(b)** positively charged black hole is shown. $E_0^{\pm}$ is in units of the particle mass m

from Fig. 21.6. Therefore, vacuum decay leads to a limiting charge-to-mass ratio of a black hole of 8×10^{-35} Cb/g. This limiting ratio can be expressed as $(Ze/M) = (2Gm^2/e^2)(e/m) = 0.5 \times 10^{42}(e/m)$, describing just the ratio between the gravitational ($\gamma = Gm/\hbar c$) and the electromagnetic coupling constant ($\alpha = e^2/\hbar c$).

For a macroscopic black hole, where the event horizon is much larger than the Compton wavelength of the produced particles, the discharge of the black hole due to pair formation may be described in the framework of Schwinger's formalism, Sect. 10.5. Since pair creation essentially occurs near the event horizon of the charged black hole, we consider the electrostatic field to be constant over the region of pair creation. The probability for pair formation per unit time interval and volume is then given by the imaginary part of the Heisenberg-Euler effective Lagrangian (10.155):

$$w \approx \alpha^2 E^2 \exp\left(-\frac{\pi m^2}{eE}\right), \tag{21.52}$$

where $E = Qr_+^{-2}$ denotes the field strength of the black hole in the vicinity of the event horizon. The exponent may be written as $-\pi mc^2/eE\lambda_c$, indicating that pair production occurs if a virtual pair gains energy equal to the rest mass from the Coulomb field over the distance of a Compton wavelength λ_c. The time evolution of the discharge process can be described by the charge loss equation

$$-\frac{\dot{Q}}{e} = -\alpha^2 E^2 \exp\left(\frac{-\pi m^2}{eE}\right) \cdot r_+^3, \tag{21.53}$$

leading to a characteristic time constant t_c for discharge of

$$t_c \approx \frac{r_+/c}{Qe\alpha^2} \exp\left[\left(\frac{\pi}{Qe}\right)\left(\frac{r_+}{\lambda_c}\right)^2\right]. \tag{21.54}$$

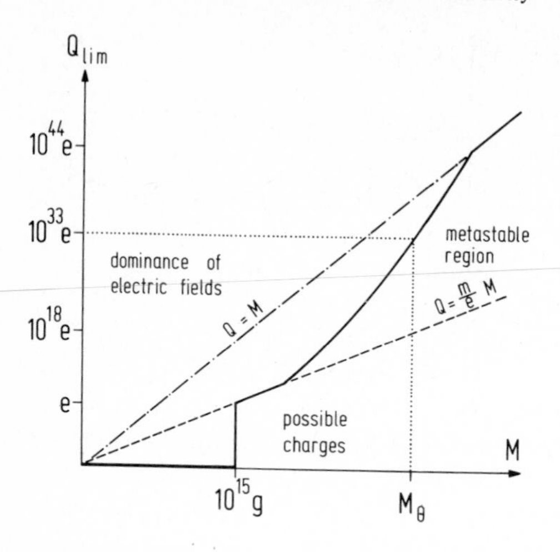

Fig. 21.7. Regime where a charged black hole spontaneously produces pairs of particles out of the vacuum. In principle, pair creation occurs if $Q > GMm/e$, but with observable likelihood only if $Q > (GmM)^2/e\hbar c$

For black holes of solar mass ($M = M_{\odot}$) decaying by electron–positron pair creation

$$(r_+/\lambda_c)^2 \sim 10^{31}\,, \qquad r_+/c \sim 5 \times 10^{-6}\,\mathrm{s}\,,$$

and thus

$$t_c \sim \frac{10}{Z^2} \exp\left(\frac{10^{33}}{Z}\right)$$

in seconds ($Q \equiv Ze$). We conclude that discharge effectively occurs only for $Q \gtrsim 10^{33}e$. In general, a charged black hole in the vacuum (effectively) discharges itself only as long as $Q > G^2m^2M^2/e\hbar c$.

Figure 21.7 shows the various regimes for pair formation in the vicinity of a charged black hole in a (M, Q) plot. Although pair formation is possible if $Q < 2GMm/e$, the process is efficient only when $Q > (m^2/e)(GM)^2$. For $Q > G^{1/2}M$ the event horizon disappears in the Reissner-Nordstrøm field and the field is considered to be unphysical.

It would be interesting to know whether such a discharge of a Reissner-Nordstrøm black hole is likely to occur in nature or not. Large quantities of excess charge should play no role in the universe; for $eQ > GMm$ the electromagnetic forces dominate the gravitational forces and a charged black hole will discharge itself by collecting particles with opposite charge from its surroundings. However, particle creation occurs only if $eQ > GMm$. Therefore, the probability that one observes the decay of the Dirac vacuum in a Reissner-Nordstrøm black hole in the universe is extremely small. Angular momentum of the black hole helps to lower the threshold of supercriticality, and sudden energy release ($\sim 10^{20}$ eV) due to pair formation in a Kerr-Newman geometry has been connected [De 77] with observed x-ray and γ-ray bursters like 3U 1820 – 30 situated in the globular cluster NGC 6624.

21.3 Uniform Acceleration and Rindler Space

One of the most intriguing phenomena of black hole physics is the breakdown of the gap between particle and antiparticle states encountered when the radius of a star approaches the Schwarzschild horizon r_s and that also showed up in the effective potential of the Schwarzschild metric (Fig. 21.4). To study its physical consequences it is useful to investigate the implications of the presence of an event horizon, i.e. a surface where $g_{00}(x) = 0$, on a "toy model". This model is supplied by the geometry seen by a uniformly accelerated observer in Minkowski space. In many respects it corresponds to the Heisenberg-Euler model of a constant electric field in QED, but the crucial difference is that a uniform acceleration field can be transformed away in general relativity by an appropriate coordinate transformation (this is just Einstein's principle of equivalence). Physically, a constant gravitational acceleration field would be produced by an infinite, planar mass distribution, just as a constant electric field is produced by a capacitor with infinitely extending charged plates, Fig. 21.8.

For simplicity, we restrict our discussion to one space and one time dimension in Minkowski space with coordinates z and t. The equations of motion of a uniformly accelerated observer may be derived from

$$\begin{aligned} u_\mu u^\mu &= -1 = (u^0)^2 - (u^3)^2 , \\ u^\mu a_\mu &= 0 = u^0 a^0 - u^3 a^3 , \\ a^\mu a_\mu &= g^2 = (a^0)^2 - (a^3)^2 , \end{aligned} \tag{21.55}$$

where $u^\mu = (dt/d\tau, dz/d\tau)$ is the four-velocity of the observer, a^μ is its four-acceleration and τ its proper time. Then g is the constant acceleration as measured in an instantaneous rest frame of the observer. When (21.55) are solved for a^0 and a^3

$$a^0 = g u^3 = \frac{du^0}{d\tau} , \qquad a^3 = g u^0 = \frac{du^3}{d\tau} , \tag{21.56}$$

the equations of motion arise

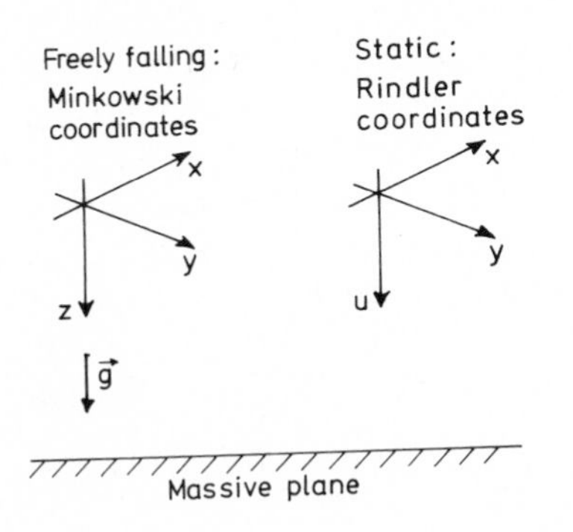

Fig. 21.8. Minkowski and Rindler coordinates in a uniform acceleration field

$$t = g^{-1} \sinh (g\tau), \qquad z = g^{-1} \cosh (g\tau). \tag{21.57}$$

Thus the uniformly accelerated observer moves on a hyperbola in Minkowski space (Fig. 21.9), the trajectory being given by

$$z^2 - t^2 = g^2. \tag{21.58}$$

As can be seen from Fig. 21.9, Minkowski space as seen by the accelerated observer is divided into four sectors: I (right), II (left), F (future) and P (past). One may define comoving coordinates (v, u) in the four sectors according to $(v = g\tau)$:

$$\left.\begin{array}{ll} t = u \sinh v & z = u \cosh v \\ v = \operatorname{artanh}(t/z) & u = \operatorname{sgn}(z)\sqrt{z^2 - t^2} \end{array}\right\} \text{ in I, II}$$
$$\left.\begin{array}{ll} t = u \cosh v & z = u \sinh v \\ v = \operatorname{artanh}(z/t) & u = \operatorname{sgn}(t)\sqrt{t^2 - z^2} \end{array}\right\} \text{ in F, P}. \tag{21.59}$$

The observer moves on a $u = \text{const}$ trajectory in sector I. Curves with $u = \text{const}$ are hyperbolae, those with $v = \text{const}$ are straight lines through the origin. Notice that in I and II the coordinate v is time-like, while u is the space-like coordinate, whereas it is the other way around in F and P. Sectors I and II are called *Rindler* space [Ri 66].

The Minkowski line element $ds^2 = dt^2 - dz^2$ in the new coordinates (v, u) is

$$\begin{aligned} ds^2 &= u^2 dv^2 - du^2 \quad \text{in} \quad \text{I, II}, \\ ds^2 &= du^2 - u^2 dv^2 \quad \text{in} \quad \text{F, P}. \end{aligned} \tag{21.60}$$

As v is the time-like coordinate in sectors I and II, the line $g_{00} = u^2 = 0$ defines the event horizon of the accelerated observer, e.g. the line $v = +\infty$, $u = 0$, separating sectors I and F, forms the borderline of the observable events, so the uniformly accelerated observer travelling in sector I cannot obtain information about any event in sector F of Minkowski space.

We now consider the Dirac equation (21.8)

$$\left[\mathrm{i}\gamma^\mu \left(\frac{\partial}{\partial x^\mu} + \Gamma_\mu\right) - m\right]\psi = 0 \tag{21.61}$$

in these four sectors of Minkowski space.

Beginning with sector I, the affine connection here is given by

$$\Gamma_\mu = (-\tfrac{1}{2}\tilde{\gamma}_0\tilde{\gamma}_3, 0) \tag{21.62}$$

and the Dirac equation takes the form

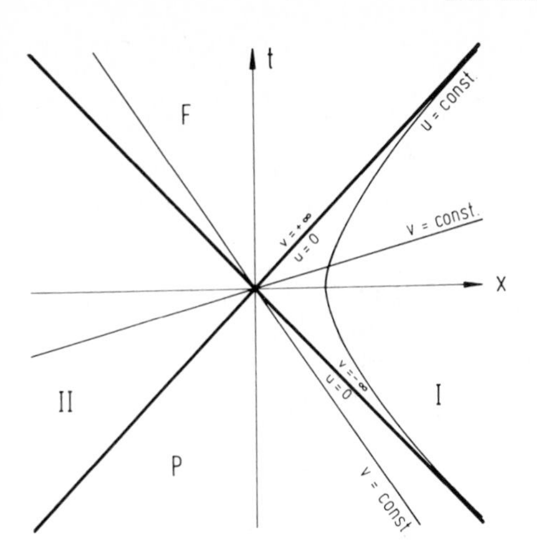

Fig. 21.9. Owing to the dynamics of a uniformly accelerated observer, two-dimensional Minkowski space is divided into four sectors: *right* (I), *left* (II), *future* (F) and *past* (P)

$$\mathrm{i}\frac{\partial}{\partial v}\psi = u\left[-\mathrm{i}\alpha_3\left(\frac{\partial}{\partial u}+\frac{1}{2u}\right)+\beta m\right]\psi \equiv \hat{H}_R\psi\,. \tag{21.63}$$

As the time-like coordinate v does not enter the operator on the rhs, we can search for stationary states in the Rindler coordinates by separating

$$\psi(v,u) = \mathrm{e}^{-\mathrm{i}\omega v}\psi(u)\,. \tag{21.64}$$

For spin-up states the only normalizable solution is

$$\psi(u) = \begin{pmatrix} \Phi_\omega^- + \Phi_\omega^+ \\ 0 \\ \Phi_\omega^- - \Phi_\omega^+ \\ 0 \end{pmatrix}, \qquad \Phi_\omega^\pm = H^{(1)}_{\mathrm{i}\omega\pm\frac{1}{2}}(\mathrm{i}mu)\,, \tag{21.65}$$

where ω can take all real values.

Note that the $\Phi_\omega^\pm$ satisfy the Bessel differential equation

$$\left(u\frac{d}{du}u\frac{d}{du}\right)\Phi_\omega^\pm(u) = \left(m^2u^2-\left(\omega\mp\frac{\mathrm{i}}{2}\right)^2\right)\Phi_\omega^\pm(u)\,, \tag{21.66}$$

which differs from the Klein-Gordon equation in Rindler space by the term $\pm\mathrm{i}/2$ from the affine bispinor connection Γ_μ. The norm of ψ_ω may be obtained by using (21.63):

$$\langle\psi_\omega|\psi_{\omega'}\rangle = \int\limits_0^\infty du\,\overset{+}{\psi}_\omega\psi_{\omega'} = \frac{\mathrm{i}u}{\omega-\omega'}\,\overset{+}{\psi}_\omega\alpha_3\psi_{\omega'}\Bigg|_0^\infty\,. \tag{21.67}$$

The surface term at infinity ($|u|\to\infty$) does not contribute and one obtains the orthogonality relation

$$\langle\psi_\omega|\psi_{\omega'}\rangle = \frac{8e^{\pi\omega}}{m\cosh\pi\omega}\,\delta(\omega-\omega')\,, \tag{21.68}$$

where $\sin(nx)/x \xrightarrow{n\to\infty} \pi\delta(x)$ was used. In the following the wave function ψ_ω will be given an index I to indicate that it was obtained as a solution of the Dirac equation in sector I.

As just shown, the spectrum of the Dirac operator in sector I covers all real numbers ω. The accelerated observer will classify particles and antiparticles according to his proper time τ, i.e. according to the sign of ω in (21.64).

Then the following picture is obtained. Owing to the dynamics the observer experiences an event horizon (the curve $u=0$ in Fig. 21.9). This leads to an effective red shift of $\sqrt{g_{00}}=u$ with $u\to 0$ (the event horizon) of all spectral lines an inertial observer would measure, (21.63). In other words, *the apparent event horizon breaks down the energy gap* between "particle" and "antiparticle" states for the accelerated observer.

In the future sector (F), u and v interchange their meaning as space- or time-like coordinates and the affine connection is given by

$$\Gamma_\mu = (0, -\tfrac{1}{2}\tilde{\gamma}_0\tilde{\gamma}_3)\,. \tag{21.69}$$

Therefore, the Dirac equation (21.61) takes the form

$$\omega\psi_\omega = \left[-\mathrm{i}\alpha_3 u^{1/2}\frac{\partial}{\partial u}u^{1/2} + \tilde{\gamma}_3 m\right]\psi_\omega\,. \tag{21.70}$$

The following two independent solutions arise for (21.70) and spin-up states:

$$^{\mathrm{F}}\psi_\omega^{(-)} = \mathrm{e}^{-\mathrm{i}\omega v}\begin{pmatrix}\varphi_\omega^- - \mathrm{i}\varphi_\omega^+\\ 0\\ \varphi_\omega^- + \mathrm{i}\varphi_\omega^+\\ 0\end{pmatrix}\,,\qquad \varphi_\omega^\pm = H^{(1)}_{i\omega\pm\frac{1}{2}}(mu)\,, \tag{21.71}$$

$$^{\mathrm{F}}\psi_\omega^{(+)} = \mathrm{e}^{-\mathrm{i}\omega v}\begin{pmatrix}\tilde{\varphi}_\omega^- - \mathrm{i}\tilde{\varphi}_\omega^+\\ 0\\ \tilde{\varphi}_\omega^- + \mathrm{i}\tilde{\varphi}_\omega^+\\ 0\end{pmatrix}\,,\qquad \tilde{\varphi}_\omega^\pm = H^{(2)}_{i\omega\pm\frac{1}{2}}(mu)\,. \tag{21.72}$$

Since v is space-like in sector F, ω has the physical meaning of a momentum. Therefore the sign of ω cannot be used to distinguish between particle and antiparticle states. Hence in F two independent normalizable wave functions can be found in contrast to the single wave function (21.65) in sector I. The normalization for $^{\mathrm{F}}\psi_\omega^{(\pm)}$ is found to be [So 80b]

$$\langle {}^{\mathrm{F}}\psi_\omega^{(i)} | {}^{\mathrm{F}}\psi_{\omega'}^{(k)}\rangle = \frac{16\exp(-k\pi\omega)}{m}\delta(\omega-\omega')\,\delta_{ik}\quad (i,k=\pm 1)\,. \tag{21.73}$$

The analogous wave function in regions II and P are similarly obtained as

$$^{\mathrm{II}}\psi_\omega(t,z) = \alpha_3\,{}^{\mathrm{I}}\psi_\omega(-t,-z)\,,\qquad {}^{\mathrm{P}}\psi_\omega^{(\pm)}(t,z) = \alpha_3\,{}^{\mathrm{F}}\psi_\omega^{(\mp)}(-t,-z)\,. \tag{21.74}$$

We now have a complete set of solutions of the Dirac equation in Rindler coordinates. In the following section these solutions become a basis for a second quantized theory of fermions in Rindler space, and we compare this theory with the corresponding theory in Minkowski space.

21.4 Event Horizon and Thermal Particle Spectrum

We may now build a quantum field theory on the whole Minkowski manifold by constructing an orthonormal basis of particle (antiparticle) wave functions $\psi_\omega^{(\pm)}$ of the Dirac equation in all Minkowski space by joining ${}^{\mathrm{I,II}}\psi_\omega$ and ${}^{\mathrm{F,P}}\psi_\omega^{(\pm)}$. In the four sectors $\psi_\omega^{(\pm)}$ may be expressed

$$\psi_\omega^+ = N(a_1{}^{\mathrm{I}}\psi_\omega + a_2{}^{\mathrm{F}}\psi_\omega^{(+)} + a_3{}^{\mathrm{P}}\psi_\omega^{(+)} + a_4{}^{\mathrm{II}}\psi_\omega)\,, \tag{21.75a}$$

$$\psi_\omega^- = N'(b_1{}^{\mathrm{I}}\psi_\omega + b_2{}^{\mathrm{F}}\psi_\omega^{(-)} + b_3{}^{\mathrm{P}}\psi_\omega^{(-)} + b_4{}^{\mathrm{II}}\psi_\omega)\,. \tag{21.75b}$$

The asymptotic expansion of the Hankel functions

$$H_\nu^{(1)}(z) \xrightarrow{|z|\to\infty} \sqrt{\frac{2}{\pi z}}\exp\left[\mathrm{i}\left(z - \frac{\nu\pi}{2} - \frac{\pi}{4}\right)\right]$$

gives for the function of (21.71) in sector F

$$\varphi_\omega^\pm \xrightarrow{u\to\infty} \sqrt{\frac{2}{\pi m u}}\,\mathrm{e}^{imu}\exp\left[-\mathrm{i}\left(\frac{\pi}{4} + \frac{\pi}{2}\left(\mathrm{i}\omega \pm \frac{1}{2}\right)\right)\right]. \tag{21.76}$$

For large times u, $\varphi_\omega^\pm$ behaves like a wave with frequency (energy) $-m$, thus ${}^{\mathrm{F}}\psi_\omega^{(-)}$ is part of an antiparticle mode. The constants $a_i(b_i)$ in (21.75) are determined by the requirement that $\psi_\omega^\pm$ should obey the wave equation in the full Minkowski space, including the light cone through the origin. The various wave functions ${}^{\mathrm{I,II}}\psi_\omega$, ${}^{\mathrm{F,P}}\psi_\omega$ are not solutions to Dirac's equation in all Minkowski space, but instead they possess distributions as source terms on the light cone through the origin. This is evident from the form

$$\begin{aligned} {}^{\mathrm{I}}\psi &= {}^{\mathrm{I}}\psi\,\Theta(z+t)\,\Theta(z-t)\,, & {}^{\mathrm{II}}\psi &= {}^{\mathrm{II}}\psi\,\Theta(-z-t)\,\Theta(t-z)\,, \\ {}^{\mathrm{F}}\psi &= {}^{\mathrm{F}}\psi\,\Theta(z+t)\,\Theta(t-z)\,, & {}^{\mathrm{P}}\psi &= {}^{\mathrm{P}}\psi\,\Theta(-z-t)\,\Theta(z-t)\,. \end{aligned} \tag{21.77}$$

The constants a_i, (b_i) in (21.75) can now be determined by requiring that source terms on the light cone through the origin are cancelled in the wave functions of the two adjoining sectors. The following orthonormal set of wave functions $\psi_\omega^{(\pm)}$ is thus obtained [So 80b]:

$$\psi_\omega^+ = \sqrt{\frac{m}{48}}\,({}^{\mathrm{I}}\psi_\omega - \mathrm{e}^{\pi\omega/2}\,{}^{\mathrm{F}}\psi_\omega^{(+)} + \mathrm{e}^{-\pi\omega/2}\,{}^{\mathrm{P}}\psi_\omega^{(+)} - \mathrm{i}\mathrm{e}^{-\pi\omega}\,{}^{\mathrm{II}}\psi_\omega)\,, \tag{21.78a}$$

$$\psi_\omega^- = \sqrt{\frac{m}{48}}\,(\mathrm{e}^{-\pi\omega}\,{}^{\mathrm{I}}\psi_\omega + \mathrm{e}^{-\pi\omega/2}\,{}^{\mathrm{F}}\psi_\omega^{(-)} + \mathrm{e}^{\pi\omega/2}\,{}^{\mathrm{P}}\psi_\omega^{(-)} + \mathrm{i}\,{}^{\mathrm{II}}\psi_\omega)\,. \tag{21.78b}$$

The usual quantum field theory of fermions in Minkowski space is obtained via the following spectral representation of the Dirac operator $\hat{\psi}$, defining the Minkowski vacuum $|0_{\mathrm{M}}\rangle$,

$$\hat{\psi} = \sum_{\mathrm{s}} \int_{-\infty}^{\infty} d\omega\,[\hat{a}_{\omega,\mathrm{s}}\,\psi^+_{\omega,\mathrm{s}} + \hat{b}^+_{\omega,\mathrm{s}}\,\psi^-_{\omega,\mathrm{s}}]\,, \tag{21.79a}$$

$$\hat{a}_{\omega,\mathrm{s}}|0_{\mathrm{M}}\rangle = \hat{b}_{\omega,\mathrm{s}}|0_{\mathrm{M}}\rangle = 0\,, \tag{21.79b}$$

where s sums over spin states.

Suppose a uniformly accelerated observer moves on a $u = \mathrm{const}$ trajectory in sector I. She can receive information from the past (sector P), send signals into the future (sector F) but she is completely decoupled from sector II. *Fulling, Unruh* and others have shown that the observer measures only wave functions causally connected with herself [Fu 73b, Un 76]. Her world is therefore described by the (uniquely determined) normalized wave functions ψ_ω of the Dirac equation in the sectors I, F and P of Minkowski space:

$$\psi_\omega = \frac{\sqrt{m}\,\mathrm{e}^{-\pi\omega/2}}{\sqrt{96\cosh\pi\omega}}\,[2\cosh(\pi\omega)\,{}^{\mathrm{I}}\psi_\omega + \mathrm{e}^{-\pi\omega/2}\,{}^{\mathrm{F}}\psi_\omega^{(-)} - \mathrm{e}^{3\pi\omega/2}\,{}^{\mathrm{F}}\psi_\omega^{(+)} + \mathrm{e}^{\pi\omega/2}\,{}^{\mathrm{P}}\psi_\omega^{(+)} + \mathrm{e}^{\pi\omega/2}\,{}^{\mathrm{P}}\psi_\omega^{(-)}]\,. \tag{21.80}$$

Note that ψ_ω does not contain ${}^{\mathrm{II}}\psi_\omega$ because the observer is decoupled from sector II. At this point the global aspect of the space-time manifold, that is naturally connected with the observer, plays a crucial role for the interpretation of $\hat{\psi}$. The observer modes ψ_ω are related to the Minkowski modes $\psi_\omega^{(\pm)}$ by a Bogoluibov transformation

$$\psi_\omega = \alpha_\omega\,\psi_\omega^+ + \beta_\omega\,\psi_\omega^- \tag{21.81}$$

with the coefficients

$$\alpha_\omega = \frac{\mathrm{e}^{\pi\omega/2}}{\sqrt{2\cosh\pi\omega}}\,, \qquad \beta_\omega = \frac{\mathrm{e}^{-\pi\omega/2}}{\sqrt{2\cosh\pi\omega}}\,. \tag{21.82}$$

The accelerated observer experiences a spectral representation of $\hat{\psi}$ according to positive and negative values of ω

$$\hat{\psi}_{\mathrm{B}} = \sum_{\mathrm{s}} \int_{0}^{\infty} d\omega\,[\hat{c}_{\omega,\mathrm{s}}\,\psi_{\omega,\mathrm{s}} + \hat{d}^+_{\omega,\mathrm{s}}\,\psi_{-\omega,\mathrm{s}}]\,, \tag{21.83}$$

defining the B vacuum by the conditions

$$\hat{c}_{\omega,\mathrm{s}}|0_{\mathrm{B}}\rangle = \hat{d}_{\omega,\mathrm{s}}|0_{\mathrm{B}}\rangle = 0 \qquad \text{for all } \omega \text{ and s}\,. \tag{21.84}$$

We can now calculate the number of particles seen by the accelerated observer in the ordinary Minkowski vacuum as

$$\langle \mathrm{O_M}|\hat{c}^{+}_{\omega,s}\hat{c}_{\omega',s'}|\mathrm{O_M}\rangle = \beta^2_\omega \delta(\omega-\omega')\delta_{ss'} \qquad (\omega,\omega'>0)\,. \tag{21.85}$$

Because of $\delta(\omega-\omega') = 1/2\pi \int \exp[\mathrm{i}(\omega-\omega')v]\,dv$, we interpret this result similarly as in Sect. 10.7, so that the rate of observed particles per unit interval of proper time is given by an integration over energies:

$$\frac{dN}{d\tau} = 2g\int_0^\infty \frac{d\omega}{2\pi}\beta^2_\omega = 2\int_0^\infty \frac{d\tilde{\omega}}{2\pi}[\exp(2\pi\omega/g)+1]^{-1}\,, \tag{21.86}$$

with $v = g\tau$, and $\tilde{\omega} = \omega g$ set as the physically measured energy variable of the observer. The factor 2 accounts for the two spin degrees of freedom. We conclude that an accelerated observer measures a thermal flux of Dirac particles, where the effective temperature is the Fulling-Unruh temperature

$$T_\mathrm{B} = \frac{g\hbar}{2\pi c k_\mathrm{B}} = 10^{-23} g\left[k\frac{\mathrm{s}^2}{\mathrm{cm}}\right], \tag{21.87}$$

where k_B is the Boltzmann constant, and we have inserted appropriate $\hbar$ and c factors.

It is remarkable that the fermion mass m does not enter into this result. This is due to the fact that the temperature is essentially determined at the event horizon ($u = 0$) where m plays no role. As shown, the energy gap between particle and antiparticle modes breaks down to the event horizon independent of m, (21.63).

In many respects, this result is reminiscent of the results obtained in Sect. 10.5 for pair production in a uniform electric field. There, in (10.156, 157) the probability for pair production per unit time and volume was given by

$$\frac{dN_{\mathrm{e^+e^-}}}{dt\,d^3x} = 2\,\mathrm{Im}\{L_\mathrm{eff}\} = \frac{m}{4\pi^2}\int_1^\infty ds\frac{\pi(s-1)}{\mathrm{e}^{\beta s}-1} = \frac{\alpha E^2}{\pi^2}\sum_{n=1}^\infty \frac{\mathrm{e}^{-n\beta}}{n^2}\,, \tag{21.88}$$

where $\beta = \pi m^2/eE$. It is suggestive to interpret $2ms$ as the energy of the pair and to define a "temperature" [Mü 77b]

$$k_\mathrm{B}T_\mathrm{E} = \frac{2m}{\beta} = \frac{eE}{2\pi m}\,, \tag{21.89}$$

so that (21.88) takes the form

$$\frac{dN_{\mathrm{e^+e^-}}}{dt\,d^3x} = \frac{m^2}{8\pi}\int_{2m}^\infty \frac{d\omega}{2\pi}\,\frac{(\omega-2m)}{\exp(\omega/k_\mathrm{B}T_\mathrm{E})-1}\,, \tag{21.90}$$

with $\omega = 2ms$, in complete analogy to (21.86) for the uniform acceleration field. The analogy goes even further, since eE/m is the acceleration of an electron or positron by the external electric field, so that (21.89) is analogous to (21.87). Of course, interpreting the parameter T_E as a temperature in the strict thermodynamic sense is precluded by the "wrong" sign in the denominator of the integral in (21.90).

In contrast, the distribution of particles given by (21.85), as experienced by a uniformly accelerated observer, is a thermal spectrum in the strictest sense. The particles are completely uncorrelated, and T_B in (21.87) is a true thermodynamic temperature. Topological methods show that this always occurs when the gravitational acceleration field leads to the existence of an event horizon [Ch 78, Sa 81, So 82c].

We now return to our discussion of the physical significance of the vanishing of the particle – antiparticle gap at the Schwarzschild radius r_s. To bring in our results for the uniform acceleration field, we write the line element of the Schwarzschild metric (21.11) with the help of the coordinate r^* defined in (21.42) in the form ($d\Theta = d\varphi = 0$):

$$\begin{aligned} ds^2 &= \left(1 - \frac{r_s}{r}\right) dt^2 - \left(1 - \frac{r_s}{r}\right)^{-1} dr^2 \\ &= \left(1 - \frac{r_s}{r}\right)(dt^2 - dr^{*2}) . \end{aligned} \tag{21.91}$$

This must be compared with the time element (21.60) in the Rindler sector II of Minkowski space

$$ds^2 = u^2 dv^2 - du^2 = u^2(dv^2 - du^{*2}) , \tag{21.92}$$

where $u^* = \ln u$. In the vicinity of the horizon, i.e. for $r \approx r_s$, with (21.42),

$$\left(1 - \frac{r_s}{r}\right) \approx \frac{r - r_s}{r_s} \approx \exp\left(\frac{r^*}{r_s}\right),$$

which must be compared with

$$u^2 = \exp(2u^*) .$$

The Schwarzschild horizon corresponds therefore to an acceleration

$$g_s = \frac{1}{2r_s} = \frac{1}{4MG} , \tag{21.93}$$

so that an observer will find that the black hole produces particles at a rate given by (21.86) with the temperature

$$T_H = \frac{g_s}{2\pi k_B} = \frac{1}{8\pi MG k_B} . \tag{21.94}$$

This result was first obtained by *Hawking* [Ha 74, 75] and later confirmed by many other calculations. The review article by *Davies* [Da 78] also discusses the thermodynamic implications of *Hawking's* result.

Thus the gravitational field becomes supercritical at the Schwarzschild radius of a black hole. The vacuum becomes unstable, and particles are produced with a thermal spectrum. As all particles interact gravitationally in the same way, this is valid for all known particles equally, but the production occurs most efficiently for massless particles, such as photons and neutrinos (for a black hole of solar mass $T_H \approx 10^{-7}$ K). By radiating off particles, the black hole would lose some of its mass, thereby increasing its temperature according to (21.94). What happens, when all its mass has been radiated away, and what the final state looks like, is an unsolved problem.

Bibliographical Notes

An introduction to general relativity can be found in [Ad 65, We 72, Mi 73a].

Schmutzer extensively treats the coupling of particle fields to a classical gravitational field [Schm 68]. The theory of quantum fields in curved space-time is set forth in the textbook of *Birrell* and *Davies* [Bi 82].

Various aspects of the stability and decay of the Dirac vacuum in general relativity are discussed in detail in the review article [So 82c].

References

Ab 65 M. Abramowitz, I. A. Stegun: *Handbook of Mathematical Functions* (Dover, New York 1965)

Ac 79 P. Achutan, T. Chandramohan, K. Venkatesan: J. Phys. **A12**, 2521 (1979)

Ad 65 R. Adler, M. Bazin, M. Schiffer: *Introduction to General Relativity* (Mc Graw-Hill, New York 1965)

Ad 70 S. L. Adler, J. N. Bahcall, C. G. Callan, M. N. Rosenbluth: Phys. Rev. Lett. **25**, 1061 (1970)

Ad 81 S. Adler: Phys. Rev. **D23**, 2905 (1981)

Ag 77 D. Agassi, C. M. Ko, H. A. Weidenmüller: Ann. Phys. (N.Y.) **107**, 140 (1977)

Ag 79 D. Agassi, C. M. Ko, H. A. Weidenmüller: Ann. Phys. (N.Y.) **117**, 407 (1979)

Ai 82 I. J. R. Aitchison, A. J. G. Hey: *Gauge Theories in Particle Physics* (Hilger, Bristol 1982)

Al 60 E. Almquist, D. A. Bromley, J. A. Kuehner: Phys. Rev. Lett. **4**, 515 (1960)

Am 80 J. Ambjorn, P. Olesen: Nucl. Phys. **B170**, 60 (1980)

An 78a R. Anholt: Z. Phys. **A288**, 257 (1978)

An 78b R. Anholt, W. E. Meyerhof: Phys. Lett. **64A**, 381 (1978)

An 84 R. Anholt: "X-ray and bremsstrahlung production in nuclear reactions", in *Atomic Inner-Shell Physics*, ed. by B. Crasemann (Plenum, New York 1984, in print)

Ar 78 P. Armbruster, H. H. Behncke, S. Hagmann, D. Liesen, F. Folkmann, P. H. Mokler: Z. Phys. **A288**, 277 (1978)

Ar 79 P. Armbruster, P. Kienle: Z. Phys. **A291**, 399 (1979)

Ba 53 D. R. Bates, H. S. W. Massey, A. L. Stewart: Proc. R. Soc. **A216**, 437 (1953)

Ba 58 D. R. Bates, R. McCarroll: Proc. R. Soc. **A245**, 175 (1958)

Ba 59 J. Bang, J. M. Hansteen: K. Dan. Vidensk. Selsk. Mat. Fys. Medd. **31**, No. 13 (1959)

Ba 75 M. Bawin, J. P. Lavine: Phys. Rev. **D12**, 1192 (1975)

Ba 77 I. A. Batalin, S. G. Matinyan, G. K. Savvidy: Sov. J. Nucl. Phys. **26**, 214 (1977)

Ba 78 H. Backe, L. Handschug, F. Hessberger, E. Kankeleit, L. Richter, F. Weik, R. Willwater, H. Bokemeyer, P. Vincent, Y. Nakayama, J. S. Greenberg: Phys. Rev. Lett. **40**, 1443 (1978)

Ba 79a H. Backe: In *Trends in Physics 1978* (Hilger, Bristol 1979) p. 445

Ba 79b J. Bang, J. M. Hansteen: Phys. Lett. **72A**, 218 (1979)

Ba 79c M. Bawin, J.-P. Lavine: Lett. Nuovo Cim. **26**, 586 (1979)

Ba 80a A. Balanda, H. J. Beeskow, K. Bethge, H. Bokemeyer, H. Folger, J. S. Greenberg, H. Grein, A. Gruppe, S. Ito, S. Matsuki, R. Schulé, R. Schulz, D. Schwalm, J. Schweppe, R. Steiner, P. Vincent, M. Waldschmidt: GSI Scientific Report **80-3** (Darmstadt 1980) p. 161

Ba 80b A. O. Barut: *Foundations of Radiation Theory and Quantum Electrodynamics* (Plenum, New York 1980)

Ba 81a M. Baker, J. S. Ball, F. Zachariasen: Nucl. Phys. **B186**, 531, 560 (1981)

Ba 81b J. Bang, J. M. Hansteen: Phys. Scr. **22**, 609 (1981)

Ba 81c M. Bawin, J. Cugnon: Phys. Lett. **107B**, 257 (1981)

Ba 83a H. Backe, W. Bonin, E. Kankeleit, M. Krämer, R. Krieg, V. Metag, P. Senger, N. Trautmann, F. Weik, J. B. Wilhelmy: In [Gr 83a] p. 107

Ba 83b H. Backe, P. Senger, W. Bonin, E. Kankeleit, M. Krämer, R. Krieg, V. Metag, N. Trautmann, J. B. Wilhelmy: Phys. Rev. Lett. **50**, 1838 (1983)

Ba 84 H. Backe, B. Müller: In *Atomic Inner-Shell Physics*, ed. by B. Crasemann (Plenum, New York 1984, in print)

Be 21 G. Bertrand: Comp. Rend. **172**, 1458 (1921)
Be 34 H. A. Bethe, W. Heitler: Proc. R. Soc. **A146**, 83 (1934)
Be 47 H. A. Bethe: Phys. Rev. **72**, 339 (1947)
Be 63 F. Beck, H. Steinwedel, G. Süssmann: Z. Phys. **171**, 189 (1963)
Be 71 V. B. Berestetskii, E. M. Lifshitz, L. P. Piatevskii: *Relativistic Quantum Theory* (Pergamon, Oxford 1971)
Be 74 J. Bernstein: Rev. Mod. Phys. **46**, 7 (1974)
Be 75 W. Betz, G. Heiligenthal, J. Reinhardt, R. K. Smith, W. Greiner: In *The Physics of Electronic and Atomic Collisions*, ed. by J. S. Risley, R. Geballe (Univ. of Washington Press, Seattle 1975) p. 531
Be 76a W. Betz: Diplomarbeit, Frankfurt University (1976)
Be 76b W. Betz, G. Soff, B. Müller, W. Greiner: Phys. Rev. Lett. **37**, 1046 (1976)
Be 77 G. Bertsch, J. Borysowicz, H. McManus, W. G. Love: Nucl. Phys. **A284**, 399 (1977)
Be 78 H. H. Behncke, D. Liesen, S. Hagmann, P. H. Mokler, P. Armbruster: Z. Phys. **A288**, 35 (1978)
Be 79 H. H. Behncke, P. Armbruster, F. Folkmann, S. Hagmann, J. R. MacDonald, P. Mokler: Z. Phys. **A289**, 333 (1979)
Be 80a E. Berdermann, F. Bosch, M. Clemente, F. Güttner, P. Kienle, W. Koenig, C. Kozhuharov, B. Martin, W. Potzel, E. Povh, C. Tsertos, W. Wagner, T. Walcher: GSI Scientific Report **80-3**, p. 103
Be 80b W. Betz: Dissertation, Frankfurt University (1980)
Be 81 E. Berdermann, F. Bosch, M. Clemente, F. Güttner, P. Kienle, W. Koenig, C. Kozhuharov, B. Martin, B. Povh, H. Tsertos, W. Wagner, T. Walcher: GSI Scientific Report **81-2**, p. 128
Be 82 E. Berdermann, F. Bosch, M. Clemente, P. Kienle, W. Koenig, C. Kozhuharov, H. Tsertos, W. Wagner: GSI Scientific Report **82-1**, p. 138
Bi 70 Z. Bialynicka-Birula, I. Bialynicki-Birula: Phys. Rev. **D2**, 2341 (1970)
Bi 75 I. Bialynicki-Birula, Z. Bialynicka-Birula: *Quantum Electrodynamics* (Pergamon, Oxford 1975)
Bi 82 N. D. Birrell, P. C. W. Davies: *Quantum Fields in Curved Space* (Cambridge University Press, Cambridge 1982)
Bj 64 J. D. Bjorken, S. D. Drell: *Relativistic Quantum Mechanics* (McGraw-Hill, New York 1964)
Bj 65 J. D. Bjorken, S. D. Drell: *Relativistic Quantum Field Theory* (McGraw-Hill, New York 1965)
Bl 82 J. S. Blair, R. Anholt: Phys. Rev. **A25**, 907 (1982)
Bo 51 D. Bohm: *Quantum Theory* (Prentice-Hall, New York 1951) pp. 257 – 261
Bo 59 N. N. Bogoliubov, D. V. Shirkov: *Introduction to the Theory of Quantized Fields* (Wiley-Interscience, New York 1959)
Bo 67 P. N. Bogolioubov: Ann. Inst. Henri Poincaré **8**, 163 (1967)
Bo 78 F. Bosch, H. Krimm, B. Martin, B. Povh, T. Walcher, K. Traxel: Phys. Lett. **78B**, 568 (1978)
Bo 80 F. Bosch, D. Liesen, P. Armbruster, D. Maor, P. H. Mokler, H. Schmidt-Böcking, R. Schuch: Z. Phys. **A296**, 11 (1980)
Bo 81 H. Bokemeyer, H. Folger, H. Grein, S. Ito, D. Schwalm, P. Vincent, K. Bethge, A. Gruppe, R. Schulé, M. Waldschmidt, J. S. Greenberg, J. Schweppe, N. Trautmann: GSI Scientific Report **81-2**, p. 127
Bo 82 F. Bosch: Phys. Bl. **38**, 205 (1982)
Bo 83a H. Bokemeyer, K. Bethge, H. Folger, J. S. Greenberg, H. Grein, A. Gruppe, S. Ito, R. Schulé, D. Schwalm, J. Schweppe, N. Trautmann, P. Vincent, M. Waldschmidt: In [Gr 83a], p. 273
Bo 83b E. Bozek, U. Gollerthan, E. Kankeleit, G. Klotz, M. Krämer, R. Krieg, U. Meyer, H. Oeschler, P. Senger: GSI Scientific Report **81-1**, p. 148
Bo 84 H. Bokemeyer, H. Folger, H. Grein, Y. Kido, T. Cowan, J. S. Greenberg, J. Schweppe, K. Bethge, A. Gruppe, D. Schwalm, P. Vincent, R. Fonte, H. Backe, M. Begemann, M. Klüner, N. Trautmann: "On the origin of the structure in positron spectra from U + Cm collisions", GSI Annual Report (1983)
Br 59a G. E. Brown: Rev. Mod. Phys. **31**, 893 (1969)
Br 59b G. E. Brown, J. S. Langer, G. W. Schaefer: Proc. R. Soc. London, Ser. **A251**, 92 (1959)

Br 70a B. H. Bransden: *Atomic Collision Theory* (Benjamin, New York 1970)
Br 70b E. Brezin, C. Itzykson: Phys. Rev. **D2**, 1191 (1970)
Br 71 E. Brezin, C. Itzykson: Phys. Rev. **D3**, 618 (1971)
Br 74 D. A. Bromley: In *Reactions between Complex Nuclei*, Vol. 2, ed. by R. L. Robinson, F. K. McGowan, J. B. Bell, J. H. Hamilton (North-Holland, Amsterdam 1974) p. 603
Br 76 G. E. Brown, W. Weise: Phys. Rep. **27C**, 2 (1976)
Br 78 S. J. Brodsky, P. J. Mohr: In *Structure and Collisions of Ions and Atoms*, ed. by I. A. Sellin (Springer, Berlin, Heidelberg, New York 1978) p. 3
Bu 66 G. A. Burginyon, J. S. Greenberg: Nucl. Inst. Meth. **41**, 109 (1966)
Bu 74 D. Burch, P. Ingalls, H. Wieman, R. Vandenbosch: Phys. Rev. **A10**, 1245 (1974)

Ca 48 H. B. G. Casimir: Proc. K. Ned. Akad. Wet. **51**, 793 (1948)
Ca 50 K. M. Case: Phys. Rev. **80**, 797 (1950)
Ca 83 L. J. Carson, R. Goldflam, L. Wilets: Phys. Rev. **D28**, 385 (1983)
Ch 68 H. Y. Chiu, V. Canuto, I. Canuto-Fassio: Phys. Rev. **176**, 1438 (1968)
Ch 71 D. Christodoulou, R. Ruffini: Phys. Rev. **D4**, 3552 (1971)
Ch 74a M. S. Child: *Molecular Collision Theory* (Academic, New York 1974)
Ch 74b A. Chodos, R. L. Jaffe, K. Johnson, C. B. Thorn, V. F. Weisskopf: Phys. Rev. **D9**, 3471 (1974)
Ch 76 K. T. Cheng, W. R. Johnson: Phys. Rev. **A14**, 1943 (1976)
Ch 78 S. Christensen, M. Duff: Nucl. Phys. **B146**, 11 (1978)
Ch 79 C. B. Chiu, S. Nussinov: Phys. Rev. **D20**, 945 (1979)
Ci 65 G. Ciocchetti, A. Molinari: Nuovo Cimento **29**, 1262 (1965)
Ci 81 N. Cindro, R. A. Ricci, W. Greiner (eds.): *Dynamics of Heavy-Ion Collisions* (North-Holland, Amsterdam 1981)
Cl 83 M. Clemente: Dissertation, Technical University, Munich (1983)
Cl 84 M. Clemente, E. Berdermann, P. Kienle, H. Tsertos, W. Wagner, C. Kozhuharov, F. Bosch, W. Koenig: Phys. Lett. **137B**, 41 (1984)
Co 34 W. M. Coates: Phys. Rev. **46**, 542 (1934)
Co 67 C. A. Coulson, A. Joseph: J. Quant. Chem. **1**, 337 (1967)
Co 73 S. Coleman, E. Weinberg: Phys. Rev. **D7**, 1888 (1973)
Co 85 T. Cowan, H. Backe, M. Begemann, K. Bethge, H. Bokemeyer, H. Folger, J. S. Greenberg, H. Grein, A. Gruppe, Y. Kido, M. Klüver, D. Schwalm, J. Schweppe, K. E. Stiebing, N. Trautmann, P. Vincent: Phys. Rev. Lett. **54**, 1761 (1985)

Da 65 M. Danos, W. Greiner: Phys. Rev. **B138**, 876 (1965)
Da 72 M. Danos, W. Greiner, J. Rafelski: Phys. Rev. **D6**, 3476 (1972)
Da 74a M. Danos, J. Rafelski: Lett. Nuovo Cimento **10**, 106 (1974)
Da 74b C. K. Davies, J. S. Greenberg: Phys. Rev. Lett. **32**, 1215 (1974)
Da 78 P. C. W. Davies: Rep. Prog. Phys. **41**, 1314 (1978)
Da 79 M. Danos, J. Rafelski: Nuovo Cimento **49A**, 326 (1979)
De 53 M. Demeur: Acad. R. Belg. Cl. Sci. Mem. Collect. **28**, 1643 (1953)
De 56 M. Deutsch, J. S. Greenberg: Phys. Rev. **102**, 415 (1956)
De 71 A. M. Desiderio, W. R. Johnson: Phys. Rev. **A3**, 1267 (1971)
De 74 N. Deruelle, R. Ruffini: Phys. Lett. **52B**, 437 (1974)
De 77 N. Deruelle: In *Proceedings of the First Marcel Grossmann Meeting on General Relativity*, ed. by R. Ruffini (North-Holland, Amsterdam 1977)
deD 26 T. de Donder, H. van Dungen: C. R. (July 1926)
deG 75 T. deGrand, R. L. Jaffe, K. Johnson, J. Kiskis: Phys. Rev. **D12**, 2060 (1975)
deR 81 T. de Reus: Diploma Thesis, Franfurt University (1981);
see also G. Soff, T. de Reus, U. Müller, J. Reinhardt, P. Schlüter, K. H. Wietschorke, B. Müller, W. Greiner: "Spectroscopy of superheavy quasimolecules", in [Gr 83a] p. 233
deR 84 T. de Reus, J. Reinhardt, B. Müller, W. Greiner, G. Soff, U. Müller: J. Phys. **B17**, 615 (1984)
DeW 67 B. S. De Witt: Phys. Rev. **162**, 1195, 1239 (1967)
DeW 75 B. S. De Witt: Phys. Rep. **19**, 295 (1975)

Di 28 P. A. M. Dirac: Proc. R. Soc. **A117**, 610 (1928)
Di 30a P. A. M. Dirac: Proc. R. Soc. **A126**, 360 (1930)
Di 30b P. A. M. Dirac: Proc. Cambridge Philos. Soc. **26**, 361 (1930)
Di 76 W. Dittrich: J. Phys. **A9**, 1171 (1976)
Do 71 H. G. Dosch, J. H. D. Jensen, V. F. Müller: Phys. Norv. **5**, 2 (1971)
Dy 49 F. J. Dyson: Phys. Rev. **75**, 486, 1736 (1949)
Dy 67 F. S. Dyson: J. Math. Phys. **8**, 1538 (1967)

Ei 48 L. Eisenbud: Dissertation, Princeton University (1948)
Ei 60 R. M. Eisberg, D. R. Yennie, D. H. Wilkinson: Nucl. Phys. **18**, 338 (1960)
Ei 70 J. M. Eisenberg, W. Greiner: *Nuclear Theory,* Vol. 2, (North-Holland, Amsterdam 1970)
Er 49 H. A. Erickson, E. L. Hill: Phys. Rev. **75**, 29 (1949)

Fa 61 U. Fano: Phys. Rev. **124**, 1866 (1961)
Fa 65 U. Fano, W. Lichten: Phys. Rev. Lett. **14**, 627 (1965)
Fa 67 L. D. Faddeev, V. N. Popov: Phys. Lett. **25B**, 291 (1967)
Fä 83 A. Fäßler: In [Gr 83a] p. 701
Fe 32 E. Fermi: Rev. Mod. Phys. **4**, 87 (1932)
Fe 39 R. P. Feynman: Phys. Rev. **56**, 340 (1939)
Fe 48 R. P. Feynman: Phys. Rev. **74**, 939 (1948)
Fe 58 H. Feshbach, F. Villars: Rev. Mod. Phys. **30**, 24 (1958)
Fe 62 H. Feshbach: Ann. Phys. (N.Y.) **19**, 287 (1962)
Fe 65 R. P. Feynman, A. R. Hibbs: *Quantum Mechanics and Path Integrals* (McGraw-Hill, New York 1965)
Fe 71 A. L. Fetter, J. D. Walecka: *Quantum Theory of Many-Particle Systems* (McGraw-Hill, New York 1971)
Fe 81 J. M. Feagin, L. Kocbach: J. Phys. **B14**, 4349 (1981)
Fi 80 J. Finger, J. E. Mandula, J. Weyers: Phys. Lett. **96B**, 367 (1980)
Fl 80 H. Flyvvbjerg: Nucl. Phys. **B176**, 379 (1980)
Fl 84 M. Flensburg, C. Peterson, L. Sköld: Z. Phys. **C22**, 293 (1984)
Fo 26 V. Fock: Z. Phys. **38**, 242 (1926);
ibid **39**, 226 (1926)
Fo 29a V. Fock: Z. Phys. **57**, 261 (1929)
Fo 29b V. Fock, D. Iwanenko: Phys. Z. **30**, 648 (1929)
Fo 76 C. Foster, T. P. Hooghamer, P. Woerlee, F. W. Saris: J. Phys. **B9**, 1943 (1976)
Fr 52 J. Friedel: Philos. Mag. **43**, 153 (1952)
Fr 65 N. Fröman, P. O. Fröman: *JWKB Approximation* (North-Holland, Amsterdam 1965)
Fr 72 H. Fritzsch, M. Gell-Mann: Proc. 16th Int. Conf. on High-Energy Physics, Chicago 1972, Vol. 2
Fr 76a W. Frank, P. Gippner, K. H. Kaun, P. Manfraß, Y. P. Tretyakov: Z. Phys. **A277**, 333 (1976)
Fr 76b B. Fricke, T. Morovic, W. D. Sepp, A. Rosen, D. E. Ellis: Phys. Lett. **59A**, 375 (1976)
Fu 37 W. H. Furry: Phys. Rev. **51**, 125 (1937)
Fu 51 W. H. Furry: Phys. Rev. **81**, 115 (1951)
Fu 73a L. Fulcher, A. Klein: Phys. Rev. **D8**, 2455 (1973)
Fu 73b S. Fulling: Phys. Rev. **D7**, 2850 (1973)
Fu 74 L. Fulcher, A. Klein: Ann. Phys. (N.Y.) **84**, 335 (1974)
Fu 79 L. P. Fulcher, J. Rafelski, A. Klein: Sci. Am. **241**, 150 (1979)
Fu 80 R. Fukuda: Phys. Rev. **D21**, 485 (1980)

Ga 74 S. Gasiorowicz: *Quantum Physics* (Wiley, New York 1974)
Gä 79 P. Gärtner: Diploma Thesis, Frankfurt University (1979)
Gä 80 P. Gärtner, B. Müller, J. Reinhardt, W. Greiner: Phys. Lett. **95B**, 181 (1980)
Gä 81 P. Gärtner, U. Heinz, B. Müller, W. Greiner: Z. Phys. **A300**, 143 (1981)
Ge 26 C. Gerthsen: Z. Phys. **36**, 540 (1926)

Ge 51 M. Gell-Mann, F. Low: Phys. Rev. **84**, 350 (1951)
Ge 60 M. Gell-Mann, H. Lévy: Nuovo Cimento **16**, 705 (1960)
Ge 62 M. Gell-Mann: Phys. Rev. **125**, 1067 (1962)
Ge 70 S. S. Gershtein, Y. B. Zeldovich: Sov. Phys. JETP **30**, 358 (1970)
Gi 74 P. Gippner, K. H. Kaun, F. Stary, W. Schulze, Y. P. Tretyakov: Nucl. Phys. **A230**, 509 (1974)
Go 26 H. Gordon: Z. Phys. **40**, 117 (1926)
Go 55 R. H. Good: Rev. Mod. Phys. **27**, 187 (1955)
Go 71 L. Gomberoff, V. Tolmachev: Phys. Rev. **D3**, 1796 (1971)
Go 80 A. Gobbi, W. Nörenberg: In *Heavy Ion Collisions*, Vol. 2, ed. by R. Bock (North-Holland, Amsterdam 1980) p. 127
Go 81a T. J. Goldman, R. W. Haymaker: Phys. Lett. **100B**, 276 (1981)
Go 81b J. Goldstone, F. Wilczek: Phys. Rev. Lett. **49**, 986 (1981)
Go 82 R. Goldflam, L. Wilets: Phys. Rev. **D25**, 1951 (1982)
Go 83 J. Goldstone, R. L. Jaffe: Phys. Rev. Lett. **51**, 1518 (1983)
Gr 65 I. S. Gradshteyn, I. W. Ryzhik: *Table of Integrals, Series and Products* (Academic, New York 1965)
Gr 69a W. Greiner: Panel Discussion, Proc. Int. Conf. on the Properties of Nuclear States, Montreal (1969)
Gr 69b W. Greiner: Discussion at the Welsh-Foundation Conference (Organizer: G. T. Seaborg) Houston, Texas (1969)
Gr 73 D. J. Gross, F. Wilczek: Phys. Rev. Lett. **30**, 1343 (1973); Phys. Rev. **D8**, 3633 (1973)
Gr 74 J. S. Greenberg, C. K. Davies, P. Vincent: Phys. Rev. Lett. **33**, 473 (1974)
Gr 76 M. Gros, B. Müller, W. Greiner: J. Phys. **B9**, 1849 (1976)
Gr 77 J. S. Greenberg, H. Bokemeyer, E. Emling, E. Grosse, D. Schwalm, F. Bosch: Phys. Rev. Lett. **39**, 1404 (1977)
Gr 79a W. Greiner: *Vorlesungen über Theoretische Physik V: Quantenmechanik II – Symmetrien* (Harri Deutsch, Frankfurt 1979)
Gr 79b E. K. V. Gross, R. M. Dreizler: Phys. Rev. **A20**, 1798 (1979)
Gr 80 W. Greiner, J. Hamilton: Am. Sci. **68**, 145 (1980)
Gr 81a T. A. Green: Phys. Rev. **A23**, 519, 532 (1981)
Gr 81b W. Greiner: *Vorlesungen über Theoretische Physik VI: Relativistische Wellengleichungen* (Harri Deutsch, Frankfurt 1981)
Gr 82a J. S. Greenberg, W. Greiner: Phys. Today **35**, 24 (Aug. 1982)
Gr 82b J. S. Greenberg: "In search of spontaneous positron creation", in: *X-Ray and Atomic Inner-Shell Physics – 1982*, ed. B. Crasemann (American Institute of Physics, New York 1982) p. 173
Gr 83a W. Greiner (ed.): *Quantum Electrodynamics of Strong Fields*, Proc. of NATO Advanced Study Institute at Lahnstein 1983 (Plenum, New York 1983)
Gr 83b J. S. Greenberg: In [Gr 83a] p. 853
Gr 83c W. Greiner: The Decay of the Vacuum in Supercritical Fields of Giant Nuclear Systems, Proc. Int. Conf. on Nuclear Physics at Florence (1983)
Gr 83d E. K. V. Gross, R. M. Dreizler: In [Gr 83a] p. 383
Gr 83e B. Grossman: Phys. Rev. Lett. **50**, 464 (1983); ibid. **51**, 959 (1983)
Gr 84a W. Greiner, B. Müller: *Quantenmechanik II: Symmetrien* (Harri Deutsch, Frankfurt 1984)
Gr 84b W. Greiner, J. Reinhardt: *Vorlesungen über Theoretische Physik VII: Quantenelektrodynamik* (Harri Deutsch, Frankfurt 1984)
Gr 85 J. S. Greenberg, P. Vincent: "High energy atomic physics (experiment)", in *Treatise on Heavy Ion Science*, Vol. 5, ed. by D. A. Bromley (Plenum, New York, in print)
Gu 62 P. C. Gugelot: In *Direct Reactions and Nuclear Reaction Mechanisms*, ed. by E. Clementel, C. Villi (Gordon and Breach, New York 1962) p. 382
Gü 82 F. Güttner, W. Koenig, B. Martin, B. Povh, H. Skapa, J. Soltani, T. Walcher, F. Bosch, C. Kozhuharov: Z. Phys. **A304**, 207 (1982)
Gy 75 M. Gyulassy: Nucl. Phys. **A244**, 497 (1975)

Ha 74 S. Hawking: Nature **248**, 30 (1974)
Ha 75 S. Hawking: Commun. Math. Phys. **43**, 199 (1975)
Ha 77 C. K. Hargrove, E. P. Hincks, R. J. McKee, H. Mes, A. L. Carter, M. S. Dixit, D. Kessler, J. S. Wadden, H. L. Anderson, A. Zehnder: Phys. Rev. Lett. **39**, 307 (1977)
Ha 78 P. Hasenfratz, J. Kuti: Phys. Rep. **40**, 75 (1978)
Ha 81a A. Hansen, F. Ravndal: Phys. Scr. **23**, 1036 (1981)
Ha 81b P. Hasenfratz, R. R. Horgan, J. Kuti, J. M. Richard: Phys. Scr. **23**, 917 (1981)
Ha 82 T. H. Hansson, K. Johnson, C. Peterson: Phys. Rev. **D26**, 2069 (1982)
Ha 83 N. D. Hari Dass: In [Gr 83a] p. 783
He 27 W. Heitler, F. London: Z. Phys. **44**, 455 (1927)
He 36 W. Heisenberg, H. Euler: Z. Phys. **98**, 718 (1936)
He 37 H. Hellmann: *Einführung in die Quantenchemie* (Deuticke, Leipzig 1937)
He 46 W. Heisenberg: Z. Naturforsch. **1**, 608 (1946)
He 81a U. Heinz, B. Müller, W. Greiner: Phys. Rev. **A23**, 562 (1981)
He 81b H. Herold, H. Ruder, G. Wunner: J. Phys. **B14**, 751 (1981)
He 83a U. Heinz, B. Müller, W. Greiner: Ann. Phys. (N.Y.) **151**, 227 (1983)
He 83b U. Heinz, J. Reinhardt, B. Müller, W. Greiner, U. Müller: Z. Phys. **A314**, 125 (1983)
He 84a U. Heinz, U. Müller, J. Reinhardt, B. Müller, W. Greiner: Ann. Phys. (N.Y.) **158**, 476 (1984)
He 84b U. Heinz, J. Reinhardt, B. Müller, W. Greiner, W. T. Pinkston: Z. Phys. **316**, 341 (1984)
He 84c P. O. Hess, W. Greiner, W. T. Pinkston: Phys. Rev. Lett. **53**, 1535 (1984); Il Nuovo Cim. **83A**, 76 (1984)
He 84d M. A. Herath-Banda: Private communication
Hi 53 D. L. Hill, J. A. Wheeler: Phys. Rev. **89**, 1102 (1953)
Ho 69 P. Holzer, U. Mosel, W. Greiner: Nucl. Phys. **A138**, 241 (1969)
Hu 27 F. Hund: Z. Phys. **40**, 742 (1927)
Hu 41 F. Hund: Z. Phys. **117**, 1 (1941)
Hu 54 F. Hund: *Materie als Feld* (Springer, Berlin,Heidelberg, New York 1954)
Hu 78 R. J. Hughes: Nucl. Phys. **B161**, 156 (1978)
Hy 31 F. A. Hylleraas: Z. Phys. **71**, 739 (1931)

Ia 79 E. Iacopini, E. Zavattini: Phys. Lett. **85B**, 151 (1979)
Im 69 B. Imanishi: Nucl. Phys. **A125**, 33 (1969)
It 80 C. Itzykson, J. B. Zuber: *Quantum Field Theory* (McGraw-Hill, New York 1980)
It 82 S. Ito, P. Armbruster, H. Bokemeyer, F. Bosch, H. Emling, H. Folger, E. Grosse, R. Külessa, D. Liesen, D. Maar, D. Schwalm, J. S. Greenberg, R. Schulé, N. Trautmann: In GSI Scientific Report **82-1**, Darmstadt (1982) p. 141

Ja 55 J. M. Jauch, F. Rohrlich: *The Theory of Photons and Electrons* (Springer, Berlin, Heidelberg, New York 1955, new ed. 1976)
Ja 69 B. Jancovici: Phys. Rev. **187**, 2275 (1969)
Ja 76a R. Jackiw, C. Rebbi: Phys. Rev. **D13**, 986 (1976)
Ja 76b D. Jakubassa: Phys. Lett. **58A**, 163 (1976)
Ja 81 R. Jackiw, J. R. Schrieffer: Nucl. Phys. **B190**, 253 (1981)
Ja 82 M. Jacob, H. Satz (eds.): *Quark Matter Formation and Heavy Ion Collisions* (World Scientific, Singapore 1982)
Jo 76 K. Johnson, C. B. Thorn: Phys. Rev. **D13**, 1934 (1976)

Ka 80 E. Kankeleit: Nukleonika **25**, 253 (1980)
Ka 81a E. Kankeleit, R. Köhler, M. Kollatz, M. Krämer, R. Krieg, P. Senger, H. Backe: GSI Scientific Report **81-2**, P. 195
Ka 81b J. I. Kapusta: Nucl. Phys. **B190**, 425 (1981)
Kä 58 G. Källén: *Quantenelektrodynamik*, Handbuch der Physik V/1 (Springer, Berlin, Heidelberg, New York 1958)
Ki 78 J. Kirsch, W. Betz, J. Reinhardt, G. Soff, B. Müller, W. Greiner: Phys. Lett. **72B**, 298 (1978)

Ki 80 J. Kirsch, B. Müller, W. Greiner: Z. Naturforsch. **35a**, 579 (1980)
Ki 83 P. Kienle: In [Gr 83a] p. 293
Kl 26 O. Klein: Z. Phys. **37**, 895 (1926)
Kl 29 O. Klein: Z. Phys. **53**, 157 (1929); see also ibid **41**, 407 (1927)
Kl 75a A. Klein, J. Rafelski: Phys. Rev. **D11**, 300 (1975)
Kl 75b A. Klein, J. Rafelski: Phys. Rev. **D12**, 1194 (1975)
Kl 77 A. Klein, J. Rafelski: Z. Phys. **A284**, 71 (1977)
Kl 80 M. Klaus: Preprint, Dept. Mathematics, University of Virginia, Charlottesville, USA (1980)
Ko 75 J. Kogut, L. Susskind: Phys. Rev. Lett. **34**, 767 (1975)
Ko 77 C. Kozhuharov, P. Kienle, D. Jakubaßa, M. Kleber: Phys. Rev. Lett. **39**, 540 (1977)
Ko 78 D. Kolb, W. D. Sepp, B. Fricke, T. Morovic: Z. Phys. **A286**, 169 (1978)
Ko 79a C. M. Ko, D. Agassi, H. A. Weidenmüller: Ann. Phys. (N.Y.) **117**, 237 (1979)
Ko 79b C. Kozhuharov, P. Kienle, E. Berdermann, H. Bokemeyer, J. S. Greenberg, Y. Nakayama, P. Vincent, H. Backe, L. Handschug, E. Kankeleit: Phys. Rev. Lett. **42**, 237 (1979)
Ko 82 C. Kozhuharov: In *Physics of Electronic and Atomic Collisions*, ed. by S. Datz (North-Holland, Amsterdam 1982) p. 179
Kr 74 G. Kraft, P. H. Mokler, H. J. Stein: Phys. Rev. Lett. **33**, 475 (1974)
Ku 26 J. Kudor: Ann. Phys. **81**, 632 (1926)

La 37 R. E. Langer: Phys. Rev. **51**, 669 (1937)
La 79a The relevant characteristics of the ANTARES laser fusion project at Los Alamos can be found, e.g., in publication LASL-79-29 (10^{14} W on a spot of less than 1 mm diameter)
La 79b P. Langacker, H. Pagels: Phys. Rev. **D19**, 2070 (1979)
La 80 M. A. Lampert, R. S. Crandall: Phys. Rev. **A21**, 362 (1980)
Le 49 N. Levinson: K. Dan. Vidensk. Selsk. Mat. Fys. Medd. **25**, No. 9 (1949)
Le 74 T. D. Lee, G. C. Wick: Phys. Rev. **D9**, 2291 (1974)
Le 81 T. D. Lee: *Particle Physics and Introduction to Field Theory* (Chur, Harwood 1981)
Li 73 E. M. Lifschitz, L. P. Pitajewski: *Lehrbuch der Theoretischen Physik*, Band 4b, Relativistische Quantenfeldtheorie, § 110 (Akademie-Verlag, Berlin 1973)
Li 77 V. I. Lisin, M. S. Marinov, V. S. Popov: Phys. Lett. **69B**, 141 (1977)
Li 78 D. Liesen, P. Armbruster, H. H. Behncke, S. Hagmann: Z. Phys. **A288**, 417 (1978)
Li 80 D. Liesen, P. Armbruster, F. Bosch, S. Hagman, P. H. Mokler, H. J. Wollersheim, H. Schmidt-Boecking, R. Schuch, J. B. Wilhelmy: Phys. Rev. Lett. **44**, 983 (1980)
Li 82 D. Liesen: Comments At. Mol. Phys. **12**, 39 (1982)
Lo 62 G. J. Lockwood, E. Everhart: Phys. Rev. **125**, 567 (1962)
Lu 68 D. Lurié: *Particles and Fields* (Wiley, New York 1968)

Ma 33 H. S. W. Massey, R. A. Smith: Proc. R. Soc. **A142**, 142 (1933)
Ma 71 J. B. Mann, W. R. Johnson: Phys. Rev. **A4**, 41 (1971)
Ma 72 J. A. Maruhn, W. Greiner: Z. Phys. **251**, 431 (1972)
Ma 74 M. S. Marinov, V. S. Popov, V. L. Stolin: JETP Lett. **19**, 49 (1974)
Ma 75a M. S. Marinov, V. S. Popov, V. L. Stolin: Sov. Phys. JETP **41**, 205 (1975)
Ma 75b M. S. Marinov, V. S. Popov, V. L. Stolin: J. Comput. Phys. **19**, 241 (1975)
Ma 76 J. E. Mandula: Phys. Rev. **D14**, 3497 (1976)
Ma 77a J. E. Mandula: Phys. Lett. **67B**, 175 (1977)
Ma 77b J. E. Mandula: Phys. Lett. **69B**, 495 (1977)
Ma 77c M. S. Marinov, V. S. Popov: Fortschr. Phys. **25**, 373 (1977)
Ma 78a J. R. MacDonald, P. Armbruster, H. H. Behncke, F. Folkmann, S. Hagmann, D. Liesen, P. H. Mokler, A. Warczak: Z. Phys. **A284**, 57 (1978)
Ma 78b M. Magg: Phys. Lett. **74B**, 246 (1978)
Ma 78c A. N. Makhlin, V. P. Oleinik: Sov. J. Nucl. Phys. **27**, 867 (1978)
Ma 79a M. Magg: Nucl. Phys. **B158**, 154 (1979)
Ma 79b C. Mahaux, H. A. Weidenmüller: Ann. Rev. Nucl. Part. Sci. **29**, 1 (1979)
Ma 79c S. G. Mamayev, N. N. Frolov: Sov. J. Nucl. Phys. **30**, 677 (1979)

Ma 85 Z. Q. Ma, G. J. Ni: Phys. Rev. **D31**, 1482 (1985)
McV 82 K. W. McVoy, H. A. Weidenmüller: Phys. Rev. **A25**, 1462 (1982)
Me 66 A. Messiah: *Quantum Mechanics* (North-Holland, Amsterdam 1966)
Me 67 H. Meldner: Ark. Fys. **36**, No. 1 – 77, Paper No. 66, 593 (1967)
Me 70 E. Merzbacher: *Quantum Mechanics*, 2nd ed. (Wiley, New York 1970) p. 107
Me 71 R. Mehlig: Dissertation, Frankfurt University (1971)
Me 73a W. E. Meyerhof: Phys. Rev. Lett. **31**, 1341 (1973)
Me 73b W. E. Meyerhof, T. K. Saylor, S. M. Lazarus, W. A. Little, B. B. Triplett, L. F. Chase: Phys. Rev. Lett. **30**, 1279 (1973)
Me 74 W. E. Meyerhof: Phys. Rev. **A10**, 1005 (1974)
Me 75 W. E. Meyerhof, T. K. Saylor, R. Anholt: Phys. Rev. **A12**, 2641 (1975)
Me 77 W. E. Meyerhof, R. Anholt, Y. El Masri, D. Cline, F. S. Stephens, R. Diamond: Phys. Lett. **B69**, 41 (1977)
Me 79 W. E. Meyerhof, D. L. Clark, C. Stoller, E. Morenzoni, W. Wölfli, F. Folkmann, P. Vincent, P. H. Mokler, P. Armbruster: Phys. Lett. **70A**, 303 (1979)
Me 84a G. Mehler, T. de Reus, J. Reinhardt, G. Soff, U. Müller: Z. Phys. **A320**, 355 (1985)
Me 84b W. E. Meyerhof, J. F. Chemin: "Nuclear-reaction effects on atomic inner-shell ionization", in *Advances in Atomic and Molecular Physics*, ed. by D. R. Bates, B. Bedersen (Academic, New York 1984, in print)
Mi 72 A. B. Migdal: Sov. Phys. JETP **34**, 1184 (1972)
Mi 73a C. W. Misner, K. S. Thorne, J. A. Wheeler: *Gravitation* (Freeman, San Francisco 1973)
Mi 73b M. H. Mittleman, H. Tai: Phys. Rev. **A8**, 1880 (1973)
Mi 73c A. B. Migdal: Nucl. Phys. **A210**, 421 (1973)
Mi 76 A. B. Migdal: Sov. Phys. JETP **43**, 211 (1976)
Mi 77 A. B. Migdal, V. S. Popov, D. N. Voskresenskij: Sov. Phys. JETP **45**, 436 (1977)
Mi 78 A. B. Migdal: Pis'ma Zh. Eksp. Tear. Fiz. **28**, 37 (1978)
Mo 31 N. F. Mott: Proc. Cambridge Philos. Soc. **27**, 553 (1931)
Mo 33 N. F. Mott, H. S. W. Massey: *Theory of Atomic Collisions* (Oxford University Press, Oxford 1933)
Mo 65 E. Moll, E. Kankeleit: Nukleonik **7**, 180 (1965)
Mo 68 U. Mosel, W. Greiner: Z. Phys. **217**, 256 (1968)
Mo 69 U. Mosel, W. Greiner: Z. Phys. **222**, 261 (1969)
Mo 72 P. H. Mokler, H. J. Stein, P. Armbruster: Phys. Rev. Lett. **29**, 827 (1972)
Mo 74 P. J. Mohr: Ann. Phys. (N.Y.) **88**, 26, 52 (1974)
Mo 77 P. J. Moffa, C. B. Dover, J. P. Vary: Phys. Rev. **C16**, 1857 (1977)
Mo 79 V. M. Mostepanenko: Phys. Lett. **75A**, 11 (1979)
Mo 80 L. G. Moretto, R. P. Schmitt: Phys. Rev. **C21**, 204 (1980)
Mo 83 J. Morishita, M. Oka, M. Kaburagi, H. Munakata, T. Kitazoe: Z. Phys. **C19**, 167 (1983)
Mo 84 P. H. Mokler, D. Liesen: "X-rays from superheavy collision systems", in *Progress in Atomic Spectroscopy, Part C*, ed. by H. J. Beyer, H. Kleinpoppen (Plenum, New York 1984) p. 321
Mø 45 C. Møller, K. Dan. Vidensk. Selsk. Mat. Fys. Medd. **23**, No. 1 (1945)
Mu 28 R. S. Mulliken: Phys. Rev. **32**, 186 (1928)
Mu 78 V. D. Mur, V. S. Popov, D. N. Voskresenskij: JETP Lett. **28**, 129 (1978)
Mü 72a B. Müller, H. Peitz, J. Rafelski, W. Greiner: Phys. Rev. Lett. **28**, 1235 (1972)
Mü 72b B. Müller, J. Rafelski, W. Greiner: Z. Phys. **257**, 62 (1972)
Mü 72c B. Müller, J. Rafelski, W. Greiner: Z. Phys. **257**, 183 (1972)
Mü 73a B. Müller, J. Rafelski, W. Greiner: Phys. Lett. **47B**, 5 (1973)
Mü 73b B. Müller, J. Rafelski, W. Greiner: Nuovo Cimento **18A**, 551 (1973)
Mü 74a B. Müller, W. Greiner: Phys. Rev. Lett. **33**, 469 (1974)
Mü 74b B. Müller, R. K. Smith, W. Greiner: Phys. Lett. **49B**, 219 (1974)
Mü 75 B. Müller, J. Rafelski: Phys. Rev. Lett. **34**, 349 (1975)
Mü 76a B. Müller: Ann. Rev. Nucl. Sci. **26**, 351 (1976)
Mü 76b B. Müller, W. Greiner: Z. Naturforsch. **31a**, 1 (1976)
Mü 77a B. Müller, W. Greiner: Acta Phys. Austr. (Suppl.) XVIII, 163 (1977)
Mü 77b B. Müller, W. Greiner, J. Rafelski: Phys. Lett. **63A**, 181 (1977)

Mü 78 B. Müller, G. Soff, W. Greiner, V. Ceausescu: Z. Phys. **A285**, 27 (1978)
Mü 81 B. Müller, J. Rafelski: Phys. Lett. **101B**, 111 (1981)
Mü 82 B. Müller, J. Rafelski: Phys. Rev. **D25**, 566 (1982)
Mü 83a B. Müller: In [Gr 83a] p. 41
Mü 83b B. Müller, J. Reinhardt, W. Greiner, G. Soff: Z. Phys. **A311**, 151 (1983)
Mü 83c U. Müller, G. Soff, T. de Reus, J. Reinhardt, B. Müller, W. Greiner: Z. Phys. **A313**, 263 (1983)
Mü 85 B. Müller: *The Physics of the Quark-Gluon Plasma*, Lecture Notes in Physics, Vol. 225 (Springer, Berlin, Heidelberg 1985)

Ne 29 J. von Neumann, E. Wigner: Z. Phys. **30**, 467 (1929)
Ne 54 R. G. Newton: Phys. Rev. **96**, 523 (1954)
Ne 71 R. G. Newton: Phys. Rev. **D3**, 626 (1971)
Ni 65 J. R. Nix, W. J. Swiatecki: Nucl. Phys. **71**, 1 (1965)
Ni 69a S. G. Nilsson, S. G. Thompson, C. F. Tsang: Phys. Lett. **B28**, 458 (1969)
Ni 69b K. Nishijima: *Fields and Particles* (Benjamin, Reading 1969)
Ni 70 A. I. Nikishov: Nucl. Phys. **B21**, 346 (1970)
Ni 78 N. K. Nielsen, P. Olesen: Nucl. Phys. **B144**, 376 (1978); Phys. Lett. **79B**, 304 (1978)
Ni 79a H. B. Nielsen, M. Ninomiya: Nucl. Phys. **B156**, 1 (1979)
Ni 79b H. B. Nielsen, P. Olesen: Nucl. Phys. **B160**, 380 (1979)
Ni 83 A. J. Niemi, G. W. Semenoff: Phys. Rev. Lett. **51**, 2077 (1983)
No 18 E. Noether: Nachr. Akad. Wiss. Göttingen, Math.-Phys. Kl. 2a (Math.-Phys.-Chem. Abt.) (1918) p. 235

Ob 76 V. Oberacker, G. Soff, W. Greiner: Nucl. Phys. **A259**, 324 (1976)
O'Br 80 J. J. O'Brian, E. Liarokapis, J. S. Greenberg: Phys. Rev. Lett. **44**, 386 (1980)
O'Co 68 R. F. O'Connell: Phys. Rev. Lett. **21**, 397 (1968)
Op 30 R. Oppenheimer: Phys. Rev. **35**, 939 (1930)
Or 77 V. N. Oraevskii, A. I. Rex, V. B. Semikoz: Sov. Phys. JETP **45**, 428 (1977)
Os 82 E. Oset, H. Toki, W. Weise: Phys. Rep. **83**, 281 (1982)

Pa 34 W. Pauli, V. F. Weisskopf: Helv. Phys. Acta **1**, 709 (1934)
Pa 56 A. Papapetrou: Ann. Phys. (Leipzig) **17**, 214 (1956)
Pa 72 V. O. Papanyan, V. I. Ritus: Sov. Phys. JETP **34**, 1195 (1972)
Pa 78 H. Pagels, E. Tomboulis: Nucl. Phys. **B143**, 485 (1978)
Pa 80 L. Parker: Phys. Rev. Lett. **44**, 1559 (1980); Phys. Rev. **D22**, 1922 (1980)
Pa 81 J. Y. Park, W. Scheid, W. Greiner: In [Ci 81] p. 53
Pa 83 M. B. Paranjape, G. W. Semenoff: Phys. Lett. **132B**, 369 (1983)
Pa 84 J. Y. Park, W. Scheid, W. Greiner: Rep. Prog. Phys., in print
Pi 69 W. Pieper, W. Greiner: Z. Phys. **218**, 327 (1969)
Pl 83 G. Plunien: Diploma Thesis, Frankfurt University (1983), to be published in Physics Reports
Po 10 H. Poincaré: Leçons de Méchanique Céleste, Vol. III, Chapt. X, Paris (1910)
Po 45 I. Pomeranchuk, J. Smorodinsky: J. Phys. (Moscow) **9**, 97 (1945)
Po 71 V. S. Popov: Sov. Phys. JETP **32**, 526 (1971)
Po 73a V. S. Popov: Sov. J. Nucl. Phys. **17**, 322 (1973)
Po 73b V. S. Popov, M. S. Marinov: Sov. J. Nucl. Phys. **16**, 449 (1973)
Pr 63 G. W. Pratt, Jr.: Rev. Mod. Phys. **35**, 502 (1963)
Pr 70 K. Prüß, W. Greiner: Phys. Lett. **33B**, 197 (1970)

Ra 71 J. Rafelski, L. P. Fulcher, W. Greiner: Phys. Rev. Lett. **27**, 958 (1971); see also Nuovo Cimento **B7**, 137 (1972)
Ra 72 J. Rafelski, B. Müller, W. Greiner: Lett. Nuovo Cimento **4**, 469 (1972)
Ra 74 J. Rafelski, B. Müller, W. Greiner: Nucl. Phys. **B68**, 585 (1974)

Ra 76a J. Rafelski, B. Müller: Argonne Nat. Lab. Phys. Div. Internal Report (1976)
Ra 76b J. Rafelski, B. Müller: Phys. Rev. Lett. **36**, 517 (1976)
Ra 76c J. Rafelski, B. Müller: Phys. Lett. **65B**, 205 (1976)
Ra 76d J. Rafelski, B. Müller: Phys. Rev. **D8**, 3532 (1976)
Ra 78a J. Rafelski: Phys. Lett. **79B**, 419 (1978)
Ra 78b J. Rafelski, L. P. Fulcher, A. Klein: Phys. Rep. **38C**, 228 (1978)
Ra 78c J. Rafelski, B. Müller, W. Greiner: Z. Phys. **A285**, 49 (1978)
Ra 78d J. Rafelski, in: *Nonlinear Equations in Physics and Mathematics*, ed. A. O. Barnt, (Reidel, Dordrecht 1978) p. 399
Ra 81 P. Ramond: *Field Theory* (Benjamin/Cummings, Reading 1981)
Ra 83 J. Rafelski: "Particle condensates in strongly coupled quantum field theory", in [Gr 83a] p. 539
Re 70 P. G. Reinhard: Lett. Nuovo Cimento **3**, 313 (1970)
Re 71 P. G. Reinhard, W. Greiner, H. Arenhövel: Nucl. Phys. **A166**, 173 (1971)
Re 73 J. F. Reading: Phys. Rev. **A8**, 3262 (1973)
Re 76a J. Reinhardt, W. Greiner: Phys. Unserer Zeit **7**, 171 (1976)
Re 76b J. Reinhardt, G. Soff, W. Greiner: Z. Phys. **A276**, 285 (1976); see also J. Reinhardt: Diploma Thesis, Institut für Theoretische Physik, Frankfurt University (1975)
Re 77 J. Reinhardt, W. Greiner: Rep. Progr. Phys. **40**, 219 (1977)
Re 79 J. Reinhardt, B. Müller, W. Greiner, G. Soff: Phys. Rev. Lett. **43**, 1307 (1979)
Re 80a J. F. Reading, A. L. Ford: Phys. Rev. **A21**, 124 (1980)
Re 80b J. Reinhardt, B. Müller, W. Greiner: In *Coherence and Correlation in Atomic Collisions*, ed. by H. Kleinpoppen, J. F. Williams (Plenum, New York 1980) p. 331
Re 81a J. Reinhardt, U. Müller, B. Müller, W. Greiner: Z. Phys. **A303**, 173 (1981)
Re 81b J. Reinhardt, B. Müller, W. Greiner: Phys. Rev. **A24**, 103 (1981)
Re 83 J. Reinhardt, B. Müller, W. Greiner, U. Müller: Phys. Rev. **A28**, 2558 (1983)
Re 84 J. Reinhardt: "Heavy ion atomic physics (theory)", in *Treatise on Heavy Ion Science*, Vol. 5, ed. by D. A. Bromley (Plenum, New York, in print)
Re 85 J. Reinhardt, W. Greiner: Phys. Bl. **41**, 38 (1985)
Rh 83 M. J. Rhoades-Brown, V. E. Oberacker, M. Seiwert, W. Greiner: Z. Phys. **A310**, 287 (1983)
Ri 66 W. Rindler: Am. J. Phys. **34**, 1174 (1966)
Ri 75 G. A. Rinker, L. Wilets: Phys. Rev. **A12**, 748 (1975)
Ri 78a R. D. Richtmyer: *Principles of Advanced Mathematical Physics, Vol. 1,* Texts and Monographs in Physics (Springer, Berlin, Heidelberg, New York 1978)
Ri 78b T. H. Rihan, N. S. Aly, E. Merzbacher, B. Müller, W. Greiner: Z. Phys. **A285**, 397 (1978)
Ro 60 P. Roman: *Theory of Elementary Particles* (North-Holland, Amsterdam 1960)
Ro 61 M. E. Rose: *Relativistic Electron Theory* (Wiley, New York 1961)
Ro 69 P. Roman: *Introduction to Quantum Field Theory* (Wiley, New York 1969)
Ro 82 M. Roos et al.: Phys. Lett. **111B**, 1 (1982)
Ru 52 M. Rudkjøbling: K. Dan. Vidensk. Selsk. Mat. Fys. Medd. **27**, 5 (1952)
Ru 76 H. Rumpf: Phys. Lett. **61B**, 272 (1976)
Ru 78 H. Rumpf, H. Urbantke: Ann. Phys. (N.Y.) **114**, 332 (1978)
Ru 79 H. Rumpf: Gen. Rel. Grav. **10**, 509, 525, 647 (1979)

Sa 31 F. Sauter: Z. Phys. **69**, 742 (1931); ibid **73**, 547 (1931)
Sa 72 F. W. Saris, W. F. van der Weg, H. Tawara, W. A. Laubert: Phys. Rev. Lett. **28**, 717 (1972)
Sa 73 R. F. Sawyer, D. J. Scalapino: Phys. Rev. **D7**, 953 (1973)
Sa 77 G. K. Savvidy: Phys. Lett. **71B**, 133 (1977)
Sa 79 G. R. Satchler, W. G. Love: Phys. Rep. **55**, 183 (1979)
Sa 81 N. Sanchez: Phys. Rev. **D24**, 2100 (1981)
Sa 82 H. Satz: Phys. Rep. **88**, 349 (1982)
Sc 79 M. D. Scadron: *Advanced Quantum Theory*, Texts and Monographs in Physics (Springer, Berlin, Heidelberg, New York 1979)

Scha 84 A. Schäfer, B. Müller, W. Greiner: Phys. Lett. **149B**, 455 (1984)
Scha 85 A. Schäfer, J. Reinhardt, B. Müller, W. Greiner, G. Soff: J. Phys. G (in print)
Sche 70 W. Scheid, W. Greiner, R. Lemmer: Phys. Rev. Lett. **25**, 176 (1970)
Schi 40 L. I. Schiff, H. Snyder, J. Weinberg: Phys. Rev. **57**, 315 (1940)
Schl 81 P. Schlüter, G. Soff, W. Greiner: Phys. Rep. **75**, 327 (1981)
Schl 83 P. Schlüter, T. de Reus, J. Reinhardt, B. Müller, G. Soff: Z. Phys. **A314**, 297 (1983)
Schl 84a P. Schlüter: Ph. D. Thesis, Frankfurt University (1984)
Schl 84b P. Schlüter, G. Soff, K. H. Wietschorke, W. Greiner: J. Phys. **B18**, 1685 (1985)
Schm 68 E. Schmutzer: *Relativistische Physik* (Teubner, Leipzig 1968)
Schm 78 R. Schmidt, V. D. Toneev, G. Wolschin: Nucl. Phys. **A311**, 247 (1978)
Schn 69 S. B. Schneidermann, A. Russek: Phys. Rev. **181**, 311 (1969)
Schr 26 E. Schrödinger: Ann. Phys. **81**, 109 (1926)
Schr 77 W. U. Schröder, J. R. Huizenga: Ann. Rev. Nucl. Sci. **27**, 465 (1977)
Schr 85 S. Schramm, J. Reinhardt, B. Müller, W. Greiner: Z. Phys. A (to be published)
Schw 48 J. Schwinger: Phys. Rev. **73**, 4162 (1948)
Schw 49 J. Schwinger: Phys. Rev. **74**, 1439 (1949)
Schw 51 J. Schwinger: Phys. Rev. **82**, 664 (1951)
Schw 53 J. Schwinger: Phys. Rev. **91**, 713 (1953)
Schw 54 J. Schwinger: Phys. Rev. **94**, 1362 (1954)
Schw 58 J. Schwinger (ed.): *Quantum Electrodynamics* (Dover, New York 1958)
Schw 61 S. S. Schweber: *Relativistic Quantum Field Theory* (Harper and Row, New York 1961)
Schw 62 J. Schwinger: Phys. Rev. **128**, 2425 (1962)
Schw 73 J. Schwinger: *Particles, Sources and Fields*, Vol. 2, (Addison-Wesley, Reading 1973)
Schw 83 J. Schweppe, A. Gruppe, K. Bethge, H. Bokemeyer, T. Cowan, H. Folger, J. S. Greenberg, H. Grein, S. Ito, R. Schulé, D. Schwalm, K. E. Stiebing, N. Trautmann, P. Vincent, M. Waldschmidt: Phys. Rev. Lett. **51**, 2261 (1983)
Schw 84 D. Schwalm: In *Electronic and Atomic Collisions* ed. by J. Eichler, I. V. Hertel, N. Stolteroht (Elsevier, 1984) p. 295
Se 84 M. Seiwert, W. Greiner, T. W. Pinkston: J. Phys. **G11**, L21 (1985)
Sl 53 L. A. Sliv: Sov. Phys. JETP **25**, 7 (1953); J. Phys. Rad. **16**, 589 (1953)
Sl 60 L. J. Slater: *Confluent Hypergeometric Functions* (Cambridge University Press, Cambridge 1960)
Sm 60 F. T. Smith: Phys. Rev. **118**, 349 (1960)
Sm 74 K. Smith, H. Peitz, B. Müller, W. Greiner: Phys. Rev. Lett. **32**, 554 (1974)
Sm 75 R. K. Smith, B. Müller, W.Greiner: J. Phys. **B8**, 75 (1975)
Sn 40 H. Synder, J. Weinberg: Phys. Rev. **57**, 307 (1940)
So 73a G. Soff, B. Müller, J. Rafelski, W. Greiner: Z. Naturforsch. **28a**, 1389 (1973)
So 73b G. Soff, J. Rafelski, W. Greiner: Phys. Rev. **A7**, 903 (1973)
So 77a G. Soff, J. Reinhardt, B. Müller, W. Greiner: Phys. Rev. Lett. **38**, 592 (1977)
So 77b M. Soffel, B. Müller, W. Greiner: J. Phys. A: Math. Nucl. Gen. **10**, 551 (1977)
So 78a G. Soff, B. Müller, W. Greiner, E. Merzbacher: Phys. Lett. **65A**, 19 (1978)
So 78b G. Soff, B. Müller, W. Greiner: Phys. Rev. Lett. **40**, 540 (1978)
So 79a G. Soff, W. Greiner, W. Betz, B. Müller: Phys. Rev. **A20**, 169 (1979)
So 79b G. Soff, J. Reinhardt, B. Müller, W. Greiner: Phys. Rev. Lett. **43**, 1981 (1979)
So 79c M. Soffel, B. Müller, W. Greiner: Phys. Lett. **70A**, 167 (1979)
So 80a G. Soff, J. Reinhardt, B. Müller, W. Greiner: Z. Phys. **A294**, 137 (1980)
So 80b M. Soffel, B. Müller, W. Greiner: Phys. Rev. **D22**, 1935 (1980)
So 81a G. Soff, B. Müller, W. Greiner: Z. Phys. **A299**, 189 (1981)
So 81b G. Soff, J. Reinhardt, W. Greiner: Phys. Rev. **A23**, 701 (1981)
So 82a G. Soff, T. de Reus, B. Müller, W. Greiner: Phys. Lett. **88A**, 398 (1982)
So 82b G. Soff, P. Schlüter, B. Müller, W. Greiner: Phys. Rev. Lett. **48**, 1465 (1982)
So 82c M. Soffel, B. Müller, W. Greiner: Phys. Rep. **85**, 51 (1982)
Sp 82 B. L. Spokoiny: Sov. J. Nucl. Phys. **36**, 277 (1982)
St 41 E. C. G. Stückelberg: Helv. Phys. Acta **14**, 588 (1941)

St 64 R. F. Streater, A. S. Wightman: *PCT, Spin and Statistics, and All That* (Benjamin, New York 1964)
St 77 C. Stoller, W. Wölfli, G. Bonani, M. Stöckli, M. Suter: J. Phys. **B10**, L347 (1977)
St 78 C. Stoller, W. Wölfli, G. Bonani, E. Morenzoni, M. Stöckli: Z. Phys. **A287**, 33 (1978)
St 81 P. M. Stevenson: Phys. Rev. **D23**, 2916 (1981)
St 84a J. D. Stack: Phys. Rev. **D29**, 1213 (1984)
St 84b C. Stoller, M. Nessi, E. Morenzoni, W. Wölfli, W. E. Meyerhof, J. D. Molitoris, E. Grosse, C. Michel: Phys. Rev. Lett. **53**, 1329 (1984)

Ta 30 I. Tamm: Z. Phys. **62**, 7 (1930)
Te 30 E. Teller: Z. Phys. **61**, 458 (1930)
Th 65 W. R. Thorson: J. Chem. Phys. **42**, 3878 (1965)
tH 71 G. t'Hooft: Nucl. Phys. **B33**, 173 (1971); ibid **B35**, 167 (1971)
Th 78 W. R. Thorson, J. B. Delos: Phys. Rev. **A18**, 117, 135 (1978)
Th 79 J. Theis, J. Reinhardt, B. Müller: J. Phys. **B12**, L479 (1979)
Th 80 S. Théberge, A. W. Thomas, G. A. Miller: Phys. Rev. **D22**, 2838 (1980)
Th 81 J. Theis, B. Müller: Phys. Rev. **A24**, 89 (1981)
Th 84 A. W. Thomas: Adv. Nucl. Part. Phys. **13**, 1 (1984)
To 46 S., Tomonaga: Prog. Theor. Phys. **1**, 27 (1946)
To 83 T. Tomoda, H. A. Weidenmüller: Phys. Rev. **C28**, 739 (1983)
To 84 T. Tomoda: Phys. Rev. **A29**, 536 (1984)
Tr 75 V. Trimble: Rev. Mod. Phys. **47**, 877 (1975)
Tr 78 J. Trümper, W. Pietsch, C. Reppin, W. Voges, R. Staubert, E. Kendziorra: Astrophys. Lett. **219**, L105 (1978)
Ts 74 W. Y. Tsai: Phys. Rev. **D10**, 1342 (1974)
Ts 83 T. Tserruya, R. Schuch, H. Schmidt-Böcking, J. Barrette, Wang Da-Hai, B. M. Johnson, M. Meron, K. W. Jones: Phys. Rev. Lett. **50**, 30 (1983)

Ue 35 E. A. Uehling: Phys. Rev. **48**, 55 (1935)
Un 76 W. Unruh: Phys. Rev. **D14**, 870 (1976)
Ut 56 R. Utiyama: Phys. Rev. **101**, 1597 (1956)

Va 65 V. S. Vanyashin, M. V. Terent'ev: Sov. Phys. JETP **21**, 375 (1965)
Va 80 J. Vaaben, K. Taulbjerg: J. Phys. **B14**, 1815 (1980)
Va 83 D. Vasak, K. H. Wietschorke, B. Müller, W. Greiner: Z. Phys. **C21**, 119 (1983)
Va 85 D. Vasak: Ph.D. Thesis, Frankfurt University (1985)
Vi 80 R. D. Viollier, J. Rafelski: Helv. Phys. Acta **53**, 352 (1980)
Vi 84 P. Vincent: "X-Ray processes in heavy ion collisions", in *Atomic Inner-Shell Physics*, ed. by B. Crasemann (Plenum, New York 1984, in print)
Vo 61 V. V. Voronkov, N. N. Koleznikov: Sov. Phys. JETP **12**, 136 (1961)
Vo 73 P. Vogel: Phys. Rev. **A7**, 63 (1973)

Wa 70 W. L. Wang, C. M. Shakin: Phys. Lett. **32B**, 421 (1970)
Wa 74 J. Waldeck: Diploma thesis, Institut für Theoretische Physik, Frankfurt University (1974)
Wa 81 Wang Gia-Zhan, Bi Pin-Zhen, Yin Pong-Cheng: Phys. Lett. **107B**, 381 (1981)
We 29 H. Weyl: Z. Phys. **56**, 330 (1929)
We 36 V. Weisskopf: K. Dan. Vidensk. Selsk. Mat. Fys. Medd. **14**, No. 6 (1936)
We 49 G. Wentzel: *Quantum Theory of Fields* (Interscience, New York 1949), Chap. 21
We 72 S. Weinberg: *Gravitation and Cosmology* (Wiley, New York 1972)
We 73 S. Weinberg: Phys. Rev. Lett. **31**, 494 (1973)
Wh 37 J. A. Wheeler: Phys. Rev. **52**, 1107 (1937)
Wi 53 R. R. Wilson: Phys. Rev. **90**, 720 (1953)
Wi 55 E. Wigner: Phys. Rev. **98**, 145 (1955)
Wi 56 E. H. Wichmann, N. M. Kroll: Phys. Rev. **101**, 843 (1956)
Wi 79 K. H. Wietschorke, B. Müller, W. Greiner, G. Soff: J. Phys. **B12**, L31 (1979)

Wi 83 K. H. Wietschorke, P. Schlüter, W. Greiner: J. Phys. **A16**, 2017 (1983)
Wi 84 K. H. Wietschorke, P. Schlüter: To be published
Wo 82 H. J. Wollersheim, W. W. Wilcke, J. R. Birkelund, J. R. Huizenga: Phys. Rev. **C25**, 338 (1982)

Ya 54 C. N. Yang, R. L. Mills: Phys. Rev. **96**, 191 (1954)
Ya 83 H. Yamagishi: Phys. Rev. **D27**, 2383 (1983)
Yo 74 S. Yoshida: Ann. Rev. Nucl. Part. Sci. **24**, 1 (1974)

Ze 72 Y. B. Zeldovich, V. S. Popov: Sov. Phys. Usp. **14**, 673 (1972)

Subject Index

Texts and Monographs in Physics

W. Rindler: **Essential Relativity: Special, General, and Cosmological,** Revised printing of the second edition (1980)

K. Chadan and P. C. Sabatier: **Inverse Problems in Quantum Scattering Theory** (1977)

C. Truesdell and S. Bharatha: **The Concepts and Logic of Classical Thermodynamics as a Theory of Heat Engines: Rigourously Constructed upon the Foundation Laid by S. Carnot and F. Reech** (1977)

R. D. Richtmyer: **Principles of Advanced Mathematical Physics.** Volume I (1978). Volume II (1981)

R. M. Santilli: **Foundations of Theoretical Mechanics.** Volume I: The Inverse Problem in Newtonian Mechanics. Corrected 2nd printing (1984). Volume II: Birkhoffian Generalization of Hamiltonian Mechanics (1983)

A. Böhm: **Quantum Mechanics.** Second edition (1985)

H. Pilkuhn: **Relativistic Particle Physics** (1979)

M. D. Scadron: **Advanced Quantum Theory and Its Applications Through Feynman Diagrams.** Corrected 2nd printing (1981)